PRACTICAL COOLING TECHNOLOGY

OVERSIZE

ACPL ITEM DISCARDED

697.93 J63p
Johnson, William M.
Practical cooling technology

ALLEN COUNTY PUBLIC LIBRARY
FORT WAYNE, INDIANA 46802

You may return this book to any location of
the Allen County Public Library.

DEMCO

PRACTICAL COOLING TECHNOLOGY

WILLIAM M. JOHNSON

Central Piedmont Community College
Charlotte, North Carolina

Delmar Publishers
I(T)P An International Thomson Publishing Company

Albany • Bonn • Boston • Cincinnati • Detroit • London • Madrid • Melbourne
Mexico City • New York • Pacific Grove • Paris • San Francisco • Singapore • Tokyo
Toronto • Washington

NOTICE TO THE READER

Publisher does not warrant or guarantee any of the products described herein or perform any independent analysis in connection with any of the product information contained herein. Publisher does not assume, and expressly disclaims, any obligation to obtain and include information other than that provided to it by the manufacturer.

The reader is expressly warned to consider and adopt all safety precautions that might be indicated by the activities herein and to avoid all potential hazards. By following the instructions contained herein, the reader willingly assumes all risks in connections with such instructions.

The publisher makes no representation or warranties of any kind, including but not limited to, the warranties of fitness for particular purpose or merchantability, nor are any such representations implied with respect to the material set forth herein, and the publisher takes no responsibility with respect to such material. The publisher shall not be liable for any special, consequential, or exemplary damages resulting, in whole or part, from the readers' use of, or reliance upon, this material.

Cover photos courtesy of: Robinair Division, SPX Corporation, Beckman Industrial Corporation, and Bill Johnson
Cover Design: Charles Cummings Advertising/Art, Inc.

Delmar Staff
Publisher: Robert Lynch
Acquisitions Editor: Vernon Anthony
Developmental Editor: Denise Denisoff
Project Editor: Thomas Smith
Production Coordinator: Karen Smith
Art and Design Coordinator: Cheri Plasse

COPYRIGHT © 1997
By Delmar Publishers
an International Thomson Publishing Company

The ITP logo is a trademark under license

Printed in the United States of America

For more information, contact:

Delmar Publishers
3 Columbia Circle, Box 15015
Albany, New York 12212-5015

International Thomson Publishing Europe
Berkshire House 168 - 173
High Holborn
London WC1V 7AA
England

Thomas Nelson Australia
102 Dodds Street
South Melbourne, 3205
Victoria, Australia

Nelson Canada
1120 Birchmount Road
Scarborough, Ontario
Canada M1K 5G4

Delmar Publishers' Online Services
To access Delmar on the World Wide Web, point your browser to:
http://www.delmar.com/delmar.html
To access through Gopher: gopher://gopher.delmar.com
(Delmar Online is part of "thomson.com", an Internet site with information on more than 30 publishers of the International Thomson Publishing organization.)
For information on our products and services:
email: info@delmar.com
or call 800-347-7707

International Thomson Editores
Campos Eliseos 385, Piso 7
Col Polanco
11560 Mexico D F Mexico

International Thomson Publishing GmbH
Königswinterer Strasse 418
53227 Bonn
Germany

International Thomson Publishing Asia
221 Henderson Road
#05 - 10 Henderson Building
Singapore 0315

International Thomson Publishing Japan
Hirakawacho Kyowa Building, 3F
2-2-1 Hirakawacho
Chiyoda-ku, Tokyo 102
Japan

All rights reserved. No part of this work covered by the copyright hereon may be reproduced or used in any form or by any means—graphic, electronic, or mechanical, including photocopying, recording, taping, or information storage and retrieval systems—without the written permission of the publisher.

1 2 3 4 5 6 7 8 9 10 XXX 02 01 00 99 98 97 96

Library of Congress Cataloging-in-Publication Data

Johnson, William M.
   Practical cooling technology / William M. Johnson.
     p.   cm.
   Includes index.
   ISBN 0-8273-6814-3 (case)
   1. Air conditioning.   I. Title.
TH7687.J69   1996
697.9'33—dc20

96-41093
CIP

BRIEF CONTENTS

Preface		xi
Unit 1	General Safety Practices	1
Unit 2	Heat, Pressure, Matter, and Energy	12
Unit 3	Refrigeration for Air Conditioning	26
Unit 4	Tubing and Piping	49
Unit 5	System Evacuation	68
Unit 6	Refrigerant Management—Recovery, Recycling, and Reclamation	86
Unit 7	System Charging	98
Unit 8	Basic Electricity and Magnetism	105
Unit 9	Introduction to Automatic Controls	129
Unit 10	Automatic Control Components and Applications	136
Unit 11	Troubleshooting Basic Controls	154
Unit 12	Electronic and Programmable Controls	170
Unit 13	Electric Motors	186
Unit 14	Motor Applications	202
Unit 15	Motor Starting	215
Unit 16	Troubleshooting Electric Motors	225
Unit 17	Comfort and Psychrometrics	235
Unit 18	Room Air Conditioners	251
Unit 19	Evaporators	277
Unit 20	Compressors	289
Unit 21	Condensers	314
Unit 22	Expansion Devices	327
Unit 23	Heat Gain Calculations	345
Unit 24	Application of Air Conditioning Equipment	376
Unit 25	Air Distribution and Balance	411
Unit 26	Installation	458
Unit 27	Typical Operating Conditions	478
Unit 28	Troubleshooting	491
Unit 29	Chilled-Water Air Conditioning Systems	511
Unit 30	Cooling Towers and Pumps	553
Unit 31	Chiller Operation, Maintenance, and Troubleshooting	572
Appendices		591
Glossary		613
Index		641

CONTENTS

PREFACE — xi

UNIT 1 GENERAL SAFETY PRACTICES — 1
1.1 Introduction   1.2 Pressure Vessels and Piping   1.3 Electrical Hazards   1.4 Heat
1.5 Cold   1.6 Mechanical Equipment   1.7 Moving Heavy Objects   1.8 Refrigerants in
Your Breathing Space   1.9 Using Chemicals   Summary   Review Questions

UNIT 2 HEAT, PRESSURE, MATTER, AND ENERGY — 12
2.1 Temperature   2.2 Heat Transfer   2.3 Sensible Heat   2.4 Latent Heat
2.5 Specific Heat   2.6 Pressure   2.7 Matter   2.8 Mass   2.9 Density
2.10 Specific Gravity   2.11 Specific Volume   2.12 Gas Laws   2.13 Energy
2.14 Horsepower   2.15 Electrical Power—The Watt   Summary   Review Questions

UNIT 3 REFRIGERATION FOR AIR CONDITIONING — 26
3.1 Refrigeration for Air Conditioning   3.2 Rating Air Conditioning Equipment
3.3 Pressure/Temperature Relationship   3.4 Refrigeration Components
3.5 The Refrigeration Cycle   3.6 Refrigerant-22 (R-22)   3.7 Refrigerant Recovery, Recycling,
and Reclamation   3.8 Plotting the Refrigeration Cycle   Summary   Review Questions

UNIT 4 TUBING AND PIPING — 49
4.1 Purpose of Tubing and Piping in A/C Systems   4.2 Types and Sizes of Tubing
4.3 Tubing Insulation   4.4 Line Sets   4.5 Cutting Tubing   4.6 Bending Tubing
4.7 Soldering and Brazing Processes   4.8 Heat Sources for Soldering and Brazing
4.9 Soldering Techniques   4.10 Brazing Techniques   4.11 Practical Soldering and Brazing Tips
4.12 Making Flare Joints   4.13 Swaging   4.14 Steel and Wrought Iron Pipe
4.15 Joining Steel Pipe   4.16 Installing Steel Pipe   4.17 Plastic Pipe   Summary
Review Questions

UNIT 5 SYSTEM EVACUATION — 68
5.1 Evacuating A/C Systems   5.2 Understanding Evacuation   5.3 Measuring a Vacuum
5.4 Recovering Refrigerant   5.5 Vacuum Pumps   5.6 Vacuum Pump Oil
5.7 Achieving a Deep Vacuum   5.8 Multiple Evacuation   5.9 Leak Detection While
in a Vacuum   5.10 Leak Detection—Standing Pressure Test   5.11 Removing Moisture
with a Vacuum   5.12 General Evacuation Tips   5.13 Cleanup After Compressor Burnout
Summary   Review Questions

UNIT 6 REFRIGERANT MANAGEMENT—RECOVERY, RECYCLING, AND RECLAMATION — 86
6.1 Refrigerants and the Environment   6.2 Refrigerant Classifications
6.3 Refrigerant Regulations   6.4 Refrigerant Recovery, Recycling, or Reclamation
6.5 EPA Evacuation Requirements   6.6 Technician Certification
6.7 Mechanical Recovery Systems   6.8 Reclaiming Refrigerant
6.9 Recovering Refrigerant from Window Air Conditioners   6.10 Technician Information
Summary   Review Questions

viii CONTENTS

UNIT 7 SYSTEM CHARGING — 98
7.1 Charging a Refrigeration System   7.2 Vapor Refrigerant Charging
7.3 Liquid Refrigerant Charging   7.4 Measuring Refrigerant   Summary
Review Questions

UNIT 8 BASIC ELECTRICITY AND MAGNETISM — 105
8.1 Structure of Matter   8.2 Movement of Electrons   8.3 Conductors   8.4 Insulators
8.5 Electricity Produced from Magnetism   8.6 Direct Current   8.7 Alternating Current
8.8 Electrical Units of Measurement   8.9 The Electrical Circuit   8.10 Making Electrical
Measurements   8.11 Ohm's Law   8.12 Series Circuits   8.13 Parallel Circuits
8.14 Electrical Power   8.15 Magnetism   8.16 Inductance   8.17 Transformers
8.18 Capacitance   8.19 Impedance   8.20 Electrical Measuring Instruments   8.21 Wire Sizes
8.22 Circuit Protection Devices   8.23 Semiconductors   Summary   Review Questions

UNIT 9 INTRODUCTION TO AUTOMATIC CONTROLS — 129
9.1 Types of Automatic Controls   9.2 Automatic Controls Used in A/C Systems
9.3 Temperature-Sensitive Devices   Summary   Review Questions

UNIT 10 AUTOMATIC CONTROL COMPONENTS AND APPLICATIONS — 136
10.1 Recognition of Control Components   10.2 Temperature Controls   10.3 Temperature
Measurements   10.4 Pressure-Sensitive Devices   10.5 Electromechanical Controls
10.6 Maintenance of Controls   10.7 Technician Service Call   Summary   Review Questions

UNIT 11 TROUBLESHOOTING BASIC CONTROLS — 154
11.1 Introduction to Troubleshooting   11.2 Troubleshooting a Simple Circuit
11.3 Troubleshooting an Air Conditioning Circuit   11.4 Troubleshooting a Thermostat
11.5 Troubleshooting Amperage in Low-Voltage Circuits   11.6 Troubleshooting Voltage in
Low-Voltage Circuits   11.7 Pictorial and Line Diagrams   11.8 Technician Service Calls
Summary   Review Questions

UNIT 12 ELECTRONIC AND PROGRAMMABLE CONTROLS — 170
12.1 Electronic Controls   12.2 Air Conditioning Applications   12.3 Electronic Thermostats
12.4 Diagnostic Thermostats   12.5 Troubleshooting Electronic Controls
12.6 Technician Service Calls   Summary   Review Questions

UNIT 13 ELECTRIC MOTORS — 186
13.1 Electric Motor Applications   13.2 Motor Components and Operation
13.3 Motors Used in A/C Applications   13.4 Motor Circuit Control Devices
13.5 Motor Capacity Control   13.6 Motor Cooling   Summary   Review Questions

UNIT 14 MOTOR APPLICATIONS — 202
14.1 Motor Selection   14.2 Power Supply   14.3 Working Conditions
14.4 Temperature Classifications   14.5 Bearings   14.6 Motor Mounting Characteristics
14.7 Motor Drives   Summary   Review Questions

UNIT 15 MOTOR STARTING — 215
15.1 Motor Control Devices   15.2 Amperage Ratings   15.3 Relays   15.4 Contactors
15.5 Motor Starters   15.6 Motor Protection   15.7 Service Factor   Summary
Review Questions

UNIT 16 TROUBLESHOOTING ELECTRIC MOTORS — 225
16.1 Electric Motor Troubleshooting   16.2 Mechanical Motor Problems
16.3 Electrical Problems   16.4 Motor Starting Problems   16.5 Checking Capacitors
16.6 Wiring and Connectors   16.7 Troubleshooting Hermetic Motors
16.8 Technician Service Calls   Summary   Review Questions

UNIT 17	COMFORT AND PSYCHROMETRICS	235

17.1 Comfort **17.2** Food Energy and the Body **17.3** Psychrometrics
17.4 The Psychrometric Chart **17.5** Ventilation and Infiltration Summary
Review Questions

UNIT 18	ROOM AIR CONDITIONERS	251

18.1 Air Conditioning and Heating with Room Units **18.2** Cooling-Only Units
18.3 Combination Cooling/Heating Units **18.4** Installation **18.5** Controls for Cooling-Only
Units **18.6** Controls for Cooling/Heating Units **18.7** Maintenance and Service
18.8 Technician Service Calls Summary Review Questions

UNIT 19	EVAPORATORS	277

19.1 Evaporator Functions **19.2** Evaporator Construction and Operation **19.3** Types of
Evaporator Coils **19.4** Condensate and the Evaporator **19.5** Sensible and Latent Heat
Removal Summary Review Questions

UNIT 20	COMPRESSORS	289

20.1 Compressor Functions **20.2** Reciprocating Compressors **20.3** Hermetic Compressors
20.4 Semi-Hermetic Compressors **20.5** Open-Drive Compressors **20.6** Reciprocating
Compressor Efficiency **20.7** Compressor Capacity Control **20.8** Rotary Compressors
20.9 Scroll Compressors **20.10** Motors for Hermetic and Semi-Hermetic Compressors
Summary Review Questions

UNIT 21	CONDENSERS	314

21.1 Condenser Functions **21.2** Refrigerant-to-Earth Heat Exchangers
21.3 Refrigerant-to-Water Heat Exchangers **21.4** Refrigerant-to-Air Heat Exchangers
Summary Review Questions

UNIT 22	EXPANSION DEVICES	327

22.1 Expansion Device Functions **22.2** Thermostatic Expansion Valves
22.3 Bulb Charges **22.4** TXV Operation **22.5** TXV Installation and Service Considerations
22.6 Balanced Port Expansion Valves **22.7** Solid-State Controlled Expansion Valves
22.8 Receivers **22.9** Fixed-Bore Metering Devices **22.10** Operating Charge for Fixed-
Bore Devices Summary Review Questions

UNIT 23	HEAT GAIN CALCULATIONS	345

23.1 System Design **23.2** Design Conditions **23.3** Heat Transfer through Building Materials
23.4 Solar Heat Gain **23.5** Heat Gain due to Infiltration and Ventilation
23.6 Heat Gain through Ductwork **23.7** Internal Heat Gains Summary Review Questions

UNIT 24	APPLICATION OF AIR CONDITIONING EQUIPMENT	376

24.1 Application of Equipment **24.2** Types of Equipment **24.3** Piping Practices
24.4 Refrigerant Line Sizing **24.5** Special-Application Cooling Equipment
Summary Review Questions

UNIT 25	AIR DISTRIBUTION AND BALANCE	411

25.1 Forced-Air Systems **25.2** Duct System Pressures **25.3** Types of Fans
25.4 Supply Duct System **25.5** Duct System Standards **25.6** Duct Materials
25.7 Combination Duct Systems **25.8** Installation Considerations **25.9** Blending
the Conditioned Air with Room Air **25.10** Return Air Duct System **25.11** Friction Loss
25.12 Measuring Air Movement for Balancing **25.13** Commercial Duct Applications
Summary Review Questions

UNIT 26	INSTALLATION	458

26.1 Equipment Installation **26.2** Duct Installation **26.3** Electrical Installation
26.4 Refrigeration System Installation **26.5** Equipment Start-Up Summary
Review Questions

CONTENTS

UNIT 27 TYPICAL OPERATING CONDITIONS — 478
  27.1 Measuring Operating Conditions   **27.2** Mechanical Operating Conditions
  27.3 Equipment Grades   **27.4** Manufacturer's Literature   **27.5** Establishing a Reference Point on Unknown Equipment   **27.6** Equipment Efficiency Ratings
  27.7 Typical Electrical Operating Conditions   Summary   Review Questions

UNIT 28 TROUBLESHOOTING — 491
  28.1 Troubleshooting A/C Systems   **28.2** Mechanical Troubleshooting
  28.3 Gage Manifold Usage   **28.4** Temperature Readings   **28.5** Charging Procedures
  28.6 Electrical Troubleshooting   **28.7** Preventive Maintenance   **28.8** Technician Service Calls
  Summary   Review Questions

UNIT 29 CHILLED-WATER AIR CONDITIONING SYSTEMS — 511
  29.1 Chilled-Water Applications   **29.2** High-Pressure Compreession Cycle Chillers
  29.3 Low-Pressure Compression Cycle Chillers   **29.4** Absorption Air Conditioning Chillers
  29.5 Motors and Drives   Summary   Review Questions

UNIT 30 COOLING TOWERS AND PUMPS — 553
  30.1 Cooling Tower Functions   **30.2** Types of Cooling Towers   **30.3** Flow Patterns
  30.4 Tower Construction   **30.5** Fan Section   **30.6** Water Control and Distribution
  30.7 Water Pumps Used in Cooling Towers   Summary   Review Questions

UNIT 31 CHILLER OPERATION, MAINTENANCE, AND TROUBLESHOOTING — 572
  31.1 Chiller Systems   **31.2** Start-Up Procedures for Chilled-Water Systems
  31.3 Operating Procedures for Chilled-Water Systems   **31.4** Maintenance Procedures for Chilled-Water Systems   **31.5** Procedures for Absorption System Start-Up
  31.6 Operation and Maintenance Procedures for Absorption Systems
  31.7 General Maintenance Tips for All Chillers   **31.8** Technician Service Calls
  Summary   Review Questions

APPENDICES — 591
GLOSSARY — 613
INDEX — 641

PREFACE

Practical Cooling Technology was written to provide students and technicians extended coverage of the practical aspects of cooling equipment and systems as well as more professional installation and service techniques. Installation and service techniques that are more professional result in fewer problems and help to create a more satisfying profession. The more professional technicians also have the best career opportunities.

The fundamentals of air conditioning are discussed in the beginning of the text to prepare the student for a more detailed explanation of the air conditioning process. Each type of equipment is discussed in detail. This text discusses how the equipment and its components are constructed and what the manufacturers' intentions are when the various features are designed into the equipment.

Refrigerant pipe sizing and system design is discussed to broaden the technician's knowledge of subjects that are very seldom discussed in other textbooks.

An effort has been made to address the more common systems, whether old or new. Many older types of cooling systems that are still in use as well as many of the newer types of systems are included; since literature for older systems is out of print, the technician may benefit from this information when troubleshooting these systems. Newer types of systems may not be discussed in other textbooks for many years.

ORGANIZATION OF THE TEXT

The text is organized so that the instructor can cover Units 1, 2, and 3 and then move on to the desired next subject; for example, room air conditioners (Unit 18) or refrigerant management (Unit 6).

Every effort has been made to keep the material at a reading level that most students can easily understand. Everyday language has been used along with much trade terminology so the student will be using the same terms that the industry uses. In addition, special effort has been made to provide enough terminology so that the student is able to converse intelligently with designers and engineers.

This text is illustrated with many easy-to-follow illustrations so the student or technician can get a visual picture of how all of the components function. The illustrations are often described in the text with additional explanations in the figure captions and callouts in the illustrations.

FEATURES OF THE TEXT

Objectives are listed at the beginning of each unit. The objective statements are kept clear and simple to give the students direction. A *Summary* and *Review Questions* appear at the end of each unit. Students should answer the questions while reviewing what they have read; instructors can use the summaries and review questions to stimulate class discussion and for classroom unit review.

Practical troubleshooting procedures are an important feature of this text. There are practical component and system troubleshooting suggestions and techniques. In many units, practical examples of service technician calls are presented in a down-to-earth situational format.

Safety precautions and techniques are discussed throughout the text. Each unit cautions the technician about the particular safety aspects that should be known and followed at this point of study.

Illustrations and *photos* are used generously throughout the text. An effort has been made to illustrate each component and its relationship to the system function in a manner that allows the student to picture each component or event.

SUPPLEMENTS TO THIS TEXT

To further enhance the training of the cooling technician, the following supplements are also available to accompany the text.

Practical Cooling Technology Lab Manual   A series of practical exercises to help guide the student through many types of equipment used in the field. When the student has successfully completed these exercises, a better understanding of how to accomplish various service functions will be accomplished. The student will have handled all of the components and performed many tasks on these components for the purpose of diagnostics and repair.

Practical Cooling Technology Instructor's Guide A guide of suggested activities and methods for presenting the text material in each unit, including material from other sources that may be used. It includes the answers to all questions in the text and the lab manual.

GENERAL SAFETY PRACTICES

OBJECTIVES

Upon completion of this unit, you should be able to

- **Describe proper procedures for working with pressurized systems and vessels, electrical energy, heat, cold, rotating machinery, chemicals, and moving heavy objects.**
- **Work safely, avoiding safety hazards.**

1.1 INTRODUCTION

As an air conditioning technician, you will work close to many potentially dangerous situations: liquids and gases under pressure, electrical energy, heat, cold, chemicals, rotating machinery, and so on. Each job must be completed in a manner that is safe both for you and for those around you.

This unit describes general safety practices and procedures. Whether in a school laboratory, shop, or manufacturing plant, always be familiar with the location of emergency exits and first aid and eye wash stations. Determine how you would get out of a building or particular location should an emergency occur. Many other more specific safety practices and tips are given in the units in which they may be applied. Always use common sense and be prepared.

1.2 PRESSURE VESSELS AND PIPING

Pressure vessels and piping are part of many systems that are serviced by air conditioning technicians. These components are under pressure and must be treated with great caution.

Refrigerant Cylinders and Piping

Refrigerant cylinders are under pressure and can be dangerous if handled improperly. For example, a cylinder of R-22 in the back of an open truck with the sun shining on it may have a cylinder temperature of 110°F on a summer day. For this refrigerant, the pressure inside the cylinder at 110°F is 226 psig (pounds per square inch gage). This pressure reading means that the cylinder has a pressure of 226 pounds for each square inch of surface area (Figure 1-1). This pressure is well contained and will be safe if the cylinder is protected. Do not drop it.

Warning: Move the cylinder only while the protective cap is on it, if it is designed for one (Figure 1-2). Cylinders that are too large to be carried should be moved chained to an approved cart (Figure 1-3).

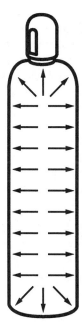

Figure 1-1. Pressure is exerted across the entire surface area of a refrigerant cylinder.

Figure 1-2. Refrigerant cylinder with protective cap. *Photo by Bill Johnson.*

2 UNIT 1 GENERAL SAFETY PRACTICES

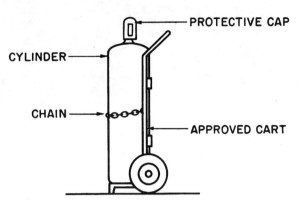

Figure 1-3. Pressurized cylinders should be moved on an approved cart. The protective cap must be secured.

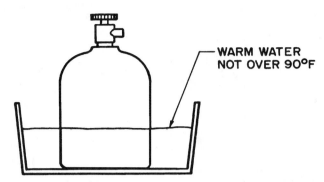

Figure 1-5. Refrigerant cylinder in tepid water (no warmer than 90°F).

The cylinder pressure can be thought of as a *potential* danger. It will not become dangerous unless it is allowed to escape in an uncontrolled manner. The cylinder has a relief valve in the top in the vapor space. If the pressure builds up to the relief valve setting, it will start relieving vapor. As the vapor pressure is relieved, the liquid in the cylinder will begin to cool and reduce the pressure (see Figure 1-4). Relief valves are generally set at values above the worst typical operating conditions and are usually over 400 psig.

The refrigerant cylinder has a fusible plug made of a material with a low melting temperature. The plug will melt and blow out if the cylinder gets too hot. This prevents the cylinder from bursting and injuring personnel and property.

Warning: Never apply direct heat to refrigerant cylinders. To keep the pressure from dropping in the cylinder while charging the system, place the cylinder in a container of tepid water with a temperature no higher than 90°F (Figure 1-5).

Taking pressure readings on air conditioning systems can be dangerous. Liquid R-22 boils at -41°F when released to the atmosphere. If you are careless and get this refrigerant on your skin or in your eyes, it will quickly cause frostbite. Protect your skin and eyes from any refrigerant. When gages are attached to take refrigerant pressure readings or to transfer refrigerant into or out of a system, wear gloves and side shield goggles. Figure 1-6 shows a pair of protective eye goggles. These goggles are vented to keep them from fogging over with condensation and also to help keep the technician cool.

Warning: If a leak develops and refrigerant is escaping, step back and look for a valve with which to shut it off. Do not try to stop it with your hands.

Caution: All cylinders into which refrigerant is transferred must be approved by the United States Department of Transportation as recovery cylinders. These cylinders are color coded (yellow tops with grey bodies) to indicate that they are approved for this purpose. It is against the law to transport refrigerant in nonapproved cylinders.

Nitrogen and Oxygen Cylinders

In addition to the pressure potential inside a refrigerant cylinder, there is tremendous pressure potential inside nitrogen and oxygen cylinders. They are shipped at pressures of up to 2500 psig. These cylinders must not be

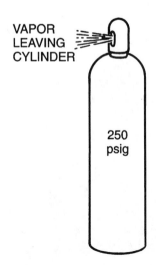

Figure 1-4. Pressure relief valve will reduce pressure in the cylinder.

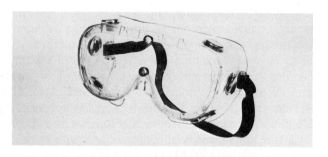

Figure 1-6. Protective goggles. *Photo by Bill Johnson.*

ELECTRICAL HAZARDS 3

This can be particularly dangerous if it occurs at the compressor shell.

Because of this high pressure, oxygen also must be regulated to a lower pressure for use. In addition, all oxygen lines must be kept absolutely oil free. Oil residue in an oxygen regulator connection may cause an explosion strong enough to blow the regulator apart in your hand.

Acetylene Cylinders

Oxygen is often used with acetylene. Acetylene cylinders are not under the same pressure as nitrogen and oxygen, but must be treated with the same respect because acetylene is highly explosive. A pressure-reducing regulator must be used. Always use an approved cart, chain the cylinder to the cart, and secure the protective cap. All stored cylinders must be supported so they will not fall over (Figure 1-8) and they must be separated as required by code.

1.3 ELECTRICAL HAZARDS

Electrical shocks and burns are ever-present hazards. It is impossible to troubleshoot all circuits with the power off, so you must learn safe methods for troubleshooting "live" circuits. As long as electricity is contained in the conductors and the devices where it is supposed to function, there is nothing to fear. When uncontrolled electrical flow occurs (e.g., if you touch two live wires), you are very likely to get hurt.

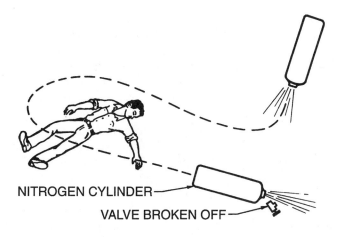

Figure 1-7. When a cylinder valve is broken off, the cylinder becomes a projectile until the pressure is exhausted.

moved unless the protective cap is in place and they are chained to an approved cart. If a cylinder is dropped and the protective cap is not in place, the cylinder valve may break off. The pressure inside the cylinder can propel the cylinder like a balloon full of air that is turned loose (Figure 1-7).

Nitrogen must have its pressure regulated before it can be used. The pressure in the cylinder is too great to be connected to a system. If nitrogen under cylinder pressure is allowed to enter a refrigeration system, the pressure can burst any weak points in the system.

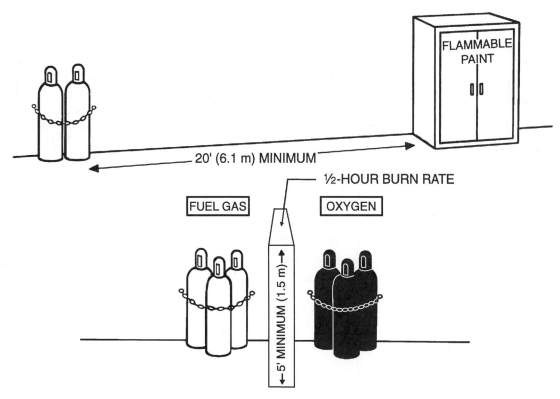

Figure 1-8. Pressurized cylinders must be chained when stored. The minimum safe distance between stored fuel gas cylinders and any flammable materials is 20 feet or a separating wall that is 5 feet high. From Jeffus & Johnson, *Welding: Principles and Applications, 3E,* © 1992 by Delmar Publishers, Inc.

4 UNIT 1 GENERAL SAFETY PRACTICES

Warning: Electrical power should always be shut off at the distribution or entrance panel, tagged, and locked out in an approved manner when installing equipment. Whenever possible, the power should also be shut off, tagged, and locked out when servicing the equipment. Electrical panels are furnished with a place for a lock for the purpose of lock out. To avoid having someone inadvertently turn the power back on, keep the panel locked when you are out of sight of it and keep the only key in your possession. Don't ever think that you are good enough or smart enough to work with live electrical power when it is not necessary.

Certain tests have to be made with the power on. Extreme care should be taken when making these tests. Ensure that your hands touch only the meter probes and that your arms and the rest of your body remain clear of all electrical terminals and connections. Know the voltage in the circuit you are checking. Make sure that the range selector on the test instrument is set properly before using it. Don't stand in a wet or damp area when making these checks. Use only proper test equipment, and make sure it is in good condition. Wear heavy shoes with an insulating sole and heel. Intelligent and competent technicians take all precautions.

Electrical Shock

Electrical shock occurs when you become part of the circuit. Electricity flows through your body and can cause your heart to stop pumping, resulting in death if it is not restarted quickly. It is a good idea to take a first aid course that includes cardiopulmonary resuscitation.

To prevent electrical shock, don't become a conductor between two live (hot) wires or a live wire and ground. The electricity must have a path to flow through. Don't let your body be the path. Figure 1-9 shows situations in which the technician has become part of a circuit.

Use only properly-grounded power tools connected to properly-grounded circuits. Caution should be taken when using portable electric tools. These are handheld devices with electrical energy inside just waiting for a path to flow through. Some portable electric tools are constructed with metal frames. They should have a grounding wire in the power cord. The grounding wire protects the operator from electric shocks. If the motor inside the tool develops a loose connection and the frame of the tool becomes electrically hot, the third wire, rather than your body, will carry the current, and a fuse or breaker will interrupt the circuit (see Figure 1-10).

In some instances, you may need to use a three- to two-prong adapter at job sites if the wall receptacle has only two connections and your portable electric tool has a three-wire plug. The adapter has a third wire that must be connected to a ground for the operator to be protected.

Warning: If this wire is fastened under the wall plate screw and the screw terminates in an ungrounded box nailed to a wooden wall, you are *not* protected (Figure 1-11). Ensure that the third or ground wire is properly connected to a ground.

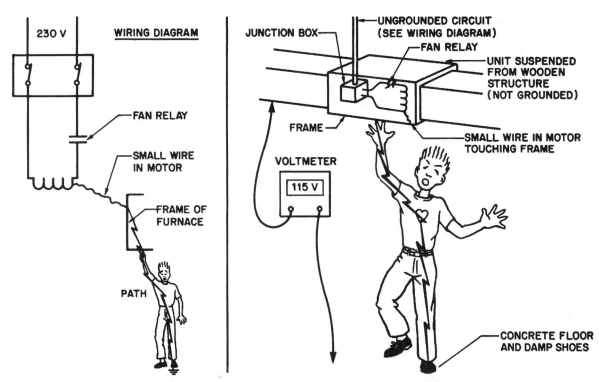

Figure 1-9. Ways for the technician to become part of the circuit and receive an electric shock.

ELECTRICAL HAZARDS 5

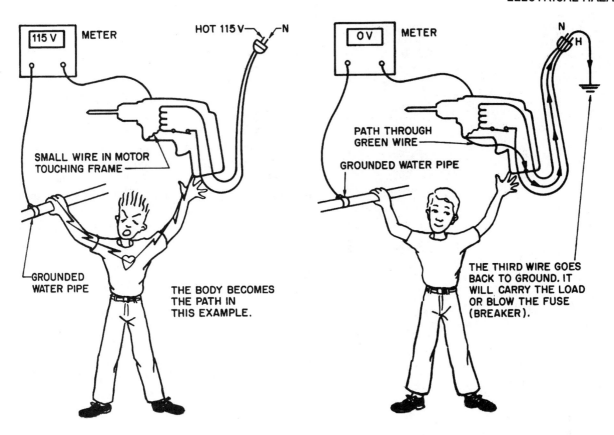

Figure 1-10. Electrical circuit to ground from metal frame of drill.

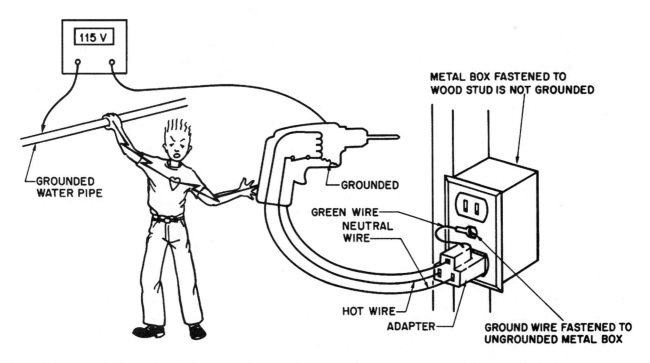

Figure 1-11. The wire from the adapter is intended to be fastened under the screw in the duplex wall plate. However, this will provide no protection if the outlet box is not grounded.

6 UNIT 1 GENERAL SAFETY PRACTICES

Figure 1-12. Extension cord with ground fault circuit interrupter (GFCI) receptacle. *Photo by Bill Johnson.*

Alternatives to traditional metal-cased electric tools are modern plastic-cased and battery-operated tools. In the plastic-cased tool, the motor and electrical connections are insulated within the tool. Plastic-cased tools are called *double-insulated* and are considered safe. Battery-operated tools use rechargeable batteries and provide both safety and convenience.

An extension cord with a ground fault circuit interrupter (GFCI) or an extension cord used with a GFCI receptacle is recommended for use with portable electric tools (Figure 1-12). These receptacles monitor the circuit and detect very small electrical leaks to ground. A small electrical current leak will cause the GFCI to open the circuit, preventing further current flow.

Nonconducting Ladders

Nonconducting ladders are made from wood or fiberglass rather than a conductive metal. They should be used on all jobs and should be of the type furnished with service trucks. Nonconducting ladders are much safer than aluminum ladders. They also work as well as aluminum ladders; they are just heavier.

Warning: A ladder may inadvertently be raised into a power line or placed against a live electrical hazard. When you are standing on a nonconducting ladder, it will provide protection from electrical shock to the ground; however, it will not always provide protection between two or more electrical conductors. Before climbing any ladder, make sure that it is secure and is not touching any electrical source. Never climb a ladder unless you are working with a second technician to steady the ladder and to spot potential problems.

Electrical Burns

Do not wear jewelry (rings and watches) while working on live electrical circuits because they can cause shocks and possible burns.

Never use a screwdriver or other tools in an electrical panel when the power is on. Electrical burns can result from an electrical arc, such as in a short circuit to ground when uncontrolled electrical energy flows. For example, if a screwdriver slips while you are working in a panel and the blade completes a circuit to ground, the potential flow of electrical energy is tremendous. *Ohm's Law* can be used to calculate the current flow in a circuit [Current (I) = Voltage (E) ÷ Resistance (R)]. When a circuit has a resistance of 10 ohms (Ω) and is operated on 120 volts (V), it would, using Ohm's Law, have a current flow or amperage (A) of:

$$I = \frac{E}{R} = \frac{120 \text{ V}}{10 \text{ }\Omega} = 12 \text{ A}$$

If this example is calculated again with less resistance, the current will be greater because the voltage is divided by a smaller number. If the resistance is lowered to 1 ohm; the current flow is then:

$$I = \frac{E}{R} = \frac{120 \text{ V}}{1 \text{ }\Omega} = 120 \text{ A}$$

If the resistance is reduced to 0.1 ohm, the current flow will be 1200 A. By this time, the circuit breaker would trip, but you would have already incurred burns or an electric shock (Figure 1-13). Current flow through the body of .015 amps or less can prove fatal.

1.4 HEAT

The use of heat requires special care. A high concentration of heat comes from the torches used for soldering and brazing. Many combustible materials may be in the

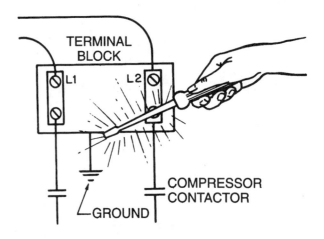

Figure 1-13. Wiring illustration showing short circuit caused by a slip of the screwdriver.

Figure 1-14. A shield used when soldering. *Courtesy Wingaersheek.*

area where soldering is required. When soldering or using concentrated heat, a fire extinguisher should always be close by, and you should know exactly where it is and how to use it. Learn how to use a fire extinguisher *before* a fire occurs. A fire extinguisher should always be included as part of the standard tools and equipment on a service truck.

When a solder connection must be made next to combustible materials or a finished surface, use a noncombustible shield for insulation (Figure 1-14). A fire-resistant spray may also be used to decrease the flammability of wood if a torch must be used nearby. This spray retardant should be used with an appropriate shield. It is also important to use a shield when soldering within an equipment cabinet, such as in a condensing unit.

Caution: Never solder tubing lines that are sealed. Service valves or Schrader ports should be opened before soldering is attempted.

Warning: Hot motors and the hot discharge line from the compressor to the condenser can burn your skin and leave a permanent scar. Care should be used while working around them.

1.5 COLD

Cold can be as harmful as heat. Liquid refrigerant can freeze your skin or eyes instantaneously. Long exposure to cold is also harmful. Working in cold weather can cause frostbite. Wear proper clothing. Dress in several loose layers and wear waterproof work boots. Waterproof boots not only protect your feet from water and cold, but also help to protect you from electric shocks when working in wet weather. Remember, a cold, wet technician will not always make decisions based on logic. Make it a point to stay warm.

1.6 MECHANICAL EQUIPMENT

Rotating equipment can cause serious injuries and property damage. Motors that drive fans, compressors, and pumps are among the most dangerous because they have so much power. Loose clothing should never be worn around rotating machinery. If a shirt sleeve or coat is caught in a motor drive pulley or coupling, it can result in a catastrophic crushing injury or dismemberment. Even a small electric motor can wind a necktie up before it can be shut off (Figure 1-15).

When starting an open motor, stand well to the side of the motor drive mechanism. If the coupling or belt flies off the drive, it will fly outward in the direction of rotation of the motor. All set screws or holding mechanisms must be tight before a motor is started, even if the motor is not connected to a load. Do not leave wrenches near a coupling or pulley. A wrench or nut thrown from a coupling can become a lethal projectile (Figure 1-16).

Figure 1-15. Never wear a necktie or loose clothing when using or working around rotating equipment.

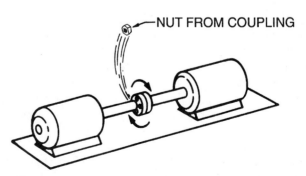

Figure 1-16. Ensure that all nuts are tight on couplings and other components.

8 UNIT 1 GENERAL SAFETY PRACTICES

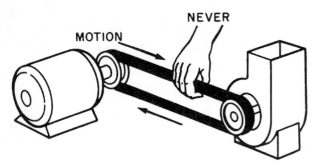

Figure 1-17. Never attempt to stop a motor or other mechanism by gripping the belt.

Figure 1-18. Jewelry can catch on nails or other objects and cause injury.

Warning: When any motor, such as a fan motor, is coasting to a stop, don't try to stop it by interfering with its mechanical components. For example, if you try to stop a motor by gripping the belts, the momentum of the equipment may pull your hand into the pulley and under the belt (see Figure 1-17).

Never wear jewelry while working on a job that requires much movement. A ring may be caught on a nail head, or a watchband or bracelet may be caught on the tailgate of a truck as you jump down (Figure 1-18).

When using a grinder to sharpen tools, remove burrs, or for other reasons, use a face shield (Figure 1-19). Most grinding stones are made for grinding ferrous metals, such as cast iron or steel. However, other stones are made for nonferrous metals such as aluminum, copper, or brass. Use the correct grinding stone for the metal you are grinding. The tool rest should be positioned approximately 1/16 inch from the grinding stone, as shown in Figure 1-20. As the stone wears down, keep the tool rest adjusted to this setting. Don't use a grinding stone on a grinder that turns faster than the stone's rated maximum revolutions per minute (rpm) or the stone may explode (Figure 1-21).

Figure 1-19. Use a face shield when grinding. *Courtesy Jackson Products.*

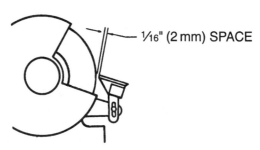

Figure 1-20. Keep the tool rest on a grinder adjusted properly.

1.7 MOVING HEAVY OBJECTS

Heavy objects must be moved from time to time. To avoid injury and equipment damage, take the time to plan each move. Don't just use muscle power. Special tools can help you to move equipment. When equipment must be installed on top of a building, a crane or even a helicopter can be used. Don't take a chance by trying to lift heavy equipment by yourself; get help from another person and use the tools and equipment designed for this purpose. A technician without proper equipment is limited.

When you must lift, use your legs rather than your back and wear an approved back brace belt (Figure 1-22). Available lifting tools include a pry bar, lever truck, refrigerator hand truck, the lift gate on a pickup truck, and a portable dolly (Figure 1-23).

MOVING HEAVY OBJECTS 9

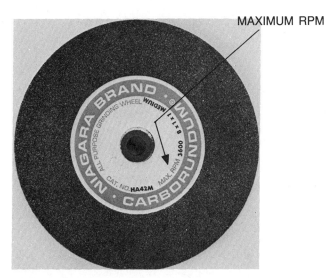

Figure 1-21. When installing a grinding stone, ensure that it is compatible with the grinder. From Jeffus & Johnson, *Welding: Principles and Applications, 3E*, © 1992 by Delmar Publishers, Inc.

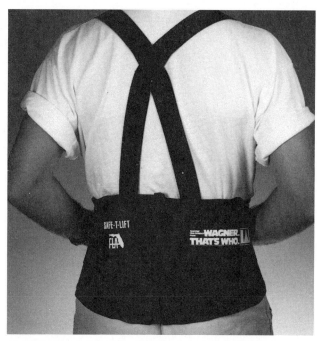

Figure 1-22. Use legs, not back, to lift objects. Keep back straight. Use back brace belt. *Courtesy Wagner Products Corp.*

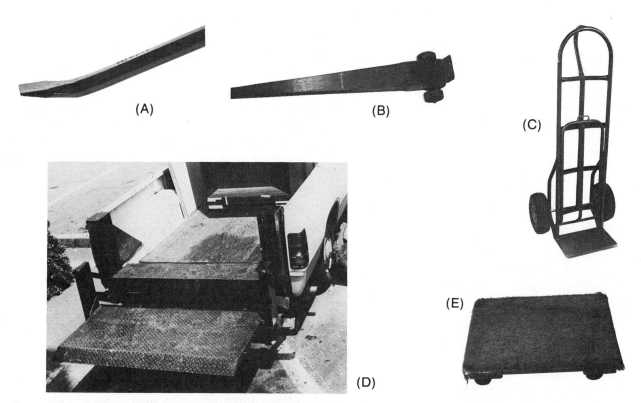

Figure 1-23. (A) Pry bar. (B) Lever truck. (C) Hand truck. (D) Lift gate on a pickup truck. (E) Portable dolly. *Photos by Bill Johnson.*

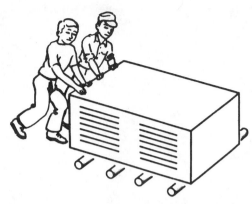

Figure 1-24. Moving equipment using short lengths of pipe as rollers.

When moving large equipment across a carpeted or tiled floor, or across a gravel-coated roof, first lay down some plywood. Keep the plywood in front of the equipment as it is moved along. When equipment has a flat bottom, such as a package air conditioner, short lengths of pipe may be used to move the equipment across a solid floor (Figure 1-24).

1.8 REFRIGERANTS IN YOUR BREATHING SPACE

Refrigerant vapors and many other gases are heavier than air and can displace the oxygen in a closed space. Proper ventilation must be used at all times to prevent suffocation due to the lack of oxygen. If you are in a confined area and the concentration of refrigerant becomes too great, you may not notice it until it is too late. Symptoms may include a dizzy feeling and numbing of the lips and other mucous membranes. If you experience these symptoms, leave the area immediately and get some fresh air.

Proper ventilation should be set up before starting a job. Fans may be used to push or pull fresh air into a confined space where work must be performed. Cross ventilation can help prevent a buildup of fumes (Figure 1-25).

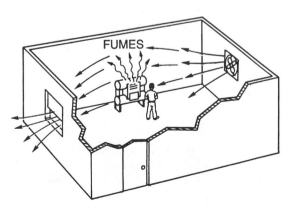

Figure 1-25. Cross ventilation with fresh air will help prevent fumes from accumulating.

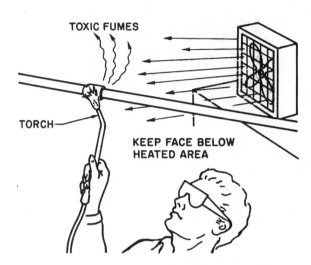

Figure 1-26. Keep your head below the heated area and ensure that the area is well ventilated.

Heated refrigerant vapor is especially toxic. It has a strong odor when exposed to heat or when passing through an open flame, such as a leak in the presence of a fire. If you are soldering in a confined area, keep your head below the rising fumes and provide plenty of ventilation (see Figure 1-26). It is a good idea to perform a leak test prior to doing any soldering in a refrigeration system. Refrigerant leak detectors are discussed later in this book.

1.9 USING CHEMICALS

Chemicals are often used to clean equipment such as air-cooled condensers and evaporators. They are also used for water treatment. These chemicals are normally simple and mild, except for some harsh cleaning products used for water treatment.

Caution: All chemicals should be handled according to the manufacturer's directions. Do not get careless. If you spill chemicals on your skin or splash them in your eyes, follow the manufacturer's directions and go to a doctor. Always read the entire label before starting a job. It's hard to read the first aid treatment for eyes after they've been damaged.

Refrigerant and oil from a motor burnout can be harmful. The contaminated refrigerant and oil may be hazardous to your skin, eyes, and lungs because it contains acid. Keep your distance if a line is subject to rupture or any amount of refrigerant is allowed to escape.

SUMMARY

- The technician must use every precaution when working with pressures, electrical energy, heat, cold, rotating machinery, chemicals, and moving heavy objects.

- Safety situations involving pressure are encountered while working with pressurized systems and vessels.
- Electrical energy is present while troubleshooting energized electrical circuits. Be careful. If possible, turn off electrical power when working on an electrical component. Lock the panel or disconnect box and keep the only key in your possession.
- Heat is encountered while soldering and working near the compressor discharge line.
- Liquid R-22 refrigerant boils at -41°F at atmospheric pressure and will cause frostbite.
- Rotating equipment such as fans and pumps can be dangerous and should be treated with caution.
- When moving heavy equipment, use correct techniques, appropriate tools and equipment, and wear a back brace belt.
- Chemicals are used for cleaning and water treatment and must be handled with care.

REVIEW QUESTIONS

1. Where would a technician encounter freezing temperatures with liquid refrigerant?
2. Should a technician ever refill a refrigerant cylinder painted green?
3. What can happen to an oxygen or nitrogen cylinder if the valve is broken off?
4. Why must nitrogen be used with a regulator when charging a system?
5. What can happen when oil is mixed with oxygen under pressure?
6. Why should technicians not use their hands to stop liquid refrigerant from escaping?
7. To what part of the body does electric shock pose the greatest threat?
8. What two types of injury are caused by electrical energy?
9. How can the technician prevent electric shocks?
10. How can the technician prevent damage to adjacent surfaces while soldering?
11. What safety precaution should be taken before starting an open motor?
12. What is the third wire used for on a portable electric drill?
13. How can heavy equipment be moved across a rooftop?
14. What precautions should be taken before using chemicals to clean an air-cooled condenser?
15. How can refrigerant from a motor burnout be harmful?

2 HEAT, PRESSURE, MATTER, AND ENERGY

OBJECTIVES

Upon completion of this unit, you should be able to

- Define temperature.
- Describe molecular motion at absolute zero.
- Make conversions between Fahrenheit and Celsius.
- Define the British thermal unit.
- Describe heat flow between substances of different temperatures.
- Explain the transfer of heat via conduction, convection, and radiation.
- Discuss sensible heat, latent heat, and specific heat.
- State the value of atmospheric pressure at sea level and explain why atmospheric pressure varies at different elevations.
- Explain the terms *psig* and *psia* as they apply to pressure gages.
- Define matter.
- List three states in which matter is commonly found.
- Define density.
- Discuss Boyle's Law.
- State Charles' Law.
- Discuss Dalton's Law as it relates to the pressure of different gases.
- Define specific gravity and specific volume.
- Describe work.
- Define horsepower.
- Convert horsepower to watts.
- Convert watts to British thermal units.

2.1 TEMPERATURE

Temperature can be thought of as a measurement of the level of heat, and *heat* can be thought of as energy in the form of molecules in motion. The higher the heat level on the temperature scale, the faster the molecules in a substance move. For example, imagine a pot of water on a stove. As the water is heated, the molecules begin to move more quickly and to collide at an increasingly higher rate, until eventually, the water boils. The starting point of the temperature scale is the point at which molecules begin to move. It is thought to be -460°F, which is known as *absolute zero*. Other reference points on the temperature scale include the freezing point of water, at 32°F, and the boiling point of water, at 212°F.

Glass-stem thermometers are often used to measure temperatures. They operate on the principle that when the substance in the bulb is heated, it will expand and rise in the tube. Mercury and alcohol are commonly used for this application.

We must qualify the statement that water boils at 212°F. Pure water boils at precisely 212°F at sea level when the temperature of the atmosphere is 70°F. That is to say, the boiling point of water (and other fluids) is related to both the altitude and the surrounding temperature. The statement that water boils at 212°F at sea level when the atmosphere is 70°F is important because these are *standard conditions* that will be applied to actual practice. The relationship between pressure and temperature is covered later in this unit.

Fahrenheit is a system of temperature measurement used in the English measurement system. The United States is one of the few countries in the world that uses this system. Celsius is the temperature scale used in the metric measurement system. It will be increasingly important as world trade opportunities develop for the businesses and individuals in the United States to change to the metric system of measurement.

Figure 2-1 illustrates a thermometer with some important Fahrenheit and Celsius equivalent temperatures.

Up to this point, we have discussed temperature in everyday terms. It is equally important for you to understand the more technical temperature scale known as the *absolute temperature scale*. Performance ratings of equipment are established in terms of absolute temperature. The Fahrenheit absolute scale is called the *Rankine (R) scale* (named for its inventor, W. J. M. Rankine), and the Celsius absolute scale is known as the *Kelvin (K) scale* (named for the scientist, Lord Kelvin). Absolute temperature scales begin where molecular motion starts; they use zero as the starting point. For example, zero on the Fahrenheit absolute scale is called *absolute zero* or *0°R*. Similarly, zero on the Celsius absolute scale is called *absolute zero* or *0 K*.

Note: The degree sign is not used when expressing temperatures on the Kelvin scale.

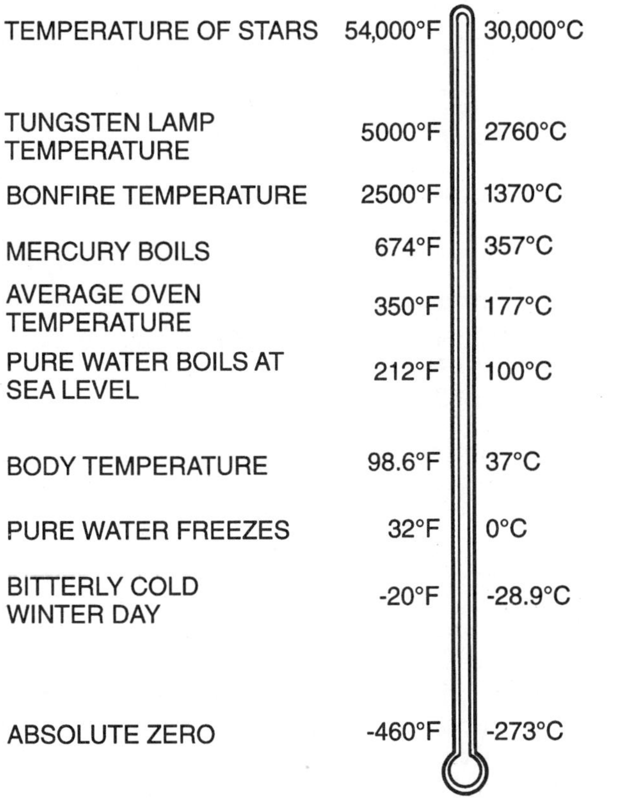

TEMPERATURE

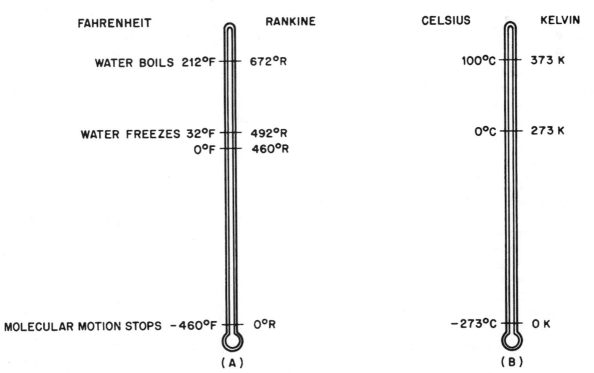

Figure 2-1. Equivalent temperatures on the Fahrenheit and Celsius scales.

The Fahrenheit/Celsius and Rankine/Kelvin scales are used interchangeably to describe air conditioning equipment and processes. Figure 2-2 shows the relationship between these four scales.

Temperature Conversions – Fahrenheit and Celsius

You may find it necessary to convert specific temperatures from Fahrenheit to Celsius or vice versa. This conversion can be done using either conversion tables or the following formulas.

$$°C = \frac{°F - 32°}{1.8} \qquad °F = 1.8\,(°C) + 32°$$

or

$$°C = \frac{5}{9}\,(°F - 32°) \qquad °F = \frac{9}{5}\,(°C + 32°)$$

To convert a room temperature of 75°F to Celsius:

$$°C = \frac{75° - 32°}{1.8} = 23.9$$

75°F = 23.9°C

To convert a room temperature of 25°C to Fahrenheit:

°F = 1.8 × 25° + 32° = 77°

25°C = 77°F

Figure 2-2. (A) Fahrenheit and Rankine thermometer. (B) Celsius and Kelvin thermometer.

ROOM TEMPERATURE (70°F)

10-POUND TANK OF WATER
IS HEATED TO 200°F
(ABOUT 1 GALLON)

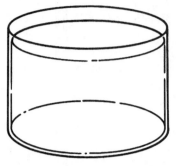

100,000-POUND TANK OF WATER
IS HEATED TO 175°F
(ABOUT 12,000 GALLONS)

Figure 2-3. The smaller tank will cool to room temperature first because it contains a smaller quantity of heat.

2.2 HEAT TRANSFER

A law of thermodynamics states that *heat can neither be created nor destroyed*. This means that all of the heat that the world experiences is not created but is energy available in one form or another and then converted to usable heat. Heat can also be accounted for when it is transferred from one substance to another.

As indicated previously, temperature describes the level of heat with reference to no heat. The term used to describe the quantity of heat is known as the *British thermal unit (Btu)*.

A Btu is defined as the amount of heat required to raise the temperature of one pound of water 1°F. For example, when one pound of water (about one pint) is heated from 68°F to 69°F, one Btu of heat energy is absorbed into the water. To actually measure how much heat is absorbed in this process, we need a special instrument known as a *calorimeter*. Notice the similarity to the word *calorie*, the term for energy as it relates to food.

When there is a temperature difference between two substances, heat transfer will take place. The rapidly moving molecules in the warmer substance give up some of their energy to the slower molecules in the cooler substance. The warmer substance cools because the molecules have slowed. The cooler substance becomes warmer because the molecules are moving faster.

The following example illustrates the difference in the quantity of heat compared to the level of heat (Figure 2-3). One tank of water weighing 10 pounds (slightly more than one gallon) is heated to a temperature level of 200°F. A second tank of water weighing 100,000 pounds (slightly more than 12,000 gallons) is heated to 175°F. The 10-pound tank will cool to room temperature much faster than the 100,000-pound tank. Even though the initial temperature of the 100,000-pound tank is 25°F lower, the cool-down time is much longer due to the larger quantity of water.

In practical terms, each piece of heating or cooling equipment is rated according to the amount of heat it will produce or absorb. If the equipment had no such rating, it would be difficult for a buyer to intelligently choose the correct appliance.

In the metric or SI (Systems International) system of measurement, the term *joule (J)* is used to express the quantity of heat. Because a joule is very small, metric units of heat are usually expressed in *kilojoules (kJ)*. The prefix *kilo* means 1000, so one kilojoule is equal to 1000 joules. One Btu equals 1.055 kJ.

The term *gram (g)* is used to express weight in the metric system. Again, this is a very small quantity, so the term *kilogram (kg)* is often used. One pound equals 0.45359 kg. The amount of heat required to raise the temperature of one kilogram of water 1°C is equal to 4.187 kJ.

Cold is a comparative term used to describe lower temperature levels. Because all heat has a positive value in relation to no heat, cold is not a true value. When a person says it is "cold" outside, it is an expression of the difference between the indoor and outdoor temperatures or the actual outdoor temperature and the expected temperature for that time of year.

Conduction

Conduction is the transfer of energy from one molecule to another. As a molecule moves faster, it causes others to do the same. For example, if one end of a copper rod is placed in a flame, the other end may get too hot to handle. The heat travels up the rod from molecule to molecule.

Conduction is used in many heat transfer applications. For example, heat is transferred via conduction from a hot electric stove burner to a tea kettle. It is then transferred from the metal kettle into the water.

Heat does not conduct at the same rate in all materials. For example, copper is an excellent conductor of heat, while glass is a very poor conductor of heat.

Touching a wooden fence post or another piece of wood on a cold morning does not give the same sensation as touching a car fender or another piece of steel. The steel feels much colder. Actually, the steel is not colder, it just conducts heat out of the hand faster.

The different rates at which various materials conduct heat have an interesting similarity to the conduction of electricity. Frequently, substances that are poor conductors of heat are also poor conductors of electricity. For example, copper is one of the best conductors of electricity and heat, and glass is one of the poorest conductors of both. Glass is actually used as an insulator of electrical current flow.

Convection

Convection is the transfer of heat from one location to another. When heat is moved, it is normally transferred via a fluid, such as air or water. Many large buildings have a central heating and/or cooling plant where water is heated or cooled and pumped throughout the building to the final conditioned space.

Warm air rises and cool air falls, producing natural convection currents. It is best if air conditioning systems are designed to take advantage of this process. For example, in a building in a southern climate where the cooling load is greater than the heating load, it is better if the air distribution registers are high on the wall or in the ceiling so that the cold air will drop, replacing the warm air that rises (Figure 2-4). The opposite is true of a building in a northern climate, where the greatest load is usually the heating load.

Radiation

Heat transferred via *radiation* travels through the air, the atmosphere, or space without heating the area in between. The heat is absorbed by any solid objects it encounters. Heat from the sun is an example of radiant heat. The sun is approximately 93,000,000 miles from the Earth's surface, yet its intensity can be felt. The Earth does not

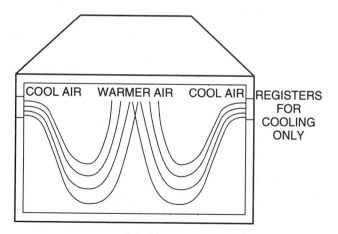

Figure 2-4. Cool air registers are located high on the wall. The cold air will fall, replacing the warmer air that rises.

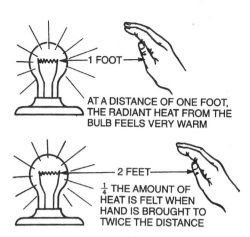

Figure 2-5. The intensity of the heat diminishes by the square of the distance.

experience the total heat from the sun because the intensity of radiant heat diminishes by the square of the distance traveled. This means that every time the distance is doubled, the heat intensity decreases by one-fourth. For example, if you hold your hand close to a light bulb, it feels quite warm, but if you move your hand twice as far away, you feel only one-quarter of the heat intensity (Figure 2-5). Keep in mind that, because of the square-of-the-distance explanation, radiant heat does not transfer the full value of the heat. If it did, the Earth would be as hot as the sun.

2.3 SENSIBLE HEAT

A change in heat level can be measured on a thermometer when the temperature of a substance has been changed. When a change in heat level results in a measurable change in temperature, it is known as *sensible heat*. If sensible heat is added to a substance, it is referred to as *sensible heat gain*, and if sensible heat is taken away from a substance, it is known as *sensible heat loss*.

2.4 LATENT HEAT

Heat can be added to (or removed from) a substance without registering a change on a thermometer. When this occurs, the temperature of the substance does not change. This is called *latent heat* or *hidden heat*. An example of this is when heat is added to water while it is boiling in an open container. Adding more heat only makes it boil faster, it does not raise the temperature. When heat is added to a substance and the temperature remains the same, the substance can be changing from a solid to a liquid or from a liquid to a vapor. When heat is removed from a substance and the temperature remains the same, the substance may be changing from a vapor to a liquid or from a liquid to a solid. When the addition of heat to a substance causes a change in state, it is called the *latent heat of melting* (solid to liquid) or the *latent heat of vaporization* (liquid to vapor). When the removal

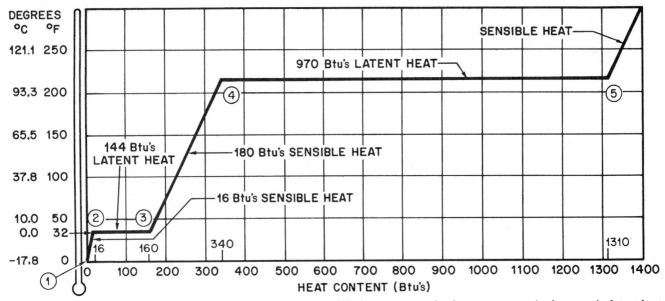

Figure 2-6. Heat/temperature graph for water. An increase in sensible heat causes a rise in temperature. An increase in latent heat causes a change in state.

of heat from a substance causes a change in state, it is called the *latent heat of condensation* (liquid to vapor) or the *latent heat of fusion* (liquid to solid).

The following example describes the sensible heat and latent heat characteristics of water. Examine the graph in Figure 2-6. Temperature is plotted on the left margin, and heat content is plotted along the bottom. Note that as heat is added, the temperature rises, except during the latent heat process, when the temperature does not change.

The following statements refer to the graph in Figure 2-6.

1. Water is in the form of ice at point 1. Note that point 1 is *not* absolute 0, it is 0°F.
2. Heat added from point 1 to point 2 is sensible heat. This is a registered rise in temperature. Note that it only takes 0.5 Btu of heat to raise the temperature of one pound of ice 1°F.
3. When point 2 is reached, the ice is saturated with heat. This means that if more heat is added, it will be latent heat and will melt the ice but not raise the temperature. Adding 144 Btu's of heat will change the pound of ice to a pound of water. Removing any heat will cool the ice below 32°F.
4. When point 3 is reached, the substance is now water and is known as a *saturated liquid*. Adding more heat causes a rise in temperature (this is sensible heat). Removal of any heat at point 3 would result in some of the water changing back to ice (this would be latent heat).
5. Heat added from point 3 to point 4 is sensible heat. When point 4 is reached, 180 Btu's of heat will have been added: 1 Btu/lb./°F.
6. Point 4 represents another saturated point. The water is saturated with heat to the point where the removal of any heat causes the liquid to cool off below the boiling point. Heat added is identified as latent heat and causes the water to boil and start changing into a vapor (steam). Adding 970 Btu's makes the pound of liquid boil to point 5 and become a vapor.
7. Point 5 represents another saturated point. The water is now in the vapor state. Heat removed would be latent heat and would change some of the vapor back to a liquid. This is called *condensing the vapor*. Any heat added at point 5 is sensible heat; it raises the vapor temperature above the boiling point. Heating the vapor above the boiling point is called *superheat*. Superheat will be important in future studies. Note that in the vapor state, it only takes 0.5 Btu to heat the water vapor (steam) 1°F.

Warning: Exercise caution if examining these principles. The water and steam are well above body temperature and can cause burns.

2.5 SPECIFIC HEAT

Every substance responds differently to heat. For example, when 1 Btu of heat energy is added to a pound of water, it increases the temperature by 1°F. However, this only holds true for water.

Specific heat is the amount of heat necessary to raise the temperature of one pound of a substance 1°F. Every substance has a different specific heat. The specific heat of water is 1 Btu/lb./°F. See Figure 2-7 for the specific heat values of other substances.

PRESSURE

SUBSTANCE	SPECIFIC HEAT Btu/lb/°F	SUBSTANCE	SPECIFIC HEAT Btu/lb/°F
ALUMINUM	0.224	BEETS	0.90
BRICK	0.22	CUCUMBERS	0.97
CONCRETE	0.156	SPINACH	0.94
COPPER	0.092	BEEF, FRESH	
ICE	0.504	LEAN	0.77
IRON	0.129	FISH	0.76
MARBLE	0.21	PORK, FRESH	0.68
STEEL	0.116	SHRIMP	0.83
WATER	1.00	EGGS	0.76
SEA WATER	0.94	FLOUR	0.38
AIR	0.24 (AVERAGE)		

Figure 2-7. Specific heat table.

Specific heat also varies depending on the state of the substance. For example, when heat energy is added to water while it is in a different state (either ice or steam) it requires only 0.5 Btu to raise the temperature by 1°F. In these states, water heats at twice the rate.

2.6 PRESSURE

Pressure may be defined as force per unit of area. This is normally expressed in *pounds per square inch (psi)*. Simply stated, when a one-pound weight rests on an area of one square inch (1 in.^{2}), the pressure exerted downward is one pound per square inch (psi). Similarly, when a 100-pound weight rests on a 1 in.^{2} area, 100 psi of pressure is exerted (Figure 2-8).

When you swim below the surface of water, you feel a pressure pushing inward on your body. This pressure is caused by the weight of the water. A different sensation is felt when flying in an airplane without a pressurized cabin. Your body is subjected to less pressure instead of more, and you still feel uncomfortable. When swimming, the weight of the water pushes in; in an airplane, the pressure inside your body pushes out.

Water weighs 62.4 pounds per cubic foot (lb./ft.^{3}). A cubic foot (7.48 gallons) exerts a downward pressure of 62.4 lbs./sq. ft. when it is in its actual cube shape. How much weight is then resting on 1 in.^{2}? The bottom of the cube has an area of 144 in.^{2} (12 inches × 12 inches) sharing the weight. Each square inch has a total pressure of 0.433 lb. (62.4 ÷ 144) resting on it. Thus, the pressure at the bottom of the cube is 0.433 psi (Figure 2-9).

Atmospheric Pressure

The Earth's atmosphere is like an ocean of air instead of water. It has weight and exerts pressure. The Earth's surface can be thought of as being at the bottom of this ocean of air. Different locations are at different depths. There are sea level locations such as Miami, Florida, and mountainous locations such as Denver, Colorado. The atmospheric pressures at these two locations are different. For now, we will assume that we live at the bottom of this ocean of air (sea level).

As mentioned previously, the Earth's atmosphere exerts a pressure of 14.696 psi at sea level when the surrounding temperature is 70°F.

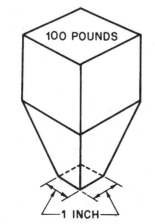

Figure 2-8. Both weights are resting on a 1 in.^{2} surface. One weight exerts a pressure of 1 psi, the other exerts a pressure of 100 psi.

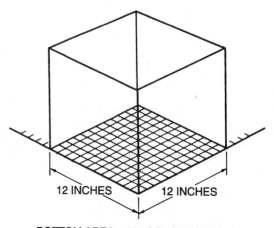

Figure 2-9. One cubic foot of water exerts a downward pressure of 0.433 psi. (62.4 lbs./sq. ft. ÷ 144 = 0.433 psi)

18 UNIT 2 HEAT, PRESSURE, MATTER, AND ENERGY

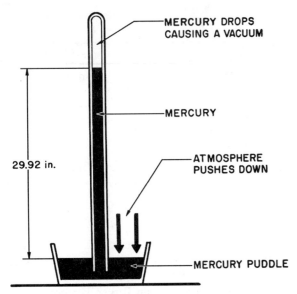

Figure 2-10. Mercury (Hg) barometer.

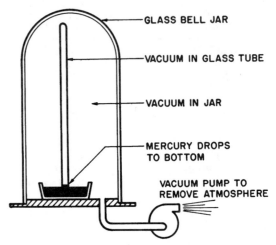

Figure 2-11. When the mercury barometer is placed in a closed bell jar and the atmosphere evacuated, the column of mercury drops to the level of the puddle.

Atmospheric pressure can be measured with an instrument called a *barometer*. A barometer is a glass tube about 36 inches long that is closed on one end and filled with mercury (Hg). It is then inserted open-side down into a puddle of mercury and held upright. The mercury will try to run down into the puddle, but it will not all run out. The atmosphere pushes down on the puddle, and a vacuum is formed in the top of the tube. The mercury in the tube will fall to 29.92 inches at sea level when the surrounding atmospheric temperature is 70°F (Figure 2-10). If the barometer is moved to a higher elevation, such as on a mountain, the mercury column will start to fall. It will fall about one inch per 1000 feet of altitude. When the barometer is at standard conditions and the mercury drops, it is called a *low-pressure system*.

Warning: Mercury is an extremely toxic substance. When using a mercury barometer, do not touch the mercury.

If the barometer is placed inside a closed bell jar and the atmosphere evacuated, the mercury column falls to the same level as the puddle in the bottom (Figure 2-11). When the atmosphere is allowed back into the jar, the mercury again rises because a vacuum exists above the mercury column in the tube.

The mercury in the column has weight and counteracts the atmospheric pressure of 14.696 psi at standard conditions. Therefore, a pressure of 14.696 psi is equal to the weight of a column of mercury that is 29.92 inches high. The term *inches of mercury* thus becomes an expression of pressure and can be converted to pounds per square inch. The conversion factor is 1 psi = 29.92 ÷ 14.696 or 2.036 in. Hg. (This is often rounded off to 2.)

Another type of barometer is the *aneroid barometer* (Figure 2-12). This is a more practical instrument for field use.

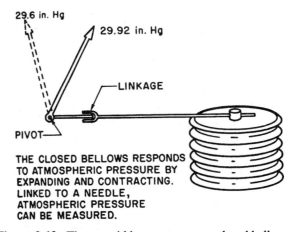

Figure 2-12. The aneroid barometer uses a closed bellows that expands and contracts with atmospheric pressure changes.

Pressure Gages

Measuring pressures in a closed system requires different instruments. The *Bourdon tube* is a device that can be linked to a needle to measure pressures above and below atmospheric pressure (Figure 2-13). A common device used in the air conditioning industry to take readings in the field or shop is the *gage manifold* (Figure 2-14). It is a combination of a low-pressure gage (called the *low-side gage*) and a high-pressure gage (called the *high-side gage*). The gage on the left reads pressures above and below atmospheric pressure. It is called a *compound gage*. The gage on the right reads pressures up to 500 psi and is called the *high-pressure gage*.

These gages read 0 psi when opened to the atmosphere. If they do not, they should be calibrated to 0 psi. These gages are designed to read pounds-per-square-inch gage pressure (psig). Atmospheric pressure is used as the starting or reference point. To obtain the absolute pressure, you must add the atmospheric pressure to the gage

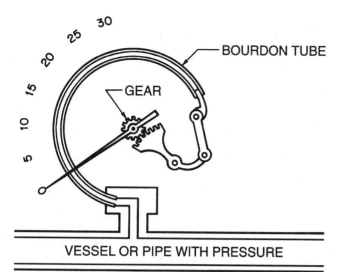

Figure 2-13. The Bourdon tube is made of a thin piece of metal, such as brass. It is closed on one end. When the pressure increases, the tube tends to straighten out. When attached to a needle linkage, pressure changes are indicated.

Figure 2-14. The gage on the left is called a *compound gage*. It reads below-atmospheric pressures in inches of mercury (in. Hg) and above-atmospheric pressures in psig. The gage on the right is called the *high-pressure gage*. It reads pressures up to 500 psi. *Photo by Bill Johnson.*

reading. For example, to convert a gage reading of 50 psig to absolute pressure, you must add the atmospheric pressure of 14.696 psi to the gage reading. Fifty psig + 14.696 = 64.696 psia (pounds-per-square-inch absolute). (This can be rounded off to 65.)

Warning: Pressures that are above or below atmospheric pressure can cause bodily injury. A vacuum can cause a blood blister on the skin. Pressures above atmospheric pressure can pierce the skin or inflict damage through flying objects.

Pressure Measured in Metric Terms

Like temperature, pressure can also be expressed in metric terms. While several terms have been used in the past, the present standard metric expression for pressure is newtons per square meter (N/m^2). Pressure in the English measurement system is expressed in pounds per square inch (psi). It is difficult to compare pounds per square inch with newtons per square meter. To make this comparison easier, newtons per square meter has been given the name *pascal* in honor of the scientist and mathematician Blaise Pascal. The standard metric term for pressure is the *kilopascal (kPa)*. One psi is equal to 6890 pascals, or 6.89 kPa. To convert psi to kPa, simply multiply the number of psi by 6.89.

2.7 MATTER

Matter is commonly explained as a substance that occupies space and has weight. The weight comes from the Earth's gravitational pull. Matter is made up of atoms. Atoms are the smallest unit of an element. Atoms of one element may be combined chemically with those of another element to form molecules of a new substance. When molecules are formed, they cannot be broken down any further without changing the chemical content of the substance.

Matter exists in three states: *solids*, *liquids*, and *gases*. Both heat content and pressure have an effect on the state of matter.

Water is made up of molecules containing atoms of hydrogen (H) and oxygen (O). There are two atoms of hydrogen and one atom of oxygen in each molecule of water. The chemical expression for water is H_2O.

Water in the solid state is known as *ice*. It exerts all of its force downward (Figure 2-15). The molecules in ice are highly attracted to one another.

When water is heated above the freezing point, it begins to change into a liquid. The molecular activity is higher, and the water molecules have less attraction for each other. Water in the liquid state exerts a pressure both outward and downward and seeks its own level (Figure 2-16).

When water is heated above the boiling point, 212°F at standard conditions, it becomes a vapor. In the vapor state, the molecules have even less attraction for each other and are said to travel at random. The vapor exerts pressure more or less in all directions (Figure 2-17).

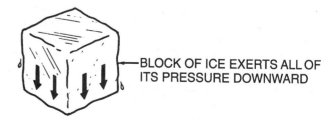

Figure 2-15. Solids exert all their pressure downward.

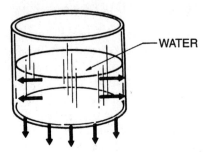

Figure 2-16. Water exerts pressure both outward and downward.

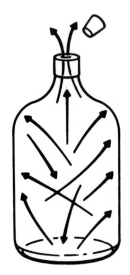

Figure 2-17. Gas molecules travel at random and exert pressure in all directions.

2.8 MASS

Mass is a term that is used along with weight to describe matter. The universe is made up of matter that has weight or mass and takes up space. All solid matter has mass. A liquid such as water is said to have mass. The air in the atmosphere has weight or mass. When the atmosphere is evacuated from a container or vessel, the mass is removed and a vacuum is created.

2.9 DENSITY

The *density* of a substance describes its mass-to-volume relationship. The mass contained in a particular volume is the density of that substance. In the English system of units, volume is measured in cubic feet. It is sometimes advantageous to compare different substances according to weight per unit volume. Water, for example, has a density of 62.4 lbs./ft.^{3}. Wood floats on water because the density (weight per volume) of wood is less than the density of water. In other words, it weighs less per cubic foot. Iron, on the other hand, sinks because it is denser than water. Figure 2-18 lists the densities of some common substances.

SUBSTANCE	DENSITY lb/ft^3	SPECIFIC GRAVITY
ALUMINUM	168	2.7
BRASS	536	8.7
COPPER	555	8.92
GOLD	1204	19.3
ICE	57	0.92
IRON	490	7.86
LITHIUM	33	0.53
TUNGSTEN	1186	19.0
MERCURY	845	13.54
WATER	62.4	1

Figure 2-18. Table of density and specific gravity.

2.10 SPECIFIC GRAVITY

Specific gravity compares the densities of various substances. The specific gravity of water is 1.0. The specific gravity of iron is 7.86. This means that a volume of iron is 7.86 times heavier than an equal volume of water. Since water weighs 62.4 lbs./ft.^{3}, a cubic foot of iron would weigh 62.4 × 7.86 or 490 pounds. Aluminum has a specific gravity of 2.7, so it has a density or weight per cubic foot of 2.7 × 62.4 = 168 pounds.

2.11 SPECIFIC VOLUME

Specific volume compares the densities of gases. It indicates the space (volume) a weight of gas will occupy. One pound of clean, dry air has a volume of 13.33 ft.^{3} at standard conditions. Hydrogen has a density of 179 ft.^{3}/lb. under the same conditions. Because there are more cubic feet of hydrogen per pound, it is lighter than air. Although both are gases, the hydrogen has a tendency to rise when mixed with air.

The specific volumes of various gases is valuable information that enables the engineer to choose the size of the compressor or vapor pump to do a particular job. The specific volumes for vapors vary according to pressure. An example is R-22, which is a common refrigerant used in residential air conditioning units. At 3 psig, about 2.5 ft.^{3} of gas must be pumped to move one pound of refrigerant. At the standard design condition of 70 psig, only 0.48 ft.^{3} of gas needs to be pumped to move a pound of the same refrigerant. The values of specific volume for various pressures can be found in the properties of liquid and saturated vapor tables in engineering manuals for any refrigerant.

2.12 GAS LAWS

It is necessary to have a working knowledge of gases and how they respond to pressure and temperature changes. Several scientific formulas known as the *gas laws* will help you to understand the reaction of gases and the pressure/temperature relationships in various parts of a refrigeration system.

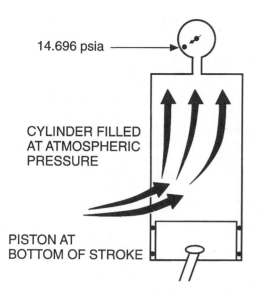 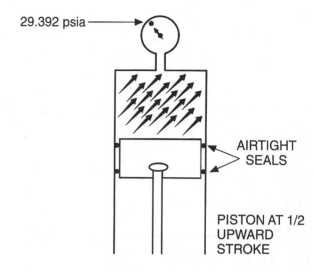

Figure 2-19. The absolute pressure in a cylinder doubles when the volume is reduced by half.

Boyle's Law

In the early 1600s, Robert Boyle developed what has come to be known as *Boyle's Law*. He discovered that when pressure is applied to a volume of air that is contained, the volume of air becomes smaller and the pressure greater. Boyle's Law states that *the volume of a gas varies inversely with the absolute pressure, provided the temperature remains constant*. See Figure 2-19. If the enclosed cylinder shown is filled with air, and the piston moves halfway up the cylinder, the pressure of the air will double.

The formula for Boyle's Law is:

$$P_1 \times V_1 = P_2 \times V_2$$

where: P_1 = original absolute pressure

V_1 = original volume

P_2 = new pressure

V_2 = new volume

For example, if the original pressure was 40 psia and the original volume was 30 in.³, what would the new volume be if the pressure is increased to 50 psia?

First, the formula must be rearranged to solve for the new volume.

$$V_2 = \frac{P_1 \times V_1}{P_2}$$

$$V_2 = \frac{40 \times 30}{50}$$

$$V_2 = 24 \text{ in.}^3$$

Charles' Law

In the 1800s, a French scientist named Jacques Charles made discoveries regarding the effect of temperature on gases. Charles' Law states that *at a constant pressure, the volume of a gas varies directly with the absolute temperature, and at a constant volume, the pressure of a gas varies directly with the absolute temperature*. Stated in a different form, when an unconfined gas is heated, it will expand, and the volume will vary directly with the absolute temperature. If a gas is confined in a container that will not expand as it is heated, the pressure will vary directly with the absolute temperature.

Each part of this law can also be stated algebraically. The first formula pertains to volume and temperature:

$$\frac{V_1}{T_1} = \frac{V_2}{T_2}$$

where: V_1 = original volume

V_2 = new volume

T_1 = original temperature

T_2 = new temperature

For example, if 2000 ft.³ of air is passed through an air conditioning unit and its temperature drops from 75°F to 55°F, what is the volume of the air leaving the unit? See Figure 2-20.

V_1 = 2000 ft.³

T_1 = 75°F + 460° = 535°R (absolute)

V_2 = unknown

T_2 = 55°F + 460° = 515°R

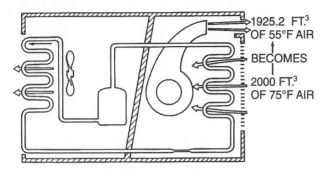

Figure 2-20. Air contracts when cooled.

First, we must rearrange the formula to solve for V_2.

$$V_2 = \frac{V_1 \times T_2}{T_1}$$

$$V_2 = \frac{2000 \text{ ft.}^3 \times 515°R}{535°R}$$

$$V_2 = 1925.2 \text{ ft.}^3$$

The second formula pertains to pressure and temperature:

$$\frac{P_1}{T_1} = \frac{P_2}{T_2}$$

where: P_1 = original pressure

T_1 = original temperature

P_2 = new pressure

T_2 = new temperature

For example, if a large natural gas tank holding 500,000 ft.^{3} of gas is stored at 70°F in the spring and the temperature rises to 95°F in the summer, what would the pressure be if the original pressure was 25 psig?

P_1 = 25 psig + 14.696 (atmospheric pressure) or 39.696 psia

T_1 = 70°F + 460°R or 530°R (absolute)

P_2 = unknown

T_2 = 95°F + 460° or 555°R (absolute)

Again, the formula must be rearranged so that the unknown is on one side of the equation by itself.

$$P_2 = \frac{P_1 \times T_2}{T_1}$$

$$P_2 = \frac{39.696 \text{ psia} \times 555°R}{530°R}$$

$$P_2 = 41.57 \text{ psia} - 14.696 = 26.87 \text{ psig}$$

General Law of Perfect Gas

Under ordinary conditions, when a gas is compressed, some heat is transferred to the gas from the mechanical compression, and when gas is expanded, heat is given up. Because of this, Boyle's Law cannot be directly applied to practical use because it requires that the temperature remain constant. However, Boyle's Law, when combined with Charles' Law, provides an accurate method of calculating changes in pressure, temperature, and volume.

A general gas law, often called the *General Law of Perfect Gas*, is a combination of Boyle's and Charles' Laws. The formula for this law can be stated as follows:

$$\frac{P_1 \times V_1}{T_1} = \frac{P_2 \times V_2}{T_2}$$

where: P_1 = original pressure

V_1 = original volume

T_1 = original temperature

P_2 = new pressure

V_2 = new volume

T_2 = new temperature

For example, 20 ft.^{3} of gas is being stored in a container at 100°F and a pressure of 50 psig. This container is connected by a pipe to another container that will hold 30 ft.^{3} (a total of 50 ft.^{3}). The gas is allowed to equalize between the two containers, then the temperature of the gas is lowered to 80°F. What is the pressure in the combined containers?

P_1 = 50 psig + 14.696 or 64.696

V_1 = 20 ft.^{3}

T_1 = 100° + 560°R

P_2 = unknown

V_2 = 50 ft.^{3}

T_2 = 80°F + 460° or 540°R

The formula must be rearranged so that P_2 is on one side of the equation by itself.

$$P_2 = \frac{P_1 \times V_1 \times T_2}{T_1 \times V_2}$$

$$P_2 = \frac{64.696 \times 20 \times 540}{560 \times 50}$$

$$P_2 = 24.95 - 14.696 = 10.26 \text{ psig}$$

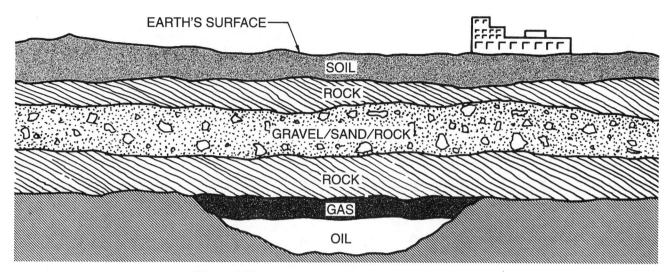

Figure 2-21. Gas and oil deposits settle into depressions.

Dalton's Law

In the early 1800s, John Dalton made the discovery that the atmosphere is made up of several different gases. He found that each gas created its own pressure and that the total pressure was the sum of the individual pressures. Dalton's Law states that *the total pressure of a confined mixture of gases is the sum of the pressures of each of the gases in the mixture.* For example, when nitrogen and oxygen are placed in a closed container, the pressure on the container will be the pressure of the nitrogen as if it were in the container by itself plus the pressure of the oxygen as if it were in the container by itself.

2.13 ENERGY

Energy in the form of electricity drives the pump, fan, and compressor motors in air conditioning systems. Most of the energy we use to produce the heat for generating electricity comes from the fossil fuels: natural gas, oil, and coal. These fossil fuels are a result of the decomposition of plant and animal growth over thousands of years. This decayed matter is in various states, such as gas, oil, or coal, depending on the pressure and temperature conditions it was subjected to in the past (Figure 2-21).

We indicated previously that temperature is a measure of the level of heat and that heat is a form of energy because of the motion of molecules. Because molecular motion does not stop until -460°F, energy is still available in a substance even at very low temperatures. For example, if two substances at very low temperatures are moved close together, heat will transfer from the warmer substance to the cooler substance. In Figure 2-22, a substance at -200°F is placed next to a substance at -350°F. As discussed earlier, the warmer substance gives up heat energy to the cooler substance.

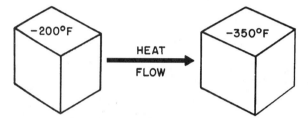

Figure 2-22. Heat energy is still available even at very low temperatures and will transfer from the warmer substance to the cooler substance.

Heat energy is converted to electrical energy through the use of magnetism. The process of producing electrical energy will be discussed in detail in a later unit.

Electrical energy must be transferred from the utility company to the consumer. This electricity is transferred through electrical lines across the country through a meter to the residence, business, or industry.

Energy Used as Work

Energy purchased from electrical utilities is known as *electric power. Power* is the rate of doing work. *Work* can be explained as a force moving an object in the direction of the force. It is expressed by the following formula:

$$\text{work} = \text{force} \times \text{distance}$$

For example, when a 150 pound man climbs a flight of stairs 100 feet high (about the height of a 10 story building), he performs work. The amount of work is equivalent to the amount of work necessary to lift this man the same height.

$$\text{work} = 150 \text{ lbs.} \times 100 \text{ ft.}$$

$$\text{work} = 15{,}000 \text{ ft.-lbs.}$$

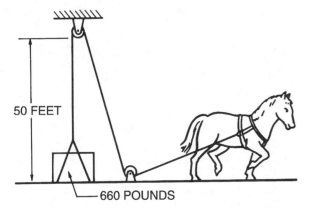

Figure 2-23. When a horse lifts 660 pounds a height of 50 feet in one minute, it has done the equivalent of 33,000 ft.-lbs. of work in one minute or one horsepower (hp).

Notice that no time limit has been included. This task can be accomplished by a healthy person in a few minutes. But if it were to be accomplished by a machine such as an elevator, we would need to know how much time it would take. The quicker the task is completed, the more power it will take.

2.14 HORSEPOWER

Power is the rate of doing work. An expression of power is *horsepower (hp)*. Many years ago, it was determined that an average horse could lift the equivalent of 33,000 pounds a height of one foot in one minute, which is the same as 33,000 ft.-lbs./min. or 1 hp (Figure 2-23). This is the same amount of work as lifting 330 pounds a height of 100 feet in one minute or lifting 660 pounds 50 feet in one minute. As a point of reference, the fan motor in the average air conditioning unit can be rated at ½ hp.

When horsepower is related to the man climbing the stairs, the man would have to climb the 100 feet in less than 30 seconds to equal one horsepower. A ½ hp motor could lift the 150 pound man 100 feet in one minute if only the man were lifted. The reason is that 15,000 ft.-lbs. of work is required.

2.15 ELECTRICAL POWER – THE WATT

The unit of measurement for electrical power is the *watt (W)*. When converted to electrical energy, 1 hp = 746 W; that is, when 746 W of electrical energy is properly used, the equivalent of 1 hp of work has been accomplished. 1000 watts equals 1 kW. 1 kW equals 3413 Btu's. Utility companies measure electricity by the kilowatt-hour (kWh).

SUMMARY

- Four temperature scales are Fahrenheit, Celsius, Fahrenheit absolute (Rankine), and Celsius absolute (Kelvin).
- Molecules in matter are constantly moving. The higher the temperature, the faster they move.
- The British thermal unit (Btu) describes the quantity of heat in a substance. One Btu is equal to the amount of heat necessary to raise the temperature of one pound of water 1°F.
- The transfer of heat through conduction is the transfer of heat from molecule to molecule. As molecules in a substance move faster and with more energy, they cause others near them to do the same.
- The transfer of heat through convection is the actual movement of heat in a fluid (vapor or liquid state) from one place to another.
- Radiant heat is a form of energy transmitted through a medium, such as air, without heating it. Solid objects absorb the energy, become heated, and again transfer the heat to the air.
- Sensible heat gains or losses can be measured with a thermometer.
- Latent (or hidden) heat is the heat added to a substance that causes a change in state rather than a change in temperature. For example, heat added to ice causes the ice to melt but does not increase its temperature.
- Specific heat is the amount of heat (measured in Btu's) required to raise the temperature of one pound of a substance 1°F. Different substances have different specific heat values.
- Pressure is the force applied to a specific unit of area. The atmosphere around the earth has weight and therefore exerts pressure. This weight or pressure is greater at sea level (14.696 psi or 29.92 in. Hg at 70°F) than it is at higher elevations.
- Barometers measure atmospheric pressure in inches of mercury. Two common barometers are the mercury barometer and the aneroid barometer.
- Gages have been developed to measure pressures in enclosed systems. Two common gages used in the air conditioning industry are the compound gage and the high-pressure gage.
- In the English system of units, pressure is expressed in pounds per square inch (psi); in the metric system, pressure is expressed in kilopascals (kPa). One psi equals 6.89 kPa.
- Matter takes up space, has weight, and can be in the form of a solid, a liquid, or a gas.
- In the English system of units, density is the weight of a substance per cubic foot.
- Specific gravity is the term used to compare the densities of various substances.
- Specific volume is the amount of space occupied by a pound of a vapor.
- Boyle's Law states that the volume of a gas varies inversely with the absolute pressure, provided the temperature remains constant.
- Charles' Law states that at a constant pressure, the volume of a gas varies directly with the absolute temperature, and at a constant volume, the pressure of a gas varies directly with the absolute temperature.

- The General Law of Perfect Gas combines both Boyle's and Charles' Laws to calculate changes in pressure, temperature, and volume.
- Dalton's Law states that the total pressure of a confined mixture of gases is the sum of the pressures of each of the gases in the mixture.
- Electrical energy and heat energy are two forms of energy used in this industry.
- Work is the amount of force necessary to move an object: work = force × distance.
- One horsepower is the equivalent of lifting 33,000 pounds a height of one foot in one minute.
- Watts are a measurement of electrical power. One horsepower equals 746 W.
- 1 kW (1000 W) = 3413 Btu's.
- Utility companies measure electricity by the kilowatt-hour (kWh).

REVIEW QUESTIONS

1. Define temperature.
2. Describe how a glass-stem thermometer works.
3. List the standard conditions necessary for water to boil at 212°F.
4. List four types of temperature scales.
5. Under standard conditions, at what point on the Celsius scale will water freeze?
6. At what Fahrenheit temperature is it assumed that all molecular motion stops?
7. Convert 80°F to degrees Celsius.
8. Convert 22°C to degrees Fahrenheit.
9. Define the British thermal unit.
10. Describe the direction of heat flow from one substance to a substance of a different temperature.
11. Describe heat transfer via conduction.
12. Describe heat transfer via convection.
13. Describe heat transfer via radiation.
14. State the difference between sensible heat and latent heat.
15. What is the atmospheric pressure at sea level under standard conditions?
16. Why is the atmospheric pressure less when measured on a mountaintop?
17. Approximately how many inches of mercury are equal to atmospheric pressure at sea level under standard conditions?
18. Explain the operation of a pressure gage using a Bourdon tube.
19. Explain the difference between psig and psia.
20. What modern metric unit is used to express pressure?
21. Define matter.
22. What are the three states in which matter is commonly found?
23. In what direction does a solid exert force?
24. Describe how a liquid exerts pressure.
25. Describe how a vapor exerts pressure.
26. Define density.
27. Define specific gravity.
28. Describe specific volume.
29. Why is information regarding the specific volume of gases important to the designer of air conditioning equipment?
30. Describe Boyle's Law.
31. At a constant pressure, how does a volume of gas vary with respect to the absolute temperature?
32. Describe Dalton's Law as it relates to a confined mixture of gases.
33. What are the two types of energy most frequently used or considered in this industry?
34. What is work?
35. State the formula for determining the amount of work accomplished in a particular task.
36. If an air conditioning compressor weighing 300 pounds had to be lifted four feet to be mounted on a base, how many ft.-lbs. of work must be accomplished?
37. Describe horsepower and list the three quantities needed to determine horsepower.
38. How many watts of electrical energy are equal to one horsepower?
39. How many Btu's are in 4000 W (4 kW)?
40. What unit of energy does the power company charge the consumer for?

3 REFRIGERATION FOR AIR CONDITIONING

OBJECTIVES

Upon completion of this unit, you should be able to

- Describe the term *ton of refrigeration*.
- Describe the basic refrigeration cycle.
- Explain the relationship between pressure and the boiling point of water or other liquids.
- Describe the function of the evaporator or cooling coil.
- Explain the purpose of the compressor.
- List the compressors normally used in residential and light commercial buildings.
- Discuss the function of the condensing coil.
- State the purpose of the metering device.
- Describe various methods of leak testing a refrigeration system.
- Describe how a refrigerant can be stored or processed while a refrigeration system is being serviced.

3.1 REFRIGERATION FOR AIR CONDITIONING

One of the most common applications of refrigeration is air conditioning (*comfort cooling*). This type of refrigeration is also known as *high-temperature refrigeration*. High-temperature refrigeration involves cooling a space to temperatures between 45°F and 75°F.

To help you understand this refrigeration process, we will examine a window air conditioner. The purpose of residential air conditioning is to pump the heat from inside the house to outside the house. Room air at approximately 75°F enters the unit, where the refrigeration components extract the excess heat. The air leaves the unit at approximately 55°F, Figure 3-1. This cooled air is recirculated into the conditioned space, and excess heat is exhausted outdoors.

For this example, our coil and inside/outside temperatures are listed below. These conditions are also guidelines to some of the design data used throughout the air conditioning field.

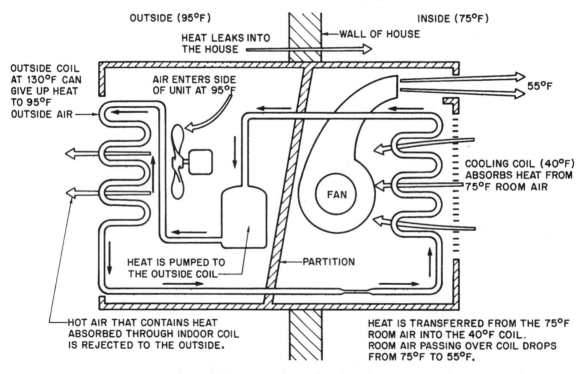

Figure 3-1. Window air conditioning unit.

- The outside design temperature is 95°F.
- The inside temperature is 75°F.
- The cooling coil temperature is 40°F. This coil transfers heat from the room into the refrigeration system. Note that with a 75°F room temperature and a 40°F cooling coil temperature, heat will transfer from the room air into the coil. This heat transfer cools the air leaving the coil to about 55°F.
- The outside coil temperature is 125°F. This coil transfers heat from the system to the outside air. Note that when the outside air temperature is 95°F and the coil temperature is 125°F, heat will be transferred from the system to the outside air.

Figure 3-1 shows that heat from the house is transferred into the system through the indoor coil and transferred from the system to the outdoors through the outdoor coil. The system is pumping the heat out of the house. The system capacity must be large enough to maintain the indoor temperature within the desired range.

3.2 RATING AIR CONDITIONING EQUIPMENT

Refrigeration equipment must have a rating system so that properly-sized equipment can be selected and so that equipment from various manufacturers can be compared. The method for rating this equipment goes back to the days of using ice as the method of removing heat. It takes 144 Btu's of heat energy to melt a pound of ice at 32°F. This same figure is also used in the rating of air conditioning equipment.

The term used for this rating is the ton. One ton of refrigeration is the amount of heat required to melt one ton of ice in a 24-hour period. If it takes 144 Btu's of heat to melt a pound of ice, it would take 2000 times that amount to melt a ton of ice (2000 pound = 1 ton).

$$144 \text{ Btu's/lb.} \times 2000 \text{ lbs.} = 288,000 \text{ Btu's}$$

When accomplished in a 24-hour period, it is known as one ton of refrigeration. The same applies when removing heat from a substance. For example, an air conditioner having a one-ton capacity will remove 288,000 Btu's per day or 12,000 Btu's per hour or 200 Btu's per minute.

Humidity Control

All types of comfort cooling, as well as most types of commercial refrigeration, are concerned with controlling both temperature and humidity. For example, a typical air conditioning system operating at an indoor temperature of 75°F with a humidity level (moisture content) of 50%. These conditions are to be maintained inside the house. The air in the house gives up heat to the refrigerant. The humidity factor must be considered because the indoor coil is also responsible for removing some of the moisture from the air to keep the humidity at an acceptable level. This is known as *dehumidifying*.

Moisture removal requires considerable energy. Approximately the same amount (970 Btu's) of latent heat removal is required to condense a pound of water from the air as it takes to condense a pound of steam. All air conditioning systems must also have a method for dealing with this moisture after it has become a liquid. Some units drip, some drain the liquid into plumbing waste drains, and some use the liquid at the outdoor coil to increase the system capacity.

3.3 PRESSURE/TEMPERATURE RELATIONSHIP

The *pressure/temperature relationship* determines the boiling or condensing point of water and other liquids. Pure water boils at 212°F at sea level when the air temperature is 70°F (standard conditions) because this condition exerts a pressure on the water's surface of 29.92 in. Hg (14.696 psi). This suggests that water has other boiling temperatures under different pressures (Figure 3-2). The boiling point of water can be changed and controlled by controlling the vapor pressure above the water.

BOILING POINT OF WATER (°F)	ABSOLUTE PRESSURE	
	psia	in. Hg
10	0.031	0.063
20	0.050	0.103
30	0.081	0.165
32	0.089	0.180
34	0.096	0.195
36	0.104	0.212
38	0.112	0.229
40	0.122	0.248
42	0.131	0.268
44	0.142	0.289
46	0.153	0.312
48	0.165	0.336
50	0.178	0.362
60	0.256	0.522
70	0.363	0.739
80	0.507	1.032
90	0.698	1.422
100	0.950	1.933
110	1.275	2.597
120	1.693	3.448
130	2.224	4.527
140	2.890	5.881
150	3.719	7.573
160	4.742	9.656
170	5.994	12.203
180	7.512	15.295
190	9.340	19.017
200	11.526	23.468
210	14.123	28.754
212	14.696	29.921

Figure 3-2. Pressure/temperature chart for water.

For example, in Denver, Colorado, which is at an altitude of 5000 feet with an atmospheric pressure of 24.92 in. Hg, water boils at about 203°F. The thinner atmosphere at the 5000-foot level causes a reduction in pressure, about 1 in. Hg/1000 ft. of altitude. This makes cooking foods such as potatoes and dried beans more difficult because they need a higher temperature. By placing the food in a closed container that can be pressurized, such as a pressure cooker, and allowing the pressure to go up to about 15 psi above atmosphere (30 psia), the boiling point can be raised to 250°F.

Studying the water pressure/temperature chart reveals that when the pressure is increased, the boiling point increases, and when the pressure is decreased, the boiling point decreases. For example, if a pan of pure water at 70°F is placed in a bell jar with a barometer and thermometer, and the atmosphere within the is jar evacuated with a vacuum pump, the water will begin to boil and vaporize when the pressure reaches 0.739 in. Hg or 0.363 psia (Figure 3-3). This can be checked using the water pressure/temperature chart in Figure 3-2.

If the pressure is reduced to 0.248 in. Hg (0.122 psia) the water will boil at 40°F. If this boiling water at 40°F is circulated through a cooling coil, and room air is passed over it, the coil will absorb heat from the room air. Because this room air is giving up heat to the coil, the air leaving the coil is cold (Figure 3-4).

When water is used in this way it is considered a *refrigerant*. A refrigerant is a substance that can be changed readily to a vapor by boiling it and then to a liquid by condensing it. The refrigerant must be able to

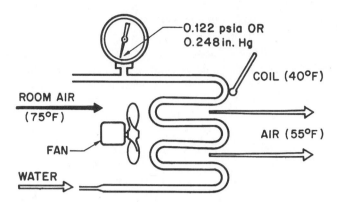

Figure 3-4. The water is boiling at 40°F because the pressure is 0.122 psia or 0.248 in. Hg. The room air is 75°F and gives up heat to the 40°F coil.

make this change repeatedly without altering its characteristics. Refrigerants are usually designated with an "R" and a number for identification. When a manufacturer selects the refrigerant for a particular system, four characteristics must be considered. A refrigerant must be safe, detectable, have a low boiling point, and good pumping characteristics. Water is not normally used as a refrigerant in small applications for reasons that will be discussed later. We used it in this example because most people are familiar with its characteristics.

To illustrate how a typical refrigeration system works, we will use Refrigerant-22 (R-22) in the following examples because it is commonly used in residential and light commercial air conditioning systems. Figure 3-5 shows the pressure/temperature chart for R-22 and other refrigerants. This chart is like that for water but at different temperature and pressure levels. Find 40°F in the left column, read to the right, and note that the gage reading is 68.5 psig for R-22. What does this mean in usable terms?

The pressure and temperature of a refrigerant will correspond when both liquid and vapor are present under two conditions:

1. When the change of state (boiling or condensing) is occurring.
2. When the refrigerant is at *equilibrium* (i.e., no heat is added or removed).

Caution: Refrigerants are stored in pressurized containers and should be handled with care. Goggles with side shields and gloves should be worn when working with refrigerants.

If a cylinder of R-22 is allowed to remain in a room until it reaches the room temperature of 75°F it will be in equilibrium because no outside forces are acting on it.

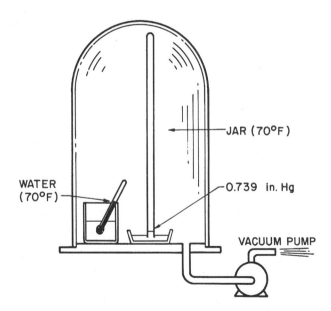

Figure 3-3. Pressure in the bell jar is reduced to 0.739 in. Hg, and the boiling temperature of the water is reduced to 70°F because the pressure is 0.739 in. Hg (0.363 psia).

PRESSURE/TEMPERATURE RELATIONSHIP

TEMPERATURE °F	REFRIGERANT				TEMPERATURE °F	REFRIGERANT				TEMPERATURE °F	REFRIGERANT			
	12	22	134a	502		12	22	134a	502		12	22	134a	502
−60	19.0	12.0		7.2	12	15.8	34.7	13.2	43.2	42	38.8	71.4	37.0	83.8
−55	17.3	9.2		3.8	13	16.4	35.7	13.8	44.3	43	39.8	73.0	38.0	85.4
−50	15.4	6.2		0.2	14	17.1	36.7	14.4	45.4	44	40.7	74.5	39.0	87.0
−45	13.3	2.7		1.9	15	17.7	37.7	15.1	46.5	45	41.7	76.0	40.1	88.7
−40	11.0	0.5	14.7	4.1	16	18.4	38.7	15.7	47.7	46	42.6	77.6	41.1	90.4
−35	8.4	2.6	12.4	6.5	17	19.0	39.8	16.4	48.8	47	43.6	79.2	42.2	92.1
−30	5.5	4.9	9.7	9.2	18	19.7	40.8	17.1	50.0	48	44.6	80.8	43.3	93.9
−25	2.3	7.4	6.8	12.1	19	20.4	41.9	17.7	51.2	49	45.7	82.4	44.4	95.6
−20	0.6	10.1	3.6	15.3	20	21.0	43.0	18.4	52.4	50	46.7	84.0	45.5	97.4
−18	1.3	11.3	2.2	16.7	21	21.7	44.1	19.2	53.7	55	52.0	92.6	51.3	106.6
−16	2.0	12.5	0.7	18.1	22	22.4	45.3	19.9	54.9	60	57.7	101.6	57.3	116.4
−14	2.8	13.8	0.3	19.5	23	23.2	46.4	20.6	56.2	65	63.8	111.2	64.1	126.7
−12	3.6	15.1	1.2	21.0	24	23.9	47.6	21.4	57.5	70	70.2	121.4	71.2	137.6
−10	4.5	16.5	2.0	22.6	25	24.6	48.8	22.0	58.8	75	77.0	132.2	78.7	149.1
−8	5.4	17.9	2.8	24.2	26	25.4	49.9	22.9	60.1	80	84.2	143.6	86.8	161.2
−6	6.3	19.3	3.7	25.8	27	26.1	51.2	23.7	61.5	85	91.8	155.7	95.3	174.0
−4	7.2	20.8	4.6	27.5	28	26.9	52.4	24.5	62.8	90	99.8	168.4	104.4	187.4
−2	8.2	22.4	5.5	29.3	29	27.7	53.6	25.3	64.2	95	108.2	181.8	114.0	201.4
0	9.2	24.0	6.5	31.1	30	28.4	54.9	26.1	65.6	100	117.2	195.9	124.2	216.2
1	9.7	24.8	7.0	32.0	31	29.2	56.2	26.9	67.0	105	126.6	210.8	135.0	231.7
2	10.2	25.6	7.5	32.9	32	30.1	57.5	27.8	68.4	110	136.4	226.4	146.4	247.9
3	10.7	26.4	8.0	33.9	33	30.9	58.8	28.7	69.9	115	146.8	242.7	158.5	264.9
4	11.2	27.3	8.6	34.9	34	31.7	60.1	29.5	71.3	120	157.6	259.9	171.2	282.7
5	11.8	28.2	9.1	35.8	35	32.6	61.5	30.4	72.8	125	169.1	277.9	184.6	301.4
6	12.3	29.1	9.7	36.8	36	33.4	62.8	31.3	74.3	130	181.0	296.8	198.7	320.8
7	12.9	30.0	10.2	37.9	37	34.3	64.2	32.2	75.8	135	193.5	316.6	213.5	341.2
8	13.5	30.9	10.8	38.9	38	35.2	65.6	33.2	77.4	140	206.6	337.2	229.1	362.6
9	14.0	31.8	11.4	39.9	39	36.1	67.1	34.1	79.0	145	220.3	358.9	245.5	385.0
10	14.6	32.8	11.9	41.0	40	37.0	68.5	35.1	80.5	150	234.6	381.5	262.7	408.4
11	15.2	33.7	12.5	42.1	41	37.9	70.0	36.0	82.1	155	249.5	405.1	280.7	432.9

Figure 3-5. Pressure/temperature chart for various refrigerants.

The cylinder and its partial liquid/partial vapor contents will be at the room temperature of 75°F. The pressure/temperature chart indicates a pressure of 132.2 psig (Figure 3-6).

If the same cylinder of R-22 is moved into a walk-in cooler and allowed to reach the room temperature of 35°F and attain equilibrium, the cylinder will reach a new pressure of 61.5 psig (Figure 3-7). This is because while it is cooling off to 35°F, the vapor inside the cylinder is reacting to the cooling effect by partially condensing; therefore, the pressure drops.

If this cylinder (now at 35°F) is moved back into the warmer room (75°F) and allowed to warm up, the liquid inside it reacts to the warming effect by boiling slightly and creating a vapor. Thus, the pressure gradually increases to 132.2 psig, which corresponds to 75°F.

If the cylinder (now at 75°F) is moved into a room at 100°F, the liquid will again respond to the temperature change by slightly boiling and creating more vapor. As the liquid boils and makes vapor, the pressure steadily increases (according to the pressure/temperature chart) until it corresponds to the liquid temperature. This continues until the contents of the cylinder reach the pressure 195.9 psig, corresponding to 100°F (Figure 3-8).

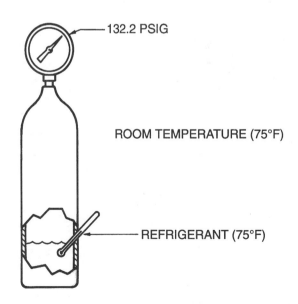

Figure 3-6. The cylinder of R-22 is left in a 75°F room until it and its contents are at room temperature. The cylinder contains a partial liquid, partial vapor mixture; when both reach room temperature, they are in equilibrium; no more temperature changes will be occurring. At this time, the cylinder pressure, 132.2 psig, will correspond to its temperature of 75°F.

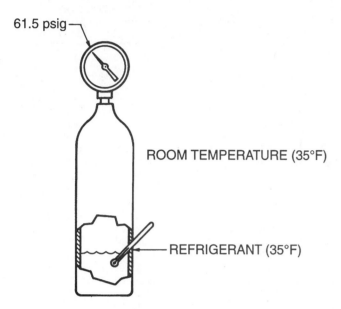

Figure 3-7. The cylinder is moved into a walk-in cooler and left until it and its contents become the same temperature as the inside of the cooler, 35°F. Until the cylinder and its contents reach 35°F some of the vapor will be changing to a liquid, reducing the pressure. When it reaches the temperature of the cooler, the cylinder pressure will correspond to the 35°F temperature (61.5 psig).

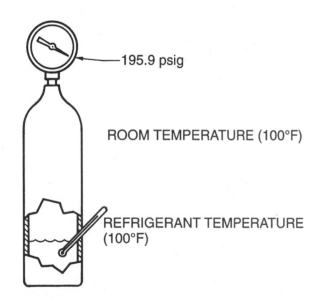

Figure 3-8. The cylinder is moved into a 100°F room and allowed to reach the point of equilibrium at 100°F, 196 psig. The pressure rise is due to some of the liquid refrigerant boiling to a vapor and increasing the total cylinder pressure.

3.4 REFRIGERATION COMPONENTS

Four major components of mechanical refrigeration systems are covered in this book:

1. Evaporators
2. Compressors
3. Condensers
4. Refrigerant metering devices

Evaporators

The *evaporator* absorbs heat into the air conditioning system. When the refrigerant is boiled at a lower temperature than that of the surrounding air, it absorbs heat from the air. The boiling temperature of 40°F was chosen in the previous examples because it is the design temperature normally used for air conditioning systems. The reason is that the room temperature is 75°F, which readily gives up heat to a 40°F coil. The 40°F temperature is also well above the freezing point of the coil. See Figure 3-9 for the coil-to-air relationships.

As R-22 refrigerant enters the evaporator coil, it is a mixture of about 75% liquid and 25% vapor. The mixture is tumbling and boiling down the tube, with the liquid being turned to vapor all along the coil because heat is being added to the coil from the air, Figure 3-10. About halfway down the coil, the mixture becomes more vapor than liquid. The purpose of the evaporator is to boil all of the liquid into a vapor just before the end of the coil. This occurs approximately 90% of the way through the coil, when all of the liquid is gone, leaving pure vapor. At this precise point, we have a saturated vapor. This is the point at which the vapor will start to condense if heat is removed, or become *superheated* if any heat is added. When a vapor is superheated, it no longer corresponds to the pressure/temperature chart; it will take on sensible heat and its temperature will rise. Superheat is considered "refrigeration insurance," because when there is superheat, there is no liquid leaving the evaporator.

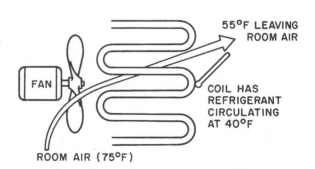

Figure 3-9. The evaporator is operated at 40°F to absorb heat from the 75°F air.

REFRIGERATION COMPONENTS   31

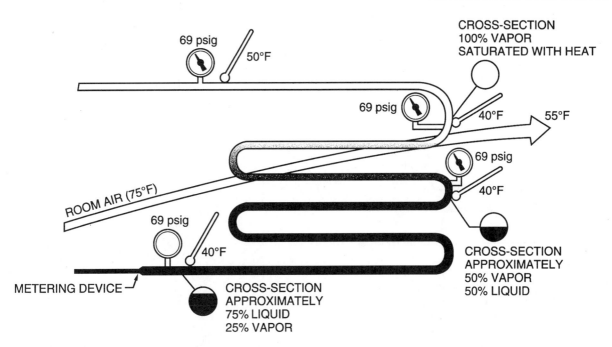

Figure 3-10. The evaporator absorbs heat into the air conditioning system by boiling the refrigerant at a temperature lower than the surrounding room air. The 75°F room air readily gives up heat to the 40°F evaporator.

Evaporators have many design configurations. For now, just remember that they absorb the heat into the system from the surrounding air. See Figure 3-11 for a typical evaporator.

Compressors

The compressor can be thought of as the "heart" of the refrigeration system. It compresses refrigerant vapor and pumps heat-laden refrigerant vapor and refrigerant liquid through the system. It reduces the pressure in the *low-pressure* side of the system, which runs from the metering device to the compressor inlet, and increases the pressure in the *high-pressure side* of the system, which runs from the compressor discharge back to the metering device. The compression of the refrigerant vapor can be accomplished in several ways depending on the type of compressor. The most common compressors used in residential and light commercial air conditioning are the *reciprocating* and the *scroll compressor*. The *rotary compressor* is used in some smaller applications such as in window air conditioners and a few small central air conditioning systems.

The reciprocating compressor uses a piston in a cylinder to compress the refrigerant, Figure 3-12. Valves, usually reed or flapper valves, ensure that the refrigerant flows in the correct direction, Figure 3-13. This type of compressor is known as a *positive displacement compressor*. When the cylinder is filled with vapor, it must be emptied as the compressor turns or damage will occur. For many years, it was the most common compressor in systems up to 100 hp. Today, newer and more efficient compressor designs are also being used.

Figure 3-11. Typical air conditioning evaporator coil. *Reproduced courtesy of Carrier Corporation.*

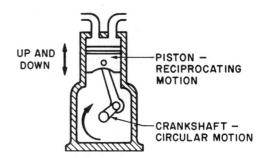

Figure 3-12. The crankshaft converts the circular motion of the motor to the reciprocating or back and forth motion of the piston.

32 UNIT 3 REFRIGERATION FOR AIR CONDITIONING

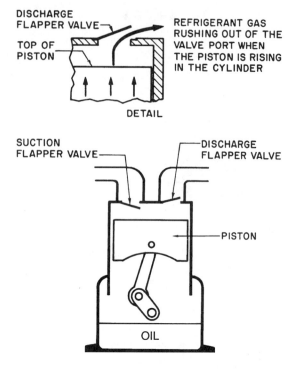

Figure 3-13. Flapper valves and compressor components.

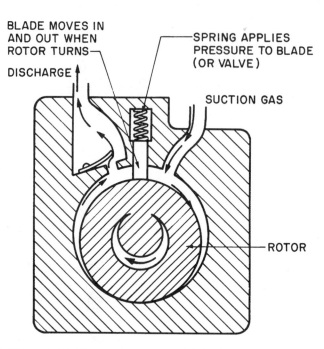

Figure 3-14. Rotary compressor with motion in one direction.

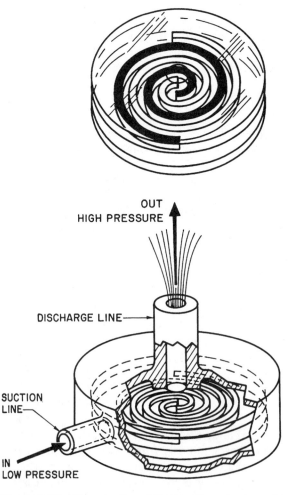

Figure 3-15. The operation of the scroll mechanism of a scroll compressor.

Rotary compressors are also positive displacement compressors, are extremely efficient, and have few moving parts. This type of compressor uses a rotating, drum-like piston that squeezes the vapor refrigerant out the discharge port, Figure 3-14. Rotary compressors are typically smaller than reciprocating compressors with the same capacity.

The scroll compressor is one of the latest designs to be developed and has an entirely different working mechanism. It has a stationary scroll and a moving scroll that matches and meshes with the stationary part, Figure 3-15. The movable scroll orbits inside the stationary scroll and squeezes the vapor between the movable and stationary parts. Several states of compression are taking place in the scrolls at the same time, making it a very smooth-running compressor with few moving parts. The scroll compressor is sealed on the bottom and top with the rubbing action and at the tip with a tip seal. These sealing surfaces prevent refrigerant from the high-pressure side from pushing back to the low-pressure side while running. It is also a positive displacement compressor to a certain degree. It uses positive displacement until too much pressure differential builds up, then the scrolls are capable of moving apart and high-pressure refrigerant can flow back through the compressor and prevent overload.

Reciprocating and scroll compressors are also used in medium to large systems. Larger commercial systems use other types of compressors because they must move much more refrigerant through the system. The centrifugal compressor is used in large air conditioning systems.

It is similar to a large fan and is not a positive displacement compressor (Figure 3-16). The screw compressor is also used in larger systems and is a positive displacement compressor (Figure 3-17). Compressors for larger systems are discussed in detail later in this text.

All compressors perform the same function. They increase the pressure in the system and move the vapor refrigerant from the low-pressure side to the high-pressure side into the condenser.

Condensers

The *condenser* rejects heat from the refrigeration system. It receives the hot gas after it leaves the compressor through the short pipe between the compressor and the condenser known as the *hot gas line* or *discharge line* (Figure 3-18). The hot gas is forced into the top of the condenser coil by the compressor. The gas is pushed along at high speeds and high temperatures (about 200°F). The gas does not correspond to the pressure/temperature chart because the head pressure (high-side pressure) is 278 psig for R-22. The head pressure for 200°F would be off the pressure/temperature chart. The temperature at which the refrigerant changes state from a gas to a liquid is 125°F. This temperature establishes the head pressure of 278 psig.

The gas entering the condenser is so hot compared to the surrounding air that heat exchange begins to occur immediately. The surrounding air that is being passed over the condenser is 95°F as compared to the near 200°F of the gas entering the condenser. As the gas moves through the condenser, it begins to give up heat to the surrounding air. This causes a drop in gas temperature. The gas keeps cooling off until it reaches the condensing temperature of 125°F, and the change in state begins to occur. This happens slowly at first, with small amounts of vapor changing to liquid, and becomes faster as the combination gas/liquid mixture moves toward the end of the condenser.

When the condensing refrigerant moves about 90% of the way through the condenser, it is almost all liquid, or all liquid. Now, more heat can be taken from the liquid. The liquid at the end of the condenser is at the condensing temperature of 125°F and can still give up some heat to the surrounding 95°F air. When the liquid at the end of the condenser goes below 125°F, it is said to be *subcooled* (Figure 3-19).

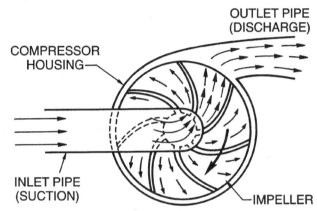

Figure 3-16. The operation of a centrifugal compressor mechanism.

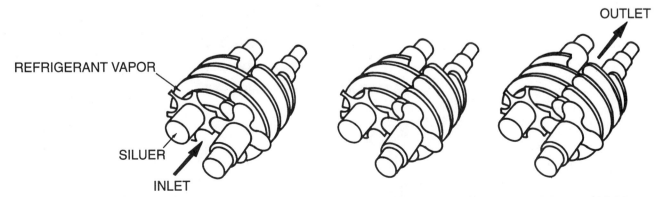

Figure 3-17. The internal working mechanism of a screw compressor.

34 UNIT 3 REFRIGERATION FOR AIR CONDITIONING

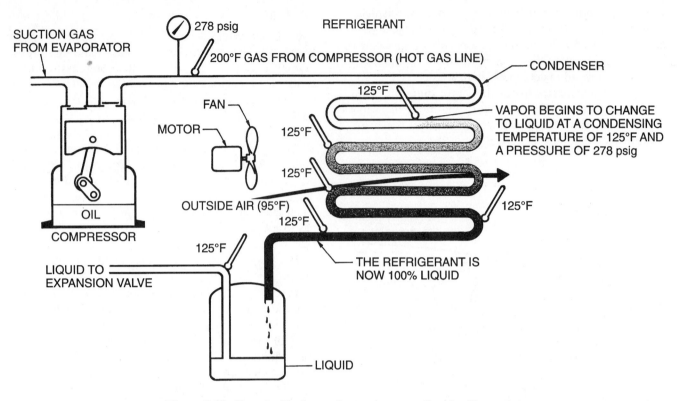

Figure 3-18. Vapor inside the condenser changes to liquid refrigerant.

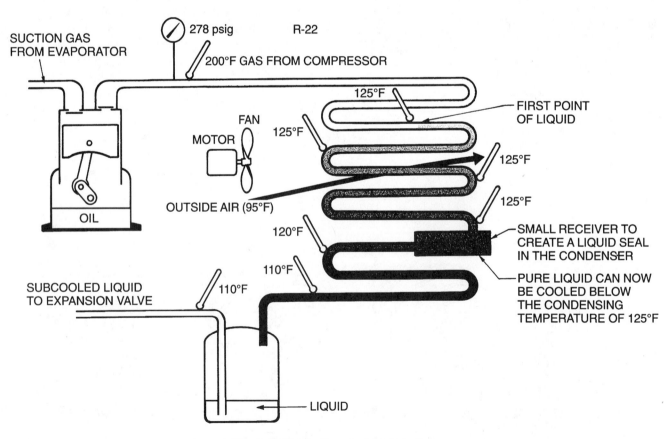

Figure 3-19. Condenser with subcooling.

Three important things happen to the refrigerant in the condenser:

1. The hot gas from the compressor is de-superheated from the hot discharge temperature to the condensing temperature. Remember, the condensing temperature determines the head pressure.
2. The refrigerant is condensed from a vapor into a liquid.
3. The liquid refrigerant temperature may then be lowered below the condensing temperature (subcooled). The refrigerant can usually be subcooled to between 10°F and 20°F below the condensing temperature, Figure 3-19.

Many types of condensing devices are available, but they all have the same purpose, which is to remove heat from the refrigeration system. The heat may have to be rejected into a liquid or gas substance. Figure 3-20 shows a typical condensing unit. The compressor is generally included in the condensing unit assembly.

Metering Devices

The warm liquid moves from the condenser through what is known as the *liquid line* in the direction of the *metering device*. The liquid temperature is about 115°F and may still give up some heat to the surrounding air before reaching the metering device. This line may be routed under a house or through a wall where it may easily reach a new temperature of about 110°F. Any excess heat rejected in this way will improve the system capacity.

One of the simplest metering devices is the *orifice*. An orifice is a small restriction of a fixed size in the refrigerant line (Figure 3-21). This device holds back the full flow of refrigerant and is the dividing point between the high-pressure and low-pressure sides of the system. The pipe leading to the orifice may be the size of a pencil, and the precision-drilled hole in the orifice may be the size of a very fine sewing needle. As you can see from the figure, the gas flow is greatly restricted by this device. The liquid refrigerant entering the orifice is at a pressure of 278 psig; the refrigerant leaving the orifice is a mixture of about 75% liquid and 25% vapor at a new pressure of 69 psig and a new temperature of 40°F, Figure 3-21. Two questions may arise at this time:

1. Why did 25% of the liquid change into a vapor?
2. How did the refrigerant drop from 110°F to 40°F in such a short space?

To understand this process, picture a garden hose under pressure. If you put your thumb over the opening to the hose, the water coming out feels cooler. The water actually is cooler because some of it evaporates and turns to mist. This evaporation removes heat out from the rest of the water and cools it down. When the high-pressure refrigerant passes through the orifice, it is similar to what happens to the water in the hose; it changes pressure (278 psig to 69 psig), which cools the remaining refrigerant to the pressure/temperature relationship of 69 psig and 40°F. This quick drop in pressure in the metering device lowers the boiling point of the liquid leaving the metering device.

One of the most common types of fixed-orifice metering devices is the capillary tube (Figure 3-22). Many other types of metering devices are available for different applications. Metering devices will be covered in detail in later units.

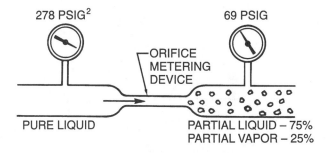

Figure 3-21. Orifice metering device.

Figure 3-20. Air conditioning condensing unit. *Courtesy York International.*

Figure 3-22. Capillary tube. *Courtesy Parker Hannifin Corporation.*

UNIT 3 REFRIGERATION FOR AIR CONDITIONING

At this point, the basic components of the mechanical refrigeration system have been described according to function. These components must be properly matched for each specific application. For instance, a low-temperature compressor cannot be applied to a high-temperature application because of the pumping characteristics of the compressor.

The next section describes a correctly-matched system operating at design conditions. Later, we will explain malfunctions and adverse operating conditions.

3.5 THE REFRIGERATION CYCLE

This section describes the complete refrigeration cycle of the window air conditioner shown earlier. Remember that part of the system is inside the house and part of it is outside. The following numbers correspond to the circled numbers in Figure 3-23.

1. A mixture of 75% liquid and 25% vapor leaves the metering device and enters the evaporator.
2. The mixture is R-22 at a pressure of about 69 psig, which corresponds to a 40°F boiling point. It is important to remember that the pressure is 69 psig because the evaporating refrigerant is boiling at 40°F.
3. The mixture tumbles down the tube in the evaporator with the liquid evaporating as it moves along.
4. When the mixture is about halfway through the coil, it is composed of 50% liquid and 50% vapor and is still at the same pressure/temperature relationship because a change in state is taking place.
5. The refrigerant is now 100% vapor. In other words, it has reached the *saturation point* of the vapor. It is saturated with heat because if any heat is removed at this point, some of the vapor changes back to a liquid, and if any heat is added, the vapor rises in temperature. This is called *superheat*. (Superheat is sensible heat.) The saturated vapor is still at 40°F and is able to absorb heat from the 75°F room air.
6. Pure vapor now exists that is normally superheated about 10°F above the saturation point. Examine the line in Figure 3-23 at this point and you will see that the temperature is about 50°F. To calculate the superheat reading, perform the following steps:

   a. Note the suction pressure reading from the suction gage (69 psig).
   b. Convert the suction pressure reading to suction temperature using the pressure/temperature chart for R-22 (40°F).
   c. Use a suitable thermometer to record the actual temperature of the suction line (50°F).
   d. Subtract the saturated suction temperature from the actual suction line temperature (50°F - 40°F = 10°F of superheat).

This vapor is said to be *heat-laden* because it contains the heat removed from the room air.

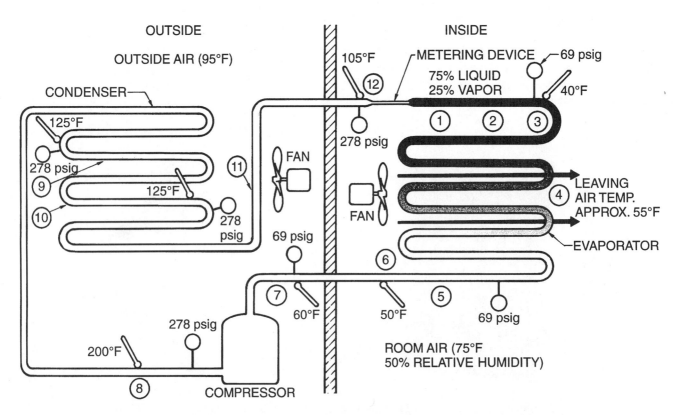

Figure 3-23. Typical air conditioning system for R-22 showing temperatures and airflow.

7. The vapor is drawn toward the compressor by its pumping action, which creates a low-pressure suction. As the vapor moves along toward the compressor, it is contained in the suction line. This line is usually copper and insulated to keep it from drawing heat into the system from the surroundings; however, it still picks up some heat. Because the suction line carries vapor, any heat that it picks up will quickly raise the temperature. It does not take much sensible heat to raise the temperature of a vapor. Depending on the length of the line and the quality of the insulation, the suction line temperature may be 60°F at the compressor inlet.

8. The compressor accepts the heat-laden refrigerant vapor and compresses it, increasing its pressure and temperature and pumping it through the system.

9. Highly superheated gas leaves the compressor through the hot gas line on the high-pressure side of the system. This line is normally very short because the condenser is usually close to the compressor. On a hot day, the hot gas line may be close to 200°F with a pressure of 278 psig. Because the saturated temperature corresponding to 278 psig is 125°F, the hot gas line has about 75°F of superheat that must be removed before condensing can occur. Because the line is so hot and a vapor is present, the line will give up heat readily to the surrounding air. The surrounding (ambient) air temperature is 95°F.

10. The superheat has been removed down to the 125°F condensing temperature, and liquid refrigerant is beginning to form. The coil temperature corresponds to the high-side pressure of 278 psig and 125°F. The high-pressure reading of 278 psig is due to the refrigerant condensing at 125°F. The condensing conditions are arrived at by knowing the efficiency of the condenser. This example uses a standard condenser, which has a condensing temperature that is about 30°F higher than the surrounding air. The outdoor air temperature of 95°F is used to absorb the heat, so 95°F + 30°F = 125°F condensing temperature. Some condensers will condense at 25°F above the surrounding air; these are high-efficiency condensers, and the high-pressure side of the system will be operating under less pressure.

11. The refrigerant is now 100% liquid at the saturated temperature of 125°F. As the liquid continues along the coil, the air continues to cool the liquid to below the actual condensing temperature. The liquid may go as much as 20°F below the condensing temperature of 125°F before it reaches the metering device.

12. The liquid refrigerant reaches the metering device through a pipe, usually copper, from the condenser. This liquid line is often field-installed and is not insulated. The distance between the two may be long, and the line may give up heat along the way. Heat given up here is leaving the system, which improves the system capacity. The refrigerant entering the metering device may be as much as 20°F cooler than the condensing temperature of 125°F, so the liquid line entering the metering device may be 105°F. The refrigerant entering the metering device is 100% liquid. In the short distance of the metering device orifice (a pinhole about the size of a small sewing needle), the above liquid is changed to a mixture of about 75% liquid and 25% vapor. The 25% vapor is known as *flash gas* and is used to cool the remaining 75% of the liquid down to 40°F, the boiling temperature of the evaporator.

The refrigerant has now completed the refrigeration cycle and is ready to go around again. It should be evident that a refrigerant does the same thing over and over, changing from a liquid to a vapor in the evaporator and back to liquid form in the condenser. The expansion device (metering device) meters the flow to the evaporator, and the compressor pumps the refrigerant through the system. The following statements briefly summarize the refrigeration cycle.

1. The evaporator absorbs heat into the system.
2. The condenser rejects heat from the system.
3. The compressor pumps the heat-laden vapor through the system.
4. The expansion device meters the flow of refrigerant.

3.6 REFRIGERANT-22 (R-22)

R-22 is used in most residential and light commercial air conditioning systems. The chemical name for R-22 is monochlorodifluoromethane. The chemical formula is $CHClF_2$.

The boiling point of R-22 is very suitable for high-temperature refrigeration (air conditioning), and the pumping characteristics meet the requirements for air conditioning systems.

Each refrigerant has a color code to identify it and cylinders containing the refrigerant are painted accordingly. The color code for R-22 is green.

R-22 is currently considered safe, although it is heavier than air and will displace the oxygen in the air, which can lead to suffocation.

Warning: Areas where there are potential refrigerant leaks must be well ventilated to avoid suffocation.

A large leak can be detected by listening for a hiss of the escaping refrigerant, Figure 3-24(A). This method would not be acceptable for small leaks.

Soap bubbles also provide a practical and inexpensive leak detector. Commercially-prepared products that blow large, elastic bubbles are used by many service technicians. These are helpful when it is known that a leak is in a certain area. Soap bubble solution can be applied with a brush to the tubing joint to see exactly where the leak is, Figure 3-24(B).

38 UNIT 3 REFRIGERATION FOR AIR CONDITIONING

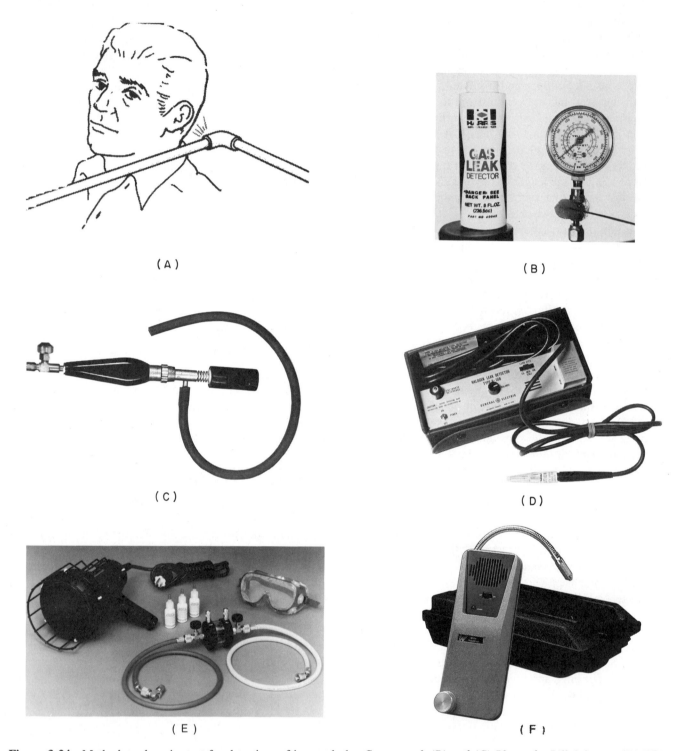

Figure 3-24. Methods and equipment for detecting refrigerant leaks. *Courtesy of: (B) and (C) Photos by Bill Johnson, (D) White Industries, (E) Yokogawa, (F) Spectronics Corporation.*

A halide leak detector, Figure 3-24(C), is available for use with acetylene and propane gas. It operates on the principle that when refrigerant is exposed to an open flame in the presence of glowing copper, the flame will change color. Halide leak detectors were very popular in the past but their use has diminished with the development of safer and more reliable detection methods.

All of the leak detection devices mentioned above are capable of finding fairly large leaks. More accurate methods are required on some newer equipment because refrigerant charges have been reduced to the bare working minimum. Some residential air conditioning equipment has refrigerant charge specifications that call for half-ounce accuracy.

The leak detector in Figure 3-24(D) uses a flexible probe attached to a sensing device. It is battery-operated and small enough to be easily carried.

The electronic leak detector shown in Figure 3-24(E) is capable of detecting leak rates down to one-half ounce per year. Another system utilizes a high-intensity ultraviolet lamp, Figure 3-24(F). An additive is introduced into the refrigeration system. This additive will show as a bright yellow-green glow under the ultraviolet light at the source of the leak. After the leak has been repaired, the area can be wiped clean with a general purpose cleaner and then reinspected. The additive can remain in the system. Should a new leak be suspected at a later date, it will still show the yellow-green color under the ultraviolet light. This system will detect leaks as small as a quarter-ounce per year.

3.7 REFRIGERANT RECOVERY, RECYCLING, AND RECLAMATION

Warning: Technicians must recover and/or recycle refrigerants during installation and servicing operations to help reduce emissions of chlorofluorocarbons (CFCs) to the atmosphere. Failure to do so can result in stiff fines and/or imprisonment.

Examples of recovery equipment are shown in Figure 3-25. Many larger systems can be fitted with receivers or dump tanks into which the refrigerant can be pumped and stored while the system is serviced. However, in smaller systems, it is not often feasible to provide these components. Recovery units or other storage devices may be necessary. Most recovery and recycle units that have been developed to date vary in their technology and capabilities, so manufacturers' instructions must be followed carefully when using this equipment. Used refrigerants that cannot be recycled for use in the same system (or another system with the same owner) must be sent back to the manufacturer for reclamation. These processes are described in detail later in this text.

3.8 PLOTTING THE REFRIGERATION CYCLE

A graphic picture of the refrigeration cycle may be plotted on a pressure/enthalpy diagram. *Enthalpy* describes how much heat a substance contains from a specific starting point. Refer to Figure 2-6, the heat/temperature graph for water. 0°F was used as the starting point of heat for water, knowing that more heat can be removed from water (ice) lowering the temperature below 0°F. The process was described as the amount of heat added starting at 0°F. This heat is called *enthalpy*. A similar diagram is available for all refrigerants called a pressure/enthalpy diagram and used to plot the refrigerant cycle in a complete loop. Figure 3-26 is a pressure/enthalpy diagram for R-22.

The pressure/enthalpy diagram plots pressure on the left hand column and enthalpy or total heat on the bottom

(A)

(B)

Figure 3-25. Refrigerant recovery systems. *Courtesy Robinair Division, SPX Corporation.*

of the diagram. The enthalpy readings start at -40°F saturated liquid with a heat content of 0 Btu/lb. Readings below -40°F saturated liquid show as minus values. Notice that the temperature corresponds to pressure and is plotted on the inside of the chart along the left and right of the horseshoe-shaped curve.

This horseshoe-shaped curve is the saturation curve with the temperature corresponding to the absolute pressure. Whenever a plot falls on this curve, the refrigerant is saturated with heat. There are two saturation curves: the one on the left is the saturated liquid curve; the one on the right is the saturated vapor curve. On the liquid curve, if heat is added, the refrigerant will start changing state to a vapor. If heat is removed, the liquid is subcooled. On the vapor curve, if heat is added, the vapor will superheat. If heat is removed, the vapor will start changing state to a liquid. Notice that the saturated liquid

40 UNIT 3 REFRIGERATION FOR AIR CONDITIONING

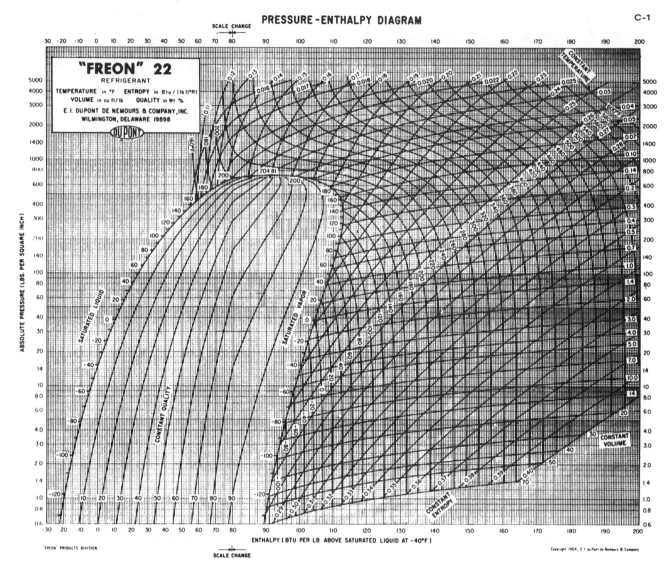

Figure 3-26. Pressure/enthalpy diagram for R-22 expressed in Btu/lb. of refrigerant circulated. The diagram uses -40°F saturated liquid as the starting point for heat content. To find this point, find 0 Btu/lb. on the bottom line and follow it straight up to the saturated liquid curve at -40°F. *Courtesy E.I. DuPont.*

and vapor curves touch at the top. This is called the *critical temperature* (or pressure). Above this point, the refrigerant will not condense. It is a vapor regardless of how much pressure is applied.

The area between the saturated liquid and saturated vapor curve, inside the horseshoe-shaped curve, is where the change of state occurs. Whenever a plot falls between the saturation curves, the refrigerant is in the partial liquid, partial vapor state. The slanted, near vertical lines between the saturated liquid and saturated vapor lines are the constant quality lines and describe the percent of liquid to vapor in the mixture between the saturation points. If the plot is closer to the saturated liquid curve, there is more liquid than vapor. If the plot is closer to the saturated vapor curve, there is more vapor than liquid. For example, find a point on the chart at 40 psia (from the left column) and 30 Btu/lb. (along the bottom), Figure 3-27. This point is inside the horseshoe-shaped curve and the refrigerant is 90% liquid and 10% vapor.

A refrigeration cycle is plotted in Figure 3-28. The system to be plotted is an air conditioning system using R-22. The system is operating at 130°F condensing temperature (296.8 psig or 311.5 psia discharge pressure) and an evaporating temperature of 40°F (68.5 psig or 83.2 psia suction pressure). The cycle plotted has no subcooling, and 10°F superheat as the refrigerant leaves the evaporator with another 10°F superheat absorbed by the suction line in route to the compressor. The compressor is an air-cooled compressor and the suction gas enters the suction valve adjacent to the cylinders.

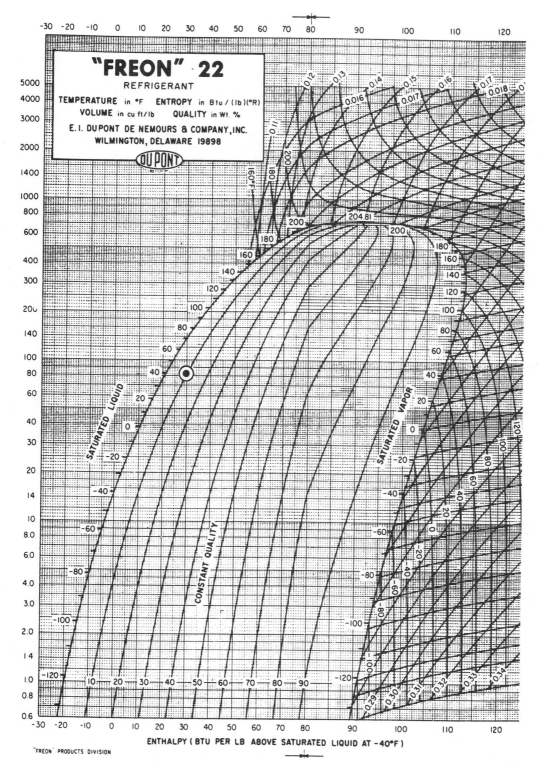

Figure 3-27. Follow the 40°F line to the right until it intersects the vertical 30 Btu/lb. line and we find the refrigerant mixture to be 90% liquid and 10% vapor. *Courtesy E.I. DuPont.*

42 UNIT 3 REFRIGERATION FOR AIR CONDITIONING

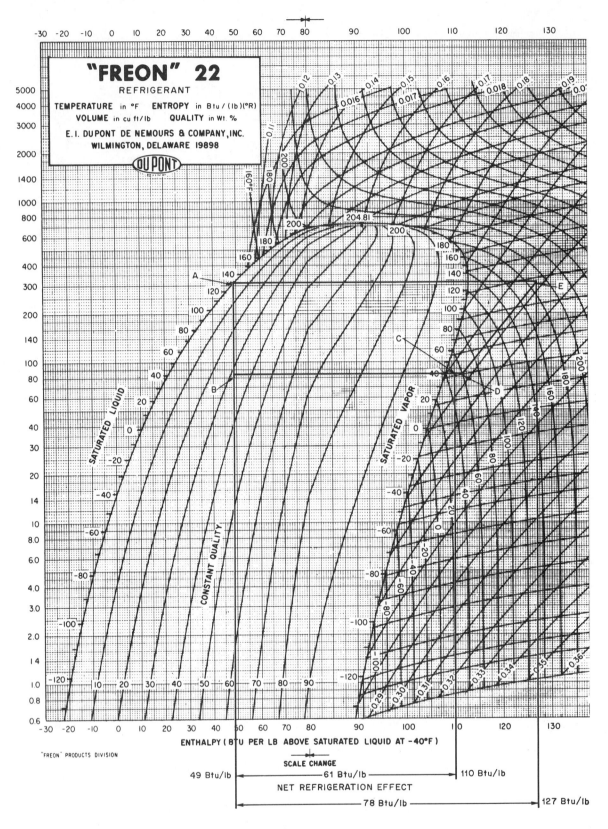

Figure 3-28. Refrigeration cycle plotted on the pressure/enthalpy diagram. *Courtesy E.I. DuPont.*

PLOTTING THE REFRIGERATION CYCLE   43

1. Refrigerant R-22 enters the expansion device as a saturated liquid at 311.5 psia (296 psig) and 130°F, Point A. The heat content is 49 Btu/lb. entering the expansion valve and 49 Btu/lb. leaving the expansion valve. The temperature of the liquid refrigerant before the valve is 130°F and the temperature leaving the valve is 40°F. The temperature drop can be accounted for by observing that we have 100% liquid entering the valve and about 67% liquid leaving the valve. About 33% of the liquid changed to a vapor (called flash gas), lowering the remaining liquid temperature to 40°F.

2. Usable refrigeration starts at Point B where the refrigerant has a heat content of 49 Btu's/lb. As heat is added in the evaporator, the refrigerant changes the state to a vapor. All liquid changes to a vapor when it reaches the saturated vapor curve, and a small amount of heat is added to the refrigerant in the form of superheat (10°F). The refrigerant leaves the evaporator at Point C, with a heat content of about 110 Btu's/lb. This is a net refrigeration effect of 61 Btu's/lb. (110 Btu/lb. - 49 Btu/lb. = 61 Btu's/lb.) of refrigerant circulated. The net refrigeration effect is the same as usable refrigeration, the heat actually extracted from the conditioned space. About 10 more Btu's/lb. are absorbed into the suction line before reaching the compressor inlet at Point D. This is not usable refrigeration because the heat does not come from the conditioned space, but it is heat that must be pumped by the compressor and rejected by the condenser.

3. The refrigerant enters the compressor at Point D and leaves the compressor at Point E. No heat has been added in the compressor except heat of compression, because the compressor is air cooled. The refrigerant enters the compressor cylinder from the suction line. (A fully-hermetic compressor with a suction-cooled motor would not plot out in this way. We have no way of knowing how much heat is added by the suction-cooled motor so we cannot know the temperature of the suction gas entering the compressor cylinder. Manufacturers obtain their own figures for this using internal thermometers during testing.) The compression process in this compressor is called adiabatic compression, where no heat is added or taken away.

4. The refrigerant leaves the compressor at Point E and contains about 127 Btu's/lb. This condenser must reject 78 Btu's/lb. (127 Btu's/lb. - 49 Btu's/lb. = 78 Btu's/lb.), called the heat of rejection. The temperature of the discharge gas is about 190°F (see the constant temperature lines for temperature of superheated gas). When the hot gas leaves the compressor it contains the maximum amount of heat that must be rejected by the condenser.

5. The refrigerant enters the condenser at point E as a highly superheated gas. The refrigerant condensing temperature is 130°F and the hot gas leaving the compressor is 190°F, so it contains 60°F (190°F - 130°F = 60°F) of superheat. The condenser will first remove the superheat down to the condensing temperature, then it will condense the refrigerant to a liquid of 130°F for reentering the expansion device at point A for another trip around the cycle.

The refrigerant cycle in the previous example can be improved by removing some heat from the condensed liquid by subcooling it. This can be seen in Figure 3-29, a scaled up diagram. The same conditions are used in this figure as in Figure 3-28, except the liquid is subcooled 15°F (from 130°F condensing temperature to 115° liquid). The system then has a net refrigeration effect of 68 Btu's/lb. instead of 61 Btu's/lb. This is an increase in capacity of about 11%. Notice the liquid leaving the expansion valve is about 23% vapor instead of the 33% vapor in the first example. This is where the capacity is gained. Less capacity is lost to flash gas.

Other conditions may be plotted on the pressure/enthalpy diagram. For example, if the head pressure is higher due to a dirty condenser, Figure 3-30, and using the first example of saturated liquid, raising the condensing temperature to 140°F (337.2 psig or 351.9 psia), the percent of liquid leaving the expansion valve is about 64% with a heat content of 53 Btu's/lb. Using the same heat content leaving the evaporator, 110 Btu's/lb., there is a net refrigeration effect of about 7% from the original example, which contained 49 Btu's/lb. at the same point. This shows the importance of keeping condensers clean.

Figure 3-31 shows how increased superheat affects the first system in Figure 3-28. The suction line has not been insulated and absorbs heat. The suction gas may leave the evaporator at 50°F and rise to 80°F before entering the compressor. Notice the high discharge temperature (about 210°F). This is approaching the temperature that will cause oil to break down and form acids in the system. Most compressors must not exceed 250°F. The compressor must pump more refrigerant to accomplish the same refrigeration effect and the condenser must reject more heat.

Pressure/enthalpy diagrams are useful in showing the refrigeration cycle for the purpose of establishing the various conditions around the system. They are partially constructed from properties of refrigerant tables. Figure 3-32 is a page from a typical table for R-22. Column 1 is the temperature corresponding to the pressure columns for the saturation temperature.

Columns 4 and 5 list the specific volume for the saturated refrigerant in cubic feet per pound. For example, at 60°F, the compressor must pump 0.4727 cubic feet of refrigerant to circulate one pound of refrigerant in the system. The specific volume along with the net refrigeration effect help the engineer determine the compressor pumping capacity. The example in Figure 3-28 using R-22 has a net refrigeration effect of 61 Btu's/lb. of refrigerant circulated. If we had a system needing to circulate enough refrigerant to absorb 36,000 Btu's per

44 UNIT 3 REFRIGERATION FOR AIR CONDITIONING

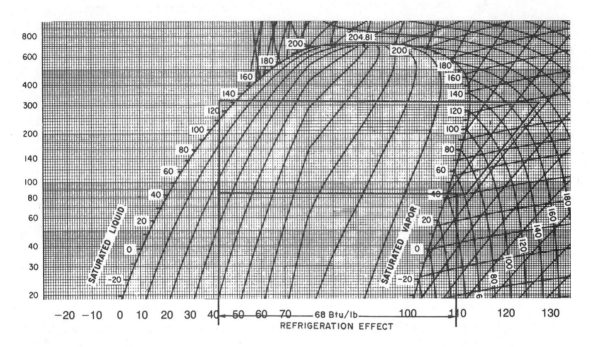

Figure 3-29. Subcooling the refrigerant 15°F (compared to Figure 3-28) increases the net refrigeration effect from 61 to 68 Btu's/lb. This increases the capacity by about 11%. Only a very slight amount of extra heat must be rejected by the condenser with no added load to the compressor. *Courtesy E.I. DuPont.*

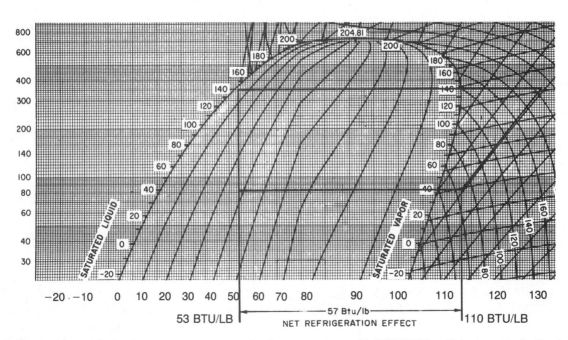

Figure 3-30. An increase in head pressure to a new condensing temperature of 140°F (337.2 psig) causes a reduction in capacity of about 7%. *Courtesy E.I. DuPont.*

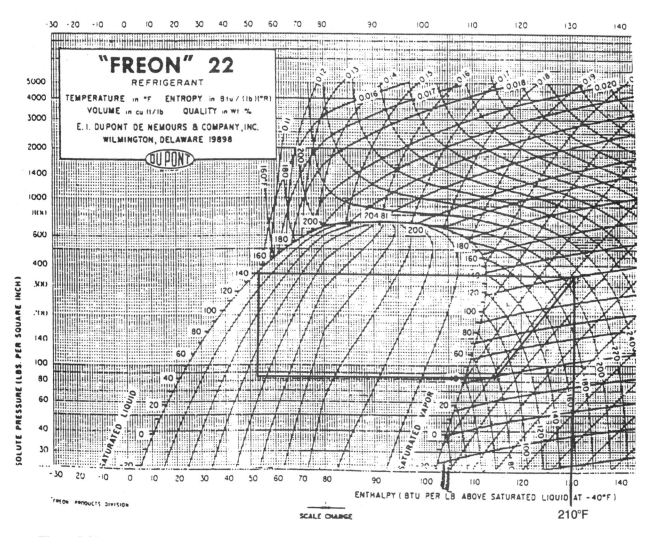

Figure 3-31. An increase in superheat causes an increase in compressor discharge temperature. *Courtesy E.I. DuPont.*

hour (Btuh) (3 tons of refrigeration) we would need to circulate 590.2 pounds of refrigerant per hour (36,000 Btuh divided by 61 Btu's/lb. = 590.2 pounds per hour). If the refrigerant entered the compressor at 60°F, the compressor must move 279 cubic feet of refrigerant per hour (590.2 pounds per hour × 0.4727 cubic feet per pound = 279 cubic feet per hour). There is a slight error in this calculation because the 0.4727 cubic feet per hour is for saturated refrigerant and the vapor is superheated entering the compressor. Superheat tables are available, but will only complicate this calculation more and there is very little error. Many compressors are rated in cubic feet per minute so this compressor would need to pump 4.65 cubic feet per minute (279 cubic feet per hour divided by 60 minutes per hour = 4.65 cubic feet per minute).

The density section of the table tells the engineer how much a particular volume of liquid refrigerant will weigh at the rated temperature. For example, R-22 weighs 76.773 pounds per cubic foot when the liquid temperature is 60°F. This is important in determining the weight of refrigerant in components, such as evaporators, condensers, and receivers.

The enthalpy section of the table (total heat) lists the heat content of the liquid and vapor and the amount of latent heat required to boil 1 pound of liquid to a vapor. For example, at 60°F, saturated liquid refrigerant would contain 27.172 Btu's/lb. compared to 0 Btu's/lb. at -40°F. It would require 82.54 Btu's/lb. to boil one pound of 60°F saturated liquid to a vapor. The saturated vapor would then contain 109.712 Btu's/lb. total heat (27.172 + 82.54 = 109.712).

"FREON" 22 SATURATION PROPERTIES—TEMPERATURE TABLE

TEMP. °F	PRESSURE		VOLUME cu ft/lb		DENSITY lb/cu ft		ENTHALPY Btu/lb			ENTROPY Btu/(lb)(°R)		TEMP. °F
	PSIA	PSIG	LIQUID v_f	VAPOR v_g	LIQUID $1/v_f$	VAPOR $1/v_g$	LIQUID h_f	LATENT h_{fg}	VAPOR h_g	LIQUID s_f	VAPOR s_g	
10	47.464	32.768	0.012088	1.1290	82.724	0.88571	13.104	92.338	105.442	0.02932	0.22592	10
11	48.423	33.727	0.012105	1.1077	82.612	0.90275	13.376	92.162	105.538	0.02990	0.22570	11
12	49.396	34.700	0.012121	1.0869	82.501	0.92005	13.648	91.986	105.633	0.03047	0.22548	12
13	50.384	35.688	0.012138	1.0665	82.389	0.93761	13.920	91.808	105.728	0.03104	0.22527	13
14	51.387	36.691	0.012154	1.0466	82.276	0.95544	14.193	91.630	105.823	0.03161	0.22505	14
15	52.405	37.709	0.012171	1.0272	82.164	0.97352	14.466	91.451	105.917	0.03218	0.22484	15
16	53.438	38.742	0.012188	1.0082	82.051	0.99188	14.739	91.272	106.011	0.03275	0.22463	16
17	54.487	39.791	0.012204	0.98961	81.938	1.0105	15.013	91.091	106.105	0.03332	0.22442	17
18	55.551	40.855	0.012221	0.97144	81.825	1.0294	15.288	90.910	106.198	0.03389	0.22421	18
19	56.631	41.935	0.012238	0.95368	81.711	1.0486	15.562	90.728	106.290	0.03446	0.22400	19
20	57.727	43.031	0.012255	0.93631	81.597	1.0680	15.837	90.545	106.383	0.03503	0.22379	20
21	58.839	44.143	0.012273	0.91932	81.483	1.0878	16.113	90.362	106.475	0.03560	0.22358	21
22	59.967	45.271	0.012290	0.90270	81.368	1.1078	16.389	90.178	106.566	0.03617	0.22338	22
23	61.111	46.415	0.012307	0.88645	81.253	1.1281	16.665	89.993	106.657	0.03674	0.22318	23
24	62.272	47.576	0.012325	0.87055	81.138	1.1487	16.942	89.807	106.748	0.03730	0.22297	24
25	63.450	48.754	0.012342	0.85500	81.023	1.1696	17.219	89.620	106.839	0.03787	0.22277	25
26	64.644	49.948	0.012360	0.83978	80.907	1.1908	17.496	89.433	106.928	0.03844	0.22257	26
27	65.855	51.159	0.012378	0.82488	80.791	1.2123	17.774	89.244	107.018	0.03900	0.22237	27
28	67.083	52.387	0.012395	0.81031	80.675	1.2341	18.052	89.055	107.107	0.03958	0.22217	28
29	68.328	53.632	0.012413	0.79604	80.558	1.2562	18.330	88.865	107.196	0.04013	0.22198	29
30	69.591	54.895	0.012431	0.78208	80.441	1.2786	18.609	88.674	107.284	0.04070	0.22178	30
31	70.871	56.175	0.012450	0.76842	80.324	1.3014	18.889	88.483	107.372	0.04126	0.22158	31
32	72.169	57.473	0.012468	0.75503	80.207	1.3244	19.169	88.290	107.459	0.04182	0.22139	32
33	73.485	58.789	0.012486	0.74194	80.089	1.3478	19.449	88.097	107.546	0.04239	0.22119	33
34	74.818	60.122	0.012505	0.72911	79.971	1.3715	19.729	87.903	107.632	0.04295	0.22100	34
35	76.170	61.474	0.012523	0.71655	79.852	1.3956	20.010	87.708	107.719	0.04351	0.22081	35
36	77.540	62.844	0.012542	0.70425	79.733	1.4199	20.292	87.512	107.804	0.04407	0.22062	36
37	78.929	64.233	0.012561	0.69221	79.614	1.4447	20.574	87.316	107.889	0.04464	0.22043	37
38	80.336	65.640	0.012579	0.68041	79.495	1.4697	20.856	87.118	107.974	0.04520	0.22024	38
39	81.761	67.065	0.012598	0.66885	79.375	1.4951	21.138	86.920	108.058	0.04576	0.22005	39
40	83.206	68.510	0.012618	0.65753	79.255	1.5208	21.422	86.720	108.142	0.04632	0.21986	40
41	84.670	69.974	0.012637	0.64643	79.134	1.5469	21.705	86.520	108.225	0.04688	0.21968	41
42	86.153	71.457	0.012656	0.63557	79.013	1.5734	21.989	86.319	108.308	0.04744	0.21949	42
43	87.655	72.959	0.012676	0.62492	78.892	1.6002	22.273	86.117	108.390	0.04800	0.21931	43
44	89.177	74.481	0.012695	0.61448	78.770	1.6274	22.558	85.914	108.472	0.04855	0.21912	44
45	90.719	76.023	0.012715	0.60425	78.648	1.6549	22.843	85.710	108.553	0.04911	0.21894	45
46	92.280	77.584	0.012735	0.59422	78.526	1.6829	23.129	85.506	108.634	0.04967	0.21876	46
47	93.861	79.165	0.012755	0.58440	78.403	1.7112	23.415	85.300	108.715	0.05023	0.21858	47
48	95.463	80.767	0.012775	0.57476	78.280	1.7398	23.701	85.094	108.795	0.05079	0.21839	48
49	97.085	82.389	0.012795	0.56532	78.157	1.7689	23.988	84.886	108.874	0.05134	0.21821	49
50	98.727	84.031	0.012815	0.55606	78.033	1.7984	24.275	84.678	108.953	0.05190	0.21803	50
51	100.39	85.69	0.012836	0.54698	77.909	1.8282	24.563	84.468	109.031	0.05245	0.21785	51
52	102.07	87.38	0.012856	0.53808	77.784	1.8585	24.851	84.258	109.109	0.05301	0.21768	52
53	103.78	89.08	0.012877	0.52934	77.659	1.8891	25.139	84.047	109.186	0.05357	0.21750	53
54	105.50	90.81	0.012898	0.52078	77.534	1.9202	25.429	83.834	109.263	0.05412	0.21732	54
55	107.25	92.56	0.012919	0.51238	77.408	1.9517	25.718	83.621	109.339	0.05468	0.21714	55
56	109.02	94.32	0.012940	0.50414	77.282	1.9836	26.008	83.407	109.415	0.05523	0.21697	56
57	110.81	96.11	0.012961	0.49606	77.155	2.0159	26.298	83.191	109.490	0.05579	0.21679	57
58	112.62	97.93	0.012982	0.48813	77.028	2.0486	26.589	82.975	109.564	0.05634	0.21662	58
59	114.46	99.76	0.013004	0.48035	76.900	2.0818	26.880	82.758	109.638	0.05689	0.21644	59
60	116.31	101.62	0.013025	0.47272	76.773	2.1154	27.172	82.540	109.712	0.05745	0.21627	60
61	118.19	103.49	0.013047	0.46523	76.644	2.1495	27.464	82.320	109.785	0.05800	0.21610	61
62	120.09	105.39	0.013069	0.45788	76.515	2.1840	27.757	82.100	109.857	0.05855	0.21592	62
63	122.01	107.32	0.013091	0.45066	76.386	2.2190	28.050	81.878	109.929	0.05910	0.21575	63
64	123.96	109.26	0.013114	0.44358	76.257	2.2544	28.344	81.656	110.000	0.05966	0.21558	64

Figure 3-32. Portion of table showing properties of R-22. *Courtesy E.I. DuPont.*

STANDARD REFRIGERANT DESIGNATION	CHEMICAL NAME	BOILING POINT °F	CHEMICAL FORMULA	MOLECULAR WEIGHT
11	Trichlorofluoromethane	74.8	CCl_3F	137.4
12	Dichlorodifluoromethane	−21.6	CCl_2F_2	120.9
13	Chlorotrifluoromethane	−114.6	$CClF_3$	104.5
22	Chlorodifluoromethane	−41.4	$CHClF_2$	86.5
30	Methylene Chloride	105.2	CH_2Cl_2	84.9
40	Methyl Chloride	−10.8	CH_3Cl	50.5
50	Methane	−259.	CH_4	16.0
113	Trichlorotrifluoroethane	117.6	CCl_2FCClF_2	187.4
114	Dichlorotetrafluoroethane	38.4	$CClF_2CClF_2$	170.9
123	Dichlorotrifluoroethane	82.17	CCl_2HCF_3	152.93
134a	Tetrafluoroethane	−15.08	CF_3CFH_2	102.03
170	Ethane	−127.5	CH_3CH_3	30.0
290	Propane	−44.0	$CH_3CH_2CH_3$	44.0
500*	Refrigerants 12/152a; 73.8/26.2% (Wt)	−28.0	CCl_2F_2/CH_3CHF_2	99.3
502*	Refrigerants 22/115; 48.8/51.2% (Wt)	−50.1	$CHClF_2/CClF_2CF_3$	112.0
601	Isobutane	14.0	$CH(CH_3)_3$	58.1
717	Ammonia	−28.0	NH_3	17.0
718	Water	212.0	H_2O	18.0

*Denotes Azeotropic Mixture

Figure 3-33. Common refrigerants and their characteristics.

The entropy column is of no practical value except on the pressure/enthalpy chart where it is used to plot the compressor discharge temperature.

These charts and tables are not normally used in the field for troubleshooting but are for engineers to use to design equipment. They help the technician understand the refrigerants and the refrigerant cycle. Different refrigerants have different pressure/temperature and enthalpy relationships. These all must be considered by the engineer when choosing the correct refrigerant for a particular application.

Although most residential and light commercial air conditioning systems use R-22, we have included a list of several other refrigerants and some of their characteristics in Figure 3-33, and a pressure/temperature comparison graph for various refrigerants in Figure 3-34.

SUMMARY

- Residential and light commercial air conditioning is sometimes called high-temperature refrigeration. It is used to cool a conditioned space to temperatures between 45°F and 75°F.
- One ton of refrigeration is the amount of heat necessary to melt one ton of ice in a 24-hour period. It takes 288,000 Btu's to melt one ton of ice in a 24-hour period.
- The boiling point of a liquid can be changed and controlled by controlling the pressure on the liquid. The relationship between vapor pressure and boiling point is called the pressure/temperature relationship. When the pressure increases, the boiling point increases; when the pressure decreases, the boiling point decreases.
- Refrigerant is boiled in the evaporator at a lower temperature than the air to be cooled. Heat from the air is absorbed into the refrigerant cooling the air.
- A compressor lowers the pressure in the evaporator to the desired temperature and increases the pressure in the condenser to a level at which the vapor may be condensed to a liquid. It also pumps the refrigerant through the system.
- The condenser rejects the heat absorbed by the evaporator. As the heat leaves the condenser, the vapor condenses to a liquid.
- The liquid refrigerant moves from the condenser to the metering device. The metering device causes part of the liquid to vaporize, and the pressure and temperature are greatly reduced as the refrigerant enters the evaporator and starts the cycle over again.
- Refrigerants are usually designated with an R and a number for identification. R-22 is used in most residential and light commercial air conditioning systems.
- A refrigerant must be safe, detectable, have a low boiling point, and good pumping characteristics.
- Refrigerant cylinders are color-coded to indicate the type of refrigerant they contain. The color for R-22 is green.
- Refrigerants must not be vented to the atmosphere. If they need to be removed from a system, they must be recovered or stored while the system is being serviced, then recycled if appropriate, or sent to a manufacturer to be reclaimed.
- Pressure/enthalpy diagrams may be used to plot refrigeration cycles.

PRESSURE–TEMPERATURE RELATIONSHIPS OF REFRIGERANTS

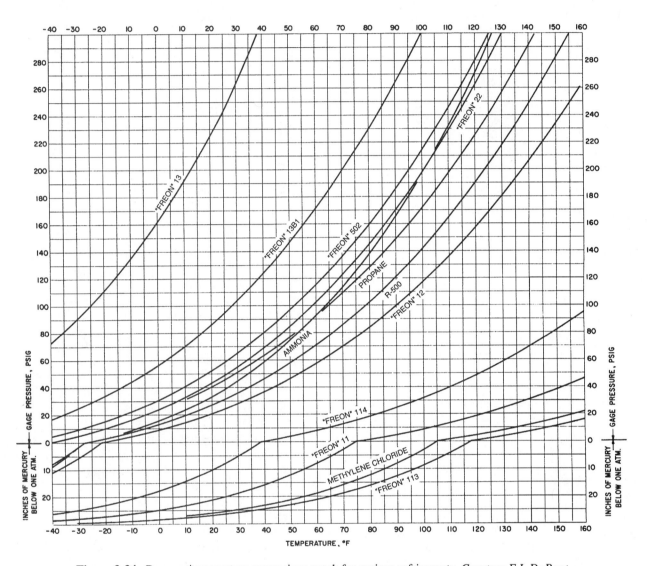

Figure 3-34. Pressure/temperature comparison graph for various refrigerants. *Courtesy E.I. DuPont.*

REVIEW QUESTIONS

1. What is a ton of refrigeration?
2. What is the relationship between pressure and the boiling point of liquids?
3. What is the function of the evaporator in an air conditioning system?
4. What does the compressor do in an air conditioning system?
5. What happens to the refrigerant in the condenser?
6. What is the function of the metering device?
7. Describe the difference between a reciprocating compressor and a scroll compressor.
8. What refrigerant is used extensively in residential and light commercial air conditioning systems?
9. List four characteristics a manufacturer must consider when choosing a refrigerant.
10. State two methods for storing or recovering refrigerant while an air conditioning system is being serviced.

TUBING AND PIPING 4

OBJECTIVES

After studying this unit, you should be able to

- List the different types of tubing used in air conditioning applications.
- Describe two common methods for cutting copper tubing.
- List the procedures for bending tubing.
- Discuss the procedures for soldering and brazing tubing.
- Describe two methods for making flared joints.
- Explain the procedures for making swaged joints.
- Describe the procedures for preparing and threading steel pipe ends.
- List four types of plastic pipe and describe their applications.

4.1 PURPOSE OF TUBING AND PIPING IN A/C SYSTEMS

The correct size, layout, type, and installation of tubing, piping, and fittings helps to keep air conditioning systems operating properly and prevents refrigerant loss. The piping system carries refrigerant to the evaporator, compressor, condenser, and expansion valve. It also provides the pathway for oil return to the compressor. Tubing, piping, and fittings are also used to transport water to (and from) chiller terminal units and between cooling towers and condensers.

Caution: Careless handling of the tubing and poor soldering or brazing techniques may cause serious damage to system components. You must keep contaminants, including moisture, from air conditioning systems.

4.2 TYPES AND SIZES OF TUBING

Copper tubing is generally used for plumbing and refrigerant piping. Plastic pipe is used for waste drains, condensate drains, and some water supplies.

Copper tubing is available as either soft copper or hard-drawn copper. Soft copper may be bent or used with elbows, tees, and other fittings. Hard-drawn copper is not intended to be bent; use it only with fittings to obtain the necessary configurations. There are two basic types of copper tubing used in this industry: air conditioning and refrigeration (ACR) tubing and plumbing tubing. ACR tubing is used for refrigerant piping. It is measured and sold by its outside diameter (OD); e.g., ½ inch ACR tubing has an OD of ½ inch. Plumbing tubing is measured and sold by its inside diameter (ID) rather than its outside diameter. This means that the OD of plumbing tubing is approximately ⅛ inch larger than the size indicated; e.g., ½ inch plumbing tubing has an OD of ⅝ inch. (See Figure 4-1.)

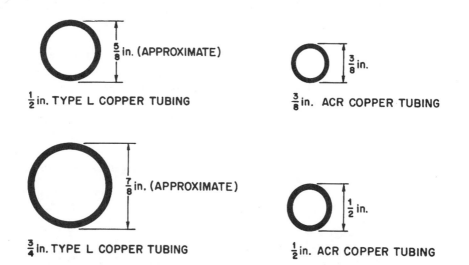

Figure 4-1. Tubing used for plumbing is sized by its inside diameter; ACR tubing is sized by its outside diameter.

49

Both types of tubing are available in several standard weights: Type K is heavy duty; Type L is the standard size and is used most frequently; the other types are not used extensively in this industry. Copper tubing is normally available in diameters from $3/16$ of an inch to greater than six inches.

Soft Copper Tubing

Soft copper tubing is normally available in 25-foot or 50-foot rolls and in diameters from $3/16$ inch to $3/4$ inch. It can be special ordered in 100-foot lengths. ACR tubing is capped on each end to keep it dry and clean inside and often has a charge of nitrogen to keep it free of contaminants. Proper practice should be used to remove tubing from the coil. Never uncoil the tubing from the side of the roll. Place it on a flat surface and unroll it, slowly. Cut only what you need and recap the ends.

Do not bend or straighten the tubing more than necessary because it will harden. This is called *work hardening*. Work-hardened tubing can be softened by heating and allowing it to cool slowly. This is called *annealing*. When annealing, don't use a high concentrated heat in one area. Instead, use a flared flame over one foot at a time. Heat the tubing to a cherry red and allow it to cool slowly.

Hard-Drawn Copper Tubing

Hard-drawn copper tubing is available in 20-foot lengths and in larger diameters than soft copper tubing. Be as careful with hard-drawn copper as with soft copper, and recap the ends after removing the desired length.

4.3 TUBING INSULATION

ACR tubing is often insulated on the low-pressure side of an air conditioning system between the evaporator and compressor to keep the refrigerant from absorbing heat (Figure 4-2). Insulation also prevents condensation from forming on the lines. The closed-cell structure of this insulation eliminates the need for a vapor barrier.

Figure 4-2. ACR tubing with insulation. *Photo by Bill Johnson.*

Figure 4-3. A typical line set. *Photo by Bill Johnson.*

The insulation may be purchased separately from the tubing, or it may be factory installed. If you install the insulation, it is easier, where practical, to apply it to the tubing before assembling the line. The inner surface of the insulation is usually powdered to allow it to slide easily over the tubing, even around most bends. A special adhesive is used to seal the ends of the insulation together.

For existing lines, or when it is impractical to insulate before installing the tubing, the insulation can be slit with a sharp utility knife and snapped over the tubing. All seams must be sealed with insulation adhesive. Do not use tape.

Do not stretch tubing insulation because it reduces the wall thickness of the insulation and therefore reduces its effectiveness. It may also cause the adhesive to fail.

4.4 LINE SETS

Tubing can be purchased as line sets. These sets are precharged with refrigerant, sealed on both ends, and may be obtained with the insulation installed. These line sets will normally have fittings on each end for quicker and cleaner field installation (Figure 4-3). Precharging helps to eliminate improper field charging. It also reduces the possibility of contamination in the system, eliminating clogged systems and compressor damage.

4.5 CUTTING TUBING

Tubing is normally cut with a tube cutter or a hacksaw. The tube cutter is most often used with soft tubing and smaller diameter hard-drawn tubing. A hacksaw may be used with larger diameter hard-drawn tubing.

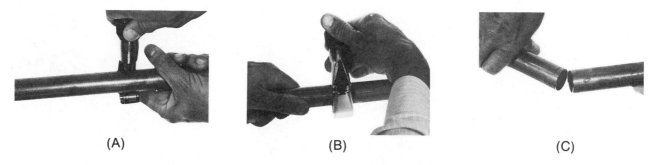

(A) (B) (C)

Figure 4-4. Proper procedure for using a tube cutter. *Photos by Bill Johnson.*

Warning: Always wear eye protection when performing any task that may release small particles into the air.

To cut tubing with a tube cutter, follow the steps listed below and shown in Figure 4-4:

1. Place the tubing in the cutter and align the cutting wheel with the cutting mark on the tube. Tighten the adjusting screw until a moderate pressure is applied to the tubing.
2. Revolve the cutter around the tubing, keeping a moderate pressure applied to the tubing by gradually turning the adjusting screw.
3. Continue until the tubing is cut.

Caution: Do not apply excessive pressure because it may break the cutter wheel and constrict the opening in the tubing.

When the cut is finished, the excess material (called a *burr*) pushed into the pipe by the cutter wheel must be removed (Figure 4-5). Burrs cause turbulence and restrict the fluid or vapor passing through the pipe.

Warning: Burrs are very sharp and can cause cuts or become lodged in the skin. Use care when reaming tubing to avoid injury.

To cut the tubing with a hacksaw, make the cut at a 90° angle to the tubing. A fixture may be used to ensure an accurate cut (Figure 4-6). After cutting, ream the tubing and file the end. Remove all chips and filings, making sure that no debris or metal particles get into the tubing.

Warning: Use caution when making cuts with a hacksaw.

4.6 BENDING TUBING

Only soft tubing should be bent. Use as large a radius bend as possible, Figure 4-7A. All areas of the tubing must remain round. *Do not allow it to flatten or kink*, Figure 4-7B. Carefully bend the tubing, gradually working around the radius.

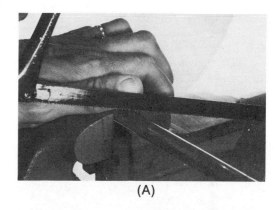

(A)

(B)

Figure 4-6. Proper procedure for using a hacksaw. *Photos by Bill Johnson.*

Figure 4-5. Removing a burr. *Photo by Bill Johnson.*

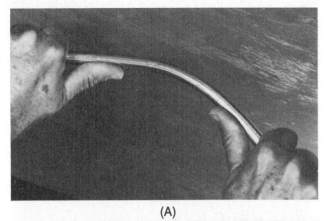

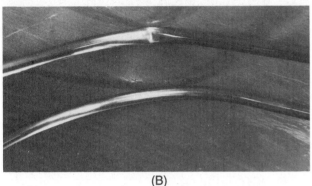

Figure 4-7. (A) Tubing bent by hand. Use as large a radius as possible. (B) Do not allow the tubing to flatten or kink when bending. *Photos by Bill Johnson.*

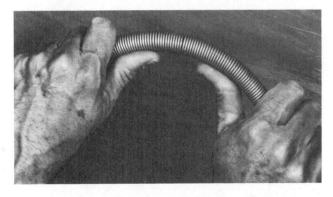

Figure 4-8. Tube bending springs used inside or outside the tubing. Be sure to use the proper size. *Photo by Bill Johnson.*

Tube bending springs may be used to help make the bend (Figure 4-8). They can be used either inside or outside the tube. They are available in different sizes for various tubing diameters. To remove the spring after the bend, you may have to twist it. If you use a spring on the outside, bend the tube before flaring so that the spring may be removed.

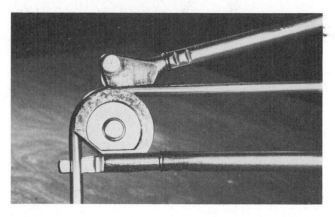

Figure 4-9. Use of a lever-type tube bender. *Photo by Bill Johnson.*

Lever-type tube benders (Figure 4-9) are also available in different sizes. They are used to bend soft copper and thin-walled steel tubing.

4.7 SOLDERING AND BRAZING PROCESSES

Soldering is a process used to join piping and tubing to fittings. It is used primarily in systems utilizing hard copper and brass piping and fittings. Large refrigeration systems use hard tubing and fittings. Soldering, often called soft soldering, is done at temperatures under 800°F, usually in the 375°F to 500°F range.

For moderate pressures and temperatures, 50/50 tin-lead solder is suitable. For higher pressures, or where greater joint strength is required, 95/5 tin-antimony solder can be used.

Brazing requires higher temperatures. It is often called *silver brazing* and is similar to soldering. It is used to join tubing and piping in air conditioning and refrigeration systems. Don't confuse this with welding brazing. In brazing processes, temperatures over 800°F are used. The differences in temperature are necessary due to the different combinations of alloys used in the filler metals.

The brazing filler metals that are used for joining copper tubing are alloys containing 15% to 60% silver (BAg), or copper alloys containing phosphorus (BCuP). In soldering and brazing, the base metal (the piping or tubing) is heated to the melting point of the filler material. *The piping or tubing must not melt.* When two close-fitting, clean, smooth metals are heated to the point where the filler metal melts, the molten metal is drawn into the close-fitting space. This is known as *capillary attraction.* If the soldering is done properly, the molten solder will be absorbed into the pores of the base metal, adhere to all surfaces, and form a strong bond (Figure 4-10).

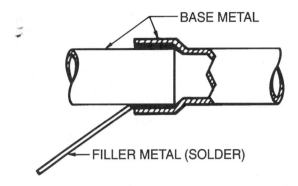

Figure 4-10. The molten solder in a soldered joint will be absorbed into the surface pores of the base metal.

4.8 HEAT SOURCES FOR SOLDERING AND BRAZING

Common heat sources for soldering and brazing include propane, butane, air-acetylene, and oxyacetylene torches. A *propane* or *butane* torch can be easily ignited and adjusted to the type and size of the joint being soldered. Various tips are available.

An *air-acetylene* unit is a type of heat source used often by air conditioning technicians. It usually consists of a tank of acetylene gas, regulator, hose, and torch. Various sizes of standard tips are available. The smaller tips are used for small-diameter tubing, while larger tips are used for larger tubing and for high-temperature applications. A high-velocity tip may be used to provide a greater concentration of heat.

Warning: The torches used for soldering and brazing can operate in excess of 5000°F and may cause serious burns if the operator is careless. Always wear appropriate eye protection and avoid close contact with the flame. Protect adjacent surfaces with fire-resistant material to avoid accidental combustion.

Caution: Poor soldering or brazing techniques can result in system damage due to the introduction of contaminants. Always exercise caution when working with an open system.

To use an air-acetylene unit, follow the procedure listed as follows and shown in Figure 4-11.

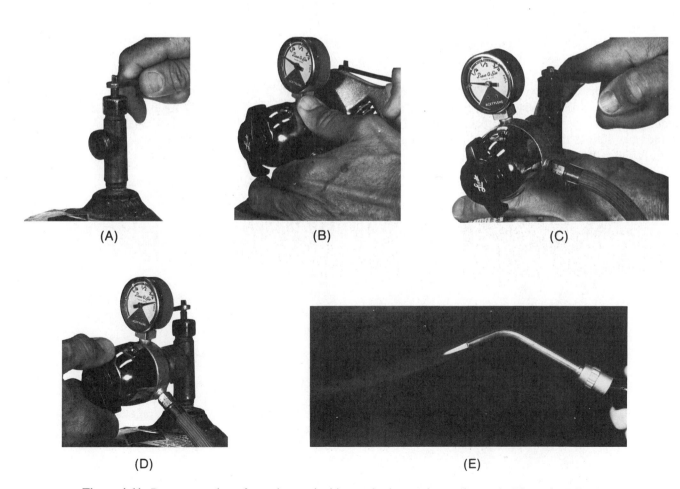

Figure 4-11. Proper procedures for setting up, igniting, and using an air-acetylene unit. *Photos by Bill Johnson.*

54 UNIT 4 TUBING AND PIPING

1. Before connecting the regulator to the tank, open the tank valve slightly to blow out any dirt that may be lodged at the valve.
2. Connect the regulator assembly to the tank. Be sure that all connections are tight.
3. Open the tank valve one-half turn.
4. Adjust the regulator valve to about midrange.
5. Open the needle valve on the torch slightly, and ignite the gas with a spark lighter.

Warning: Do not use matches or cigarette lighters.

Adjust the flame using the needle valve at the handle so that there will be a sharp inner flame and a blue outer flame. After each use, shut off the valve on the tank and bleed off the acetylene in the hose by opening the valve on the torch handle. Bleeding the acetylene from the hoses relieves the pressure when not in use.

Oxyacetylene torches are be preferred by some technicians, particularly when brazing large-diameter tubing or other applications requiring higher temperatures.

Warning: Oxyacetylene equipment can be extremely dangerous if not used properly. Before attempting to use this equipment, you must receive specialized instruction under the close supervision of a qualified technician.

Oxyacetylene brazing and welding processes use a high-temperature flame. Oxygen is mixed with acetylene gas to produce this high heat. The equipment includes oxygen and acetylene cylinders, oxygen and acetylene pressure regulators, hoses, fittings, safety valves, torches, and tips (Figure 4-12).

The regulators each have two gages, one to register tank pressure and the other to register pressure to the torch. The pressures indicated on these regulators are in pounds per square inch gage (psig). These regulators must be used only for the gases and service for which they are intended.

Warning: All connections must be free of dirt, dust, grease, and oil. Oxygen can produce an explosion when in contact with grease or oil. A reverse flow valve should be used. These valves allow the gas to flow in one direction only. Always follow the manufacturer's instructions when attaching these valves; because they are designed to attach to the hose connection on the torch body while others attach to the hose connection on the regulator.

The red hose is attached to the acetylene regulator with left-handed threads and the green hose to the oxygen regulator with right-handed threads. The torch body is then attached to the hoses and the appropriate tip to the torch (Figure 4-13). Many tip sizes and styles are available, depending on the application.

A neutral flame should be used for these operations. Figure 4-14 shows a neutral flame, carbonizing flame (too much acetylene), and oxidizing flame (too much oxygen).

Warning: As stated previously, there are many safety precautions that should be followed when using oxyacetylene equipment. Be sure that you follow all of these precautions. You should become familiar with them during your specialized training for this equipment.

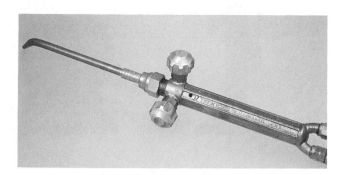

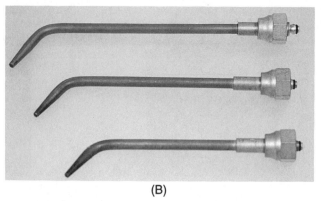

Figure 4-12. Oxyacetylene equipment. *Photo by Bill Johnson.*

Figure 4-13. (A) Oxyacetylene torch with tip. (B) Assortment of oxyacetylene torch tips. *Photos by Bill Johnson.*

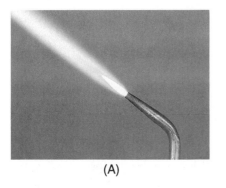 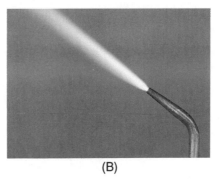 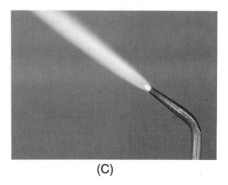

Figure 4-14. (A) Neutral flame. (B) Carbonizing flame. (C) Oxidizing flame. *Photos by Bill Johnson.*

4.9 SOLDERING TECHNIQUES

The mating diameters of tubing and fittings are designed or sized to fit together properly. For good capillary attraction, there should be a space between the metals of approximately 0.003 inch. After the tubing has been cut to size and deburred, you must perform the following procedures:

1. Clean mating parts of the joint.
2. Apply a flux to the male connection.
3. Assemble the tubing and fitting.
4. Heat the joint and apply the solder.
5. Wipe the joint clean.

Cleaning

The end of the copper tubing and the inside of the fitting must be absolutely clean. Even though these surfaces may look clean, they may contain fingerprints, dust, or oxidation. A fine sand cloth, cleaning pad, or special wire brush may be used to clean the surfaces. When the piping system is for a hermetic compressor, a nonconducting, approved sand cloth should be used (Figure 4-15).

Fluxing

Apply the flux immediately after the surfaces have been cleaned. For soft soldering, the flux may be a paste, jelly, or liquid. Apply the flux with a *clean* brush or applicator.

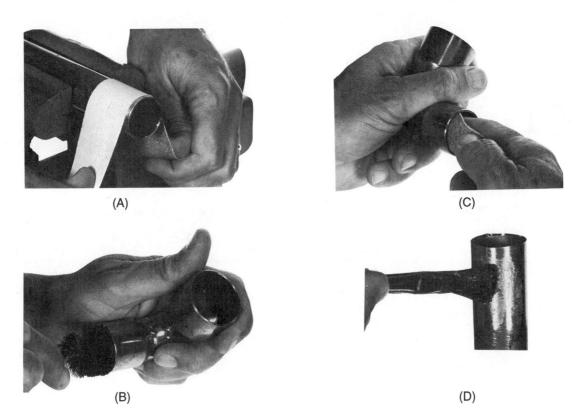

Figure 4-15. Cleaning and fluxing. (A) Clean tubing with sand cloth. (B) Clean fitting with a brush. (C) Clean fitting with sand cloth. (D) Apply flux. *Photos by Bill Johnson.*

Do *not* use a brush that has been used for any other purpose. Apply the flux only to the area to be joined, and avoid getting it into the piping system. The flux minimizes oxidation while the joint is being heated. It also helps to float dirt or dust out of the joint.

Assembly

Soon after the flux has been applied, assemble and support the joint so that it is straight and will not move while being soldered (Figure 4-15).

Heating and Applying Solder

First, heat the tubing near the fitting for a short time. Then, move the torch from the tubing to the fitting. Keep moving the torch back and forth between the tubing and the fitting. This will spread the heat evenly and avoid overheating any one area. Do not point the flame into the fitting socket. Hold the torch so that the inner cone of the flame just touches the metal. After briefly heating the joint, touch the solder to the joint. If it does not readily melt, remove it and continue heating the joint. Continue to test the heat of the metal with the solder. Do *not* melt the solder with the flame; use the heat in the metal. When the solder flows freely from the heat of the metal, feed enough solder in to fill the joint. Do not use excessive solder. Figure 4-16 shows the step-by-step procedure for heating the joint and applying solder.

For horizontal joints, it is preferable to apply the filler metal first to the bottom, then to the sides, and finally to the top, making sure the operations overlap. On vertical joints, it does not matter where the filler is first applied.

Wiping

While the joint is still hot, you may wipe it with a rag to remove excess solder. This is not necessary for producing a good bond, but it improves the appearance of the joint.

4.10 BRAZING TECHNIQUES

Cleaning and Fluxing

The cleaning and fluxing procedures for brazing are similar to those for soldering. The brazing flux is applied with a brush to the cleaned area of the tube end. Avoid getting flux inside the piping system. Some types of silver-copper-phosphorus alloy do not require extensive cleaning when brazing copper to copper. Always follow the filler-material manufacturer's instructions.

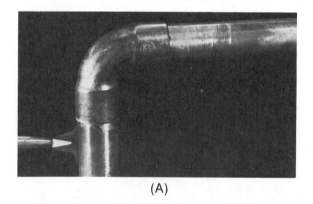

(A)

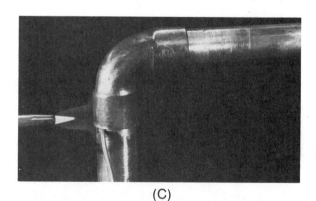

(C)

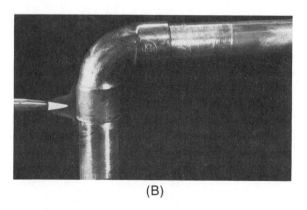

(B)

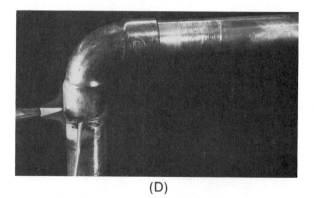

(D)

Figure 4-16. Proper procedures for heating a joint and applying solder. (A) Start by heating the tubing. (B) Keep moving the flame. Do not point flame into edge of fitting. (C) Touch solder to joint to check for proper heat. Do not melt solder with flame. (D) When joint is hot enough, solder will flow. *Photos by Bill Johnson.*

Heating and Applying Solder

Before you heat the joint, it is best to force nitrogen or carbon dioxide into the system to purge the air and reduce the possibility of oxidation. Apply heat to the parts to be joined using an air-acetylene or oxyacetylene torch. Heat the tube first, beginning about one inch from the edge of the fitting, and sweeping the flame around the tube. It is very important to keep the flame in motion and to avoid overheating any one area. Next, switch the flame to the fitting at the base of the cup. Heat uniformly, sweeping the flame from the fitting to the tube. Apply the filler rod or wire at a point where the tube enters the socket. When the proper temperature is reached, the filler metal will flow readily into the space between the tube and the fitting. As in soldering, do not heat the filler metal itself. The temperature of the metal at the joint should be hot enough to melt the filler metal. When the joint is at the correct temperature, it will be cherry red in color. The procedures are the same as with soldering except for the materials used and the higher heat applied.

Note: The flux used in the brazing process will cause oxidation. After the joint has cooled, it should be washed with soap and water.

4.11 PRACTICAL SOLDERING AND BRAZING TIPS

Low-Temperature Soldering

The adjoining surfaces must be absolutely clean before soldering. Clean surfaces are necessary to ensure leak-free connections. It takes much longer to prepare surfaces for soldering than it does to actually make the soldered joint. It is best to solder the joint immediately after cleaning it. If this is not possible, perform a touch-up cleaning before the joint is soldered. Copper oxidizes and iron or steel begin to rust as soon as they are exposed to the air. Certain fluxes may be applied after cleaning to prevent oxidation and rust until the tubing and fittings are ready for soldering.

Only the best solders should be used for low-temperature solder connections in refrigeration and air conditioning systems. For many years, systems were soldered successfully with 95/5 solder. If the soldered connection is completed in the correct manner, 95/5 can still be used. However, it should never be used on the high-pressure side of the system close to the compressor. The high temperature of the discharge line and the vibration will very likely cause it to leak.

Modern alternatives to 95/5 include low-temperature solders with a silver content. They offer greater strength with low melting temperatures.

One of the problems with most low-temperature solders is that the melting and flow points are too close together; i.e., it often flows too fast and you have a hard time keeping it in the clearance in the joint. Some of the silver-type, low-temperature solders have a wider melt and flowpoint and are easier to use. They also have the advantage of being more elastic during the soldering procedure. This allows gaps between fittings to be filled more easily.

High-Temperature Soldering (Brazing)

There are several types of high-temperature brazing materials available. Some include a high silver content (45% silver) and must always be used with a flux. Others do not have a high silver content (15% silver) and may not require flux when making copper-to-copper connections. Still, others have been developed with no silver content. The choice of material is largely a matter of personal preference.

Different Joints

The type of joint dictates the solder or brazing materials used. Not all connections are copper-to-copper. Some may be copper-to-steel, copper-to-brass, or brass-to-steel. These are called *dissimilar metal connections*. Examples include:

1. *A copper suction line to a steel compressor or connection.* The logical choice is to braze the joint using 45% silver filler material. This provides a strong connection with a high melting temperature.
2. *A copper suction line to a brass accessory valve.* From a strength standpoint, the best choice would be to braze the joint using 45% silver. Another option would be to use a low-temperature solder with a silver filler material. It still provides a strong joint, and the valve body will not have to be heated to the high melting temperature of the 45% silver filler material.
3. *A copper liquid line to a steel filter drier.* Brazing with 45% silver filler material is a good choice but requires a lot of heat. A low-temperature, silver-type solder may be the best choice. It also provides the option of easily replacing the drier at a later date.
4. *A large copper suction line connection using hard-drawn tubing.* A high-temperature brazing material with low silver content would be the choice of many technicians because it can be used without flux, but the high temperature may affect the temper of the hard-drawn tubing. A low-temperature solder with a silver content will provide the needed strength, but will not remove the temper from the pipe.

Heat Sources for Soldering and Brazing

Many technicians prefer air-acetylene torches because oxyacetylene units are heavy and difficult to use. In addition, the air-acetylene torches that use the twist-tip method for mixing the air and acetylene may be used for both low-temperature and high-temperature soldering. For general piping work, there are practically no instances where oxyacetylene is necessary for brazing unless you are installing systems over 15 tons.

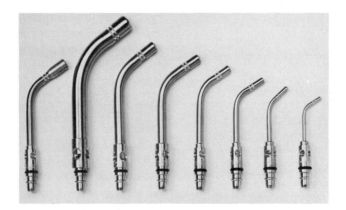

ACETYLENE TORCH TIPS

Tip No.	Tip Size in.	Tip Size mm	Gas Flow @ 14 psi ft³/hr	Gas Flow (0.9 Bar) m³/hr	Copper Tubing Size Capacity — Soft Solder in.	Soft Solder mm	Silver Solder in.	Silver Solder mm
A-2	3/16	4.8	2.0	.17	1/8-1/2	3-15	1/8-1/4	3-10
A-3	1/4	6.4	3.6	.31	1/4-1	5-25	1/8-1/2	3-12
A-5	5/16	7.9	5.7	.48	3/4-1 1/2	20-40	1/4-3/4	10-20
A-8	3/8	9.5	8.3	.71	1-2	25-50	1/2-1	15-30
A-11	7/16	11.1	11.0	.94	1 1/2-3	40-75	7/8-1 5/8	20-40
A-14	1/2	12.7	14.5	1.23	2-3 1/2	50-90	1-2	30-50
A-32*	3/4	19.0	33.2	2.82	4-6	100-150	1 1/2-4	40-100
MSA-8	3/8	9.5	5.8	.50	3/4-3	20-40	1/4-3/4	10-20

*Use with large tank only.
NOTE: For air conditioning, add 1/8 inch for type L tubing.

Figure 4-17. Different tip sizes may be used for different pipe sizes and solder combinations. *Courtesy Thermadyne Industries, Inc.*

An air-acetylene torch has a flame temperature of 5589°F. The correct tip must be used for the type of filler material and pipe size (Figure 4-17).

Another heat source is MAPP gas. It is a composite gas that is similar in nature to propane and may be used with air. The flame temperature of MAPP gas is 5301°F. It does not get as hot as air-acetylene, but is supplied in larger and possibly lighter containers (Figure 4-18).

General Soldering and Brazing Tips

Developing effective soldering and brazing techniques takes time and practice. Keep the following in mind as you learn these procedures:

1. Always clean all surfaces to be soldered or brazed, and keep all filings and flux from entering the pipe.
2. When making horizontal soldered joints, apply heat to the top of the fitting first.
3. When soldering or brazing fittings of different weights, such as soldering a copper line to a large brass valve body, most of the heat should be applied to the large mass of metal (the valve).

Figure 4-18. MAPP gas container. *Courtesy Thermadyne Industries, Inc.*

4. Do not overheat the connections. The heat may be varied by moving the torch closer to or further from the joint. Once heat is applied to a connection, do not allow it to cool because air will move in and oxidize the metal before you have a chance to solder the joint.
5. Do not apply excess solder. It is a good idea to mark the desired length of solder with a bend (Figure 4-19). When you get to the bend, stop or you will overfill the joint and the excess may end up in the system.
6. When using flux with high-temperature brazing material, always chip the flux away when finished. *Wear eye protection.* The flux appears as a hard, glassy material on the brazed connection (Figure 4-20). This substance may cover a leak and be blown out later.
7. When working with corrosive fluxes, such as those used with low-temperature solders, wash the flux off the connection or corrosion will occur. A corroded connection reflects poor workmanship.
8. Often, the best source of information is the expert at the solder supply house.

4.12 MAKING FLARE JOINTS

Another method for joining tubing and fittings is the *flare joint*. This joint uses a flare on the end of the tubing against an angle on a fitting, and is secured with a flare nut behind the flare on the tubing (Figure 4-21).

Making a Single-Thickness Flare

Standard tubing flares are made using a screw-type flaring tool. To make a flare, use the following procedure:

1. Cut the tubing to the desired length.
2. Ream the tubing to remove all burrs and clean all residue from the tubing.
3. Slip the flare nut or coupling nut over the tubing with the threaded end facing the end of the tubing.
4. Clamp the tube in the flaring block, Figure 4-22(A). Adjust it so that the tube is slightly above the block (about one third of the total height of the flare).
5. Place the yoke on the block with the tapered cone over the end of the tube. Many technicians use a drop or two of refrigerant oil to lubricate the inside of the flare while it is being made, Figure 4-22(B).
6. Turn the screw down firmly, Figure 4-22(C). Continue to turn the screw until the flare is completed.
7. Remove the tubing from the block, Figure 4-22(D). Inspect it for defects. If you find any, cut off the flare and start over.
8. Assemble the joint.

Making a Double-Thickness Flare

A double-thickness flare provides additional strength at the flare end of the tube. Double-thickness flares are sometimes used for joints between the liquid line and receiver on the compressor discharge line on small systems. Making the flare is a two-step operation.

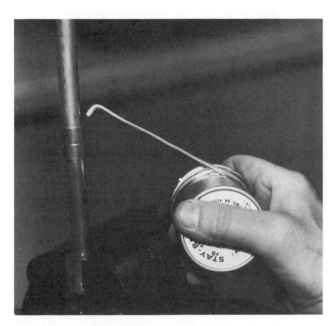

Figure 4-19. Make a bend in the end of the solder so you will know when to stop. *Photo by Bill Johnson.*

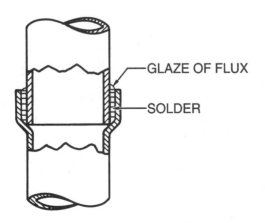

Figure 4-20. Flux used with high-temperature brazing materials will form a glaze that looks like glass.

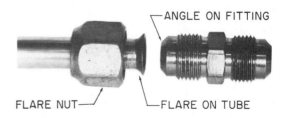

Figure 4-21. Components of a flare joint. *Photo by Bill Johnson.*

UNIT 4 TUBING AND PIPING

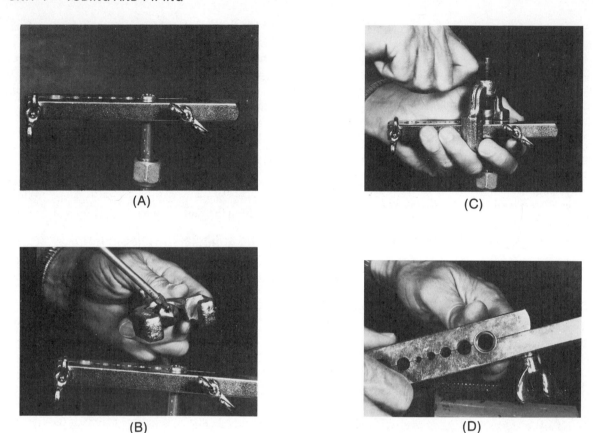

Figure 4-22. Proper procedure for making a flare joint using a screw-type flaring tool. *Photos by Bill Johnson.*

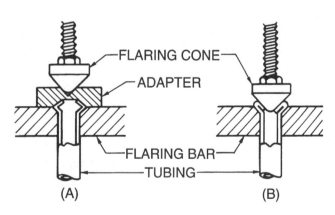

Figure 4-23. Procedure for making a double-thickness flare. (A) Place adapter of combination flaring tool over the tube extended above flaring block. Screw down to bell out tubing. (B) Remove adapter, place cone over tube, and screw down to form a double flare.

It requires a punch and block or a combination flaring tool. Figure 4-23 illustrates the procedure for making double-thickness flares with the combination flaring tool.

Figure 4-24. Examples of flare fittings. *Photo by Bill Johnson.*

Flare Fittings

Many fittings are available for use with a flare joint. Each of the fittings has a 45° angle on the end that fits against the flare on the end of the tube (Figure 4-24).

4.13 SWAGING

Swaging is less common than flaring or simply soldering or brazing tubing to fittings. It takes longer, but it can be used if a coupling is unavailable. Swaging is the joining of two pieces of copper tubing of the same diameter by expanding or stretching the end of one piece to fit over the other so the joint may be soldered or brazed (Figure 4-25).

SWAGING 61

Figure 4-25. Joining tubing using a swaged joint. *Photo by Bill Johnson.*

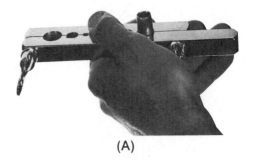

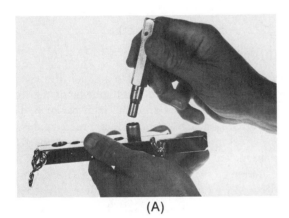

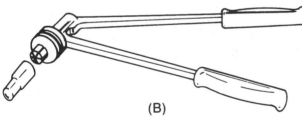

Figure 4-26. (A) Swaging punch. (B) Lever-type swaging tool. *Courtesy of: (A) Photo by Bill Johnson, (B) From Lang, Air Conditioning: Procedures and Installation,* © 1982 by Delmar Publishers, Inc.

Figure 4-27. Making a swaged joint. (A) Tube placed in block for swaging. (B) Swaging punch expanding metal. *Photos by Bill Johnson.*

Swaged joints are made using either a punch or a lever-type tool to expand the end of the tubing (Figure 4-26). To make a swaged joint using a punch, perform the following procedure:

1. Place the tubing in a flare block or an anvil block that has a hole equal to the outside diameter of the tubing. The tube should extend above the block by an amount equal to the outside diameter of the tubing plus approximately 1/8 inch, Figure 4-27(A).

2. Place the correct size swaging punch in the tube and strike it with a hammer until the proper shape and length of the joint has been obtained, Figure 4-27(B). Follow the same procedure with screw-type or lever-type tools. A drop or two of refrigerant oil on the swaging tool will help but must be cleaned off before soldering.
3. Assemble the joint. The tubing should fit together smoothly.
4. Always inspect the tubing after swaging to see if there are cracks or other defects. If any are seen or suspected, cut off the swage and start over.

Caution: Field fabrication of tubing is not done under factory-clean conditions, so you need to be extra careful that *no* foreign material enters the piping. Any foreign material in an air conditioning system *will* cause problems.

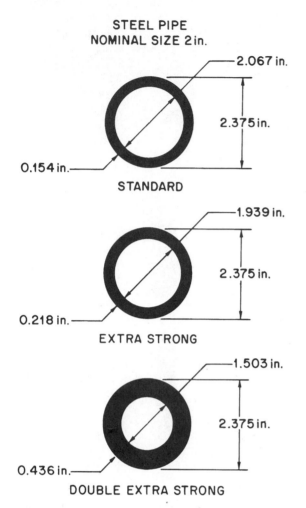

Figure 4-28. Cross-section of standard, extra strong, and double extra strong steel pipe.

4.14 STEEL AND WROUGHT IRON PIPE

The terms *steel pipe, wrought steel,* and *wrought iron pipe* are often used interchangeably and incorrectly. When you want wrought iron pipe, specify "genuine wrought iron" to avoid confusion.

When manufactured, steel pipe is either welded or produced without a seam by drawing hot steel through a forming machine. This pipe may be painted, left black, or coated with zinc *(galvanized)* to help resist rusting.

Steel pipe is often used in plumbing and chilled water cooling applications. The size of the pipe is referred to as the *nominal size*. For pipe sizes less than 12 inches in diameter, the nominal size is approximately equal to the size of the inside diameter of the pipe. For sizes larger than 12 inches in diameter the outside diameter is considered the nominal size. Steel pipe comes in many wall thicknesses, but is normally furnished in standard, extra strong, and double extra strong sizes. Figure 4-28 is a cross-section of these wall thicknesses for two-inch pipe. For most common pipe diameters, this information is supplied in the pipe schedules available from the pipe manufacturer.

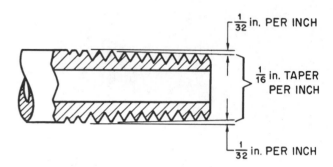

Figure 4-29. Cross-section of a pipe thread.

4.15 JOINING STEEL PIPE

Steel pipe is joined using either welded or threaded fittings. There are two types of American National Standard pipe threads: tapered pipe and straight pipe. In this industry, tapered pipe threads are used because they produce a tighter joint and help prevent leakage.

Pipe Threads

Pipe threads have been standardized. Each thread is V-shaped with an angle of 60°. The diameter of the thread has a taper of 3/4 inch per foot or 1/16 inch per inch. There should be approximately seven perfect threads and two or three imperfect threads for each joint (Figure 4-29). If the threads are nicked or broken, leaks may occur.

Thread diameters refer to the approximate inside diameter of the steel pipe. The nominal size will then be smaller than the actual diameter of the thread. Figure 4-30 shows the number of threads per inch for common pipe sizes. A thread dimension is written as follows: first the diameter, followed by the number of threads per inch, and then the letters NPT (Figure 4-31).

PIPE SIZE (INCHES)	THREADS PER INCH
$\frac{1}{8}$	27
$\frac{1}{4}, \frac{3}{8}$	18
$\frac{1}{2}, \frac{3}{4}$	14
1 to 2	$11\frac{1}{2}$
$2\frac{1}{2}$ to 12	8

Figure 4-30. Threads per inch for common pipe sizes.

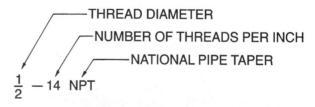

Figure 4-31. Thread specification.

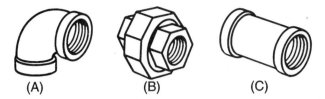

Figure 4-32. Steel pipe fittings. (A) 90° elbow. (B) Union. (C) Coupling.

Pipe Fittings

Various fittings are used to connect steel pipe. Three common threaded fittings are illustrated in Figure 4-32.

Tools

The following tools are needed to cut and thread pipe:

- A *hacksaw* (use one with 18 to 24 teeth per inch) or *pipe cutter* is generally used to cut the pipe (Figure 4-33). A pipe cutter is generally preferred because it makes a square cut, but there must be room to swing the cutter around the pipe.
- A *reamer* removes burrs from the inside of the pipe after it has been cut. The burrs must be removed because they restrict the flow of the fluid or gas (Figure 4-34).
- A *threader*, also known as a *die*, is used to cut the pipe threads. Most threading devices used in this field are fixed-die threaders (Figure 4-35).

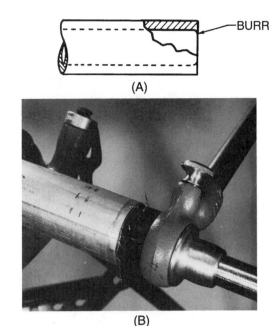

Figure 4-34. (A) Burr inside a pipe. (B) Using a reamer to remove a burr. *Photo by Bill Johnson.*

Figure 4-35. Fixed die-type pipe threader. *Courtesy Ridge Tool Company, Elyria, Ohio.*

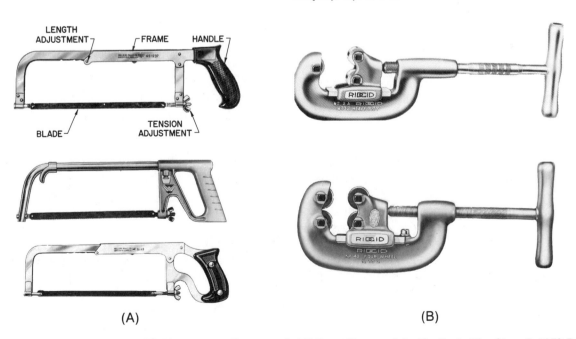

Figure 4-33. (A) Standard hacksaws. (B) Pipe cutters. *Courtesy of: (A) From Slater and Smith, Basic Plumbing, © 1979 Delmar Publishers, Inc., (B) Ridge Tool Company, Elyria, Ohio.*

64 UNIT 4 TUBING AND PIPING

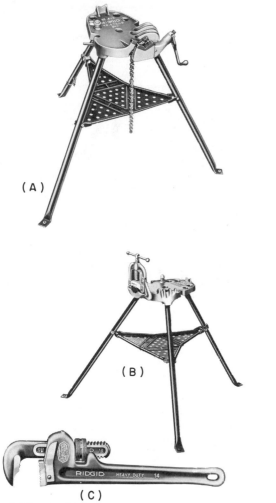

Figure 4-36. (A) Chain vise. (B) Yoke vise. (C) Pipe wrench. *Courtesy Ridge Tool Company, Elyria, Ohio.*

Figure 4-37. Pipe cutter. *Courtesy Ridge Tool Company, Elyria, Ohio.*

Figure 4-38. Cutting steel pipe with a hacksaw and holding fixture. *Photo by Bill Johnson.*

- Tools such as the *chain vise*, *yoke vise*, and *pipe wrench* (Figure 4-36) are used to hold the pipe in place during various procedures.

When large quantities of pipe are cut and threaded regularly, special machines can be used. These machines are not covered in this text.

Cutting

The pipe must be cut square to be threaded properly. If there is room to revolve a pipe cutter around the pipe, you can use a one-wheel cutter (Figure 4-37). Otherwise, use one with more than one cutting wheel. If the pipe has not be installed, secure it in a chain vise or yoke vise. Place the cutting wheel directly over the place where the pipe is to be cut. Adjust the cutter with the T handle until all the rollers or cutters contact the pipe. Apply moderate pressure with the T handle and rotate the cutter around the pipe. Turn the handle about one-quarter turn for each revolution around the pipe. Do not apply too much pressure because it will cause a large burr inside the pipe and excessive wear of the cutting wheel.

To use a hacksaw, start the cut gently. Use your thumb to guide the blade or use a holding fixture (Figure 4-38).

Warning: Keep your thumb away from the teeth.

A hacksaw will only cut on the forward stroke. Do not apply pressure on the backstroke. Do not force the hacksaw or apply excessive pressure. Let the saw do the work.

Reaming

After the pipe has been cut, insert a reamer in the end of the pipe. Apply pressure against the reamer and turn clockwise. Ream only until the burr is removed.

Threading

To thread the pipe, place the die over the end and make sure it lines up square with the pipe. Apply cutting oil on the pipe and turn the die once or twice. Then reverse the die approximately one-quarter turn. Rotate the die one or two more turns, then reverse it again. Continue this procedure and apply cutting oil liberally until the end of the pipe is flush with the far side of the die (Figure 4-39).

PLASTIC PIPE 65

Figure 4-39. Threading pipe. *Photos by Bill Johnson.*

4.16 INSTALLING STEEL PIPE

When installing steel pipe, use pipe wrenches to hold or turn the fittings and pipe. These wrenches have teeth set at an angle so that the fitting or pipe will be held securely when pressure is applied. Position the wrenches in opposite directions on the pipe and fitting (Figure 4-40).

When assembling the pipe, use the *correct* pipe thread dope on the male threads. Do not apply the dope closer than two threads from the end of the pipe (Figure 4-41), or it may get into the piping system.

Figure 4-40. Holding pipe and turning fitting with pipe wrenches. *Photo by Bill Johnson.*

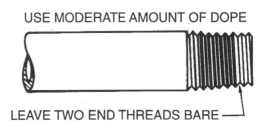

Figure 4-41. Applying pipe dope.

Caution: Never substitute pipe sizes without consulting the system designer. The designer has carefully studied the entire system and has indicated the size that will deliver the correct amount of gas or fluid. Any substitutions may have a significant impact on system operation and efficiency. In addition, all state and local codes must be followed when piping any system. You should continually familiarize yourself with all applicable codes.

4.17 PLASTIC PIPE

Plastic pipe is used for many plumbing, venting, and condensate applications. The four most common types are explained below.

ABS (Acrylonitrilebutadiene Styrene)

ABS is used for water drains, waste, and venting. It can withstand heat up to 180°F without pressure. Use a solvent cement to join ABS with ABS; use a transition fitting to join ABS with metal piping. ABS is rigid and has good impact strength at low temperatures.

PE (Polyethylene)

PE is used for water, gas, and irrigation systems. It can be used for water supply and sprinkler systems and water-source heat pumps. PE is not used for hot water supply systems; although it can stand heat with no pressure, it does not offer good impact strength at higher temperatures. It is flexible and does provide good impact strength at low temperatures.

PVC (Polyvinyl Chloride)

PVC can be used in high-pressure applications at low temperatures. It can be used for water, gas, sewage, certain industrial processes, and irrigation systems. It is a rigid pipe and has a high impact strength. PVC pipe can be joined to PVC fittings with a solvent cement, or it can be threaded and used with a transition fitting for joining to metal pipe.

CPVC (Chlorinated Polyvinyl Chloride)

CPVC is similar to PVC except that it can be used with temperatures up to 180°F at 100 psig. It is used for both hot and cold water supplies and is joined to fittings in the same manner as PVC.

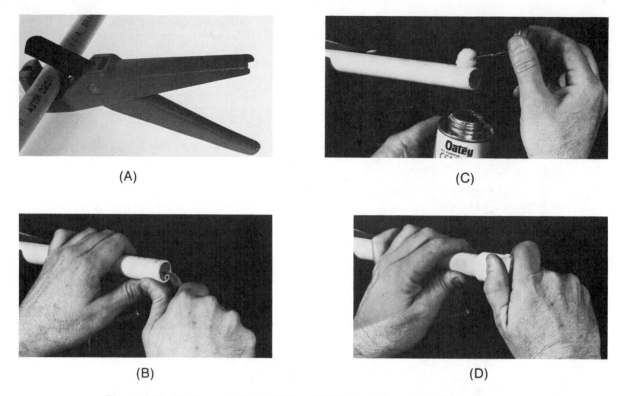

Figure 4-42. Cutting and joining PVC or CPVC pipe. *Photos by Bill Johnson.*

Joining PVC or CPVC Piping

To prepare PVC or CPVC for joining, perform the steps listed below and shown in Figure 4-42.

1. Cut the end square with a plastic tubing shear, hacksaw, or tube cutter. The tube cutter should have a special wheel for plastic pipe.
2. Deburr the pipe inside and out using a knife or a half-round file.
3. Clean the pipe end. Apply the cement primer and solvent to both the outside of the pipe and the inside of the fitting.

Warning: Avoid inhaling the pipe cement vapor. Prolonged exposure poses a health risk.

4. Insert the pipe all the way into the fitting. Turn approximately one-quarter turn to spread the cement and allow it to set (dry) for about one minute.

Schedule #80 PVC and CPVC can be threaded. A regular pipe thread die can be used.

Caution: Do not use the same die for metal and plastic pipe. The die used for metal will become too dull to be used for plastic. The plastic pipe die must be kept very sharp. Always follow the manufacturer's instructions when using any plastic pipe or cement.

Joining ABS Piping

ABS is prepared for joining in a similar way except that a primer must be applied before applying the solvent cement. ABS does not use the same cement as PVC and CPVC.

Joining PE Piping

PE piping is normally attached to fittings with two hose clamps. Place the screws of the clamps on opposite sides of the pipe (Figure 4-43).

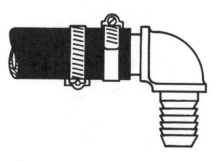

Figure 4-43. Position of clamps on PE pipe.

SUMMARY

- The use of correct tubing, piping, and fittings along with proper installation is necessary for an air conditioning system to operate properly. Careless handling of the tubing and poor soldering or brazing techniques may cause serious damage to the system components.
- Copper tubing is generally used for air conditioning piping.
- Copper tubing is available in soft-drawn or hard-drawn copper. ACR copper tubing is used in air conditioning systems.
- The nominal size of ACR tubing refers to the outside diameter.
- A proper adhesive should be used to fasten tubing insulation together.
- ACR tubing can be purchased as line sets that are precharged with refrigerant and sealed on both ends.
- Tubing may be cut with a hacksaw or tube cutter.
- Soft tubing may be bent. Tube bending springs or lever-type benders may be used, or the bend can be made by hand.
- Soldering and brazing are used to fasten tubing and fittings together. Temperatures below 800°F are used for soldering; temperatures above 800°F are used for brazing.
- Air-acetylene units are frequently used for soldering and brazing.
- Oxyacetylene equipment is also used, particularly when brazing at higher temperatures.
- The flare joint is another method for joining tubing and fittings.
- The soldered swaged joint is a method used to fasten two pieces of copper tubing together. It is not used as often as soldering or brazing fittings to tubing or using flared joints.
- ABS, PE, PVC, and CPVC are four types of plastic pipe; each has different uses.
- ABS, PVC, and CPVC are joined to fittings using a solvent cement. PE is joined using metal clamps.

REVIEW QUESTIONS

1. Which type of copper tubing may be bent?
2. What would the size $1/2$ inch refer to with regard to ACR tubing?
3. In what size rolls is soft copper tubing normally available?
4. In what lengths is hard-drawn copper normally available?
5. Why are some ACR tubing lines insulated?
6. What sealant is best for all tubing insulation seams?
7. Describe a line set.
8. Describe the procedure for cutting tubing with a tube cutter.
9. Describe the different procedures for bending soft copper tubing.
10. What are the approximate temperature ranges used when soldering?
11. What is considered the minimum temperature for brazing?
12. What type of solder is suitable for moderate pressures and temperatures?
13. What are the elements that make up brazing filler metal alloys?
14. Describe how to make a good soldered joint.
15. What type of equipment is often used in soldering and brazing applications?
16. List the proper procedures for setting up, igniting, and using an air-acetylene unit.
17. Describe the procedure for making a brazed joint.
18. Describe how a flare joint seals.
19. Explain why a drop of oil is used when making a flare joint.
20. What would be an application for a double-thickness flare?
21. What is swaging?
22. What should you do if you see or suspect a crack in a flare joint?
23. Describe the thread dimension $1/4$-18 NPT.
24. List four types of plastic pipe.
25. Describe the joining procedures for joining each type of plastic pipe.

5 SYSTEM EVACUATION

OBJECTIVES

Upon completion of this unit, you should be able to:

- Define a deep vacuum.
- Explain two different types of evacuation.
- Describe two different types of vacuum measurement instruments.
- Select a high-vacuum pump.
- List proper evacuation practices.
- Describe a high-vacuum single evacuation.
- Describe a triple evacuation.

5.1 EVACUATING A/C SYSTEMS

Air conditioning systems are designed to operate with only refrigerant and oil circulating inside them. Air enters the system during installation and service procedures. Air contains oxygen, nitrogen, hydrogen, and water vapor, all of which are detrimental to the system. First, nitrogen, hydrogen, and oxygen are generally considered to be *noncondensable*. This means that they will not condense in the condenser and will occupy condenser space that would normally be used for condensing the refrigerant. This will cause a rise in head pressure. Second, these contaminants cause chemical reactions that produce acids in the system. Acids cause deterioration of the system components, copper plating of the running gear, and breakdown of motor insulation. Air contains 13% oxygen, which will react with the various chemicals that make up the refrigerant. Some of these chemical combinations result in the formation of hydrofluoric or hydrochloric acid. These acids may remain in the system for years without showing signs of a problem. Then, the motor will burn out or copper will be deposited on the crankshaft from the copper in the system, causing the crankshaft to become slightly oversized and binding will occur. This will cause the rubbing surfaces to score and become worn prematurely.

These contaminants must be removed from the system if it is to have a normal life expectancy. Noncondensable gases are removed using a vacuum pump after the system is leak checked. The pressure inside the system is reduced to an almost perfect vacuum. Moisture is removed using a combination of evacuation and dehydration (discussed later).

5.2 UNDERSTANDING EVACUATION

To *pull a vacuum* means to lower the pressure in a system below atmospheric pressure. The atmosphere exerts a pressure of 14.696 psia (29.92 in. Hg) at sea level at 70°F. Vacuum is sometimes expressed in millimeters of mercury (mm Hg). The atmosphere will support a column of mercury 760 mm (29.92 in. Hg) high.

To pull a *perfect* vacuum in an air conditioning system, all of the atmosphere must be removed. If this were possible, the pressure inside the system would be reduced to 0 psia (29.92 in. Hg vacuum).

A compound gage is often used to indicate the vacuum level. A compound gage measures pressures in the range from 0 in. Hg vacuum down to 30 in. Hg vacuum. When the term "vacuum" is used, it is applied to the compound gage. When the term "in. Hg" is used, it is applied to a manometer or barometer.

The typical bell jar shown in Figure 5-1 can be used to describe a typical system evacuation. A refrigeration system contains a volume of gas like the bell jar. The only difference is that an air conditioning system is composed of many small chambers connected by piping. These chambers include the cylinders of the compressor, which may have a reed valve partially sealing it from the system.

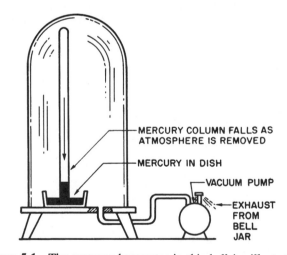

Figure 5-1. The mercury barometer in this bell jar illustrates how the atmosphere will support a column of mercury. As the atmosphere is removed from the jar, the column of mercury will begin to fall. If all of the atmosphere could be removed, the mercury would be at the bottom of the column.

As the atmosphere is pulled out of the bell jar, the barometer inside the jar changes (Figure 5-1). The standing column of mercury begins to drop. When the column drops down to 1 mm Hg, only a small amount of atmosphere is still in the jar ($1/760$ of the original volume, which was 760 mm Hg).

5.3 MEASURING A VACUUM

When the pressure in the bell jar is reduced to 1 mm Hg, the mercury column becomes difficult to read. In order to produce an accurate measurement, microns are used (1000 microns = 1 mm Hg). Microns are measured with electronic instruments, such as a thermocouple or thermistor vacuum gage. Figure 5-2 shows a typical electronic vacuum gage.

The electronic vacuum gage consists of a vacuum indicator, interconnecting wire, and a sensor with a threaded connection for attachment to the system (Figure 5-3). The sensor section of the instrument must never be exposed to system pressure, so it is advisable to install a valve between it and the system. The sensor may also be installed using gage lines, as shown in Figure 5-4.

Figure 5-2. An electronic vacuum (micron) gage used to measure vacuums in the very low range. *Courtesy Robinair Division, SPX Corporation.*

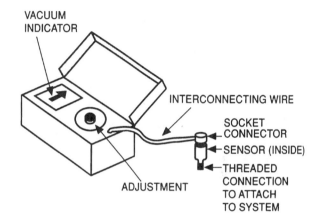

Figure 5-3. Components of electronic vacuum gage.

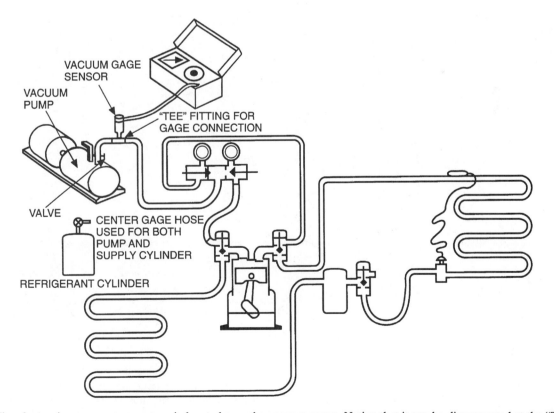

Figure 5-4. The electronic vacuum gage sensor is located near the vacuum pump. Notice that it can be disconnected at the "Tee" fitting to prevent pressure from entering the sensor.

Caution: The sensor should always be located in an upright position so that any oil in the system will not drain into the sensor.

When an electronic sensor is used, the vacuum pump should be operated until the gage manifold indicates a deep vacuum. Then, the valve to the sensor should be opened and the instrument turned on. It is pointless to turn the instrument on until a fairly deep vacuum is achieved (below 25 in. Hg) as indicated on the gage manifold.

When the vacuum gage reaches the desired level, usually about 250 microns, the vacuum pump should then be valved off and the reading marked. The instrument reading may rise for a short time (about one minute) before it stabilizes. At this point, the true system reading can be recorded. If it continues to rise, it indicates one of two problems: moisture is present and is boiling, which creases pressure; or a leak exists in the system.

One of the advantages of the electronic vacuum gage is its rapid response to very small pressure rises. The smaller the system, the faster the rise. A very large system will take longer to reach new pressure levels. These pressure differences may be seen instantly on the electronic instrument.

Caution: Be sure you allow for the first rise in pressure mentioned above assuming that a leak is present.

Another common vacuum gage is the U-tube manometer, Figure 5-5(A). This instrument consists of glass tubing formed in a U shape and closed on one side. It uses mercury as an indicator. It has two columns of mercury which balance each other. The atmosphere has been removed from one side of the manometer column so that the instrument has a standing column of about 5 in. Hg. Because of this difference in the height of the mercury, this gage starts indicating at about 25 in. Hg vacuum. When the gage is attached to the system and the vacuum pump is started, the gage will not read until the vacuum reaches about 25 in. Hg vacuum, Figure 5-5(B). Then, the gage will gradually fall until the two columns of mercury are equal. At this time, the vacuum in the system is between 1 mm Hg and a perfect vacuum. The instrument cannot be read much closer than that, Figure 5-5(C).

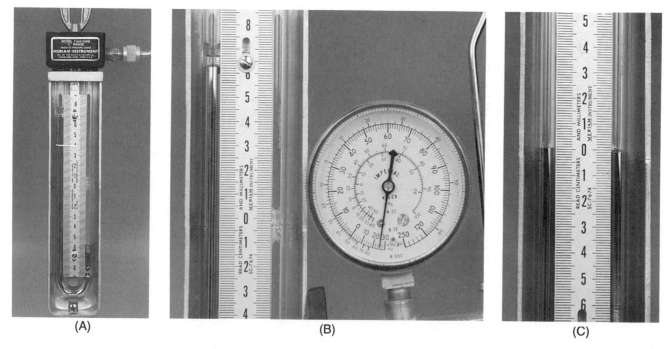

Figure 5-5. This mercury U-tube manometer is shown at various stages of evacuation. (A) The column on the left is a closed column with no atmosphere above it. When the column on the right is connected to a system that is at a very low vacuum, the column on the left will fall and the column on the right will rise. (B) This will not start until the vacuum on the right is at about 25 in. Hg vacuum or 5 in. Hg absolute. (C) As the atmosphere is pulled out of the right hand column, the column rises. At a perfect vacuum, it will rise to a position that is exactly equal to the left hand column. *Photos by Bill Johnson.*

VACUUM PUMPS 71

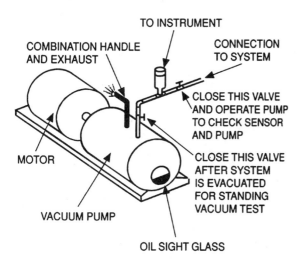

Figure 5-6. The valve arrangements used to check the vacuum pump and sensor, and to perform a standing vacuum test.

Figure 5-7. Two-stage rotary vacuum pump. *Photo by Bill Johnson.*

The system vacuum level and vacuum pump operation can both be checked using special valve arrangements (Figure 5-6). To check the sensor and vacuum pump, isolate the vacuum pump and the sensor by closing the valve closest to the system so that only the sensor is evacuated. If the vacuum pump cannot develop enough vacuum, it can be shown at this point. To check the system pressure, close the valve closest to the vacuum pump so that both the sensor and the system are evacuated. This tests can be used to determine the system pressure for a standing vacuum test.

5.4 RECOVERING REFRIGERANT

Before evacuating a system to remove contaminants or noncondensable gases, any refrigerant in the system must be removed using EPA-approved recovery equipment. The amount of vacuum to be achieved when removing the refrigerant depends on the size of the system, the type of refrigerant, and whether or not the recovery equipment was manufactured before November 15, 1993. Refrigerant recovery is discussed in more detail in Unit 6.

5.5 VACUUM PUMPS

A vacuum pump must be capable of removing the atmosphere down to a very low vacuum. The vacuum pumps used in the air conditioning field are normally manufactured with rotary compressors. The pumps that produce the lowest vacuums are two-stage rotary vacuum pumps (Figure 5-7). These pumps are capable of reducing the pressure in a leak-free vessel down to 0.1 micron. It is not practical to produce a vacuum this low in a field-installed system because the refrigerant oil in the system will boil slightly and create a vapor. The usual vacuum required by most manufacturers is approximately 250 microns, although some may require a vacuum as low as 50 microns.

Manufacturers generally have a system or formula for classifying vacuum pumps. These pumps are normally sized by the pump displacement in cubic feet per minute (cfm). The following information may be used as a general guide for selecting a vacuum pump.

Pump Displacement	Application
2 cfm	Window air conditioners
3 & 4 cfm	Residential central A/C systems
	Light commercial A/C systems
5 & 6 cfm	Commercial air conditioners
	Heat pumps
8 & 10 cfm	Industrial and larger commercial installations

System Dehydration Using a Vacuum Pump

To prevent damage, water vapor must be removed from an air conditioning system. Moisture by itself can cause corrosion over a period of time. In addition, refrigerant oil and moisture can combine to form a sludge. This sludge can cause many problems, including reducing the lubricating ability of the oil, and clogging strainers and capillary tubes.

A vacuum pump does not pull out liquid moisture. Instead, it reduces the system pressure to the point at which the moisture boils and the vapor is removed from the system by the vacuum pump.

A two-stage pump has a second pumping chamber that enables the pump to reach a greater vacuum. The exhaust of the first stage is discharged into the intake of the second stage (Figure 5-8). The second stage begins pumping at a lower pressure and pulls a higher vacuum on the system. Most pumps used with refrigerants have a vented exhaust feature called a gas ballast. This feature allows drier air from the atmosphere to enter the second stage.

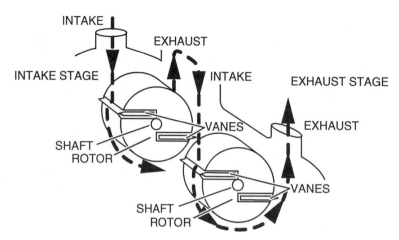

Figure 5-8. Diagram of the path from the intake to the exhaust in a rotary two-stage vacuum pump. *Courtesy Ritchie Engineering Company, Inc.*

This air helps to prevent the moisture vapor from condensing in the second stage of the pump and combining with the oil.

Small amounts of moisture can be removed with a vacuum pump and exhausted to the atmosphere, but it is not practical to remove large amounts of moisture in this manner because of the large amount of vapor produced by the boiling water. For example, one pound of water (about a pint) will produce 867 cubic feet of vapor if boiled at 70°F. The procedure for removing large amounts of moisture is described later in this unit.

5.6 VACUUM PUMP OIL

A vacuum pump cannot create a pressure lower than the vacuum that will boil its lubricating and sealing oil. The oil must seal the clearances between the parts of the pump at the normal working temperature of the pump. The specifications of a particular brand of oil will often indicate the boiling pressure in microns. This oil is normally processed from a paraffin-based crude oil.

Vacuum pumps must operate under very stressful conditions for long periods of time. It is important that they be maintained properly on a continuing basis. The oil must be changed regularly. If used with large systems or if the oil is dirty and contains considerable moisture, it may be necessary to change the oil during an evacuation. Most manufacturers recommend that it be changed at least after each use. Contaminated oil will affect the pump's performance. It may be difficult to determine if the oil is contaminated as it will not necessarily change color. A micron gage may be used to indicate whether or not the pump is performing to the manufacturer's specifications. To perform this test, isolate the pump from the air conditioning system during the evacuation and while the pump is hot, change the oil. Start the pump and run it until the oil is hot. Take a gage reading and if the pump is not performing to specifications, change the oil again. Each reading must be taken with the pump running and the oil hot. This may have to be done up to three times. The clean oil actually washes the pump's parts—by changing the oil, the contaminants are removed from the pump.

5.7 ACHIEVING A DEEP VACUUM

Achieving a deep vacuum involves reducing the pressure in the system to about 50 to 250 microns. When the vacuum reaches the desired level, the vacuum pump is valved off and the system is allowed to stand for a certain length of time to see if the pressure rises. If the pressure rises and then stops, a liquid substance such as water is boiling in the system. If this occurs, continue evacuating. If the pressure continues to rise, there is a leak, and the atmosphere is seeping into the system. In this case, the system should be pressurized and leak checked again.

When the system pressure is reduced to 50 to 250 microns and the pressure remains constant, no moisture or noncondensable gases are left in the system. Reducing the system pressure to 250 microns is a slow process because when the vacuum pump pulls the system pressure to below about 5 mm (5000 microns), the pumping process slows down. Most technicians plan to start the vacuum pump as early as possible and finish other work while the vacuum pump runs.

Some technicians leave the vacuum pump running all night because of the time involved in reaching a deep vacuum. This is an acceptable practice if some precautions are taken. When the vacuum pump pulls a vacuum, the system becomes a large volume of low pressure with the vacuum pump between this volume and the atmosphere. If the vacuum pump shuts off during the night from a power failure, its lubricating oil may be pulled out of the pump and into the system because of the vacuum in the system. If the power is restored and it starts back up, it will lack the necessary lubrication and could be damaged. This can be prevented by installing a large solenoid valve in the vacuum line entering the vacuum pump and wiring the solenoid valve coil in parallel with the vacuum pump motor.

MULTIPLE EVACUATION   73

5.8 MULTIPLE EVACUATION

Multiple evacuation is often used to reduce the system pressure down to the lowest level of contamination. Multiple evacuation is normally accomplished by evacuating a system to a low vacuum (about 1 or 2 mm Hg) and then allowing a small amount of refrigerant to bleed into the system. The system is then evacuated until the vacuum is again reduced to 1 mm Hg. The following is a detailed description of a multiple evacuation. This one is performed three times and is called a *triple evacuation*. Figure 5-10 is a diagram of the valve arrangements.

Warning: Be extremely careful when using a U-tube manometer. Mercury is a toxic substance; do not handle it.

1. Attach a U-tube mercury manometer to the system. The best place is as far from the vacuum port as possible. For example, on a refrigeration system, the pump may be attached to the suction and discharge service valves and the U-tube manometer to the liquid receiver valve port.

Warning: Do not place your hand over any opening that is under a vacuum because the vacuum may cause a blood blister on the skin.

2. Start the vacuum pump, and let it run until the manometer reaches 1 mm Hg. The mercury manometer should be positioned vertically to take accurate readings. Lay a straightedge across the manometer to help determine the column heights.

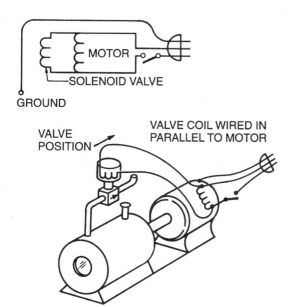

Figure 5-9. Vacuum pump with a solenoid valve in the inlet line. Note the direction of the arrow on the solenoid valve. It is installed to prevent flow from the pump. It must be installed in this direction.

The solenoid valve should have a large port to keep it from restricting the flow (Figure 5-9). Now, if the power fails, or if someone disconnects the vacuum pump (a good possibility at a construction site), the vacuum will not be lost, and the vacuum pump will not lose its lubrication. The use of solenoid valves is discussed in more detail later in this unit.

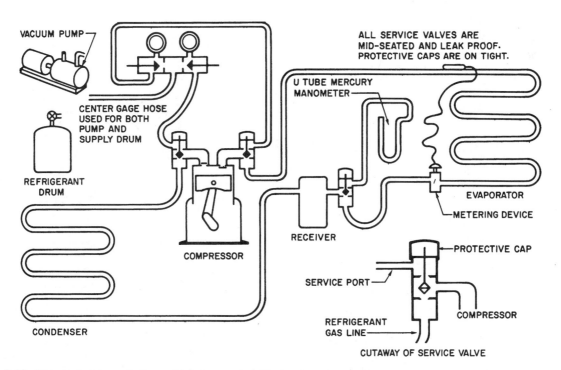

Figure 5-10. This system is ready for multiple evacuation. Notice the valve arrangements and the location of the U-tube manometer.

3. Allow a small amount of refrigerant to enter the system until the vacuum is about 20 in. Hg. This must be indicated on the manifold gage because the mercury in the mercury gage will rise to the top and give no indication. This small amount of refrigerant vapor will fill the system and absorb and mix with other vapors.
4. Open the vacuum pump valve and let the vacuum pump run until the vacuum is again reduced to 1 mm Hg. Then repeat Step 3.
5. When the refrigerant has been added to the system the second time, open the vacuum pump valve and again remove the vapor. Operate the vacuum pump for a long time during this third pull down. It is best to operate the vacuum pump until the manometer columns are equal. This is sometimes referred to as flat out.
6. When the vacuum has been pulled the third time, allow refrigerant to enter the system until the system is about 5 psig above the atmosphere. Now remove the mercury manometer (it cannot withstand the system pressure) and charge the system.

An electronic vacuum gage may be used during a triple evacuation by using the above-mentioned valve arrangements to isolate the system between evacuations. Again, remember that the advantage of the electronic vacuum gage is its rapid response to pressure changes in the system. If a leak is present, the electronic gage will show it much faster than the U-tube manometer.

5.9 LEAK DETECTION WHILE IN A VACUUM

As mentioned previously, if a leak is present in a system under vacuum, the vacuum gage will start to rise, indicating a pressure rise in the system. The vacuum gage will rise very quickly. Many technicians use this as an indicator that a leak is still in the system, but this is not a recommended leak test procedure for several reasons:

1. It allows air to enter the system.
2. The technician cannot determine from the vacuum where the leak is located.
3. When a vacuum is used for leak checking, it is only proving that the system will not leak under a pressure difference of 14.696 psi (the atmosphere trying to get into the system). Under normal conditions, the system will be under a much higher pressure. For example, an R-22 air-cooled condenser operating on a very hot day may have an operating pressure of 350 psig + 14.696 psi = 364.696 psia (Figure 5-11).
4. Using a vacuum for leak checking may also hide a leak. For example, if there is a pin-sized hole in a solder connection that has a flux buildup over it, the vacuum will tend to pull the flux into the pinhole and may even hide it to the point where a deep vacuum can be achieved. Then, when pressure is applied to the system, the flux will blow out of the pinhole, and a leak will appear.

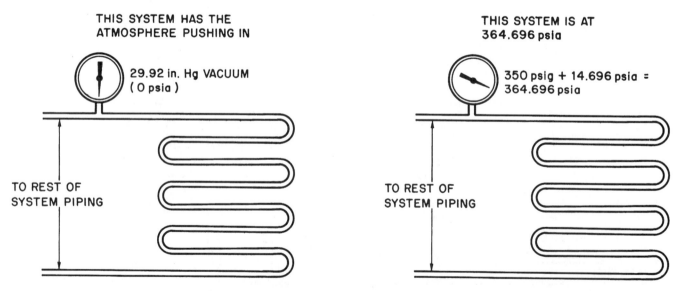

Figure 5-11. These two systems are being compared to each other under different pressure situations. One is evacuated and has the atmosphere under atmospheric pressure trying to get into the system. The other has 350 psig + 14.696 = 364.696 psia. The one under the most pressure is under the most stress. Using a vacuum as a leak test does not give the system a proper leak test.

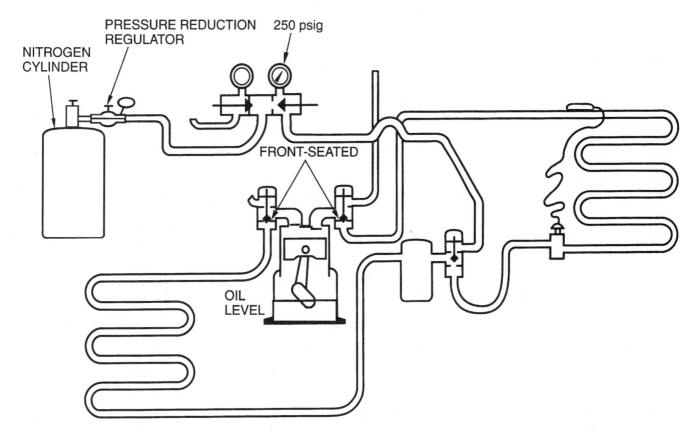

Figure 5-12. Isolating the compressor to pressurize the condenser and evaporator.

5.10 LEAK DETECTION—STANDING PRESSURE TEST

The best leak checking procedure is a *standing pressure test* using a pressure source that will not be affected by temperature changes. Nitrogen is often used for this purpose.

Warning: Never use air to pressurize a system. Air mixed with certain refrigerants under pressure can create an explosive mixture.

To perform a standing pressure test, add a small amount of refrigerant to the system for leak-checking purposes (up to about 10 psig). Using R-22 as a trace refrigerant for leak detection is acceptable by the EPA. The small amount of refrigerant can be detected with any common leak detector. After adding the refrigerant, pressurize the system with nitrogen.

Warning: Do not pressurize any system above the working pressure written on the equipment label. No manufacturer uses a working pressure lower than 150 psig with the high-pressure refrigerants discussed in this book (R-12, R-22, and R-502). Therefore, we can assume that 150 psig is safe for any system using these refrigerants.

The compressor has the lowest working pressure of any component in the system (150 psig). If more pressure is desired for a pressure test, the compressor may be isolated and other components may be pressure tested to a higher value (Figure 5-12). It is often practical to pressure test the piping, condenser, and evaporator to a higher pressure than the compressor.

When the system pressure reaches 150 psig, tap the gage slightly to make sure the needle is free, then make a mark on the gage. Let the system stand at this pressure while leak checking. (Leak detection instruments were discussed in Unit 3.) When the leak check is complete, observe the gage reading again. If it has fallen, there is a leak. Do not forget that the gage manifold and connections may also leak. When the system is leak checked and no drop is found in the gage, let the system stand for a while. The smaller the system, the shorter the standing time. For example, a small beverage cooler may need to stand for only an hour to be sure that the system is leak free, whereas a 20-ton system may need to stand under pressure for 12 hours or more. The longer the standing time, the more assurance you will have that there is no leak.

5.11 REMOVING MOISTURE WITH A VACUUM

There are two kinds of moisture in the system: vapor and liquid. When the moisture is in the vapor state, it is easy to remove. When it is in the liquid state, it is much more dif-

UNIT 5   SYSTEM EVACUATION

ficult to remove. An example earlier in this unit indicated that 867 cubic feet of vapor at 70°F must be pumped to remove one pound of water. This is not a complete explanation because as the vacuum pump begins to remove the moisture, the water will boil and the temperature of the remaining water will drop. This process will increase the volume of the resulting vapor. For example, if the water temperature drops to 50°F, one pound of water will then boil into 1702 cubic feet of vapor that must be removed. This is at a pressure level in the system of 0.362 in. Hg or 9.2 mm Hg (0.362 in. Hg × 25.4 mm/in.). At this point, the vacuum level is just reaching the low ranges. As the vacuum pump pulls lower, the water will continue to boil (if the vacuum pump has the capacity to pump this much vapor), and the temperature will decrease to 36°F. One pound of water will now create a vapor volume of 2837 cubic feet. This is at a system vapor pressure of 0.212 in. Hg or 5.4 mm Hg (0.212 in. Hg × 25.4 mm/in.). Figure 5-13 shows the relationships between temperature, pressure, and volume.

If the system pressure is reduced further, the water will turn to ice and be even more difficult to remove. If large amounts of moisture must be removed from a system with a vacuum pump, the following procedure will help.

1. Use a large vacuum pump. If the system is flooded (for example, if a water-cooled condenser pipe ruptures from freezing) a 5 cfm vacuum pump is recommended for systems up to 10 tons. If the system is larger, a larger pump or a second pump should be used.
2. Drain the system in as many low places as possible. Remove the compressor and pour the water and oil from the system. Do not add the oil back into the system until it is ready to be started after evacuation. If you add it earlier, the oil may become wet and hard to evacuate.
3. Apply as much heat as possible without damaging the system. If the system is in a heated room, the room may be heated to 90°F without fear of damaging the room and its furnishings or the system (Figure 5-14). If part of the system is outside, use a heat lamp (Figure 5-15). The entire system, including the interconnecting piping, must be heated to a warm temperature, or the water will boil into a vapor where the heat is applied and condense where the system is cool. For example, if you know water is in the evaporator inside the structure and you apply heat to the evaporator, the water will boil into a vapor. Then, if it is cool outside, the water vapor may condense outside in the condenser piping. In this case, the water is only being moved around; it is not being removed.
4. Start the vacuum pump and observe the oil level in it. As moisture is removed, some of it will condense in the vacuum pump crankcase. As mentioned earlier, some vacuum pumps have a gas ballast feature that introduces a certain amount of drier atmospheric air between the first and second stages of the two-stage pump. This prevents some of the moisture from condensing in the crankcase.

Caution: Regardless of the type of vacuum pump, watch the oil level. The water will displace the oil and raise the oil out of the pump. Soon, water may be the only lubricant in the vacuum pump crankcase and damage may occur to the vacuum pump. Vacuum pumps are very expensive and should be protected.

TEMPERATURE		SPECIFIC VOLUME OF WATER VAPOR	ABSOLUTE PRESSURE		
°C	°F	ft³/lb	lb/in.²	kPa	in. Hg
−12.2	10	9054	0.031	0.214	0.063
−6.7	20	5657	0.050	0.345	0.103
−1.1	30	3606	0.081	0.558	0.165
0.0	32	3302	0.089	0.613	0.180
1.1	34	3059	0.096	0.661	0.195
2.2	36	2837	0.104	0.717	0.212
3.3	38	2632	0.112	0.772	0.229
4.4	40	2444	0.122	0.841	0.248
5.6	42	2270	0.131	0.903	0.268
6.7	44	2111	0.142	0.978	0.289
7.8	46	1964	0.153	1.054	0.312
8.9	48	1828	0.165	1.137	0.336
10.0	50	1702	0.178	1.266	0.362
15.6	60	1206	0.256	1.764	0.522
21.1	70	867	0.363	2.501	0.739
26.7	80	633	0.507	3.493	1.032
32.2	90	468	0.698	4.809	1.422
37.8	100	350	0.950	6.546	1.933
43.3	110	265	1.275	8.785	2.597
48.9	120	203	1.693	11.665	3.448
54.4	130	157	2.224	15.323	4.527
60.0	140	123	2.890	19.912	5.881
65.6	150	97	3.719	25.624	7.573
71.1	160	77	4.742	32.672	9.656
76.7	170	62	5.994	41.299	12.203
82.2	180	50	7.512	51.758	15.295
87.8	190	41	9.340	64.353	19.017
93.3	200	34	11.526	79.414	23.468
98.9	210	28	14.123	97.307	28.754
100.0	212	27	14.696	101.255	29.921

Figure 5-13. This partial pressure/temperature relationship table for water shows the specific volume of water vapor that must be removed to remove a pound of water from a system.

REMOVING MOISTURE WITH A VACUUM 77

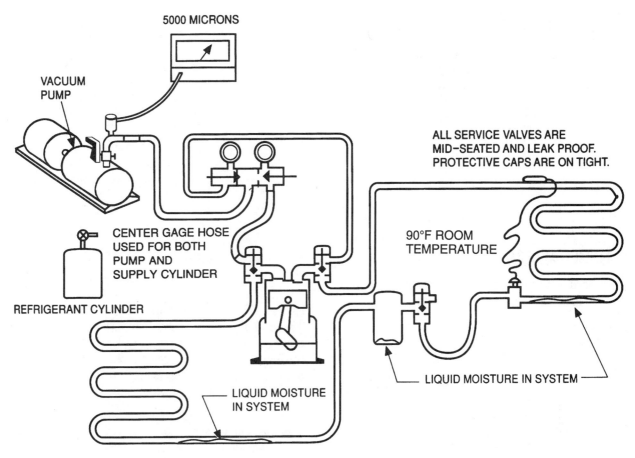

Figure 5-14. When a system containing moisture is being evacuated, heat may be applied to the system to vaporize the water so that the vacuum pump can remove it.

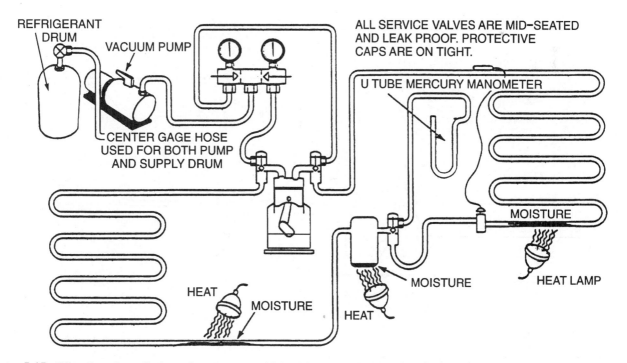

Figure 5-15. When heat is applied to a large system with both indoor and outdoor components, the entire system must be heated. If not, moisture will condense where the system is cool.

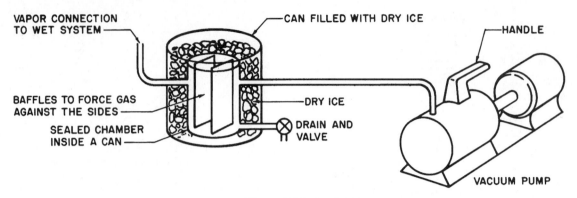

Figure 5-16. A cold trap.

5.12 GENERAL EVACUATION TIPS

This section discusses general rules that apply to deep vacuum and multiple evacuation procedures.

Cold Traps

If the system is large enough or if you must evacuate the moisture from several systems, you can construct a cold trap to use in the field. A cold trap is a refrigerated volume in the vacuum line between the wet system and the vacuum pump. When the water vapor passes through the cold trap, the moisture freezes to the walls of the trap, which is normally refrigerated with dry ice (CO_2), a commercially available product.

The trap is heated, pressurized, and drained periodically to remove the moisture (Figure 5-16). Using a cold trap can save a vacuum pump.

Removing Moisture from the Compressor

Compressors contain small chambers, such as cylinders, that may trap air or moisture. At times, it is advisable to start the compressor after a vacuum has been tried. This is easy to do when using the triple evacuation method. When the first vacuum has been reached, refrigerant or nitrogen can be charged into the system until it reaches atmospheric pressure. The compressor can then be started for a few seconds. All chambers should be flushed at this time.

Caution: Do not start a hermetic compressor while it is in a deep vacuum because motor damage may occur.

Figure 5-17 shows an example of vapor trapped in the cylinder of a compressor.

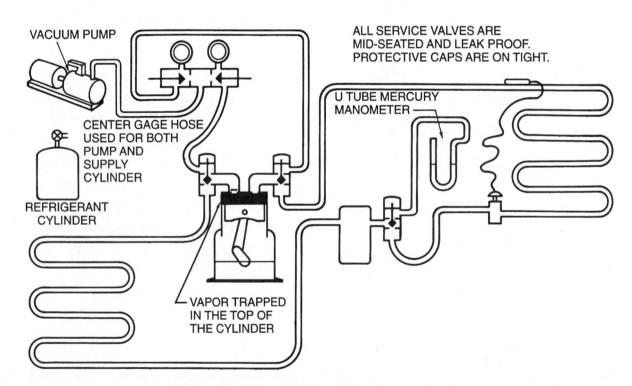

Figure 5-17. Vapor trapped in the compressor cylinder.

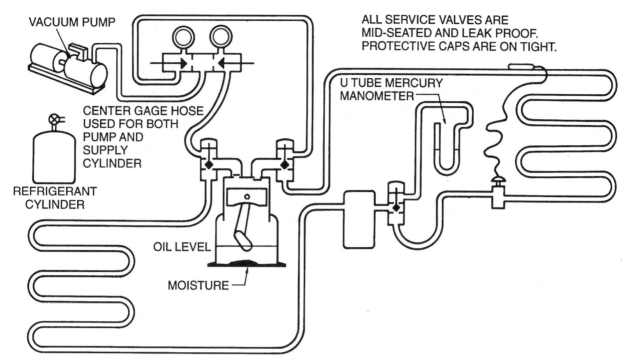

Figure 5-18. Compressor with water trapped under the oil in the crankcase.

Water can also be trapped in a compressor under the oil (Figure 5-18). The oil has surface tension, and the moisture may stay beneath it even under a deep vacuum. During a deep vacuum, the oil surface tension can be broken by vibrating the compressor. This can be done by striking the housing with a soft face hammer, but any kind of movement that causes the oil's surface to shake will work. Applying heat to the compressor crankcase will also release the water (Figure 5-19).

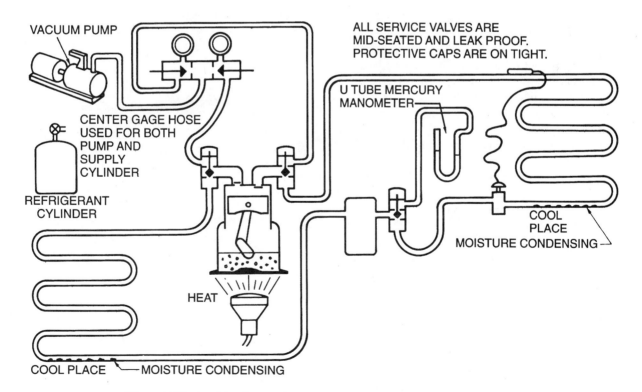

Figure 5-19. Applying heat to the compressor to boil the water under the oil.

UNIT 5 SYSTEM EVACUATION

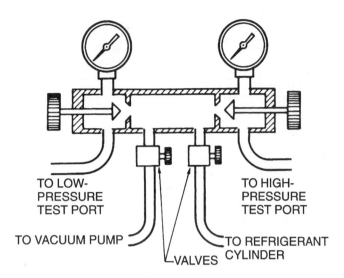

Figure 5-20. Gage manifold with four valves and four gage hoses.

Selecting and Using a Gage Manifold

A typical gage manifold may not be the best choice for system evacuation because it has very small valve ports that slow the evacuation process. Some gage manifolds are manufactured with large valve ports and a large hose for the vacuum pump connection. The gage manifold in Figure 5-20 has four valves and four hoses. The extra two valves are used to control the refrigerant and the vacuum pump lines. When using this manifold, you don't need to disconnect the vacuum pump and switch the hose line to the refrigerant cylinder to charge refrigerant into the system. When it is time to stop the evacuation and charge refrigerant into the system, you simply close one valve and open the other. This is a much easier and cleaner method than stopping to change the hose connections.

When a gage line is disconnected from the vacuum pump, air is drawn into the gage hose. This air must be purged from the gage hose at the top, near the manifold. However, it is impossible to get all of the air out of the manifold; some of it will always be trapped and pushed into the system (Figure 5-21).

Most gage manifolds have valve stem depressors in the ends of the gage hoses. These depressors are used for servicing systems with Schrader access valves, which are similar to the valve and stem on an automobile tire. When a vacuum pump pulls down to the very low ranges (1 mm Hg), these valve depressors slow the vacuum process considerably. To speed this process, the valve depressors can be removed from the ends of the gage hoses, and adapters can be used when valve depression is needed. Figure 5-22 shows one of these adapters.

Another alternative is to remove the valve stems during evacuation and replace them when the evacuation is completed. Schrader valve stems are designed to be removed either for replacement or during evacuation (Figure 5-23).

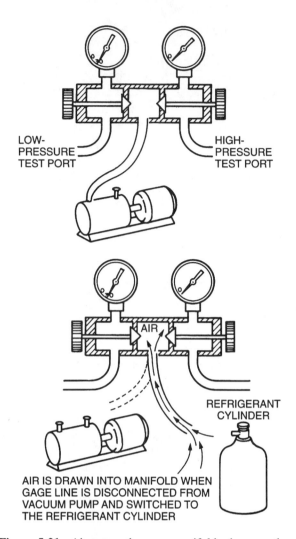

Figure 5-21. Air enters the gage manifold whenever the gage lines are disconnected.

Figure 5-22. This gage adapter can be used instead of the gage depressors that are normally found in the end of the gage lines. This adapter may be used for gage readings. *Photo by Bill Johnson.*

Figure 5-23. Schrader valve assembly. *Courtesy J/B Industries.*

A special tool, called a *field service valve*, can be used to replace Schrader valve stems under pressure, and it can also be used as a control valve during evacuation. This allows the technician to evacuate a system through the field service tool with the stem backed out of the Schrader valve. The stem is replaced when the evacuation is completed.

Schrader valves are shipped with a special cap, which is used to cover the valve when it is not in use. This cap has a soft gasket and should be the only cover used for Schrader valves. If a standard brass flare cap is used and overtightened, the Schrader valve top will be distorted and valve stem service will be very difficult, if not impossible.

Gage Manifold Hose Leaks

The standard gage manifold uses flexible hoses with connectors on the ends. These hoses sometimes get pinhole leaks, usually around the connectors, that may leak while under a vacuum but not be evident when the hose has pressure inside it. The reason is that the hose swells when pressurized. If you have trouble pulling a vacuum and you can't find a leak, substitute soft copper tubing for the gage line, (Figure 5-24).

System Valves

Larger systems may have many valves and piping runs, and perhaps even multiple evaporators. Before evacuating these systems, check the system's valves to make sure

Figure 5-24. Gage manifold with copper gage lines used for evacuation. *Photo by Bill Johnson.*

that they are open. For example, a system may have a closed solenoid valve, which can trap air in the liquid line between the expansion valve and the solenoid valve (Figure 5-25). This valve must be open for complete evacuation. It may even need a temporary power supply to operate its magnetic coil. Some solenoid valves have a screw on the bottom to manually jack the valve open (Figure 5-26).

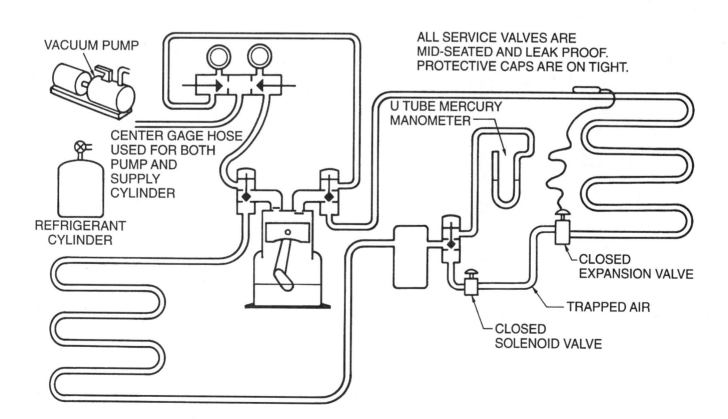

Figure 5-25. A closed solenoid valve may trap air in the system liquid line.

82 UNIT 5 SYSTEM EVACUATION

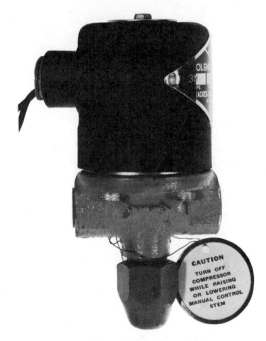

Figure 5-26. This solenoid valve has a manual opening stem on the bottom. *Photo by Bill Johnson.*

Using Dry Nitrogen

When piping is field installed, sweeping dry nitrogen through the refrigerant lines will help to push the atmosphere out and clean the system. A dry nitrogen setup is relatively inexpensive and using it will save time and money.

Whenever a system has been open to the atmosphere for a significant period of time, it needs to be evacuated. The task can be hastened by sweeping the system with dry nitrogen before evacuation. See Figure 5-27.

5.13 CLEANUP AFTER COMPRESSOR BURNOUT

In addition to the moisture and noncondensible gases that may enter a system during installation or servicing, other contaminants may be produced during a mild or severe compressor burnout. During a compressor burnout, the hermetic motor inside the sealed system produces a tremendous amount of heat. This heat can break down the oil and refrigerants to form acids, soot (carbon), and sludge that cannot be removed with a vacuum pump.

Suppose a 5-ton air conditioning system with a fully hermetic compressor experiences a severe motor burnout while the system is running. When a motor burnout occurs while the compressor is running, the soot and sludge from the hot oil will be carried to the condenser. The following procedure should be used to clean the system.

Warning: Care should be taken when handling any of the products from a contaminated system. The acids produced can cause severe burns. Always wear goggles and gloves during system cleanup.

1. Recover the refrigerant from the system. This process is discussed in detail in Unit 6.

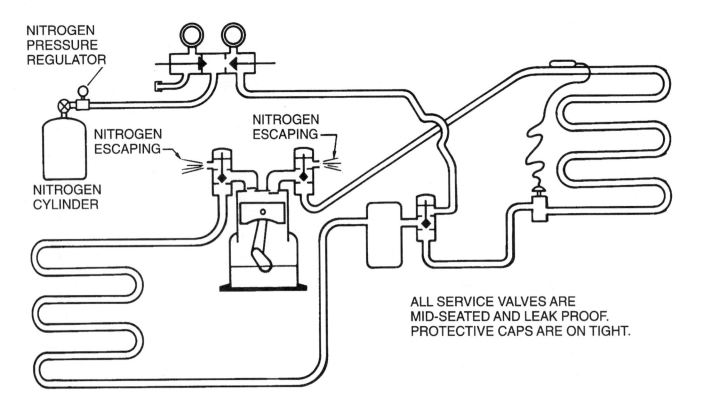

Figure 5-27. Dry nitrogen can be used to sweep a system before evacuation.

2. After the refrigerant has been removed, the compressor can be replaced, but should not be connected until after the system has been purged with nitrogen.
3. Dry nitrogen may be used to push some of the contaminants out of the system. This is done by attaching the nitrogen regulator to one of the lines and allowing it to blow out the other.

Warning: Do not exceed the system working pressure with the nitrogen. Without the compressor in the system, you can safely use 250 psig for high-pressure refrigerant systems.

Because the contaminants are known to be in the condenser, disconnect the line before the expansion device. Purge the nitrogen through the liquid line toward the compressor. It will discharge out the compressor discharge line (Figure 5-28). This will sweep all loose contaminants from this side of the system.

4. Connect the nitrogen cylinder to the expansion device side of the liquid line, and purge this line toward the compressor suction line. The velocity of the refrigerant will be reduced because of the expansion device, but in many cases, it is not practical to remove it. For example, if it is a capillary tube system, there will be several connections and it would be too time-consuming to remove them.

5. At this point, the system has been purged as much as possible, but both solid and liquid contaminants are still in the system (soot and contaminated oil). Now, connect the compressor to the system with a suction line filter drier installed just before the compressor. Filter drier manufacturers have done a good job of developing filter media that will remove acid, moisture, and carbon sludge. They claim that nothing can be created inside a system that has only oil and refrigerant that the filter driers will not remove. This will prevent any contamination from entering the compressor on start-up. Purging the system with dry nitrogen also ensures the maximum capacity from the filter drier because some of the contaminants have already been pushed out of the system.
6. Install a liquid line filter drier just before the expansion device to prevent contamination from restricting the refrigerant flow.
7. Leak check the system before evacuation.

Note: If you have both an old and a new vacuum pump, use the old one, because contamination may be pulled through the pump.

8. Either evacuate the system to a low vacuum of 250 microns, or triple evacuate, then charge the system with refrigerant.

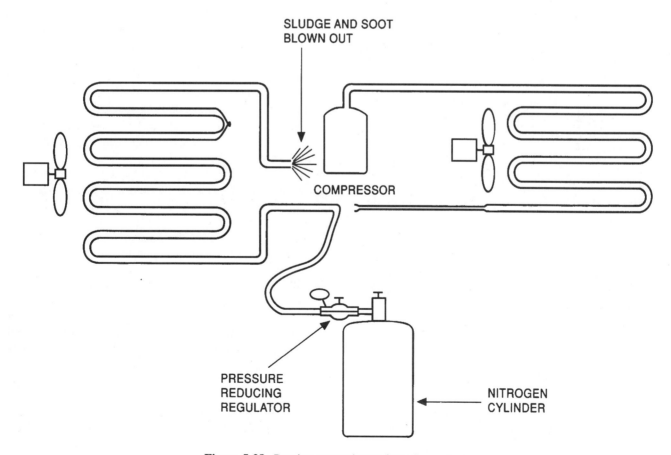

Figure 5-28. Purging contaminants from the condenser.

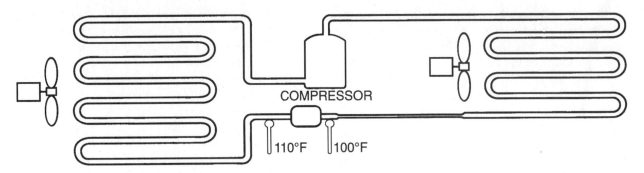

Figure 5-29. The 30°F temperature differential indicates a pressure drop; the drier should be changed.

9. Start the system and keep it running as long as practical to circulate the refrigerant through the filter driers. The refrigerant in the system is an excellent solvent. It will break loose the contaminants and they will be trapped in the filter driers. The pressure drop across the suction line drier may be monitored to see if it is gathering particles and beginning to become restricted. The manufacturer's literature will list the maximum pressure drop. If the filter drier becomes restricted beyond the manufacturer's recommendation, pump the system down and change the drier. If it does not become restricted, you may leave it in the system; it will not affect system operation.
10. Change the oil in the vacuum pump while it is hot. Run it for 30 minutes and change it again to be sure all contamination has been removed from the pump crankcase.

Caution: A technician may do a great job of cleaning the refrigeration system and neglect the vacuum pump, which is very expensive and just as important as the system compressor.

11. Check the pressure drop across the liquid line filter drier by measuring the line temperature before and after of the drier. If there is a temperature drop, there is a pressure drop (Figure 5-29).

Filter drier manufacturers have done a good job of developing filter media that will remove acid, moisture, and carbon sludge. They claim that nothing can be created inside a system that has only oil and refrigerant that the filter driers will not remove.

SUMMARY

- Only two products should be circulating in a refrigeration system: refrigerant and oil. Anything else is considered a contaminant.
- Noncondensable gases and moisture are common contaminants that enter systems during installation and servicing. They must be removed to prevent system damage.
- Evacuation using low vacuum levels removes noncondensable gases and involves pumping the system to below-atmospheric pressure.
- Vapors will be pumped out by the vacuum pump. Liquids must be boiled to be removed with a vacuum.
- Water creates a large volume of vapor when boiled at low pressure levels. If possible, it should be drained from a system before evacuation.
- Two common vacuum gages are the U-tube manometer and the electronic vacuum gage.
- Pumping a vacuum will be faster when using large unrestricted lines.
- When noncondensables are left in a system, acids (hydrochloric and hydrofluoric) will slowly form and deteriorate the system by attacking the motor windings and causing copper plating on the crankshaft.
- If nitrogen is left in a system following a standing pressure test, it will cause excess head pressure because it takes up condensing space.
- The best system pressure test is a standing pressure test. For this test, the system is normally pressurized to 150 psig.
- The only advantage in using a vacuum to test for leaks is the rapid response of vacuum instruments. The vacuum leak test only proves that the system piping will prevent the atmosphere from entering the system; it does not pinpoint the leak or represent what will happen under the stress of the much higher operating pressure.
- When a vacuum pump is allowed to run unattended, the system becomes a large vacuum reservoir. If the vacuum pump loses power, the pump oil will be pulled into the system. Then, if the vacuum pump is restarted, it will be operating without lubrication and may be damaged.
- Special valve arrangements allow the technician to check the operation of the vacuum sensor and pump and also to check the system pressure.
- System Schrader valve cores may be removed to speed the evacuation process; and then replaced before the system is put back in operation.

- Using dry nitrogen to sweep a refrigeration system will aid the evacuation process. This can be helpful when the system is first installed, when it has been open for any length of time, and following a compressor burnout.
- Do not forget to clean the vacuum pump after evacuating a contaminated system.

REVIEW QUESTIONS

1. Name the common contaminants in a refrigeration system that must be removed?
2. Define the term *system evacuation*.
3. What are the only two products that should be circulating in a refrigeration system?
4. Why is it difficult to remove water from a refrigeration system using a vacuum pump?
5. Name some things that can be done to help remove water from a refrigeration system.
6. What can be done to dislodge the water from under the oil in a compressor crankcase?
7. Name two evacuation processes.
8. What must be done to solenoid valves and other valves in a system to ensure proper evacuation?
9. Name two common vacuum gages.
10. How many microns are there in 1 mm Hg?

6 REFRIGERANT MANAGEMENT – RECOVERY, RECYCLING, AND RECLAMATION

OBJECTIVES

After studying this unit, you should be able to

- Describe ozone.
- Discuss how CFCs deplete the Earth's ozone layer.
- Differentiate between CFCs, HCFCs, and HFCs.
- Discuss EPA regulations as they relate to refrigerants.
- Define the terms *recover, recycle,* and *reclaim.*
- Describe methods for recovering refrigerants.
- Identify a Department of Transportation (DOT) approved recovery cylinder.

Note: There are several references to refrigerants and procedures in this unit that pertain to refrigeration systems other than those found in air conditioning systems. This information is included for those technicians who intend to take the certification examinations.

6.1 REFRIGERANTS AND THE ENVIRONMENT

As the world population grows and demands for comfort and newer technologies increase, more and more chemicals are produced and used in various combinations. Many of these chemicals reach the Earth's atmosphere and produce different types of pollution. These chemicals include the refrigerants that are used to cool or freeze food, and condition the homes, office buildings, stores, and other buildings in which we live and work. Many of these refrigerants were developed in the 1930s.

While contained within a system, these chemicals are stable and do not have an adverse effect on the environment. Over the years, however, quantities have been released due to system leaks or through the willful purging or venting of systems during routine service procedures.

Over time, these refrigerants are thought to slowly rise to the stratosphere (7 to 30 miles above the Earth), mostly with the help of wind currents. It is believed that certain refrigerants react with a form of oxygen called *ozone*. When this happens, new chemical compounds are formed and the ozone is depleted.

Each molecule of ozone (O_3) consists of three atoms of oxygen. The oxygen molecules we breathe consist of two atoms of oxygen (O_2).

The ozone layer in the stratosphere is formed by ultraviolet rays from the sun acting on oxygen (O_2) molecules. This ozone layer acts as a shield that prevents excess amounts of the sun's ultraviolet rays from reaching the Earth. The depletion of this layer will allow more of these harmful rays to reach the Earth.

Ultraviolet rays in sufficient quantities can cause serious burns, an increase in skin cancer, and other health problems. Research has also shown that excess ultraviolet radiation may have an adverse effect on crops and other plant growth.

When released to the Earth's atmosphere, refrigerants rise to the stratosphere very slowly, and then remain there for a long time before they break down. The refrigerants released many months or even years ago are what is suspected to be affecting the ozone layer today. The refrigerants released today may take months or possibly years to affect the ozone layer.

Ozone is also present near the Earth's surface in the troposphere (up to 7 miles above the Earth). This type of ozone is harmful and is not to be confused with the helpful ozone in the stratosphere. Tropospheric ozone is caused by ultraviolet radiation from the sun acting on smog and other air pollution. Much of this pollution can be attributed to the automobile exhaust and industrial wastes released into the lower atmosphere. This ozone is harmful to breathe and is present in greater quantities during certain weather conditions.

6.2 REFRIGERANT CLASSIFICATIONS

Many refrigerants contain chlorine atoms. When these refrigerants are bombarded by the ultraviolet radiation in the atmosphere, they release free chlorine atoms, which combine with ozone to form chlorine monoxide and oxygen. This results in ozone depletion. Most refrigerants in use today are classified according to whether or not they contain chlorine and how readily they release it into the atmosphere. The three classifications in current use are: chlorofluorocarbons (CFCs), hydrochlorofluorocarbons (HCFCs) and hydrofluorocarbons (HFCs).

CFC Refrigerants

CFCs contain chlorine, fluorine, and carbon and are considered the most damaging because they do not decompose before reaching the stratosphere. Once in the

REFRIGERANT CLASSIFICATIONS

stratosphere, the chlorine reacts with the ozone and sunlight, causing the ozone to break down.

CFC refrigerants include the following:

Refrig. No.	Chem. Name	Chem. Formula
R-11	Trichlorofluoromethane	CCl_3F
R-12	Dichlorodifluoromethane	CCl_2F_2
R-113	Trichlorotrifluoroethane	CCl_2FCClF_2
R-114	Dichlorotetrafluoroethane	$CClF_2CClF_2$
R-115	Chloropentafluoroethane	$CClF_2CF_3$

All of the refrigerants in the CFC group were phased out of production in 1995. This has had a significant impact on our industry because R-12 is commonly used for residential and light commercial refrigeration and for centrifugal chillers in some larger commercial applications. R-11 is used for many centrifugal chillers in office buildings and also as an industrial solvent to clean parts. R-113 is used in smaller commercial chillers in office buildings and also as a cleaning solvent. R-114 has been used in some household refrigerators in the past and in centrifugal chillers for marine applications. R-115 is one of the refrigerants that make up the blend R-502, which is used in low-temperature refrigeration systems. As the current supply of these refrigerants diminishes, it will become necessary to produce alternative refrigerants for these applications.

HCFC Refrigerants

HCFCs contain hydrogen, chlorine, fluorine, and carbon. The hydrogen content in these refrigerants makes them less stable in the atmosphere. Therefore, HCFCs have much less potential for ozone depletion because they tend to break down in the atmosphere, releasing the chlorine before it reaches and reacts with the ozone in the stratosphere. However, because some of the refrigerant reaches the stratosphere intact, HCFCs are scheduled to be phased out of production by the year 2030.

HCFC refrigerants include the following:

Refrig. No.	Chem. Name	Chem. Formula
R-22	Chlorodifluoromethane	CHC_lF_2
R-123	Dichlorotrifluoroethane	$CHCl_2CF_3$
R-124	Chlorotetrafluoroethane	$CHClFCF_3$

R-22 is very important because it is used in most modern air conditioning systems in this country. R-123 is used in centrifugal chillers. R-124 is used in certain marine chiller applications.

HFC Refrigerants

HFCs contain hydrogen, fluorine, and carbon. Because these refrigerants do not contain chlorine, they are considered safe for the environment. They are currently being examined as possible replacement refrigerants for CFCs and HCFCs. Originally, it was believed that R-12 could be replaced with R-134a in existing installations, but R-134a is not compatible with any oil left in an R-12 system. It is also not compatible with some of the gaskets used in the R-12 systems. To convert an R-12 system to R-134a, a complete oil change is necessary. The residual oil left in the system pipes must be removed and all system gaskets must be checked for compatibility. The compressor materials must also be checked for compatibility by contacting the manufacturer. Due to their complexity, refrigerant conversions should only be performed by an experienced technician.

HFC refrigerants include the following:

Refrig. No.	Chem. Name	Chem. Formula
R-125	Pentafluoroethane	CHF_2CF_3
R-134a	Tetrafluoroethane	CH_2FCF_3

R-134a is currently being used as a replacement refrigerant in automotive air conditioning and certain other R-12 systems. R-125 is used for low-temperature refrigeration and commercial food transportation.

Ozone Depletion Potential

Each refrigerant has been assigned a number to help identify the rate at which it is thought to deplete the ozone layer. This number is called its *ozone depletion potential (ODP)*. These values are based on R-11, which has an ODP of 1. The lower the ODP, the less harmful the refrigerant is to the environment. Other ODP numbers are as follows:

Refrigerant No.	Ozone Depletion Potential
R-11	1
R-12	1
R-22	0.05
R-113	0.8
R-114	1
R-115	0.8
R-123	0.016
R-124	not rated
R-125	not rated
R-134a	0
R-500	not rated
R-502	not rated

Note that R-11 and R-12 are major contributors to ozone depletion and will be among the first to be phased out.

Refrigerant Blends

A mixture of two or more refrigerants is known as a *blend*. Blends may be classified as either azeotropic or zeotropic. In a zeotropic blend, the refrigerants never mix chemically and may boil at different temperatures. This is known as *temperature glide*. Zeotropic blends are also subject to *fractionation*, which means that each component may leak at a different rate. Some zeotropic blends contain methane, which is natural gas and is flammable; the concern is that the refrigerant could become explosive as a result of severe fractionation.

Over time, more and more zeotropic blends will be developed as replacements for existing CFC and HCFC refrigerants, and the problems of temperature glide and fractionation may be eliminated. For example, R-410a, which is being used to replace R-22 in certain applications, has very little glide (1°F) and exhibits no fractionation.

Azeotropic refrigerants are blends that are chemically combined in such a manner that they will not separate into their individual components, even when they leak into the atmosphere. Unfortunately, most azeotropes are also CFCs and have been phased out of production. R-500 is one example of an azeotropic blend. It consists of 73.8% R-12 and 26.2% R-152 by weight. This refrigerant was used in some air conditioning (cooling) systems many years ago.

6.3 REFRIGERANT REGULATIONS

The United States, Canada, and more than 30 other countries met in Montreal, Canada in September 1987 to address the problem of ozone depletion. This conference became known as the *Montreal Protocol* and its participants agreed to reduce the production of chlorine-based refrigerants by half by the year 1999. Additional meetings have been held since 1987 and further production limits have been agreed upon. The use of refrigerants as we have known them has changed to some extent and will change dramatically in the near future.

The United States Clean Air Act Amendments of 1990 regulate the use and disposal of refrigerants. The United States Environmental Protection Agency (EPA) is charged with implementing the provisions of this legislation.

According to the Clean Air Act of 1990, technicians may not "knowingly vent or otherwise release or dispose of any substance used as a refrigerant in such an appliance in a manner which permits such substance to enter the environment. De minimus releases associated with good faith attempts to recapture and dispose of any such substance shall not be subject to the prohibition set forth in the preceding sentence." *De minimus* means that the minimum amount possible under the circumstances. To the air conditioning technician, this means that all refrigerant must be contained, except that used to purge lines and tools of the trade, such as the gage manifold or any device used to capture and save the refrigerant. This prohibition became effective July 1, 1992. Severe fines and penalties are provided for, including prison terms. The first level allows of the EPA to obtain an injunction against the offending party, prohibiting the discharge of refrigerant to the atmosphere. The second level is a $25,000 fine per day and a prison term of up to five years.

To help police the illegal discharge of refrigerants, a reward is offered to any person who furnishes information that leads to the arrest and conviction of anyone who willfully vents refrigerant to the environment.

Federal Excise Tax

The United States budget for 1990 contained provisions for an excise tax on CFC refrigerants. The CFCs included in this tax are R-11, R-12, R-113, R-114, and R-115. This tax is intended to encourage the transition to the use of new refrigerants, and applies only to newly-manufactured refrigerants. Recycled and reclaimed refrigerants are exempt from the tax. As shown below, the rate of taxation has increased substantially since its inception.

Federal Excise Tax on CFC Refrigerants
(per pound of refrigerant)

1990	$1.37	1995	$3.10
1991	1.37	1996	3.55
1992	1.67	1997	4.00
1993	2.65	1998	4.45
1994	2.65	1999	4.90

6.4 REFRIGERANT RECOVERY, RECYCLING, OR RECLAMATION

You should become very familiar with these three terms: *recover, recycle* or *reclaim*. These are the three terms the industry and its technicians must understand.

Determining When Recovery is Necessary

The air conditioning technician is responsible for determining when a refrigerant charge must be recovered. Situations that require the recovery of refrigerant from a component or system include:

1. Following a compressor motor burnout.
2. Prior to system disposal. A refrigerant system cannot be disposed of at a salvage yard if it contains any refrigerant.
3. Whenever a repair must be made to a system and the refrigerant cannot be pumped into the condenser or receiver using the system compressor.

Warning: Gloves and goggles should be worn anytime a technician transfers refrigerant from one container to another. Refrigerant can only be legally transferred into DOT-approved containers.

When refrigerant must be recovered, the technician must study the system and determine the best procedure for removal. For example, the system may have several different valve configurations, or the compressor may or may not be operable. Sometimes, the system compressor may be used to assist in refrigerant recovery. Any of the following situations may be encountered when refrigerant must be removed from a system:

1. The compressor will *not* run and the system has no service valves or access ports, such as when a window air conditioner is being discarded.
2. The compressor will *not* run and the system has no access ports, such as a compressor burnout in a window air conditioner.

3. The compressor will run and the unit has service ports (Schrader ports) only, as in a central air conditioner.
4. The compressor will *not* run and the unit has service ports (Schrader ports) only, as in a central air conditioner.
5. The compressor will run and the system has some service valves, such as a residential air conditioning system with liquid and suction line isolation valves.
6. The compressor will *not* run and the system has some service valves, such as a residential air conditioning system with liquid and suction line isolation valves.
7. The compressor will run and the system has a complete set of service valves.
8. The compressor will *not* run and the system has a complete set of service valves.

The condition of the refrigerant in the system must also be considered. It may be contaminated with air, another refrigerant, nitrogen, acid, water, or motor burnout particulates. Cross contamination to other systems must never be allowed because it can cause damage to the system into which the contaminated refrigerant is transferred. This damage may occur slowly and may not be known for long periods of time. The warranties of compressors and other system components may be voided if contamination can be proven.

Recovery and Refrigerant Oil

The refrigerant may be removed from the system as either a vapor or liquid or in the partial liquid/partial vapor state. It must be remembered that lubricating oil is also circulating in the system with the refrigerant. If the refrigerant is removed in the vapor state, the oil is more likely to stay in the system. This is desirable from two standpoints:

1. If the oil is contaminated, it may have to be handled as a hazardous waste. Much more consideration is necessary and a certified hazardous waste technician must be available.
2. If the oil remains in the system, it will not have to be measured and returned back to the system. In either case, time is saved and time is money. Always pay close attention to the management of the oil from any system.

Recovery Cylinders

Refrigerant must be transferred only into DOT-approved refrigerant cylinders. Never use DOT 39 disposable cylinders. The approved cylinders are recognizable by the color and valve arrangement. They have yellow tops with a special valve that allows liquid or vapor to be added or removed from the cylinder (Figure 6-1).

The refrigerant cylinder must be clean and in a deep vacuum before the recovery or recycle process is started. It is suggested that the cylinder be evacuated to 1000 microns (1 mm Hg) before recovery is started. This will

Figure 6-1. Department of Transportation (DOT) approved cylinders. *Photo by Bill Johnson.*

remove any refrigerant from the previous job. All lines to the cylinder must be purged of air before connections are made.

The terms *recover, recycle,* and *reclaim* are often used interchangeably, but they represent three entirely different procedures, each with its own EPA requirements.

Recovery

To recover refrigerant is "to remove refrigerant in any condition from a system and store it in an external container without necessarily testing or processing it in any way." This refrigerant may be contaminated with air, another refrigerant, nitrogen, acid, water, or motor burnout particulates. Recovered refrigerant must not be used in another system unless it is known to be clean or has been processed to meet the Air Conditioning & Refrigeration Institute (ARI) standard entitled ARI 700. Analysis of the refrigerant to determine if it meets the ARI 700 standard may only be performed by a qualified chemical lab. This test is expensive and is usually only performed on large volumes of refrigerant.

If the refrigerant charge has been removed to service a system and the system is operable, the recovered refrigerant may be put back into the system. For example, assume a system has no service valves and a leak occurs. The remaining refrigerant may be recovered and reused in this system only. It may not be sold to another customer without meeting the ARI 700 standard. However, owners of multiple equipment are permitted to reuse the refrigerant in another unit they own if they are willing to take the chance with their equipment. Others may only be willing to take this chance after the equipment is out of warranty.

It is up to the technician to use an approved, clean refrigerant cylinder for recovery to avoid cross contamination.

Recycling

To recycle refrigerant is "to clean the refrigerant by oil separation and single or multiple passes through devices, such as replaceable core filter-driers, which reduce moisture, acidity, and particulate matter. This term usually applies to procedures implemented at the job site or at a local service shop." If it is suspected that the refrigerant is dirty, it may be recycled, which means that it is filtered and cleaned. Filter-driers only remove acid, particulates, and moisture from a refrigerant. In some cases, the refrigerant will need to pass through the driers several times before complete cleanup is accomplished. Purging air from the system may only be done when the refrigerant is contained in a separate container or recycling unit.

Refrigerant recycling is sometimes performed in the system itself. When a motor burnout occurs and the compressor has service valves, the valves are front-seated and the compressor changed. The only refrigerant lost is the small amount of vapor in the compressor. Driers are used to clean the refrigerant while running the new system compressor. Often, the system is started and pumped down into the condenser and receiver, and additional acid-removing filter-driers are added to the suction line to further protect the compressor (Figure 6-2). The entire process occurs within the equipment and normal operation can be resumed after adjusting the charge.

Today special recycling units are also used to clean the contaminated refrigerant. A recycling unit removes the refrigerant from the system, cleans it, and returns it to the system. This is accomplished by transferring the refrigerant to a recovery cylinder. Then, the two valves on the cylinder are used to route the refrigerant to the recycle unit and its filtration system. Some units have an air purge that allows any air introduced into the system to be purged before transferring the refrigerant back into the system (Figure 6-3).

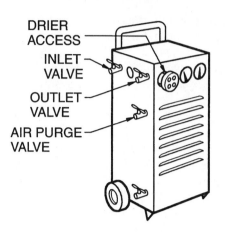

Figure 6-3. This recovery/recycle unit has an air purge valve from which accumulated air may be vented.

Reclamation

To reclaim refrigerant is "to process refrigerant to new product specifications by means which may include distillation. It will require chemical analysis of the refrigerant to determine that appropriate product specifications are met. This term usually implies the use of processes or procedures available only at a reprocessing or manufacturing facility." When refrigerant is reclaimed, it is recovered at the job site, stored in approved cylinders, and then shipped to the reprocess site where it is chemically analyzed to determine whether or not it can be reprocessed (reclaimed). It is then reprocessed to meet the ARI 700 standard.

6.5 EPA EVACUATION REQUIREMENTS

Most of the refrigerants mentioned in this unit are thought of as high-pressure refrigerants. They are also

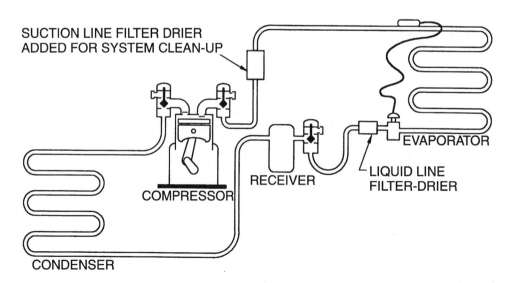

Figure 6-2. Filter-driers can be used to recycle the refrigerant within the system.

known as *low boiling point refrigerants*. Some of the CFC refrigerants, such as R-11, are low-pressure refrigerants (they have a high boiling point). R-11 boils at 74.9°F at atmospheric pressure. This refrigerant is difficult to remove from a system because of its high boiling point. A very low vacuum must be pulled to boil it all from the system. This is very different from R-12, which has a much lower boiling point of -21.62°F at atmospheric pressure. That means that if the temperature of the system is above -21.62°F, R-12 will boil and can easily be removed from the system.

The boiling points and vapor pressures of common refrigerants at atmospheric pressure are listed below.

Refrigerant	Boiling Point/ Vapor Pressure at 70°F
R-12	-21.62°F/70 psig
R-22	-41.36°F/121 psig
R-125 (to replace R-502)	-55.3°F/158 psig
R-134a (to replace R-12)	-15.7°F/71 psig
R-500	-28.3°F/85 psig
R-502	-49.8°F/137 psig

EPA Evacuation Requirements for Residential and Commercial Refrigeration Systems

The boiling point is important because the EPA has set forth pressure requirements for each recovery category. Effective August 12, 1993, all HVAC service companies were required to certify to the EPA that they own the necessary equipment to perform adequate refrigerant recovery for the systems they service.

Recovery equipment manufactured prior to November 15, 1993 must meet the qualifications set for this time period. Equipment manufactured after November 15, 1993 must be approved by an ARI-approved third-party laboratory to meet ARI Standard 740-1993.

As of November 15, 1993, the EPA has required the following minimum evacuation levels for all refrigeration systems except small appliances and motor vehicle air conditioning:

1. HCFC-22 appliances (or isolated component of such appliances) that normally contain less than 200 pounds of refrigerant, must be evacuated to 0 psig.
2. HCFC-22 appliances (or isolated components of such appliances) that normally contain more than 200 pounds of refrigerant must be evacuated to 10 in. Hg vacuum.
3. Other high-pressure appliances (or isolated components for such appliances) that normally contain less than 200 pounds of refrigerant (R-12, R-500, R-502, R-114) must be evacuated to 15 in. Hg vacuum.
4. Other high-pressure appliances (or isolated components of such appliances) that normally contain 200 pounds or more of refrigerant (R-12, R-500, R-502, R-114) must be evacuated to 15 in. Hg vacuum.
5. Very high-pressure appliances (R-13, R-503) must be evacuated to 0 psig.
6. Low-pressure appliances (R-11, R-113, R-123) must be evacuated to 25 mm Hg absolute (29 in. Hg vacuum).

One exception to the above requirements would be when the system cannot be reduced to these pressure levels, such as when the system has a large leak. In this case, there is no need to pull the system full of air.

EPA Evacuation Requirements for Small Appliances

The EPA regulations are different for small appliance repair technicians. Effective August 12, 1993, they must meet the following requirements. If their recovery equipment was manufactured prior to November 15, 1993, it must be capable of recovering 80% of the refrigerant in a system or achieving a 4 in. Hg vacuum. If the equipment was manufactured after November 15, 1993, it must be capable of recovering 90% of the refrigerant or achieving a 4 in. Hg vacuum if the compressor in the appliance is operable. If the compressor is not operable, it must be capable of recovering 80% of the refrigerant.

6.6 TECHNICIAN CERTIFICATION

Technician certification has been required to purchase refrigerant since November 14, 1994. There are four different certification areas, depending on the type of service work:

1. Type I Certification (small appliance)
2. Type II Certification (high-pressure appliances)
3. Type III Certification (low pressure appliances)
4. Universal Certification (all of the above)

Type I Certification

Type I certification covers small appliances that are manufactured, charged, and hermetically sealed with five pounds or less of refrigerant. This includes refrigerators, freezers, room air conditioners, package terminal heat pumps, dehumidifiers, under-the-counter ice makers, vending machines, and drinking water coolers.

Type II Certification

Type II certification covers high-pressure appliances that use refrigerant with a boiling point between -58°F (-50°C) and 50°F (10°C) at atmospheric pressure. This includes R-12, R-22, R-114, R-500, and R-502.

Type III Certification

Type III certification covers low-pressure appliances that use refrigerants with a boiling point above 50°F (10°C) at atmospheric pressure. This includes R-11, R-113, and R-123 refrigerants.

Universal Certification

Universal certification covers all of the above: Type I, Type II, and Type III.

Technician certification should ensure that the technician knows how to handle the refrigerant in a safe manner without exhausting it to the atmosphere.

6.7 MECHANICAL RECOVERY SYSTEMS

Many mechanical recovery systems are available. Some are sold for the purpose of recovering refrigerant only, while others are designed for both recovery and recycling. A few units claim to have reclaim capabilities. ARI is responsible for certifying the equipment specifications of recovery systems and these units are expected to perform to the 1993 ARI 740 standard.

For recovery and recycling equipment to be useful in the field, it must be portable. It must be easy to move to the rooftops where many systems are located, and must also be small enough to be hauled in the technician's truck. Units for use on rooftops should not weigh much more than 50 pounds or they may be too heavy for the technician to carry up a ladder. Some technicians may overcome this problem by using long hoses to reach from the rooftop to the unit, which may be left on the truck or on the top floor of a building. However, this is not an efficient practice because the technician must recover the refrigerant from the long hose. The long hose also slows the recovery process.

Manufacturers are meeting this portability challenge in several ways. Some have developed modular units that may be carried to the rooftops or remote jobs as separate components (Figure 6-4). Others are reducing the size of the units by limiting the internal components and using them for recovery only. In this case, the refrigerant is transferred to a recovery cylinder and taken to the shop for recycling, then returned to the job. Obviously, this process is more time consuming, and it is also less efficient, because every time the refrigerant is transferred from one container to another, refrigerant is lost.

Manufacturers are attempting to provide a unit that is as small as practical, yet will still perform properly. These units must be able to remove vapor, liquid, and vapor/liquid refrigerant mixtures and also deal with any oil that is suspended in the liquid refrigerant. Many are accomplishing this task by using a small, hermetic reciprocating compressor in the unit to pump the refrigerant vapor. However, these compressors are limited in the amount of refrigerant they can transfer, and as the compressors become larger, they also become heavier. Larger condensers are also needed, which adds to the weight. Some manufacturers are using rotary compressors to reduce the weight. Rotary compressors may also pull a deeper vacuum on the system with less effort than a reciprocating compressor. Future regulations may require deeper vacuums to remove more of the system refrigerant. All compressors in recovery units also pump oil, so some method of oil return is needed to return the oil to the recovery unit crankcase.

The fastest method of removing refrigerant from a system is to take it out in the liquid state because it occupies a smaller volume per pound of refrigerant. If the system is large enough, it may have a liquid receiver where most of the charge is collected. However, many systems are small and if the compressor is not operable, it cannot pump the refrigerant.

Figure 6-5 illustrates liquid refrigerant recovery. The recovery unit pumps refrigerant vapor to the air conditioning unit vapor side. The pressure differential between the air conditioning unit and the recovery cylinder forces the liquid refrigerant from the unit to the cylinder. Note the two valves on the cylinder, one for vapor and one for liquid. The liquid refrigerant must not go through the recovery unit or the compressor will be damaged. The level switch in the recovery cylinder will turn the recovery unit off when the cylinder is 80% full.

The slowest method of removing refrigerant is to remove it in the vapor state. When the liquid refrigerant has been recovered, the piping arrangement can be changed, as shown in Figure 6-6. This provides for the recovery of the refrigerant vapor remaining in the air conditioning unit. When removed in the vapor state using a recovery/recycle unit, the unit will remove the vapor faster if the hoses and valve ports are not restricted and if a greater pressure difference can be created. When the system is warmer, the vapor is under a greater pressure, and the compressor in the recovery/recycle unit will be able to pump more pounds per minute. As the vapor pressure in the system is reduced, the unit capacity is also reduced. The recovery rate will slow down as the refrigerant pressure drops because the compressor pumping rate is constant.

Recovery/recycle equipment must be able to operate under various conditions. For example, it may be 100°F on the rooftop. The system pressure may be very high due to the ambient temperature, which may tend to overload the unit compressor. A crankcase pressure regulator is sometimes used in recovery/recycle units to prevent overloading the compressor with high suction pressure. The unit compressor will have to condense the refrigerant using the 100°F air across the condenser Figure 6-7. The condenser must have enough capacity to condense this refrigerant without overloading the compressor.

Figure 6-4. This modular system reduces a heavy unit into two manageable pieces. *Courtesy of Robinair Division, SPX Corporation.*

MECHANICAL RECOVERY SYSTEMS 93

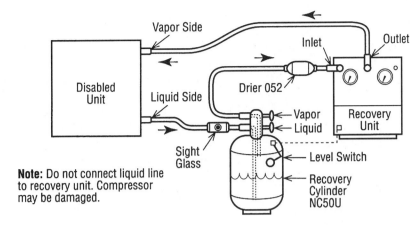

Figure 6-5. Liquid recovery of refrigerant. *Courtesy National Refrigeration Products.*

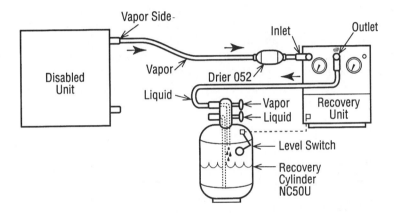

Figure 6-6. Vapor recovery of refrigerant. *Courtesy National Refrigeration Products.*

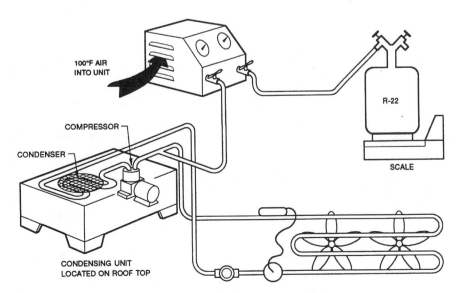

Figure 6-7. This unit will have to condense the refrigerant using 100°F air on the rooftop. The condenser must be large enough to keep the pressure down so that the compressor is not overloaded.

Manufacturers of recovery/recycle units list the typical pumping rate in their equipment specifications. This is the rate at which the equipment is capable of removing refrigerant (generally two pounds/minute for equipment under ten tons). This rate is fairly constant and rapid while removing refrigerant in the liquid state. It tends to vary somewhat when removing vapor. When the system pressure is high, the rate of removal is relatively fast. As the system pressure is lowered during the recovery process, the rate of vapor removal will slow down. With a small compressor, when the system vapor pressure reaches about 20 psig, the rate of removal will become very slow.

The compressor in the recovery/recycle unit is a vapor pump. Liquid refrigerant cannot be allowed to enter the compressor because it will cause damage. Different methods may be used to prevent this from occurring. Some manufacturers use a push-pull method of removing refrigerant. This is accomplished by connecting the liquid line fitting on the unit to the liquid line fitting on the cylinder. The suction line on the recovery unit is connected to the vapor fitting on the cylinder. A connection is then made from the unit discharge back to the system. When the unit is started, vapor is pulled out of the recovery cylinder from the vapor port and condensed by the recovery unit. A very small amount of liquid is pushed into the system where it flashes to a vapor to build pressure and push liquid into the receiving cylinder (refer to Figure 6-5). A sight glass is used to monitor the liquid and the recovery cylinder is weighed to prevent overfilling.

When it is determined that no more liquid may be removed, the suction line from the recovery unit is reconnected to the vapor portion of the system and the unit discharge is fastened to the refrigerant cylinder (Figure 6-6). The unit is started and the remaining vapor is removed from the system and condensed in the cylinder. A crankcase pressure regulator (CPR) valve is often installed in the suction line to the compressor in the recovery/recycle unit. This helps to protect it from overloading and offers some liquid refrigerant protection (Figure 6-8).

When liquid refrigerant is removed, it contains oil. Normally, this refrigerant and oil are transferred directly to the refrigerant cylinder. When this liquid refrigerant is charged back into the system, the oil will go with it. Oil that may be carried through the recovery unit with any small amount of liquid will be stopped by the oil separator. Any oil removed must be accounted for and returned to the system. Oil acid test kits are available to check the oil if there is any question as to its quality, such as after a motor burnout.

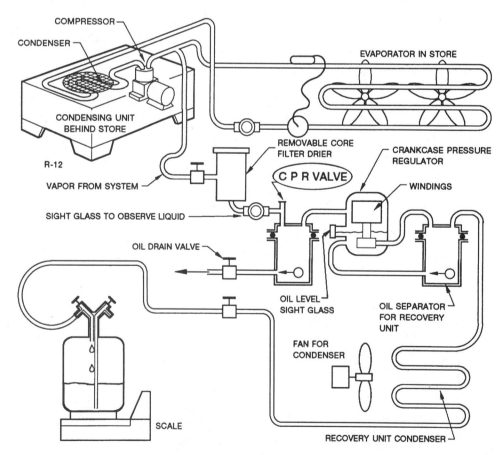

Figure 6-8. This unit uses a push-pull method of removing liquid refrigerant from a system. Vapor is pulled out of the recovery cylinder, creating a low pressure in the cylinder. A small amount of condensed liquid is allowed back into the system to build pressure and push liquid into the cylinder.

Some DOT-approved cylinders include a float and switch to shut the unit off when the cylinder is 80% full. This helps to prevent overfilling the cylinder with refrigerant. It also helps to compensate for any oil that may be mixed with the refrigerant. Oil is much lighter than refrigerant. Therefore, if a cylinder contains several pounds of oil due to poor practices, it can easily be overfilled if the cylinder is filled only by weight. The float switch prevents this from happening (Figure 6-9).

Some manufacturers remove liquid from the system through a metering device and an evaporator that is used to boil the liquid to a vapor. The compressor in the recovery/recycle unit moves the liquid refrigerant and some of the oil from the system and evaporates the refrigerant from the oil (Figure 6-10). The oil is usually removed by an oil separator.

All recovery/recycle unit compressors will pump oil out the discharge line and should have an oil separator to direct the oil back to the unit compressor crankcase.

Note: Each manufacturer may use different features to accomplish the same processes. Always follow the manufacturer's instructions when using any type of recovery or recovery/recycle unit.

Selecting Recovery or Recovery/Recycle Equipment

These units should be chosen carefully. It may be helpful to ask the following questions when determining which unit to purchase.

Recovery unit only:

1. Will it run under the conditions needed for the majority of the jobs encountered? For example, will it operate properly on a rooftop in hot or cold weather?
2. Will it recover both liquid and vapor if you need these features?
3. What is the pumping rate for the refrigerants you need to recover?
4. Are the drier cores standard and can they be purchased locally?
5. Is the unit portable enough to be handled easily during your service jobs?

Recovery/recycle unit:

1. Will it run under the conditions needed for the majority of the jobs encountered?
2. Will it recycle enough refrigerant for the type of jobs encountered?

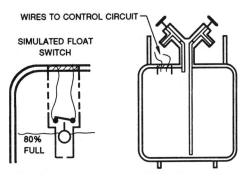

Figure 6-9. The float switch prevents overfilling the cylinder.

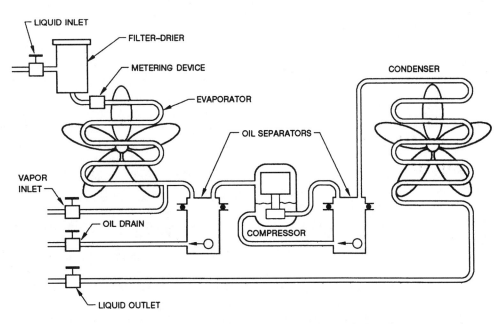

Figure 6-10. This system includes a crankcase pressure regulator (CPR) valve to protect the compressor.

3. What is the recycle rate and does it correspond to the rate you may need for the size systems you service?
4. Are the drier cores standard and can they be purchased locally?
5. Is the unit portable enough to be handled easily during your service jobs and will it fit conveniently on your truck?

6.8 RECLAIMING REFRIGERANT

Used refrigerant must meet the ARI 700 standard before it is resold or reused in another customer's equipment. This refrigerant is usually sent to a reclaim company where it can be checked and reprocessed (provided it is in good enough condition). Most service companies will not try to recycle refrigerant in the field because of the complications and the possibility of cross contamination. Reclaim companies will only accept refrigerant in cylinders of 100 pounds and larger. Some companies are buying recovered refrigerant and accumulating it to be reclaimed.

The refrigerant in the reclaim cylinders must not be mixed or the reclaim company cannot reclaim it. The service shop will need a cylinder of the correct size for each of the refrigerants that will be shipped. When the reclaim company receives refrigerant that cannot be reclaimed, it must be destroyed. Refrigerant is destroyed by incinerating it in such a manner that the fluorine in the refrigerant is captured. This is an expensive process, so the technicians must be careful not to mix the refrigerants.

6.9 RECOVERING REFRIGERANT FROM WINDOW AIR CONDITIONERS

Recovering refrigerant from window units may be an easier job than from larger systems because not as much refrigerant is involved. Most small appliances contain less than two pounds of refrigerant. This can usually be removed into an evacuated, approved cylinder by reducing the pressure in the cylinder to 1000 microns, connecting it to the appliance, and opening the valves. In small appliances, 80 to 90% of the charge needs to be removed. The procedure is as follows:

1. Evacuate the recovery cylinder to 1000 microns.
2. Fasten a line tap valve on the appliance liquid line leaving the condenser.
3. Fasten the other end of the line to the approved recovery cylinder and purge the line of air.
4. Start the compressor if it will run and open the valve to the cylinder.
5. The refrigerant will transfer to the cylinder. Place the line to your ear and listen for the flow to stop.

The compressor will pull the refrigerant from the low side of the system and move it to the high side of the system. This will transfer most of the refrigerant charge into the cylinder. If the compressor will not run, the cylinder may be cooled with ice to increase the pressure differential and pull more refrigerant into the cylinder.

Refrigerant bags are available for the recovery of refrigerant from small units. These bags are plastic and will hold the charge of several window air conditioners. When the bag is full, it may be taken to the shop and the refrigerant transferred into the reclaim cylinder for delivery to the reclaim company.

6.10 TECHNICIAN INFORMATION

It always pays to stay abreast of new developments in the industry. Participate in local trade organizations that monitor and report industry trends. For example, in the future there will be changes in the tools used to handle refrigerant. Vacuum pumps will soon be required to pump refrigerant into recovery cylinders so the refrigerant that is removed from a system during evacuation can be captured. Leak-free gage lines may soon be mandatory. Many older gage lines seep refrigerant while connected. Gage lines may also be required to incorporate valves that reduce the amount of refrigerant lost when they are disconnected from Schrader connections. These valves will be located in the end of the hoses and will hold the refrigerant in the hose when disconnected.

Make an effort to be aware of the latest information and technology and always practice the best service techniques.

SUMMARY

- It is illegal to vent refrigerant into the atmosphere.
- Ozone is a form of oxygen in which each molecule consists of three atoms of oxygen (O_3). The oxygen we breathe contains two atoms of oxygen (O_2).
- The ozone layer in the stratosphere protects the earth from harmful ultraviolet rays which can cause damage to human beings, animals, and plants.
- Chlorofluorocarbons (CFCs) contain chlorine, fluorine, and carbon atoms.
- Hydrochlorofluorocarbons (HCFCs) contain hydrogen, chlorine, fluorine, and carbon atoms.
- Hydrofluorocarbons (HFCs) contain hydrogen, fluorine, and carbon atoms.
- The hydrogen content of HCFCs helps them to decompose before they reach the stratosphere and react with the ozone.
- The EPA has established many regulations governing the handling of refrigerants.
- To recover refrigerant is "to remove refrigerant in any condition from a system and store it in an external container without necessarily testing or processing it in any way."
- To recycle refrigerant is "to clean the refrigerant by oil separation and single or multiple passes through devices, such as replaceable core filter-driers, which reduce moisture, acidity, and particulate matter. This term usually applies to procedures implemented at the job site or at a local service shop."

- To reclaim refrigerant is "to process refrigerant to new product specifications by means which may include distillation. It will require chemical analysis of the refrigerant to determine that appropriate product specifications are met. This term usually implies the use of processes or procedures available only at a reprocessing or manufacturing facility."
- Refrigerant can be transferred only into DOT-approved cylinders and tanks.
- Manufacturers have developed many types of equipment to recover and recycle refrigerant.

REVIEW QUESTIONS

1. What is the difference between the oxygen we breathe and ozone?
2. Why are CFCs more harmful to the ozone layer than HCFCs?
3. What is the name of the conference that was held in Canada in 1987 to address the problem of ozone depletion?
4. What agency of the Federal government is charged with implementing the Clean Air Act Amendments of 1990?
5. Are Federal excise taxes on reclaimed refrigerants the same as those for new refrigerants?
6. To recover refrigerant means to _____.
7. To recycle refrigerant means to _____.
8. To reclaim refrigerant means to _____.
9. What is the name of the Federal agency that must approve containers into which refrigerant is transferred?
10. What is the standard that a reclaimed refrigerant must meet before it can be sold to a different customer?

7 SYSTEM CHARGING

OBJECTIVES

Upon completion of this unit, you should be able to

- Describe how refrigerant is charged into systems in the vapor and liquid states.
- Describe system charging using two different weighing methods.
- State the advantage of using electronic scales for weighing refrigerant into a system.
- Describe various charging devices.

7.1 CHARGING A REFRIGERATION SYSTEM

Charging a system refers to the addition of refrigerant to the system. An accurate refrigerant charge is critical to ensuring proper system operation. Each component in the system must have the correct amount of refrigerant. The refrigerant may be added to the system in the vapor or liquid states by weighing, measuring, or using operating pressure charts.

7.2 VAPOR REFRIGERANT CHARGING

Vapor refrigerant charging is accomplished by allowing vapor to move out of the vapor space of a refrigerant cylinder and into the low-pressure side of the refrigeration system.

Vapor Charging an Empty System

When the system is not operating—for example, when a vacuum has just been pulled, or when the system is out of refrigerant—you can add vapor to either the low-pressure (low side) or high-pressure (high side) of the system. When the system is running, refrigerant may normally be added only to the low side of the system, because the high side is under more pressure than the refrigerant in the cylinder. For example, an R-22 system may have a head pressure of 278 psig on a 95°F day (Figure 7-1). As we studied in Unit 3, this pressure is determined by taking the outside temperature of 95°F and adding 30°F; this creates a condensing temperature of 125°F, which corresponds to 278 psig for R-22. The refrigerant cylinder is exposed to the same ambient

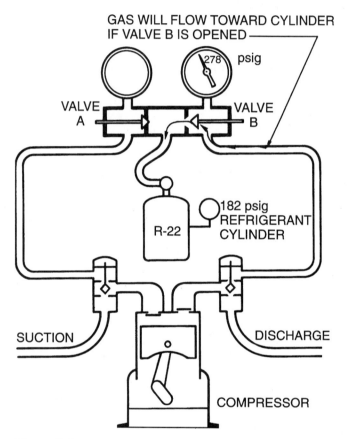

Figure 7-1. This refrigerant cylinder has a pressure of 182 psig. The high-side of the system has a pressure of 278 psig. The pressure in the system will prevent the refrigerant in the cylinder from moving into the system.

temperature of 95°F but only has a pressure of 182 psig, according to its pressure-temperature chart.

Vapor Charging an Operating System

When a refrigerant cylinder is warm, its pressure will be much higher than the low-side pressure in an operating system. For example, on a 95°F day, the cylinder will have a pressure of 182 psig, but the evaporator pressure may be only 39 psig. Refrigerant will easily move into the system from the cylinder (Figure 7-2).

VAPOR REFRIGERANT CHARGING 99

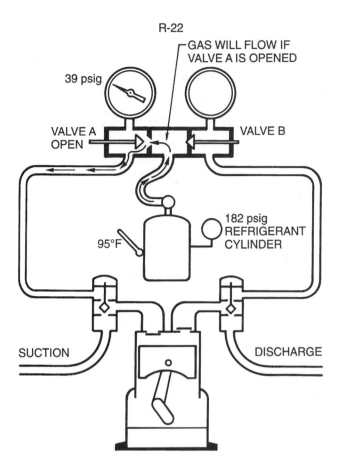

Figure 7-2. The temperature of the cylinder is 95°F. The pressure inside the cylinder is 182 psig. The low-side pressure is 39 psig.

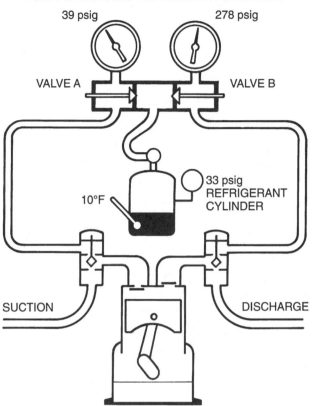

Figure 7-3. The refrigerant in this cylinder is at a low temperature and pressure because it has been in the back of a truck all night in cold weather. The cylinder pressure is 33 psig, which corresponds to 10°F. The pressure in the system is 39 psig.

In cold weather, however, the cylinder may have been in the back of the truck all night and its pressure may be lower than the low side of the system (Figure 7-3). This happens more often than you might imagine because many large installations, such as manufacturing plants, have core cooling loads even in the coldest weather, and require service year-round. In this case, the cylinder may have to be warmed to get refrigerant to move from the cylinder to the system. It is a good idea to store a cylinder of refrigerant in the equipment room of large installations. The cylinder will always be there in case you have no refrigerant in the truck, and the cylinder will be at room temperature even in cold weather.

When vapor refrigerant moves from a refrigerant cylinder, the liquid refrigerant boils to replace the vapor. As more and more vapor is released from the cylinder, the liquid in the bottom of the cylinder continues to boil, and its temperature decreases. If enough refrigerant is released, the cylinder pressure will equalize with the low-side pressure of the system. Heat will have to be added to the liquid refrigerant to keep the pressure up.

Warning: Never use concentrated heat, such as a torch, to warm a refrigerant cylinder. Instead, place the cylinder in a warm tub of water. The water temperature should not exceed 90°F.

This water temperature should maintain a cylinder pressure of 168 psig for R-22. Move the cylinder around in the tub of water to keep the liquid in the center of the cylinder in touch with the warm outside of the cylinder, and add additional warm water as necessary.

Caution: If the water begins to feel warm to the hand, it is getting too hot.

The larger the volume of liquid refrigerant in the bottom of the cylinder, the longer the cylinder will maintain the pressure. When large amounts of refrigerant must be charged into a system, use the largest cylinder available. For example, don't use a 25-pound cylinder to charge 20 pounds of refrigerant into a system if a 125-pound cylinder is available.

7.3 LIQUID REFRIGERANT CHARGING

Liquid refrigerant charging is normally accomplished by adding refrigerant to the liquid line. To dispense liquid, a liquid valve is provided on some cylinders. On others, the cylinder is simply inverted to put the cylinder valve in touch with the liquid portion of the refrigerant.

Liquid Charging an Empty System

When a system contains no refrigerant or is under a vacuum, liquid refrigerant may be charged into the king valve on the liquid line. To do this, attach the gage manifold to the liquid connection of a refrigerant cylinder and allow refrigerant to enter the system until it has nearly stopped. The liquid entering the system will move towards the evaporator and condenser. When the system is started, the refrigerant is divided between the evaporator and condenser, and there is no danger of liquid flooding into the compressor (Figure 7-4). When charging with liquid refrigerant, the cylinder pressure is not reduced, and warming is unnecessary. When large amounts of refrigerant are needed, the liquid method is preferable because it saves time.

Liquid Charging an Operating System

To charge an operating system with a king valve, front-seat the valve while the system is running (Figure 7-5). This will cause the low-side pressure of the system to drop. Liquid from the cylinder may then be charged into the system through an extra charging port. In this process, the liquid from the cylinder is actually feeding the expansion device.

Caution: Be careful not to overcharge the system.

The low-pressure control may have to be bypassed during charging to keep it from shutting the system down. Be sure to remove the bypass when charging is completed.

Warning: Do not charge liquid refrigerant directly into the suction line of a compressor. It may cause irreparable damage to the equipment.

Some charging devices allow the cylinder liquid line to be attached to the suction line for charging a system while it is operating. These are orifice-metering devices that create a restriction between the gage manifold and the system's suction line (Figure 7-6). They meter liquid refrigerant into the suction line where it flashes into a vapor. The same thing may be accomplished using the gage manifold valve (Figure 7-7). The suction line pressure is maintained at no more than 10 pounds above the system suction pressure. This will meter the liquid refrigerant into the suction line as a vapor.

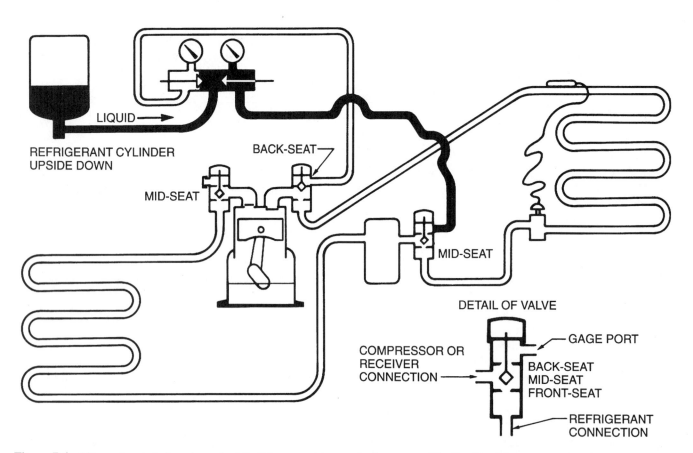

Figure 7-4. This system is being charged while it has no refrigerant in the system. The liquid refrigerant entering the system moves toward the evaporator and condenser. This prevents liquid refrigerant from entering the compressor.

LIQUID REFRIGERANT CHARGING 101

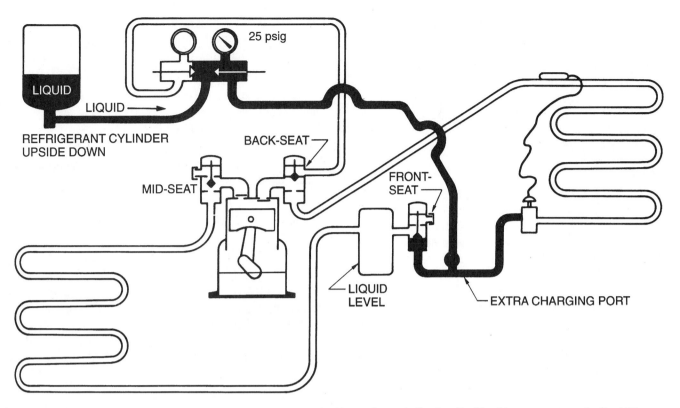

Figure 7-5. This system is being charged by front-seating the king valve and allowing liquid refrigerant to enter the liquid line.

Caution: If a charging device is used to flash liquid refrigerant into the suction line, it should only be used on compressors in which the suction gas passes over the motor windings. This will boil any small amounts of liquid refrigerant that may reach the compressor. If the lower compressor housing becomes cold, stop adding liquid. This method should only be performed under the supervision of an experienced person.

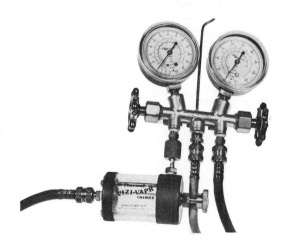

Figure 7-6. Charging device in the gage line between the liquid refrigerant in the cylinder and the suction line of the system. *Photo by Bill Johnson.*

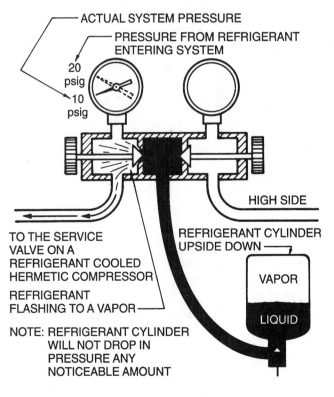

Figure 7-7. A gage manifold can be used to serve the same purpose as a charging device.

7.4 MEASURING REFRIGERANT

When a measured amount of refrigerant must be charged into a system, it may be weighed using a charging scale, measured using a graduated charging cylinder, or calculated using charging charts. Package air conditioning systems will have the recommended charge printed on the nameplate. This charge must be added to the system from a deep vacuum or it will not be correct.

Weighing Refrigerant

Weighing refrigerant may be accomplished using special charging scales. Bathroom scales and other inaccurate devices should not be used. Select portable scales and always secure them in the truck to keep the mechanism from shaking and changing the calibration. Figure 7-8 shows an accurate analog (dial) scale graduated in pounds and ounces. Analog scales are less expensive than electronic (digital) models, but they are also more difficult to use.

For example, suppose 28 ounces of refrigerant is needed, and the cylinder weight on a dial scale reads 24 pounds 4 ounces (Figure 7-9). As refrigerant transfers into the system, the cylinder weight decreases. Determining the final cylinder weight requires several calculations.

The calculated final cylinder weight is as follows:

$$24 \text{ lbs. } 4 \text{ oz. } - 28 \text{ oz. } = 22 \text{ lbs. } 8 \text{ oz.}$$

To arrive at this weight, 24 lbs. 4 oz. is first converted into ounces:

$$24 \text{ lbs.} \times 16 \text{ oz./lb.} = 384 \text{ oz.} +$$
$$\text{the remaining 4 oz.} = 388 \text{ oz.}$$

Now, subtract 28 oz. from the 388 oz. to determine the final cylinder weight.

$$388 \text{ oz. } - 28 \text{ oz. } = 360 \text{ oz.}$$

Because the scales do not read in ounces, you must convert back to pounds and ounces:

$$360 \text{ oz.} \div 16 \text{ oz./lb.} = 22.5 \text{ lbs.} = 22 \text{ lbs. } 8 \text{ oz.}$$

Electronic scales are very accurate and easy to use (Figure 7-10). Many of these scales can be adjusted back to zero with a full cylinder of refrigerant on the platform, so that as refrigerant is added to the system, the scales read a positive value. For example, if 28 ounces of refrigerant is needed, put the refrigerant cylinder on the scale and set the scale to zero.

Figure 7-8. Refrigerant charging scale. *Photo by Bill Johnson.*

Figure 7-10. Electronic scale with adjustable zero feature. *Photo by Bill Johnson.*

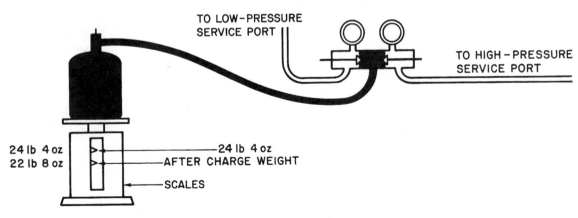

Figure 7-9. Using an analog charging scale to measure refrigerant into a system.

MEASURING REFRIGERANT 103

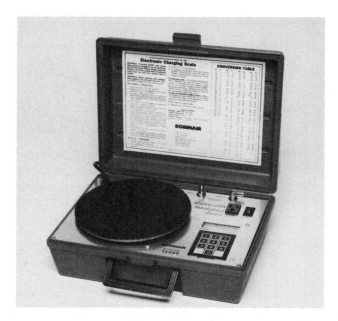

Figure 7-11. Programmable electronic scale. *Photo by Bill Johnson.*

Figure 7-12. Graduated cylinder used to measure refrigerant into a system using volume. *Photo by Bill Johnson.*

As the refrigerant leaves the cylinder, the scale counts upward. When it reaches 28 ounces, stop the refrigerant flow. This is a time-saving feature that avoids the cumbersome calculations involved with analog scales. Figure 7-11 shows an electronic scale that can be programmed for the correct amount of charge. A solid state microprocessor controls a solenoid which automatically stops the charging process when the programmed weight has been dispensed.

Measuring Refrigerant Using Charging Cylinders

Graduated charging cylinders are also used to measure refrigerant (Figure 7-12). These cylinders have a transparent column that allows you to observe the liquid level in the cylinder. A pressure gage at the top of the cylinder is used to determine the temperature of the refrigerant. Liquid refrigerant has different volumes at different temperatures, so the temperature of the refrigerant must be known. The pressure corresponding to this temperature is dialed on the graduated cylinder. The final liquid level inside the cylinder must be calculated as with a dial-type scale.

For example, suppose a graduated cylinder has 4 lbs. 4 oz. of R-22 in the cylinder at 120 psig and 28 ounces of refrigerant is required. Turn the dial to 120 psig and record the level of 4 lbs. 4 oz. The calculated final cylinder weight is as follows:

4 lbs. × 16 oz./lb. = 64 oz. + the remaining 4 oz. = 68 oz.

Next subtract 28 oz. from the cylinder weight of 68 oz. to determine the final cylinder weight.

68 oz. - 28 oz. = 40 oz.

Because charging cylinders do not read in ounces, you must convert back to pounds and ounces:

40 oz. ÷ 16 oz./lb. = 2.5 lbs. = 2 lbs. 8 oz.

One advantage of the graduated cylinder is that the refrigerant can be seen as the level drops (Figure 7-13).

Some graduated cylinders have heaters in the bottom to keep the refrigerant temperature from dropping as vapor is removed from the cylinder.

When selecting a graduated charging cylinder, be sure to choose one that is large enough for the systems you will be servicing. It is difficult to use a cylinder twice for one accurate charge. In addition, when charging systems with more than one type of refrigerant, use a separate charging cylinder for each type of refrigerant.

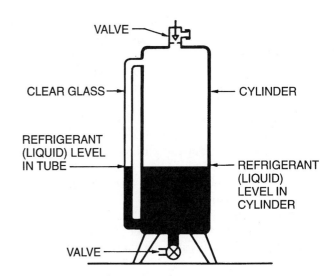

Figure 7-13. Cross-section of a graduated cylinder. The refrigerant may be seen in the tube as it moves into the system.

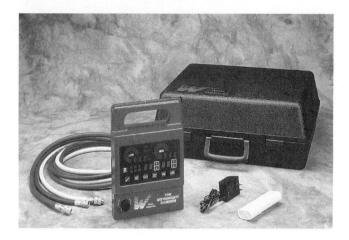

Figure 7-14. Refrigerant charger. *Courtesy White Industries.*

Refrigerant Chargers

Other refrigerant charging devices are now available which may make it more convenient to charge refrigerant into a system. Figure 7-14 shows one example. This device can be set to charge a system using either pressure or weight. A predetermined amount of refrigerant can be charged in pounds and ounces in one-ounce increments. This type of device also has many other features. Be sure to follow the manufacturer's instructions.

Measuring Refrigerant Using Charging Charts

Some system manuals provide typical operating pressures that may be compared to the gage readings for determining the correct charge. These are called charging charts and are plotted individually for each type of system (Figure 7-15). Always follow manufacturer's instructions when using charging charts.

SUMMARY

- Under proper conditions, refrigerant may be added to an air conditioning system in either the vapor or liquid state.
- When refrigerant is added in the vapor state, the refrigerant cylinder will lose pressure as the vapor is transferred out of the cylinder.
- Liquid refrigerant is normally added in the liquid line and only under the proper conditions.
- Liquid refrigerant must never be allowed to enter the compressor.
- Refrigerant is measured into systems using weight, volume, or pressure.
- Dial scales can be more difficult to use than electronic devices because the final cylinder weight must be calculated. The scales are graduated in pounds and ounces.
- Many electronic scales offer an adjustable zero feature that allows the scales to be adjusted back to zero with a full cylinder of refrigerant on the platform.

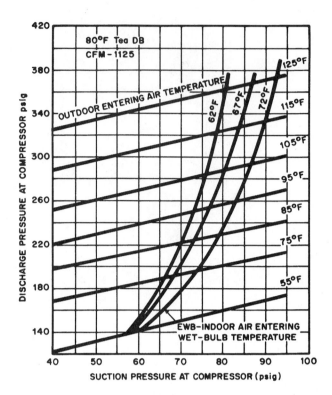

Figure 7-15. Typical system charging chart. Pressures and temperatures are plotted to arrive at the correct charge. *Reproduced courtesy of Carrier Corporation.*

- Graduated cylinders measure the volume of liquid refrigerant. This volume varies at different temperatures. The refrigerant pressure is dialed onto the cylinder for accuracy.

REVIEW QUESTIONS

1. How is liquid refrigerant added to an empty refrigeration system?
2. How is the refrigerant cylinder pressure kept above the system pressure when charging vapor from a cylinder?
3. Does the refrigerant pressure decrease in a refrigerant cylinder while charging with vapor? Can you explain why?
4. What is the main disadvantage of dial scales?
5. What type of equipment normally has the refrigerant charge printed on the nameplate?
6. Explain the use of the adjustable zero electronic charging feature found on some scales.
7. How is refrigerant pressure maintained in certain graduated charging cylinders?
8. How does a charging cylinder account for the volume change due to temperature changes?
9. What must you consider when purchasing a charging cylinder?
10. What methods besides charging scales and charging cylinders can be used for charging systems?

BASIC ELECTRICITY AND MAGNETISM 8

OBJECTIVES

Upon completion of this unit, you should be able to

- Describe the structure of an atom.
- Explain why some atoms have a positive charge and others have a negative charge.
- Explain the characteristics that make certain materials good conductors.
- Describe how magnetism is used to produce electricity.
- State the differences between alternating current and direct current.
- List the units of measurement for electricity.
- Explain the differences between series and parallel circuits.
- State the three formulas derived from Ohm's law.
- State the formula for determining electrical power.
- Describe a solenoid.
- Explain inductance.
- Describe the construction of a transformer and indicate how current is induced in a secondary circuit.
- Describe how a capacitor works.
- State the reasons for using proper wire sizes.
- Describe the physical characteristics and functions of several semiconductors.
- Describe the general procedures for making electrical measurements.

8.1 STRUCTURE OF MATTER

To understand the theory of how an electric current flows, you must understand something about the structure of matter. Matter is made up of atoms. Atoms consist of protons, neutrons, and electrons. Protons have a positive charge. Neutrons have no charge and have little or no effect as far as electrical characteristics are concerned. Electrons have a negative charge and travel around the nucleus in orbits. The number of electrons in an atom is the same as the number of protons.

The hydrogen atom contains only one proton and one electron (Figure 8-1). Very few atoms are as simple as the hydrogen atom. For example, most wiring used to conduct electric current is made of copper. Figure 8-2 illustrates a copper atom. It has 29 protons and 29 electrons.

Some electron orbits are further away from the nucleus than others. As you can see, two travel in an inner orbit, eight in the second, 18 in the third, and one in the outer orbit. It is this single electron in the outer orbit that makes copper a good conductor.

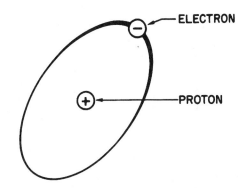

Figure 8-1. Hydrogen atom with one electron and one proton.

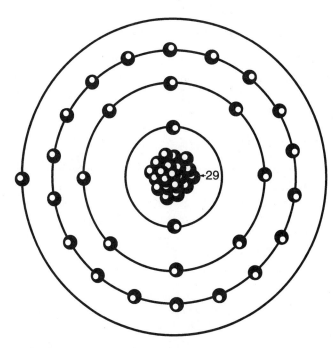

Figure 8-2. Copper atom with 29 protons and 29 electrons.

105

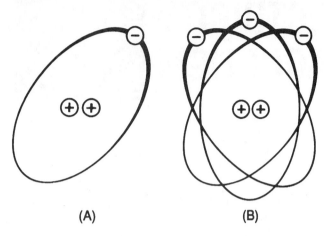

Figure 8-3. (A) This atom has two protons and one electron. It has a shortage of electrons and therefore a positive charge. (B) This atom has two protons and three electrons. It has an excess of electrons and therefore a negative charge.

8.2 MOVEMENT OF ELECTRONS

When sufficient energy or force is applied to an atom, the outer electron (or electrons) can be knocked out of orbit. If an electron leaves the atom, the atom will contain more protons than electrons. Because protons have a positive charge, the atom will now have a positive charge as shown in Figure 8-3(A). If the free electron joins another atom, this atom will contain more electrons than protons, so it will have a negative charge as shown in Figure 8-3(B).

A basic law of magnetism is that like charges repel each other, and unlike charges attract each other. Therefore, an atom with a surplus of electrons (negative charge) will be attracted to an atom with a shortage of electrons (positive charge). When a free electron approaches an orbit with a surplus of electrons, it will tend to repel any electron in that orbit and cause it to become a free electron. This transfer of energy from one atom to another is known as *conduction*.

8.3 CONDUCTORS

The most efficient conductors are those with few electrons in the outer orbit. Copper, silver, and gold are all good conductors. Each has one electron in the outer orbit. These are considered free electrons because they move easily from one atom to another.

8.4 INSULATORS

Atoms with several electrons in the outer orbit are poor conductors. These electrons are difficult to knock out of orbit, and materials made with these atoms are known as *insulators*. Glass, rubber, and plastic are examples of good insulators.

8.5 ELECTRICITY PRODUCED FROM MAGNETISM

Electricity can be produced in many ways; e.g., from chemicals, pressure, light, heat, and magnetism. The electricity in most air conditioning systems is produced by a generator using magnetism.

Magnets have opposing poles that are usually designated as the north (N) pole and the south (S) pole. They also have lines or fields of force. Figure 8-4 shows the lines of force around a permanent bar magnet. This force causes the like poles of two magnets to repel each other and the unlike poles to attract each other.

If a conductor, such as a copper wire, is passed through this magnetic field and cuts through these lines of force, the outer electrons in the copper atoms are freed and begin to travel from atom to atom (Figure 8-5). They travel in a single direction. It does not matter if the wire moves or the magnetic field moves for the wire to cut these lines of force.

This movement of electrons in one direction produces an electric current. Current is an electrical impulse transferred from one electron to the next.

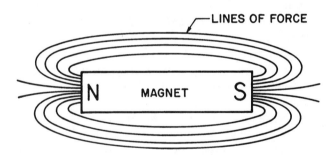

Figure 8-4. Permanent magnet with lines of force.

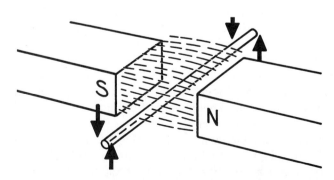

Figure 8-5. The movement of wire up and down cuts the lines of force and causes an electric current to flow in the wire.

THE ELECTRICAL CIRCUIT 107

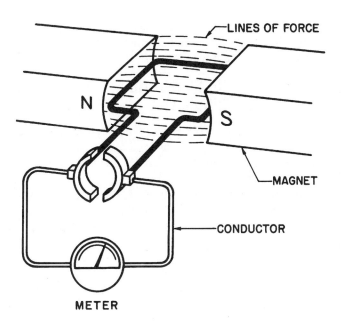

Figure 8-6. Simple generator.

An electrical generator generally has a large magnetic field and many turns of wire cutting through the lines of force. A large magnetic field or one with many turns of wire produces more electricity than a smaller field or one with fewer turns of wire. The magnetic force field for generators is usually produced by electromagnets. Electromagnets have many of the same characteristics as permanent magnets and are discussed in more detail later in this unit. Figure 8-6 shows a simple generator.

8.6 DIRECT CURRENT

Direct current (DC) travels in one direction. Because electrons have a negative charge and travel to atoms with a positive charge, DC is considered to flow from negative to positive.

8.7 ALTERNATING CURRENT

In alternating current (AC), the charge at the power source (generator) is continually changing direction; thus the current continually reverses itself. This produces voltage peaks and valleys known as waveforms. The number of waveforms per second is typically referred to as cycles per second or Hertz (Hz). Household electricity in the United States is delivered at 60 cycles per second or 60 Hz. For several reasons, most electrical energy generated for public use is AC. It is much more economical to transmit electrical energy long distances in the form of AC. The voltage of this type of electrical current flow can be readily changed so that it has many more uses. DC still has many applications, but it is usually obtained by changing AC to DC using a rectifier or by producing the DC locally, such as with a battery.

8.8 ELECTRICAL UNITS OF MEASUREMENT

Electromotive force (emf) or *voltage (V)* are both used to represent the difference in potential between two charges. When an electron surplus builds up on one side of a circuit and a shortage of electrons exists on the other side, a difference in potential or *emf* is created. The unit used to measure this force is the *volt*.

The *ampere (A)* is the unit used to measure the rate of electron flow.

All materials resist the flow of electrical current to some extent. In conductors, the resistance is low. In insulators, the resistance is high. The unit used to measure resistance is the *ohm (Ω)*. A conductor has a resistance of one ohm when a force of one volt causes a current of one ampere to flow.

Volt = Electromotive force (emf) or voltage (V)
Ampere = Current or rate of electron flow (A)
Ohm = Resistance to electron flow (Ω)

8.9 THE ELECTRICAL CIRCUIT

An electrical circuit must have a power source, a conductor to carry the current, and a load or device to use the current. Most circuits also include a switch or other device for turning the current flow on and off. Figure 8-7 shows an electrical generator for the source, a wire for the conductor, a light bulb for the load, and a switch for opening and closing the circuit.

The generator produces the current by passing many turns of wire through a magnetic field. If it is a DC generator, the current will flow in one direction. If it is an AC generator, the current will continually reverse itself. However, the effect on the circuit will generally be the same whether it is AC or DC.

The wire or conductor provides the means for the electricity to flow to the bulb and complete the circuit. The electrical energy is converted to both heat and light energy at the bulb element.

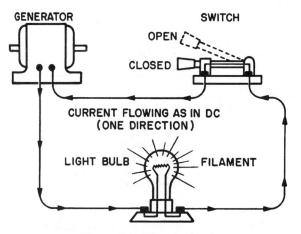

Figure 8-7. Electric circuit.

The switch is used to open and close the circuit. When the switch is open, no current will flow. When it is closed, the bulb element will produce heat and light.

Many circuits contain more than one resistance or load. These resistances may be wired in series or in parallel, depending on the application. In a series circuit, there is only one electrical pathway for the current to follow; in a parallel circuit, there are two or more pathways. Figure 8-8 shows three loads wired in series. This is shown both pictorially and using symbols. Figure 8-9 illustrates three loads wired in parallel.

Power-passing devices such as switches are wired in series. Most loads (power-consuming devices) in air conditioning systems are wired in parallel.

8.10 MAKING ELECTRICAL MEASUREMENTS

In the circuit illustrated in Figure 8-7, electrical measurements can be made to determine the voltage (emf) and amperes (current). As shown in Figure 8-10, the voltmeter is connected across the terminals of the bulb without interrupting the circuit. The ammeter is connected directly into the circuit so that the current flows through it. Figure 8-11 illustrates the same circuit using symbols.

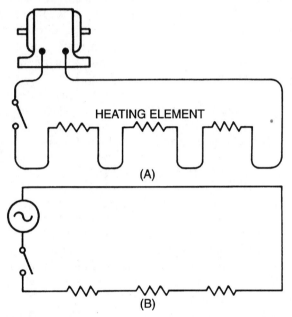

Figure 8-8. Multiple resistances (small heating elements) in series.

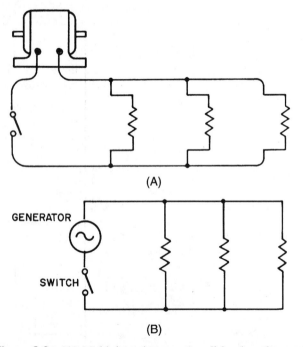

Figure 8-9. (A) Multiple resistances (small heating elements) in parallel. (B) Three resistances in parallel using symbols.

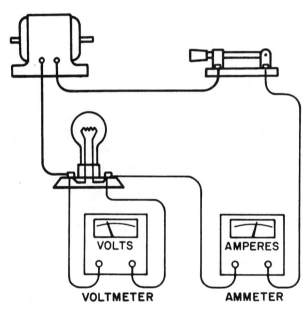

Figure 8-10. Voltage is measured across the resistance (in parallel). Amperage is measured in series.

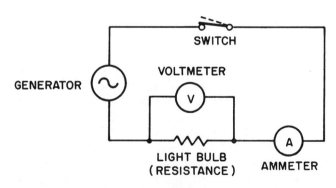

Figure 8-11. This circuit is the same circuit shown in Figure 8-10, except that it is now illustrated using electrical symbols.

Figure 8-12 illustrates how a voltmeter is connected in parallel with each of the loads. The ammeter is also shown and is connected in series with the circuit. One type of ammeter can be clamped around a single conductor to measure amperes (Figure 8-13). This is convenient because it is time consuming and often difficult to disconnect a circuit to connect the ammeter in series. This type of ammeter is often called a *clamp-on ammeter*. Meters and their uses are described in more detail later in this unit.

8.11 OHM'S LAW

During the early 1800s, the German scientist Georg S. Ohm determined that there is a relationship between each of the factors in an electrical circuit. This relationship is known as Ohm's Law. The basic equation is as follows:

$$E = I \times R$$

where: E or V = voltage (emf)
I = amperage (current)
R = resistance (load)

This equation can be rearranged to solve for resistance:

$$I = \frac{E}{R}$$

This equation can also be rearranged to solve for amperage:

$$R = \frac{E}{I}$$

Figure 8-14 shows a convenient way to remember three equations.

In Figure 8-15, the resistance of the heating element can be determined as follows:

$$R = \frac{E}{I} = \frac{120}{1} = 120\ \Omega$$

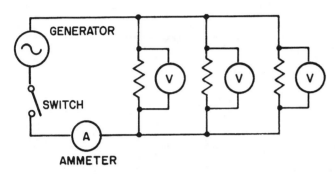

Figure 8-12. Voltage readings are taken across the resistances in the circuit.

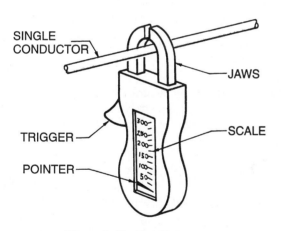

Figure 8-13. Clamp-on ammeter.

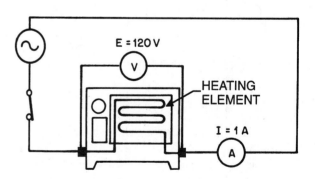

Figure 8-15. Determining the resistance of a heating element.

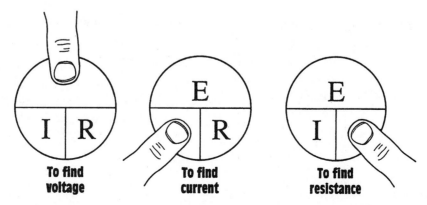

Figure 8-14. To determine the formula for the unknown quantity, cover the letter representing the unknown.

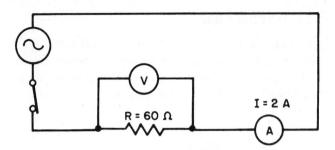

Figure 8-16. Determining the voltage across the resistance.

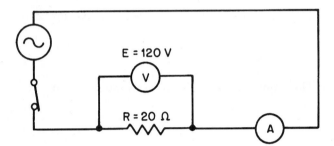

Figure 8-17. Determining the current flow.

In Figure 8-16, the voltage across the resistance can be calculated as follows:

$$E = I \times R = 2 \times 60 = 120 \text{ V}$$

Figure 8-17 indicates that the voltage is 120 V and the resistance is 20 Ω. The formula for determining the current flow is:

$$I = \frac{E}{R} = \frac{120}{20} = 6A$$

8.12 SERIES CIRCUITS

Series circuits have the following characteristics:

- The voltage is divided across the different resistances.
- The total current flows through each resistance or load.
- The resistances are added together to obtain the total resistance.

Calculating Resistance in a Series Circuit

To determine the total resistance in a series circuit, simply add all of the resistances together as if they were a single load. For example, in the circuit shown in Figure 8-18, the total resistance is 40 Ω (20 Ω + 10 Ω + 10 Ω). The amperage in this circuit can be determined as follows:

$$I = \frac{E}{R} = \frac{120}{40} = 3A$$

The individual resistances can be measured by disconnecting each load from the circuit and reading the ohms using an ohmmeter (Figure 8-19).

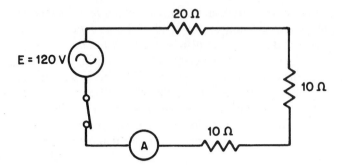

Figure 8-18. Determining the amperage in a series circuit.

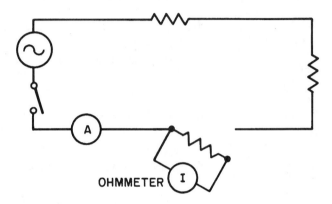

Figure 8-19. To determine the resistance, disconnect the resistance from the circuit and check with an ohmmeter.

Caution: Do not use any electrical measuring instruments without specific instructions from a qualified person. The use of electrical measuring instruments is discussed later in this unit.

8.13 PARALLEL CIRCUITS

Parallel circuits have the following characteristics:

- The total voltage is applied across each resistance.
- The current is divided between the different loads, or the total current is equal to the sum of the currents in each branch.
- The total resistance is less than the value of the smallest resistance.

Calculating Resistance in a Parallel Circuit

Calculating the total resistance in a parallel circuit requires a different procedure than simply totaling the individual resistances as in a series circuit. The general formula used to determine total resistance in a parallel circuit is as follows:

$$R_{total} = \frac{1}{\frac{1}{R_1} + \frac{1}{R_2} + \frac{1}{R_3} + \text{etc.}}$$

The total resistance of the circuit in Figure 8-20 can be calculated as follows:

$$R_{total} = \frac{1}{\frac{1}{10} + \frac{1}{20} + \frac{1}{30}}$$

$$= \frac{1}{0.1 + 0.05 + 0.033}$$

$$= \frac{1}{0.183}$$

$$= 5.46\ \Omega$$

To determine the total current draw, use Ohm's law:

$$I = \frac{E}{R} = \frac{120}{5.46} = 22\ A$$

8.14 ELECTRICAL POWER

Electrical power (P) is measured in watts. A *watt (W)* is the power used when one ampere flows at a potential difference of one volt. Therefore, power can be determined by multiplying the voltage times the amperes flowing in a circuit.

$$\text{watts} = \text{volts} \times \text{amperes}\ or$$
$$P = E \times I$$

For example, the power supplied to the circuit in Figure 8-20 can be calculated as follows:

$$P = 120 \times 22 = 2640\ W$$

The consumer of electrical power pays the electrical utility company according to the number of kilowatts (kW) used. A kW is equal to 1000 W. To determine the kilowatts being consumed, divide the number of watts by 1000:

$$P\ (\text{in kW}) = \frac{E \times I}{1000}$$

For our example:

$$P = \frac{2640}{1000} = 2.64\ kW$$

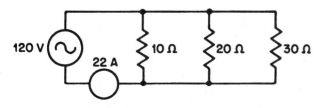

Figure 8-20. Determining the amperage in a parallel circuit.

8.15 MAGNETISM

As discussed previously, magnets are classified as either permanent or temporary. Permanent magnets are used in only a few applications in air conditioning systems. Electromagnets, a form of temporary magnet, are used in many components of air conditioning equipment.

A magnetic field exists around a wire carrying an electric current. If the wire or conductor is formed in a loop, the strength of the magnetic field will be increased (Figure 8-21). If the wire is wound into a coil, an even stronger magnetic field will be created (Figure 8-22). A coil of wire carrying an electric current is called a *solenoid*. A solenoid or electromagnet will attract or pull an iron bar into the coil (Figure 8-23).

Figure 8-21. Magnetic field around loop of wire. This is a stronger field than that around a straight wire.

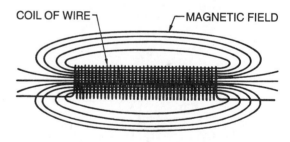

Figure 8-22. There is a stronger magnetic field surrounding wire formed into a coil.

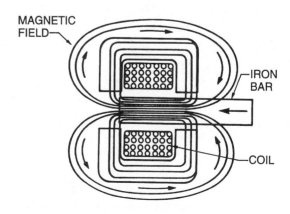

Figure 8-23. When current flows through the coil, the iron bar will be attracted into the coil.

112 UNIT 8 BASIC ELECTRICITY AND MAGNETISM

If an iron bar is inserted permanently in the coil, the strength of the magnetic field will be increased even further.

This magnetic field can be used to generate electricity and operate electric motors. Magnetic attraction can also be used to operate controls and switching devices, such as solenoids, relays, and contactors (Figure 8-24).

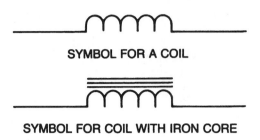

Figure 8-25. Symbols for a coil.

8.16 INDUCTANCE

As mentioned above, when voltage is applied to a conductor and current flows, a magnetic field is produced around the conductor. In an AC circuit, the current is continually changing direction. This causes the magnetic field to continually build up and then collapse. When these lines of force build up and collapse, they cut through the wire or conductor and produce an emf or voltage. This voltage opposes the existing voltage in the conductor.

In a straight conductor, this induced voltage is too small to have any effect on the circuit. However, if a conductor is wound into a coil, these lines of force overlap and reinforce each other. This develops a voltage that is strong enough to provide opposition to the existing voltage. This opposition is called *inductive reactance* and is a type of resistance in an AC circuit. Coils, chokes, and transformers are examples of components that produce inductive reactance. Figure 8-25 shows symbols for the electrical coil.

8.17 TRANSFORMERS

Transformers are electrical devices that produce a current in a second circuit through electromagnetic induction. For instance, transformers are rated in volt-amperes (VA). A 40 VA transformer is often used in the control circuit of combination heating and cooling systems. These transformers produce 24 V and can carry a maximum of 1.66 A. This is determined as follows:

$$\text{output in amperes} = \frac{\text{VA rating}}{\text{voltage}}$$

$$I = \frac{40}{24} = 1.66$$

In Figure 8-26(A), a voltage applied across terminals A-A will produce a magnetic field around the steel or iron core. The alternating current causes the magnetic field to continually build up and collapse as the current reverses. This is known as a *sine wave*. When this occurs, the magnetic field around the core in the secondary winding will cut across the conductor wound around it. This induces a current in the second winding.

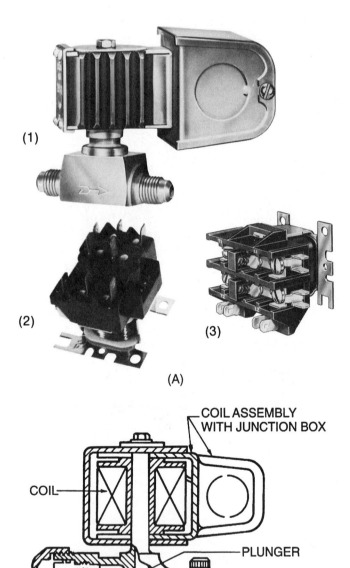

Figure 8-24. (A) (1) Solenoid. (2) Relay. (3) Contactor. (B) Cutaway view of solenoid. *Courtesy of: (A) (1) Sporlan Valve Company, (2) Photo by Bill Johnson, (3) Honeywell, Inc., Residential Division, (B) Parker Hannifin Corporation.*

TRANSFORMERS 113

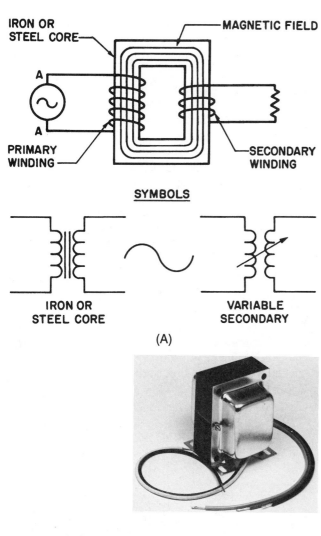

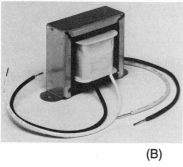

Figure 8-26. (A) Voltage supplied across the terminals produces a magentic field around an iron or steel core. (B) Transformers. *Photos by Bill Johnson.*

Transformers include a primary winding, a core that is usually made of thin plates of steel laminated together, and a secondary winding. Figure 8-26(B) shows a typical transformer. There are two types of transformers: step-down transformers and step-up transformers. A step-down transformer contains more turns of wire in the primary winding than in the secondary winding. The voltage in the secondary winding is directly related to the

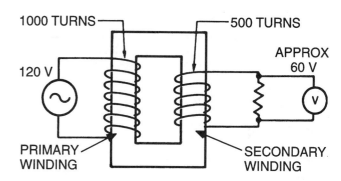

Figure 8-27. Step-down transformer.

difference between the number of turns of wire in the two windings. For example, Figure 8-27 is a transformer with 1000 turns in the primary winding and 500 turns in the secondary winding. A voltage of 115 V is applied, and the voltage induced in the secondary winding is approximately 60 V. Actually, the voltage is slightly less due to some loss of the magnetic field into the air and because of the resistance in the wire.

A step-up transformer has more turns of wire in the secondary winding than in the primary winding. This causes a larger voltage to be induced into the secondary winding. Figure 8-28 shows a transformer with 1000 turns in the primary winding, 2000 in the secondary winding, and an applied voltage of 120 V. The voltage induced in the secondary winding is doubled, or approximately 240 V.

In a transformer, the same power (watts) is available at both windings (except for a slight loss in the secondary winding). If the voltage is doubled in the secondary winding, the current capacity is reduced by half. If the voltage is reduced by half in the secondary winding, the current capacity nearly doubles.

When electricity is transmitted over long distances, it incurs a loss of voltage known as *transmission loss*. Step-up transformers are used at generating stations to provide more efficient delivery of the electrical energy over long distances to substations or other distribution centers.

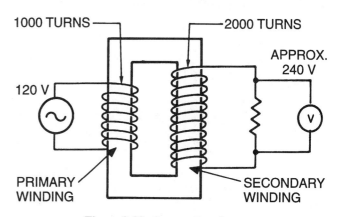

Figure 8-28. Step-up transformer.

At the substation, step-down transformers reduce the voltage for further distribution. At a residence, the voltage may be further reduced to 240 V or 120 V. Step-down transformers may also be used with air conditioning equipment to produce the 24 V commonly used in thermostats and other control devices.

8.18 CAPACITANCE

A *capacitor* is a device that allows electrical energy to be stored in the circuit for later use. A simple capacitor is composed of two plates with insulating material (dielectric) between the plates (Figure 8-29). The capacitor can store a charge of electrons on one plate (Figure 8-30). When the plate is fully charged in a DC circuit, no current will flow until there is a path back to the positive plate (Figure 8-31). When this path is available, the electrons will flow to the positive plate until the negative plate no longer has a charge (Figure 8-32). After the capacitor has discharged, both plates are neutral.

In an AC circuit, the voltage and current are continuously changing direction. As the electrons flow in one direction, one plate of the capacitor becomes charged. When the current and voltage reverse, the charge on the capacitor becomes greater than the source voltage, and the capacitor discharges. This current travels through the circuit and charges the opposite plate. This continues through each AC cycle.

Capacitance is the amount of charge that can be stored in a particular capacitor. Capacitance is determined by the following physical characteristics of the capacitor:

- The distance between the plates
- The surface area of the plates
- The type of dielectric material between the plates

Capacitors are rated in farads (F). However, farads represent such a large amount of capacitance that the term microfarad (µF) is normally used. The symbol for micro is the Greek letter µ (mu). A microfarad is one millionth (0.000001) of a farad. Capacitors can be purchased in ranges up to several hundred microfarads.

The opposition to current flow of a capacitor is similar to that of a pure resistor or the process of inductive reactance. The opposition of a capacitor is called *capacitive reactance*. The degree of capacitive reactance depends on the frequency of the voltage and the capacitance of the capacitor.

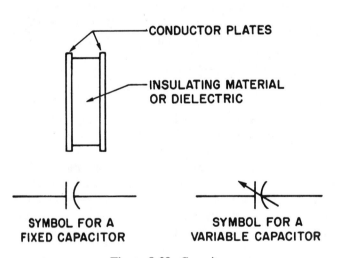

Figure 8-29. Capacitor.

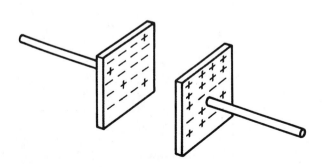

Figure 8-30. Charged capacitor.

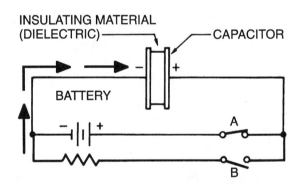

Figure 8-31. Electrons will flow from the battery to one side of the capacitor. The negative plate will charge until the capacitor has the same potential difference as the battery. At this point, no current will flow until there is a path back to the positive plate.

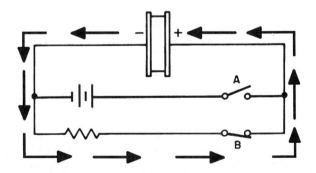

Figure 8-32. After the capacitor is charged, switch A opens and switch B closes. The capacitor then discharges through the resistor to the positive plate. After the capacitor has discharged, both plates are neutral.

ELECTRICAL MEASURING INSTRUMENTS 115

Figure 8-33. Start capacitor and run capacitor. *Photo by Bill Johnson.*

Two types of capacitors used frequently in the air conditioning industry are the start and run capacitors used to start-up electric motors (Figure 8-33). Start capacitors provide extra power to overcome the motor starting torque and run capacitors improve efficiency by supplying additional power while the motor is operating.

8.19 IMPEDANCE

We have learned that there are three types of opposition to current flow in an AC circuit: pure resistance, inductive reactance, and capacitive reactance. The total effect of all opposition in a circuit is known as *impedance*. To understand this process, we must return briefly to the topic of waveforms. As discussed previously, voltage produces a characteristic sine waveform pattern, as shown in Figure 8-34(A). As you can see, current also produces a similar waveform. In a circuit that has only resistance, the voltage and current are in phase with one another. The voltage leads the current across an inductor, as shown in Figure 8-34(B), and lags behind the current across a capacitor, as shown in Figure 8-34(C). Inductive reactance and capacitive reactance can cancel each other out. Impedance is a combination of the opposition to current flow produced by these characteristics in a circuit.

8.20 ELECTRICAL MEASURING INSTRUMENTS

Warning: When servicing equipment, shut off the electrical service at the disconnect panel whenever possible, lock and tag the disconnect panel, and keep the only key in your possession at all times.

Multimeters are combination instruments that measure voltage, milliamperes, resistance, and, on some models, temperature. They can be used for making AC or DC measurements in several ranges. A multimeter is the instrument used most often by air conditioning technicians.

A common multimeter is the volt-ohm-milliammeter (VOM). This meter is used to measure AC and DC voltages, direct current, milliamperes, resistance, and alternating current when used with an AC clamp-on ammeter adapter. The meter has two main switches: the function switch and the range switch (Figure 8-35). The function switch has –DC, +DC, and AC positions.

The range switch may be turned in either direction to obtain the desired range. It is also used to select the proper position for making alternating current measurements when using the AC clamp-on adapter.

The zero ohms control is used to adjust the meter to compensate for weak batteries. Before using the meter, always make sure that the pointer is set precisely to zero. If the pointer is off, use a screwdriver to turn the control screw clockwise or counterclockwise until the pointer is set exactly at zero.

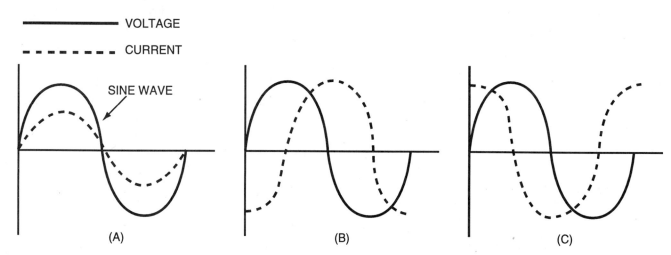

Figure 8-34. (A) In a pure resistive circuit, current and voltage are in phase with one another. (B) Voltage leads current by 90° in a pure inductive circuit. (C) Current leads voltage by a 90° in a pure capacitive circuit.

116   UNIT 8   BASIC ELECTRICITY AND MAGNETISM

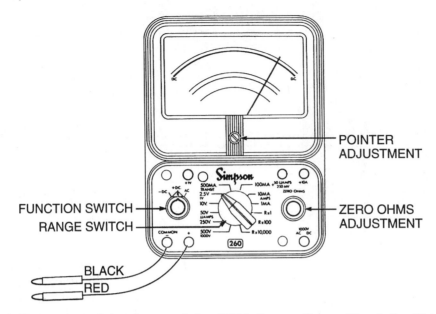

Figure 8-35. Function switch and range switch on VOM. *Courtesy Simpson Electric Co., Elgin, Illinois.*

The test leads may be plugged into any of eight jacks. In the unit shown in Figure 8-35, the common (−) and positive (+) jacks will be the only ones used. Only a few of the basic measurements will be discussed. Other measurements will be described in detail in other units.

Caution: The following instructions are for familiarization only. Do not make any measurements without approval from an instructor or supervisor, and always read and follow the manufacturer's instructions when using any type of meter.

Using a VOM to Measure DC Voltage

Figure 8-36 shows a DC circuit with 15 V from the battery power source. To check this voltage with a VOM, follow the procedure outlined below.

1. Set the function switch to +DC.
2. Set the range switch to 50 V.

Note: If you are not sure about the magnitude of the voltage, always set the range switch to the highest setting to avoid damaging the instrument. After measuring, you can switch to a lower range if necessary to obtain a more accurate reading.

3. Plug the black test lead into the common (−) jack and the red test lead into the + jack.
4. Be sure the switch in the circuit is open.
5. Connect the black test lead to the negative side of the circuit, and the red test lead to the positive side, as shown in Figure 8-36. Note that the meter is connected across the load (in parallel).
6. Close the switch and read the voltage from the DC scale.

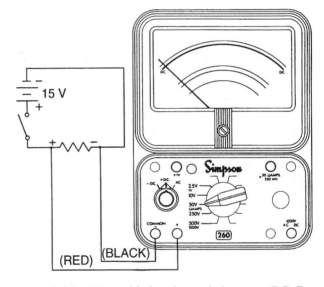

Figure 8-36. VOM with function switch set at +DC. Range switch set at 50 V. *Courtesy Simpson Electric Co., Elgin, Illinois.*

Using a VOM to Measure AC Voltage

To check the voltage in the 120 V AC circuit shown in Figure 8-37, follow the procedure outlined below.

1. Turn off the power.
2. Set the function switch to AC.
3. Set the range switch to 250 V.
4. Plug the black test lead into the common (−) jack and the red test lead into the + jack.
5. Connect the test leads across the load as shown in Figure 8-37.
6. Turn on the power. Find the scale on the meter that is marked AC. Read the voltage.

ELECTRICAL MEASURING INSTRUMENTS 117

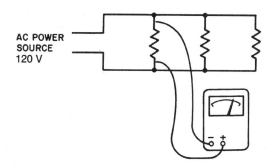

Figure 8-37. Connecting test leads across a load.

Using a VOM to Measure Resistance

To determine the resistance of a load, follow the procedure outlined below.

1. Turn off the power.
2. Make sure all power is off while doing this.
3. Make the zero ohms adjustment as follows:
   a. Turn the range switch to the desired ohms range.
      - Use R x 1 for 0–200 Ω.
      - Use R x 100 for 200–20,000 Ω.
      - Use R x 10,000 for over 20,000 Ω.
   b. Plug the black test lead into the (-) jack and the red test lead into the (+) jack.
   c. Connect the test leads to each other.
   d. Rotate the zero ohms control until the pointer indicates zero ohms. (If the pointer cannot be adjusted to zero, replace the batteries.)
4. Disconnect the ends of the test leads and connect them to the load being tested.
5. Set the function switch to either –DC or +DC (it doesn't matter which).
6. Observe the reading on the ohms scale at the top of the dial. (Note that the ohms scale reads from right to left.)
7. To determine the actual resistance, multiply the reading by the factor at the range switch position.

Using a Clamp-On Ammeter

To avoid disconnecting the load from the circuit to measure the amperage, a clamp-on ammeter can be used. This device is placed around a single wire in the circuit and the current flowing through the wire is read as amperage on the meter (Figure 8-38). Most clamp-on type ammeters do not provide accurate readings below one ampere. However, a standard clamp-on ammeter can be modified to produce an accurate reading.

Figure 8-39 illustrates how to amplify the current reading by using ten wraps of wire around the jaws of the meter. To determine the actual amperage, divide the measured amperage by ten.

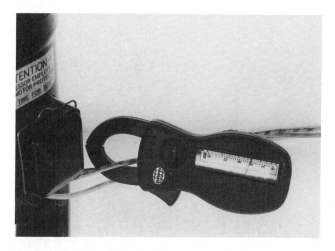

Figure 8-38. Measuring amperage by clamping meter around conductor. *Photo by Bill Johnson.*

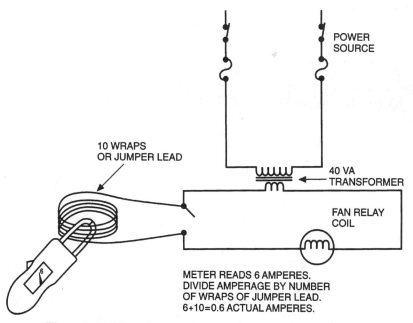

Figure 8-39. Wrapping the wire around the ammeter provides a more accurate reading.

118 UNIT 8 BASIC ELECTRICITY AND MAGNETISM

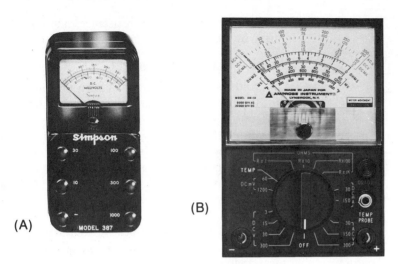

Figure 8-40. Meters used for electrical measurements. (A) DC millivoltmeter. (B) Multimeter (VOM). (C) Digital clamp-on ammeter. *Courtesy of: (A) Simpson Electric Co., Elgin, Illinois, (B and C) Amprobe Instrument Division, Core Industries.*

It is often necessary to determine voltage or amperage readings to a fraction of a volt or ampere.

Common units of voltage and amperage are listed below:

Voltage
1 volt = 1000 millivolts (mV)
1 volt = 1,000,000 microvolts (µV)

Amperage
1 ampere = 1000 milliamperes (mA)
1 ampere = 1,000,000 microamperes (µA)
(Note that the symbol for micro or millions is µ.)

Many styles and types of meters are available for making electrical measurements. Figure 8-40 shows some of these meters. Many meters now provide digital readouts (Figure 8-41).

Figure 8-41. Typical VOM with digital readout. *Courtesy Simpson Electric Co., Elgin, Illinois.*

Electrical Troubleshooting

To avoid unnecessary parts changing and ensure customer satisfaction, electrical troubleshooting must be taken one step at a time. All operable components must be eliminated as the source of the problem before any specific component is replaced or condemned. For example, Figure 8-42 shows a simplified wiring diagram of an air conditioning fan. In this case, the fan motor will not operate because of an open motor winding. The following voltage checks will help to determine where the problem is located:

1. This is a 115 V circuit, so the VOM range selector switch is set to 250 V AC. The neutral (– or common) meter lead is connected to a neutral terminal at the power source.
2. Connect the positive (+) meter probe to a line-side power source terminal, as shown in Figure 8-42, point 1. The meter should read 115 V, indicating that there is power at the source.

Note: Most household appliances are rated at 115 V. However, the actual voltage measured depends upon many factors, including the home's proximity to the transformer and the present load on the transformer. The actual voltage reading may vary by ±10%. This text uses both 115 V and 120 V for illustrative purposes.

3. Connect the positive lead to the line-side terminal of the disconnect switch (Figure 8-42, point 2). There is power.
4. Connect the positive lead to the line-side of the fuse (Figure 8-42, point 3). There is power.
5. Connect the positive lead to the load side of the fuse (Figure 8-42, point 4). There is power.
6. Connect the positive lead to the load side of the fan switch (Figure 8-42, point 5). There is power.
7. Connect the positive lead o the line side of the motor terminal (Figure 8-42, point 6). There is power.

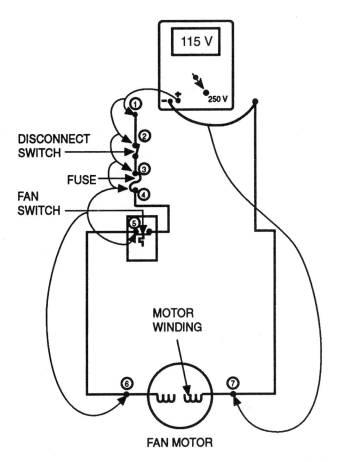
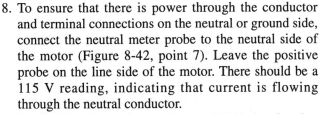

Figure 8-42. Partial diagram of air conditioning fan motor circuit.

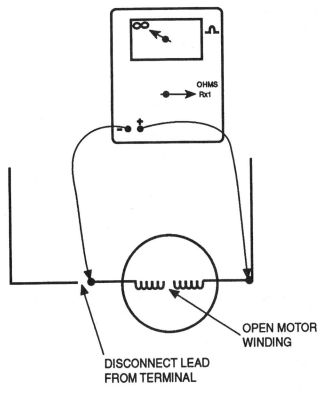

Figure 8-43. Checking an electric motor winding using an ohmmeter.

8. To ensure that there is power through the conductor and terminal connections on the neutral or ground side, connect the neutral meter probe to the neutral side of the motor (Figure 8-42, point 7). Leave the positive probe on the line side of the motor. There should be a 115 V reading, indicating that current is flowing through the neutral conductor.

In all of the above checks, there is 115 V showing, but the motor does not run. This leads us to the conclusion that the motor is defective.

The motor winding can be checked with an ohmmeter, as shown in Figure 8-43. The motor winding must have a measurable resistance for it to function properly. To check this resistance, perform the following procedure:

1. Turn off the power source to the circuit.
2. Disconnect one terminal on the motor from the circuit.
3. Set the meter selector switch to ohms (R × 1).
4. Touch the meter probes together and adjust the meter to zero ohms.
5. Touch one meter probe to one motor terminal and the other to the other terminal. The meter reads infinity (∞). This is the same reading you would get by holding the meter probes apart in the air; in other words, there is no circuit through the winding, which indicates that it is open.

8.21 WIRE SIZES

All conductors have some resistance. The degree of resistance depends on the conductor material, the cross-sectional area of the conductor, and the length of the conductor. A conductor with low resistance carries current more easily than one with high resistance.

Warning: If a wire is too small for the current passing through it, it may overheat enough to burn the insulation and cause a fire.

Always use the proper wire (conductor) size. The size of a wire is determined by its diameter or cross-section. A large diameter wire has more current-carrying capacity than a smaller diameter wire. The conductor must be sized large enough so that it does not operate beyond its rated temperature, typically 140°F (60°C) at an ambient temperature of 86°F (30°C).

Standard copper wire sizes are identified by American Standard Wire Gauge numbers and measured in circular mils. A circular mil is the area of a circle that is 1/1000 inch in diameter. Temperature is also considered because resistance increases as temperature increases. Larger wire size numbers indicate smaller wire diameters and greater resistances. Check the tables in the National Electrical Code to determine the proper wire size for each job.

8.22 CIRCUIT PROTECTION DEVICES

Circuit protection is essential to prevent the conductors and/or loads in a circuit from becoming overheated and causing equipment damage or fires. If one of the power-consuming devices in a circuit has a short circuit in its coil, the circuit protector stops the current flow before the conductor becomes overloaded and hot. For example, a circuit may be designed to carry a load of 20 amperes. As long as the circuit is carrying up to this amperage, overheating should not occur. If the amperage in the circuit is gradually increased, the conductor will begin to overheat (Figure 8-44). For specific information regarding circuit protection, refer to the National Electrical Code.

Fuses

A fuse is a simple device that usually contains a strip of metal with a higher resistance than the conductors in the circuit. This strip also has a relatively low melting point. Because of its higher resistance, it will heat up faster than the conductor. When the current exceeds the rating of the fuse, the strip melts and opens the circuit.

Plug Fuses Plug fuses have either an Edison base or a Type S base (Figure 8-45). Edison-base fuses are found in older installations and can be used for replacement only. Type S fuses are currently recommended as replacements for Edison base fuses. However, Type S fuses can be used only in a Type S fuse holder specifically designed for the fuse; otherwise, an adapter must be used. Each adapter is designed for a specific amperage rating, and cannot be interchanged. The amperage rating determines the size of the adapter. Plug fuses are rated up to 125 V and 30 A.

Figure 8-45. (A) Edison base plug fuse. (B) Type S fuse and adapter to Edison base socket.

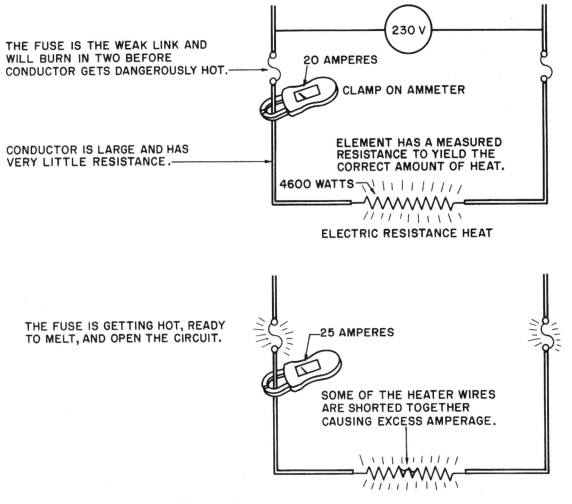

Figure 8-44. Fuses protect the circuit.

Dual-Element Plug Fuses Many circuits include electric motors. Motors draw extra current at start-up and can sometimes cause a plain (single element) fuse to burn out or open the circuit. Dual-element fuses are frequently used in this situation. One element in the fuse will melt when there is a large overload, such as a short circuit. The other element will melt and open the circuit when there is a smaller current overload lasting more than a few seconds; this protects the circuit from momentary surge of inrush current on start-up.

Cartridge Fuses For 230 V to 600 V service up to 60 A, a ferrule cartridge fuse is used, Figure 8-46(A). From 60 A to 600 A, a knife-blade cartridge fuse is used, Figure 8-46(B). A cartridge fuse is sized according to its amperage rating to prevent a fuse with an inadequate rating from being used. Many cartridge fuses have an arc-quenching material around the element to prevent damage from arcing in severe short-circuit situations (Figure 8-47).

Circuit Breakers

A circuit breaker can function as a switch as well as a means for opening a circuit when a current overload occurs. Most commercial and industrial installations and newer residences use circuit breakers rather than fuses for circuit protection. The advantage of a circuit breaker is that it can be reset after it trips, unlike fuses, which must be replaced.

Circuit breakers use two methods to protect the circuit. One is a bimetal strip that heats up with a current overload and trips the breaker, opening the circuit. The other is a magnetic coil that causes the breaker to trip and open the circuit when there is a short circuit or other excessive current overload in a very short period of time. Figure 8-48 shows a typical circuit breaker.

Ground Fault Circuit Interrupters

Ground fault circuit interrupters (GFCI) help protect against shock hazards, in addition to providing current overload protection. A GFCI, Figure 8-49, detects even a very small current leak to ground. Under certain conditions, this leak may cause an electrical shock. This small leak, which may not be detected by a conventional circuit breaker, will cause the GFCI to open the circuit.

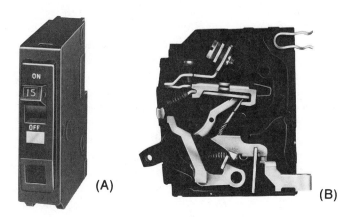

Figure 8-48. (A) Circuit breaker. (B) Cutaway. *Courtesy Square D Company.*

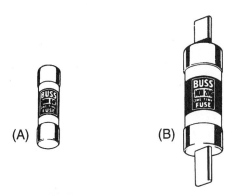

Figure 8-46. (A) Ferrule-type cartridge fuses. (B) Knife-blade cartridge fuse. *Reprinted with permission of Bussmann Division, McGraw-Edison Company.*

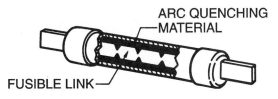

Figure 8-47. Knife-blade cartridge fuse with arc-quenching material.

Figure 8-49. Ground fault circuit interrupter. *Courtesy Square D Company.*

Figure 8-50. Circuit board with semiconductors. *Photo by Bill Johnson.*

8.23 SEMICONDUCTORS

The development of semiconductors or solid-state components has caused major changes in the design of electrical devices and controls.

Semiconductors are generally small and lightweight and can be mounted on circuit boards (Figure 8-50). This unit describes some of the individual solid-state devices and their applications. Air conditioning technicians do not normally replace solid-state components on circuit boards. However, you should have some knowledge of these components and should be able to determine when they are defective. In most cases, when a component is defective, the entire board is replaced. Circuit boards can be returned to the manufacturer or sent to a company that specializes in repairing or rebuilding them.

Semiconductors are usually made of silicon or germanium. In their pure form, semiconductors do not conduct electricity well. For semiconductors to be of value, they must conduct electricity in a controlled manner. To accomplish this, an additional substance, often called an *impurity*, is added to the crystal-like structures of the silicon or germanium. This is called *doping*. One type of impurity produces a hole in the material where an electron should be. Because the hole replaces an electron (which has a negative charge) it results in the material having fewer electrons or a net positive charge. This is called a *P-type material*. If a material of a different type is added to the semiconductor, an excess of electrons is produced, and the material has a negative charge. This is called an *N-type material*.

When voltage is applied to a P-type material, electrons fill the holes and move from one hole to the next, still moving from negative to positive. However, this makes it appear that the holes are moving in the opposite direction (from positive to negative) as the electrons move from hole to hole.

N-type material has an excess of electrons which move from negative to positive when voltage is applied.

Solid-state components are made from a combination of N-type and P-type materials. The type of solid-state component and its electronic characteristics are determined by the manner in which the materials are joined together, the thickness of the materials and various other factors.

Diodes

Diodes are simple solid-state devices. They consist of P-type and N-type material connected together. When this combination is connected to a power source one way, it will allow current to flow and is said to have forward bias. When reversed, it is said to have reverse bias, and no current will flow. Figure 8-51 is a drawing of a simple diode, and Figure 8-52 shows a photo of typical diodes. One of the connections on the diode is called the cathode and the other is called the anode (Figure 8-53). If the diode is to be connected to a battery to produce forward bias (current flow) the negative terminal on the battery should be connected to the cathode (Figure 8-54). Connecting the negative terminal of the battery to the anode will produce reverse bias (no current flow) (Figure 8-55).

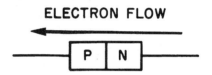

Figure 8-51. Pictorial drawing of a diode.

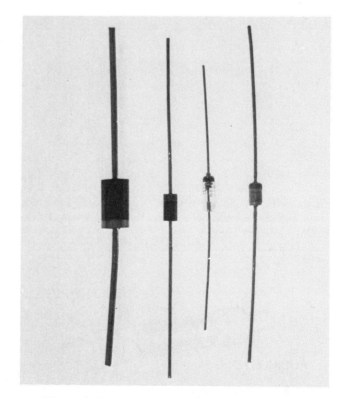

Figure 8-52. Typical diodes. *Photo by Bill Johnson.*

SEMICONDUCTORS **123**

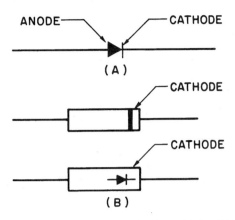

Figure 8-53. (A) Schematic symbol for a diode. (B) Identifying markings on a diode.

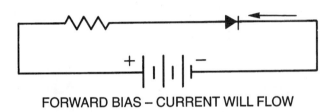

Figure 8-54. Simple diagram with diode indicating forward bias.

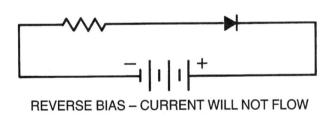

Figure 8-55. Circuit with diode indicating reverse bias.

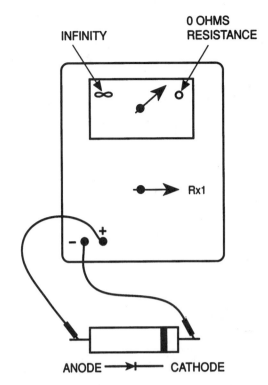

Figure 8-56. Checking a diode.

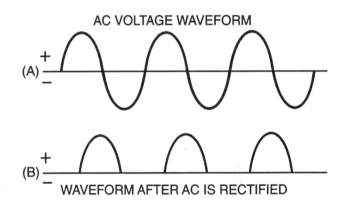

Figure 8-57. (A) Full-wave AC waveform. (B) Half-wave DC waveform.

Checking a Diode A diode may be tested by connecting an ohmmeter across it. The basic procedure is as follows:

1. Remove the diode from the circuit.
2. Set the selector switch on the meter to the R × 1 scale.
3. Zero the meter.
4. Touch the negative probe to the cathode and the positive probe to the anode. The meter should show a small resistance, indicating that there is continuity (Figure 8-56).
5. Reverse the leads. The meter should show infinity, indicating there is no continuity. A diode should show continuity in one direction and not in the other. If it shows continuity in both directions, or it does not show continuity in either direction, it is defective.

Rectifiers

A diode can be used as a solid-state rectifier to convert AC to DC. The term *diode* is normally used when the device is rated for less than one ampere. A similar component rated above one ampere is called a *rectifier*. A rectifier allows current to flow in one direction only. AC flows in first one direction and then reverses, as shown in Figure 8-57(A). The rectifier allows the AC to flow in one direction but blocks it from reversing, Figure 8-57(B). Therefore, the output of a rectifier circuit is in one direction or direct current. Figure 8-58 illustrates a rectifier circuit. This is called a *half-wave rectifier* because it allows only that part of the AC moving in one direction to pass through. Figure 8-58, point A shows the

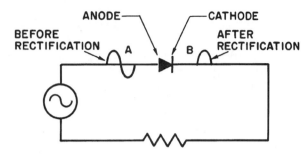

Figure 8-58. Diode rectifier circuit.

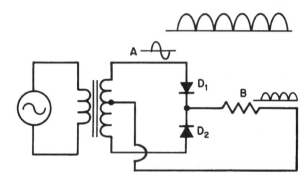

Figure 8-59. Full-wave rectifier.

AC before it is rectified and Figure 8-58, point B shows the AC after it is rectified. *Full-wave rectification* can be achieved using a more complicated circuit such as the one shown in Figure 8-59. During half of the AC cycle, D1 will conduct. Current flows through D1, then through the resistor, and finally back to the center tap of the transformer. When the voltage reverses, D2 conducts; current flows through the resistor and back to the center tap.

Silicon-Controlled Rectifiers

Silicon-controlled rectifiers (SCRs) consist of four semiconductor materials bonded together. These form a PNPN junction, Figure 8-60(A). Figure 8-60(B) illustrates the schematic symbol for this device. Notice that the schematic is similar to the diode except for the gate. The SCR is used to control devices which may use large amounts of power. The gate is the control for the SCR. These devices may be used to control the speed of motors or to control the brightness of lights. Figure 8-61 shows a photo of a typical SCR.

Checking a Silicon-Controlled Rectifier The SCR can also be checked with an ohmmeter. The basic procedure is as follows:

1. Ensure that the SCR is removed from the circuit.
2. Set the selector switch on the meter to R × 1.
3. Zero the meter.
4. Fasten the negative lead from the meter to the cathode terminal of the SCR and the positive lead to the anode terminal (Figure 8-62). If the SCR is good, the needle should not move. This is because the SCR has not fired to complete the circuit.

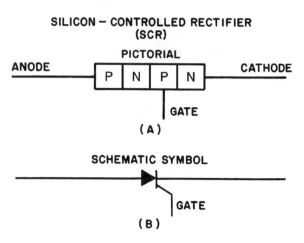

Figure 8-60. Pictorial and schematic drawings of a silicon-controlled rectifier (SCR).

Figure 8-61. Typical silicon-controlled rectifier. *Photo by Bill Johnson.*

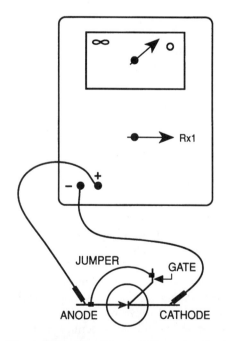

Figure 8-62. Checking a silicon-controlled rectifier.

5. Use a jumper to connect the gate terminal to the anode. The meter needle should now show continuity. If it does not, you may not have the cathode and anode properly identified.
6. Reverse the leads and change the jumper to the new suspected anode. If it fires, you had the anode and cathode reversed. When the jumper is removed, the SCR will continue to conduct if the meter has enough capacity to keep the gate closed. If the meter shows current flow without firing the gate, or if the gate will not fire after the jumper and leads are attached correctly, the SCR is defective.

Transistors

Transistors are also made of N-type and P-type semiconductor materials. Three pieces of these materials are sandwiched together. Transistors are either NPN or PNP types; Figure 8-63 shows pictorial diagrams for each type, and Figure 8-64 shows their schematic symbols. As the symbols show, each transistor has a base, collector, and emitter. In the NPN type, the collector and base are connected to the positive; the emitter is connected to the negative. The PNP transistor has a negative base and collector connection and a positive emitter. The base must be connected to the same polarity as the collector to provide forward bias. Figure 8-65 is a photo of typical transistors.

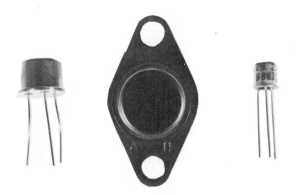

Figure 8-65. Typical transistors. *Photo by Bill Johnson.*

The transistor may be used as a switch or a device to amplify or increase an electrical signal. One application of a transistor used in an air conditioning control circuit would be to amplify a small signal to provide enough current to operate a switch or relay.

Current will flow through the base emitter and collector emitter. The base emitter current is the control and the collector emitter current produces the action. A very small current passing through the base emitter may allow a much larger current to pass through the collector emitter junction. A small increase in the base emitter junction can allow a much larger increase in current flow through the collector emitter.

Thermistors

A thermistor is a temperature-sensitive resistor (Figure 8-66). The resistance of a thermistor changes with a change in temperature. There are two types of thermistors.

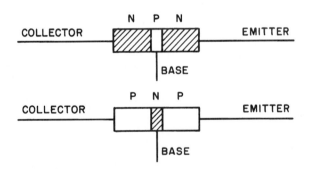

Figure 8-63. Pictorial drawings of NPN and PNP transistors.

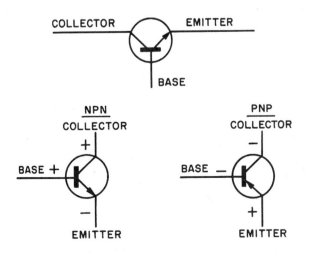

Figure 8-64. Schematic drawings of NPN and PNP transistors.

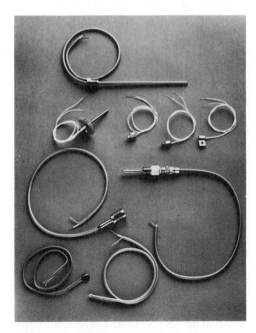

Figure 8-66. Typical thermistors. *Photo by Bill Johnson.*

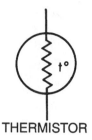

Figure 8-67. Schematic symbol for a thermistor.

A positive temperature coefficient (PTC) thermistor causes the resistance of the thermistor to increase when the temperature increases. A negative temperature coefficient (NTC) thermistor causes the resistance to decrease with an increase in temperature. Figure 8-67 illustrates the schematic symbol for a thermistor.

One application of a thermistor is to provide motor overload protection. The thermistor is imbedded in the motor windings. When the winding temperature exceeds a predetermined amount, the thermistor changes its resistance. This change in resistance is detected by an electronic circuit, which causes the motor circuit to open.

Another thermistor application is to provide start assistance in a PSC (permanent split capacitor) motor. This thermistor is a PTC thermistor. It allows full voltage to reach the start windings during motor start-up. The thermistor heats during start-up and creates resistance, turning off power to the start winding at the appropriate time. This does not provide the starting torque of a start capacitor, but is advantageous in some applications because of its simple construction and lack of moving parts.

Diacs

The diac is a bi-directional electronic device. It can operate in an AC circuit and its output is AC. It is a voltage-sensitive switch that operates in both halves of the AC waveform. When voltage is applied, it will not conduct (or will remain off) until the voltage reaches a predetermined level. Let's assume that this level is 24 V. When the voltage in the circuit reaches 24 V, the diac will begin to conduct or fire. Once it fires, it will continue to conduct even at a lower voltage. Diacs are designed to have a higher cut-in voltage and lower cut-out voltage. If the cut-in voltage is 24 V, and the cut-out voltage is 12 V, the diac will continue to operate until the voltage drops below 12 V, at which time it will cut off.

Figure 8-68 shows two schematic symbols for a diac. Figure 8-69 illustrates a diac in a simple AC circuit. Diacs are often used as switching or control devices for triacs.

Triacs

A triac is a switching device that will conduct on both halves of the AC waveform. The output of the triac is AC. Figure 8-70 shows the schematic symbol for a triac. Notice that it is very similar to the diac but has a gate lead. A pulse supplied to the gate lead can cause the triac

Figure 8-68. Schematic symbols for a diac.

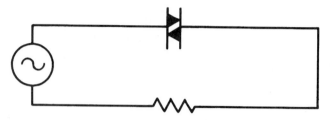

Figure 8-69. Simple diac circuit.

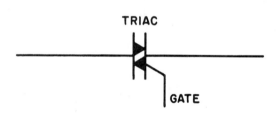

Figure 8-70. Schematic symbol for a triac.

to fire or conduct. Triacs were developed to provide better AC switching. Diacs often provide the pulse to the gate of the triac.

Heat Sinks

Some solid-state devices produce much more heat than others and will only operate properly if kept within a certain temperature range. Excess heat must be dissipated. This is done by adhering the solid-state component to an object called a heat sink (Figure 8-71). Heat will travel from the protected device to the heat sink, which has a large enough surface area to dissipate the excess heat into the surrounding air.

This has been a very brief introduction into semiconductors. As an air conditioning technician, you will be more involved with checking circuit board inputs and outputs than you will with checking individual electronic components. The preceding information has been provided so that you will have some idea of the purpose of these control devices. You are encouraged to pursue the study of these components as more and more solid-state electronics will be used in the future to control air conditioning systems. Each manufacturer will have a control sequence procedure that you must follow to successfully troubleshoot their controls. Make it a practice to attend seminars and factory schools in your area to increase your knowledge in all segments of this ever-changing field.

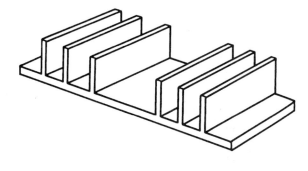

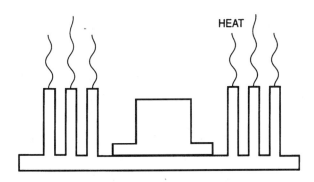

Figure 8-71. One type of heat sink.

SUMMARY

- Matter is made up of atoms.
- Atoms consist of protons, neutrons, and electrons.
- Protons have a positive charge, neutrons have a neutral or no charge, and electrons have a negative charge.
- Electrons travel in orbits around the protons and neutrons.
- Electrons in outer orbits may travel from one atom to another.
- When there is a surplus of electrons in an atom, it has a negative charge. When there is a deficiency of electrons, the atom has a positive charge.
- Conductors have few electrons in their outer orbits. They conduct electricity by allowing electrons to move easily from atom to atom.
- Insulators have several electrons in their outer orbits, which makes it difficult for the electrons to move from atom to atom. Insulators are poor conductors of electricity.
- Electricity can be produced using magnetism. A conductor cutting magnetic lines of force produces electricity.
- Direct current is an electrical current moving in one direction.
- Alternating current is an electrical current that is continually reversing.
- A volt is a measure of electrical force or pressure.
- An ampere is a measure of the quantity of electron flow.
- Ohm is a measure of the resistance to electron flow.
- An electrical circuit must have an electrical source, a conductor to carry the current, and a resistance or load to use the current.
- Resistances or loads may be wired in series or in parallel.
- Ohm's Law is represented by the equation: voltage (E) = amperage (I) × resistance (R).
- In series circuits, the voltage is divided across the resistances, the total current flows through each resistance, and the resistances are added together to obtain the total resistance.
- In parallel circuits, the total voltage is applied across each resistance, the current is divided between the resistances, and the total resistance is less than that of the smallest resistance.
- Electrical power is measured in watts. The power equation is watts or power (P) = voltage (E) × amperage (I).
- Inductive reactance is the resistance caused by the magnetic field surrounding a coil in an AC circuit.
- A coil with an electric current flowing through the loops of wire will cause an iron bar to be attracted into the coil. Electrical switching devices can be designed to take advantage of this process. These switches include solenoids, relays, and contactors.
- Transformers use a magnetic field to increase or decrease voltage. A step-up transformer increases the voltage and decreases the current. A step-down transformer decreases the voltage and increases the current.
- A capacitor in a DC circuit collects electrons on one plate. These collect until they are equal to the source voltage. When a path is provided for these electrons to discharge, they will do so until the capacitor becomes neutral.
- A capacitor in an AC circuit will continually charge and discharge as the current in the circuit reverses.
- Capacitance is the amount of charge that can be stored in a particular capacitor.
- Impedance is the opposition to current flow in an AC circuit from the combination of resistance, inductive reactance, and capacitive reactance.
- A multimeter is an electrical measuring instrument used often by air conditioning technicians. A common multimeter is the VOM (volt-ohm-milliammeter).
- Properly-sized conductors must be used. Larger wire sizes can carry more current without overheating.
- Fuses and circuit breakers are used to interrupt the current flow in a circuit when it becomes excessive.
- Semiconductors are usually made from silicon or germanium. In their pure state, they do not conduct electricity well, but when doped with an impurity, they form an N-type or P-type material that will conduct current in one direction.
- Diodes, rectifiers, transistors, thermistors, diacs, and triacs are all examples of semiconductors.

UNIT 8 BASIC ELECTRICITY AND MAGNETISM

REVIEW QUESTIONS

1. Describe the structure of an atom.
2. Name the three major components of an atom and provide their respective charges.
3. Describe the component of an atom that moves from one atom to another.
4. What effect does this movement have on the losing atom? What is the effect on the gaining atom?
5. Describe the electron structure of a conductor.
6. Describe the electron structure of an insulator.
7. Describe how electricity is generated through the use of magnetism.
8. State the differences between direct current and alternating current.
9. State the electrical units of measurement and describe each.
10. What components make up an electrical circuit?
11. Describe how a meter would be connected in a circuit to measure the voltage at a light bulb.
12. Describe how a meter would be connected in a circuit to measure the amperage draw of a light bulb.
13. Describe how an amperage reading would be made using a clamp-on ammeter.
14. Describe how the resistance in a DC circuit is determined.
15. List the three formulas derived from Ohm's Law and explain the purpose of each formula.
16. Sketch three loads wired in parallel in a circuit.
17. Illustrate how three loads would be wired in series.
18. Describe the characteristics of the voltage, amperage, and resistances when there is more than one load in a series circuit.
19. Describe the characteristics of the voltage, amperage, and resistances when there is more than one load in a parallel circuit.
20. What is the formula for the total resistance of three loads in a parallel circuit?
21. What is the unit of measurement for electrical power?
22. What is the formula for determining electrical power?
23. Explain inductance.
24. Explain how a solenoid operates.
25. Describe how a transformer operates.
26. Sketch a step-up transformer.
27. How does a step-down transformer differ from a step-up transformer?
28. Describe how a transformer is constructed.
29. Describe a capacitor.
30. How does a capacitor work in a DC circuit?
31. How does a capacitor work in an AC circuit?
32. What two types of capacitors are used frequently in the air conditioning industry?
33. What types of opposition to current flow are represented by impedance?
34. What electrical measurements can be made using a multimeter?
35. What do the letters VOM stand for when referring to an electrical measuring instrument?
36. What are the two main switches on a VOM?
37. Why is there a zero ohms adjustment on a VOM?
38. Why is it important to use a properly-sized wire in a particular circuit?
39. What is a circular mil?
40. Describe two kinds of plug fuses.
41. Describe two reasons for using a circuit breaker.
42. What force opens a circuit breaker?
43. What are the two letters used to represent the types of material in most semiconductors?
44. Briefly describe a diode.
45. What is meant by forward bias?
46. What is meant by reverse bias?
47. List four types of semiconductors and their applications.

INTRODUCTION TO AUTOMATIC CONTROLS 9

OBJECTIVES

Upon completion of this unit, you should be able to

- **Describe how bimetal controls respond to temperature changes.**
- **Make general comparisons between different bimetal applications.**
- **Describe fluid-filled controls.**
- **Describe partial liquid/partial vapor mixture controls.**
- **Distinguish between the bellows, diaphragm, and Bourdon tube.**
- **Discuss the operation of a thermistor.**

9.1 TYPES OF AUTOMATIC CONTROLS

Air conditioning systems require many types of automatic controls to operate equipment and provide various safety features.

Controls can be classified as electrical, mechanical, electromechanical, electronic, pneumatic, or hydraulic.

- Electrical controls are electrically operated and normally control electrical devices.
- Mechanical controls are operated by pressure and temperature to control fluid flow.
- Electromechanical controls are driven by pressure or temperature to provide electrical functions, or they are driven by electricity to control fluid flow.
- Electronic controls use electronic circuits and devices to perform the same functions as electrical and electromechanical controls.
- Pneumatic controls use compressed air to control, stop, and start electrical circuits and control fluid flow. These controls are discussed in more detail in applicable units.

9.2 AUTOMATIC CONTROLS USED IN A/C SYSTEMS

The automatic control of an air conditioning system is intended to maintain stable or constant operating conditions and to provide protection to individuals and/or equipment. The system must regulate itself within the design boundaries of the equipment. If it is allowed to operate outside of these boundaries, it may result in equipment damage or failure.

Temperature Control

Air conditioning systems usually provide some method of temperature control. Temperature control is used to maintain space temperature and prevent personnel and equipment damage. When used to control space temperature, the control is called a *thermostat*; when used to protect equipment, it is known as a *safety device*. A good example of both of these applications can be found in a residential air conditioning system. For example, the air conditioning system thermostat shown in Figure 9-1(A) maintains the space temperature in the residence at about 72°F. Without it, someone would have to monitor the ambient temperature 24 hours a day and manually turn the system on and off to maintain the desired conditions. A typical thermostat will normally operate without a malfunction for a number of years, automatically controlling system operation to maintain the specified (setpoint) temperature.

The air conditioning compressor protective overload device shown in Figure 9-1(B) prevents the equipment from overloading and causing damage. This control is designed to prevent compressor damage due to current overloads or problems resulting from a low refrigerant charge. For example, if the power goes off and comes right back on while the air conditioning system is running, the overload device will automatically stop the compressor for a cool-down period.

Common applications of automatic temperature control include the following:

- Residential and commercial cooling systems
- Household refrigerators
- Commercial food storage and preservation systems

9.3 TEMPERATURE-SENSITIVE DEVICES

Many automatic controls used in the air conditioning industry monitor temperature and its changes. Some controls are also used as on/off switches, such as the thermostat in Figure 9-1(A). Other controls, such as the compressor overload device in Figure 9-1(B), monitor electrical overloads by sensing temperature changes in the wiring circuits. The response of most temperature-sensitive devices is represented by a change either in the dimension of the device or electrical characteristics of its sensing element.

129

UNIT 9 INTRODUCTION TO AUTOMATIC CONTROLS

Bimetal Devices

The *bimetal device* is probably the most common device used to detect thermal change. In its simplest form, the device consists of two unlike metal strips that are attached back to back as shown in Figure 9-2(A). Each strip has a different rate of expansion. Brass and steel are commonly used. When the device is heated, the brass expands faster than the steel, and the device is warped out of shape. This warping action is a known dimensional change and the device can be attached to an electrical component or valve to stop, start, or modulate electrical current or fluid flow.

For example, when the bimetal is fixed on one end and heated, the other end moves a certain amount per degree of temperature change. Figure 9-2(B) shows a bimetal strip used as a temperature control in a cooling application. On a temperature rise, the strip warps to the right, closing the contacts and energizing the cooling equipment. This control is limited to its application by the amount of warp it can accomplish with a temperature change.

To obtain enough travel to make the bimetal practical over a wider temperature range, length is added to the bimetal strip. To conserve space, the bimetal strip may be coiled in a circle, shaped like a hair pin, wound in a helix fashion, or formed into a worm shape (Figure 9-3). The movable end of these shapes can be attached to any of the following:

- A pointer to indicate temperature
- A switch to stop or start current flow
- A valve to modulate fluid flow

One of the basic control applications is shown in Figure 9-4.

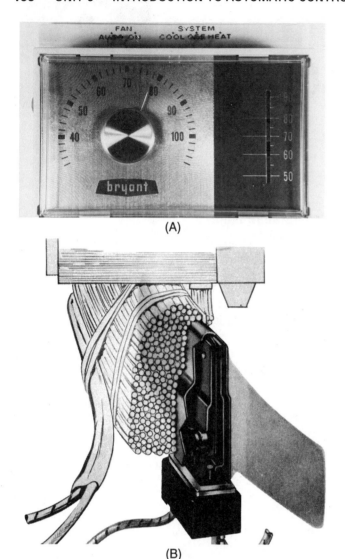

Figure 9-1. (A) Thermostat for controlling space temperature. (B) Compressor protective overload device. *Courtesy of: (A) Photo by Bill Johnson, (B) Tecumseh Products Company.*

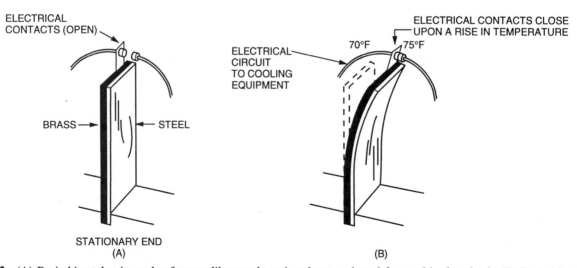

Figure 9-2. (A) Basic bimetal strip made of two unlike metals such as brass and steel fastened back to back. (B) Basic bimetal strip used in a cooling thermostat. This bimetal strip is straight at 70°F. The brass side expands faster than the steel side on a temperature rise, causing a bend to the right. This bend is a predictable amount per degree of temperature change. The longer the strip, the more the bend.

TEMPERATURE-SENSITIVE DEVICES

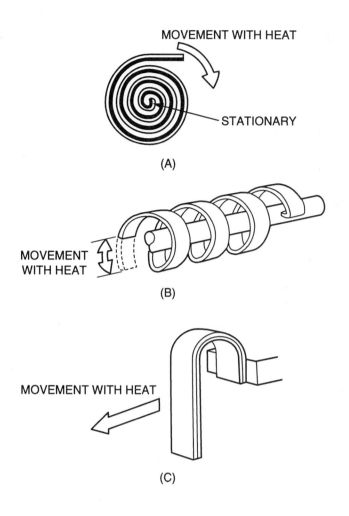

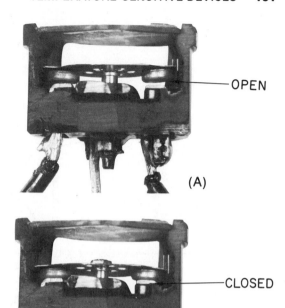

Figure 9-3. Adding length to the bimetal. (A) Coiled. (B) Wound into helix. (C) Hairpin shape. (D) Worm shape.

Figure 9-5. Snap-disc is another variation of the bimetal concept. The snap-disc is usually round and fastened on the outside. When heated, the disc snaps to a different position. (A) Open circuit. (B) Closed circuit. *Photo by Bill Johnson.*

The *snap-disc* is another type of bimetal device that is used to sense temperature changes (Figure 9-5). This control has a snap characteristic that gives it a quick open-and-close feature. Some sort of snap-action feature is incorporated into all controls that stop and start electrical loads.

Fluid Expansion Devices

Fluid expansion is another method of sensing temperature change. In Unit 2, a mercury thermometer was described as a bulb with a thin tube of mercury rising up a glass stem. As the mercury in the bulb is heated or cooled, it expands or contracts and either rises or falls in

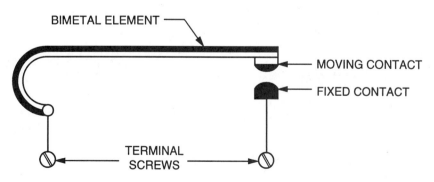

Figure 9-4. Movement of the bimetal due to changes in temperature opens or closes electrical contacts.

132 UNIT 9 INTRODUCTION TO AUTOMATIC CONTROLS

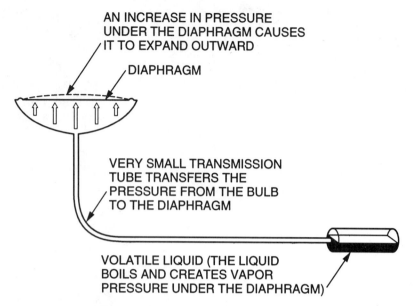

Figure 9-6. The diaphragm is a thin, flexible, movable membrane (brass/steel or other metal) used to convert pressure changes into movement. This movement can stop, start, or modulate controls.

the stem of the thermometer. The level of the mercury in the stem is based on the temperature of the mercury in the bulb. This same idea can be used to signal a control that a temperature change is taking place.

A fluid expansion device normally consists of a component in which the mechanical action takes place, a transmitting tube, and a sensing component. Figure 9-6 illustrates a type of fluid expansion device. The liquid moving through the transmitting tube has to act on some component to produce usable motion. One type of component used is the diaphragm. A *diaphragm* is a thin, flexible metal disc with a large area. It moves in and out in response to pressure changes underneath it (Figure 9-6).

A diaphragm has very little travel (range) but considerable power.

When a bulb is filled with a liquid and connected to a diaphragm with a transmitting tube, the bulb temperature can be transmitted to the diaphragm by the volatile liquid (Figure 9-7).

More accurate control is obtained by using a bulb which contains a partial liquid/partial vapor mixture. In response to temperature changes, the liquid boils into a vapor, which is then transmitted to the diaphragm at the control point (Figure 9-8). The liquid will respond to temperature change more readily than the vapor, which is used to transmit pressure.

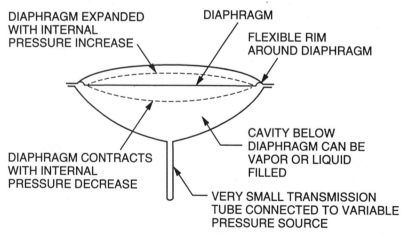

Figure 9-7. A large bulb is partially filled with a volatile liquid that boils and creates vapor pressure when heated. This causes an increase in vapor pressure, which forces the diaphragm to move outward. When cooled, the vapor condenses, and the diaphragm moves inward.

TEMPERATURE-SENSITIVE DEVICES 133

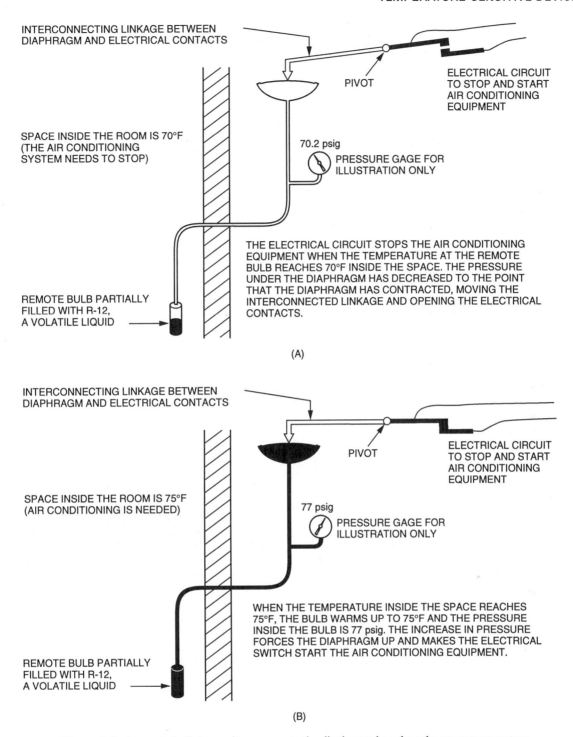

Figure 9-8. A remote bulb transmits pressure to the diaphragm based on the space temperature.

The following example using an air conditioner with R-12 describes how this can work. Refer back to Unit 3 for the temperature/pressure chart for R-12. The inside temperature is maintained by stopping the system when the room temperature reaches 70°F and starting it when the temperature reaches 75°F. A control with a remote bulb is used to regulate the space temperature. The bulb is located inside the unit, and the control is located outside the unit so it can be adjusted.

For illustration purposes a pressure gage is installed in the bulb to monitor the pressures inside the bulb as the temperature changes. Figure 9-8 presents a progressive explanation of this example. At the point that the unit needs to be cycled off, the bulb temperature is 70°F. This corresponds to a pressure of 70.2 psig for R-12. A control mechanism can be designed to open an electrical circuit and stop it at this point. When the room temperature rises to 75°F, it is time to restart the unit. At 75°F for R-12, the pressure inside the control is 77 psig, and the same mechanism can be designed to close the electrical circuit and start the air conditioning system.

134 UNIT 9 INTRODUCTION TO AUTOMATIC CONTROLS

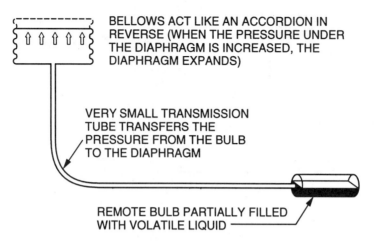

Figure 9-9. The bellows is applied where more movement per degree is desirable. This control would normally have a partially-filled bulb with vapor pushing up in the bellows section.

When more travel is needed, another device known as a *bellows*, can be used. The bellows is very much like an accordion (Figure 9-9). It has a large internal volume with a lot of travel. A bellows normally contains a vapor rather than a liquid.

A remote bulb that is partially filled with liquid may also be used with a Bourdon tube to drive a needle on a calibrated dial (Figure 9-10). Partially filled bulb controls are widely used in this industry because they are reliable, simple, and economical. This type of control is available in many configurations. The Bourdon tube is often used in the same manner as the diaphragm and the bellows to monitor fluid expansion.

Thermistors

A *thermistor* is a type of semiconductor and requires an electronic circuit to utilize its capabilities. It varies its resistance to current flow based on temperature.

The thermistor can be very small and will respond to small temperature changes (Figure 9-11). The changes in current flow in the device are monitored by special electronic circuits that can stop, start, and modulate machines or provide a temperature readout.

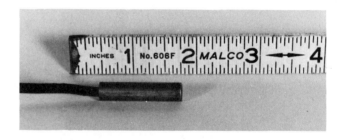

Figure 9-11. A thermometer probe using a thermistor to measure temperature. *Photo by Bill Johnson.*

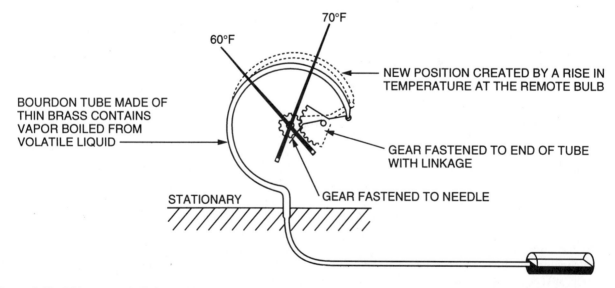

Figure 9-10. This remote bulb is partially filled with liquid. When heated, the expanded vapor is transmitted to a Bourdon tube that straightens out with an increase in vapor pressure. A decrease in pressure causes the Bourdon tube to curl inward.

SUMMARY

- A bimetal element consists of two unlike metal strips such as brass and steel. The strips fastened back to back.
- Bimetal strips warp with temperature changes and can be used to stop, start, or modulate electrical current flow and fluid flow when used with different mechanical, electrical, and electronic devices.
- The travel of the bimetal can be extended by coiling it. The helix, worm shape, hairpin shape coil are all common configurations for extending bimetal elements.
- Fluid expansion is used in the thermometer to indicate temperature and to operate controls that are totally liquid filled.
- The pressure-temperature relationship studied earlier is applied to some controls that are partially filled with liquid.
- The diaphragm is used to move the control mechanism when either liquid or vapor pressure is applied to it.
- The diaphragm has very little travel but considerable power.
- The bellows is used to provide additional travel and is normally filled with a vapor rather than a liquid.
- The Bourdon tube is sometimes used like the diaphragm or bellows.
- A thermistor is an electronic device that varies its resistance to electrical current flow based on temperature changes.

REVIEW QUESTIONS

1. Describe a bimetal strip.
2. What are some applications of the bimetal strip?
3. Name some of the configurations of an extended bimetal device.
4. What two metals can be used in a bimetal application?
5. Describe a diaphragm.
6. Name two characteristics of a diaphragm.
7. What type of expansion is used to operate a mercury thermometer?
8. Describe a bellows.
9. What is the difference between a bellows and a diaphragm?
10. Does a bellows normally contain a liquid or a vapor?
11. Describe the operation of a thermistor.
12. What must be used with a thermistor to control a mechanical device or machine?

10 AUTOMATIC CONTROL COMPONENTS AND APPLICATIONS

OBJECTIVES

Upon completion of this unit, you should be able to

- Explain the operation of a space temperature control.
- Describe a mercury control bulb.
- Describe the difference between low-voltage and high-voltage space temperature controls.
- Name components of low-voltage and high-voltage controls.
- Name two ways motors are protected from excessive temperatures.
- Describe the difference between a diaphragm and a bellows control.
- State the uses of pressure-sensitive controls.
- Describe a high-pressure control.
- Describe a low-pressure control.
- Explain the purpose of a pressure relief valve.

10.1 RECOGNITION OF CONTROL COMPONENTS

The ability to recognize various types of controls and understand their function(s) are vital skills that will eliminate confusion when reading a diagram or troubleshooting. While there are many types of controls, there are also some similarities and categories that can make control recognition an easier task.

10.2 TEMPERATURE CONTROLS

As discussed in Unit 9, temperature controls are used to satisfy both operating functions (such as a thermostat space temperature) and safety functions. Both types of controls require a sensing element to sense temperature changes and provide a known response.

The thermostat application has two different actions depending on whether heating or cooling is needed. In the cooling mode, the control must make (complete) a circuit to start cooling based on a temperature rise. In the heating mode, the control must break (disconnect) a circuit to stop heating when conditions are satisfied. The cooling thermostat closes on a rise in temperature; the heating thermostat opens on a rise in temperature.

Note: In both cases, control is described as functioning on a rise.

An overload device (motor winding thermostat) has the same circuit action as a space temperature thermostat. It opens the circuit on a rise in motor temperature and stops the motor. The space temperature thermostat and motor winding thermostat may operate in the same way, but they do not physically resemble each other (Figure 10-1).

Another difference between the two thermostats is the medium to be detected. The motor winding thermostat must be in close contact with the motor winding. It is fastened to the winding itself.

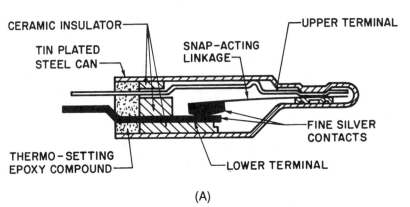

(A) (B)

Figure 10-1. Both thermostats open on a rise in temperature, but they serve two different purposes. (A) The motor winding thermostat measures the temperature of iron and copper while in close contact with the motor windings. (B) The space thermostat measures air temperature from random air currents. *(B) Courtesy Robertshaw Controls Company.*

The space temperature thermostat is mounted on a wall with the control components suspended in air under the decorative cover. This control relies on random air currents passing over it.

Current-Carrying Capacity

There is no firm rule that specifies the type of control voltage for each specific application. However, most high-voltage components use low-voltage controls. For example, the stopping and starting of a 3 hp compressor requires a relatively large switching mechanism. A 3 hp compressor might require a running current of 18 A and a starting current of 90 A. If a bimetal device had to carry the current for a 3 hp compressor, the control would be so large that it would be slow to respond to air temperature changes.

Residential air conditioning systems usually incorporate low-voltage control circuits. There are four reasons for this:

1. Low-voltage circuits are less expensive.
2. 24 V is far less dangerous than line voltage (115 V or 230 V).
3. Low-voltage circuits offer a more precise response to minor temperature changes in relatively still air.
4. In many states, a technician does not need an electrician's license to install and service low-voltage wiring.

The low-voltage thermostat is energized from the residential power supply and reduced to 24 V with a small transformer that is usually furnished with the equipment (Figure 10-2).

Low-Voltage Space Temperature Controls

The low-voltage space temperature control (thermostat) normally regulates other controls and does not carry much current—seldom more than 2 A. A cooling thermostat consists of the electrical contacts, cold anticipator, cover, thermostat assembly, and subbase.

Electrical Contacts The mercury bulb is probably the most common type of switching device used to make and break the electrical circuit in low-voltage thermostats. The mercury bulb is located inside the thermostat (Figure 10-3).

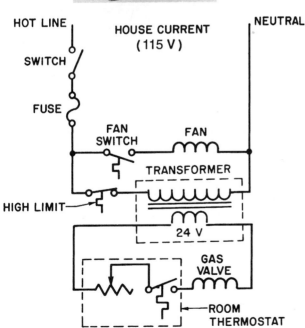

Figure 10-2. The typical transformer used in residences and light commercial buildings to convert 115 V to 24 V (control voltage). *Photo by Bill Johnson.*

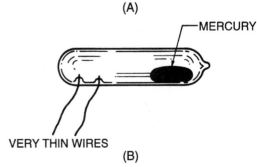

Figure 10-3. (A) A wall thermostat with the over off and the mercury bulb exposed. (B) A detail of the mercury bulb. Note the very fine wire that connects the bulb to the circuit. The bulb is attached to the movable end of a bimetal coil. When the bimetal tips the bulb, the mercury flows to the other end and closes the circuit by providing the contact between the two wires. The wires are fastened to contacts inside the glass bulb, which is filled with an inert gas. This inert gas helps to prevent the contacts from pitting and burning up (oxidizing). *Photo by Bill Johnson.*

It consists of a glass bulb filled with an inert gas and a small puddle of mercury that is free to move from one end to the other. When an electrical current is either made or broken, a small arc is present. The arc is hot enough to cause oxidation. When this arc takes place inside a bulb filled with an inert gas, there is no oxygen and, therefore no oxidation.

The mercury bulb is fastened to the movable end of the bimetal device, and rotates when the bimetal changes dimension. The wire that connects the mercury bulb to the electrical circuit is very fine to prevent drag on the movement of the bulb. The mercury cannot be in both ends of the bulb at the same time, so when the bimetal device rolls the mercury bulb to a new position, the mercury rapidly makes or breaks the electrical current flow. This is called *snap action* or *detent action*.

Some low-voltage thermostats incorporate conventional contact made of silver-coated steel (Figure 10-4). One type is simply an open set of contacts, usually with a small protective dust cover; the other is a set of contacts enclosed in a glass bulb. Both devices use a magnet mounted close to the contacts to achieve snap action.

Cold Anticipator A cooling system must anticipate the need for air conditioning and energize a few minutes early to allow the system to get up to capacity when needed. If the system is not started until it is needed, it may take as long as 15 minutes before it will be producing to capacity. This is enough time to cause a temperature rise and discomfort in the conditioned space.

The cold anticipator is normally a fixed resistor that is not adjustable in the field (Figure 10-5). It is wired in parallel with the mercury bulb cooling contacts and is energized during the off cycle (Figure 10-6).

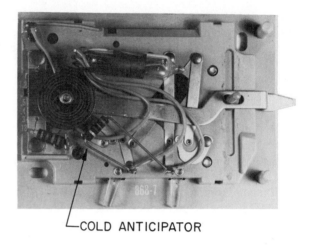

Figure 10-5. The cold anticipator is normally a fixed resistor similar to resistors found in electronic circuitry. It is a small, cylindrical device with color bands to denote the resistance and wattage. *Photo by Bill Johnson.*

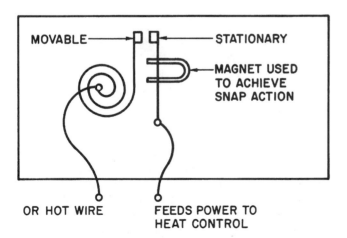

Figure 10-4. Low voltage thermostat illustrated with open contacts.

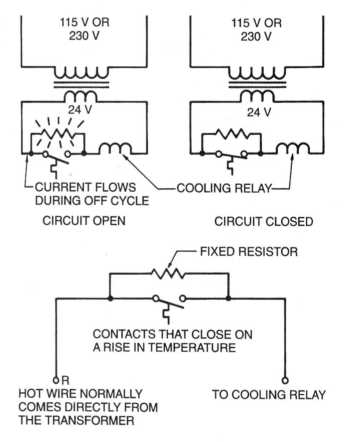

Figure 10-6. The cold anticipator is wired in parallel with the cooling contacts on the thermostat. This allows the current to flow through it during the off cycle. This means that the cold anticipator and the relay to start the cooling cycle are both in the circuit when the thermostat is open.

TEMPERATURE CONTROLS 139

Cover The thermostat cover is both decorative and protective. A thermometer is usually mounted on the cover to indicate the surrounding (ambient) temperature. This thermometer is functionally separate from any of the controls and would serve the same purpose if it were hung on the wall next to the thermostat. In other words, it is not the temperature sensor that drives the thermostat—that device is the bimetal element that operates the electrical contacts.

Thermostats come in many shapes. Each manufacturer tries to have a distinctive design.

Thermostat Assembly The thermostat assembly houses the thermostat components already mentioned and is normally mounted on a subbase fastened to the wall. In addition to a mercury bulb and anticipator, the thermostat assembly includes the movable lever that adjusts the temperature. This lever or indicator normally points to the setpoint or desired temperature (Figure 10-7). When the system is functioning correctly, the thermometer on the front of the thermostat should be at or near the setpoint temperature.

Subbase The subbase, which is usually separate from the thermostat assembly, contains the selector switching levers, such as the FAN ON/AUTO switch or the HEAT/OFF/COOL switch (Figure 10-8). The subbase is first mounted on the wall, then the interconnecting wiring for the thermostat is attached to the subbase. When the thermostat is attached to the subbase, the electrical connections are made between the components.

High-Voltage Space Temperature Controls

Some cooling units incorporate high-voltage (line-voltage) thermostats. For example, a window air conditioner is self-contained and does not require a remote thermostat. The thermostat in a window air conditioner normally uses a sensing bulb located in the return air stream (Figure 10-9). The fan usually runs continuously and

Figure 10-8. The subbase assembly normally mounts on the wall and the wiring is fastened inside the subbase on the terminals. These terminals are designed in such a way as to allow easy wire makeup. When the thermostat is screwed down onto the subbase, electrical connections are made between the two. The subbase normally contains the selector switches, such as FAN ON/AUTO and HEAT/OFF/COOL. *Photo by Bill Johnson.*

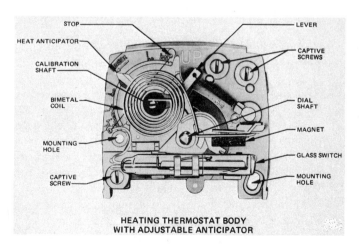

Figure 10-7. Thermostat assembly. *Courtesy Robertshaw Controls Company.*

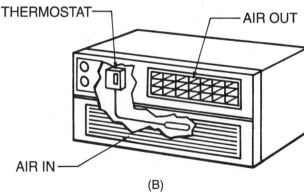

Figure 10-9. The window air conditioner has a line-voltage thermostat that stops and starts the compressor. When the selector switch is turned to COOL, the fan comes on and stays on. The thermostat cycles the compressor only. *(A) Courtesy Whirlpool Corporation.*

keeps a steady stream of room return air passing over the bulb. This provides more sensitivity in this type of application. Line-voltage thermostats are also used in freestanding package air conditioning equipment.

Line-voltage thermostats are generally heavy duty type and may not be as sensitive as the low-voltage thermostats. When replacing line-voltage thermostats, always use an exact replacement or one recommended by the equipment manufacturer.

The line-voltage thermostat must be matched to the voltage and current that the circuit is expected to use. Usually, thermostats are not rated at more than 25 A because the size becomes prohibitive. This limits the size of the compressor that can be started directly with a line-voltage thermostat to about 1½ hp on 115 V or 3 hp on 230 V. The same motor will draw exactly half the current when the voltage is doubled.

When larger current-carrying capacities are needed, a motor starter is normally used with a line-voltage thermostat or a low-voltage thermostat is used.

Line-voltage thermostats usually contain a switching mechanism, sensing element, thermostat assembly, cover, and subbase.

Switching Mechanism Some line-voltage thermostats use mercury as the contact surface, but most switches use silver-coated metal contacts (Figure 10-10). The silver helps conduct current at the contact point. The contact point takes the real load in the circuit and is the control component that usually wears first.

Sensing Element The sensing element is normally a bimetal device, bellows, or remote bulb (Figure 10-11). It is located where it can sense the space temperature. If sensitivity is important, a slight air velocity should cross the element.

Thermostat Assembly The thermostat assembly contains the sensing element, switching mechanism, and the levers or knobs used to adjust the thermostat. It is usually mounted onto a subbase beneath a protective cover.

Cover Because of the line voltage inside, the cover is usually attached with some sort of fastener to discourage easy entrance. A thermometer may be mounted on the cover to read the room temperature.

Subbase The subbase fits on the electrical outlet box. The wire leading into a line-voltage control is normally a high-voltage wire routed in conduit. The conduit is connected to the box that the thermostat is mounted on. This way, if an electrical arc due to overload or short circuit occurs, it is enclosed in the conduit box. This reduces fire hazard (Figure 10-12).

10.3 TEMPERATURE MEASUREMENTS

Accurate temperature measurements are essential to system diagnosis and troubleshooting. The method used to

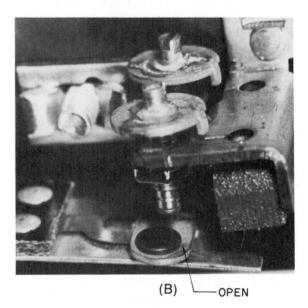

Figure 10-10. Silver-coated line voltage contacts. (A) Closed. (B) Open. *Photos by Bill Johnson.*

determine the temperature varies depending upon the state of the substance to be measured.

Motor Temperature Controls

The protection of electric motors is very important because motors are the prime movers of refrigerant, air, and water. Motors, especially the compressor motor, are the most expensive components in the system. Motors are normally made of steel and copper and need to be protected from heat and from any overload that will cause heat.

All motors build up heat as a normal function of work. The electrical energy that passes to the motor is intended to be converted to magnetism and work, but some of it is converted to heat. All motors have some means of detecting and dissipating this heat under normal or design conditions.

TEMPERATURE MEASUREMENTS   141

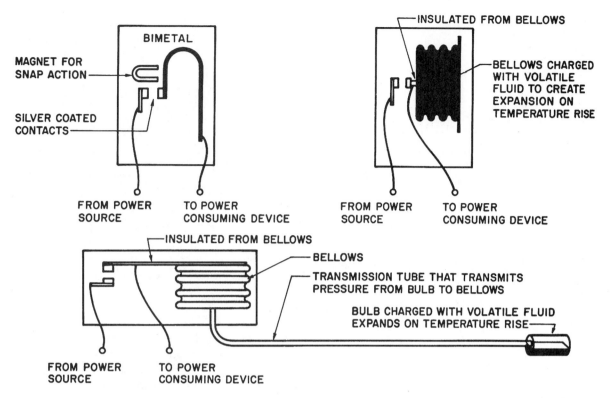

Figure 10-11. Bimetal, bellows, and remote bulb sensing elements used in line-voltage thermostats. The levers or knobs used to adjust these controls are arranged to apply more or less pressure to the sensing element to vary the temperature range.

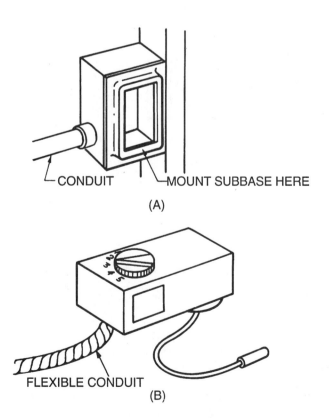

Figure 10-12. (A) Box for a line-voltage, wall-mounted thermostat. (B) Remote bulb thermostat with flexible conduit. In both cases, the interconnecting wiring is covered and protected.

Motor high-temperature protection is usually accomplished with a thermistor or bimetal device. The bimetal device (Figure 10-13) can either be mounted on the outside of the motor (usually in the terminal box) or embedded in the windings themselves.

The thermistor device is normally embedded in the windings. This close contact with the windings provides a fast, accurate response, but it also means that the wires must be brought to the outside of the compressor. This involves extra terminals in the compressor terminal box (Figure 10-14). Figures 10-15 and 10-16 show wiring diagrams of motor overload devices.

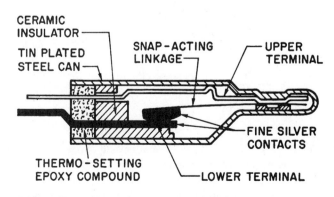

Figure 10-13. Bimetal motor temperature protection device.

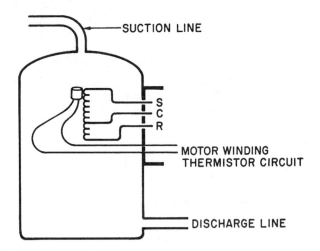

Figure 10-14. Terminal box on the side of the compressor. The extra terminals are for the internal motor protection.

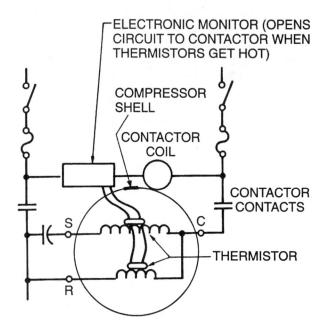

Figure 10-16. A thermistor uses an electronic monitoring circuit to measure temperature. When the temperature reaches a predetermined level, the monitor interrupts the circuit to the contactor coil and stops the compressor.

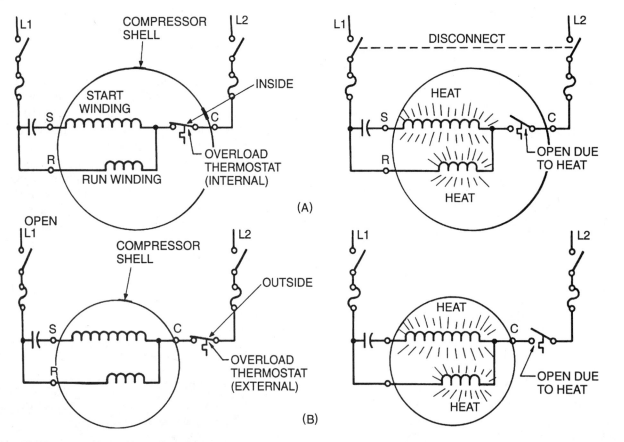

Figure 10-15. Line-voltage bimetal sensing device under hot and normal operating conditions. (A) Permanent split capacitor motor with internal protection. Note that a meter would indicate an open circuit if the ohm reading were taken at the C (common) terminal to either S (start) or R (run) if the overload thermostat were to open. There is still a measurable resistance between start and run. (B) The same motor, except the motor protection is on the outside. Note that it is easier to troubleshoot the outside overload, but it will not provide the same fast response as the internal device because it is not as close to the windings.

TEMPERATURE MEASUREMENTS 143

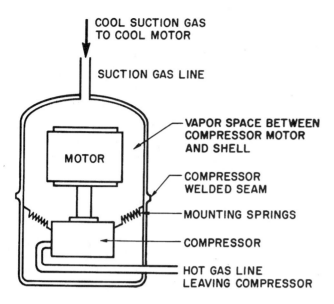

Figure 10-17. Compressor and motor suspended in vapor space in compressor shell. The vapor conducts heat slowly.

Because of its size and weight, a motor can take a long time to cool after overheating. If the motor is an open type, a fan or moving airstream can be devised to cool it more quickly. If the motor is inside a compressor shell, it may be suspended from springs inside the shell (Figure 10-17). This means that the actual motor and compressor may be hot and hard to cool even though the shell does not feel hot.

The best procedure when a hot compressor is encountered is to shut off the compressor at the space temperature thermostat and return later or even the next day, if practical. This gives it ample time to cool.

Caution: It is impossible to make an accurate diagnosis until the compressor has cooled down.

Many service technicians have diagnosed an open winding and replaced a hot compressor; then later discovered that the winding was only open because of internal thermal protection. Figure 10-18 shows a common method used to cool a hot compressor.

If you are in a hurry, set up a fan, or even cool the compressor with water, but don't allow water to get on the electrical components or into the electrical circuits. Turn the power off and cover electrical circuits with plastic before using water to cool a hot compressor. Be careful when restarting. When servicing a compressor that has tripped on a thermal overload, always use an ammeter to check for overcurrent conditions, and gages to determine the charge level of the equipment. Most hermetic compressors are cooled by the suction gas. If there is an undercharge, there will be undercooling (or overheating) of the motor.

Measuring the Temperature of Solids

Any sensing device indicates or reacts to the temperature of the sensing element. For example, a mercury thermometer indicates the temperature of the bulb at the end of the thermometer, not the temperature of the substance it is in contact with. To achieve an accurate reading, the thermometer must stay in contact with the substance long enough to attain the same temperature as the substance.

It can be quite difficult to get a round mercury bulb close enough to a flat piece of metal so that it senses only the temperature of the metal. Only a fractional part of the bulb will touch the flat metal at any time.

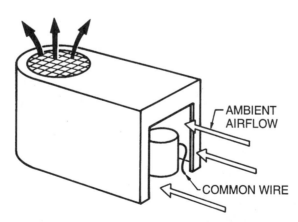

THE COMPRESSOR COMPARTMENT DOOR IS REMOVED AND THE FAN IS STARTED. THIS MAY BE ACCOMPLISHED BY REMOVING THE COMPRESSOR COMMON WIRE AND SETTING THE THERMOSTAT TO CALL FOR COOLING.

Figure 10-18. Ambient air can be used to cool the compressor motor when the compressor is located in the fan compartment.

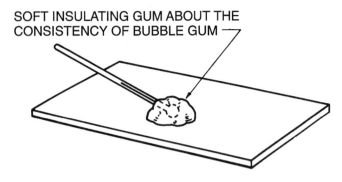

Figure 10-19. The insulating gum holds the bulb against the metal and insulates it from the ambient air.

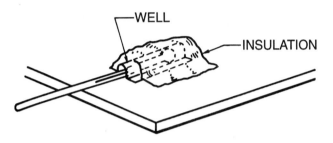

Figure 10-20. The thermometer well is fastened to the metal plate so that heat will conduct both into and out of the bulb.

This leaves most of the area of the bulb exposed to the surrounding (ambient) air. A gum-type substance can be used to hold the thermometer bulb against the surface and insulate it from the ambient air (Figure 10-19). A well in which to insert the thermometer may be used for a more permanent installation (Figure 10-20).

Some sensing elements are designed to fit the surface to be sensed. The external motor temperature sensing element is a good example. It is manufactured flat to fit close to the motor housing (Figure 10-21). One reason it can be made flat is because it is a bimetal device. This control is normally mounted inside the terminal box of the motor or compressor to shield it from the ambient temperature.

Figure 10-21. Motor temperature-sensing thermostat. *Photo by Bill Johnson.*

Measuring the Temperature of Fluids

The term fluid applies to both the liquid and vapor states of matter. Liquids are heavy and change temperature very slowly. The sensing element must be able to quickly assume the temperature of the fluid to be measured. Because liquids are contained in vessels (or pipes), the measurement can be made either by contact with the outside of the vessel or by immersion in the liquid itself. When the temperature is measured at the outside of the vessel, care must be taken that the ambient temperature does not affect the reading.

When regular temperature readings will be required at a larger pipe, a well may be welded into the pipe during installation. A thermometer or sensing bulb can be permanently mounted in this well. The well must be carefully matched to the size of the thermometer or sensing bulb or it will not produce an accurate reading.

At times, it may be desirable to remove the thermometer from the well and insert an electronic thermometer with a small probe for troubleshooting purposes. If the well is much larger than the probe, it can be packed so that the probe is held firm against the well for an accurate reading (Figure 10-22).

Another method for obtaining an accurate temperature reading in a water circuit is to bleed the system using one of the valves in the water line. For example, if the leaving-water temperature of a water-cooled condenser is needed and the thermometer in the well is questionable, or a well is not provided, try the following procedure.

Place a small container under the drain valve in the leaving-water line. Allow a small amount of water from the system to run continuously into the container.

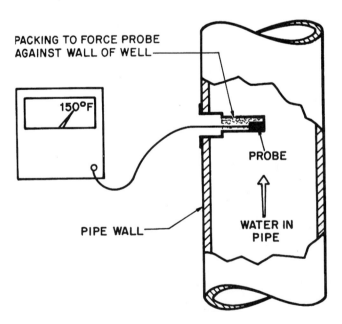

Figure 10-22. The well is packed to push the probe against the wall of the well.

PRESSURE-SENSITIVE DEVICES   145

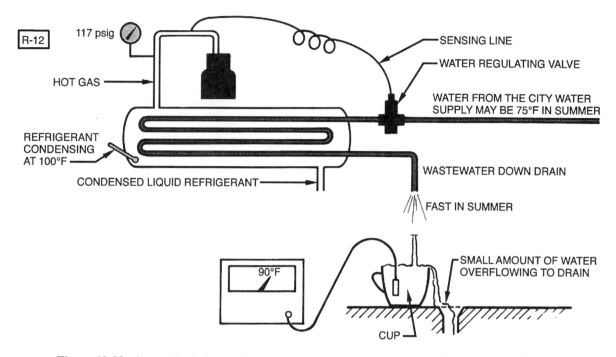

Figure 10-23. One method of obtaining a leaving-water temperature reading from a water-cooled condenser.

An accurate reading can be obtained from this water. This is not a long-term solution because it requires a constant bleed, but it is an effective field method (Figure 10-23).

Sensing Temperature in an Airstream

Sensing temperature in fast-moving airstreams such as in duct work is usually done by inserting the sensing element into the actual airstream. Generally, the sensing element is a bimetal device, such as a flat snap-disc (Figure 10-24) or a helix coil.

If you examine temperature controls carefully, you will usually be able to determine the way they operate. If you are still undecided, consult a catalog or the supplier of the control.

Figure 10-24. A snap-disc which can be used to sense air temperatures in duct work. *Photo by Bill Johnson.*

10.4 PRESSURE-SENSITIVE DEVICES

Pressure-sensitive devices are normally used when measuring or controlling the pressure of refrigerants, air, and water. These are sometimes strictly pressure controls or they can be used to operate electrical switching devices. The terms *pressure control* and *pressure switch* are often used interchangeably in the field. The actual application of the component should indicate if the device is used to control a fluid or to operate an electrical switching device.

Applications of pressure-sensitive controls include the following:

1. Pressure switches are used to stop and start electrical loads, such as motors (Figure 10-25).
2. Pressure controls contain a bellows, diaphragm, or Bourdon tube to create movement when the pressure inside it is changed. Pressure controls may be attached to either switches or valves (Figure 10-26).
3. When used as a switch, the bellows, Bourdon tube, or diaphragm is attached to the linkage that operates the electrical contacts. When used as a valve, the movement device is normally attached directly to the valve.
4. When used as a valve, the pressure control can either open or close on a rise in pressure. This opening and closing action can control water or other fluids.
5. When used as a switch, the electrical contacts associated with the control either open or close with snap action on a rise in pressure (Figures 10-27 and 10-28). The electrical contacts are the component that actually open and close the electrical circuit.

146   UNIT 10   AUTOMATIC CONTROL COMPONENTS AND APPLICATIONS

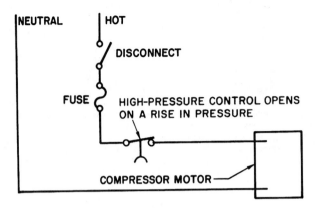

Figure 10-25. Electrical circuit of a refrigeration compressor with a high-pressure control. This control has normally closed contacts that opens on a rise in pressure.

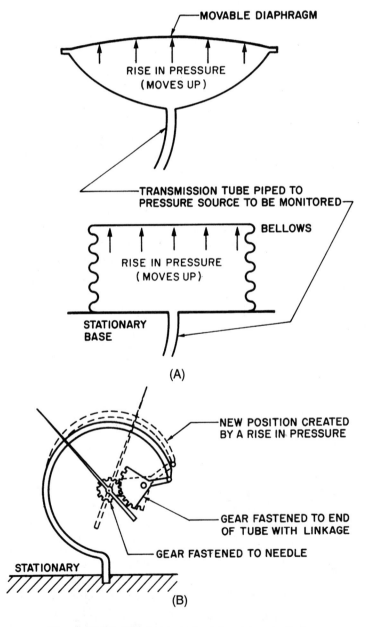

Figure 10-26. Moving part of most pressure controls.

PRESSURE-SENSITIVE DEVICES 147

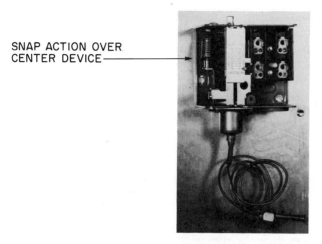

Figure 10-27. The snap action device is shown attached to the bottom of the spring. *Photo by Bill Johnson.*

Figure 10-29. The control on the left is a manual reset control, the one on the right is an automatic reset. Note the push lever on the manual control. *Photo by Bill Johnson.*

6. The pressure controls can be used as operating controls or safety controls.
7. Pressure controls can either operate at low pressures (even below atmospheric pressure) or high pressures, depending on the design of the control mechanism.
8. A pressure control can sometimes be recognized by the small pipe running to the control. This type of device is used to measure fluid pressures.
9. Pressure switches are manufactured to handle control voltages to start a compressor up to about 3 hp maximum.
10. Some pressure switches are adjustable, and others are not.

Warning: A high-pressure control is an important safety device. Do not tamper with it.

11. Some controls are automatic reset, and others are manual reset (Figure 10-29).
12. In some pressure controls, the high-pressure and low-pressure controls are built into one housing. These are known as *dual-pressure controls* (Figure 10-30).
13. Pressure controls are usually located near the compressor on air conditioning equipment.

Caution: When used as safety controls, pressure controls should be installed in such a manner that they are not subject to being valved off by the service valves.

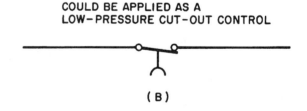

Figure 10-28. These symbols show how pressure controls connected to switches appear on control diagrams. (A) and (B) show switches that will make (complete) the circuit on a rise in pressure. (A) indicates the switch is normally open when the machine does not have power to the electrical circuit. (B) shows that the switch is normally closed without power. (C) and (D) show switches that will open the circuit on a rise in pressure. (C) is normally open and (D) is normally closed.

Figure 10-30. The control has two bellows (left and right) acting on one set of contacts (center). Either can stop the compressor. *Photo by Bill Johnson.*

14. The point or pressure setting at which the control interrupts the electrical circuit is known as the *cut-out*. The point or pressure setting at which the electrical circuit is made is known as the *cut-in*. The difference in the two settings is known as the *differential*.

High-Pressure Controls

The high-pressure control (switch) on an air conditioner stops the compressor if the pressure on the high-pressure side becomes excessive. This control appears in the wiring diagram as a normally closed control that opens on a rise in pressure. The manufacturer may determine the upper limit of operation for a particular piece of equipment and furnish a high-pressure cutout control to ensure that the equipment does not operate above these limits.

A compressor is known as a *positive displacement device*. When it has a cylinder full of vapor, it will either pump out the vapor or stall. If a condenser fan motor on an air-cooled piece of equipment burns out and the compressor continues to operate, very high pressures will occur. The high-pressure control is a safety device used to protect for the equipment and surrounding area. Some compressors are strong enough to burst a pipe or a container. The compressor overload device offers some protection against this type of problem, but it is really a secondary device because it is not directly responding to the pressures. It may also be a little slow to respond.

Low-Pressure Controls

The low-pressure control (switch) is used in the air conditioning field as a low charge protection. When the system loses some of its refrigerant, the pressure on the low-pressure side of the system will fall. The manufacturer may have a minimum pressure under which the equipment is not allowed to operate. This is the point at which the low-pressure control cuts off the compressor.

Note: Pressure control points are not standardized. Always follow the manufacturer's recommendations when adjusting any pressure control.

The recent popularity of the orifice or capillary tube as a metering device has caused the low-pressure control to be reconsidered as a standard control on all equipment. This type of metering device equalizes pressures during the off cycle and may cause the low-pressure control to short cycle the compressor. The capillary tube is a fixed-bore metering device and has no shutoff valve action. To prevent short cycling, some equipment includes a time delay circuit that will not allow the compressor to restart for a predetermined time period.

There are two reasons why it is undesirable to operate a system without an adequate charge:

1. Most compressor motors used in air conditioning applications are cooled by the refrigerant. Without this cooling action, the motor will build up heat when the charge is low. A motor overload device is used to detect this condition. It often takes the place of the low-pressure cutout by sensing motor temperature.
2. If refrigerant escapes through a leak in the low side of the system, the system can operate until it goes into a vacuum. When a vessel is in a vacuum, the atmospheric pressure is greater than the vessel pressure. This causes the atmosphere to be pushed into the system. This air in the system is sometimes hard to detect, but if it is not removed, it causes acids to form and can result in a compressor burnout.

Oil Pressure Safety Controls

The oil pressure safety control (switch) is used to ensure that the compressor has sufficient oil pressure when operating (Figure 10-31). This control is used on larger compressors and has a different sensing arrangement than the high-pressure and low-pressure controls. The high-pressure and low-pressure controls are single diaphragm or single bellows controls because they are comparing atmospheric pressure to the pressures inside the system. Atmospheric pressure can be considered a constant for any particular locality because it does not vary by more than a small amount.

The oil pressure safety control is a pressure differential control. This control measures a difference in pressure to establish that positive oil pressure is present. A study of the compressor will show that the compressor crankcase pressure (this is where the oil pump suction inlet is located) is the same as the compressor suction pressure (Figure 10-32). The suction pressure will vary from the off or standing reading to the actual running reading. For

PRESSURE-SENSITIVE DEVICES 149

example, when a system is using R-22 as the refrigerant, the pressures may be similar to the following: 125 psig while standing, 70 psig while operating, and 20 psig during a low-charge situation.

A plain low-pressure cutout control would not function at all of these levels, so a control had to be devised that would sensibly monitor pressures at all of these conditions.

Most compressors need at least 30 psig of actual oil pressure for proper lubrication. This means that whatever the suction pressure is, the actual oil pressure has to be at least 30 pounds above the oil pump inlet pressure, because the oil pump inlet pressure is the same as the suction pressure. For example, if the suction pressure is 70 psig, the oil pump outlet pressure must be 100 psig for the bearings to have a net oil pressure of 30 psig. This difference in the suction pressure and the oil pump outlet pressure is called the *net oil pressure*.

In the basic low-pressure control, the pressure is under the diaphragm or bellows and the atmospheric pressure is on the other side of the diaphragm or bellows. Remember, the atmospheric pressure is considered a constant. The oil pressure control uses a double bellows—one bellows opposing the other—to detect the net oil pressure. The pump inlet pressure is under one bellows, and the pump outlet pressure is under the other bellows. These bellows are positioned opposite of one another either physically or using a linkage. The bellows with the most pressure is the oil pump outlet, and it overrides the bellows with the least amount of pressure. This override reads out in net pressure and is attached to a linkage that can stop the compressor when the net oil pressure drops for a predetermined amount of time.

Because the control needs a differential in pressure to allow power to pass to the compressor, it must have some means of allowing the compressor to energize. There is

Figure 10-31. Two views of an oil pressure safety control. This control satisfies two requirements: how to measure net oil pressure effectively and how to get the compressor started to build oil pressure. *Photos by Bill Johnson.*

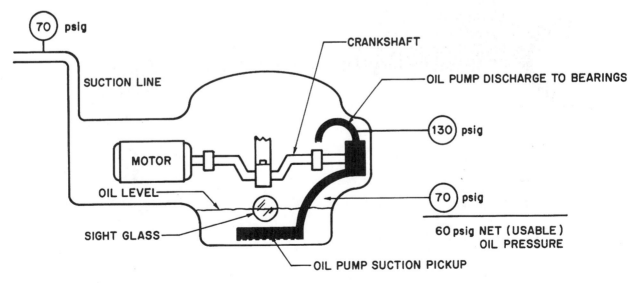

Figure 10-32. The oil pump suction (inlet) pressure is actually the suction pressure of the compressor. This means that the true oil pump pressure is the oil pump discharge (outlet) pressure less the compressor suction pressure. For example, if the coil pump discharge pressure is 130 psig and the compressor suction pressure is 70 psig, the net oil pressure is 60 psig. This is the usable oil pressure.

150 UNIT 10 AUTOMATIC CONTROL COMPONENTS AND APPLICATIONS

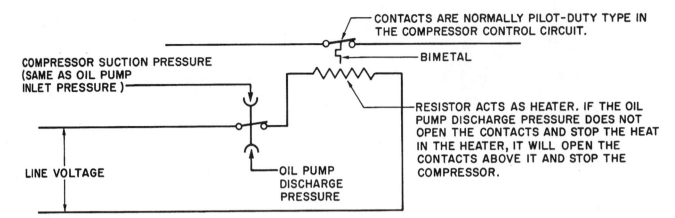

Figure 10-33. The oil pressure control is used for lubrication protection for the compressor. The oil pump that lubricates the compressor is driven by the compressor crankshaft. Therefore, a time delay is necessary to allow the compressor to start up and build oil pressure. This time delay is normally 90 seconds. The time delay is accomplished either with a bimetal device or an electronic circuit.

no pressure differential until the compressor starts to turn, because the oil pump is attached to the compressor crankshaft. There is a time delay built into the control that allows the compressor to start and prevents cutouts when the oil pressure varies momentarily. This time delay is normally about 90 seconds. It is accomplished with either a heater circuit and a bimetal device or electronically. Figure 10-33 shows a wiring diagram for a typical oil pressure control.

Caution: Always consult the manufacturer's instructions when working with any compressor that has an oil safety control.

Pressure Relief Valves

The pressure relief valve can be considered a pressure-sensitive device. It is used to detect excess pressure in any system that contains fluids (water or refrigerant). It then relieves this pressure, usually into the atmosphere. Figure 10-34 illustrates a refrigerant pressure relief valve.

Warning: A pressure relief valve is a factory-set safety device and should not be changed or tampered with.

Water Pressure Regulators

Water pressure regulators are used to control water. Water-cooled air conditioning and refrigeration systems operate at lower pressures and temperatures than air-cooled systems, but the water circuit requires special maintenance to prevent corrosion and scaling. Air-cooled equipment has now become dominant in residential and light commercial installations because it is easy to maintain. Water pressure regulating valves are used primarily on larger or specialized equipment.

The water regulating valve controls water flow to the condenser on water-cooled equipment to maintain a constant head pressure. This valve detects the pressure on the high-pressure side of the system and uses this to control the water flow to establish a predetermined head pressure (Figure 10-35).

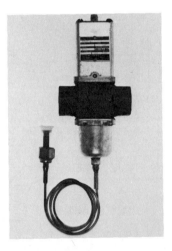

Figure 10-35. This valve maintains a constant head pressure for a water-cooled system during changing water temperatures and pressures. An adjustment screw is located on the top of the control. *Photo by Bill Johnson.*

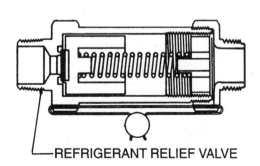

Figure 10-34. Pressure relief valve. *Courtesy Superior Valve Company.*

10.5 ELECTROMECHANICAL CONTROLS

Electromechanical controls convert a mechanical movement into some type of electrical activity. A high-pressure switch is an example of an electromechanical control. The switch contacts are the electrical part, and the bellows or diaphragm is the mechanical part. The mechanical action of the bellows is transferred to the switch to stop a motor when high pressures occur (Figure 10-36). Electromechanical controls normally appear as standardized symbols on electrical diagrams. By studying the symbol, you should be able to recognize the purpose of the control (Figure 10-37). However, some older equipment is labeled with non-standard symbols. In some cases, you will need imagination and experience to understand the purpose of the control.

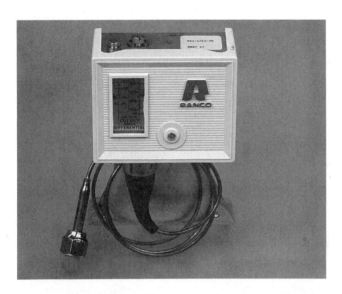

Figure 10-36. High-pressure control. It is considered an electromechanical control. *Photo by Bill Johnson.*

Figure 10-37. Table of electromechanical control symbols. (N.O. = normally open; N.C. = normally closed.)

10.6 MAINTENANCE OF CONTROLS

Mechanical controls are used to control the flow of fluids such as water, refrigerant liquids, and vapors. These fluids flowing through the controls must be contained within the system. Controls are typically designed with diaphragms, bellows, and gaskets that are subject to leakage after being used for long periods of time. A visual check is suggested any time you are near the controls for service or maintenance. Look for any signs of a leak or physical change.

Water is a difficult substance to contain and is likely to leak through even the smallest opening. All water regulating valves should be inspected for leaks by looking for wet spots or rust streaks. Water circulating in a system will also leave mineral deposits in the piping and valve mechanisms.

Electromechanical controls have both a mechanical and an electrical action. Inspect for leaks. Electromechanical pressure controls often are connected to the system with small tubing similar to capillary tubes. This tube is usually copper and is responsible for many of the leaks in air conditioning systems because of misapplication or poor installation practices. For example, a pressure control may be mounted close to a compressor and a small control tube routed to the compressor to sense low or high pressure. The control must be mounted securely to the frame and the control tube routed in such a manner that it does not touch the frame or any other component. The vibration of the compressor will vibrate the tube and rub a hole in the control tube or one of the refrigerant lines if they are not kept isolated.

The electrical section of the control will usually have a set of contacts or a mercury switch to stop and start some component in the system. The electrical contacts are often enclosed and cannot be viewed so visual inspection is not possible. If a problem occurs, look for frayed wires or burned wire insulation adjacent to the control. If the sealed contacts inside a control are becoming burned, the excess heat produced may burn the wire insulation.

The mercury in mercury switches may be viewed through the clear glass enclosure. If the mercury becomes dark, the tube is allowing the entry of oxygen and oxidation has occurred. In this case, the switch should be replaced. It is possible that the switch will conduct across the oxidation and will then appear to be closed at all times.

10.7 TECHNICIAN SERVICE CALL

A customer calls and says the central air conditioning system in their small office building is not cooling correctly. The thermostat is set at 73°F and the space temperature is 85°F. This is a split system air conditioner that uses a thermostatic expansion valve and has a low-pressure control. The problem is the unit is low on charge (due to a flare left loose by a previous technician) and the compressor is cycling off because of low pressure.

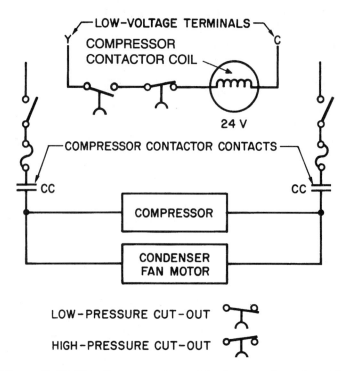

Figure 10-38. This diagram can be used for the example service call.

The technician arrives and finds the space temperature is 85°F and the thermostat is set correctly. The technician goes to the back of the building where the outdoor condensing unit is located and notes that the unit is running. When the technician reaches the unit, it shuts off. The technician removes the control panel and fastens a voltmeter lead on either side of the low-pressure control to see if this control is what is stopping the unit (Figure 10-38). When the unit is off, there is a voltage reading across the low-pressure control, verifying that the unit must be operating with a low charge. The technician fastens a low-pressure gage to the gage port and verifies that the unit is operating with a low suction pressure. The technician will have to take appropriate steps to determine the cause of the low charge condition and then make a proper repair.

SUMMARY

- Temperature controls are either operating controls or safety controls.
- Space temperature controls can either be low-voltage or line-voltage controls.
- Low voltage is normally applied to residential cooling controls.
- Temperature control devices may either use open electrical contacts that are enclosed in a bulb and use mercury as the contact surface.
- Cooling thermostats may have a cold anticipator in parallel with the thermostat contacts.
- Line-voltage thermostats are normally rated up to 25 A and they are used to switch high-voltage current.
- Low-voltage thermostats are normally rated at no more than 2 A.
- To achieve an accurate temperature reading on a flat or round surface, the sensing element (either mercury bulb, remote bulb, or bimetal device) must be in good contact with the surface and insulated from the ambient air.
- Some installations have built-in wells in which a temperature sensor can be placed.
- Motors have both internal and external types of motor-temperature sensing devices.
- The internal type of motor-temperature sensing device can be either a bimetal device or a thermistor inserted inside the motor windings.
- All temperature-sensitive elements change in some manner with a change in temperature: the bimetal warps and the thermistor changes resistance.
- Pressure-sensitive devices can be applied to all fluids.
- Pressure controls are normally operated by a diaphragm, Bourdon tube, or bellows.
- Pressure controls can operate at high pressures, low pressures, or even below-atmospheric pressures (in a vacuum). These controls can also detect differential pressures.
- Pressure-sensitive controls can be used to control or modulate fluid flow.
- Mechanical controls perform mechanical functions without electricity.

- Electromechanical controls have both electrical and mechanical functions.
- Some purely mechanical controls are water regulating valves and pressure relief valves.
- Electromechanical controls include low-pressure and high-pressure cutout controls, high limit switches, and thermostats.
- Always refer to the manufacturer's literature when servicing or troubleshooting any system.

REVIEW QUESTIONS

1. Why is it important to be able to recognize the various types of controls?
2. Why is a low-voltage thermostat normally more accurate than a high-voltage thermostat?
3. Name two kinds of switching mechanisms in a low-voltage thermostat.
4. What is an inert gas?
5. What does a cold anticipator do?
6. Describe how a bimetal element functions.
7. In a residential system, what voltage is considered low voltage?
8. What component steps down the voltage to the low-voltage value?
9. Name four reasons why low voltage is desirable for residential control voltage.
10. What is the maximum amperage usually encountered by a low-voltage thermostat?
11. What two types of switches are normally found in the subbase of a low-voltage thermostat?
12. When is a line-voltage thermostat used?
13. Name two sensing elements that can be used with a line-voltage thermostat.
14. What is the maximum amperage generally encountered in a line-voltage thermostat?
15. Why is a line-voltage thermostat mounted on an electrical box that has conduit connected to it?
16. What method is used to produce an accurate reading when using a mercury thermometer on a flat or round surface?
17. Why do motors build up heat?
18. How is motor heat dissipated?
19. What are the two types of motor-temperature sensing devices?
20. What is the principal of operation of most externally-mounted overload protection devices?
21. Name a method that can be used to speed up the cooling of an open motor.
22. Describe the best way to cool an overheated compressor.
23. Why is it important to allow a compressor to cool before making a diagnosis?
24. How can an electronic thermometer be used in a well to obtain an accurate response?
25. How can the temperature be checked in a water circuit if no temperature well is provided?
26. How does the thermistor change with a temperature change?
27. What is a pressure-sensitive device?
28. Name two methods used to convert pressure changes into action.
29. Name two actions that can be obtained with a pressure change.
30. Can pressures below atmospheric pressure be detected?
31. Name one function of a low-pressure control.
32. Name one function of a high-pressure control.
33. How can you find information about a control when there is no description with the unit?

11 TROUBLESHOOTING BASIC CONTROLS

OBJECTIVES

Upon completion of this unit, you should be able to

- Define and identify power-consuming and power-passing devices.
- Describe how a voltmeter is used to troubleshoot electrical circuits.
- Identify some typical problems in an electrical circuit.
- Describe how an ammeter is used to troubleshoot an electrical circuit.
- Recognize the components in an air conditioning electrical circuit.
- Follow the sequence of operation in an air conditioning electrical circuit.
- Differentiate between a pictorial diagram and a line or ladder diagram.

11.1 INTRODUCTION TO TROUBLESHOOTING

Before troubleshooting any electrical circuit, always familiarize yourself with its control components. Each control must be evaluated as to its function (major or minor) in the system. Studying the purpose of a control before taking action will save time and help to prevent call-backs. Look for pressure lines leading to pressure controls and temperature sensors on temperature controls. See if the control stops and starts a motor, opens and closes a valve, or provides some other function.

Electrical devices can be categorized according to whether or not they consume power in the circuit. *Power-consuming devices* use power and have either a magnetic coil or a resistance circuit. They are wired in parallel with the power source.

Devices that do not consume power are known as *power-passing devices*. They are wired in series with the power-consuming device. These terms can be understood by studying a simple light bulb circuit with a switch (Figure 11-1). The light bulb in the circuit actually consumes the power, and the switch passes the power to the light bulb. The light bulb is wired to both sides (hot and neutral) of the power supply. This places the light bulb in parallel with the power supply and establishes a complete circuit. The switch is a power-passing device and is wired in series with the light bulb.

For any power-consuming device to use power, it has to have what is known as *potential voltage*. Potential voltage is the voltage indicated on a voltmeter between two power legs (such as line 1 to line 2) of a power supply. For example, the light bulb has to be wired from the hot leg (the wire that has a fuse or breaker in it) to the neutral leg (the wire that is connected to the ground).

The following general statements should help you to understand current flow in an electrical circuit.

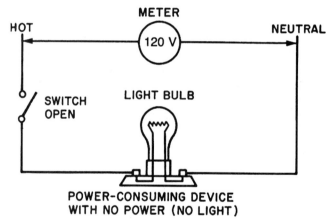

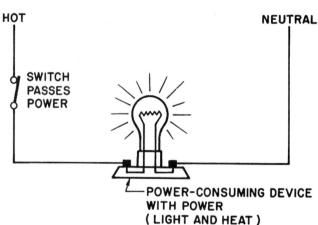

Figure 11-1. Power-consuming device (light bulb) and power-passing device (switch).

TROUBLESHOOTING A SIMPLE CIRCUIT 155

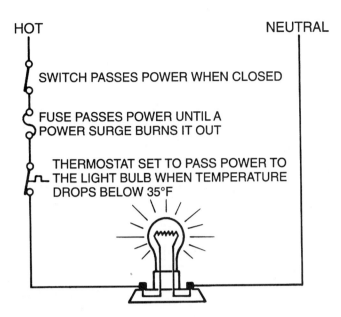

Figure 11-2. When closed, the switch passes power to the fuse.

1. It doesn't make any difference which way the electrons are flowing in an AC circuit when you keep in mind that the object is to move the electrons (current) to a power-consuming device and complete the circuit.
2. The electrons may pass through many power-passing devices before reaching the power-consuming device(s).
3. Devices that don't consume power pass power.
4. Devices that pass power can be either safety devices or operating devices.

Suppose the light bulb mentioned above is used to add heat to a well pump to keep it from freezing in the winter. To prevent the bulb from burning all the time, a thermostat is installed. A fuse must also be installed in the line to protect the circuit, and a switch must be wired to allow the circuit to be serviced (Figure 11-2). Note that the power in this circuit must pass through three power-passing devices to reach the light bulb. All three devices are on one side of the circuit. The fuse is a safety control. The thermostat is the operating control that turns the light bulb on and off. The switch is not a control but a service convenience device.

11.2 TROUBLESHOOTING A SIMPLE CIRCUIT

Suppose that the light bulb shown in Figure 11-2 does not operate. Which component is interrupting the power to the bulb: the switch, the fuse, or the thermostat? Or is the light bulb itself at fault? To determine the solution to this problem, we will use a voltmeter to check the voltage across each device until we find the point where the circuit is broken.

First, turn the voltmeter selector switch to a higher voltage setting than the voltage supply. In this case, the supply voltage is 120 V, so the 250 V scale is a good choice. To troubleshoot this circuit, follow the procedure listed below and shown in Figure 11-3.

Warning: When troubleshooting electrical malfunctions, take the following precautions:

- Disconnect the electrical power unless it is necessary to make appropriate checks.
- When checking continuity or resistance, turn off the power and disconnect at least one lead from the component being checked.
- Ensure that meter probes come in contact only with terminals or other appropriate measuring points.

1. Place the red lead of the voltmeter on the hot line and the black lead on the neutral line. The meter reads 120 V.
2. Move the red lead (the lead being used to find and detect power in the hot line) to the load side of the switch. (The load side of the switch is the side of the switch that is connected to the load. The other side of

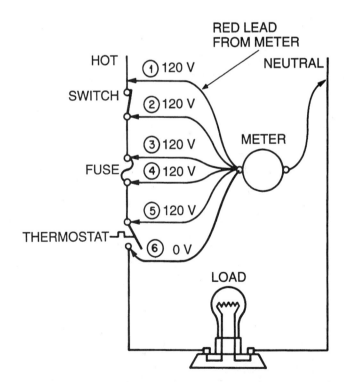

Figure 11-3. The first step in the troubleshooting procedure is to establish the main power supply of 120 V from the hot line to the neutral line. When this power supply is verified, the lead on the hot side is moved down the circuit toward the light bulb. The voltage is established at all points until point 6 is reached, where no voltage is present. The thermostat contacts are open.

the switch, where the line is connected, is known as the line side of the switch.) The black lead should remain in contact with the neutral line. The meter reads 120 V.

3. Move the red lead to the line side of the fuse. The meter reads 120 V.
4. Move the red lead to the load side of the fuse. The meter reads 120 V.
5. Move the red lead to the line side of the thermostat. The meter will read 120 V.
6. Move the red lead to the load side of the thermostat. The meter reads 0 V. There is no power available to energize the bulb because the thermostat contacts are open. At this point, we need to know if the room is cold enough to cause the thermostat contacts to make. If the room temperature is below 35°F, the circuit should be closed; the contacts should be made. Let's assume that the temperature was too warm to close the contacts, but when it drops below 35°F, the light bulb still does not operate. The meter reading across the thermostat is now 120 V.
7. Move the red meter lead to the terminal on the light bulb (Figure 11-4). Suppose it reads 120 V. Now, move the black lead to the light bulb terminal on the right. If there is no voltage, the neutral wire is open between the source and the bulb.

If there is voltage at the light bulb and it will not burn, the bulb is defective (Figure 11-5). When there is a power supply and a path to flow through, current will flow. If the light bulb filament is broken, there is no pathway for the current.

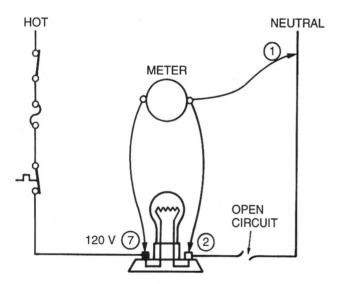

Figure 11-4. The thermostat contacts are closed, and there is a measurement of 120 V when one lead is on the neutral line and one lead is on the light bulb terminal (see position 7). When the black lead is moved to the light bulb in position 2, there is no voltage reading. The neutral line is open.

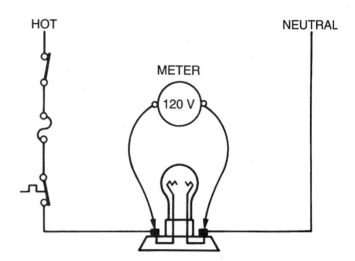

Figure 11-5. Power is available at the light bulb. The hot line on the left side and the neutral line on the right side complete the circuit. The light bulb filament is burned out, so there is no path through the bulb.

11.3 TROUBLESHOOTING AN AIR CONDITIONING CIRCUIT

The following example uses a typical air conditioning control circuit. This circuit is not standard because each manufacturer may design its circuits differently. They may vary, but will be similar in nature.

The unit selected for this example is a package air conditioner with 1½ tons of cooling. Package units resemble window air conditioners because all of the components are in the same cabinet. This unit can be installed through a wall or on a roof, and the supply and return air can be ducted to the conditioned space. This unit is available in many sizes—from 1½ tons of cooling (18,000 Btu's/hour) to very large systems. It is a popular choice in shopping centers because it allows a store to have several rooftop units to provide zone control. The unit has all of the control components, except the room thermostat, within the unit cabinet, so it can be serviced without disturbing the conditioned space. The thermostat is mounted in the conditioned space.

The first thing we will consider is the thermostat (Figure 11-6).

Note: The thermostat circuit used in this example has been simplified for illustration purposes.

1. This thermostat is equipped with a select or switch for either HEAT or COOL.
2. When the selector switch is in the cooling mode, the cooling system will start on a call for cooling. The indoor fan will start through the AUTO mode on the fan selector switch. The outdoor fan must also run when the unit is operating, so it is wired in parallel with the compressor.

TROUBLESHOOTING A THERMOSTAT 157

through this anticipator when the thermostat is satisfied (or when the thermostat's circuit is open). The cold anticipator is normally a fixed, nonadjustable device.

11.4 TROUBLESHOOTING A THERMOSTAT

The thermostat is an often misunderstood, frequently suspected component during equipment malfunctions. However, it is a fairly simple and reliable device. The thermostat monitors the room temperature and distributes the power leg of the 24 V circuit to the correct component to maintain comfortable conditions. As you approach a thermostat troubleshooting job, remember that one power leg enters the thermostat and it distributes power where required.

One way to troubleshoot a thermostat on a cooling system is to first turn the fan selector switch to the FAN ON position and see if the fan starts. The fan circuit is a switch that operates the fan regardless of the room temperature or the thermostat HEAT/COOL setting. If the fan does not start, there may be a problem with the control voltage. Control voltage has to be present for the thermostat to operate. Assume for a moment that the thermostat will operate the fan in the FAN ON mode but will not operate in the cooling mode. The next step may be to take the thermostat off the subbase and energize the circuit manually with an insulated jumper. See Figure 11-7.

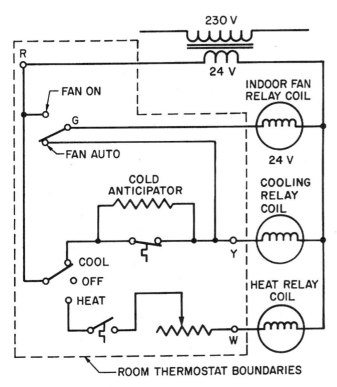

Figure 11-6. Simple thermostat with a cooling relay for starting the compressor and a fan starting circuit for the indoor fan.

Note: Only the low-voltage control circuit is shown.

As shown in Figure 11-6, the fan starting circuit is connected to the G terminal. The G terminal designation is not universal but is quite common. When the switch is in the FAN ON position, the power travels from the R terminal through the fan selector switch to the G terminal and on to the indoor fan relay coil. Power is supplied to the other side of the coil directly from the control transformer. When this coil has power (24 V), it closes a set of contacts in the high-voltage circuit (not shown) and starts the indoor fan.

Figure 11-6 also shows the Y terminal for cooling. Again, this is not the only letter used for cooling, but it is typical. Follow the power from the R terminal down to the selector switch through the COOL contacts and on to the Y terminal. When these contacts are closed, power passes through them and travels in two directions. One path is through the fan AUTO switch to the G terminal and then to the indoor fan. The indoor fan must run in the cooling cycle. The other path is straight to the cooling relay. When the cooling relay is energized (24 V), it closes a set of contacts in the high-voltage circuit (not shown) to start the cooling cycle.

Note: The cold anticipator is wired in parallel with the cooling contacts. This means that current will flow

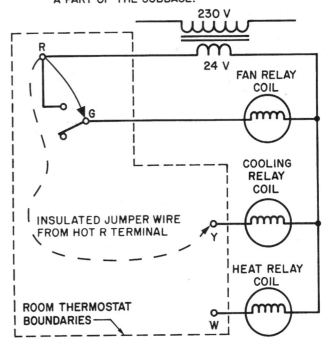

Figure 11-7. Attaching a jumper wire between the thermostat subbase terminal connections will verify the cooling circuit (R to Y) and the indoor fan circuit (R to G).

Jump from R to G, and the fan should start. Jump from R to Y, and the cooling should come on. If the circuit can be made to operate without the thermostat but not with it, the thermostat is defective.

Note: Jumping a thermostat is a safe procedure because only one leg of power is routed to the thermostat. Do not attempt this procedure unless you are under the supervision of an instructor or other qualified person. Also, do not restart the air conditioner until five minutes have passed. This allows the system pressures to equalize.

We have just covered what happens in the low-voltage circuit of a basic thermostat. The next step is to move into the high-voltage circuit and see how the thermostat actually controls the fan and cooling process.

Refer to Figure 11-8. Note the high voltage (230 V) input to the transformer. When the service disconnect is closed, the potential voltage between line 1 and line 2 is the power supply to the primary winding of the transformer. The primary winding induces power in the secondary winding. Therefore, if there is no power to the primary winding, there will be no voltage in the secondary winding.

Figure 11-8 also illustrates the high-voltage operation of the fan circuit. The power-consuming device is the fan motor in the high-voltage circuit. The fan relay contacts are in the line 1 circuit, and must be closed to pass power to the fan motor. These contacts close when the fan relay coil is energized in the low-voltage control circuit.

Figure 11-9 shows the complete wiring diagram. Three components have to operate in the cooling mode: the indoor fan, the compressor, and the outdoor fan. The compressor and outdoor fan are wired in parallel and, for our purposes, can be thought of as a single component.

The line 2 side of the circuit goes directly to the three power-consuming components. Power for the line 1 side of the circuit comes through two different relays. The cooling relay contacts start the compressor and outdoor fan motor; the indoor fan relay starts the indoor fan motor. In both cases, the relays pass power to the components upon a call from the thermostat.

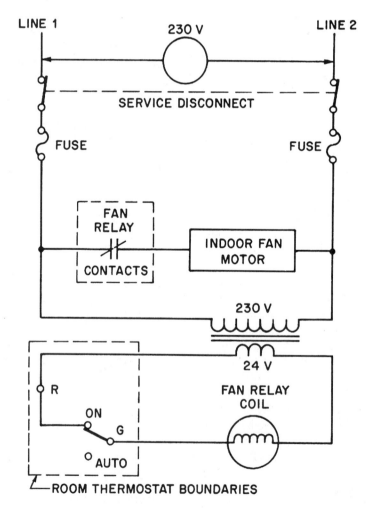

Figure 11-8. The high-voltage operation of the fan. When the fan switch is turned to the ON position, a circuit is completed to the fan relay coil. When this coil is energized, it closes the fan relay contacts and passes power to the indoor fan motor.

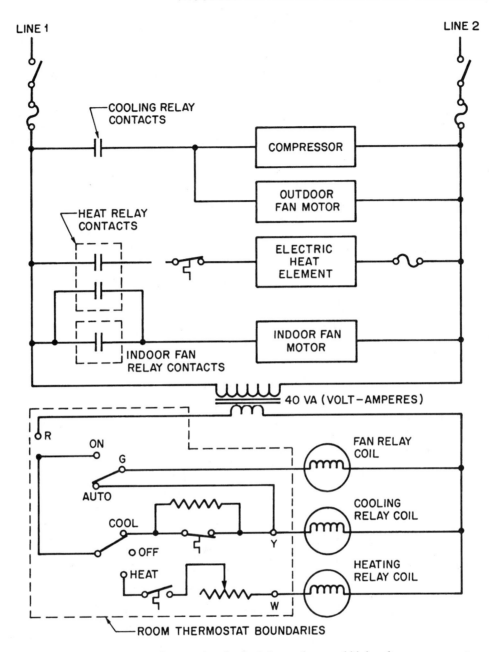

Figure 11-9. Complete wiring diagram showing both low-voltage and high-voltage components.

When the thermostat selector switch is set to the COOL position, power is passed to the contacts in the thermostat. When these contacts are closed, power passes to the cooling relay coil. When the cooling relay coil is energized, the cooling relay contacts in the high-voltage circuit are closed. This passes power to the compressor and outdoor fan motor.

The indoor fan must run whenever there is a call for cooling. When power passes through the contacts in the thermostat, a circuit is made through the AUTO side of the selector switch to energize the indoor fan relay coil and start the indoor fan. Note that if the fan selector switch is in the ON position, the fan will run all the time and will remain on during the cooling cycle.

11.5 TROUBLESHOOTING AMPERAGE IN LOW-VOLTAGE CIRCUITS

Transformers are rated in *volt-amperes (VA)*. This rating can be used to determine if the transformer is underrated or drawing too much current.

For example, it is common to use a 40 VA transformer for the low-voltage power source on combination cooling/heating equipment. At 24 V, the maximum amperage that the transformer can be expected to carry is calculated as follows:

$$\frac{40 \text{ VA}}{24 \text{ V}} = 1.66 \text{ A}$$

160 UNIT 11 TROUBLESHOOTING BASIC CONTROLS

Because clamp-on ammeters do not accurately measure low current values, a multiplier is used to amplify the current readings. Use a jumper wire and coil it ten times (called a 10-wrap amperage multiplier). Place the ammeter's jaws around the 10-wrap loop and place the jumper in series with the circuit (Figure 11-10). To obtain the actual reading, the reading on the ammeter is divided by 10. For example, if the cooling relay coil amperage reads 7 A, it is really only carrying 0.7 A.

Some ammeters have attachments to read ohms and volts. The volt attachment can be helpful for taking voltage readings, but the ohm attachment should be checked to make sure that it will read in all of the ranges that are needed. Some ohmmeter attachments will not read very high resistances. For example, the attachment may show an open circuit on a high-voltage coil that has a considerable amount of resistance.

11.6 TROUBLESHOOTING VOLTAGE IN LOW-VOLTAGE CIRCUITS

The volt-ohm-milliammeter or VOM has the capability of checking continuity, milliamps, and volts. The most common applications are voltage and continuity checks. The volt scale can be used to check for the presence of voltage, and the ohmmeter scale can be used to check for continuity. A high-quality meter should produce accurate readings for each type of measurement.

Remember that any power-consuming device in the circuit must have the correct voltage to operate. The voltmeter can be applied to any power-consuming device for the voltage check by placing one probe on one side of the coil and the other probe on the other side of the coil. When power is present at both points, the coil should function. If not, use an ohmmeter to check for continuity

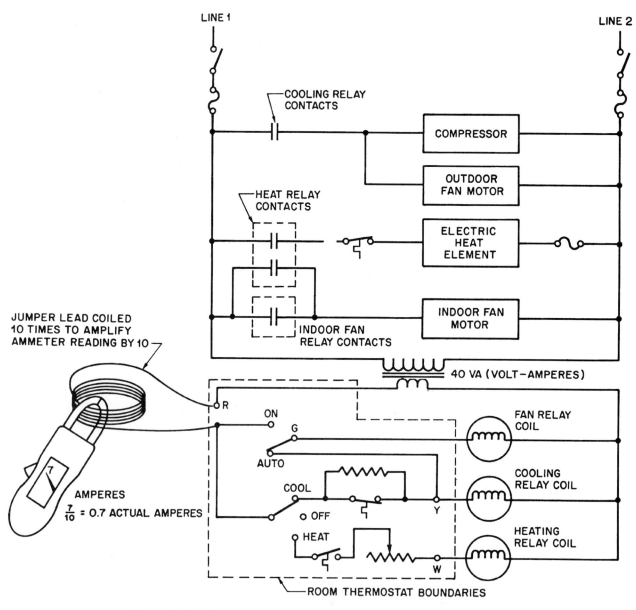

Figure 11-10. Clamp-on ammeter to measure current draw in the 24 V circuit.

PICTORIAL AND LINE DIAGRAMS   **161**

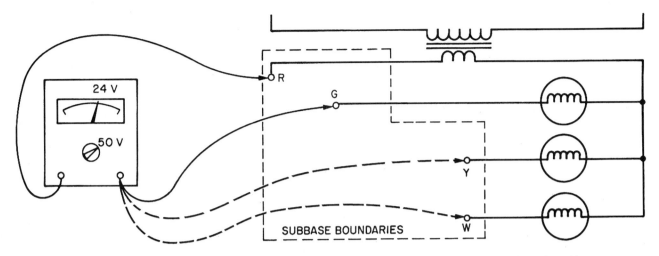

Figure 11-11. The VOM can be used at the thermostat location with the thermostat removed from the subbase.

through the coil. Again, keep in mind that when voltage is present and a path is provided, current will flow. If the flowing current does not cause the coil to do its job, the coil is defective and will have to be changed.

As discussed earlier, voltage checks can be made at the thermostat by removing the thermostat from the subbase and jumping the thermostat terminals. With the subbase terminals exposed, the voltmeter can be used in the following manner:

1. Turn the voltmeter selector switch to the scale just higher than 24 V.
2. Attach the voltmeter lead to the hot leg feeding the thermostat (Figure 11-11). This terminal is sometimes labeled R or V (there is no standard letter or number used).
3. Place the other lead on the other terminals one at a time to verify each circuits.

For example, with one lead on the R terminal and one on the G terminal (the terminal normally used for the fan circuit) the voltage reading is 24 V. The meter is reading one side of the line straight from the transformer and the other side of the line through the coil on the fan relay. When the meter probe is moved to the circuit assigned to cooling (this terminal may be lettered Y), 24 V is read. The fact that the voltage reads through the coil in the respective circuits is evidence that a complete circuit is present.

The previous examples covered troubleshooting of low-voltage circuits. The control circuits that use high voltage (115 V or 230 V) work the same way. The same rules apply with more emphasis on safety.

11.7 PICTORIAL AND LINE DIAGRAMS

The most important skills when troubleshooting electrical circuits is the ability to get a mental picture of the control circuits. With practice, you will be able to take a symptom from equipment observation and turn this symptom into a discovery that leads to repair. This starts with being able to look at the control panel and identify the various controls. Most equipment includes one or more circuit diagrams to help the technician identify the controls. There are two types of diagrams furnished with equipment: the pictorial diagram and the line diagram. Some equipment has only one diagram, while others include both.

The pictorial diagram, sometimes known as the component arrangement diagram, shows the components in their relative positions. The controls are organized just as you would see them with the panel door open. For example, if the CR (control relay) is shown in the upper left corner of the diagram (Figure 11-12), it will be in the upper left corner of the control panel when the door is open. This is helpful when you are not sure what a particular component looks like. Study the diagram until you find the control and then locate it in the corresponding place in the control panel. The diagram also provides the wire identifying characteristics to further verify each component. The wires may be colored or they may be all black and labeled with numbers. The general outline of components is the same.

Some component arrangement diagrams do not show the interconnecting wiring (Figure 11-13). The concept is for the technician to use the simplified diagram to identify the component in the unit, and then go to a working diagram that shows the wiring.

The working diagram is called a *ladder diagram* or *line diagram*. This diagram makes it easy to follow the circuit (Figure 11-14). All power-consuming devices are shown horizontally between the vertical voltage lines (like the steps on a ladder). Most manufacturers use the right side of the diagram as a common line to all power-consuming devices. The right side of the diagram will normally have few or no switches in it. This makes it easier to troubleshoot the circuit.

Pictorial and line diagrams are an example of the way most manufacturers illustrate the wiring in their equipment. However, wiring diagrams can vary significantly

162 UNIT 11 TROUBLESHOOTING BASIC CONTROLS

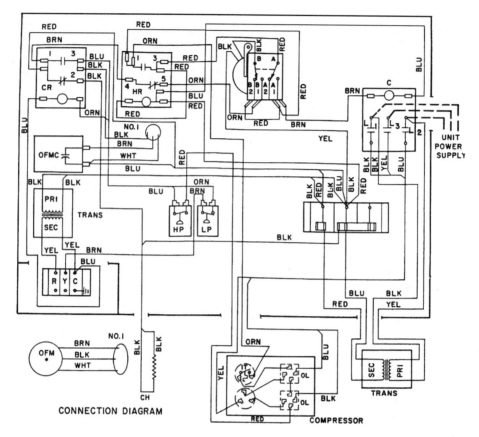

Figure 11-12. A pictorial diagram shows the relative positions of the components in the control panel. *Reproduced courtesy of Carrier Corporation.*

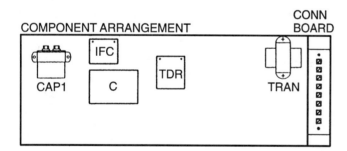

Figure 11-13. Component arrangement diagram. *Reproduced courtesy of Carrier Corporation.*

from manufacturer to manufacturer. The only items that have been standardized are the symbols used to illustrate the various components.

Note: The best method of troubleshooting complex circuits is to identify and eliminate each circuit that is working, and then concentrate on those that are not working. At first, circuit identification can appear to be a very complicated task. As you begin to work with diagrams, you may find it helpful to use colored pencils to highlight each circuit.

Using the ladder diagram and accompanying pictorial diagram in Figure 11-15, follow the sequence to start the

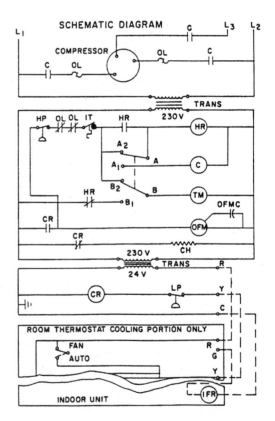

Figure 11-14. Ladder diagram of Figure 11-12. *Reproduced courtesy of Carrier Corporation.*

PICTORIAL AND LINE DIAGRAMS 163

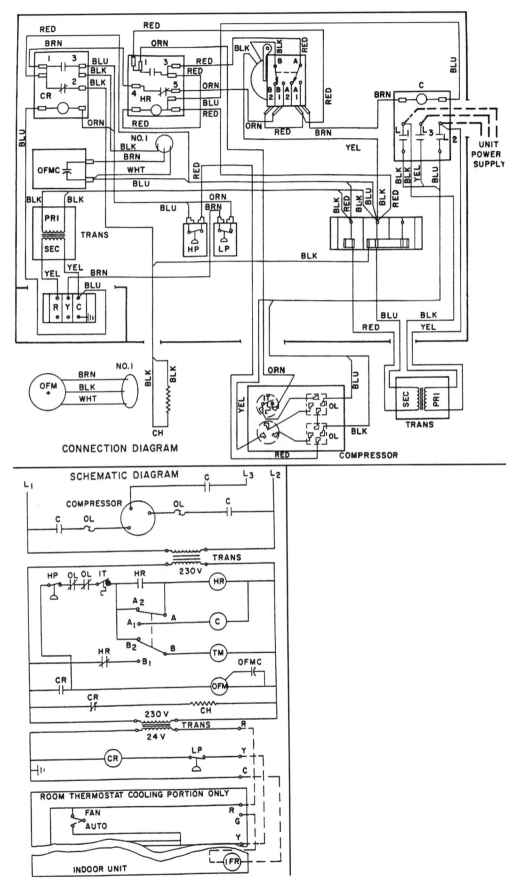

Figure 11-15. Tracing a circuit in a ladder diagram and following it in a pictorial diagram. *Reproduced courtesy of Carrier Corporation.*

compressor. This circuit has a 7-minute delay to prevent compressor short cycling. We will be starting the compressor using this time delay. Looking at the ladder diagram, the compressor is shown at the top. In order for the compressor to operate, the contacts labeled C must be closed. Let's follow the power supply as it leaves the room thermostat to close the C contacts. Follow the process in both the ladder diagram and the pictorial diagram.

1. The room thermostat contacts in the lower portion of the diagram must be closed to make a circuit from R to Y (this is not illustrated).
2. The room thermostat contacts pass power to the CR (control relay) provided that the LP (low-pressure control) contacts are closed. The CR coil is located just above the room thermostat in the ladder diagram.
3. When power passes to the CR coil, two things happen: the normally closed CR contacts open and stop the compressor crankcase heater, and the normally open CR compressor contacts close, passing power to the HP (high pressure control), two OL (overload contacts) and the IT (internal thermostat), and on to the B2 terminal in the timer.
4. The timer is used to prevent the compressor from short cycling. When power is applied to the B2 terminal, it is passed to the B terminal and the timer motor starts to run. At the same time, power is passed to the HR (holding relay) coil. The normally open HR contacts close and the normally closed HR contacts open. The holding relay is energized and will continue to stay energized by the HR contacts.
5. When the timer runs for approximately 7 minutes, the contacts shift and power passes through the A1 contacts to C (the compressor contactor). The compressor starts.

The following section includes sample service calls to help build your troubleshooting skills. The solutions are included for the first three service calls so that you will have a better understanding of the troubleshooting procedures. The last three service calls do not include solutions. Use your critical thinking abilities or class discussion to determine the answers. The solutions can be found in the Instructor's Guide.

11.8 TECHNICIAN SERVICE CALLS

Service Call 1

A customer with a package air conditioner calls and tells the dispatcher that the unit is not cooling. The problem is that the control transformer has an open circuit in the primary winding (Figure 11-16).

The technician arrives, locates the indoor thermostat, and turns the indoor fan switch to the FAN ON position. The fan will not start. This is an indication that there may

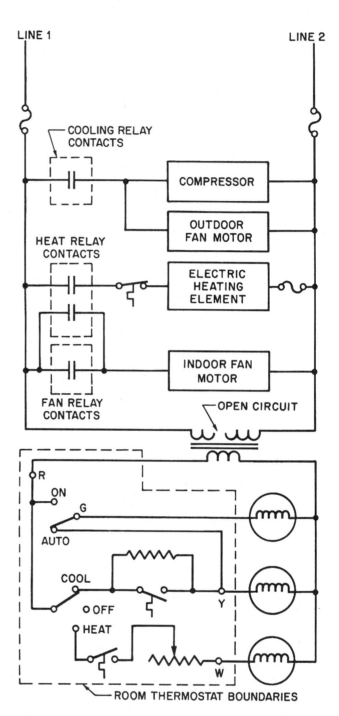

Figure 11-16. This control transformer has an open circuit.

be no control voltage. Remember, there must be high voltage before there is low voltage. To check for high voltage, the technician sets the meter range selector switch set to 250 V and check the power supply at the indoor unit. This check shows that there is power.

The cover is removed from the control compartment (usually this compartment is easy to find because the low-voltage wires enter nearby). A check for high voltage at the primary winding of the transformer shows that there is power.

A check for low voltage at the secondary winding of the transformer indicates no power. The transformer must be defective.

To prove this, the main power supply is turned off. One lead of the low-voltage transformer is removed and the ohmmeter section of the VOM is used to check the transformer for continuity. Here is where the technician finds that the primary circuit is open. A new transformer must be installed.

The technician installs the new transformer and turns the power on. The system starts and runs correctly. The paperwork is handled with the customer and the technician goes to the next call.

Service Call 2

A customer calls with a complaint of no cooling. The problem is that the cooling relay coil is open (Figure 11-17).

The technician arrives, locates the thermostat, and tries the fan circuit. When the fan switch is moved to the FAN ON position, the indoor fan starts. The thermostat is then switched to the COOL position. Again, the fan runs. This means that power is passing through the thermostat, so the problem is probably not there. The thermostat is left in position to call for cooling while the technician goes to the outdoor unit.

Since the primary winding of the low-voltage transformer receives its power from the high-voltage power supply, the power supply is established as good.

A look at the diagram shows the technician that the only requirement for the cooling relay to close its contacts is that its coil be energized. A check for 24 V at the cooling relay coil shows 24 volts. The coil must be defective. To be sure, the technician turns off the power and removes one of the leads on the coil. The ohmmeter is used to check the coil. There is no continuity. It is open.

The technician changes the whole contactor (it is less expensive to change the contactor from stock on the truck than to go to a supply house and get a coil). The technician then turns the power on and starts the unit. It runs correctly. All paperwork is handled and the technician leaves.

Service Call 3

A customer calls and complains of no cooling. The problem is that the cooling relay has a shorted coil that has overloaded the transformer and burned it out, Figure 11-18.

The technician arrives, locates the thermostat, and tries the indoor fan. It does not run. There is obviously no current flow to the fan. This could be a problem in either the high-voltage or low-voltage circuits.

The technician checks for high voltage at the outdoor unit and finds there is power. However, a check for low voltage shows there is no power. The transformer must be defective. When it is removed, the technician notices that it has a burnt smell. It is checked with the VOM and found to have an open secondary winding.

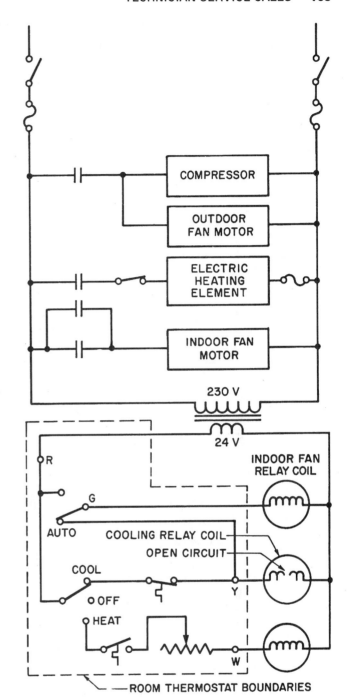

Figure 11-17. This cooling relay has an open circuit and will not close the cooling relay contacts. Note that the indoor fan is operable.

When the transformer is changed and the system is switched to COOL, the system does not come on. The technician notices that the transformer is getting hot. It appears to be overloaded. The technician turns off the power before a second transformer is damaged. The problem is to find out which circuit is overloading. Since the system was in the cooling mode, the technician assumes for the moment it is the cooling relay. The technician realizes that if too much current is allowed to pass

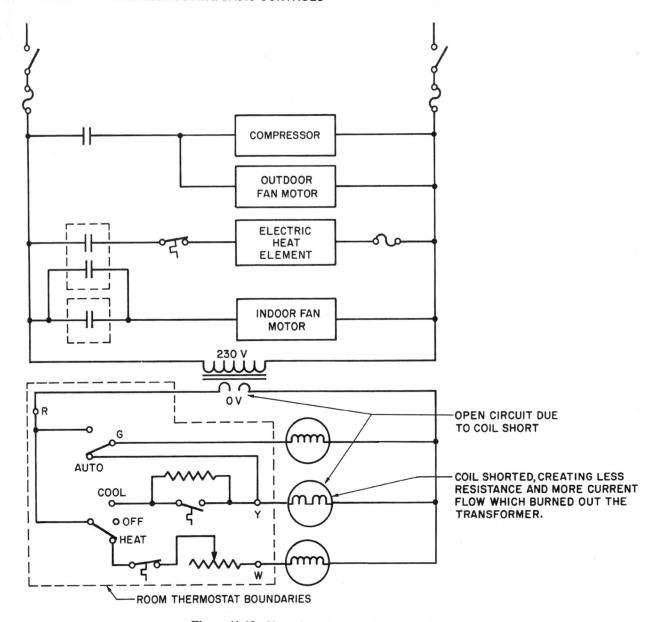

Figure 11-18. Shorted cooling relay burns transformer.

through the room thermostat, it will also burn and have to be replaced.

The technician turns off the room thermostat to take it out of the circuit, then goes to the outdoor unit where the low-voltage terminal block is located. An ammeter is applied to the transformer circuit leaving the transformer. A ten wrap coil of wire is installed to amplify the current reading. The power is turned on and no amperage is recorded. Therefore, the problem is not a short in the wiring.

The technician then jumps from R to Y to call for cooling. The measured amperage is now 25 amperes. This value is divided by 10 because of the ten wrap; the real amperage in the circuit is 2.5 amperes. This is a 40 VA transformer with a rated amperage of 1.67 amperes (40 ÷ 24 = 1.666). The cooling relay is pulling enough current to overheat the transformer and burn it up in a short period of time. It is lucky that the room thermostat has not also been damaged.

The technician changes the cooling relay and tries the circuit again. Now, the amperage reading is only 5 amperes, divided by 10, or .5 ampere. This is correct for the circuit. The jumper is removed, the panels are put back in place, and the system started with the room thermostat. It operates normally. All paperwork is handled and the technician leaves.

Note: The following service calls do not have the solution described.

Service Call 4

A customer calls and indicates that their air conditioner stopped running after the cable TV repairman was under the house running TV cable.

TECHNICIAN SERVICE CALLS   167

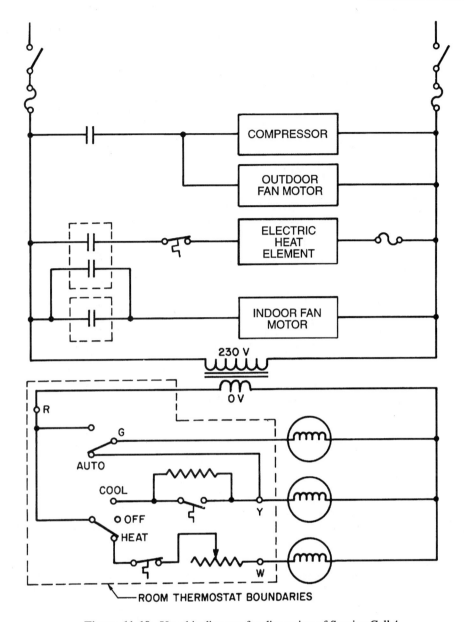

Figure 11-19. Use this diagram for discussion of Service Call 4.

The technician arrives and enters the house. It is obvious that the air conditioner is not running. The technician can hear the fan running, but it is hot in the house. A check outside shows that the condensing unit is not running. The technician then proceeds through the following checkout procedure. Use the diagram in Figure 11-19 to follow this checkout procedure.

1. A low-voltage check outside at the condenser shows zero volts.
2. Inside the house, the technician removes the room thermostat from the subbase and jumps from R to Y. The condensing unit does not start.
3. Leaving the jumper in place, the technician then goes back to the condenser and checks voltage at the contactor. Again, there is no voltage.

What is the problem and the recommended solution?

Service Call 5

A store manager calls and reports that the air conditioner is not running. This store has a package unit on the roof.

The technician arrives, talks to the customer, and uses the following checkout procedure to discover the problem. Follow the procedure using the diagram in Figure 11-20.

1. The customer says the system worked well until about an hour ago when the manager noticed the store was getting hot.
2. The technician notices that the setpoint on the room thermostat is 75°F and the store temperature is 85°F.

168 UNIT 11 TROUBLESHOOTING BASIC CONTROLS

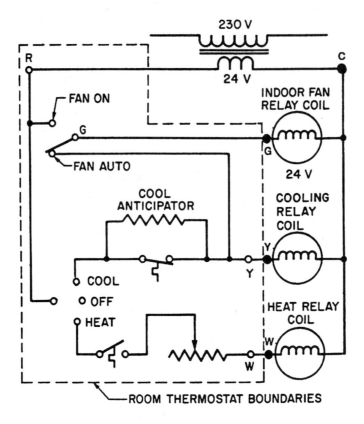

Figure 11-20. Use this diagram for discussion of Service Call 5.

3. The technician then goes to the roof and removes the low-voltage control panel. The voltage from C to R is 24 volts, and from C to Y (the cool terminal) is 0 volts. There is a 24 volt power supply, but the compressor is not operating.
4. The room thermostat is removed from the subbase and a jumper is applied from R to Y. The cooling system starts.

What is the problem and the recommended solution?

Service Call 6

A customer calls on a very hot day and explains that their air conditioner has shut off and the house is getting hot.

The technician arrives and checks the room thermostat. It is set at 73°F and the house temperature is 85°F. The thermostat is calling for cooling. The fan is running, but there is no cooling.

The technician goes to the condensing unit and removes the control compartment door. The technician checks the line voltage at the line side of the contactor and the meter reads 233 V. There are only two control wires entering the condensing unit from the room thermostat, one white and one red. The meter leads are fastened to these leads and the meter reads 24 V.

What is the problem and the recommended solution?

SUMMARY

- Each control has to be evaluated as to its purpose in a circuit.
- Electrical controls are divided into two categories: power consuming and power passing.
- One method of understanding a circuit is to remember that the potential power supply is between two different power legs of different potential.
- Devices or controls that pass power are known as safety, operating, or control devices.
- A light bulb controlled by a thermostat with a fuse in the circuit is an example of an operating device and a safety device in the same circuit.
- The voltmeter may be used to follow the circuit from the power source to the power-consuming device.
- There are two separate power-consuming circuits in the low-voltage control of a typical cooling unit: the cooling circuit and the fan circuit. The selector switch determines which function will operate.
- The fan relay and the cooling relay are power-consuming devices.
- The low-voltage relays start the high-voltage power-consuming devices.
- The voltmeter is used to trace the actual voltage at various points in a circuit.
- The ohmmeter is used to check for continuity in a circuit.

- The ammeter is used to measure current flow.
- The pictorial diagram has wire colors and destinations printed on it. It shows the actual locations of all components.
- The line or ladder diagram is used to trace the circuit to understand its purpose.

REVIEW QUESTIONS

1. Name three types of automatic controls.
2. Name two categories of electrical controls.
3. Name the circuit that is energized with a thermostat terminal designated Y.
4. Name the circuit that is energized with a thermostat terminal designated G.
5. When a thermostat terminal is designated R what name is it known by?
6. Is the cold anticipator in parallel or in series with the cooling contacts in a thermostat?
7. True or False? The indoor fan must always run in the cooling cycle.
8. True or False? When the control transformer is not working, the indoor fan will not run in the FAN ON position.
9. What switch should be turned off when working on the compressor circuit?
10. True or False? The pictorial wiring diagram is used to locate components in the unit.
11. True or False? The line diagram is used to follow and understand the intent of the circuit.
12. How can amperage be measured in a low-voltage circuit with a clamp-on ammeter?

12 ELECTRONIC AND PROGRAMMABLE CONTROLS

OBJECTIVES

Upon completion of this unit, you should be able to

- List several applications of electronic controls.
- Explain the advantages of electronic controls.
- Describe the electronic control boards used for air conditioning circuits.
- Recognize and troubleshoot a basic electronic control circuit board.
- Describe programmable thermostats.

12.1 ELECTRONIC CONTROLS

For many years, electronic controls have been used to control larger equipment. The development of these controls has now reached the point where they are feasible for use in residential and light commercial equipment. These controls are economical, reliable, and provide efficient energy management for basic or sophisticated control systems.

Electronic controls serve the same purposes as many electric and electromechanical controls. These include operating and safety functions as well as energy management applications. These controls are normally provided on circuit boards with terminal strips for external circuit connections. Regardless of the number of components on a circuit board, many manufacturers design the boards as though they were a single control. When an individual component on a circuit board is faulty, the entire board is generally removed and replaced. Most circuit boards are not field-serviceable. For troubleshooting purposes, this control board is normally treated like a switch. Recall that a switch is a power-passing device where power enters one side and exits the other side (Figure 12-1). Many circuit

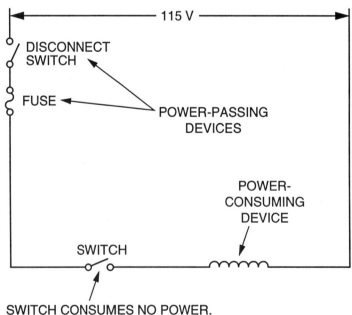

Figure 12-1. A switch is a power-passing device.

170

ELECTRONIC CONTROLS   171

boards are treated the same way when it comes to troubleshooting; under normal conditions, power in should provide power out (Figure 12-2). One difference is that a circuit board requires a small amount of power (normally 24 V or less) to operate the electronics within the board. This 24 V power supply is often converted to DC.

One advantage of electronic controls is that they often contain built-in diagnostic features. Some equipment manufacturers furnish a special module that can be plugged into the control circuit to locate problems. The module may read out in numbers that equate to a particular fault (Figure 12-3).

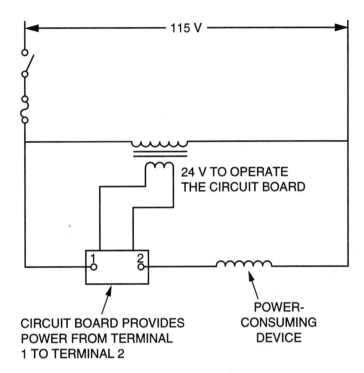

Figure 12-2. The circuit board can often be treated like a switch—power in should provide power out.

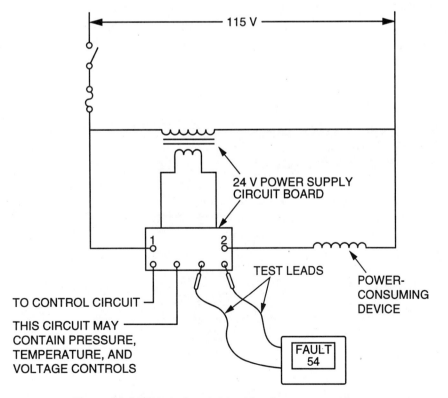

Figure 12-3. This fault code identifies the system problem.

Following are examples of electronic controls in residential, light commercial, and central system air conditioning applications.

12.2 AIR CONDITIONING APPLICATIONS

Electronic control of air conditioning systems is becoming increasingly popular. There are many functions that can be accomplished using electronics and even internal computer circuits that cannot be accomplished using electric or electromechanical controls. Even residential air conditioning equipment is now being controlled using electronic circuit boards. Figure 12-4 shows a small circuit board used in a central air conditioning system.

As equipment becomes larger, it also becomes more expensive to operate and owners are more aware of the energy savings that can be provided by using electronic controls. Energy management is very important for all commercial buildings as the air conditioning and lighting are the two most expensive utilities for the summer months. It is important that the cooling system only operate when needed and at the capacity needed. It makes no sense to operate a system at full load when part load will cool the structure. Electronics can be used to control the equipment as well as protect it from damage should problems occur. In addition, some energy management systems now offer remote building control. For example, the electronic circuits only require that small wires be routed to a remote location to control large electrical loads that may be at quite a distance (Figure 12-5). If the distance becomes too great, telephone lines may be used to control equipment from anywhere in the country. Many large buildings are controlled via telephone lines from remote locations where energy management companies monitor the buildings for the best efficiency.

Some of the applications of electronic controls include:

- Voltage monitoring
- Current monitoring and power consumption
- Phase protection
- Anti-recycle (short cycle) control
- Pressure and temperature controls
- Programmed operation
- Refrigerant flow control

Voltage Monitoring

Some manufacturers use electronic controls to monitor the voltage being supplied to a unit. Electronic monitoring can prevent the compressor from operating when the voltage is too high or too low, either of which can damage the equipment. Remember that a motor cannot be

Figure 12-4. Small circuit board used for a central air conditioning system. *Reproduced courtesy of Carrier Corporation.*

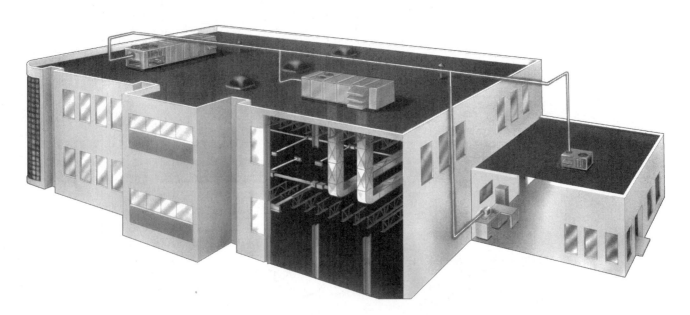

Figure 12-5. Remote control system using a computer to control the equipment. *Reproduced courtesy of Carrier Corporation.*

allowed to operate at voltages that go beyond the range of ±10% of the nameplate voltage. Electronic devices can measure these voltages very accurately.

All large systems operate using three-phase power supplies. A three-phase power supply is one that has three hot legs: Lines 1, 2, and 3 (Figure 12-6). Each line may supply a slightly different voltage. If the voltages differ markedly, it is known as a *voltage imbalance*. The voltages must balance between each of the legs in a three-phase system for the compressor or any other three-phase motor to be furnished with the correct power supply. With single-phase power, it is not noticed when there is a voltage reduction to one leg of the power supply because it only causes a reduction in the line voltage to a motor with one winding (Figure 12-7). With three-phase power, the voltage to each phase of the motor must be within certain limits. Typically, a phase imbalance of more than 2% is considered unacceptable. The % phase imbalance is calculated by taking the voltages from phase to phase and averaging them and then dividing the maximum deviation by the average voltage. For example, suppose the voltages are as follows:

L1 to L2 = 241 V
L1 to L3 = 233 V
L2 to L3 = 236 V

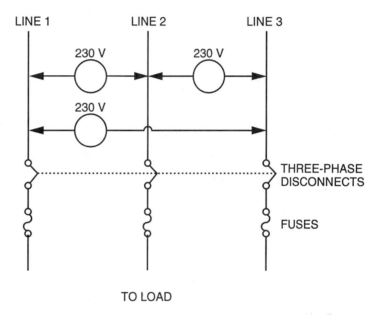

Figure 12-6. A three-phase power supply.

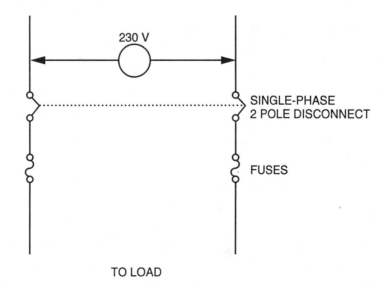

Figure 12-7. A single-phase power supply does not disclose a load imbalance.

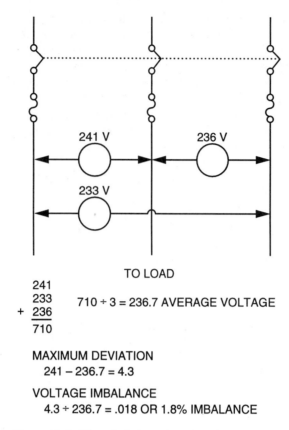

Figure 12-8. Phase imbalance in a three-phase circuit.

The average voltage is 236.7 (241 + 233 + 236 = 710 ÷ 3 = 236.7 V). The maximum deviation from the average is 4.3 volts (241 - 236.7 = 4.3 V). The voltage imbalance is 1.8% (4.3 ÷ 236.7 = .018 or 1.8%). Figure 12-8 shows the actual problem and how it is detected using field meters.

Voltage imbalance can be detected using electronic circuits and the equipment can be shut down before damage occurs.

Current Monitoring

Overload protection can be readily obtained with electromechanical devices, but low-voltage and high-voltage monitoring and protection cannot. However, electronic monitoring of current draw (amperage) can be very closely monitored using a device called a *current transformer*. This device is a type of step down transformer. It consists of a wire wound device that reduces the current down to a fraction of the actual line current. For example, a current transformer with a 100 to 1 ratio can be placed in a compressor circuit that draws 100 amps at full load and the monitoring amperage can be reduced to 1 amp for use in the electronic circuit to protect the compressor (Figure 12-9).

The current transformer can also be used to monitor the actual power consumption of the equipment. This small current can be coupled with the voltage to read out the power draw in watts. As discussed earlier, the equation for power is: Power (P) = Amperage (I) × Voltage (E).

Actually, there is more to this than just I × E because the power factor of the equipment must be considered in the equation, but this can be entered into the energy data for the equipment. This is a useful value for a large building where it is desirable to maintain close control of the building wattage (expressed in kilowatts). Most large consumers of electricity closely monitor their kW usage because they are usually billed by a process called *demand metering*. Demand metering is a process by which the power company charges a commercial customer for the month based on the highest power draw for a measured time period, usually 15 to 30 minutes. For example, a building cooling system may be operated at full load for 30 minutes on the first billing day of the month and may not require full load operation for the rest of the month. The customer's monthly bill will be based on this 30 minutes of full load operation on the first day. An energy management company with proper computer control may have the long-term weather forecast programmed into its computers and hold the building's cooling unit back to part load and reduced kW for that one day, thereby reducing the power bill for the month. This can be a great savings for the building owner.

Phase Protection

The rotation of all motors that operate using three-phase power is determined by the connections to the power supply, and it is important to replace these connections

AIR CONDITIONING APPLICATIONS

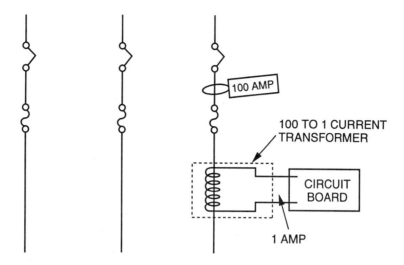

Figure 12-9. A current transformer is used to monitor current draw.

carefully to avoid reversing the phases. For example, a typical motor is connected by the following method: Line 1 is connected to the motor terminal labeled Terminal 1; Line 2 is connected to Terminal 2; and Line 3 is connected to Terminal 3 (Figure 12-10). If any two wires are reversed, the motor rotation will also be reversed (Figure 12-11). Reverse rotation can cause serious damage in some applications. For example, some large compressors can be severely damaged if they are started in reverse rotation, and scroll compressors will not function in reverse. Phase protection can be accomplished using electronic devices. These devices prevent the unit from starting if leads have been reversed, and sometimes use a light or other indicator to warn the operator of power reversal.

Anti-Recycle or Time Delay

The anti-recycle or time delay feature is easily accomplished with electronic circuits. It is used to keep the unit from short cycling. The time delay before a compressor restart (typically five minutes for a residential unit) is used to allow the pressures to equalize between the low side and the high side before the compressor is restarted. Most residential air conditioning units use fixed-bore metering devices, such as the capillary tube, that equalize during the off cycle.

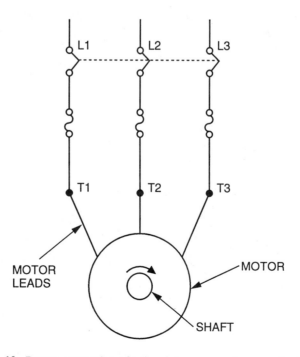

Figure 12-10. Proper connection of a three-phase motor for correct rotation.

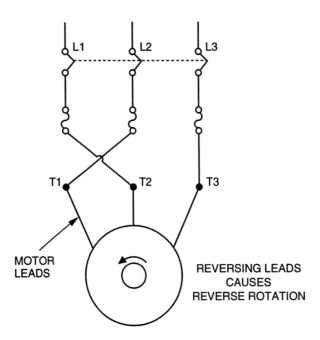

Figure 12-11. Reversing the motor leads will cause reverse rotation of the motor and may lead to motor damage.

The time delay keeps the unit from coming right back on after a low-pressure cutout or when the occupant repeatedly cycles the equipment by raising and lowering the setpoint temperature at the thermostat.

Note: Most thermostatic expansion valves do not equalize completely during the off cycle, so short cycling was not as much of a problem when that valve was used exclusively.

Larger compressors can also be damaged by short cycling. These compressors generate a large amount of heat on start-up, and the motor must run for a certain period of time to dissipate this heat. If the motor is restarted too soon, internal heat will build up and can cause motor damage. For example, it is typical for a manufacturer to only allow a large motor such as a centrifugal compressor to be started twice per hour; therefore, a 30-minute time delay between starts is required. This anti-recycling feature can be accomplished using the electronics in the control circuit. Short cycling can occur when an operator tries to re-energize the equipment too soon or from various control short cycles, such as low-pressure controls.

Pressures and Temperatures

The various pressures in the refrigeration process must be maintained for proper performance. For example, the suction pressure for an R-22 system is 68.5 psig. The refrigerant is boiling in the evaporator at 40°F at this pressure. At this time, the suction line temperature should be approximately 50°F because the evaporator should have 10°F of superheat (Figure 12-12). If conditions in the

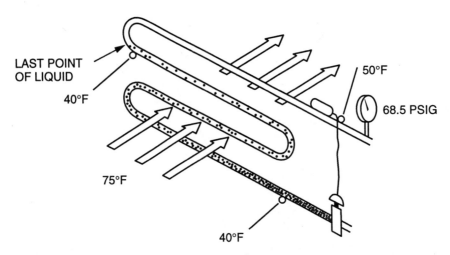

Figure 12-12. Pressures and temperatures for a properly operating evaporator.

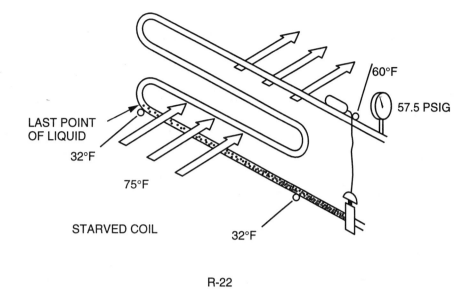

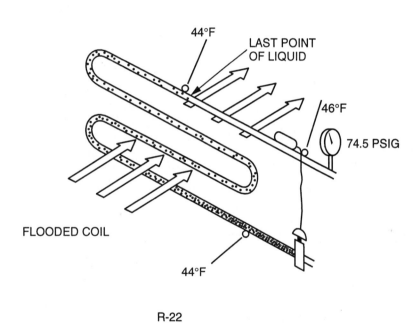

Figure 12-13. Starved and flooded evaporator conditions that may be monitored using sensing elements.

evaporator vary significantly from these design conditions, a problem has occurred. For example, if the expansion device is defective, the superheat may be 20°F instead of 10°F and the evaporator would be starving for refrigerant (Figure 12-13). On the other hand, if the superheat is 2°F, a flooded coil condition may be occurring. Electronic circuits and solid state controls can help to diagnose the above problems and many other problems.

A pressure transducer can be used to register the pressure and convert it to usable terms for an electronic circuit application. A pressure transducer is a device that changes pressure differences into electronic variables that can be monitored by an electronic circuit. This is accomplished by means of a variable capacitor that responds to pressure differences. As the pressure increases, the two plates of the capacitor move closer together, creating a higher capacitance (Figure 12-14). This difference in capacitance can be used to determine actual pressures (Figure 12-15) or to find pressure differences that would cause system problems. When the pressure is monitored using a transducer and the temperature is monitored using a thermistor, diagnostic conclusions can be arrived at for system troubleshooting. For example, if the temperature of the suction line should be 50°F when the suction

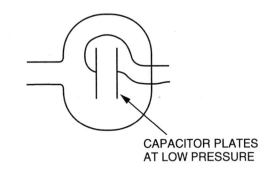

CAPACITOR PLATES AT LOW PRESSURE

PRESSURE MOVES CAPACITOR PLATES CLOSER TOGETHER, CHANGING THE CAPACITANCE

Figure 12-14. How a transducer functions.

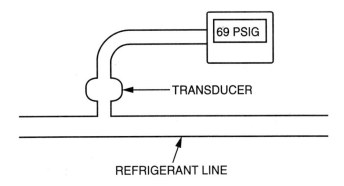

Figure 12-15. A transducer and an electronic device to read out in pressure.

pressure is 68 psig for an air conditioning system using R-22, this condition can be programmed into the computer in the unit. If these conditions are not met after a reasonable running time, the computer in the electronic circuit can declare a fault (starved evaporator or flooded evaporator) and a fault light can be used to alert the operator.

Many different conditions can be programmed into the computer circuit and the circuit board is used to execute various responses by stopping, starting, modulating, and recording the various system functions.

Programmed Operation

Commercial building owners only want their buildings to be conditioned when they are occupied, which is typically from 8:00 am to 5:00 pm during typical working hours. If it is left up to a person to turn the system on and off, it will be mishandled part of the time. The operator may not start the unit on time or may forget to turn the unit off at the end of the day. Building management must also deal with holidays and time changes from daylight to standard time. Programmed operation using electronic circuits can offer various programs that will stop, start, and even modulate the operation of the cooling system for the best efficiency. For systems with multiple stages of cooling capacity, electronic controls can provide the ability to continuously monitor the indoor and outdoor conditions and stage the sequence of operation to achieve optimum efficiency and energy savings. The building indoor and outdoor lighting may also be included in the same package. The need for outdoor lighting varies from one month to the next as the seasons change. Since energy management systems include all of the above, the air conditioning technician often gets involved in the programming.

Refrigerant Flow Control

In the past, most refrigeration equipment used a thermostatic expansion valve to control refrigerant flow to the evaporator. Today, many systems use electronic expansion valves because they offer better refrigerant flow control. It is desirable for the evaporator to have as much liquid as possible and still have superheat at the end of the coil. The less superheat the better. Since a thermostatic expansion valve has an external thermal contact with the refrigerant suction line, a sizable margin of safety must be used, typically 8°F to 10°F of superheat. The electronic expansion valve shown in Figure 12-16 is used for chilled-water applications (discussed in a later unit). It uses two thermistors to measure the refrigerant temperature. One thermistor is in the boiling refrigerant and the other is located in the compressor where the suction refrigerant gas enters the cylinders. This expansion valve has 760 steps of metering for very close control. By maintaining 20°F of superheat at the cylinders, after the suction gas has passed over the motor windings, a very low superheat is maintained in the evaporator, approximately 3 to 5°F. This allows the chiller to function at peak efficiency.

The electronic expansion device can also be used to throttle the refrigerant entering the chiller on start-up to prevent the compressor from becoming overloaded. This could be due to chiller water that is too hot and would overload the compressor.

Many applications for electronic controls have been developed and more will be developed in the future. The professional technician will stay up to date on new controls and equipment.

TROUBLESHOOTING ELECTRONIC CONTROLS 179

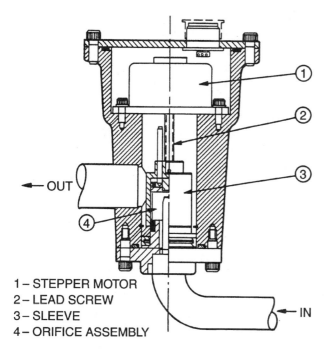

1 – STEPPER MOTOR
2 – LEAD SCREW
3 – SLEEVE
4 – ORIFICE ASSEMBLY

Figure 12-16. Electronic expansion valve. *Reproduced courtesy of Carrier Corporation.*

12.3 ELECTRONIC THERMOSTATS

Electronic thermostats provide variable control at a reasonable price. This makes them attractive to the small user who wants both comfort and economy. They can be easily programmed to stop and start the air conditioning system at predetermined times (Figure 12-17). These thermostats can cut the system back in the home during the day while people are at work, or during the night, when occupants are sleeping, and bring the home back to the desired temperature just before the residents get up in the morning or arrive home. Some thermostats have a seven-day program that allows a different schedule for each day of the week. However, most users generally need only two programs, one for weekdays and the other for weekends. The program schedule can be altered when necessary. It will automatically return to the original program at the next scheduled change.

The temperature control element in many electronic thermostats has a faster response time than the typical bimetal type because of the mass of the bimetal sensing element. Electronic thermostats often use a thermistor element, which is lighter than a bimetal device and therefore responds more quickly.

When troubleshooting electronic thermostats, follow the manufacturer's instructions to check the program to make sure that the problem is not due to operator error. If the program is correct and the thermostat does not function as designed, it will probably have to be replaced.

12.4 DIAGNOSTIC THERMOSTATS

Advanced control systems now offer thermostats that act as complete diagnostic centers. Sensors are mounted in strategic locations in the system to monitor pressures and temperatures. When the unit experiences difficulty, the computer sends a signal to the thermostat, which then displays the problem(s).

Many commercial and industrial installations are now using central system control. These systems link several thermostats to a main computer terminal, which can then provide a readout of potential problems. This data can be relayed to the technician before he leaves the shop. As this becomes more common, it will help the technician to carry the correct repair parts to the job site and also help the company owner because every service truck will not have to carry such a large inventory. For example, imagine how advantageous it would be for everyone concerned if the electronics in the system told the technician that the fan motor had an open circuit so that a fan motor could be carried on the first trip.

12.5 TROUBLESHOOTING ELECTRONIC CONTROLS

Troubleshooting electronic circuit boards is much like troubleshooting a single component in the circuit. See Figure 12-18 for an electronic thermostat and subbase. The low-voltage power supply is doing the same thing as in a conventional circuit in that it is still working to supply power to the power-consuming control circuit. The hot leg of the transformer goes directly to the thermostat, and the thermostat distributes this hot leg of power to the individual components just as in a conventional thermostat. There is still a compressor contactor and fan relay to energize.

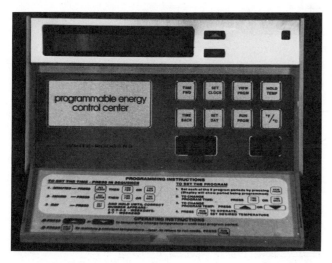

Figure 12-17. Electronic thermostat that is programmable for energy management. It can be used in any 24 volt circuit for single-stage cooling and single-stage heating. *Photo by Bill Johnson.*

180 UNIT 12 ELECTRONIC AND PROGRAMMABLE CONTROLS

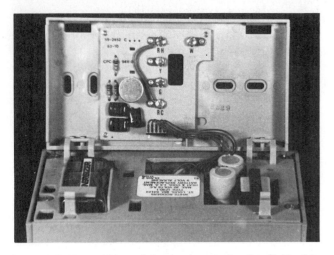

Figure 12-18. Subbase of the thermostat showing field wiring connections. *Photo by Bill Johnson.*

When a technician has a problem with a circuit and believes that it may be the circuit board, the first thing to do is to check the manufacturer's troubleshooting procedures. This may be attached to the unit or you may need to place a call to the distributor of the product. If your problem occurs at a time when research is not practical, you may have to proceed on your own. First look for any warnings on the circuit board or in the compartment. Then determine from the wiring diagram which wire may be carrying the signal to the circuit board and which wire may be used to carry the signal from the circuit board to the power-consuming device, usually the compressor contactor or fan relay. For example, on an air conditioning circuit board, the wire feeding the low-voltage signal to the board may be the Y wire, which may feed the circuit through the board's electronic circuit to the compressor contactor coil. In some cases, a jumper across this circuit may start the unit. In this case, remove the jumper immediately as you have proven the board to be defective (Figure 12-19).

Larger systems using electronic circuit boards may have very detailed checkout procedures, so always follow the manufacturer's instructions when troubleshooting. If there are no directions with the unit, a call to the manufacturer can often solve the problem. Diagrams and directions can be sent over the phone lines by means of FAX machines to remote locations. Do not hesitate to call on the manufacturer for help.

Troubleshooting the Electronic Thermostat

The following is an example of how a typical programmable thermostat for combination cooling and heating can be diagnosed for some common problems. This could be considered a replacement for a standard heat/cool thermostat.

Remember that the hot terminal feeding the thermostat provides the power to the other terminals.

The circuit signal enters the thermostat at the RH terminal (Figure 12-20). A jumper caries the power to terminal RC for cooling. If there is a problem with the circuit and the thermostat is suspected, a jumper applied between RC and any other circuit will prove the thermostat is functioning. If the system will start with the jumper in place but will not start without it, the thermostat is defective. Remember, there may be a time delay built into the thermostat, so be sure to allow five minutes for the unit to start.

When the existing unit has two different power supplies, the transformers must be in phase, or they will oppose each other (Figure 12-21). Isolation relays are

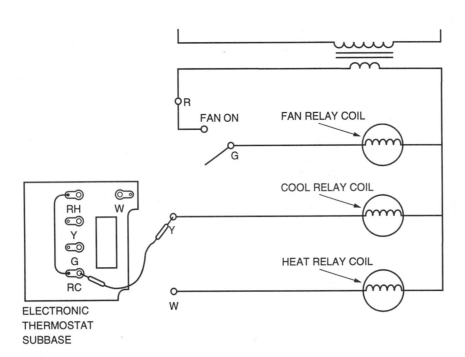

Figure 12-19. Jumping the circuit board to verify that the circuits are good and the board is defective.

TROUBLESHOOTING ELECTRONIC CONTROLS 181

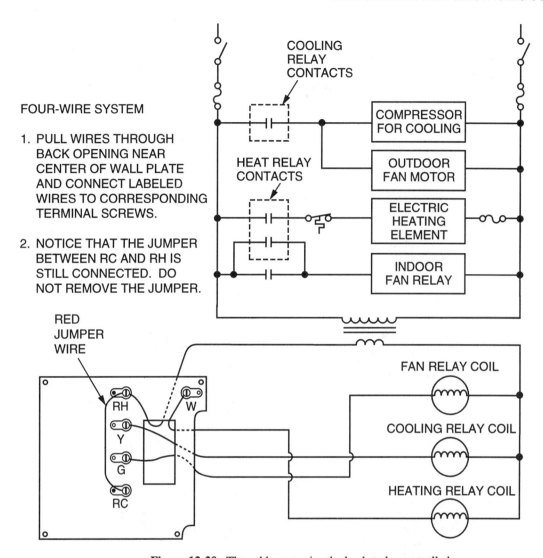

Figure 12-20. The subbase carries the loads to be controlled.

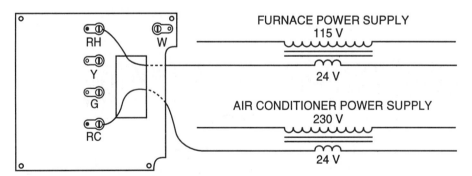

FIVE-WIRE SYSTEM WHERE SEPARATE TRANSFORMERS ARE USED ON HEATING SYSTEM AND COOLING SYSTEM.
1. REMOVE AND DISCARD THE RED JUMPER WIRE THAT CONNECTS TERMINALS RH AND RC ON WALL PLATE.
2. PULL WIRES THROUGH BACK OPENING NEAR CENTER OF WALL PLATE AND CONNECT LABELED WIRES TO CORRESPONDING TERMINAL SCREWS.

Figure 12-21. A thermostat application with two 24 V power supplies. One power supply comes to the RH (heat) terminal and the other to the RC (cool) terminal. The thermostat subbase keeps the two power supplies separate. If the two power supplies are phase checked, it is permissible to wire them together. Phase checking involves making sure that the primaries and the secondaries of the transformers are parallel (in phase).

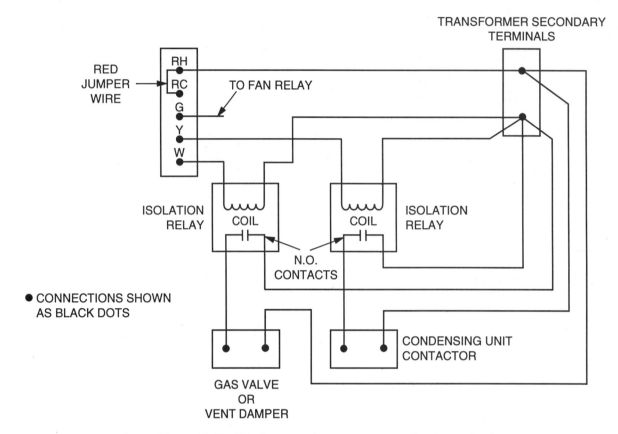

Figure 12-22. This thermostat has two power supplies that are in phase.

often used to prevent unwanted electrical feedback that can cause erratic operation (Figure 12-22).

Power Outage and the Electronic Thermostat

Manufacturers use various methods to prevent program loss due to power failures. For example, the thermostat in Figure 12-23 is backed up by a battery that prevents program loss for up to three days. When an extended power failure occurs, the program is likely to be lost. When the program is lost, the thermostat will default to a factory-set program until reprogrammed by the owner. The owner's manual describes how to program the unit. Most thermostats maintain an acceptable setpoint temperature with the default program until the program is reestablished. This protects the premises from overheating or overcooling.

12.6 TECHNICIAN SERVICE CALLS

The following section provides sample service calls to help build your troubleshooting skills.

Service Call 1

A customer calls and tells the company dispatcher that the unit in her house is not cooling. The problem is that a wire is burned off the compressor contactor due to a bad connection (Figure 12-24).

The technician arrives and talks to the customer about the problem, then goes to the indoor thermostat, which is a programmable type. The indoor fan is switched to ON. It runs. The technician then switches the thermostat to call for cooling. The indoor fan comes on again. This proves that the control voltage and thermostat are trying to operate the equipment.

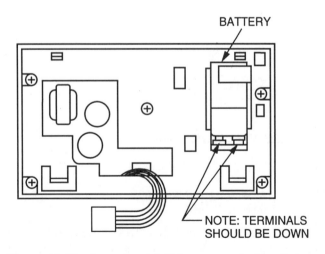

Figure 12-23. An optional 9 V battery can be installed as a backup for power failure.

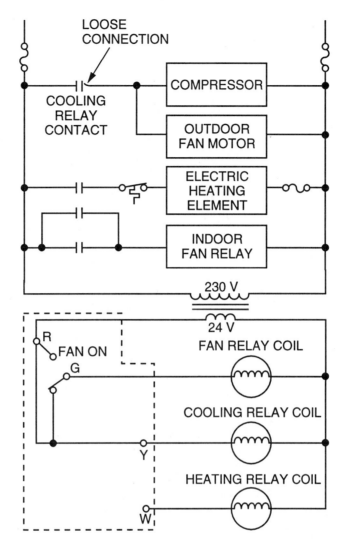

Figure 12-24. A loose connection at the cooling relay terminal does not allow power to get to the compressor and outdoor fan motor. The coil is energized because it is humming. This represents what happens to the electronic circuit on a call for cooling. R passes power to G and Y, but the compressor will not start.

The technician then goes to the outdoor unit. The cooling relay is humming. This means that the low-voltage call for cooling is reaching the outdoor unit. When the contactor is humming, it is energized and is trying to start the compressor.

The power is turned off and the cover to the compartment and controls is removed. When the cover is removed, the power is turned on. The cooling relay is energized and still humming. Line power has to be reaching the unit because it is the power source for the control transformer. Line power is checked into and out of the compressor contactor. Power is going in but is not coming out. The relay has a bad connection. The power is turned off and the connection repaired. When the power is restored, the unit starts and runs normally.

The technician explains the problem to the customer, fills out the paperwork, and leaves for another job.

Service Call 2

A customer calls and complains that the cooling is not operating in a small retail craft shop in a shopping center. The store has a split system air conditioner with the condenser located behind the store. The system will not cool because the electronic programmable heat/cool thermostat will not pass power (Figure 12-25).

The technician goes to the indoor thermostat and switches on the indoor fan. It starts. This proves that the line voltage is reaching the indoor unit because the 24 V control transformer is located in the indoor unit. The thermostat is switched to call for cooling. Neither the indoor fan nor the outdoor unit starts.

The technician goes to the outdoor unit, removes the panel to the low-voltage terminal block, and checks for voltage at the cooling relay coil. There is no voltage.

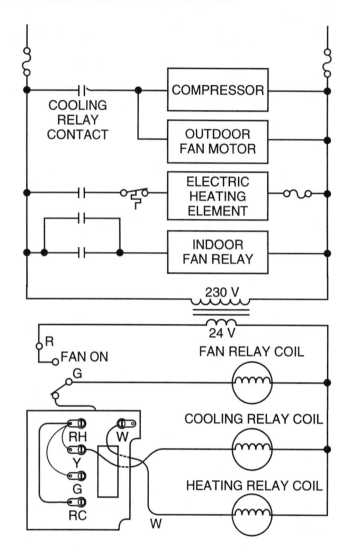

Figure 12-25. The 24 V control voltage is going into the thermostat but will not come out on a call for cooling. The thermostat is removed from the subbase, a jumper is attached from R to G and the indoor fan starts. Another jumper is attached from R to Y and the outdoor unit starts. The thermostat is defective.

The technician returns to the indoor thermostat and checks the program according to the manufacturer's instructions. The program is correct. The thermostat is removed from the subbase. A jumper is placed on the R terminal and the G terminal. The indoor fan starts. The jumper is left on the fan circuit and another is placed between the R terminal and the Y terminal. The outdoor fan and compressor start and run with the indoor fan. Obviously, the thermostat is not passing power and needs to be replaced.

The technician obtains another thermostat and changes it out. The system is then switched to cool, and after a short time delay, the condensing unit starts cooling. While the technician is there, the complete control circuit is checked, including the fan and heating cycles. It all functions correctly.

The store owner asks the technician to program the thermostat for the typical store hours, 8:00 am to 5:00 pm, Monday through Saturday and off on Sunday.

After this is done, the paperwork is completed and the technician moves to another job.

SUMMARY

- Always review the instructions provided by the manufacturer before starting any installation or service job. The distributor will also be helpful if you have questions.
- Electronic controls are being used more frequently because they are reliable and economical.
- Electronic controls are basically used to replace electrical and electromechanical controls where they are applicable.

- Electronic controls are used for safety and operating functions, as well as energy management and equipment efficiency.
- Electronic controls can monitor high and low voltages, current flow and voltage imbalance, and easily add time delays and sequence of operation to the control system.
- The residential and light commercial air conditioning control circuit may use electronics to monitor high and low voltages, time delays, and the current draw of the compressor.
- The electronic programmable thermostat is becoming a popular alternative to traditional thermostats.
- Troubleshooting electronic controls is similar to troubleshooting electromechanical controls because each circuit board is normally treated as a single component—power in should provide power out.

REVIEW QUESTIONS

1. True or False? Electronic controls are normally more economical than electromechanical controls.
2. True or False? Electronic thermostats react more quickly to temperature changes than bimetal thermostats.
3. True or False? Voltage imbalance is not as important in three-phase motors as it is in single-phase motors.
4. True or False? An anti-recycle feature prevents a motor from restarting too quickly.
5. True or False? Electronic programmable thermostats can be programmed for only one program.
6. True or False? An electronic programmable thermostat can normally be installed in place of a traditional thermostat.
7. True or False? Electronic circuits can generally be repaired in the field.
8. True or False? Electronic programmable thermostats are designed so that the owner can reprogram them.
9. True or False? A qualified technician should be called to troubleshoot electronic circuit problems.
10. A voltage imbalance is suspected in a three-phase system. The voltage to each phase is recorded as follows: L1 to L2 = 228 V, L1 to L3 = 240 V, and L2 to L3 = 235 V. What is the % voltage imbalance?

13 ELECTRIC MOTORS

OBJECTIVES

Upon completion of this unit, you should be able to

- Describe the different types of open single-phase motors used to drive fans, compressors, and pumps.
- Describe the applications of the various types of motors.
- State which motors have high starting torque.
- List the components that cause a motor to have a higher starting torque.
- Describe a multispeed PSC motor and indicate how the different speeds are obtained.
- Explain the operation of a three-phase motor.
- Describe a motor used for a hermetic compressor.
- Explain the motor terminal connections in various compressors.
- Describe the different types of compressors that use hermetic motors.
- Describe the use of variable speed motors.

13.1 ELECTRIC MOTOR APPLICATIONS

Electric motors are used to turn the fans, pumps, and compressors that are the prime movers of air, water, and refrigerant (Figure 13-1). There are several types of motors available, each with its particular application. For example, some applications require motors that will start under heavy loads and still develop their rated work horsepower at a continuous running condition, while others are used in installations that do not require a great deal of starting torque but must develop their rated horsepower under a continuous running condition. Some motors run for years in dirty operating conditions, while others operate in the relatively sterile refrigerant atmosphere. The technician must understand which motor is suitable for each job so that effective troubleshooting can be accomplished and, if necessary, the motor replaced by the proper type. We will begin by reviewing the basic operating principles of an electric motor. Although there are many types of electric motors, most motors operate on similar principles.

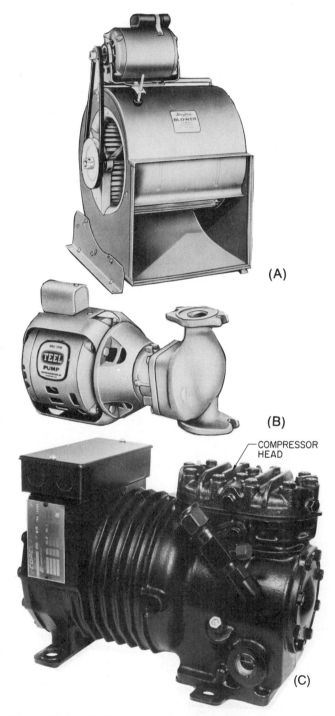

Figure 13-1. (A) Fans move air. (B) Pumps move water. (C) Compressors move vapor refrigerant. *Courtesy of: (A and B) W.W. Grainger, Inc., (C) Copeland Corporation.*

13.2 MOTOR COMPONENTS AND OPERATION

Electric motors have a stator with windings, a rotor, bearings, end bells, housing, and some means to hold these parts in the proper position (Figure 13-2).

Electric Motors and Magnetism

Most motors operate on the principle of electromagnetism. As you will recall, unlike poles of a magnet attract each other, and like poles repel each other. For example, if a stationary horseshoe magnet is placed with its two poles (north and south) at either end of a free-turning magnet, one pole of the free rotating magnet will line up with the opposite pole of the horseshoe magnet (Figure 13-3). If the horseshoe magnet were an electromagnet and the wires on the battery were reversed, the poles of this magnet would reverse and the poles on the free magnet would be repelled, causing it to rotate until the unlike poles again were lined up. This is the basic principle of electric motor operation. The horseshoe magnet is the stator, and the free rotating magnet is the rotor.

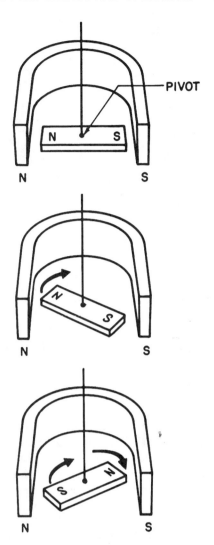

Figure 13-3. The north and south poles on the rotating magnet will line up with the opposite poles on the stationary magnet.

Run Winding

In a two-pole split-phase motor, the stator has two poles with an insulated wire winding called the *run winding*. When an electrical current is applied, these poles become an electromagnet with the polarity changing constantly. In normal 60-cycle operation, the polarity changes 60 times per second.

The rotor may be constructed of bars (Figure 13-4). This is known as a *squirrel cage rotor*. When the rotor shaft is placed in the bearings in the bell type ends, it is positioned within the coil of the run winding. When an alternating current is applied to the run winding, a magnetic field is produced in the winding and a magnetic field is also induced in the rotor. The bars in the rotor actually form a coil. This is similar to the field induced in a transformer secondary by the magnetic field in the transformer primary. The field induced in the rotor has a polarity opposite to that in the run winding.

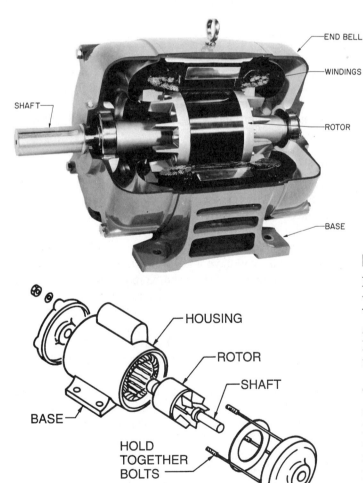

Figure 13-2. Cutaway of an electric motor. *Courtesy Century Electric, Inc.*

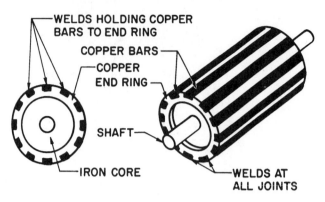

Figure 13-4. Simple sketch of a squirrel cage motor.

The attracting and repelling action between the poles of the run winding and the rotor sets up a rotating magnetic field and causes the rotor to turn. Since this is alternating current that reverses 60 times per second, the rotor turns, in effect "chasing" the changing polarity in the run winding.

Start Winding

A start winding is placed between the run windings and the stator to ensure that the rotor starts properly and that it turns in the desired direction (Figure 13-5). The start winding has more turns than the run winding and is constructed of a smaller-diameter wire. This produces a larger magnetic field and greater resistance, which helps the rotor to start turning and determines the direction of rotation. The different positions and resistances of the two windings creates a phase shift between the two currents; therefore, motors of this type are known as split-phase motors.

We have just described a two-pole split-phase induction motor, which is rated to run at 3600 revolutions per minute (rpm). When the motor reaches approximately 75% of its normal speed, a device such as a centrifugal switch is used to open the circuit to the start winding and the motor continues to operate on only the run winding.

Determining Motor Speed

The following formula can be used to determine the synchronous speed (without load) of motors.

$$\text{Speed (rpm)} = \frac{\text{Frequency} \times 120}{\text{Number of Poles}}$$

Frequency is the number of cycles per second (also called *Hertz*).

Note: The magnetic field builds and collapses twice each second (each time it changes direction); therefore, 120 is used in the formula instead of 60.

Speed of two-pole split phase motors:

$$\frac{60 \times 120}{2} = 3600$$

Speed of four-pole split phase motors:

$$\frac{60 \times 120}{4} = 1800$$

The actual speed under load of each motor will always be less than the synchronous speed. For the above motors, the actual speed will be approximately 3450 rpm and 1750 rpm, respectively. The difference between synchronous speed and actual speed is called slip. Slip is caused by the load.

Starting and Running Characteristics

Two major considerations of electric motor applications are the starting and running characteristics. A motor applied to a refrigeration compressor must have a high starting torque; that is, it must be able to start under heavy starting loads. For example, a refrigeration compressor may have a head pressure of 155 psig and a suction pressure of 5 psig and be required to start in systems where the pressures do not equalize (Figure 13-6). The pressure difference of 150 psig is the same as starting the compressor with a resistance of 150 psi of piston area. If this compressor has a 1-inch diameter piston, the area of the piston is 0.78 in.^{2} (A = πr^2 = 3.14 × 0.5 × 0.5). When this area is multiplied by the pressure difference of 150 psi, it provides the starting resistance of the motor (117 pounds). This is similar to a 117-pound weight resting on top of the piston when it tries to start.

To start a small fan, a motor does not need as much starting torque. The motor must simply overcome the friction needed to start the fan moving. There is no pressure difference because the pressures equalize when the fan is not running.

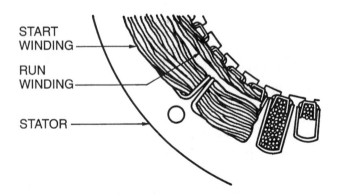

Figure 13-5. Placement of the start and run windings inside a stator.

MOTOR COMPONENTS AND OPERATION   189

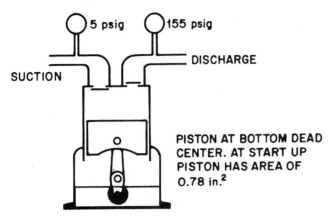

Figure 13-6. Compressor with a high-side pressure of 155 psig and a low-side pressure of 5 psig.

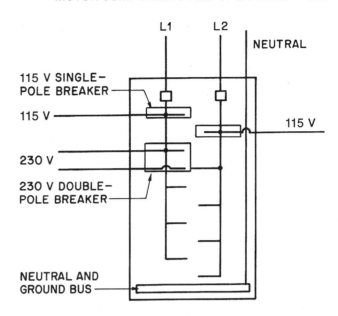

Figure 13-7. Main breaker panel for a typical residence.

Power Supplies for Electric Motors

One other basic difference in electric motors is the power supply that is used to operate them. For example, only single-phase power is available to homes; therefore, motors designed to operate on single-phase power must be used. For large loads in commercial buildings, single-phase power is inadequate, so three-phase power is used. The difference in the two power sources changes the starting and running characteristics of the motors.

The power supply for most single-phase motors is either 115 V or 208-230 V. Residences normally use both types of power (Figure 13-7). For example, a home furnace may have a power supply of 115 V, while the air conditioner uses a power supply of 230 V. A commercial building may have either 230 V or 208 V, depending on the power company. Some single-phase motors are dual voltage. In this case, the motor has two run windings and one start winding. The two run windings have the same resistance and the start winding has a higher resistance. The motor will operate with the two run windings in parallel in the low-voltage mode. When it is required to run in the high-voltage mode, the technician changes the numbered motor leads according to the manufacturer's instructions. This wires the run windings in series with one another and delivers an effective voltage of 115 V to each winding. It can be said that the motor windings are actually only 115 V because they only operate on 115 V, no matter which mode they are in. The technician can change the voltage at the motor terminal box (Figure 13-8).

Some commercial and industrial installations may use a 460 V power supply for large motors. The 460 V may then be reduced to a lower voltage to operate smaller motors. The smaller motors may be single phase and must operate from the same power supply (Figure 13-9).

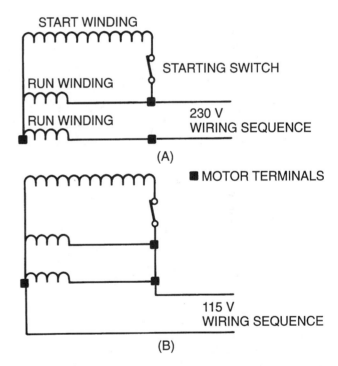

Figure 13-8. Wiring diagram of a dual-voltage motor. It is made to operate using 115 V or 230 V, depending on how the motor is wired. (A) 230 V wiring sequence. (B) 115 V wiring sequence.

A motor can rotate either clockwise or counterclockwise. Some motors are reversible from the motor terminal box (Figure 13-10).

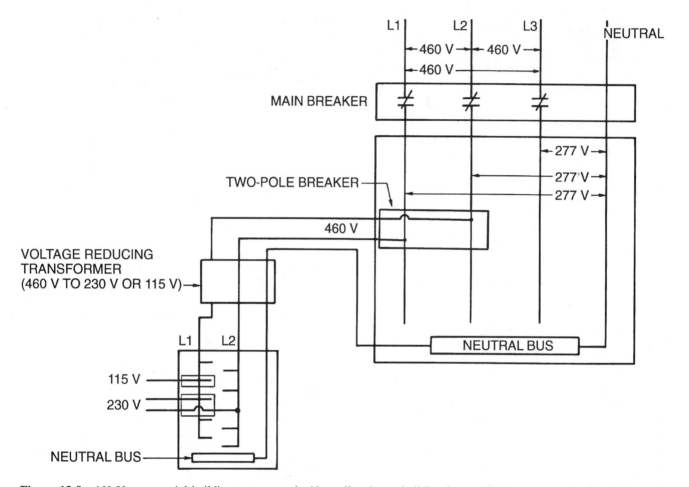

Figure 13-9. 460 V commercial building power supply. Normally when a building has a 460 V power supply, it will have a step-down transformer to produce 115 V for office machines and small appliances.

Figure 13-10. A single-phase motor can be reversed by changing the connections in the motor terminal box. The direction the motor turns is determined by the start winding. This can be shown by disconnecting the start winding leads and applying power to the motor. It will hum and will not start. The shaft can be turned in either direction and the motor will run in that direction.

13.3 MOTORS USED IN A/C APPLICATIONS

This section describes some of the motors currently used in the cooling industry. This text emphasizes the electrical characteristics rather than the working conditions. There are some older motors in operation, but they will not be discussed in this text.

Split-Phase Open Motors

Split-phase motors have two distinctly different windings (Figure 13-11). They have a medium amount of starting torque and good operating efficiency. The split-phase motor is normally used for operating fans in the fractional horsepower range. Its normal operating speeds are 1800 rpm and 3600 rpm. An 1800 rpm motor will normally slip to between 1725 and 1750 rpm under a load. If the motor is loaded to the point where the speed falls below 1725 rpm, the current draw will climb above the rated amperage. Motors rated at 3600 rpm will normally slip to between 3450 and 3500 rpm. Some of these motors are designed to operate at either speed (1750 or 3450). The speed of the motor is determined by the number of motor poles and the

MOTORS USED IN A/C APPLICATIONS 191

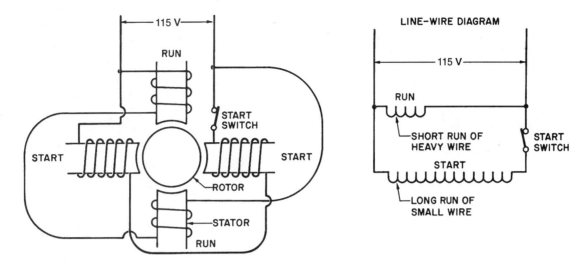

Figure 13-11. Diagram showing the difference in the resistance between the start and run windings.

method of wiring. The technician can change the speed of a two-speed motor at the motor terminal box.

Capacitor-Start Motors

The capacitor-start motor is the same basic motor as the split-phase motor (Figure 13-12). It has the two distinctly different windings for starting and running. A start capacitor is wired in series with the start winding to give the motor more starting torque. Figure 13-13 shows voltage and current cycles in an induction motor. In an inductive circuit, the current lags the voltage. In a capacitive circuit, the current leads the voltage. The amount by which the current leads or lags the voltage is called the *phase angle*. A capacitor is chosen to make the phase angle such that it is most efficient for starting the motor. Figure 13-14 shows a start capacitor. This capacitor is not designed to be used while the motor is running, and it must be switched out of the circuit soon after the motor starts. This is done when the start winding is taken out of the circuit.

Capacitor-Start, Capacitor-Run Motors

Capacitor-start, capacitor-run motors have both start and run capacitors (Figure 13-15). The run capacitor is wired into the circuit to provide the most efficient phase angle between the current and voltage when the motor is running. It is in the circuit whenever the motor is running. If a run capacitor fails because of an open circuit within the capacitor, the motor may start, but the running amperage will be about 10% too high and the motor will overheat if operated under full load. The capacitor-start, capacitor-run motor is one of the most efficient motors used in refrigeration and air conditioning equipment. It is normally used with belt-drive fans and compressors.

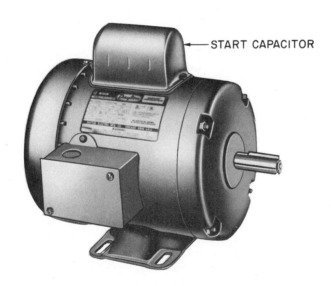

Figure 13-12. Capacitor-start motor. *Courtesy W.W. Grainger, Inc.*

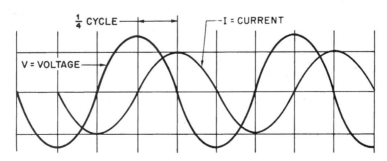

Figure 13-13. Voltage and current in an inductive circuit. The current lags the voltage.

192 UNIT 13 ELECTRIC MOTORS

Figure 13-14. Start capacitor. *Photo by Bill Johnson.*

Figure 13-16. Permanent split-capacitor motor. *Courtesy Universal Electric Company.*

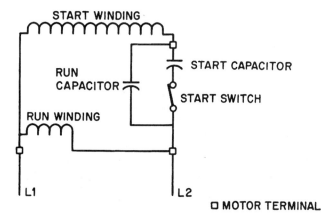

Figure 13-15. Wiring diagram of a capacitor-start, capacitor-run motor. The start capacitor is only in the circuit at start-up, and the run capacitor is in the circuit both at start-up and while the motor is running.

Figure 13-17. Multispeed PSC motor. *Photo by Bill Johnson.*

Permanent Split-Capacitor Motors

The permanent split-capacitor (PSC) motor (Figure 13-16) is similar to the capacitor-start, capacitor-run motor, but it does not have a start capacitor. Instead it uses one run capacitor. This is the simplest split-phase motor. It is very efficient and has no moving parts for the starting of the motor; however, the starting torque is very low, so the motor can only be used in low starting torque applications.

A PSC motor may have several speeds. A multispeed motor can be identified by the many wires at the motor electrical connections (Figure 13-17). As the resistance of the motor winding decreases, the speed of the motor increases. When the resistance increases, the motor speed decreases. The motor speed can be changed by switching the wires. Most manufacturers use this motor in the fan section in air conditioning and heating systems. Earlier systems used a capacitor-start, capacitor-run motor and a belt drive, and air volumes were adjusted by varying the drive pulley diameter.

The permanent split-capacitor motor may be used to obtain higher fan speeds during the summer cooling season for cooler leaving air temperatures (Figure 13-18). The fan speed can be decreased by switching to a different resistance in the winding using a relay. This will adjust for the lower airflow requirements in the heating season.

MOTORS USED IN A/C APPLICATIONS   193

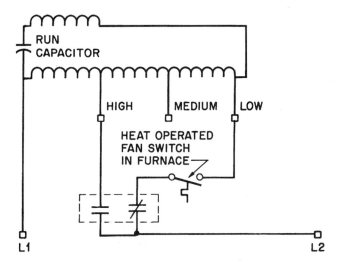

Figure 13-18. Diagram of a PSC motor showing how the motor can be applied for high air volume in the summer and low air volume in the winter.

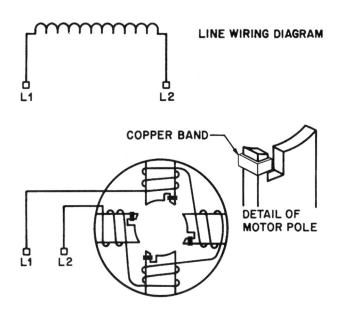

Figure 13-20. Wiring diagram of a shaded-pole motor.

Figure 13-19. Shaded-pole motor. *Courtesy Universal Electric Company.*

Shaded-Pole Motors

The shaded-pole motor (Figure 13-19) has very little starting torque and is not as efficient as the PSC motor, so it is only used for light-duty applications. These motors have small start windings at the corner of each pole that help the motor start by providing an induced current and a rotating field (Figure 13-20). These motors are very economical from a first-cost standpoint. The shaded-pole motor is normally manufactured in the fractional horsepower range, and is often used in air-cooled condensers to turn the fans.

Three-Phase Motors

Three-phase motors are normally used only in commercial and industrial equipment. (Three-phase power is seldom found in a home.) Three-phase motors have no start windings or capacitors. They can be thought of as having three single-phase power supplies (Figure 13-21). Each of the phases can have either two or four poles. A 3600 rpm motor will have three sets, each with two poles (total of six), and an 1800 rpm motor will have three sets, each with four poles (total of 12). Each phase changes the direction of current flow at different times but always in the same order. A three-phase motor has a high starting torque because of the three phases of current that operate the motor (Figure 13-22). At any given point in the rotation of the motor, one of the windings is in position for high torque. This makes it easy to start large fans and compressors.

The three-phase motor rpm also slips to about 1725 and 3450 rpm under full load. These motor are not normally available with dual speed; they are either 1800 rpm or 3600 rpm motors.

The rotation of a three-phase motor may be changed by switching any two motor leads (Figure 13-23). This rotation must be carefully observed when three-phase fans are used. If a fan rotates in the wrong direction it will move only about half as much air. If this occurs, reverse the motor leads and the fan will turn in the correct direction.

UNIT 13 ELECTRIC MOTORS

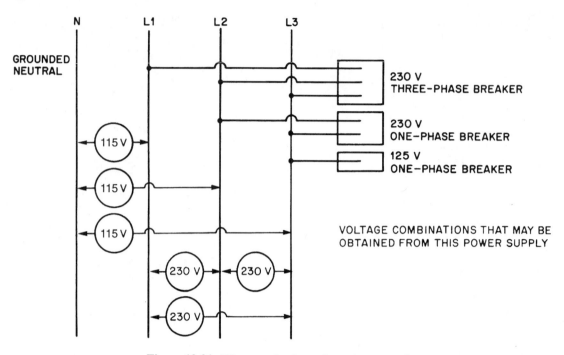

Figure 13-21. Diagram of a three-phase power supply.

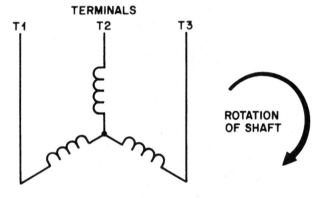

Figure 13-22. Diagram of a three-phase motor.

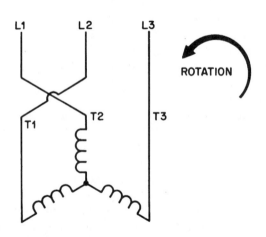

Figure 13-23. Wiring diagram of a three-phase motor. The rotation of this motor can be reversed by changing any two motor leads.

Single-Phase Hermetic Motors

The wiring in a single-phase hermetic motor is similar to that in a split-phase motor. It has both start and run windings, each with a different resistance. The motor runs with the run winding, but uses a potential relay rather than a centrifugal switch to open the circuit to the start winding. A run capacitor is often used to improve running efficiency.

A hermetic motor is designed to operate in a refrigerant (usually vapor) atmosphere. It is undesirable for liquid refrigerant to enter the shell, as by an overcharge. Single-phase hermetic compressors usually are manufactured up to 5 hp (Figure 13-24). If more capacity is needed, multiple systems or larger three-phase units are used.

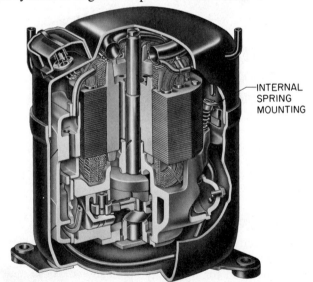

Figure 13-24. Typical motor for a hermetic compressor. *Courtesy Tecumseh Products Company.*

MOTORS USED IN A/C APPLICATIONS 195

Figure 13-25. Motor terminal box on the outside of the compressor. *Photo by Bill Johnson.*

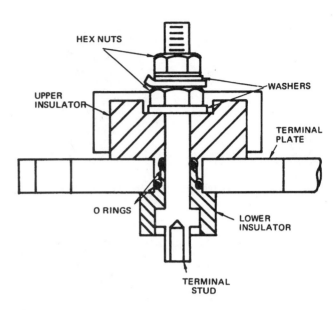

Figure 13-27. These motor terminals use neoprene O-rings as the insulator between the terminal and the compressor housing.

Hermetic compressor motor materials must be compatible with the refrigerant and oil circulating in the system. The coatings on the windings, the materials used to tie the motor windings, and the papers used as wedges, must all be of the correct material. The motor is assembled in a clean, dry atmosphere.

Because the hermetic motor is enclosed in refrigerant, the motor leads must pass through the compressor shell to the outside. A terminal box on the outside of the compressor (Figure 13-25) houses the three motor terminals: one for the run winding, one for the start winding, and one for the line common to the run and start windings. See Figure 13-26 for a wiring diagram of a three-terminal compressor. The start winding has more resistance than the run winding.

The motor leads are insulated from the steel compressor shell. For years, neoprene was the most popular insulating material used (Figure 13-27). However, if the motor terminal becomes too hot, the neoprene may eventually become brittle and possibly leak. Many compressors now use a ceramic material to insulate the motor leads.

Three-Phase Hermetic Motors

Large commercial and industrial installations will have three-phase power for the air conditioning and refrigeration equipment. Three-phase compressor motors normally have three motor terminals, but the resistance across each winding is the same (Figure 13-28). As explained earlier, three-phase motors have a high starting torque and, consequently, should experience no starting problems.

Welded hermetic compressors were limited to 7½ tons for many years, but are now being manufactured in sizes up to about 50 tons. The larger welded hermetic compressors are traded to the manufacturer when they fail and are remanufactured. These must be cut open for service.

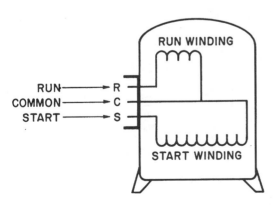

Figure 13-26. Wiring diagram showing the components behind the three terminals on a single-phase compressor.

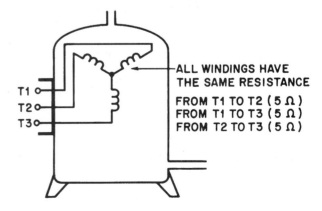

Figure 13-28. Three-phase compressor with three leads for the three windings. The windings all have the same resistance.

UNIT 13 ELECTRIC MOTORS

Figure 13-29. Serviceable hermetic compressors. *Courtesy Copeland Corporation.*

Serviceable hermetic compressors of the reciprocating type are manufactured in sizes up to about 125 tons (Figure 13-29). These compressors may have dual-voltage motors for 208/230 V or 460 V operation (Figure 13-30). These compressors are normally rebuilt or remanufactured when the motor fails, and can either be rebuilt in the field or traded for remanufactured equipment. Most large metropolitan areas will have companies that can rebuild the compressor to the proper specifications.

Note: The difference between rebuilding and remanufacturing is that one is done by an independent rebuilder and the other is done by the original manufacturer or an authorized rebuilder.

13.4 MOTOR CIRCUIT CONTROL DEVICES

Various control devices are used to disconnect the start winding from the circuit when the motor has reached 75% of its rated speed. These devices include centrifugal switches, electronic relays, potential relays, current relays, and positive temperature coefficient devices.

Centrifugal Switches

The centrifugal switch (Figure 13-31) is used to disconnect the start winding from the circuit when the motor reaches three-quarters of its rated speed. Centrifugal

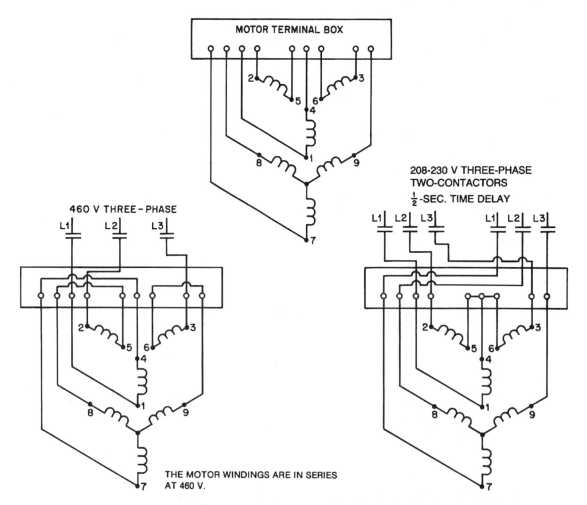

Figure 13-30. Dual-voltage compressor wiring diagrams.

MOTOR CIRCUIT CONTROL DEVICES 197

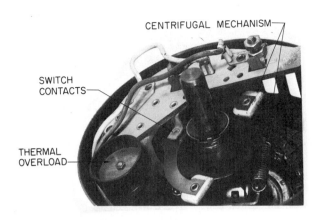

Figure 13-31. Centrifugal switch located at the end of the motor. *Photo by Bill Johnson.*

switches are used in motors that are open to the atmosphere. When a motor is started in the air, the arc from the centrifugal switch will not harm the atmosphere. (It will harm refrigerant, so there must be no arc in a refrigerant atmosphere.)

The centrifugal switch is a mechanical device attached to the end of the shaft with weights that will swing outward when the motor reaches three-quarter speed. For example, if the motor has a rated speed of 1725 rpm, at 1294 rpm (1725 × 0.75) the centrifugal weights will change position and open a switch to remove the start winding from the circuit. This switch is under a fairly large current load, so a spark will occur. If the switch fails to open its contacts and remove the start winding, the motor will draw too much current and the overload device will cause it to stop.

The more the switch is used, the more its contacts will burn from the arc. If this type of motor is started many times, the first thing that will likely fail will be the centrifugal switch. This switch makes an audible sound when the motor starts and stops.

Electronic Relays

The electronic relay is also used with some open motors to open the start winding after the design speed has been obtained.

Potential Relays

A potential relay is often used in hermetic motors to break the circuit to the start winding when the motor reaches approximately 75% of its normal speed. The start winding is not removed from the circuit in the same way as for an open motor because the windings are in a refrigerant atmosphere. Open single-phase motors are operated in air, and a spark is allowed when the start winding is disconnected. This cannot be allowed in a hermetic motor because the spark will deteriorate the refrigerant. This relay has a normally closed set of contacts. The coil is designed to operate at a slightly higher voltage than the applied (line) voltage. When the rotor begins to turn, a

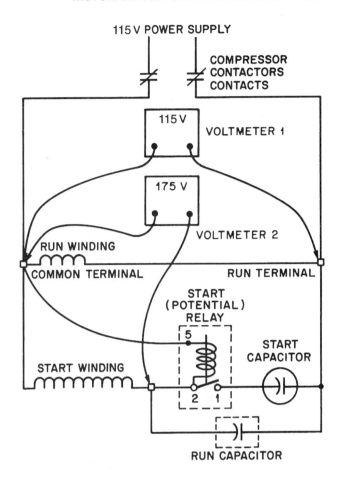

Figure 13-32. Wiring diagram illustrating the higher voltage of the start winding in a typical motor.

transformer action (called back electromotive force or back emf) takes place at the start winding; as the rotor approaches 75% of its design speed, the voltage exceeds the applied voltage and is sufficient to energize the coil. This opens the contacts, which open the start winding circuit (Figure 13-32).

Current Relays

A current relay can also be used in a hermetic compressor to break the circuit to the start winding. It is generally only used on very small motors. It uses the inrush current of the motor to determine when the motor is running up to speed. A motor draws what is known as locked-rotor current during the time when power has been applied to the windings but the motor has not yet started turning. As the motor starts turning, the current peaks; it begins to reduce as the motor continues to turn. The current relay has a set of normally open contacts that close when the inrush current flows through its coil, energizing the start winding. When the motor reaches about three-quarters of the rated rpm, the current relay opens its contacts, either by gravity or using a spring (Figure 13-33). The coil is wired in series with the run winding of the motor.

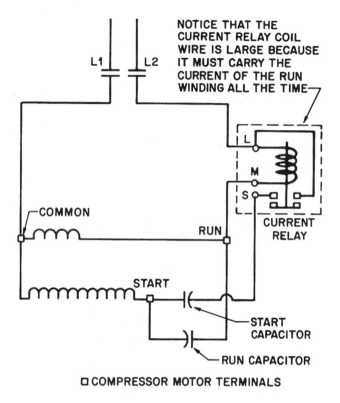

Figure 13-33. Wiring diagram of a current relay.

The full current of the motor must flow through the coil of the current relay. The current relay may always be identified by the size of the wire in the relay coil. This wire is unusually large because it must carry the full-load current of the motor (Figure 13-34).

These starting methods are used on many compressors with split-phase motors that need high starting torque. If a system has a capillary tube metering device or a fixed-orifice metering device, the pressures will equalize during the off cycle and a compressor with high starting torque may not be necessary. A PSC motor may be used for this application. PSC motors are often used in residential air conditioners and heat pumps that have a pressure-equalizing refrigeration cycle using fixed-bore metering devices.

Figure 13-34. A current relay is identified by the size of the wire in the holding coil. *Photo by Bill Johnson.*

In the PSC motor, the start and run windings are both energized whenever there is power to the motor. The motor does utilize a run capacitor wired between the run and start terminals, so line voltage is not applied directly to the start winding.

Positive Temperature Coefficient Devices

The permanent split-capacitor (PSC) motor may not need any start assistance when conditions are well within the design parameters. If it does need start assistance, a potential relay and start capacitor may be added to provide additional torque or a positive temperature coefficient (PTC) device may be used. The PTC device is a thermistor that has no resistance to current flow when the unit is off. Remember, a thermistor changes resistance with a change in temperature. When the unit is started, the current flow through the PTC causes it to heat very fast and create a high resistance in its circuit. This changes the phase angle of the start winding. It will not provide the same degree of starting torque as a start capacitor, but is advantageous because it has no moving parts. The PTC device is wired in parallel with the run capacitor and acts like a short across the run capacitor during starting. This provides full line voltage to the start winding during starting (see Figure 13-35).

13.5 MOTOR CAPACITY CONTROL

Manufacturers incorporate various features to provide greater efficiency or to adjust the motor capacity to meet the changing load.

Two-Speed Compressor Motors

Two-speed compressor motors are used by some manufacturers to control the capacity required from small compressors. For example, a residence or small office building may have a 5 ton air conditioning load at the peak of the season and a 2½ ton load as a minimum. Capacity control is desirable for this application. A two-speed compressor may be used to accomplish capacity control. Two-speed operation is obtained by wiring the compressor motor to operate as a two-pole motor or a four-pole motor. The automatic changeover is accomplished with the space-temperature thermostat and the proper compressor contactor for the proper speed. For all practical purposes, this can be considered two motors in one compressor housing. One motor turns at 1800 rpm, and the other at 3600 rpm. The compressor uses either motor, based on the capacity needs. This compressor has more than three motor terminals to accommodate the two speeds of the compressor.

Special-Application Motors

Some special-application single-phase motors have more than three motor terminals, but are not two-speed motors. Some manufacturers design an auxiliary winding in the compressor to provide additional motor efficiency. These

MOTOR CAPACITY CONTROL 199

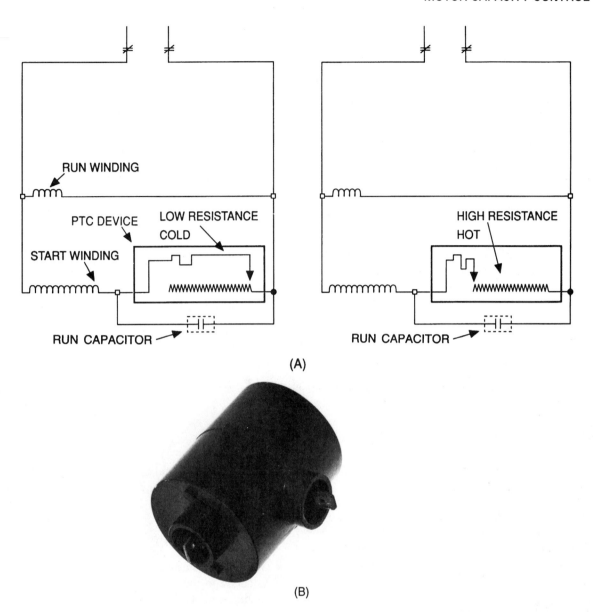

Figure 13-35. A positive temperature coefficient (PTC) device. *Photo by Bill Johnson.*

motors are normally in the 5 hp and smaller range. Other special motors may have the winding thermostat wired through the shell with two extra motor terminals. The winding thermostat can be wired out of the circuit if it should fail with its circuit open (Figure 13-36).

Variable Speed Motors

The desire to control motors to provide greater efficiency has led the industry to explore the development and use of variable speed motors. Most motors do not need to operate at full speed and load except during the peak temperature of the season and could easily satisfy the load at other times by operating at a slower speed. When the motor speed is reduced, the power to operate the motor reduces proportionately. For example, if a home or building only needs 50% of the capacity of the air conditioning unit to satisfy the space temperature, it would be advantageous to reduce the capacity of the unit rather than continuously stop and restart it. When the power consumption can be reduced in this manner, the unit becomes more efficient.

The voltage and frequency (cycles per second) of the power supply determine the speed of a conventional motor. New motors are being used that can operate at different speeds by the use of electronic circuits. There are several methods used to vary the voltage and frequency of the power supply, depending on the type of motor. The compressor and fan motors may be controlled through any number of speed combinations based on the system requirements.

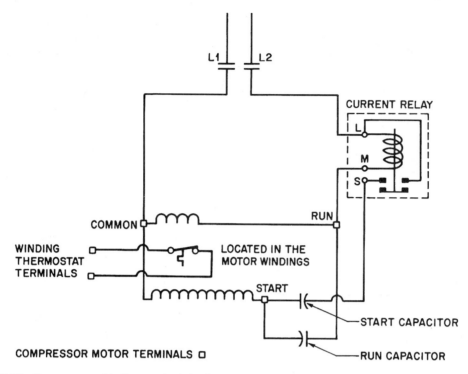

Figure 13-36. Compressor with five terminals in the terminal box, two of which are wired to the winding thermostat.

Benefits of variable speed technology include the following:

- Power savings
- Load reduction based on demand
- Soft starting of the motor (no locked-rotor amperage)
- Better space temperature and humidity control
- Solid-state motor starters do not have open contacts
- A unit may be oversized for future expansion and run at part load until the expansion occurs

13.6 MOTOR COOLING

All motors must be cooled because part of the electrical energy input to the motor is given off as heat. Most open motors are air cooled. Hermetic motors may be cooled with air, water, or refrigerant gas (Figure 13-37). Small to medium-sized motors are normally cooled with refrigerant gas. Only very large motors are water cooled. An air-cooled motor has fins on the surface to provide additional surface area for dissipating heat, Figure 13-37(B). These motors must be located in a moving airstream. To operate properly, motors cooled with refrigerant gas must have an adequate refrigerant charge.

SUMMARY

- Motors turn fans, compressors, and pumps.
- Refrigeration compressors normally require motors with high starting torque.
- Small fans normally use motors with low starting torque.
- The voltage supplied to a particular installation will determine the motor voltage. The common voltage for furnace fans is 115 V, while 230 V is the common voltage for home air conditioning systems.
- Common single-phase motors include: split phase; capacitor start; capacitor start, capacitor run; PSC; and shaded pole.
- When more starting torque is needed, a start capacitor is added to the motor, and for improved running efficiency, a run capacitor is added.
- A centrifugal switch is used in open motors to break the circuit to the start winding when the motor is up to running speed. The switch changes position with the speed of the motor.
- An electronic switch may also be used to interrupt power to the start winding.
- The common rated speed of a single-phase motor is determined by the number of poles or windings in the motor. The common speeds are 1800 rpm, which will slip to about 1725 rpm, and 3600 rpm, which will slip to about 3450 rpm.
- The difference between the rated speeds of 1800/3600 rpm and the running speeds of 1725/3450 is known as slip. Slip is due to the load imposed on the motor while operating.
- Three-phase motors are used for all large applications. They have a high starting torque and high running efficiency. Three-phase power is not available at most residences, so these motors are limited to commercial and industrial installations.

Figure 13-37. All motors must be cooled or they will overheat. (A) This compressor is cooled by the refrigerant gas passing over the motor winding. (B) This compressor is cooled by air from a fan. The air-cooled motor has fins on the compressor to help dissipate the heat. *Courtesy Copeland Corporation.*

- The power to operate hermetic motors must be conducted through the shell of the compressor by way of insulated motor terminals.
- Since the winding of a hermetic compressor is in the refrigerant atmosphere, a centrifugal switch may not be used to interrupt the power to the start winding. Instead, a potential relay is used to remove the start winding from the circuit using back emf.
- A current relay can also be used in a hermetic compressor to break the circuit to the start winding using the motor's run current.
- A PSC motor is used when high starting torque is not required. It needs no starting device other than the run capacitor.
- A PTC device is used with some PSC motors to provide small amounts of starting torque. It has no moving parts.
- When compressors are larger than 5 tons, they are normally three phase.
- Serviceable hermetic compressors come in sizes up to about 125 tons.
- Dual-voltage three-phase compressors are built with two motors wired into the housing.
- Variable speed motors operate at higher efficiencies with varying loads.

REVIEW QUESTIONS

1. Name the two typical operating voltages in residences.
2. Describe the differences between start and run windings of a split-phase motor.
3. What device may be wired into the circuit to improve the starting torque of a compressor?
4. What device may be wired into the circuit to improve the running efficiency of a compressor?
5. Why is a hermetic compressor constructed of special materials?
6. Name the two types of motors used for single-phase hermetic compressors.
7. How does the power pass through the compressor shell to the motor windings?
8. Name a device that can be used to take the start winding out of the circuit when an open motor gets up to running speed?
9. What is back emf?
10. Name the two types of relays used to start a single-phase hermetic compressor.
11. What is a PTC device?
12. How are some small compressors operated at two different voltages?
13. Why is variable speed control desirable?
14. How are some compressor motors operated at different voltages?
15. Name two methods used to cool hermetic compressors.

14 MOTOR APPLICATIONS

OBJECTIVES

Upon completion of this unit, you should be able to

- Identify the proper power supply for a motor.
- Describe the application of three-phase versus single-phase motors.
- Explain how the noise level in a motor can be isolated from the conditioned space.
- Describe the different types of motor mounts.
- Identify the various types of motor drive mechanisms.

14.1 MOTOR SELECTION

Because electric motors perform so many different jobs, choosing the proper motor is necessary for safe and effective performance. Motor selection is usually the job of the manufacturer or design engineer. However, as a technician, you will occasionally need to substitute a motor when an exact replacement is not available, so you should understand the reasons for choosing a particular motor for a job. For example, when a fan motor burns out in an air conditioning condensing unit, the correct motor must be obtained or another failure may occur.

Note: In an air-cooled condenser, the air is normally pulled through the hot condenser coil and passed over the fan motor. This air is used to cool the motor. You must be aware of this, or you will install the wrong motor. The motor must be able to withstand the operating temperatures of the condenser air, which may be as high as 130°F.

This unit concentrates on the applications and selection of open motors. The following are some of the design differences that influence the application:

- Power supply
- Work requirements
- Temperature classifications
- Bearing types
- Mounting characteristics

14.2 POWER SUPPLY

The power supply must provide the correct voltage and sufficient current. For example, the power supply in a small shop building may be capable of operating the existing lighting and heating loads, as well as a 5 hp air compressor. However, if air conditioning is installed, it may overload the breaker panel. If the air conditioning contractor prices the job expecting the electrician to use the existing power supply, the customer may be in for a surprise. The electrical service for the whole building may have to be changed. It is important to assess the existing power supply before bidding or installing an additional load. The motor nameplate (Figure 14-1) and the manufacturer's catalogs provide the needed information for the additional service, but someone must put the whole project together. The installing air conditioning contractor may have that responsibility, but it is usually preferable to have a licensed electrician make the final calculations.

The power supply data contains:

1. Voltage (115 V, 208 V, 230 V, 460 V)
2. Current capacity in amperes
3. Frequency in Hertz or cycles per second (60 Hz in the United States and 50 Hz in many foreign countries)
4. Phase (single-phase or three-phase power)

Voltage

The voltage of an installation is important because every motor operates within a specified voltage range—

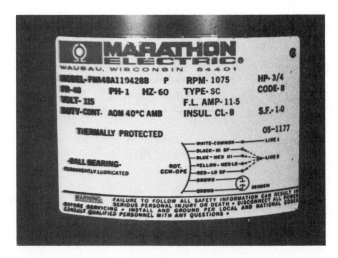

Figure 14-1. Motor nameplate. *Photo by Bill Johnson.*

208 VOLT-RATED MOTOR	+10%	228.8 VOLTS
	−10%	187.2 VOLTS
230 VOLT-RATED MOTOR	+10%	253 VOLTS
	−10%	207 VOLTS
208–230 VOLT-RATED MOTOR	+10%	253 VOLTS
	−10%	187.2 VOLTS

Figure 14-2. Table showing the maximum and minimum operating voltages of typical motors.

APPROXIMATE FULL-LOAD AMPERAGE VALUES FOR ALTERNATING CURRENT MOTORS

Motor	Single Phase		3-Phase-Squirrel Cage Induction		
HP	120 V	230 V	230 V	460 V	575 V
$\frac{1}{6}$	4.4	2.2			
$\frac{1}{4}$	5.8	2.9			
$\frac{1}{3}$	7.2	3.6			
$\frac{1}{2}$	9.8	4.9	2	1.0	0.8
$\frac{3}{4}$	13.8	6.9	2.8	1.4	1.1
1	16	8	3.6	1.8	1.4
$1\frac{1}{2}$	20	10	5.2	2.6	2.1
2	24	12	6.8	3.4	2.7
3	34	17	9.6	4.8	3.9
5	56	28	15.2	7.6	6.1
$7\frac{1}{2}$			22	11.0	9.0
10			28	14.0	11.0

Does not include shaded pole.

Figure 14-3. Chart showing approximate full-load amperage values. *Courtesy BDP Company.*

usually within ±10%. Figure 14-2 shows the upper and lower voltage limits for typical motors. If the voltage is too low, the motor will draw a high current. For example, if a motor is designed to operate on 230 V but the supply voltage is really 200 V, the motor current draw will increase. The motor is trying to do its job, but it lacks the power and will overheat.

If the applied voltage is too high, the motor may develop localized hot spots within its windings, but it will not always experience high amperage. The high voltage will actually give the motor more power than it can use. For example, a 1 hp motor with a voltage rating of 230 V that is operating at 260 V is running at more than 10% above its rated voltage. This motor may be able to develop $1\frac{1}{4}$ hp at this voltage, but the windings are not designed to operate at that level. The motor can overheat and eventually burn out if it continues to run overloaded. This can happen without drawing excessive current.

Current Capacity

There are two current ratings for a motor. The full-load amperage (FLA) is the current the motor draws while operating under a full load at the rated voltage. This is also called the rated load amperage or run-load amperage (RLA). For example, a 1 hp motor will draw approximately 16 A at 115 V or 8 A at 230 V in a single-phase circuit. Figure 14-3 shows approximate full-load amperages for some typical motors.

The other amperage rating that may be given for a motor is the locked-rotor amperage (LRA). The LRA and FLA values are available for every motor and are usually stamped on the motor nameplate for an open motor. (Some compressors do not have both ratings printed.) The LRA or FLA can be used to tell if the motor is operating outside its design parameters. Normally, the LRA is about five times the FLA. For example, a motor that has an FLA of 5 A will normally have an LRA of about 25 A. If the LRA is listed on the nameplate and the FLA is not provided, divide the LRA by 5 to find the approximate FLA. For example, if a compressor nameplate shows an LRA of 80 A, the approximate FLA is 80 ÷ 5 or 16 A.

Every motor has what is known as a *service factor*. (It is normally listed in the manufacturer's literature.) This service factor represents reserve horsepower. For example, a service factor of 1.15 means that the motor can operate at 15% over the nameplate horsepower before it exceeds its design parameters. A motor operating with a variable load and above-normal conditions for short periods of time should have a larger service factor. If the voltage varies at a particular installation, a motor with a high service factor may be chosen. The service factor is standardized by the National Electrical Manufacturer's Association (NEMA). Figure 14-4 is a typical manufacturer's chart showing service factors for various motors.

Frequency

The frequency in cycles per second (cps) or Hertz (Hz) is determined by the power company. The technician has no control over this. Most motors are rated at 60 Hz in the United States, but may be rated at 50 Hz in other countries. Most 60-Hz motors will run on 50 Hz, but they will develop only five sixths of their rated speed (50/60). If you believe that the supply voltage is not 60 Hz, contact the local power company. When motors are operated with local generators as the power supply, the generator speed will determine the frequency. A meter is normally mounted on the generator and can be checked to determine the frequency.

THREE PHASE • DRIPPROOF

Type SC Squirrel Cage • Fractional HP
- 60 Hertz
- Ball Bearing
- 40°C Ambient
- Class B Insulation

• NEMA Service Factor
  1/20 thru 1/8 HP—1.40
  1/6 thru 1/3 HP—1.32
  1/2 thru 3/4 HP—1.25
  1 thru 200 HP—1.15

• Versatile 208-430/460 volt motors available in many ratings.

HP	RPM	Volts	Full Load (5) Amps	Frame
Rigid Base				
1/4	1800	200-230/460	0.8	K48
	1200	230/460	0.6	H56
1/3	3600	200-230/460	0.7	B56
	1800	200-230/460	0.8	K48
		208-230/460	0.8	B56
	1200	200-208	1.7	J56
		230/460	0.8	J56
1/2	3600	208-230/460	0.9	B56
	1800	200-208	2.4	B56
		230/460	1.1	B56
		208-230/460	1.1	B56
	1200	200-208	2.0	J56
		230/460	1.0	J56
3/4	3600	208-230/460	1.2	J56
	1800	200-208	3.2	H56
		230/460	1.3	H56
		200-230/460	1.3	H56
	1200	200-208	3.3	J56
		200-208	3.3	M143T
		230/460	1.6	J56
		230/460	1.6	M143T
1	3600	200-208	3.2	J56
		230/460	1.5	J56
	1800	200-208	3.8	J56
		200-230/460	1.7	L143T
		200-230/460	1.7	J56
		575	1.4	L143T
	1200	200-208	3.8	N145T
		230/460	1.9	K56
		230/460	1.9	N145T

Figure 14-4. Chart showing service factors for motors. *Courtesy Century Electric, Inc.*

Phase

The number of phases of power supplied to a particular installation is also determined by the power company. This determination is based on the total electrical load and the types of equipment that must operate from the power supply. Normally, single-phase power is supplied to residences and three-phase power is supplied to commercial and industrial installations. Single-phase motors will operate on two phases of three-phase power (Figure 14-5), but three-phase motors will not operate on single-phase power. The technician must match the motor to the number of phases of the power supply (Figure 14-6).

POWER SUPPLY 205

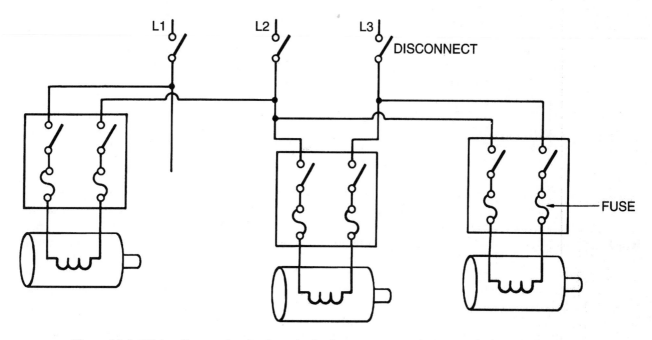

Figure 14-5. Wiring diagram showing how single-phase motors are wired into a three-phase circuit.

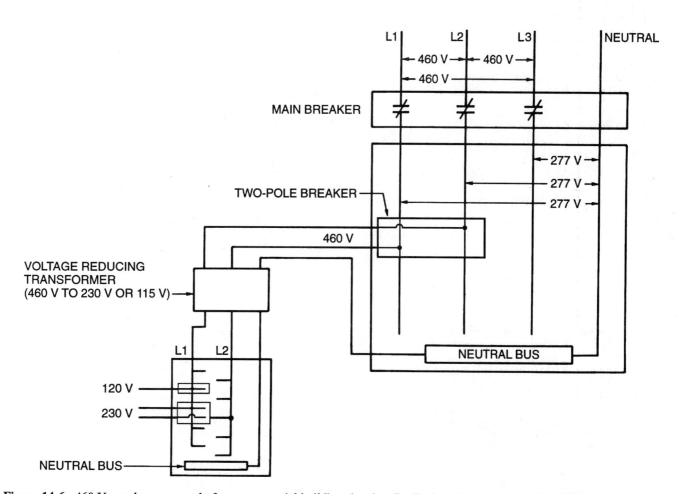

Figure 14-6. 460 V panel power supply for a commercial building showing distribution of various loads. The 460 V to 115 V step-down transformer is used to operate appliances and office machines.

14.3 WORKING CONDITIONS

The motor's working conditions determine which motor should be selected for the particular job. For example, an open motor with a centrifugal starting switch (single phase) for the air conditioning fan may not be used in a room with explosive gases. When the motor's centrifugal switch opens to interrupt the power to the start winding, the gas may ignite. An explosion-proof motor must be used (Figure 14-7). Local codes should be checked and adhered to. Economics is another consideration. For example, the explosion-proof motor is very safe, but is too expensive to be used unless conditions warrant it.

The cleanliness of the surrounding area also plays a part in motor selection. For example, a motor operated in a very dirty area may need to be enclosed, eliminating the natural ventilation for the motor windings. This motor must have an alternative method of dissipating the heat from the windings.

In areas where water may drip onto an exposed motor, a drip-proof motor should be used. It is designed to shed water (Figure 14-8).

14.4 TEMPERATURE CLASSIFICATIONS

The motor insulation class determines the maximum operating temperature that a motor can withstand at a particular ambient condition. As mentioned earlier, some motor applications require that the unit be able to withstand higher ambient temperatures, such as an air conditioning condensing unit. Motors are normally classified by the maximum allowable operating temperatures of the motor windings (Figure 14-9).

For many years, motors were rated by the allowable temperature rise of the motor above the ambient temperature, expressed in °C. Some motors still use this rating. For example, a typical motor has an allowable temperature rise of 40°C. If the maximum allowable ambient temperature is 40°C (104°F), the motor should have a winding temperature of 40°C + 40°C = 80°C (176°F). (Temperature conversions were discussed in Unit 2.) If the temperature of the motor winding can be determined, the technician can tell if a motor is running too hot for the conditions. For example, suppose a motor in a 70°F room is allowed a 40°C rise on the motor (Figure 14-10). The maximum winding temperature is determined as follows:

$$70°F = 21°C; \; 21°C + 40°C \text{ rise} = 61°C \text{ or } 142°F$$

Class A	221°F	(105°C)
Class B	266°F	(130°C)
Class F	311°F	(155°C)
Class H	356°F	(180°C)

Figure 14-9. Temperature classifications of typical motors.

Figure 14-7. Explosion-proof motor. *Courtesy W.W. Grainger, Inc.*

Figure 14-8. Drip-proof motor. *Courtesy W.W. Grainger, Inc.*

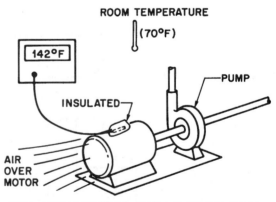

NOTE: CARE MUST BE TAKEN THAT TEMPERATURE TESTER LEAD IS TIGHT ON MOTOR AND NEXT TO MOTOR WINDINGS. THE LEAD MUST BE INSULATED FROM SURROUNDINGS. THE TEMPERATURE CHECK POINT MUST BE AS CLOSE TO THE WINDING AS POSSIBLE.

Figure 14-10. This motor is being operated in an ambient temperature that is less than the maximum allowable for its insulation.

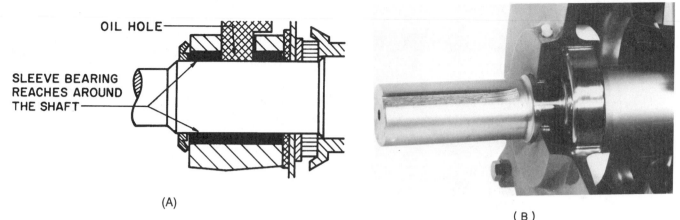

Figure 14-11. (A) Sleeve bearing. (B) Ball bearing. (B) *Courtesy Century Electric, Inc.*

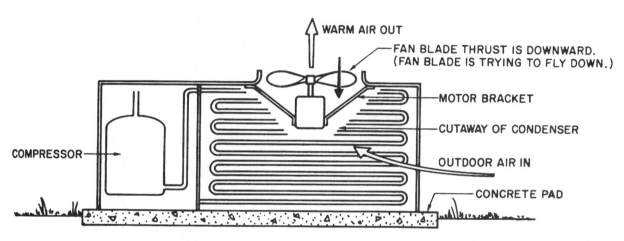

Figure 14-12. Motor working against two conditions: the normal motor load and the fact that the fan is trying to flow downward while pushing air upward.

14.5 BEARINGS

Both load characteristics and noise level determine the type of bearings selected for the motor. Two common types are sleeve bearings and ball bearings (Figure 14-11).

Sleeve Bearings

A sleeve-bearing motor is used where the load is light and system noise must be minimized (e.g., a fan motor on a residential furnace). A ball-bearing motor would probably produce excessive noise in the conditioned space. Metal ductwork is an excellent sound carrier. Any noise in the system is carried throughout the entire system through the ductwork. For this reason, sleeve bearings are preferred in smaller applications, such as residential and light commercial air conditioning. They are quiet and dependable, but cannot withstand excessive pressures (e.g., if the fan belts are too tight). These motors have either vertical or horizontal shaft applications. For example, the typical air-cooled condenser has a vertical motor shaft and pushes the air out the top of the unit. This results in a downward thrust on the motor bearing (Figure 14-12). Some furnace fans have a horizontal motor shaft (Figure 14-13). These two fans may not look very different, but they have very different bearing loads. The horizontal furnace fan has a

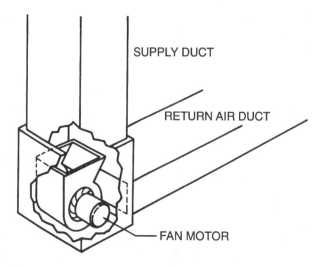

Figure 14-13. Furnace fan motor mounted in a horizontal position.

208　UNIT 14　MOTOR APPLICATIONS

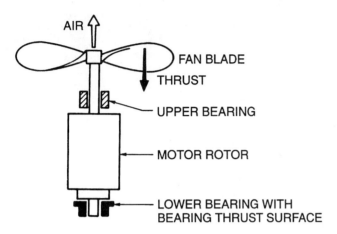

Figure 14-14. Thrust surface on a fan motor bearing.

relatively minor bearing load. On the other hand, the vertical condenser fan is trying to fly downward into the unit to push air out the top. This puts a significant load on the end of the bearing (called the thrust surface) (Figure 14-14).

Sleeve bearings are made from material that is softer than the motor shaft, and will not deteriorate if not protected by a film of lubricating oil. The shaft actually floats on this oil film and should never touch the bearing surface. Sleeve bearings are normally lubricated using either a refillable reservoir and oil port or a permanent lubrication system.

The oil port sleeve bearing has an oil reservoir that is filled from the outside by means of an access port. This bearing must be lubricated at regular intervals with the correct type of oil, usually 20-weight nondetergent motor oil or an oil specially formulated for electric motors. If the oil is too thin, it will allow the shaft to rub against the bearing surface. If the oil is too thick, it will not run into the clearance between the shaft and the bearing surface. The correct interval for lubricating a sleeve bearing depends on the design and use of the motor. Always follow manufacturer's instructions. Some motors have large reservoirs and may not require the addition of oil for several years. This is convenient when there is limited access to the motor.

The permanently lubricated sleeve bearing is constructed with a large reservoir and a wick to gradually feed the oil to the bearing. This bearing does not require lubrication unless the oil deteriorates. If the motor has been running hot for a long period of time, the oil will deteriorate and fail. These bearings are often used in shaded-pole motors.

Ball Bearings

Ball-bearing motors are not as quiet as sleeve-bearing motors and are used in locations where their noise levels will not disturb occupants. Large fan and pump motors are normally located far enough from the conditioned space to muffle the bearing noises. These bearings are made of very hard material and are usually lubricated with grease rather than oil. Ball-bearing motors generally have either permanently lubricated bearings or grease fittings.

Permanently lubricated ball-bearing motors have reservoirs of grease sealed in the bearing. They typically last for years with the lubrication furnished by the manufacturer, unless the operating conditions exceed the design parameters of the motor.

Bearings requiring periodic lubrication have grease fittings (Figure 14-15). A grease gun is used to force grease into the bearing. Only an approved grease should be used. Figure 14-16 shows the relief screw at the bottom of the bearing housing. When grease is pumped into the bearing, this screw must be removed, or the pressure of the grease may push the grease seal out and grease will leak down the motor shaft.

Large motors use a type of ball bearing called a *roller bearing*, which has cylindrical rollers instead of balls. This type of motor is lubricated in the same way as the ball-bearing motor.

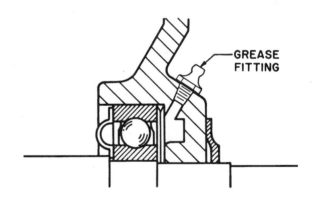

Figure 14-15. Fitting through which the bearing is greased.

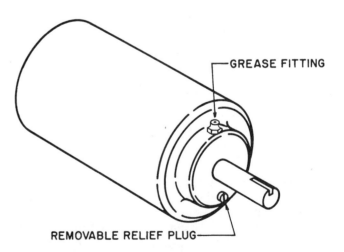

Figure 14-16. Motor bearing using grease for lubrication. Note the relief plug.

MOTOR MOUNTING CHARACTERISTICS   209

Figure 14-17. The grounding strap is used to carry current from the frame if the motor has a grounded winding. *Courtesy Universal Electric Company.*

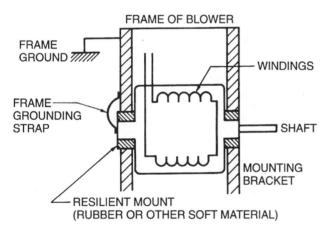

Figure 14-18. How the grounding strap works.

14.6 MOTOR MOUNTING CHARACTERISTICS

Mounting characteristics determine how a motor will be secured during operation. Noise level must be considered when mounting a motor. Two primary methods of attaching a motor are the *rigid mount* and the *resilient* or *rubber mount*. Rigid-mount motors are bolted, metal to metal, to the frame of the fan or pump and will transmit any motor noise into the piping or ductwork. Motor hum is an electrical noise, which is different from bearing noise, and must also be isolated in some installations.

Resilient-mount motors are both electrically and mechanically isolated from the metal framework of the system. Notice the grounding strap on the resilient-mount motor shown in Figure 14-17. The grounding strap prevents the motor frame from becoming electrified if there is a short in one of the motor windings (Figure 14-18). The soft mounting material insulates the surrounding area from motor vibration.

Warning: When replacing a motor, always connect the grounding strap properly or the motor could become dangerous.

The four basic mounting styles are *cradle mount*, *rigid-base mount*, *end mount*, and *bellyband mount*, all of which fit standard dimensions established by NEMA and which are distinguished from each other by frame numbers. Figure 14-19 shows some typical examples.

Cradle-Mount Motors

Cradle-mount motors are used for either direct-drive or belt-drive applications. They have a cradle that fits the motor end housing on each end (Figure 14-20). The end housing is held down with a bracket. The cradle is fastened to the equipment or pump base with machine screws (Figure 14-21). Cradle-mount motors are normally resilient-mounted and are available only in the small horsepower range. One advantage of these motors is that they can easily be removed.

MOTOR DIMENSIONS FOR NEMA FRAMES

Standardized motor dimensions as established by the National Electrical Manufacturers Assiciation (NEMA) are tabulated below and apply to all base-mounted motors listed herein which carry a NEMA frame designation.

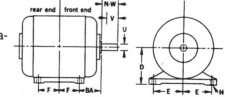

NEMA FRAME	All Dimensions in Inches							V(§) Min.	Key			NEMA FRAME
	D(*)	2E	2F	BA	H	N-W	U		Wide	Thick	Long	
42	2⅝	3½	1¹¹⁄₁₆	2¹⁄₁₆	⁹⁄₃₂ slot	1⅛	⅜	—	—	²¹⁄₆₄ flat	—	42
48	3	4¼	2¾	2½	¹¹⁄₃₂ slot	1½	½	—	—	²⁹⁄₆₄ flat	—	48
56 56H	3½	4⅞	3 3&5(‡)	2⅜	¹¹⁄₃₂ slot	1⅞(†)	⅝(†)	—	³⁄₁₆(†)	³⁄₁₆(†)	1⅜(†)	56 56H
56HZ	3½	**	**	**	**	2¼	⅞	2	³⁄₁₆	³⁄₁₆	1⅜	56HZ

Figure 14-19. Dimensions of typical motor frames. *Courtesy W.W. Grainger, Inc.*

210 UNIT 14 MOTOR APPLICATIONS

Figure 14-20. Cradle-mount motor. *Courtesy W.W. Grainger, Inc.*

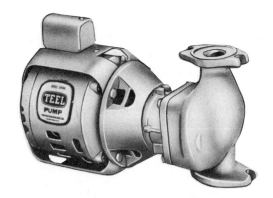

Figure 14-21. Cradle fastened to base of pump. *Courtesy W.W. Grainger, Inc.*

Figure 14-22. Rigid-mount motor. *Courtesy W.W. Grainger, Inc.*

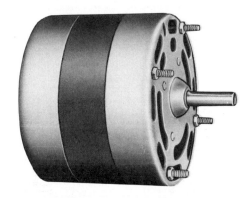

Figure 14-23. This motor is end-mounted with tabs and studs. *Courtesy W.W. Grainger, Inc.*

Figure 14-24. Bellyband-mount motor. *Courtesy W.W. Grainger, Inc.*

Rigid-Mount Motors

Rigid-mount motors are similar to cradle-mount motors except that the base is fastened to the motor body (Figure 14-22). The only sound isolation for this motor is in the belt (if one is used) that drives the prime mover. The belt is flexible and dampens motor noise. This motor is often used as a direct drive to turn a compressor or pump. A flexible coupling is used between the motor and prime mover. Flexible couplings are discussed in greater detail later in this unit.

End-Mount Motors

End-mount motors are very small motors mounted to the prime mover with tabs or studs fastened to the motor housing (Figure 14-23). These motors are normally rigid-mounted. Flange-mounted motors have a flange as a part of the motor housing.

Bellyband-Mount Motors

Bellyband-mount motors have a strap that wraps around the motor to secure it with brackets mounted to the strap (Figure 14-24). These motors are normally resilient-mounted and are often used in air conditioning air handlers. Several universal types of motor kits are belly-band-mounted and will fit many different applications. These motors are all direct drive.

14.7 MOTOR DRIVES

Motor drives are devices or systems that connect a motor to the driven load. For instance, the motor is a driving device, and a fan is a driven component. There are three main types of motor drives: belts, direct drives through couplings or gears, or the driven component may be mounted on the motor shaft. Gear drives will not be covered in this text because they are used mainly in large industrial applications.

The drive mechanism is intended to transfer the motor's rotating power or energy to the driven device. For example, a compressor motor is designed to transfer the motor's power to the compressor, which compresses the refrigerant vapor and pumps refrigerant from the low side of the system to the high side of the system. Some of the factors involved in this transfer are efficiency, the speed of the driven device, and the noise level. It takes energy to turn the belts and pulleys on a belt-drive system, in addition to pulling the compressor load. Therefore, a direct-drive motor may be better suited for this application. Figure 14-25 is an example of both direct and belt drives.

Belt-Drive Applications

Belt drives have been used for years to drive both fans and compressors. Pulley sizes can be changed to modify the speed of the driven device (Figure 14-26). This can be a great advantage if the capacity of a compressor or fan needs to be adjusted. However, the changes must be made within the capacity of the drive motor.

Belts are manufactured in different types and sizes. Some incorporate special fibers to prevent stretching.

Warning: Handle belts carefully during installation. A belt designed for minimum stretch must not be installed by forcing it over the side of the pulley because it may not stretch enough. Fibers will break and weaken the belt, Figure 14-27. Do not get your fingers between the belt and the pulley. Never touch the belt when it is moving.

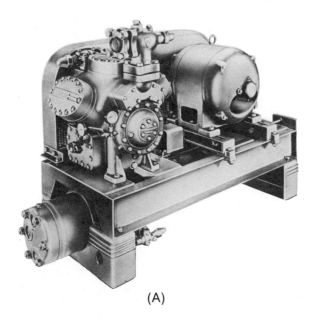

(A)

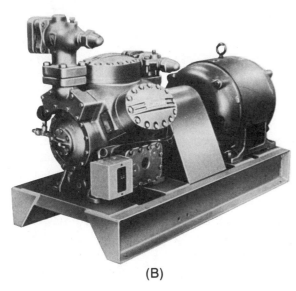

(B)

Figure 14-25. Motor drive mechanisms. (A) Belt drive. (B) Direct drive. *Reproduced courtesy of Carrier Corporation.*

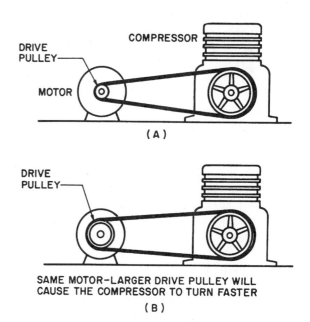

Figure 14-26. Belt drive.

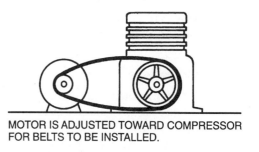

Figure 14-27. Correct method for installing belts over a pulley. The adjustment is loosened to the point where the belts may be passed over the pulley side.

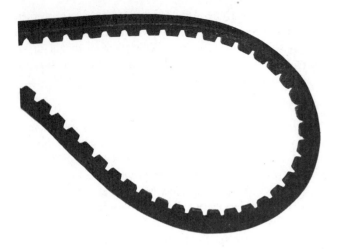

Figure 14-28. This belt has grooves with a tractor-type grip. *Photo by Bill Johnson.*

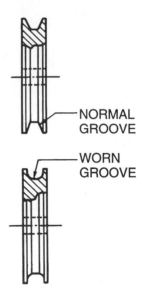

Figure 14-29. Normal and worn pulley comparison.

Belt widths are denoted by "A" ($^{17}/_{32}$ inch wide) and "B" ($^{21}/_{32}$ inch wide). An A-width belt must not be used with a B-width pulley or vice versa. Belts can also have different grips. Figure 14-28 shows a belt with a tractor-type grip.

When a drive has more than one belt, the belts must be matched. Two belts with the same length marked on the belt are not necessarily matched. They may not be exactly the same length. A matched set of belts means the belts are exactly the same length. A set of 42 inch belts marked as a matched set means each belt is exactly 42 inches. If the belts are not marked as a matched set, one may be $42^1/_4$ inches, and the other may be $41^3/_4$ inches. Thus, the belts will not pull evenly—one belt will take most of the load and will wear out first.

Belts and pulleys wear like any moving or sliding surface. When a pulley begins to wear, the surface roughens and wears out the belts. Normal pulley wear is caused by use or running time. Belt slippage will cause premature wear. Pulleys must be inspected often (Figure 14-29).

Belts must have the correct tension, or they will cause the motor to operate in an overloaded condition. A belt tension gage should be used to correctly adjust belts to the proper tension. This gage is used in conjunction with a chart that gives the correct tension for different types of belts of various lengths. Figure 14-30 shows two types of belt tension gages.

Direct-Drive Applications

Direct-drive motor applications are normally used with drive motors for fans, pumps, and compressors. Small fans and hermetic compressors are direct drive, but the motor shaft is actually an extended shaft with the fan or compressor on the end (Figure 14-31). The technician can do nothing to alter these. When this type is used in a large open-drive application, some sort of coupling must be

Figure 14-30. Belt tension gages. *Courtesy Robinair Division, SPX Corporation.*

installed between the motor and the driven device. Figure 14-32 shows a common flexible coupling. Some couplings have springs between the two coupling halves to absorb small amounts of vibration from the motor or pump.

A more complicated coupling is used between the motor and a larger pump or compressor (Figure 14-33). This coupling and shaft must be in very close alignment, or vibration will occur. The alignment must be checked to see that the motor shaft is parallel with the compressor or pump shaft. Alignment is a very precise operation and should be done only by experienced technicians. If two

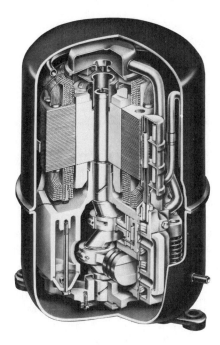

Figure 14-31. Direct-drive compressor. *Courtesy Tecumseh Products Company.*

Figure 14-33. Coupling used to connect larger motors to large compressors and pumps. *Courtesy Lovejoy, Inc.*

Figure 14-32. Small flexible coupling. *Courtesy Lovejoy, Inc.*

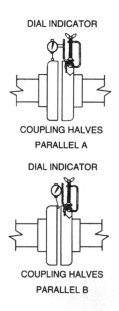

Figure 14-34. The two shafts must be aligned closely for the system to operate correctly. *Courtesy Amtrol, Inc.*

shafts are aligned to within tolerance while the motor and driven mechanism is at room temperature, the alignment must be checked again after the system has run long enough to reach operating temperature. The motor may not expand and move the same distance as the driven mechanism and the alignment may need to be adjusted to the warm value (Figure 14-34).

When a new motor must be installed, try to use an exact replacement. The replacement motor may be found at a motor supply house or it may have to be obtained from the original equipment manufacturer. When the motor is not normally stocked by the supply house, you can save time and trouble by bringing the old motor with you when you order the new one.

SUMMARY

- In many installations only one type of motor can be used.
- The power supply determines the applied voltage, current capacity, frequency, and number of phases.
- The working conditions (duty) for a motor deal with the atmosphere in which the motor must operate (i.e., wet, explosive, or dirty).
- Motors are also classified according to the motor temperature under which they operate.
- Each motor uses sleeve, ball, or roller bearings. Sleeve bearings are the quietest but will not stand heavy loads.

UNIT 14 MOTOR APPLICATIONS

- Four types of motor mounting are the cradle mount, rigid mount, bellyband mount, and end mount.
- The drive mechanism transfers the motor's energy to the driven device (the fan, pump, or compressor).
- An exact motor replacement should be obtained whenever possible.

REVIEW QUESTIONS

1. Name four electrical values that are determined by the motor's power supply.
2. What is the allowable voltage variation for most motors?
3. What are the two main power supply characteristics that the technician has some control over?
4. What is meant by the service factor of a motor?
5. What two categories of electric motors concern the power supply?
6. Name some of the typical conditions under which a motor must operate.
7. How does the temperature classification of a motor affect its use?
8. Name two types of bearings commonly used on small motors.
9. Name four types of motor mounts.
10. What does the drive mechanism do?
11. What is the best replacement motor to use for a special application?
12. What is a matched set of belts?
13. Name the different types of belts.
14. Why must direct-drive couplings be aligned?
15. Why are springs used in a small coupling?

MOTOR STARTING 15

OBJECTIVES

Upon completion of this unit, you should be able to

- **Describe the differences between relays, contactors, and starters.**
- **State how the locked-rotor current of a motor affects the type of motor starter used.**
- **List the basic components of a contactor and starter.**
- **Compare two types of external motor overload protection.**
- **Describe conditions that must be considered when resetting safety devices to restart electric motors.**

15.1 MOTOR CONTROL DEVICES

This unit covers the relays, contactors, and starters used to close or open the power supply circuit to a motor. For example, a compressor in a residential air conditioner is controlled in the following manner. The thermostat contacts close on a temperature rise in the conditioned space. The thermostat contacts in the motor circuit then close and pass low voltage to energize the coil in the compressor relay. This closes the contacts in the compressor relay, allowing the applied or line voltage to pass to the compressor motor windings. Figure 15-1 shows a diagram of a start relay. When energized, the control relay contacts close and pass the line current to the motor. Note that the relay is a major part of the motor circuit, even though it is external to the motor windings. Power is fed to the motor windings and start relay from the control relay.

The motor size and application usually determine the switching device(s) required. For example, a small manual switch can start and stop a hand-held hair drier. On the other hand, a large 100 hp motor that drives an air conditioning compressor must start, run, and stop automatically. It will also consume much more current than the hair drier. The components that start and stop large motors must be more elaborate than those that start and stop small motors.

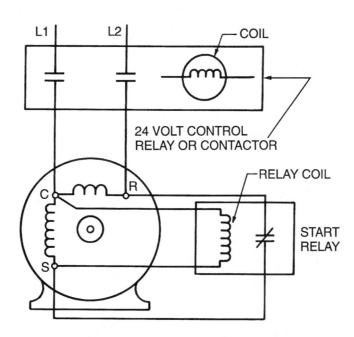

Figure 15-1. The start relay is actually part of the motor circuit. The control relay or contactor passes power to the start relay.

15.2 AMPERAGE RATINGS

Electric motors have two current (amperage) ratings: the *run-load amperage (RLA)*, sometimes referred to as the *full-load amperage (FLA)*, and the *locked-rotor amperage (LRA)*. The RLA or FLA is the current drawn while the motor is running. The LRA is the current drawn by the motor just as it begins to start. Both currents must be considered when choosing the component (relay, contactor, or starter) that passes the line voltage to the motor.

15.3 RELAYS

A *relay* has a magnetic coil that closes one or more sets of contacts (Figure 15-2). It is considered a disposable device because replacement parts are not available.

Relays are designed for light-duty applications. For example, *pilot relays* can be used to switch larger contactors or starters on and off. Pilot relays for switching circuits are very light duty and are not designed to start motors directly. Conversely, relays designed for starting motors are not suitable as switching devices because their contacts have more resistance.

Pilot relay contacts are often made of a fine silver alloy and are designed for low-level current switching. Use on a higher load would melt the contacts. Heavy-duty motor switching relays are often made of silver cadmium oxide with a higher surface resistance, and are physically larger than pilot relays.

If a relay starts the indoor fan in the cooling mode in a central air conditioning system, it must be able to withstand the inrush current of the fan motor on start-up (Figure 15-3). (Remember, the starting current is typically equal to five times the running current.) Relays are often rated in horsepower. For example, if a relay is rated for a 3 hp motor, it will be able to stand the inrush or locked-rotor current of a 3 hp motor.

A relay may have more than one type of contact configuration. For example, it may have two sets of contacts that close when the magnetic coil is energized (Figure 15-4) or it may have two sets of contacts that close when the coil is energized and one set that opens when the coil is energized (Figure 15-5). A relay with a single set of contacts that close when the coil is energized is called a *single-pole-single-throw (SPST)* normally open relay.

Figure 15-3. Fan relay that may be used to start the evaporator fan in a central air conditioning system. *Photo by Bill Johnson.*

Figure 15-2. Relay for starting a motor. This device has a magnetic coil that closes the contacts when the coil is energized. *Photo by Bill Johnson.*

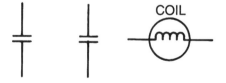

Figure 15-4. Double-pole, single-throw relay with two sets of contacts that close when the coil is energized.

Figure 15-5. Relay with two sets of contacts that close when the coil is energized and one set that opens when it is energized.

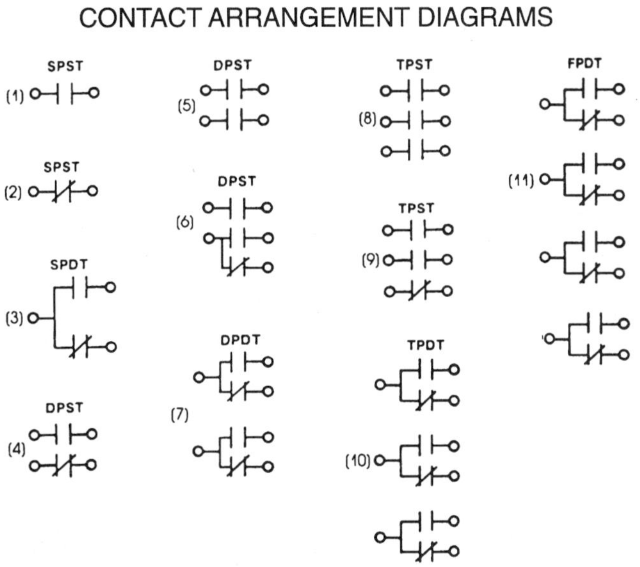

CONTACT ARRANGEMENT DIAGRAMS

Figure 15-6. Some of the more common combinations of contacts supplied with relays. *Reproduced courtesy of Carrier Corporation.*

A relay with two sets of contacts that close and one that opens is called a *triple-pole-single-throw (TPST)* relay. It has two normally open sets of contacts and one normally closed set of contacts. Figure 15-6 shows various contact arrangements.

15.4 CONTACTORS

A *contactor* is a larger version of a relay (Figure 15-7). It can be rebuilt if it fails (Figure 15-8). A contactor has movable and stationary contacts. The holding coil can be designed for various operating voltages; 24 V, 115 V, 208-230 V, or 460 V operation.

Contactors can be very small or very large, but all have similar parts. A contactor may have many configurations of contacts—from the most basic, with one contact, to more elaborate types, with several contacts. The single-contact contactor is used on many residential air conditioning units. It interrupts power to one leg of the compressor, which is all that is necessary to stop a single-phase compressor. A single-contact contactor can also be used to provide crankcase heat to the compressor. A trickle charge of electricity passes through this contact to the motor windings when the contacts are open (during the off cycle). If you used a contactor with two contacts for a replacement, the compressor the crankcase heater would not operate. Once again, an exact replacement is the best choice. Figure 15-9 shows this method of supplying crankcase (oil sump) heat during the off cycle.

Some contactors have as many as five or six sets of contacts. Larger motors usually have three heavy-duty sets to start and stop the motor; the remaining contacts, also known as the *auxiliary contacts*, can be used to switch auxiliary circuits (Figure 15-10).

Figure 15-7. Contactor. *Courtesy Honeywell, Inc.—Residential Division.*

Figure 15-8. Replaceable components in a contactor include the contacts (both movable and stationary), the springs that hold the contacts, and the holding coil. *Photo by Bill Johnson.*

218 UNIT 15 MOTOR STARTING

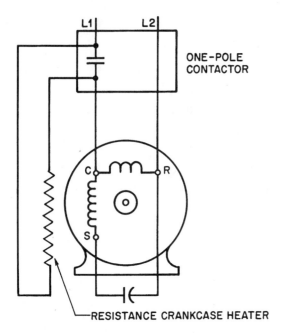

Figure 15-9. This contactor only has one set of contacts; therefore, only one line of the power needs to be broken in order to stop the motor. This is a method of supplying crankcase heat to the compressor during the off cycle. When the unit is operating, the contacts close and the current bypasses the heater and flows to the run winding.

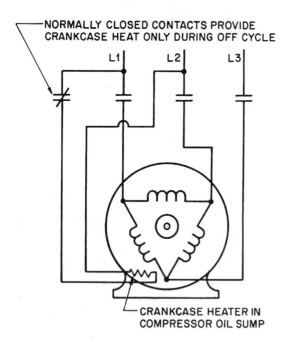

Figure 15-10. Contactor with auxiliary contacts for switching other circuits.

15.5 MOTOR STARTERS

The *starter*, or *motor starter* as it is sometimes called, is similar to the contactor. In some cases, a contactor may be converted to a motor starter. The motor starter differs from the contactor because it has motor overload protection built into its framework (Figure 15-11). Motor starters are available in a variety of sizes, and can be rebuilt when individual components fail. Figure 15-12 shows the replaceable components in a typical motor starter.

Motor overload protection protects that particular motor. See Figure 15-13 for a melting alloy-type overload heater. The fuse or circuit breaker cannot be wholly relied upon for protection because it protects the entire circuit, which may have many components. In some cases, the motor starter's protection is a better indication of a motor problem than the motor winding thermostat protection.

Figure 15-11. Motor starter. *Courtesy Square D Company.*

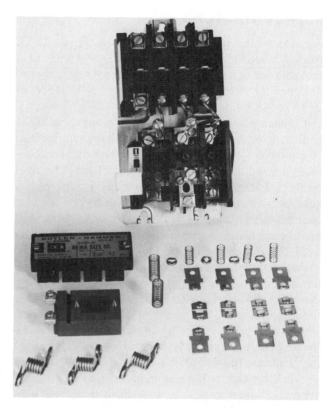

Figure 15-12. Components that may be changed on a starter include the movable and stationary contacts, springs, coil, and overload protection devices. *Photo by Bill Johnson.*

MOTOR STARTERS 219

Figure 15-13. Melting alloy-type overload heater. *Courtesy Square D Company.*

Figure 15-14. Clean contacts contrasted with dirty, pitted contacts. *Courtesy Square D Company.*

Contact surfaces become dirtier and more pitted with each motor starting sequence (Figure 15-14). Some technicians believe that these contacts may be buffed or cleaned with a file or sandpaper. However, filing or sanding exposes the base metal under the silver plating and speeds its deterioration. When contacts are dirty or pitted, they should be replaced. If the device is a relay, the complete relay must be replaced. If the device is a contactor or starter, the contacts may be replaced. There are both movable and stationary contacts and springs that hold tension on the contacts. Be sure to use an exact replacement whenever possible.

A voltmeter can be used to check the resistance across a set of contacts under full load. When the meter leads are placed on each side of a set of contacts and there is a measurable voltage, there is resistance in the contacts. When the contacts are new and have no resistance, no voltage should be read on the meter. An old set of contacts will have a slight voltage drop and will produce heat due to the resistance of the contact surfaces. Contacts may also be checked with an ohmmeter (Figure 15-15).

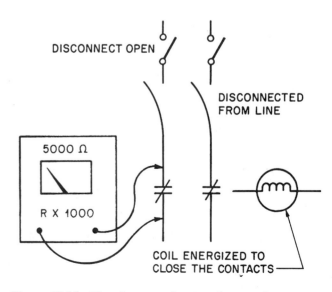

Figure 15-15. The ohmmeter shows an increase in contactor resistance caused by dirt and corrosion.

220 UNIT 15 MOTOR STARTING

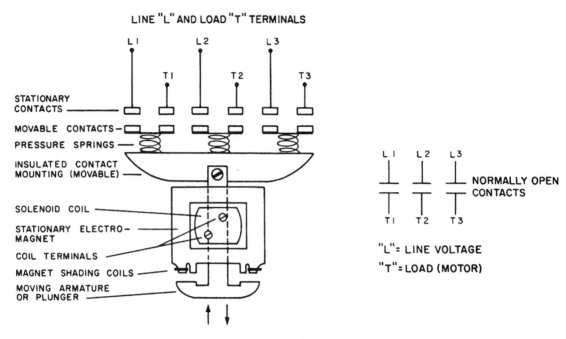

Figure 15-16. When energized, springs will hold the three movable contacts tightly against the stationary contacts. From Alerich, *Electric Motor Control*, © 1993 by Delmar Publishers, Inc.

Each time a motor starts, the contacts will be exposed to the inrush current (the same as LRA). These contacts are under a tremendous load for this moment. When the contacts open, there is an arc caused by the breaking of the electrical circuit. The contacts must make and break as quickly as possible. The contacts have a magnet that pulls them together with a tension mechanism (such as a spring) to take up any slack. For example, there may be three large movable contacts in a row that must all be held equally tight against the stationary contacts. The springs behind the movable contacts keep this pressure even (Figure 15-16). If there is resistance in the contact surfaces, the contacts may get hot enough to take the tension out of the springs. This will further reduce the contact-to-contact pressure and increase the heat. The tension in these springs must be maintained.

15.6 MOTOR PROTECTION

The electric motors used in HVAC equipment are the most expensive single components in the system. These motors consume large amounts of electrical energy and place considerable stress on the motor windings. Therefore, they deserve the best protection possible within the economic boundaries of a well-designed system. The more expensive the motor, the more expensive and elaborate the protection should be.

Fuses are usually used as circuit (not motor) protectors.

A motor may be operating at an overloaded condition that would not cause the conductor to be overloaded; hence the fuse will not open the circuit. Let's use a central air conditioning system as an example. There are two motors in the condensing unit: the compressor (the largest) and the condenser fan motor (the smallest). In a typical unit, the compressor may have a current draw of 23 A, and the fan motor may use 3 A (Figure 15-17). The fuse protects the total circuit, which means that it must be able to handle at least 23 A. Therefore, if the fan motor is overloaded, the fuse may not open the circuit.

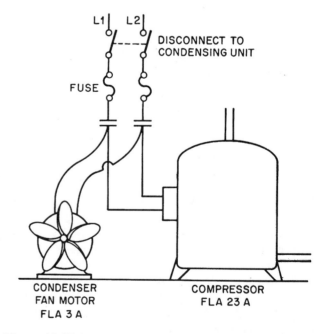

Figure 15-17. Two motors in the same circuit served by the same conductor.

Each motor should be protected within its own operating range.

Motors can operate for brief periods at a slight overcurrent condition without being damaged. The overload protection is designed to disconnect the motor at a current draw that is slightly more than the RLA, so the motor can be operated at its maximum design condition. In addition, the higher the current value above the RLA, the more quickly the overload should react. The amount of the overload and the time are both figured into the design of the particular overload device.

Overload current protection is applied to motors in different ways. For example, overload protection is not needed for many small motors because they do not consume enough power to damage the motor unless shorted from winding to winding or from the winding to the frame (to ground). Some small condenser fan motors do not draw enough amperage at the LRA condition to overheat. These motors do not have overload protection, and are described as *impedance protected*. This type of motor is inexpensive, and may only generate 50 watts of power, the same as a 50 watt light bulb. If the motor fails because of a burnout, the current draw will be interrupted by the circuit protector.

Overload protection is divided into inherent (internal) protection and external protection.

Warning: The conductor wiring in the circuit must not be allowed to pass too much current or it will overheat and cause conductor failures or fires.

Inherent Motor Protection

Inherent protection is that provided by internal thermal overloads in the motor windings or the thermally-activated snap-disc (bimetal) device (Figure 15-18). The same types of devices are used with open motors.

External Motor Protection

External protection is often applied to the device passing power to the motor contactor or starter. These devices are normally actuated by current overload and break the circuit to the contactor coil. The contactor stops the motor. When a motor is started with a relay, the motor is normally small and only has internal protection. Contactors are used to start larger motors, and either inherent protection or external protection is used. Large motors (above 5 hp in HVAC systems) use starters and have overload protection either built into the starter or in the contractor circuit.

Both the value (trip point) and the type of overload protection are normally chosen by the system design engineer or manufacturer. The technician checks the overload device when there is a problem, such as random shutdowns due to overload trips. The technician must be able to understand the designer's intent with regard to the motor's operation and the overload device operation because they are closely related in a working system.

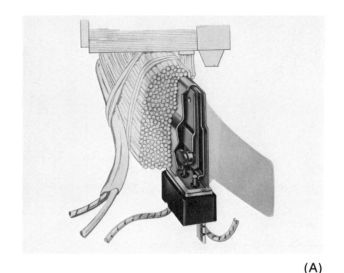

(A)

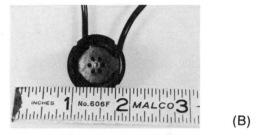

(B)

Figure 15-18. (A) Inherent overload protector. (B) Snap-disc (bimetal) overload protector. *Courtesy of: (A) Tecumseh Products Company, (B) Photo by Bill Johnson.*

15.7 SERVICE FACTOR

As discussed in the previous unit, many motors have a service factor, which is the reserve capacity of the motor. Typical service factors range between 1.15 and 1.40. The smaller the motor, the larger the service factor. For example, a motor with an FLA of 10 A and a service factor of 1.25 can operate at 12.5 A without damaging the motor. The overload protection for a particular motor takes the service factor into account.

National Electrical Code Standards

The National Electrical Code (NEC) sets the standard for all electrical installations, including motor overload protection. In addition, the code book published by NEMA should be consulted for any overload problems or misunderstandings that may occur regarding correct selection of the overload device.

Temperature-Sensing Devices

Various sensing devices are used for overcurrent situations. The most popular devices are those that respond to temperature changes.

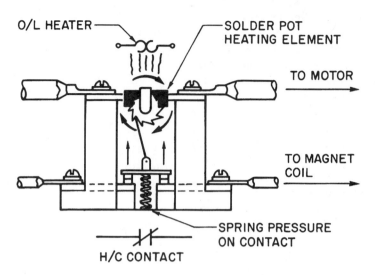

Figure 15-19. This overload device uses a resistance heater that heats a low-temperature solder. From Alerich, *Electric Motor Control,* © 1993 by Delmar Publishers, Inc.

The bimetal element is an example. The line current of the motor passes through a heater that heats a bimetal strip. (This heater can be changed to suit a particular motor amperage.) When the current is excessive, the heater warps the bimetal, opening a set of contacts that interrupt power to the contactor coil circuit. Many bimetal devices are designed as snap discs to avoid excessive arcing.

A low-melting solder may be used in place of the bimetal device (Figure 15-19). This is called a *solder pot*. The solder will melt from the heat caused by an overcurrent condition. The overload heater is sized for the particular amperage draw of the motor it is protecting. The overload control circuit will interrupt power to the motor contactor coil and stop the motor in case of overload. The solder melts and the overload mechanism turns because it is spring-loaded. It can be reset when it cools.

Both of these overload devices are sensitive to temperature. The temperature of the heater causes them to function, but heat from any source may trip these devices, even if it has nothing to do with a motor overload. For example, if the overload device is located in a metal control panel in the sun, the heat from the sun may affect the performance of the overload device. In addition, a loose connection on one of the overload device leads will cause local heat and may open the circuit to the motor even though there is actually no overload (Figure 15-20). If you are troubleshooting a circuit with an overload that cannot be attributed to the motor, external causes should be investigated.

Magnetic Overload Devices

Magnetic overload devices are separate components and are not attached to the motor starter (Figure 15-21). This component is very accurate and is not affected by the ambient temperature. The advantage of this device is that it can be located in a hot cabinet on the roof and the temperature will not affect it. It will shut the motor off at an accurate ampere rating regardless of the temperature.

Restarting the Motor

Warning: When a motor trips on a safety device, such as an overload, do not restart it immediately. Look around for the possible problem before restarting the motor.

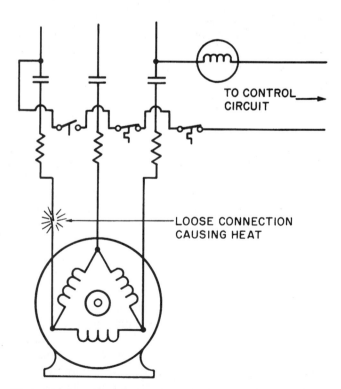

Figure 5-20. Local heat, such as from a loose connection, may open a thermal overload device.

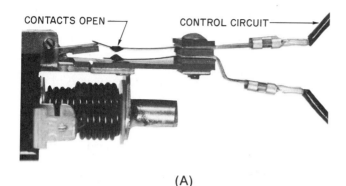

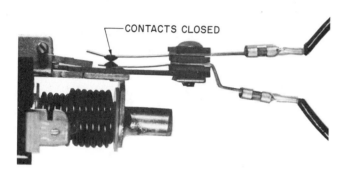

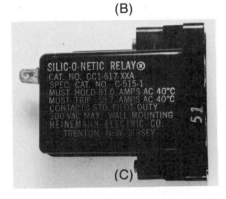

Figure 15-21. Magnetic overload device. *Photos by Bill Johnson.*

Figure 15-22. Manual-reset overload device. *Courtesy Square D Company.*

When a motor stops because it is overloaded, the overload condition is reduced to 0 A at the instant the motor is stopped. However, this does *not* mean that the motor should be immediately restarted. The cause of the overload may still exist. Also, the motor could be too hot and may need time to cool.

There are various ways of restarting a motor after an overload condition has occurred. Some manufacturers design their control circuits with a manual reset to keep the motor from restarting, while others use a time delay to keep the motor off for a predetermined amount of time. Some manufacturers use a relay that will keep the motor off until the thermostat is reset.

Units with a manual reset at the overload device may be somewhat inconvenient, especially if the unit is located in a remote area, such as on a rooftop. Figure 15-22 shows an example of a manual-reset overload device.

When the reset is in the thermostat circuit, the protection devices may be reset from the room thermostat. This is very convenient, but the reset button may operate multiple controls simultaneously, and the technician may not know which control is being reset. When the manual reset button is pushed and a restart occurs, there is no doubt which control has been reset to start the motor.

Time-delay reset devices keep the unit from short cycling but may reset even though the problem condition still exists. If the problem is serious, motor damage can result.

Many cooling systems use a combination of these types of devices, with manual resets used for more critical circuits and time delays or thermostat resets used for less critical circuits.

SUMMARY

- The relay, contactor, and motor starter are three types of motor starting and stopping devices.
- The relay is used for switching circuits and motor starting.
- Motor starting relays are used in heavy-duty applications, while switching relays are used in light-duty applications.
- Contactors are large relays that may be rebuilt.
- Starters are contactors with motor overload protection built into the framework of the contactor.
- The contacts on relays, contactors, and starters should not be filed or sanded.
- Large motors are often the most expensive component in the system and should be protected with individual motor overload devices.
- Inherent motor overload protection is provided by sensing devices within the motor.

- External motor overload protection is applied to the current-passing device: the relay, contactor, or starter.
- To effectively troubleshoot motor controls, the technician must have an understanding of both the motor operation and the overload device operation.
- The service factor is the reserve capacity of the motor.
- Bimetal and solder-pot devices are thermally operated.
- Magnetic overload devices are very accurate and are not affected by ambient temperature.
- Most motors should not be restarted immediately after tripping on a safety device because they may need time to cool. When possible, determine the reason for the overloaded condition before restarting the motor.

REVIEW QUESTIONS

1. How would you repair a defective relay?
2. What components can be changed on a contactor?
3. What components can be replaced on a starter?
4. What are the two types of relays?
5. What two amperages influence the selection of a motor starting component?
6. What is the difference between a contactor and a starter?
7. What materials are used in the contact surfaces of relays, contactors, and starters?
8. What causes an overload protection device to function?
9. What are the typical coil voltages used for contactors?
10. Why is it not a good idea to file or sand contact surfaces?
11. Why is it not a good idea to rely upon circuit protection devices to protect motors from overload conditions?
12. Under what conditions are motors allowed to operate with slightly higher than design loads?
13. Describe the difference between inherent and external overload protection.
14. What is the purpose of overload protection at the motor?
15. State the reasons why a motor should not be restarted immediately.

TROUBLESHOOTING ELECTRIC MOTORS 16

OBJECTIVES

Upon completion of this unit, you should be able to

- Describe different types of electric motor problems.
- Identify common electrical problems in electric motors.
- Identify various mechanical problems in electric motors.
- Describe a capacitor checkout procedure.
- Explain the differences between troubleshooting a hermetic motor problem and troubleshooting an open motor problem.

16.1 ELECTRIC MOTOR TROUBLESHOOTING

Electric motor problems are either mechanical or electrical. Mechanical problems may sometimes appear to be electrical. For example, a bearing dragging in a small PSC fan motor may not make any noise. The motor may not start, and it appears to be an electrical problem. The technician must know how to correctly diagnose the problem. This is particularly true with open motors, because if the driven component is stuck, a motor may be changed unnecessarily. If the stuck component is a hermetic compressor, the whole compressor must be changed; if it is a serviceable hermetic compressor, the motor can be replaced or the compressor running gear can be rebuilt.

16.2 MECHANICAL MOTOR PROBLEMS

Mechanical motor problems normally occur in the bearings or the shaft where the drive is attached. The bearings can be tight or worn due to lack of lubrication. Grit can easily get into the bearings of some open motors and cause them to wear.

Motor problems are not usually repaired by HVAC technicians. They are handled by technicians trained in rebuilding motor and other rotating equipment. Your primary responsibility is to isolate motor problems and, if necessary, replace the motor or call in a specialist for the necessary repairs. For example, a problem with motor vibration may require you to seek help from a qualified balancing technician. Before calling in a specialist, explore every possibility to ensure that the vibration is not caused by a field problem, such as a fan loaded with dirt or liquid flooding into a compressor.

Motor bearing failure with roller and ball bearings can often be identified by the bearing noise. When sleeve bearings fail, they normally lock up (will not turn) or sag to the point that the motor is out of its magnetic center. At this point, the motor will not start.

When motor bearings fail, they can be replaced. If the motor is small, the motor is normally replaced because it would cost more to disassemble the motor and change the bearings than it would to purchase and install a new motor. This is particularly true for fractional horsepower fan motors. These small motors almost always have sleeve bearings pressed into the end bells of the motor, and special tools may be needed to remove and replace the bearing (Figure 16-1).

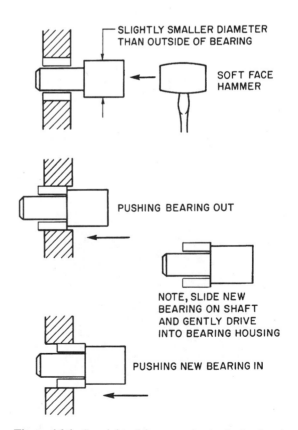

Figure 16-1. Special tool for removing/replacing bearings.

Removing Drive Assemblies

To remove the motor, you need to remove the pulley, coupling, or fan wheel from the motor shaft. The fit between the shaft and the adjoining assembly may be very tight. Removing the assembly from the motor shaft must be done with care. The assembly may have been running on this shaft for years, and it may have rust between the shaft and the assembly. You must remove the assembly without damaging it. Special pulley pullers will help (Figure 16-2), but other tools or procedures may be required. Always consult with the manufacturer or an experienced technician when attempting to disassemble an unfamiliar device.

Most assemblies are held to the motor shaft with set screws threaded through the assembly and tightened against the shaft. A flat spot is usually provided on the shaft for seating the set screw and prevent it from damaging the shaft surface (Figure 16-3). The set screw is made of very hard steel, much harder than the motor shaft. When larger motors with more torque are used, a matching keyway is normally machined in the shaft and assembly. This keyway provides a better bond between the motor and assembly (Figure 16-4). A set screw is usually tightened down on the top of the key to secure the assembly to the motor shaft.

Many technicians make the mistake of trying to drive a motor shaft out of the assembly by striking the end of the shaft with a hammer or other instrument. In doing so, they blunt or distort the end of the motor shaft. When it is distorted, the motor shaft cannot pass through the assembly without damage. If the shaft must be driven, you may need to use a similar shaft with a slightly smaller diameter as the driving tool (Figure 16-5).

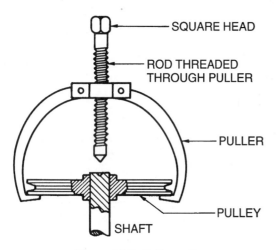

Figure 16-2. Pulley puller.

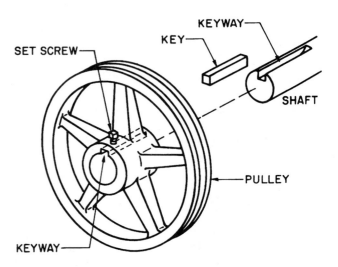

Figure 16-4. This pulley has a groove cut in it that matches a groove in the shaft. A key is placed in these grooves and a set screw is tightened down on top of the key.

Figure 16-3. Flat spot on motor shaft where pulley set screw is tightened. *Photo by Bill Johnson.*

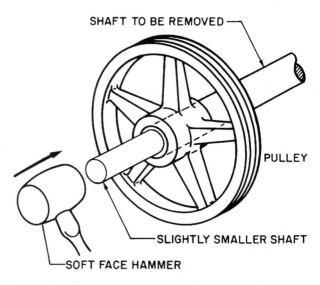

Figure 16-5. Shaft driven through the pulley with another shaft as the contact surface. The shaft used as the contact surface is smaller in diameter than the original shaft. This prevents shaft damage.

Belt Tension

Belts that are too tight strain the bearings so that they wear out prematurely. Incorrect belt tension can lead to motor failure. You should be aware of the specifications for the motor belt tension on belt-drive systems. A belt tension gage (Figure 16-6) will ensure properly adjusted belts when the gage manufacturer's directions are followed. Belts that are too loose can result in slippage and wear.

Pulley Alignment

Pulley alignment is essential to proper motor operation. If the drive pulley and driven pulley are not in line, a strain is imposed on the shaft drive mechanism. The pulleys may be aligned with the help of a straightedge such as a carpenter's level. There is a certain amount of adjustment tolerance built into the motor base on small motors, and this may be enough to allow the motor to be out of alignment if the belt becomes loose. Aligning shafts on equipment may be awkward depending on the system location, but it must be done or the motor or drive mechanism will not last.

When mechanical problems occur with a motor, the motor is normally either replaced or taken to a motor repair shop. Bearings can be replaced in the field by a competent technician, but it is generally better to leave this type of repair to motor experts. However, when the problem is in the pulley or drive mechanism, you may be responsible for the repair.

Caution: Proper tools must be used for motor repair, or shaft and motor damage may occur.

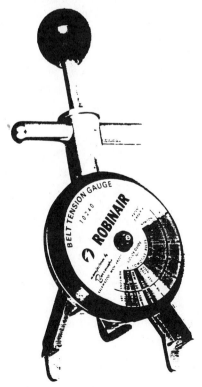

Figure 16-6. Belt tension gage. *Courtesy Robinair Division, SPX Corporation.*

16.3 ELECTRICAL PROBLEMS

Electrical problems are the same for both hermetic and open motors. Open motor problems are a little easier to diagnose because they can often be seen. For example, when an open motor burns up, the charred windings may be seen through the end bells. With a hermetic motor, instruments must be used because they are the only means of diagnosing problems inside the compressor. There are three common electrical motor problems:

1. An open winding
2. A short circuit from winding to winding
3. A short circuit from winding to ground

Warning: If you suspect that a motor has electrical problems, stop the motor immediately to prevent personnel hazards and further equipment damage.

Open Windings

Open motor windings can be diagnosed using an ohmmeter. Every motor has a known, measurable resistance from terminal to terminal. These values are often supplied with the product data for the equipment. Single-phase motors must have the applied system voltage at the run winding to run and at the start winding during start-up. Figure 16-7 is an illustration of a motor with an open start winding.

Shorted Motor Windings

Short circuits in windings occur when the conductors in the winding touch each other where the insulation is worn or otherwise defective. This creates a short circuit through which the electrical energy flows. This path has a lower resistance and increases the current flow in the winding. Although motor windings appear to be made from bare copper wire, they are coated with an insulator to keep the copper wires from touching each other.

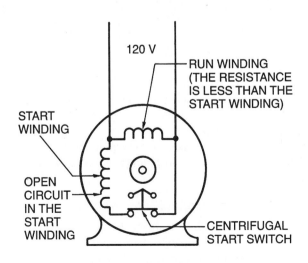

Figure 16-7. Motor with open winding.

| Compressor Model | Voltage | MOTOR AMPS | | | | FUSE SIZE | | Winding Resistance in Ohms |
| | | Full Winding | | 1/2 Winding | | Recommended Max | | |
		Rated Load	Locked Rotor	Rated Load	Locked Rotor	Fusetron	Std.	
9RA - 0500 - CFB	230/1/60	27.5	125.0			FRN-40	50	Start 1.5 Run 0.40
9RB TFC	208-230/3/60	22.0	115.0			FRN-25	40	0.51-0.61
9RJ TFD	460/3/60	12.1	53.0			FRS-15	15	2.22-2.78
9TK TFE	575/3/60	7.8	42.0			FRS-10	15	3.40-3.96
MRA FSR	200-240/3/50	17.0	90.0	8.5	58.0	FRN-25	35	0.58-0.69
MRB FSM	380-420/3/50	9.5	50.0	4.8	32.5	FRS-15	20	1.80-2.15
MRF								

Figure 16-8. Resistances for typical hermetic compressors. *Courtesy Copeland Corporation.*

The best way to check a motor for electrical soundness is to know what the resistance should be for a particular winding and verify it with a good ohmmeter. Some motors include the winding resistances in the manufacturer's product data (Figure 16-8). When a motor has short-circuit problems, the measured resistance will normally be less than the rated value. The decrease in resistance causes the current to rise, which causes motor overload devices to open the circuit and possibly even trip the circuit overload protection. If the resistance does not read within these tolerances, there is a problem with the winding.

Resistance values are particularly helpful when troubleshooting hermetic compressors. This information may not be easy to obtain for open motors, and the windings are not as easy to check because the individual windings do not all come out to terminals as they do on a hermetic compressor.

If the decrease in resistance in the windings is in the start winding, the motor may not start. If it is in the run winding, the motor may start, but will draw too much current while running. If the motor winding resistance cannot be determined, then it is hard to know whether the motor is overloaded or whether it has a defective winding.

When the motor is an open motor, the load can be removed. For example, the belts can be removed or the coupling can be taken apart, and the motor can be started without the load (Figure 16-9). If the motor starts and runs correctly without the load, the load may be too great.

Three-phase motors have three identical windings and should show the same resistance for each winding (Figure 16-10). If the resistances are not equal, there is a problem. An ohmmeter check will quickly reveal an imbalance in winding resistance.

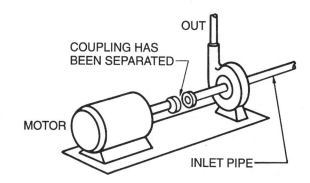

Figure 16-9. With an open motor, the load can be disconnected to determine if it is preventing the motor from operating.

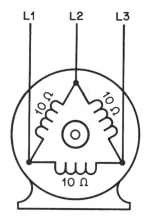

Figure 16-10. Wiring diagram of a three-phase motor. All three windings show the same resistance.

Short Circuit to Ground (Frame)

A short circuit from winding to ground may also be detected with a good ohmmeter. No circuit should be detectable from the winding to ground. The copper suction line on a compressor is a good place to check the circuit. *Ground* and *frame* are interchangeable terms because the frame should be grounded to the earth ground through the building electrical system (Figure 16-11).

ELECTRICAL PROBLEMS   229

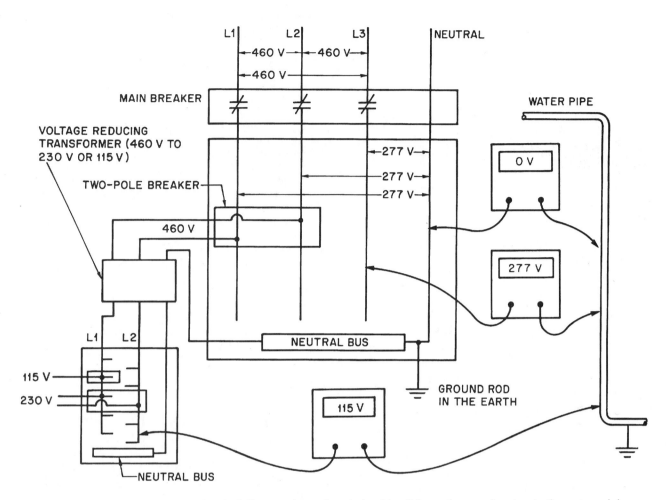

Figure 16-11. This building electrical diagram shows the relationship of the earth ground system to the system piping.

To check a motor for a ground, use an ohmmeter with an R (10,000 ohm) scale. Special instruments for finding very high resistances to ground are used for larger, more sophisticated motors, but most technicians use ohmmeters. Top-quality instruments called *Meggers*® can detect a ground in the 10,000,000 Ω and higher range. These instruments even have an internal high-voltage DC supply to help create conditions to detect the ground (Figure 16-12). The term *Megger*® refers to the meter's ability to measure resistances in megohms or 1,000,000 ohms. It has to do with the capacity of the meter to detect very high resistances.

A typical ohmmeter will detect a ground of about 1,000,000 Ω or less. The rule of thumb is that if an ohmmeter set to the R × 10,000 scale moves its needle even slightly when one lead is touching the motor terminal and the other is connected to a ground (such as a copper suction line), the motor should be started with caution.

Warning: If the meter needle moves to the midscale area, do not start the motor or damage may occur (Figure 16-13).

When a meter reads a very slight resistance to ground in an open motor, the windings may be dirty or damp.

Figure 16-12. Megger®. *Photo by Bill Johnson.*

Clean the motor and the ground will probably be eliminated. Some motors may indicate a slight circuit to ground in damp weather. Air-cooled condenser fan motors are an example. When the motor is started and allowed to run long enough to become warm and dry, the ground circuit may disappear.

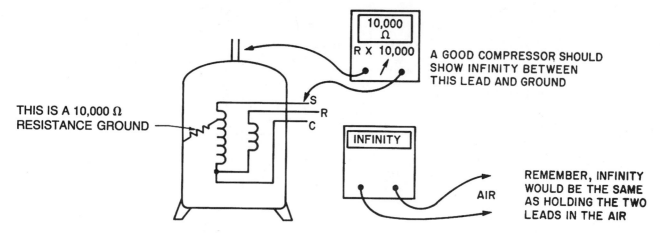

Figure 16-13. Volt-ohmmeter detecting a circuit to ground in a compressor winding.

Hermetic compressors may occasionally have a slight ground due to the oil and liquid refrigerant in the motor splashing on the windings. The oil may have dirt suspended in it and show a slight ground. Liquid refrigerant causes this condition to worsen. If the ohmmeter shows a slight ground but the motor starts, run the motor for a short time and check again. If the ground persists, the motor is likely to fail if the system is not cleaned up. A suction line filter-drier may help remove particles that are circulating in the system and causing the slight ground.

When troubleshooting electric motors, remember that for current to flow, there must be a power supply, a path (conductor), and a measurable resistance. Voltmeters, ohmmeters, and ammeters are all used to diagnose electric motor circuit problems.

16.4 MOTOR STARTING PROBLEMS

Symptoms of electric motor starting problems can include any of the following:

1. The motor hums and then shuts off.
2. The motor runs for short periods and shuts off.
3. The motor makes no sound at all.

The technician must decide whether there is a motor mechanical or electrical problem, a circuit problem, or a load problem. If the motor is an open motor, turn the power off and try to rotate the motor by hand. If it is a fan or pump, it should be easy to turn (Figure 16-14). If it is a compressor, the shaft may be difficult to turn by hand. Use a wrench to grip the coupling when trying to turn a compressor.

Warning: Be sure the power is off before trying to turn any motor using a wrench or other manual method.

If the motor and the load both turn freely, examine the motor windings and components. If the motor is humming and not starting, the starting switch may need to be replaced or the windings may be burned. If the motor is open, you may be able to visually check the windings; if not, remove the motor end bell (Figure 16-15).

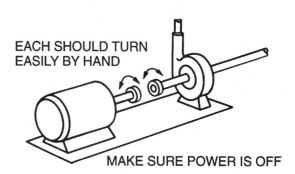

Figure 16-14. Example of how a pump or compressor coupling can be disconnected and the shaft turned by hand to check for a hard-to-turn shaft or stuck component.

Figure 16-15. The end bell has been removed from this motor to examine the start switch and the windings. *Photo by Bill Johnson.*

16.5 CHECKING CAPACITORS

An ohmmeter may be used to check for a faulty capacitor. First, turn off the power to the motor and remove one lead of the capacitor. Short from one terminal to the other with a 20,000 Ω 5 W resistor to discharge the capacitor in case it has a charge stored in it. Use insulated pliers. Some start capacitors have a resistor across these terminals to bleed off the charge during the off cycle (Figure 16-16). However, you should still short across the capacitor because the resistor may be open and may not bleed the charge. If you place ohmmeter leads across a charged capacitor, the meter movement may be damaged.

Set the ohmmeter to the R × 100 scale and touch the leads to the capacitor terminals (Figure 16-17). If the capacitor is good, the meter needle will move to zero and begin to fall back toward infinity. If the leads are left on the capacitor, the needle will eventually show infinite resistance. If the needle falls part of the way back and will drop no more, the capacitor has an internal short. If the needle will not rise at all, try the R × 1000 scale or try reversing the leads. The ohmmeter is charging the capacitor with its internal battery. This is DC voltage. When the capacitor is charged in one direction, the meter leads

Figure 16-16. Start capacitor with bleed resistor across the terminals to bleed off the charge during the off cycle. *Photo by Bill Johnson.*

must be reversed for the next check. If the capacitor has a bleed resistor, the capacitor will charge to 0 Ω, then drop back to the value of the resistor.

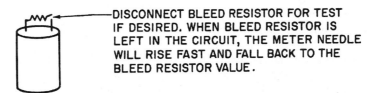

FOR BOTH RUN AND START CAPACITORS

1. FIRST SHORT THE CAPACITOR FROM POLE TO POLE USING A 20,000 OHM 5-WATT RESISTOR.

2. USING THE R × 100 OR R × 1000 SCALE TOUCH THE METER'S LEADS TO THE CAPACITOR'S TERMINALS. METER NEEDLE SHOULD RISE FAST AND FALL BACK SLOWLY. IT WILL EVENTUALLY FALL BACK TO INFINITY IF THE CAPACITOR IS GOOD. (PROVIDED THERE IS NO BLEED RESISTOR).

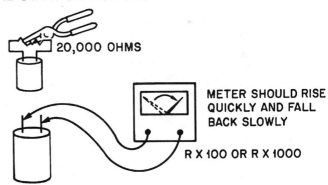

3. YOU CAN REVERSE THE LEADS FOR A REPEAT TEST, OR SHORT THE CAPACITOR TERMINALS AGAIN. IF YOU REVERSE THE LEADS, THE METER NEEDLE MAY RISE EXCESSIVELY HIGH AS THERE IS STILL A SMALL CHARGE LEFT IN THE CAPACITOR.

4. FOR RUN CAPACITORS THAT ARE IN A METAL CAN: WHEN ONE LEAD IS PLACED ON THE CAN AND THE OTHER LEAD ON A TERMINAL INFINITY SHOULD BE INDICATED ON THE METER USING THE R × 10,000 OR R × 1000 SCALE

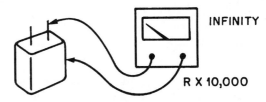

Figure 16-17. Procedure for checking a capacitor.

Figure 16-18. Capacitor analyzer. *Photo by Bill Johnson.*

This simple check will not provide the actual capacitance of a capacitor. A capacitor analyzer is used to measure capacitance (Figure 16-18). It is not often that a capacitor will change in value, so this instrument is used primarily by technicians who perform many capacitor checks.

Capacitor Identification

A run capacitor is contained in a metal can and is oil filled. If this capacitor becomes overheated due to overcurrent, it will often swell, disturbing the sides of the metal container. The capacitor should then be changed. Run capacitors have an identified terminal to which the lead that feeds power to the capacitor should be connected (Figure 16-19). When the capacitor is wired in this manner, a fuse will blow if the capacitor is shorted to the container. If the capacitor is not wired in this manner and a short occurs, current can flow through the motor winding to ground during the off cycle and overheat the motor.

Start capacitors do not use oil and may be contained in a shell of paper or plastic. Paper containers are no longer manufactured, but you may find one in an older motor. If this capacitor has been exposed to overcurrent, the rubber diaphragm at the top of the container may show a small bulge. The capacitor should then be changed.

16.6 WIRING AND CONNECTORS

The wiring and connectors that carry the power to a motor must be in good condition. When a connection becomes loose it overheats, and oxidation of the copper wire occurs. The oxidation acts as an electrical resistance and creates additional heat, which in turn causes more oxidation. This condition will only get worse. Loose connections will result in low voltages at the motor and overcurrent conditions. A voltmeter may be used to check for loose connections (Figure 16-20). If a connection is loose enough to create an overcurrent condition, it can often be located by a temperature rise at the loose connection.

16.7 TROUBLESHOOTING HERMETIC MOTORS

Diagnosing hermetic compressor motor problems differs from diagnosing open motor problems because the motor is enclosed in a shell and cannot be seen. In addition, the motor sound level may be dampened by the compressor shell. Motor noises that are obvious in an open motor may be hard to hear in a hermetic motor.

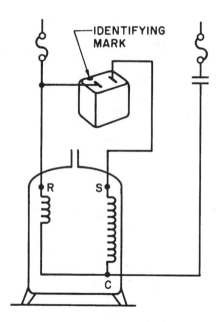

Figure 16-19. The wiring to a run capacitor should be connected to the identified terminal for proper fuse protection of the circuit.

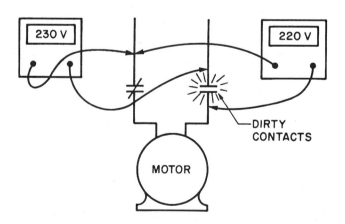

Figure 16-20. When voltmeter leads are applied to both sides of a set of dirty contacts, the meter will show a voltage drop across the contacts.

The motor inside a hermetic compressor can only be checked electrically from the outside. It will have the same problems as an open motor—open circuit, short circuit, or grounded circuit. The technician must give the motor a complete electrical checkout from the outside of the shell; if it is defective, the compressor must be changed. A motor checkout includes the starting and running components for a single-phase compressor, such as the run and start capacitors and the start relay.

As mentioned earlier, compressors operate in remote locations and are not normally attended. For example, a compressor can throw a rod, and the motor may keep running until permanently damaged. The compressor damage is not detected because it cannot be seen. When a compressor is changed because of a bad motor, you should suspect that mechanical damage is the cause.

16.8 TECHNICIAN SERVICE CALLS

Service Call 1

The manager of a retail store calls. The air conditioning is not working and the piping going into the unit is frozen solid. The air handler is in the stock room. The evaporator motor winding is open, and the outdoor unit has continued to run without air passing over the evaporator, causing it to freeze.

Upon arriving at the store and noticing that the fan is not running, the technician turns the cooling thermostat to OFF to stop the condensing unit, then turns the fan switch to ON. The technician then turns off the power and removes the fan compartment cover. The fan motor body is cool to the touch, indicating that the motor has been off for some time. The technician then turns on the power and checks for voltage at the motor terminal block. The motor has power. The power is turned off and the technician checks the motor winding with an ohmmeter. The motor windings are open.

The fan motor is changed, but this is not the end of the problem. The unit cannot be started until the coil is allowed to thaw. To save both time and money, the technician does not stand around and wait for the coil to thaw. Instead, the technician instructs the store manager to allow the fan to run until air is felt at an outlet in the store, then, 15 minutes after air flow has started, to set the thermostat to COOL and the fan switch to AUTO. After providing these instructions, the technician then leaves. A call back later in the day verifies that the system is now cooling correctly.

Service Call 2

A retail store customer calls. Their cooling system is not operating. The compressor (located outside) starts up, then shuts off. The fan motor is not running. The dispatcher tells the customer to shut the unit off until the technician arrives. The PSC fan motor for the condenser has bad bearings. The motor shaft turns freely by hand, but if power is applied to the motor when it is turning, the motor stops.

The technician arrives and speaks with the store manager. The technician then goes to the outdoor unit and disconnects the power so that the system can be controlled from outside. The room thermostat is set to call for cooling.

The technician returns to the outdoor unit, turns the disconnect on, and observes the unit operation. The compressor starts up, but the fan motor does not. The power is turned off and the compressor leads are disconnected at the load side of the contactor so that the compressor will not be damaged. The technician makes the disconnect switch again and, using a clamp-on ammeter, checks the current going to the fan motor. It is drawing current and trying to turn. It is not known at this point whether the motor is at fault or the capacitor is bad. The power is shut off. The technician spins the motor and turns the power on while the motor is turning. The motor acts like it has brakes and stops. This indicates bad bearings or an internal electrical motor problem.

The motor is changed and the new motor performs normally even with the old capacitor. The compressor is reconnected, and power is resumed. The system now cools normally. The technician changes the air filter and oils the indoor fan motor before leaving.

The following service calls do not have the solutions described. The solutions can be found in the Instructor's Guide.

Service Call 3

An insurance office customer calls because their cooling system is not operating. This system has an electric furnace with the air handler (fan section) mounted above the suspended ceiling in the stock room. The condensing unit is on the roof. The low-voltage power supply is at the air handler.

The technician arrives and goes to the room thermostat. It is set for cooling, and the indoor fan is running. The technician goes to the roof and discovers that the breaker has tripped. Before resetting the breaker, the technician must find out what caused it to trip. The cover to the electrical panel of the unit is removed. A voltage check shows that the breaker has all the voltage off. The contactor's 24 V coil is the only thing energized because its power comes from downstairs at the air handler.

The technician checks the motors for a grounded circuit by placing one lead on the load side of the contactor and the other on a ground with the meter set to R × 10,000. The meter shows a short circuit to ground. From this test, the technician does not know if the problem is the compressor or the fan motor. The wires are disconnected from the line side of the contactor to isolate the two motors. The compressor shows a normal reading (moves to 0 then returns to infinity). The fan motor shows 0 Ω resistance to ground.

What is the problem and the recommended solution?

Service Call 4

A commercial customer calls. There is no air conditioning on the second floor of their three-story office building. This building has an air handler on each floor and a chiller in the basement.

The technician arrives and goes to the fan room on the second floor. There is no need to check the chiller or the other floors because there would be complaints if the system was not operating.

The chilled water coil piping feels cold, which verifies that the chiller is running and furnishing cold water to the coil. The fan motor is not running. Since the motor may have electrical problems, the technician opens the electrical disconnect. The technician then pushes the reset button on the fan motor starter and hears the ratchet mechanism reset. The motor will not try to start while pushing the reset button because the disconnect switch is open. The unit must have been pulling too much current, causing the overload to trip.

Using an ohmmeter set to R × 10,000, the technician checks for a ground by touching one lead to a ground terminal and the other to one of the motor leads on the load side of the starter. The motor is not grounded. The resistances between each of the motor windings are also checked and found to be equal, so the motor appears to have no electrical problems. The technician then turns the motor over by hand to see if the bearings are too tight; the motor turns normally. The technician shuts the fan compartment door and then fastens a clamp-on ammeter to one of the motor leads on the load side of the starter. The electrical disconnect is then used to start the motor. When it is closed, the motor tries to start. It will not start and pulls a high amperage. The motor seems to be normal from an electrical standpoint and turns freely, so the power supply is now suspected.

The technician quickly moves the disconnect to the off position. Each fuse is checked with an ohmmeter. A blown fuse is discovered and replaced and the motor is started again. The motor starts and runs normally with normal amperage on all three phases.

Why did the fuse blow and what is the recommended solution?

Service Call 5

The owner of a truck stop calls. There is no cooling in the main dining room. The owner does some of the system maintenance, so the technician can expect some problems related to this service program.

The technician arrives and can hear the condensing unit on the roof running; it shuts off soon after the technician gets out of the truck. The owner has recently serviced the system by changing the filters, oiling the motors, and checking the belts. Since then, the fan motor has been shutting off. It can be reset by pressing the reset button at the fan contactor, but it will not run for very long.

The technician and the owner go to the stock room where the air handler is located. The fan is not running, and the condensing unit has shut off on a low-pressure cutout. The technician fastens the clamp-on ammeter to one of the motor leads and restarts the motor. The motor is pulling too much current on all three phases and seems to be making too much noise.

What is the problem and the recommended solution?

SUMMARY

- Motor problems can be divided into mechanical problems and electrical problems.
- Mechanical problems are usually either bearing or shaft problems.
- Bearing problems are often caused by belt tension.
- Shaft problems can be caused by the technician while removing pulleys or couplings.
- Motor balancing problems are normally not handled by the HVAC technicians.
- Before assuming that a vibrating motor is defective, make sure that the vibration is not caused by the system.
- Most electrical problems consist of open windings, short-circuited windings, or grounded windings.
- The laws of current flow must be used while troubleshooting motors. For current to flow, there must be a power supply, a path (conductor), and a measurable resistance.
- If the motor is receiving the correct voltage and is electrically sound, check the motor components for mechanical problems.
- Troubleshooting hermetic compressor motors is different from troubleshooting open motors because hermetic motors are enclosed and may mask motor problems that would be obvious in an open motor.

REVIEW QUESTIONS

1. What are the two main categories of motor problems?
2. In what category is a motor shaft problem?
3. Who normally works with motor balancing problems?
4. How is a stuck hermetic compressor repaired?
5. What can be done to check an open motor to determine whether the load is excessive?
6. What three electrical conditions must be met to have current flow?
7. What are the test instruments most used for diagnosing motor problems?
8. Where is a convenient place to check a motor for an electrical ground?
9. If the motor is electrically sound, what should then be checked?
10. How does checking a hermetic motor differ from checking an open motor?

COMFORT AND PSYCHROMETRICS 17

OBJECTIVES

Upon completion of this unit, you should be able to

- **Recognize the four conditions that must be controlled to achieve comfort.**
- **Explain the relationship of body temperature to room temperature.**
- **Define psychrometrics.**
- **Explain the difference between wet bulb and dry bulb temperature measurements.**
- **Define dew-point temperature.**
- **Explain the vapor pressure of water in air.**
- **Describe humidity.**
- **Plot air conditions using a psychrometric chart.**

17.1 COMFORT

Comfort describes a delicate balance of pleasant feelings in the body produced by its environment. A comfortable atmosphere describes our surroundings when we are not aware of discomfort. Providing a comfortable environment is the job of the HVAC technician. Achieving comfort involves the control of four conditions:

1. Temperature
2. Humidity
3. Air movement
4. Air cleanliness

The human body has a sophisticated control system for both protection and comfort. For example, it can move from a warm house to 0°F outside, and will immediately begin to compensate for the change in temperature. Conversely, it can move from a cool house to 95°F outside, and it will start to adjust to keep the body comfortable and prevent overheating. Body adjustments are accomplished via the circulatory and respiratory systems. When the body is exposed to a climate that is too cold, it starts to shiver (an involuntary reaction) to warm the body. In addition, the blood vessels next to the skin constrict keeping the warm blood toward the center of the body to protect vital organs. When the body gets too warm, the vessels next to the skin dilate to move the blood closer to the surrounding air in an effort to increase the heat exchange with the air. If this does not cool the body, it will perspire. When this perspiration evaporates, it takes heat from the body and cools it.

17.2 FOOD ENERGY AND THE BODY

The human body may be compared to a coal hot-water boiler. The coal is burned in the boiler to create heat. Heat is energy. Food to the human body is like the coal in a boiler. The coal in a boiler is converted to heat for space heating. Some heat goes up the flue, some escapes to the surroundings, and some is carried away in the ashes. If fuel is added to the fire and the heat cannot be dissipated, the boiler will overheat. Similarly, our bodies use food to produce energy. Some energy is stored as fatty tissue, some leaves as waste, some as heat, and some is used as energy to keep the body functioning. If the body cannot dissipate excess heat to the surroundings, it will overheat.

Body Temperature

Our body temperature is normally 98.6°F. We are comfortable when the heat level in our body due to food intake is transferring to the surroundings at the correct rate. However, certain conditions must be met for this comfortable, or balanced, condition to exist.

The body gives off and absorbs heat via the three methods of heat transfer discussed in an earlier unit: *conduction*, *convection*, and *radiation*. The evaporation of perspiration can also be considered a fourth method of heat transfer. When the indoor air is maintained within a specific range of temperatures and humidities known as the *comfort zone*, the body will release heat at a steady rate and we feel comfortable. Typically, when the body is at rest (sitting), in surroundings of 75°F and 50% relative humidity (RH) with a slight air movement, most people are comfortable. Notice that the room air at this condition is 23.6°F cooler than the human body (98.6°F - 75°F).

The following statements can be used as guidelines for comfort.

1. In winter:
   a. The comfort zone is about 72°F to 80°F with humidities between 70% and 20% RH. Lower temperatures can be offset with higher humidity levels.

b. The lower the humidity, the higher the temperature must be in order to achieve comfort.
c. Air movement is more noticeable in cooler weather.
2. In summer:
    a. The comfort zone for the cooling season is about 68°F to 78°F, with humidities between 80% and 20% RH. Higher temperatures can be offset with lower humidity levels.
    b. When the humidity is high, air movement helps to increase comfort.
3. Clothing styles vary in different parts of the country and make a slight difference in the conditioned space-temperature requirements for comfort. For example, in Maine the styles would be warmer in the winter than in Georgia, so the inside temperature of a home or office will not have the same comfort level.
4. Body metabolism varies from person to person. For example, older people and thin people often feel the cold more acutely than people with larger frames. In addition, sedentary people may feel colder than more active people at the same room temperature.

17.3 PSYCHROMETRICS

The study of air and its properties is called *psychrometrics*.

Density and Resistance

When we move through a room, we are not aware of the air inside the room, but the air has weight and takes up space like the water in a swimming pool. The water in a swimming pool is more dense than the air in the room; that is, it weighs more per unit of volume. Air weighs 0.075 lb./ft.^{3} at 70°F at sea level. The density of water under the same conditions is 62.4 lbs./ft.^{3}.

Air offers resistance to movement. To prove this, take a large piece of cardboard and try to swing it around with the flat side moving through the air. It is difficult because of the resistance of the air. The larger the area of the cardboard, the more resistance there will be.

For another example, invert an empty glass and push it down in a shallow pan of water. The air in the glass resists the water going up into the glass. Since a cubic foot of air at 70°F has a weight of 0.075 lb./ft.^{3}, the weight of air in a room can be calculated by multiplying the room volume by the weight of a cubic foot of air. In a 10 foot × 10 foot × 10 foot room, the volume is 1000 ft.^{3}, so the room air weighs 1000 ft.^{3} × 0.075 lb./ft.^{3} = 75 lbs.

The number of cubic feet of air required to make a pound of air can be obtained by calculating the reciprocal of the density. The reciprocal of a number is 1 divided by that number. The reciprocal of the density of air at 70°F is calculated as follows:

$$1 \div 0.075 \text{ lb./ft.}^3 = 13.33 \text{ cubic feet per pound of air.}$$

Superheated Gases in Air

Air is made up of nitrogen (78%), oxygen (21%), and approximately 1% other gases. These gases are highly superheated; in other words, they are far above their boiling points. At atmospheric pressure, nitrogen boils at -319°F and oxygen boils at -297°F. Each gas exerts pressure according to *Dalton's Law of Partial Pressures*. Simply stated, this law says that each gas in a mixture of gases acts independently of the other gases and the total pressure of a gas mixture is the sum of the pressures of each gas in the mixture. More than one gas can occupy a space at the same time.

Water Vapor in Air

Air is not totally dry. Surface water and rain keep moisture in the atmosphere (even in deserts) at all times. The water vapor that is suspended in the air is a gas that exerts its own pressure and occupies space with the other gases. Water at 70°F in an open dish in the atmosphere exerts a pressure of 0.7392 in. Hg (Figure 17-1). If the water vapor pressure in the air is less than the water vapor pressure in the dish, the water in the dish will evaporate slowly to the lower pressure area of the water vapor in the air. For example, the room may be at a temperature of 70°F with a humidity of 30%. The vapor pressure for the moisture suspended in the air at 30% is equal to 0.206 in. Hg. Conversely, if the water vapor pressure in the dish is less than the pressure of the vapor in the air, water from the air will condense into the water in the dish (Figure 17-2).

Note: The vapor pressures of moisture in the air can be found in some psychrometric charts and saturated water tables.

When water vapor is suspended in the air, the air is sometimes called *wet air*. If the air has a large amount of moisture, the moisture can be seen (for example, fog or a cloud). Actually, the air is not wet because the moisture is suspended in the air. This could more accurately be called a nitrogen, oxygen, and water vapor mixture.

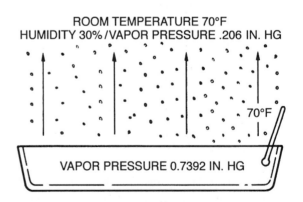

Figure 17-1. Vapor pressure of 70°F water in an open dish in a room that is also at 70°F.

PSYCHROMETRICS 237

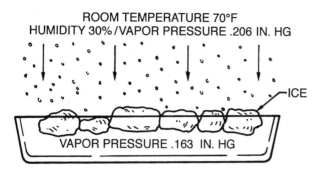

Figure 17-2. The moisture in the dish has ice on it, which lowers the vapor pressure to 0.163 in. Hg. The room temperature is still 70°F with a humidity of 30% which equates to a vapor pressure of 0.206 in. Hg.

The moisture content in air (humidity) is measured by weight, expressed in pounds or grains (7,000 grains per pound). Air can hold very little water vapor. 100% humid air at standard atmospheric pressure (29.92 in. Hg) and 70°F can hold 110.5 grains (gr) of moisture (0.01578 lb.) per pound of air (Figure 17-3). Several methods are used to calculate the percentage of moisture content in the air. *Relative humidity* is the most common and practical method. It is based on the weight of water vapor in a pound of air compared to the weight of water vapor that a pound of air could hold if it were 100% saturated (Figure 17-3).

Dry Bulb and Wet Bulb Temperatures

The moisture content of air can be checked using a combination of dry bulb (DB) and wet bulb (WB) temperatures. The dry ulb temperature represents the sensible heat level of the air and is taken with an ordinary thermometer. The wet bulb temperature is normally measured using a special thermometer known as a psychrometer. This thermometer has a wick on one end that is soaked with distilled water. The reading from a wet-bulb thermometer takes into account the moisture content of the air and reflects the total heat content of the air.

The wet bulb thermometer will be cooler than the dry-bulb thermometer due to the evaporation of the distilled water.

The difference between the dry bulb reading and the wet bulb reading is called the wet bulb depression. Figure 17-4 shows a wet bulb depression chart. As the amount of moisture suspended in the air decreases, the wet bulb depression increases and vice versa. For example, a room with a dry bulb temperature of 76°F and a wet bulb temperature of 64°F has a wet bulb depression of 12°F and a relative humidity of 52%. If the 76°F dry bulb temperature is maintained and moisture is added to the room so that the wet bulb temperature rises to 74°F, the relative humidity increases to 91% and the new wet bulb depression is 2°F. If the wet bulb depression is allowed to fall to 0°F, (e.g., 76°F dry bulb and 76°F wet bulb), the relative humidity will be 100%. The air is holding all of the moisture it can—in other words, it is saturated with moisture.

Dew-Point Temperature

The dew-point temperature is the temperature at which moisture begins to condense out of the air. For example, if you set a glass of warm water in a room with a temperature of 75°F and 50% relative humidity, the water level in the glass will slowly drop as the water evaporates. If you gradually cool the glass with ice, when the glass surface temperature falls to 55.5°F (the dew-point temperature), water will begin to form on the outside of the glass. Moisture from the room will also collect in the glass and the water level will begin to rise. Air can be dehumidified by passing it over a surface that is below the dew-point temperature of the air. When this occurs, moisture will collect on the colder surface (e.g. an air conditioning coil, as shown in Figure 17-5). This is the moisture that you see running out of the condensate line of an air conditioner.

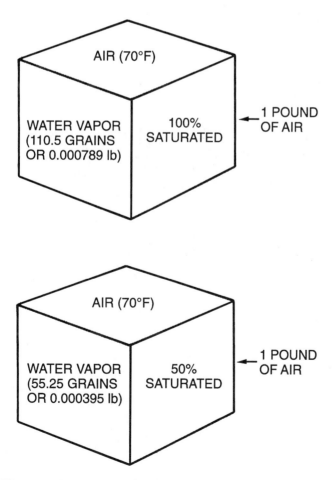

Figure 17-3. Relative humidity is based on the water vapor suspended in a pound of air compared to the weight of water vapor a pound of air could hold if 100% saturated.

DB TEMP.	1	2	3	4	5	6	7	8	9	10	11	12	13	14	15	16	17	18	19	20	21	22	23	24	25	26	27	28	29	30
32	90	79	69	60	50	41	31	22	13	4																				
36	91	82	73	65	56	48	39	31	23	14	6																			
40	92	84	76	68	61	53	46	38	31	23	16	9	2																	
44	93	85	78	71	64	57	51	44	37	31	24	18	12	5																
48	93	87	80	73	67	60	54	48	42	36	34	25	19	14	8															
52	94	88	81	75	69	63	58	52	46	41	36	30	25	20	15	10	6	0												
56	94	88	82	77	71	66	61	55	50	45	40	35	34	26	24	17	12	8	4											
60	94	89	84	78	73	68	63	58	53	49	44	40	35	31	27	22	18	14	6	2										
64	95	90	85	79	75	70	66	61	56	52	48	43	39	35	34	27	23	20	16	12	9									
68	95	90	85	81	76	72	67	63	59	55	51	47	43	39	35	31	28	24	21	17	14									
72	95	91	86	82	78	73	69	65	61	57	53	49	46	42	39	35	32	28	25	22	19									
76	96	91	87	83	78	74	70	67	63	59	55	52	48	45	42	38	35	32	29	26	23									
80	96	91	87	83	79	76	72	68	64	61	57	54	54	47	44	41	38	35	32	29	27	24	21	18	16	13	11	8	6	1
84	96	92	88	84	80	77	73	70	66	63	59	56	53	50	47	44	41	38	35	32	30	27	25	22	20	17	15	12	10	8
88	96	92	88	85	81	78	74	71	57	64	61	58	55	52	49	46	43	41	38	35	33	30	28	25	23	21	18	16	14	12
92	96	92	89	85	82	78	75	72	69	65	62	59	57	54	51	48	45	43	40	38	35	33	30	28	26	24	22	19	17	15
96	96	93	89	86	82	79	76	73	70	67	74	61	58	55	53	50	47	45	42	40	37	35	33	31	29	26	24	22	20	18
100	96	93	90	86	83	80	77	74	71	68	65	62	59	57	54	52	49	47	44	42	40	37	35	33	31	29	27	25	23	21
104	97	93	90	87	84	80	77	74	72	69	66	63	61	58	56	53	51	48	46	44	41	39	37	35	33	31	29	27	25	24
108	97	93	90	87	84	81	78	75	72	70	67	64	62	59	57	54	52	50	47	45	43	41	39	37	35	33	31	29	28	26

Header spans: WB DEPRESSION

Figure 17-4. Wet bulb depression chart. The figures in the center of the chart represent %RH values. Whenever the wet bulb depression is 0°F, the relative humidity is 100%.

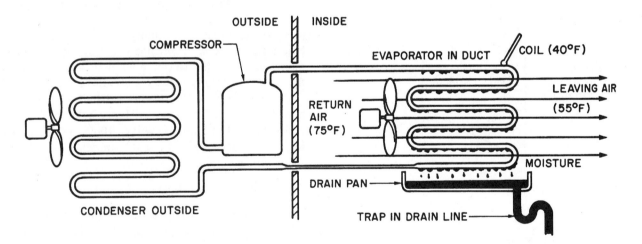

Figure 17-5. The cold surface of the evaporator coil condenses moisture from the air passed over it.

17.4 THE PSYCHROMETRIC CHART

All of the air qualities we have discussed can be plotted on a psychrometric chart (Figure 17-6A). The chart looks very complicated, but a clear plastic straightedge and a pencil will help you to understand it. See Figures 17-6B through 17-6G for some examples of plottings of the different conditions on a psychrometric chart.

If you know any two conditions previously mentioned, you can plot any of the other conditions. The easiest conditions to determine from room air are the wet bulb and dry bulb temperatures. If you do not have a psychrometer, you can make a wet bulb thermometer using the leads on an electronic thermometer. Take two leads and tape them together with one lead about two inches below the other one (Figure 17-7). A simple wick can be made from a piece of white cotton fabric. Make sure that the fabric is clean. Wet the lower bulb (the one with the wick on it) with distilled water that is warmer than the room air. Distilled water is used because the mineral content of regular tap water may effect the reading. Water from a clean condensate drain line may be used if other distilled water is not available.

THE PSYCHROMETRIC CHART 239

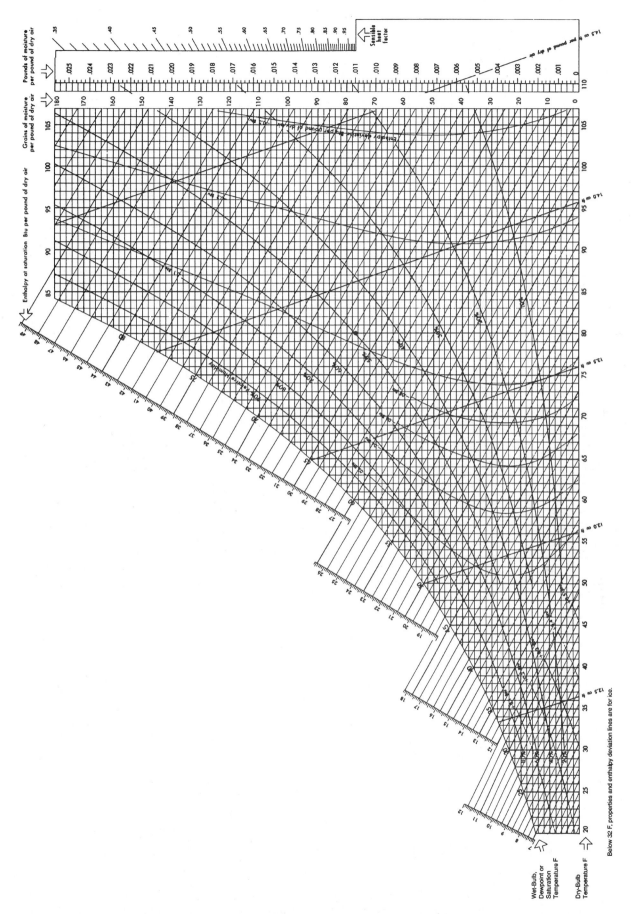

Figure 17-6A. Psychrometric chart. *Reproduced courtesy of Carrier Corporation.*

240 UNIT 17 COMFORT AND PSYCHROMETRICS

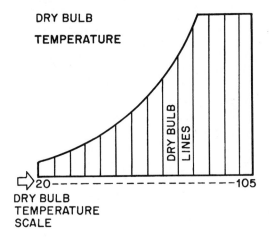

Figure 17-6B. Skeleton chart showing the dry bulb temperature lines. From Lang, *Principles of Air Conditioning*, © 1987 by Delmar Publishers Inc.

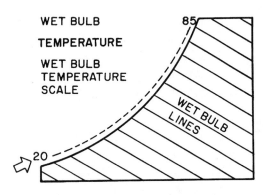

Figure 17-6C. Skeleton chart showing the wet bulb lines. From Lang, *Principles of Air Conditioning*, © 1987 by Delmar Publishers Inc.

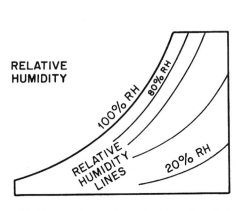

Figure 17-6D. Skeleton chart showing the relative humidity lines. From Lang, *Principles of Air Conditioning*, © 1987 by Delmar Publishers Inc.

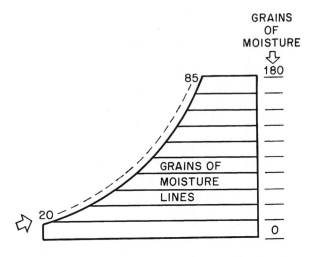

Figure 17-6E. Skeleton chart showing the moisture content of air expressed in grains per pound of air. From Lang, *Principles of Air Conditioning*, © 1987 by Delmar Publishers Inc.

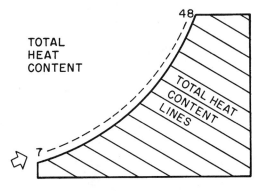

Figure 17-6F. Skeleton chart showing the total heat content of air in Btu's/lb. These lines are almost parallel to the wet bulb lines. From Lang, *Principles of Air Conditioning*, © 1987 by Delmar Publishers Inc.

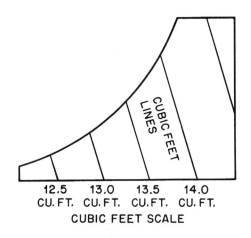

Figure 17-6G. Skeleton chart showing the specific volume of air at different conditions. From Lang, *Principles of Air Conditioning*, © 1987 by Delmar Publishers Inc.

THE PSYCHROMETRIC CHART 241

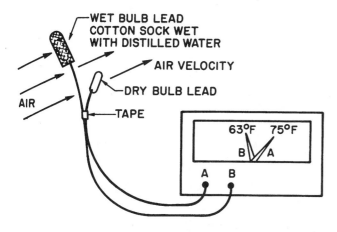

Figure 17-7. How to make a wet bulb thermometer from an electronic thermometer.

Hold the leads about three feet back from the element on the end and slowly spin them in the air. The wet lead will drop to a lower temperature than the dry lead. Keep spinning them until the lower lead stops dropping in temperature but is still damp. Figure 17-8 shows this procedure using a sling psychrometer. Quickly read the wet bulb and dry-bulb temperatures without touching the bulbs. Suppose the reading is 75°F DB and 62.5°F WB. Put your pencil point at this place on the psychrometric chart (Figure 17-6) and make a dot. Draw a light circle around it so you can find the dot again. The following information can be derived from these two measurements:

1. Dry bulb temperature   75°F
2. Wet bulb temperature   62.5°F
3. Dew-point temperature   55.5°F
4. Total heat content of 1 lb. of air   27.2 Btu's
5. Moisture content of 1 lb. of air   65 gr
6. Relative humidity   50%
7. Specific volume of air   13.6 ft.³/lb.

Figure 17-8. Technician spinning a sling psychrometer.

Plotting on the Psychrometric Chart

The condition of air can be plotted on the psychrometric chart as it is being conditioned. The following examples show how different air conditioning applications are plotted.

- Air is heated. Movement through the heating equipment can be followed as an increase in sensible heat on the chart (Figure 17-9A).
- Air is cooled. There is no moisture removal. (This is not a typical situation but is used for illustrative purposes only.) This results in a decrease in sensible heat direction on the chart (Figure 17-9B).
- Air is humidified. No heat is added or removed. This results in an increase in moisture content (latent heat) and dew-point temperature (Figure 17-9C).
- Air is dehumidified. No heat is added or removed. (This is not a typical situation but is used for illustrative purposes only). This results in a decrease in moisture content and dew-point temperature (Figure 17-9D).
- Air is cooled and humidified using an evaporative cooler. These are popular in hot, dry climates. This results in a decrease in sensible heat and an increase in latent heat (Figure 17-9E).

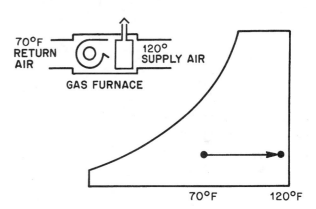

Figure 17-9A. Air passing through a sensible heat exchange furnace.

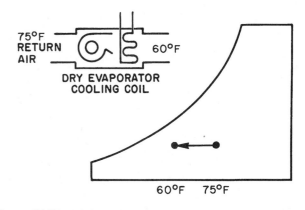

Figure 17-9B. Air is cooled with a dry evaporator coil, operating above the dew-point temperature of the air. No moisture is removed. The sensible heat level decreases.

242 UNIT 17 COMFORT AND PSYCHROMETRICS

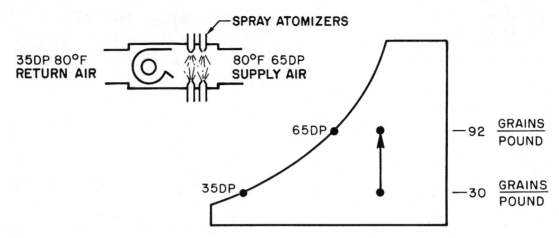

Figure 17-9C. Spray atomizers are used to add moisture to the air. The dew-point temperature and moisture content both increase.

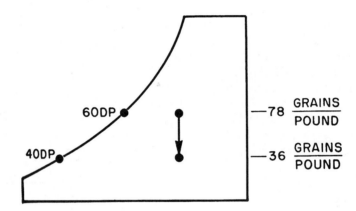

Figure 17-9D. Moisture is removed from the air without changing the air temperature. This results in a decrease in moisture content and dew-point temperature.

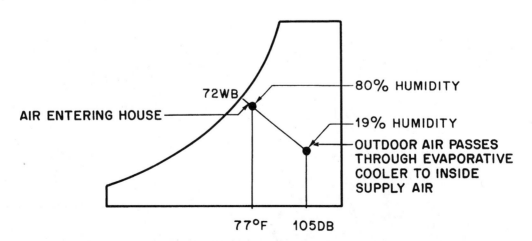

Figure 17-9E. Hot, dry air passes through the water circuit of the evaporative cooler. Heat is given up to the cooler water and the air entering the house is cooled and humidified; the water is cool via evaporation.

When air enters air conditioning equipment, it can be plotted on a chart. Figure 17-10 shows a chart indicating that from the reference point in the middle, the air may be conditioned to heat, cool, humidify, or dehumidify. Some equipment can be used to add both heat and moisture, or to remove both heat and moisture. The following examples will show what happens in the most common heating and cooling systems.

- The most common winter application is heating and humidification. This will result in a rise in temperature, moisture content, and dew-point (Figure 17-11).
- The most common summer application is cooling and dehumidification. A decrease in temperature, moisture content, and dew-point will take place (Figure 17-12).

It is important to note that any change in heat content or moisture content will cause a change in the wet bulb reading and, therefore, a change in the total heat content.

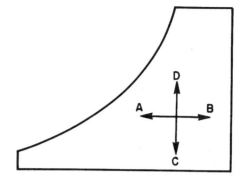

WHEN AIR IS CONDITIONED AND THE PLOT MOVES IN THE DIRECTION OF:
(A) SENSIBLE HEAT IS REMOVED.
(B) SENSIBLE HEAT IS ADDED.
(C) LATENT HEAT IS REMOVED, MOISTURE REMOVED.
(D) LATENT HEAT IS ADDED, MOISTURE ADDED.

Figure 17-10. A summation of sensible and latent heat.

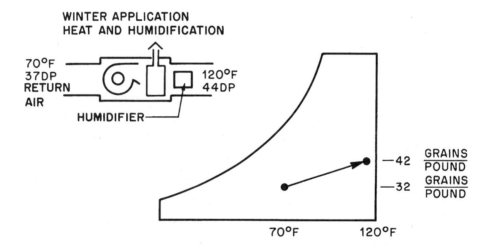

Figure 17-11. Sensible heat raises the temperature of the air from 70°F to 120°F. Moisture is evaporated and latent heat is added to the air.

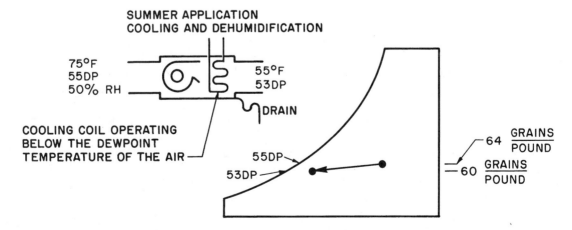

Figure 17-12. Removal of sensible heat cools the air. Removal of latent heat removes moisture from the air.

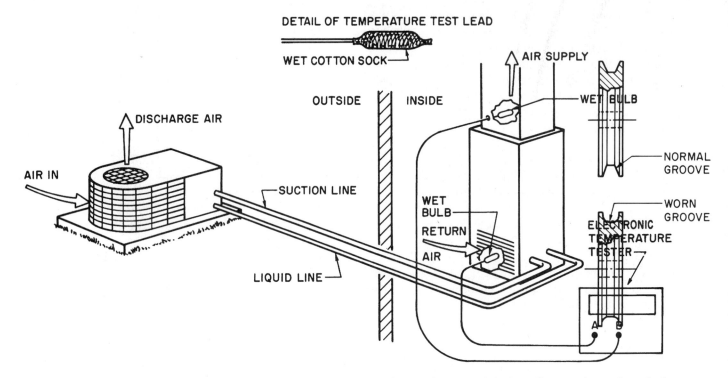

Figure 17-13. A wet bulb reading can be taken on each side of an air-to-air heat exchanger and these readings can be used to calculate the capacity of the unit using total heat.

The capacity of a heating or cooling unit may be field checked using a special formula and the psychrometric chart (discussed later). Simply stated, if the amount of air passing over a heat exchanger is known, the total heat can be checked where it enters and leaves the heat exchanger. This will provide a fairly accurate account of the performance of the heat exchanger (Figure 17-13).

17.5 VENTILATION AND INFILTRATION

The air that surrounds us has to be maintained at the correct conditions for us to be comfortable. The existing air in our homes is treated by heating it, cooling it, dehumidifying it, humidifying it, and cleaning it so that our bodies will give off the correct amount of heat for comfort. In addition, a small amount of outdoor air is induced into the air conditioner to keep the air from becoming oxygen starved and stagnant. This is called fresh-air intake or *ventilation*. If a system has no mechanical ventilation, it must rely on the air infiltrating the structure around doors and windows.

Modern, energy-efficient homes can be built so tight that infiltration does not provide enough fresh air. Studies show that indoor pollution in homes and commercial buildings has been on the increase for the past few years. As a result of the increase in awareness of energy conservation, modern buildings are constructed to allow much less air infiltration from the outside. When a home is too loose, a great deal of energy is lost due to infiltration. In this case, the homeowner may take the following measures to prevent outside air from entering the structure:

- Install storm windows and doors
- Caulk around windows and doors
- Install dampers on exhaust fans and dryer vents

All of these will reduce the amount of outside air that leaks into the structure and improve energy costs. However, care must be taken to avoid poor indoor air due to a home that is made too tight. Poor air exchange rates may lead to problems due to the following indoor air pollution sources:

- Chemicals in new carpets, drapes, and upholstered furniture
- Cooking odors
- Vapors from cleaning chemicals
- Bathroom odors and moisture
- Vapors from freshly painted rooms
- Vapors from aerosol cans, hair sprays, and room deodorizers
- Vapors from particle building board epoxy resins
- Pets and their upkeep
- Radon gas leaking into the structure from the soil

These indoor pollutants may be diluted with outdoor air in the form of ventilation. This may be accomplished with a duct from the outside to the return air side of the equipment (Figure 17-14).

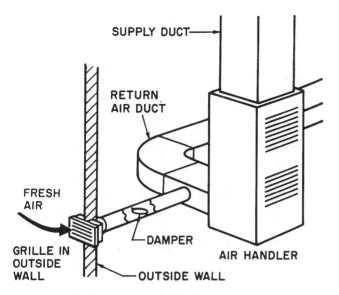

Figure 17-14. Fresh air is drawn into the return air duct to improve the air quality inside the house.

Calculating Ventilation Air for Residences

There is some discussion as to how much air should be introduced, but it is generally agreed that *at least* a 0.35 air change per hour is desirable for most residential applications. This means that 35% of the indoor air is pushed out by inducing air into the system. For example, suppose a 2000 ft.² home with an 8 foot ceiling requires ventilation because the home is very tight. How many cubic feet of air per minute must be introduced to change 35% of the air per hour?

$$2000 \text{ ft.}^2 \times 8 \text{ ft. ceiling} \times 0.35 = 5600 \text{ ft.}^3/\text{hr.}$$

$$\frac{5600 \text{ ft.}^3/\text{hr.}}{60 \text{ min./hr.}} = 93 \text{ cfm}$$

This adds a considerable load to the equipment. For example, suppose the house is located in Cleveland, Ohio, where the outdoor design temperature is 5°F in the winter. This represents the worst-case scenario for the heating system, which will occur for 1% of the time, and is found using the Design Conditions Tables located in the back of this book. If the home is to be maintained at 70°F indoors when the outdoor temperature is 5°F, there is a 65°F temperature difference. The sensible heating load (Q_s) on the HVAC equipment due to ventilation alone will be:

$$Q_s = 1.08 \times \text{cfm} \times TD$$

where: Q_s = sensible heat

1.08 = a constant

TD = temperature difference

$Q_s = 1.08 \times 93 \times 65$

$Q_s = 6528.6$ Btuh

Remember, the outside temperature of 5°F is only for 1% of the year. The fresh air will be warmer than 5°F the other 99% of the year. Because of this, many system designers will elect to use a higher design temperature for system equipment selection. In the event of a cold winter, indoor conditions may be colder than desired. This may result in initial savings if smaller equipment is selected, but will present a comfort problem during a very cold winter.

The summertime calculation is made in much the same manner, except a different formula is used. Both sensible and latent heat must be taken into account. In the summer, the design temperatures for Cleveland, Ohio are 91°F dry bulb and 73°F wet bulb. The total heat formula may be used for this calculation.

$$Q_t = 4.5 \times \text{cfm} \times \text{total heat difference}$$

where: Q_t = total heat

4.5 = a constant used to change pounds of air to cfm

total heat difference = the difference in total heat between the indoor and outdoor air

To solve the problem of fresh air, plot the indoor air and the outdoor air on the psychrometric chart. The derived values are as follows:

91°F dry bulb and 73°F wet bulb = 36.75 Btu's/lb.

75°F dry bulb and 50% RH indoors = 28.20 Btu's/lb.

total heat difference = 8.55 Btu's/lb. difference

Next, insert the total heat difference into the above formula:

$Q_t = 4.5 \times \text{cfm} \times \text{total heat difference}$

$Q_t = 4.5 \times 93 \times 8.55$

$Q_t = 3578.18$ Btu's/hr. total heat added due to ventilation

System designers would require this total heat calculation to be broken down into the sensible heat gain and the latent heat gain. Equipment must be selected for the correct sensible and latent heat capacities or space humidity will not be correct. This is done by calculating the sensible heat and the latent heat formula as separate calculations. You will not arrive at the exact same total heat, because you cannot see the lines on the psychrometric chart close enough for total accuracy.

The following exercise calculates sensible heat, then the latent heat formula uses the grains of moisture in the air to calculate the latent heat load.

$$Q_s = 1.08 \times \text{cfm} \times TD$$

plus

$$Q_l = 0.68 \times \text{cfm} \times \text{grains difference}$$

where: Q_l = latent heat

0.68 = a constant used to change cfm to pounds of air and grains per pound

grains difference = the difference in the grains per pound of air for the indoor air and outdoor air

246 UNIT 17 COMFORT AND PSYCHROMETRICS

For our Cleveland, Ohio example:

$$Q_s = 1.08 \times cfm \times TD$$
$$Q_s = 1.08 \times 93 \times 16 \ (91 - 75)$$
$$Q_s = 1607.04 \text{ Btuh sensible heat}$$

From the psychrometric chart points plotted earlier, you will find that the outdoor air contains 93 grains per pound and the indoor air contains 64 grains per pound, for a difference of 29 grains per pound of air. Therefore:

$$Q_l = 0.68 \times cfm \times \text{grains difference}$$
$$Q_l = 0.68 \times 93 \times 29$$
$$Q_l = 1833.96$$

Finally, we can combine the two values to find the total heat:

Total heat = $Q_s + Q_l$

Total heat = 1607.04 + 1833.96

Total heat = 3441 Btuh

Commercial Ventilation Requirements

Office buildings have the same pollution problems as residences, except that they are multiplied to a certain degree. For example, office buildings have a tendency to be remodeled more often, leading to more construction-type pollution. Copying machines that use liquid copy methods give off vapors. Also, there are more people per square foot in office buildings.

The national code requirements for fresh air in public buildings is based on the number of people in the building and the type of use. Different buildings may have different indoor pollution rates because of the activity in the building. For example, the requirements for a department store are much less stringent than those for a restaurant. In a department store, shoppers are generally active and are less sensitive to stagnant indoor air. In a restaurant, diners are at rest and additional ventilation is required to offset the odors of cooking and smoking and maintain a pleasant indoor environment. Figure 17-15 shows some of the fresh air requirements recommended by ASHRAE for different applications.

The design engineer is responsible for choosing the correct fresh air makeup for a building, and the service technician is responsible for regulating the airflow in the field. The technician may be given a set of specifications for a building and told to regulate the outdoor air dampers to induce the correct amount of outdoor air into the building.

The mixed air condition may be determined by calculation for the purpose of setting the outside air dampers. The condition may be plotted on the psychrometric chart for ease of understanding. The dampers may then be adjusted for the correct conditions. For example, suppose a building with a return air condition of 73°F DB and 60°F WB is taking in and mixing air from the outside of 90°F DB and 75°F WB. When these two conditions are plotted on the psychrometric chart and a line is drawn between them, the condition of the mixed air will fall on the line (Figure 17-16). If the mixture is ½ outdoor air and ½ indoor air, the condition of the mixed air will be midway between points A and B. Figure 17-17 shows some examples of air mixtures.

RECOMMENDED OUTDOOR AIR VENTILATING RATES
Abbreviated from ASHRAE Standard 62-1989

	cfm or see footnote
Dining Room	20
Bars & cocktail lounges	30
Hotel conference rooms	20
Office spaces	20
Office conference rooms	35
Retail stores	.02 to .03(a)
Beauty shops	25
Ballrooms & discos	25
Spectator areas	15
Theater auditoriums	15
Transportation waiting rooms	15
Classrooms	15
Hospital patient rooms	25
Residences	.35(b)
Smoking lounges	60

a cfm per square feet of space
b air changes per hour

Figure 17-15. Fresh air requirements for some typical applications. *Reprinted by permission of American Society of Heating, Refrigerating and Air-Conditioning Engineers.*

VENTILATION AND INFILTRATION 247

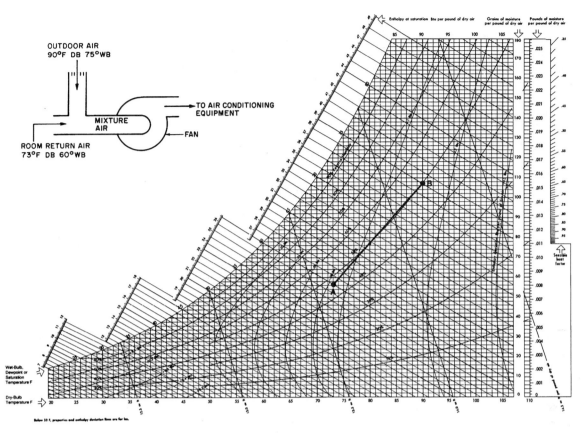

Figure 17-16. Point A is the condition of the indoor air and point B is the condition of the outdoor air. When the two are mixed, the mixed air condition will fall on the line drawn from A to B. *Adapted from Carrier Corporation Psychrometric Chart.*

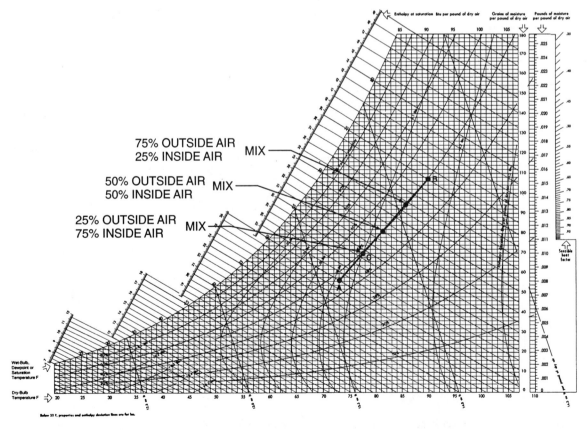

Figure 17-17. Mixture of different percentages of outside and inside air. *Adapted from Carrier Corporation Psychrometric Chart.*

UNIT 17 COMFORT AND PSYCHROMETRICS

The point on line A to B may be calculated for any air mixture if the percentage of either air, outdoor or indoor, is known. For example, if the mixture is 25% outdoor air and 75% indoor air, the following calculation can be used to find the DB point on the line. When the DB point is known and it falls on the line, you have a coordinate to find any other information for this air condition.

25% outdoor air: 0.25 × 90°F = 22.50°F
75% indoor air: 0.75 × 73°F = 54.75°F
Mixture DB: 22.5°F × 54.75°F = 77.25

Figure 17-17 shows this mixed air DB temperature at point C. Note that the mixed air temperature will always fall closest to the air temperature with the greater percentage of air.

Like any other system problem, the first step when approaching a ventilation problem is to determine how the system *should be* operating, and then compare that to how the system *is* operating. For example, suppose the building specifications are as follows:

- Total building air = 50,000 cubic feet per minute (cfm)
- Recirculated air = 37,500 cfm, 75% return air
- Makeup air = 12,500 cfm, 25% outdoor air
- Indoor design temperature = 75°F DB, 50% RH (62.5°F WB)
- Outdoor design temperature (Atlanta, Georgia) = 94°F DB, 74°F WB

The technician arrives on the job to check the outdoor air percentage and finds that the outdoor temperature is not at the design temperature. It is 93°F DB and 75°F WB and the indoor conditions are 73°F DB and 59°F WB. What should the mixed air temperature be?

A calculation for the mixed air is made. The actual conditions are plotted in Figure 17-18.

93°F × 0.25 = 23.25
73°F × 0.75 = 54.75
Mixed air DB = 23.25°F + 54.75°F = 78°F

The technician checks the mixed air temperature. If it is too high, the outdoor air dampers will be closed slightly until the mixed air stabilizes at the correct temperature. If the mixed temperature is too low, the outdoor dampers will be opened slightly until the mixed air temperature is stable and correct.

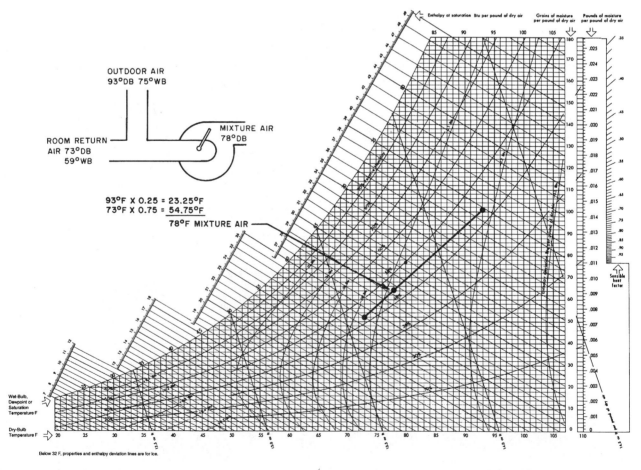

Figure 17-18. Use this psychrometric plot to find the mixed air condition for the sample problem. Point C represents the mixed air. *Adapted from Carrier Corporation Psychrometric Chart.*

Note: The outdoor and indoor conditions should be rechecked after adjustment because of weather changes and indoor condition changes. These calculations and proper damper settings also depend on correct instrumentation while checking the air temperatures.

Capacity Checks

We can also use the psychrometric chart for field checking the capacity of HVAC equipment. For example, suppose the cooling capacity of a 5 ton (60,000 Btuh) unit in a shoe store is in question and the unit uses gas as the heat source. The heat source is used to determine the cfm of the equipment. For example, most standard efficiency gas-burning appliances are 80% efficient.

It is a good idea to perform the cfm part of this problem first thing in the morning, because it is going to involve operating the heating system long enough to arrive at the cfm. The unit heat is turned on and an accurate temperature rise across the gas heat exchanger is taken at 69.5°F.

Note: Be careful not to let radiant heat from the heat exchanger influence the thermometer lead.

The unit nameplate lists an input of 187,500 Btuh. If the unit has a heating efficiency rating of 80%, the output would be 80% of the input, or 187,500 Btuh × 0.80 = 150,000 Btuh. The fan motor will add about 600 watts (3.431 Btu's per watt = 2048 Btu's for a total of 152,048 Btuh).

Note: The fan is operated using the fan on position on the room thermostat. This ensures the same fan speed in heating and cooling.

Using the sensible heat formula, we can find the unit cfm.

$$Q_s = 1.08 \times \text{cfm} \times \text{TD}$$

Solved for cfm:

$$\text{cfm} = \frac{Q_s}{108 \times \text{TD}}$$

$$\text{cfm} = \frac{152,048}{1.08 \times 69.5}$$

$$\text{cfm} = 2026 \text{ ft.}^3/\text{min.}$$

Next, switch the unit to cooling, allow it to run for 15 minutes, then check the wet bulb temperature in the entering and leaving air stream. Suppose the entering WB is 62°F and the leaving WB is 53°F (Figure 17-19).

Total heat at 62°F WB = 27.85

Total heat at 53°F WB = 22.00

Total heat difference = 5.85 Btu's/lb. of air

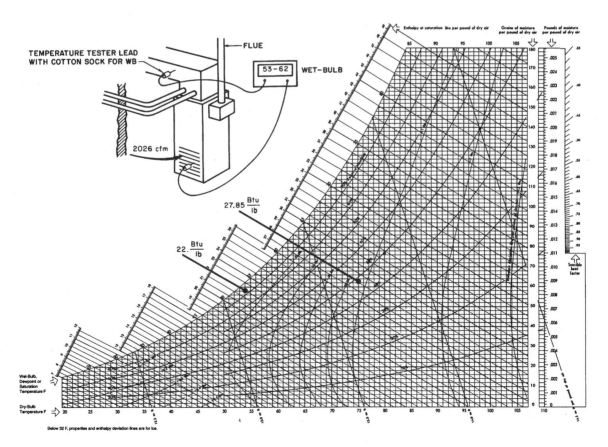

Figure 17-19. Capacity check for cooling mode. *Adapted from Carrier Corporation Psychrometric Chart.*

Using the total heat formula:

$$Q_t = 4.5 \times cfm \times \text{total heat difference}$$
$$Q_t = 4.5 \times 2026 \times 5.85$$
$$Q_t = 53{,}334 \text{ Btuh total heat}$$

The unit has a cooling capacity of 60,000 Btuh, so it is operating close to capacity. This is about as close as a technician can expect to obtain using field calculations.

SUMMARY

- *Comfort* describes a delicate balance of pleasant feelings in the body produced by its environment.
- The body burns food and converts it to energy. The body stores energy as fatty tissue, expels it as waste, consumes it in work, and releases it as heat to its surroundings.
- For the body to be comfortable, it must be warmer than its surroundings, so it can give up excess heat to the surroundings.
- The body gives off heat in three conventional ways: conduction, convection, and radiation. Evaporation may be considered a fourth way.
- Air contains 78% nitrogen, 21% oxygen, and 1% other gases, as well as suspended water vapor.
- The density of air at 70°F and 29.92 in. Hg is 0.075 lbs./ft^3.
- The specific volume of air is the reciprocal of the density: $1 \div 0.075 = 13.33$ ft.^{3}/lb.
- The moisture content of air affects the transfer of heat from the human body; therefore, different temperatures and humidities can provide the same relative comfort level.
- Slight air movement can help to offset higher temperatures in the summer.
- Dry bulb temperature is registered with a regular thermometer.
- Wet-bulb temperature is registered with a thermometer that has a wet wick. The wet bulb thermometer lead gets colder than the dry bulb thermometer lead because the moisture evaporates on the wick.
- The difference between the wet bulb reading and the dry bulb reading is the wet bulb depression. It can be used to determine the relative humidity of a conditioned space.
- The dew-point temperature of air is the point at which moisture begins to condense out of the air.
- Water vapor in the air creates its own vapor pressure.
- The psychrometric chart is used to plot various air conditions.
- The wet bulb reading on a psychrometric chart shows the total heat content of a pound of air.
- When the cfm is known, the wet bulb reading in and out of a heat exchanger can provide the total heat being exchanged. This can be used for field calculating the capacity of a unit.

REVIEW QUESTIONS

1. Name four comfort factors.
2. Name three ways the body gives off heat.
3. How does perspiration cool the body?
4. How can lower room temperatures be offset in winter?
5. What two conditions can offset higher room temperatures in summer?
6. How is the relative humidity of a conditioned space measured?
7. What is the name of the chart used to plot the various air conditions?
8. What two unknowns are the easiest to obtain in the field to plot air conditions?
9. Describe the dew-point temperature.
10. What is the density of air at 70°F?

ROOM AIR CONDITIONERS 18

OBJECTIVES

Upon completion of this unit, you should be able to

- Describe the various methods of installing window air conditioning units.
- Discuss the differences between window and through-the-wall units.
- List the major components in the refrigeration cycle of a window cooling unit.
- Explain the purpose of the heat exchange between the suction line and capillary tube.
- Describe the heating cycle in the heat pump or reverse cycle room air conditioner.
- Describe the controls for room air conditioning (cooling) units.
- Describe the controls for room air conditioning (cooling/heating) units.
- Discuss the service procedures for room air conditioners.
- List the procedures to be followed when determining whether or not to install gages.
- State the proper procedures for charging a room air conditioner.
- Explain the circumstances under which an expansion valve may be substituted for a capillary tube.
- List the components in a room unit that may require electrical service.

18.1 AIR CONDITIONING AND HEATING WITH ROOM UNITS

Room air conditioning is the process of controlling the temperature and humidity of one or more rooms using individual air conditioning units.

Room units are constructed of the same components that we have discussed in detail earlier in this book. When a particular component or service technique is mentioned, you will be referred to the appropriate section of the text for the necessary details. Situations that are specific to room units will be discussed here.

Single room air conditioning can be accomplished in several ways, but each way involves the use of package (self-contained) systems. The most common type is the room air conditioning window unit for cooling only. This type of unit may be expanded to include electric strip heaters in the air stream with controls for both heating and cooling. Adequate air circulation between rooms allows many owners to utilize a room unit for more than one room.

18.2 COOLING-ONLY UNITS

Cooling-only units may be either window or through-the-wall types (Figure 18-1). They are much the same, typically having only one fan motor for both the evaporator and condenser. The capacity of these units may range from about 4000 Btuh (1/3 ton) to 24,000 Btuh (2 tons).

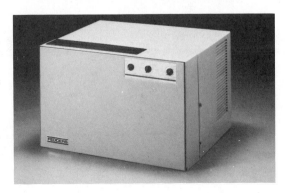

Figure 18-1. Window and through-the-wall units. *Courtesy Fedders Air Conditioning USA, Inc.*

251

Some units are front-air discharge and others are top-air discharge. Top-air discharge is more common for through-the-wall units. These controls are located on top and are commonly found in motels where individual room control is desirable. The controls are built in so the unit may be serviced in place or changed for a spare unit and repaired in the shop.

Both window and wall units are designed for easy installation and service. Window units have two types of cases. One type is fixed to the chassis of the unit and the other is a case that fastens to the window opening and the chassis slides in and out. Older units were all slide-out types, while the smaller, newer units have the case attached to the chassis.

The case design is important from a service standpoint. For units that have the case attached to the chassis, the entire unit must be removed for service. On units with a slide-out chassis, the chassis may be removed for easy access.

One special application unit manufactured by several companies is a roof-mount design for travel trailers and motor homes (Figure 18-2). Because it is up out of the way, this unit is also suited for use in certain small buildings, such as gas station attendant booths. The case on this unit lifts off for service. The controls and part of the control circuit are located next to the air discharge (Figure 18-3).

The manufacturer's primary design objectives for room units are efficiency of space and equipment and low noise level. The units are as compact as current manufacturing and design standards will allow. The intent is to get the most capacity from the smallest unit.

Cooling-Only Refrigeration Cycle

You should have a full understanding of Unit 3 before studying the following text.

The most common refrigerant used for room units is R-22. This refrigerant will be the only one discussed here. When a refrigerant of another type is encountered, the pressure and temperature chart may be consulted for

Figure 18-3. In this roof-mount unit, part of the control circuit is located under the front cover. *Photo by Bill Johnson.*

the difference in pressure. The operating temperatures will be the same.

The refrigeration cycle consists of the same four major components described in Unit 3: an evaporator to absorb heat into the system (Figure 18-4), a compressor to pump the heat-laden refrigerant through the system (Figure 18-5), a condenser to reject the heat from the system (Figure 18-6), and an expansion device (usually a capillary tube) to control the flow of refrigerant (Figure 18-7). This is done by maintaining a pressure difference between the high-pressure and low-pressure sides of the system. Most units have a heat exchange between the metering device (capillary tube) and the suction line (Figure 18-8). The complete refrigeration system is shown in Figure 18-9.

Figure 18-2. Roof-mount unit for recreation vehicles. Sometimes these are used in small standalone buildings. *Photo by Bill Johnson.*

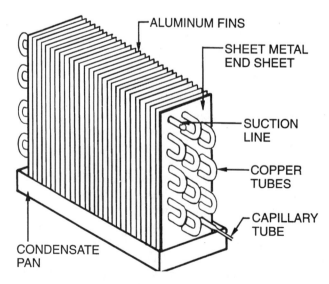

Figure 18-4. Room unit evaporator.

COOLING-ONLY UNITS 253

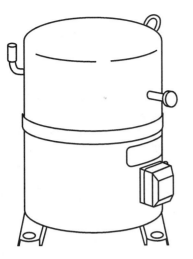

Figure 18-5. Room unit compressor.

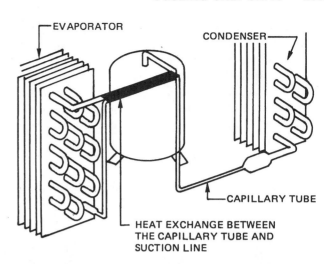

Figure 18-8. Suction-to-capillary tube heat exchange.

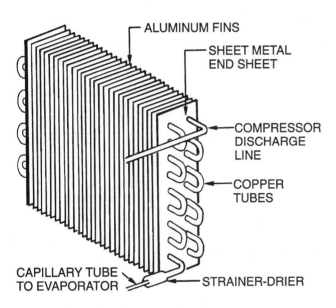

Figure 18-6. Room unit condenser.

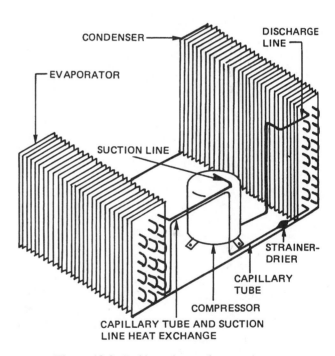

Figure 18-9. Refrigeration cycle components.

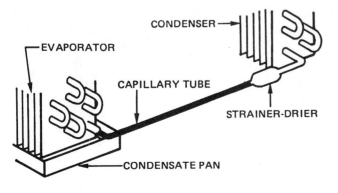

Figure 18-7. Room unit capillary tube.

Cooling units are considered high temperature refrigeration systems and must operate above 32°F to prevent condensate from freezing on the coil. Typically, the evaporators boil the refrigerant at about 35°F. Remember, central air conditioners normally boil the refrigerant at 40°F. Room units boil the refrigerant much closer to freezing, so more care must be taken to prevent freezing of the evaporator.

Room unit evaporators are small and designed for optimum heat exchange. They are usually constructed of copper or aluminum tubing with aluminum fins.

Typical designs include fins that force the air to move from side to side and tubes that are staggered to force the air to pass in contact with each tube (Figure 18-10).

The evaporator may have several refrigerant circuits. They may take many different paths, such as two, three, or four circuits that may be in line with the airflow (series) or that operate as multiple evaporators (Figure 18-11).

The evaporator typically operates below the dew-point temperature of the room air for the purpose of dehumidification. Condensate forms on the coil and drains to a pan beneath the coil. The condensate is generally drained back to the condenser section and evaporated.

Figure 18-12 shows the pressure/temperature conditions in a typical evaporator.

The compressor for room units is typically hermetically sealed. It may be either rotary or reciprocating. Compressors are described in detail in Unit 20.

Figure 18-10. Some fins force the air to move against the fins and tubes.

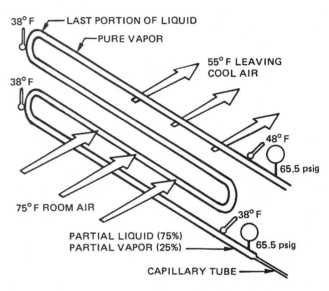

Figure 18-12. Typical evaporator pressures and temperatures.

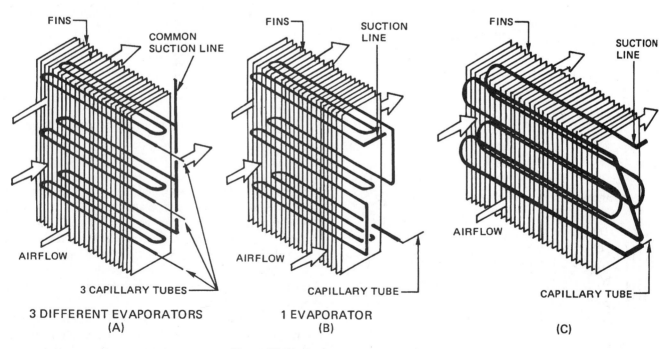

Figure 18-11. Typical evaporator circuits.

COOLING-ONLY UNITS 255

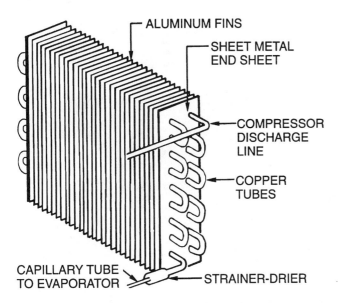

Figure 18-13. Finned-tube condenser.

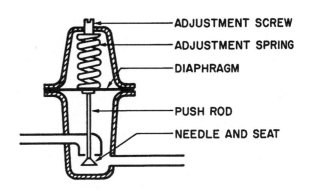

Figure 18-15. Some older room units use an automatic expansion valve rather than a capillary tube.

Room unit condensers typically have copper or aluminum tubes and aluminum fins, similar to the evaporator (Figure 18-13). The condenser serves two purposes: it condenses the heat-laden vapor refrigerant inside the tubes and evaporates the condensate from the evaporator section. This is usually accomplished by using the heat from the discharge line or a slinger ring on the condenser fan (Figure 18-14). Evaporating the condensate serves two purposes: it keeps the unit from dripping and improves the efficiency of the condenser.

The capillary tube is the metering device used in most room units manufactured during the past several years. Some early units used the automatic expansion valve (Figure 18-15). The automatic expansion valve has the advantage of controlling the pressure, which in turn controls the coil temperature. This can be used to prevent the coil from freezing. Unit 22 describes automatic expansion devices.

Most room units are designed to exchange heat between the capillary tube and the suction line. This exchange adds some superheat to the suction gas and subcools the refrigerant in the first part of the capillary tube. The pressure and temperature of the refrigerant reduces all along the tube without a heat exchange (Figure 18-16). The tube is colder at the outlet (where it enters the evaporator) than at the inlet (where it leaves

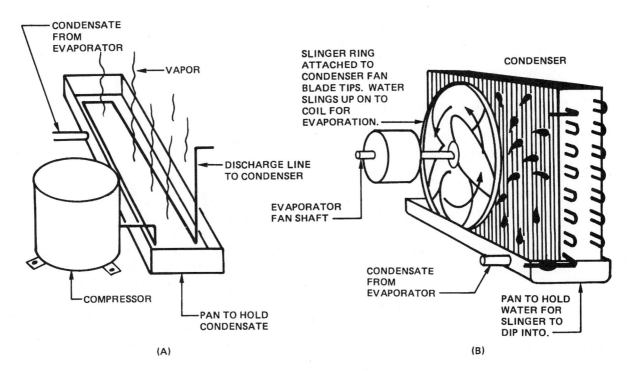

Figure 18-14. Moisture evaporated with the compressor discharge line (A) or (B) slinger.

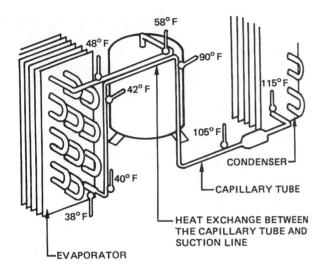

Figure 18-16. Typical pressures and temperatures along the capillary tube-to-suction line heat exchange.

Figure 18-17. The reversing valve for a heat pump. *Photo by Bill Johnson.*

the condenser). When this capillary tube is attached to a suction line, it will result in an increase in temperature at the compressor. The increase in subcooling of the capillary tube may also help the manufacturer to achieve additional capacity from the evaporator coil. It should be noted that this heat exchange improves the capillary tube function and hurts the suction line function as far as efficiency is concerned. There is no efficiency change in the overall unit.

18.3 COMBINATION COOLING/HEATING UNITS

Heating may be accomplished with either electric strip heaters or reverse cycle heat. Reverse cycle heat is more complex, but is also more economical. Some window and through-the-wall units have reverse cycle capabilities similar to heat pumps. They can absorb heat from the outdoor air in the winter and reject heat to the indoors. This is accomplished with a four-way (reversing) valve (Figure 18-17). This valve is used to redirect the suction and discharge gas at the proper time to provide heat or cooling. Check valves are used to assure flow through the correct metering device at the proper time.

Heat Pump Refrigeration Cycle

The following is a typical cycle description. There are many variations, depending on the manufacturer. Follow along in Figure 18-18 as we review the refrigeration cycle of a typical unit.

During the cooling cycle, the refrigerant leaves the compressor discharge line as a hot gas. The hot gas enters the reversing valve and is directed to the outdoor coil where the heat is rejected to the outdoors. The refrigerant is condensed into a liquid, leaves the condenser, and flows to the indoor coil through the capillary tube.

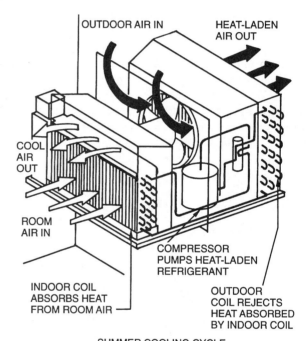

Figure 18-18. Refrigerant cycle and description (cooling).

The refrigerant is expanded in the indoor coil where it boils into a vapor, just like the cooling cycle on a regular cooling unit. The cold, heat-laden vapor leaves the evaporator and enters the reversing valve body. The piston in the reversing valve directs the refrigerant to the suction line of the compressor where the refrigerant is compressed and the cycle repeats itself.

The heating cycle may be followed in Figure 18-19. The hot gas leaves the compressor and enters the reversing valve body as in the previous example. However, the piston in the reversing valve is shifted to the heating position and the hot gas now enters the indoor coil.

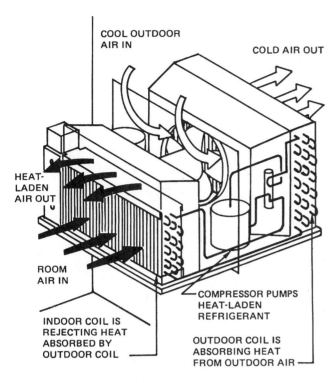

Figure 18-19. Refrigerant cycle and description (heating).

The indoor coil now acts as the condenser and rejects heat to the conditioned space. The refrigerant condenses into a liquid and flows out the indoor coil through the capillary tube and to the outdoor coil. The refrigerant is expanded and absorbs heat while boiling into a vapor. The heat-laden vapor leaves the outdoor coil and enters the reversing valve where it is directed to the compressor suction line. The vapor is compressed and moves to the compressor discharge line to repeat the cycle.

The compressor in a heat pump is not a conventional air conditioning compressor. It is a heat-pump compressor; it has a different displacement and horsepower. The compressor must have enough pumping capacity for low-temperature operation. The unit evaporator also operates at temperatures below freezing during the heating cycle and frost will build up on the evaporator. Different manufacturers handle this problem in different ways; for example, a defrost cycle may be used where the valve may reverse to the cooling cycle until defrost is accomplished. Another point to remember is that the capacity of a heat pump is lower in colder weather, so supplemental heat will probably be used. Electric strip heat is the most common form of auxiliary heat used with heat pumps. It may also be used to prevent the unit from blowing cold air during the defrost cycle.

18.4 INSTALLATION

There are two types of installation for room units: window installation and through-the-wall installation.

A window installation may be considered temporary because a unit may be removed and the window put back to its original use. A wall installation is permanent because a hole is cut in the wall and must be patched if the unit is removed. When a window air conditioner must be replaced, a new unit can usually be found to fit the window because windows are generally standard in size. A new unit may not be available to fit the hole in a wall.

Window installations are either for double-hung windows or casement-type windows. The double-hung window unit is the most popular because there are more double-hung windows. Window units may be installed in other types of windows, such as picture windows or jalousie windows, but these require considerable carpentry skills. A special unit is manufactured for casement window installation (Figure 18-20).

Every installation should follow the general guidelines provided here. The first consideration is always safety. Keep the following in mind when installing any HVAC appliance:

- Wear back brace belts when lifting or moving any appliance.
- Lift with your legs keeping your back straight. Whenever possible use appropriate equipment to move heavy objects.
- Observe proper electrical safety techniques when servicing any appliance or system.
- Wear goggles and gloves when transferring refrigerant into or out of a room air conditioning unit.

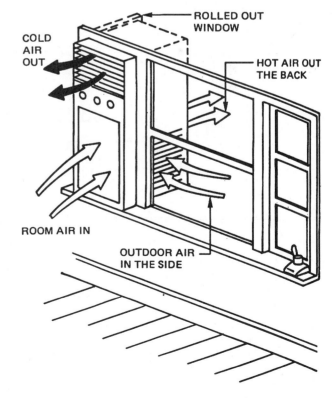

Figure 18-20. Window unit installed in a casement window.

258 UNIT 18 ROOM AIR CONDITIONERS

The inside part of the installation requires that the proper electrical outlet be located within the length of the cord furnished with the unit.

Caution: Extension cords are not recommended by any manufacturer.

The proper electrical outlet must match the plug on the end of the cord (Figure 18-21). Window units may be either 115 V or 208/230 V and should be on a circuit by themselves.

When 115 V operation is chosen, it is usually to prevent having to install an electrical circuit. Often this choice is made without investigating the existing load on the circuit. For example, a 115 V electrical outlet may be part of the lighting circuit and have a television and several lamps on the same circuit. The addition of a window air conditioner of any size may overload the circuit (Figure 18-22).

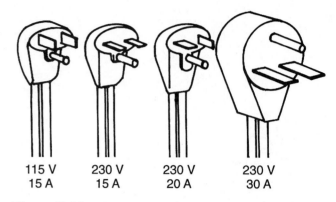

Figure 18-21. The proper outlet must be used to match the plug.

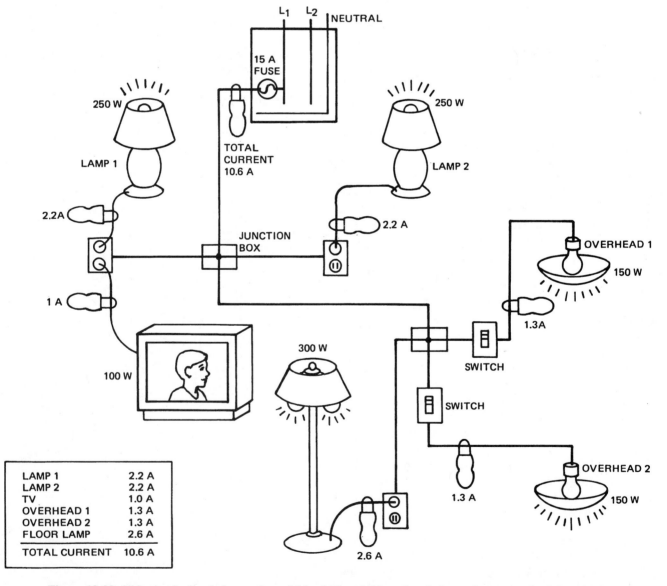

Figure 18-22. This circuit already has a substantial load. The addition of a window unit may overload the circuit.

Before adding a load to a circuit, first determine if the circuit has the extra capacity required for the unit. A typical 5000 Btuh unit will pull about 7 amperes when 115 V is used. To determine the existing load on a circuit, turn on all the lights and other appliances in the room where the unit is to be installed. The lights in adjacent rooms may also be turned on. Then turn off the breaker that controls the outlet. The current load of the lights may be totaled by using an ammeter on the wire to the circuit. For example, if the circuit is already using 10 amperes and it is a 15 ampere circuit, the addition of a window unit consuming 7 amperes will overload the circuit. A homeowner, who buys a unit at a department store and installs it without examining the load may be in trouble. If the unit will even run on an overloaded circuit, the television picture will roll over every time the compressor starts.

Other considerations on the inside of the room are the best window to use and the air direction. A window that will give the best total room circulation may not be the best choice as it may be next to the easy chair in a den, the bed in a bedroom, or behind the dining table. The blast of cold air from the unit may cause occupant discomfort.

Some people will purchase a large window unit to condition several rooms. The unit must be able to circulate the air to the adjacent rooms or the thermostat will shut the unit off and cool only one room. A floor fan may be used to move the air between rooms (Figure 18-23).

Through-the-wall units in motels are usually installed on the outside wall under the window, where the table and chairs are located. This is the best practical place for the unit, but is not necessarily the best location for occupant comfort (Figure 18-24).

A unit located where air will recirculate (e.g., behind drapes) will be a problem in any installation. Air will continually recirculate to the return air grille and be cold enough to satisfy the thermostat, shutting the unit off.

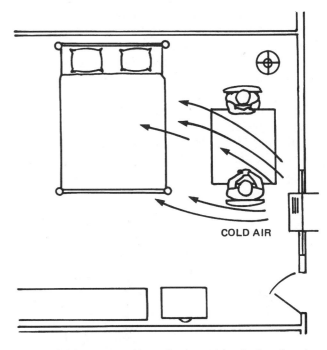

Figure 18-24. The outside wall may not be the best location for comfort, but in most rooms, it is the only place to locate an air conditioner.

Some small units have no thermostat, only an on-off switch. In this case, recirculation may freeze the evaporator.

Typically, the airflow should be directed upward, because cold air will fall (Figure 18-25). This will also keep the supply air from mixing with the return air and freezing the evaporator or satisfying the thermostat.

The window must be wide enough for the unit to be installed and open high enough for the unit to be positioned on the window ledge.

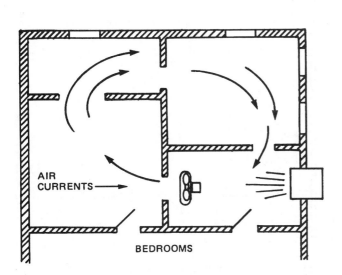

Figure 18-23. A floor fan may be used to circulate air to adjacent rooms.

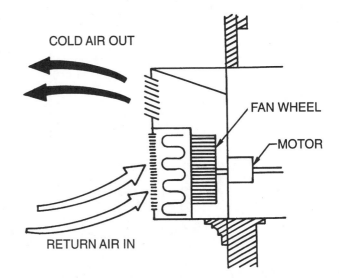

Figure 18-25. Cold air should be directed upward for best air distribution.

Smaller window units are designed so that when they are placed on the window ledge, they will almost balance. The center of gravity of the unit is near the center of the unit (Figure 18-26). In larger units, the compressor is located more to the rear, causing the unit to have a tendency to fall out the window unless half of the unit is extended into the room (Figure 18-27). This of course is not desirable. Most manufacturers provide a brace kit to support the back of the unit. The bracket takes the pressure off the upper part of the window and places the load on the bracket.

When the unit is placed in the window, it never completely fills the opening. A kit is provided with each new unit to aid in neatly filling the hole. This kit may include either telescoping side panels or straight panels that match the unit color and can be cut to fit. When the window is partly raised to accommodate the unit, there will be a space between the movable window parts that must be insulated. A foam strip is usually provided for this purpose.

Directions are included with each new unit. However, if the unit is removed (e.g. for winter storage) the braces and window kit may be misplaced. If the unit is reinstalled without braces or a proper window kit, a very poor installation may result. Units that are not properly braced will vibrate, and are sometimes called window shakers. A properly braced unit will typically be slanted slightly to the rear for proper condensate drainage (discussed later). When the weight is supported by the back, it relieves the pressure on the window and reduces the vibration.

The unit must be installed in such a manner that air can move across the condenser and does not recirculate. It must be allowed to escape the vicinity and not heat the surrounding air. For example, window units have been located with the condenser in a spare room, such as in a business (Figure 18-28). This is poor practice as recirculation will cause high head pressure and higher operating costs. The same problems will occur when the condenser is obstructed by outdoor foliage (Figure 18-29). Always check for potential obstructions when installing these units.

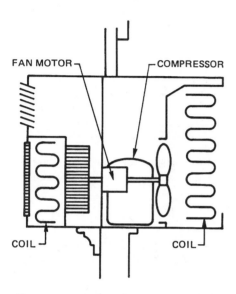

Figure 18-26. The center of gravity of a small window unit is near the center of the unit. This allows it to almost balance on the window ledge.

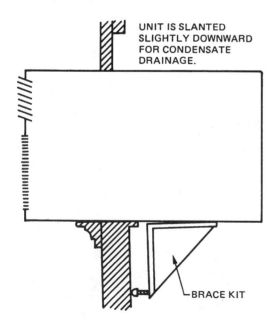

Figure 18-27. Larger units should be positioned so that they do not protrude into the room. A brace kit should be used to support the unit.

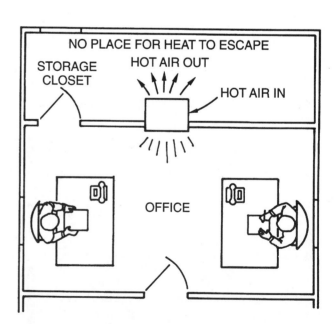

Figure 18-28. The heat must be able to escape from around the unit. Units with the condenser in an adjacent room will not operate efficiently.

INSTALLATION 261

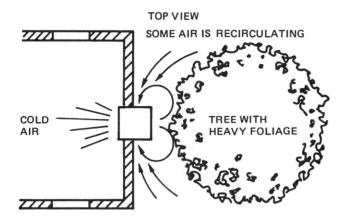

Figure 18-29. Recirculated air will cause high head pressure.

Care must be taken to ensure proper condensate drainage for room units. For example, when these units are installed in a motel, the outdoor portion protrudes into the walkway (Figure 18-31). If the condensate is not completely evaporated, it will run across the walkway and cause a hazard. Always make sure that the condensate drains to a suitable location, such as a lawn, flower bed, or gutter.

Through-the-wall units may be installed while the building is under construction. This is accomplished by installing wall sleeves and covering the openings until the unit is set in place (Figure 18-32). The wiring is run to the vicinity of the unit and the unit is connected when installed. It is very important to size the hole carefully. The wall sleeve should be purchased from the manufacturer in advance and used to determine the precise dimensions to the wall opening.

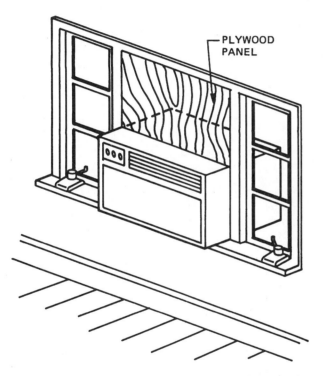

Figure 18-30. A regular window unit is sometimes installed in the fixed portion of a casement window between the roll-out sections.

Figure 18-31. Most of the through-the-wall unit is located on the outside of the wall.

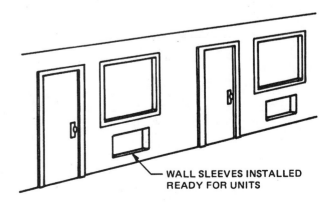

Figure 18-32. Wall sleeves may be installed in advance.

Casement window installations require a special unit. This unit is narrow and tall to fit the roll-out portion of the casement window. A regular window unit may also be installed by removing the fixed portion of the casement window and filling in the gap with a plywood panel (Figure 18-30). As mentioned previously, this requires significant carpentry skills and should be avoided, if possible. When this type of installation is performed, all stress must be on the windowsill. If the unit is removed, another repair job must be performed for the window to be utilized.

18.5 CONTROLS FOR COOLING-ONLY UNITS

Cooling-only units are typically plug-in appliances furnished with a power cord. All controls are furnished with the unit. There is no wall-mounted remote thermostat as with central air conditioning. The room thermostat sensor is located in the unit return air stream (Figure 18-33). A typical room unit may have a selector switch to control the fan speed and provide power to the compressor circuit (Figure 18-34). The selector switch may be considered a power distribution center. The power cord will usually be wired straight to the selector switch for convenience. Therefore, it may contain a hot, neutral, and ground wire for a 115 V unit. A 208/230 V unit will have two hot wires and a ground wire (Figure 18-35). The neutral wire for 115 V service is routed straight to the power-consuming devices.

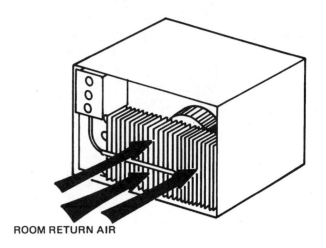

Figure 18-33. The thermostat is located in the return air stream.

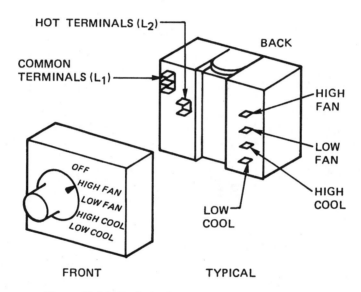

Figure 18-34. Typical selector switch for a room unit.

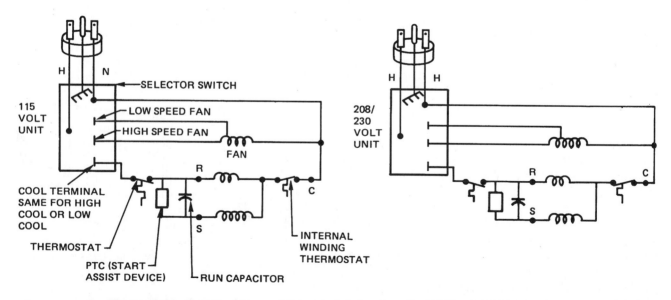

Figure 18-35. Typical selector switch terminal designations for 115 V and 208/230 V units.

CONTROLS FOR COOLING-ONLY UNITS   263

On 208/230 V units, one hot wire is wired through the selector switch to the power-consuming devices and the second is connected directly to the other side of the power-consuming devices. There are many combinations of selector switches that will allow the owner to select fan speed, high cool or low cool, and exhaust or fresh air. The level of cooling is determined by the fan speed. Since the unit has only one fan motor, a reduction in fan speed slows both the indoor and the outdoor fan. This reduces the capacity and noise level of the unit, and increases the system efficiency (Figure 18-36).

The exhaust/fresh air control is provided by a lever that positions a damper to either bring in outdoor air or exhaust indoor air (Figure 18-37).

The selector switch sends power to various circuits including the compressor start circuit. Motor starting was discussed in Unit 15. Figure 18-38 shows typical selector switch and compressor wiring diagrams with explanations.

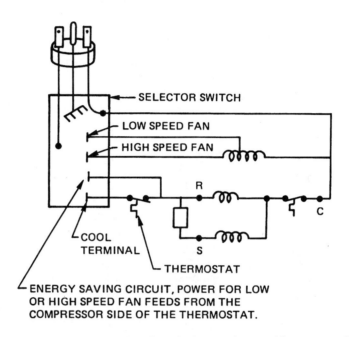

Figure 18-36. Some units automatically adjust the fan speed to provide energy savings.

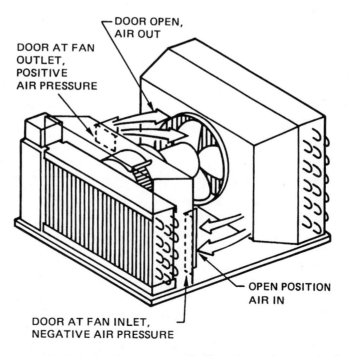

Figure 18-37. The exhaust and fresh air functions are controlled by a damper lever on the front of the unit.

UNIT 18 ROOM AIR CONDITIONERS

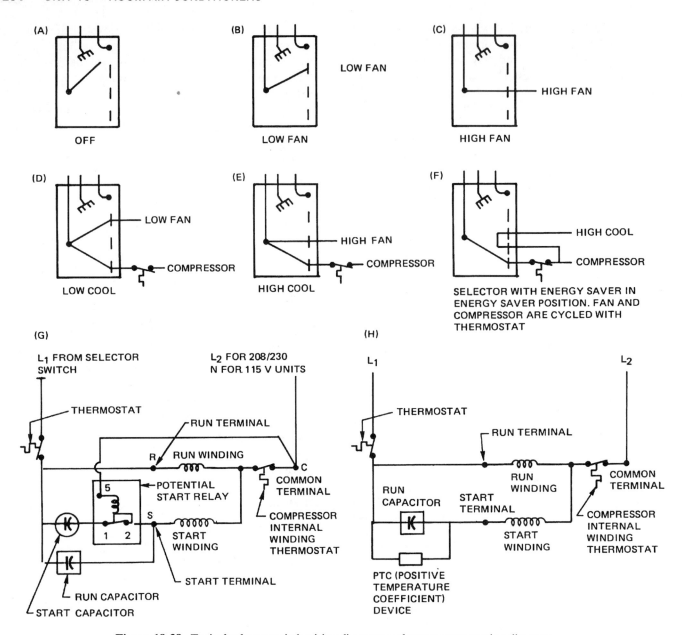

Figure 18-38. Typical selector switch wiring diagrams and compressor starting diagrams.

18.6 CONTROLS FOR COOLING/HEATING UNITS

Combination cooling/heating units may be either plug-in appliances or through-the-wall units. Through-the-wall units usually have an electrical service furnished to the unit. The cooling controls for these units are the same as those for cooling-only units. Typically, a selector switch changes the unit from cooling to heating. When the unit has electric strip heat, the selector switch merely directs the power to the heating element through the thermostat (Figure 18-39). A heat pump will have a selector switch to select heating or cooling. The cooling cycle is the same as a typical cycle with the possible addition of the reversing valve. In the heating cycle, defrost may have to be considered. A separate defrost control is sometimes provided.

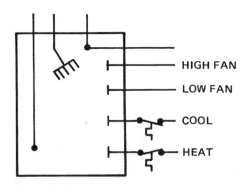

Figure 18-39. The selector switch directs the power to either the heating circuit or the cooling circuit.

18.7 MAINTENANCE AND SERVICE

Maintenance of room units basically involves keeping the filters and coils clean. Modern units typically have permanently-lubricated motors. Older models may need lubrication. If the filter is not maintained, the indoor coil will become dirty and cause the unit to operate at low suction pressures. As mentioned previously, the coil is already operating at about 38°F in the cooling mode. If the coil temperature drops more than a few degrees, ice will begin to form. Some manufacturers provide freeze protection which will shut the compressor off and allow the indoor fan to run until the the temperature rises. This feature may be controlled by a thermostat located on the indoor coil or mounted on the suction line (Figure 18-40).

Service may involve both mechanical and electrical problems. Mechanical problems usually concern the fan motor and bearings or refrigerant circuit problems. All room units have a critical refrigerant charge and gages should only be installed if absolutely necessary. The system will normally be sealed and require line tap valves or the installation of service ports on the process tubes for gages to be installed. Figure 18-41 shows an example of a line tap valve.

Note: Be sure to follow the manufacturer's instructions when installing these valves.

System Evaluation Using a Bench Test

A bench test is often performed on room units to determine whether or not to install gages when a low charge is suspected. Since the condenser and evaporator are at the same temperature during the bench test, the condenser

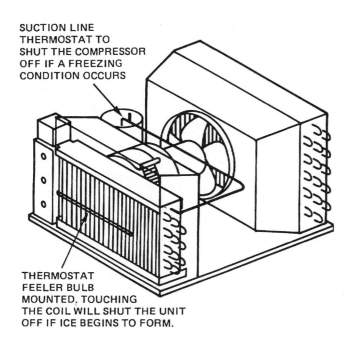

Figure 18-40. Freeze protection for room unit.

Figure 18-41. Line tap valve. *Photo by Bill Johnson.*

airflow may need to be reduced to move any refrigerant from the condenser due to low head pressure. A full explanation of condenser operation and charging under low ambient conditions is discussed in Unit 27. Refer to Figure 18-42 while reviewing the following procedure:

1. Remove the unit from its case.
2. Make sure that air flows through the coils. For temporary testing, cardboard may be positioned over places where panels force air to flow through the coils.
3. Start the unit in high cool.
4. Let the unit run for about five minutes and observe the sweat on the suction line. It should come close to the compressor. The evaporator should be cold from bottom to top.
5. Cover a portion of the condenser to force the head pressure to rise. If the room is cool, a portion of the charge may be held in the condenser. The condenser airflow may be blocked to the point that the air leaving it is hot (about 110°F).
6. With the head pressure increased, allow the unit to run for five minutes. The sweat line should move to the compressor. If the humidity is too low for the line to sweat, the suction line should be cold. The evaporator should be cold from bottom to top. If only part of the evaporator is cold (frost may form), the evaporator is starved. This may be the result of either a restriction or a low charge.
7. If the compressor is not pumping to capacity and the charge is correct, the coil will not be cold anywhere, only cool. An ammeter may be clamped on the common wire to the compressor and the amperage compared to the full-load amperage of the compressor. If it is very low, the compressor may be pumping on only one of two cylinders.

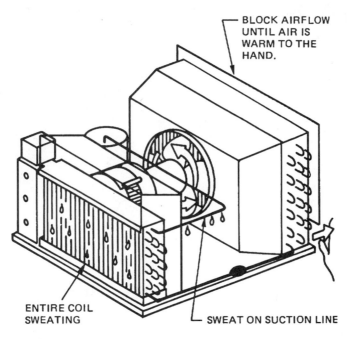

Figure 18-42. Testing for a low charge before installing gages.

When the previous test indicates the charge is low (the suction line is not cold or sweating) gages should be installed.

Leak Checking

When the charge is low, the leak must be found and repaired, if possible. If the unit has never had a line tap valve or service port installed and has operated for a long period of time without a problem, it can be assumed that the leak has occurred recently. An electronic leak detector may be used to find the leak. Turn the unit off and allow the pressures to equalize. Make sure there is enough refrigerant in the unit to perform a leak check (Figure 18-43). For example, if there is a very small amount of refrigerant in the system (if the system pressure and temperature relationship indicates there is no liquid present) then there is so little gas in the system that what is left could be considered trace refrigerant for leak checking purposes. If this is the case and the leak cannot be found, turn the unit off, then use dry nitrogen to increase the pressure to the lowest working pressure of the unit. This would be the evaporator working pressure and may be 150 psig, but check the nameplate to be sure. When the leak check is completed, the nitrogen and trace refrigerant will have to be exhausted, the system evacuated, and a new charge added.

Warning: Always wear goggles and gloves when handling refrigerant.

Adjusting the System Charge

The correct method to charge a unit is to remove and recover the remaining charge, evacuate the unit, then

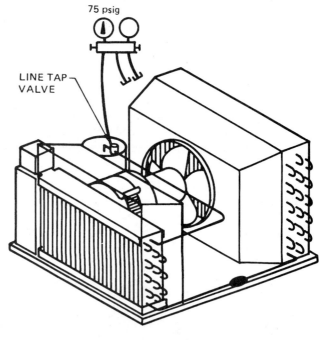

Figure 18-43. Make sure that there is enough refrigerant in the unit for leak checking purposes.

measure in the correct charge printed on the nameplate. Use accurate scales to measure the charge (Figure 18-44). There are times when you may want to simply adjust the charge rather than fully evacuate and recharge the system. A charge may be adjusted using the following method (see Figure 18-45).

1. Install suction and discharge gage lines.

MAINTENANCE AND SERVICE 267

Figure 18-44. Charging scales. *Photos by Bill Johnson.*

2. Purge the gage lines and connect to the correct refrigerant cylinder.
3. Make sure that air is passing through the evaporator and condenser coils. Cardboard may be used for temporary panels.
4. Start the unit in high cool.
5. Watch the suction pressure and do not let it fall below atmospheric pressure into a vacuum. Add refrigerant through the suction gage line if it falls too low.
6. Adjust the airflow across the condenser until the head pressure is 250 psig, corresponding to a condensing temperature of 120°F for R-22. Refrigerant may have to be added in order to obtain the correct head pressure.
7. Add refrigerant vapor at intervals with the head pressure of 250 psig until the compressor suction line is sweating to the compressor. As refrigerant is added, the airflow will have to be adjusted over the condenser or the pressure will rise too high. You should also note that the evaporator coil has become colder toward the end of the coil as refrigerant is added.
8. The correct charge is verified after about 15 minutes of operation at 250 psig and a cold (probably sweating) suction line at the compressor. The suction pressure should be about 65 psig and the compressor should be operating at near full-load amperage.

When a system must be evacuated, achieving a low vacuum will be faster and easier using a Schrader valve rather than a line tap valve. To add a Schrader valve, solder a tee fitting and service stem in the line with the Schrader valve on the end (Figure 18-46). This fitting can then be capped and used for any future service on the unit.

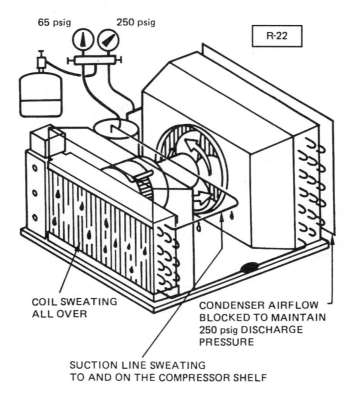

Figure 18-45. Adjusting a unit charge by sweat line.

Figure 18-46. This service stem has been added to the system for better evacuation. When the service procedure is completed, the valve is capped to prevent leaking and can be used for future service.

Capillary Tube Restrictions

If the head pressure rises and the suction pressure does not, and refrigerant will not fill the evaporator coil, suspect a capillary tube restriction. The capillary tube is very difficult to service on a room air conditioning unit. If you believe that the capillary tube is restricted, change the strainer located before the capillary tube and try the unit again. If this does not correct the problem, the unit may need to be scrapped. At times, a capillary tube may need to be changed. The capillary tube-to-suction line heat exchange for a room air conditioner presents a problem. It may not be worth the time to make the change. Some service technicians may install a thermostatic or automatic expansion valve instead of trying to change a capillary tube (Figure 18-47). In this case, you must make sure the compressor has a start relay and capacitor or it may not start. Thermostatic and automatic expansion valves do not equalize during the off cycle.

Removing the Fan Motor

Fan motor removal may be difficult in many units. The units are so compact that the fan and motor are in very close proximity to the other components. The evaporator fan is normally a centrifugal (squirrel cage) design and the condenser fan is typically a propeller type. The condenser fan shaft usually extends to a point very close to the condenser coil. Often there is not enough room to slide the propeller fan to the end of the shaft (Figure 18-48). Either the fan motor must be raised up or the condenser coil moved back (Figure 18-49). The evaporator fan wheel is normally locked in place with an Allen-head set screw that must be accessed through the fan wheel using a long Allen wrench (Figure 18-50). When a unit is reassembled after fan motor replacement, it is often difficult to align the components. For best results, the unit should be placed on a level surface such as a work bench.

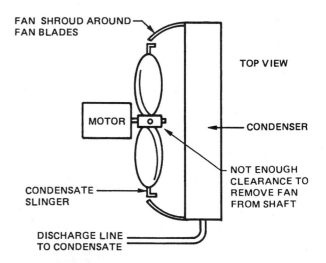

Figure 18-48. The condenser fan is sometimes difficult to remove because there may be insufficient clearance between the fan and the condenser.

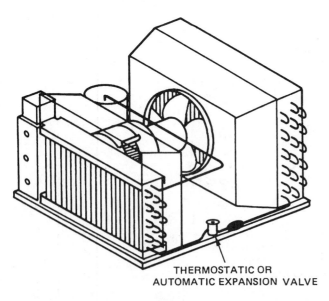

Figure 18-47. A thermostatic or automatic expansion valve may sometimes be installed instead of replacing a capillary tube.

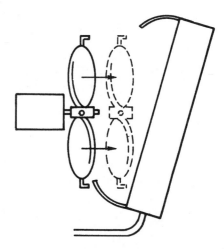

Figure 18-49. The condenser may be moved back or the fan motor may be raised for removal of the fan blade.

MAINTENANCE AND SERVICE 269

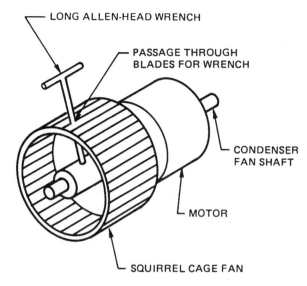

Figure 18-50. An Allen-head set screw may be loosened by sliding a long Allen wrench between the blades.

Electrical Service

Electrical service requires the technician to be familiar with the electrical power supply. All units are required to be grounded. This involves a green wire that is run to every motor in the unit from the grounded neutral bus bar in the main control panel (Figure 18-51). Many older units do not have this ground wire; these units will operate correctly because the ground wire does not alter normal system operation. Instead, it is a safety circuit to protect the service technician and others from electrical shock hazard.

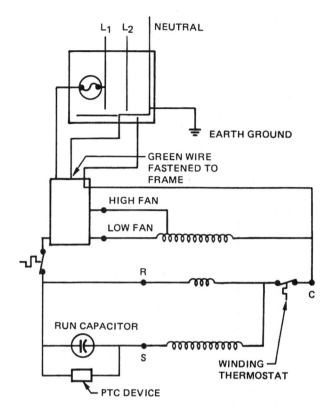

Figure 18-51. This wiring diagram shows where the green ground wire terminates in the main control panel.

The power supply must be of the correct voltage and have adequate current-carrying capacity. Electrical service may involve the fan motor, thermostat, selector switch, compressor, and power cord. The fan motor will be either a shaded pole or a permanent split capacitor, (Figure 18-52). These motors are discussed in detail in Unit 13.

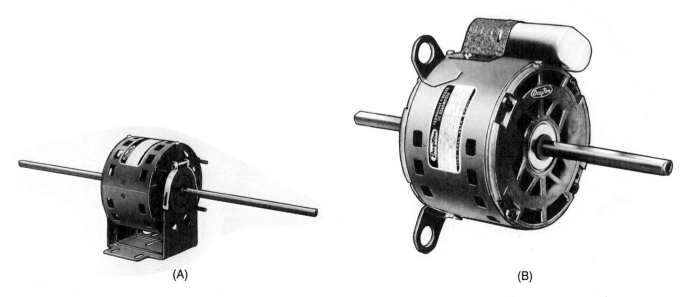

Figure 18-52. Fan motors (A) Shaded pole. (B) Permanent split capacitor. *Courtesy Dayton Electric Manufacturing Company.*

The thermostat for room units is usually mounted with the control knob in the control center and the remote bulb in the vicinity of the return air. These thermostats are the line-voltage type discussed in Unit 12. They are typically slow to respond, but this problem is overcome somewhat by the fact that most units operate with the fan running continuously anytime the selector switch is set for cooling or heating. Therefore, the thermostat has a moving air stream over it at all times. Some systems may have an energy saving feature that allows the thermostat to cycle the fan. In this case, the thermostat will not provide the same level of control.

The thermostat passes power to the compressor start circuit and may be treated as a switch for troubleshooting purposes. If it is suspected that the thermostat is not passing power to the compressor start circuit, you may unplug the unit (shut off the power if wired direct) and remove the plate holding the thermostat. The plate may be moved forward far enough to apply voltmeter leads to the thermostat terminals. Turn the voltmeter selector switch to the correct setting (higher than the voltage of the unit), and plug the unit in (resume power). Turn the unit selector switch to the high cool position and observe the voltmeter reading. If there is a voltage reading, the thermostat contacts are open (Figure 8-53).

Selector switch troubleshooting is similar to that for the thermostat. The panel that holds the thermostat usually has the selector switch mounted on it. If it is suspected that the selector switch is not passing power, turn the power off and remove the panel. If the power panel contains a protective device that removes the power when the panel is removed, you may need to bypass this device so that the unit may be started for the test. The selector switch is checked using the common terminal and the power lead. For example, attach one lead of a voltmeter to the common lead and the other to the circuit to be tested (high cool) as shown in Figure 18-54. Resume the power supply and turn the selector switch to the high cool position. Full line voltage should be indicated on the meter. If not, check the hot lead entering the selector switch. If power enters, but does not leave, the switch is defective.

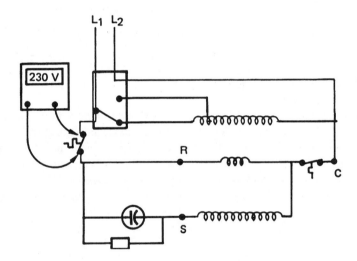

Figure 18-53. If the meter reads voltage, the thermostat contacts are open.

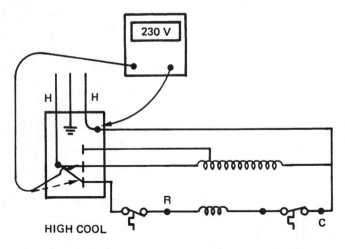

Figure 18-54. Using a voltmeter to check a selector switch.

TECHNICIAN SERVICE CALLS 271

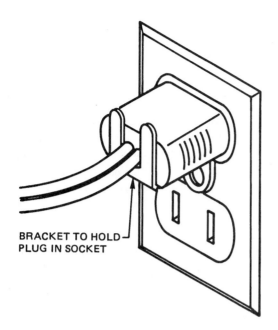

Figure 18-55. A power cord may need support.

Compressor electrical problems may either be in the starting circuit or the compressor itself. Room units generally use single-phase power, and incorporate either a potential relay or positive temperature coefficient (PTC) device for starting the compressor (see Unit 13). Electrical troubleshooting and capacitor checks are discussed in Unit 16. A review of these units will help you to understand motors and their problems.

Power cord problems usually occur at the end of the cord, where it is plugged into the wall outlet. The most common problems are loose connections because the plug does not fit well or the plug or power cord has been stepped on and partially unplugged. The power cord should make the best connection possible at the wall outlet. Sometimes the cord may need support to keep the strain off the connection (Figure 18-55). If the cord becomes hot to the point of discoloration or the plug swells, the plug should be changed. The wall receptacle may need inspection to ensure that it is not damaged. If the plastic is discolored or distorted, it should be changed or the problem will recur.

18.8 TECHNICIAN SERVICE CALLS

Service Call 1

A customer in an apartment calls and says that the window unit in their dining room is freezing solid every night. This unit is used to serve the kitchen, dining room, and living room. The tenants both work during the day. They shut the unit off in the morning and turn it on when they get home from work. The air temperature is 80°F inside and 75°F outside by the time they turn the unit on. There is a high heat load in the apartment keeping the unit running. The problem is that the head pressure is too low during the last part of the cycle, starving the evaporator and causing it to operate below freezing.

The apartment house technician arrives late in the afternoon when the tenants arrive home. The tenants show the technician how the thermostat is set each night. The technician notes that the outdoor temperature is cooler than the temperature in the apartment by the time the unit is turned on. It is explained that a better approach would be to leave the unit on low cool during the day, and raise the thermostat setting several degrees. This allows the unit to run part of the day, and relieves the large load that would occur if the unit were started only once each day. A call back in a few days verifies that the unit is performing well using this approach.

Service Call 2

A customer calls to indicate that the air conditioner in their guest bedroom is not cooling the room. The coil behind the filter has ice on it. This sounds like a low charge complaint. The problem is that the drapes partially cover the front of the unit, causing the air to recirculate. The supply air louvers are also pointing downward.

The technician arrives at the home and notices that the drapes hang in front of the unit. The technician explains that the drapes will have to be kept out of the air stream. The air discharge is also pointed down, causing the air to move toward the return air inlet. This is adjusted. Just to be sure that the unit does not also have a low charge, the technician pulls the unit out of the case and places it on a small stool so it can be started. The technician notices that when the unit is started, air blows out the top of the blower compartment and bypasses the indoor coil, so a piece of cardboard is placed over the fan compartment to force the air through the coil. The unit is started and the condenser is partially blocked to force the head pressure higher. After the unit has run for a few minutes, the suction line starts to sweat back to the compressor. This unit has never had gages installed, so it has the original charge and it is still correct. The unit is stopped and pushed back into its case in the window. The filter is cleaned, then unit is started and left operating normally.

Service Call 3

A customer calls and reports that their new unit is not cooling. The problem is a leak in the high-pressure side of the system.

The technician arrives and starts the unit. The compressor runs, but nothing else happens. The technician turns the unit off and slides it out of its case onto a stool and restarts the compressor. It is obvious that the unit has problems, so the technician uses a hand truck to remove the unit and takes it back to the shop.

At the shop, the unit is moved to a workbench. A line tap valve is installed on the suction process tube. When the gages are fastened to the line tap valve, no pressure registers on the gage. The system is out of refrigerant.

The technician allows R-22, the refrigerant for the system, to flow into the unit. A leak cannot be heard, so an electronic leak detector is used. A leak is found on the discharge line connection to the condenser. The refrigerant is recovered from the unit. The process tube is cut off below the line tap valve connection. A ¼ inch flare connection is soldered to the end of the process tube and a gage manifold connected to the low side. The leak is repaired using high-temperature silver solder.

When the torch is removed from the connection, the gage manifold valve is closed to prevent air from being drawn into the system when the vapor inside cools and shrinks. When the connection is cooled, refrigerant is allowed into the system and the leak is checked again. It is not leaking. A small amount of refrigerant and air are purged from the system. It is now ready for evacuation.

The vacuum pump is attached to the system and started. While the system is being evacuated, the electronic scales are moved to the system and prepared for charging. Since the service connection is on the low-pressure side of the system, the charge will have to be allowed into the system in the vapor state. The cylinder is a 30 pound cylinder. This should be enough refrigerant to keep the pressure from dropping during charging. The unit calls for 18.5 ounces of refrigerant. When a vacuum is reached (29 in. Hg on the low-side gage) the vacuum pump is disconnected and the hose connected to the refrigerant cylinder. The gage line is purged of air and the charge is allowed into the system. When the correct charge is established, the unit is started. It is cooling correctly.

The technician takes the unit back to the customer and places it in the window. The unit is started and operates correctly.

Service Call 4

A customer in a warehouse office calls and reports that the window unit working the office is not cooling to capacity. A quick fix must be performed until a new unit can be installed. The problem is a slow leak.

The technician arrives and examines the unit. The unit definitely has a low charge because the evaporator coil is only cold halfway to the end. The unit is pulled out of the case and placed on a stool. The technician points out that the unit is old and needs replacing. The customer asks if there is anything that can be done to get them through the day.

A line tap valve is fastened to the suction line and a gage line connected. The suction pressure is 50 psig. This unit uses R-22, so this pressure corresponds to an evaporator temperature of 26.5°F. The unit is showing signs of frost on the bottom of the coil. Refrigerant will be added to correct the charge. The unit is in a room where it is about 80°F. The condenser airflow is blocked until the leaving air feels warm. Refrigerant is added until the suction line is sweating to the compressor. The gages are removed and the unit is placed back in the case. The leak was not found. The company requested that a new unit be installed tomorrow. The refrigerant will have to be recovered from the old unit before the unit can be discarded.

Service Call 5

A new unit is returned to the shop for repair. The service repair order indicates that it is not cooling. The problem is the compressor is stuck and will not start.

The technician places the unit on a work bench. Next, the technician checks to make sure the selector switch is in the off position and plugs the unit into the correct wall outlet (230 volts). The work bench is made for servicing appliances, so it has an ammeter mounted in it. The technician turns the fan to high and the fan runs. The fan selector switch is then turned to low and the fan still runs. The amperage is normal, so the technician turns the selector switch to high cool, keeping a hand on the switch. The fan speeds up but the compressor does not start. The thermostat is not calling for cooling. The thermostat is turned to a lower setting and the compressor tries to start. The ammeter rises to a high amperage and does not fall back. The technician turns the unit off before the compressor overload reacts, and then disconnects the power cord.

The technician removes the unit from the case and looks for the compressor terminal box. The cover is removed and a test cord is attached to the common, run, and start terminals (Figure 18-56). The test cord is attached to a power supply on the bench. An ammeter is still in the circuit on the test bench. The technician tries to start the compressor with the test cord by turning the selector switch to start and then releasing the selector knob, per the test-cord instructions. It will not start. The compressor is stuck and must be changed.

The technician installs a line tap valve on the suction process tube and starts to recover the refrigerant. While

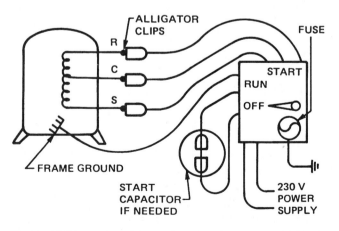

Figure 18-56. Test cord fastened to the common, run, and start terminals on the compressor.

the refrigerant is being recovered slowly, a new compressor, a ¼ inch flare fitting for the process tube, and a suction-line drier are brought from the supply room. The technician removes the compressor hold-down bolts. When the refrigerant has been recovered, the technician uses an air-acetylene torch with a high-velocity tip to remove the suction and discharge lines. The old compressor is removed and the new one installed. The suction line is altered to install the suction-line drier, but it is not installed. The compressor discharge line is soldered in place. The process tube was cut off below the line tap connection and the ¼ inch flare fitting is installed on the process tube connection. The suction line is soldered to the compressor and the suction-line drier is soldered into the line. A gage line is fastened to the ¼ inch flare connection. A small amount of R-22 is added to the system (to 5 psig), then nitrogen is used to pressurize the system to 150 psig. A leak check is performed on all new fittings. The unit is leak free.

The small amount of refrigerant vapor and nitrogen are purged from the system and the vacuum pump is fastened to the gage manifold and started. When the mercury manometer shows that the vacuum pump has obtained a deep vacuum, the correct charge is weighed into the unit. The compressor wiring is reconnected while the vacuum pump operates so the unit is ready to start. The technician plugs in the cord and turns the selector switch to high cool; the compressor starts. The unit is allowed to run for 30 minutes and operates satisfactorily. The process tube is pinched off and cut off below the ¼ inch flare fitting. The process tube is closed on the end and soldered shut.

Service Call 6

A unit is brought into the shop with a tag saying "no cooling." The problem is the start relay is defective (it has a burned coil).

The technician places the unit on the workbench and checks the selector switch. It is in the off position. The unit is plugged into the correct power supply (230 volts). The workbench has a built-in ammeter. The technician turns the selector switch to the high fan setting. The fan runs. The unit is then switched to low fan and the fan still runs. The technician then turns the selector switch to high cool, keeping a hand on the switch. The fan speeds up to high, but the compressor does not start. The ammeter reading is very high (25 amperes). The technician turns the unit off before the compressor overload device reacts.

The unit is unplugged and removed from its case. The compressor motor terminal wires are removed and a test cord connected. The test cord is plugged in. The technician turns the test cord to start, then releases the knob (as per directions) and the compressor starts. This narrows the problem to the compressor start circuit. The compressor is then turned off.

The technician removes the terminals from the start relay and checks the coil circuit from terminals 1 to 5. It has an open circuit (Figure 18-57). The relay coil is

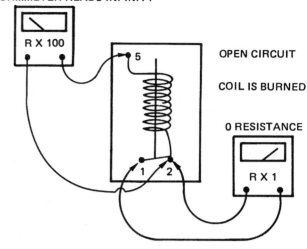

Figure 18-57. A start relay is checked using an ohmmeter.

defective, keeping the unit in start. The capacitors are checked as explained in Unit 16 and they are satisfactory.

The relay is changed and the compressor is wired back into the circuit. The unit is plugged in again and the selector switch turned to high cool. The compressor starts and runs. The amperage is normal when compared to the amperage indicated on the nameplate.

Service Call 7

A customer calls to report that the window unit servicing their den and kitchen has tripped the breaker. This is a fairly new unit (about six years old). The owner is told to shut the unit off and not reset the breaker. The problem is the unit has a grounded compressor.

The technician arrives and decides to give the unit an ohm test before trying to reset the breaker and start the unit. The technician unplugs the unit and fastens one ohmmeter lead to each of the hot wires on the 230 V plug (Figure 18-58). The ohmmeter selector switch is set to R × 1 and the meter is zeroed. The unit selector switch is turned to high cool. There is a measurable resistance, and all appears to be well. Before plugging the unit in, however, one more ohm test is necessary. This is the ground test. The ohmmeter is set to the R × 1000 scale and one lead is moved to the ground terminal. The meter reads zero resistance, which means that there is a ground somewhere in the unit (Figure 18-59). It could be in either the fan motor or the compressor. The unit is pulled from its case. The compressor terminals are disconnected. One lead of the meter is touched to the discharge line and the other to the common terminal. The meter reads zero resistance, so the compressor has the grounded circuit.

The customer is informed of the problem and given a price to repair the unit. The customer decides to have it repaired as it will cost much less than purchasing a new unit.

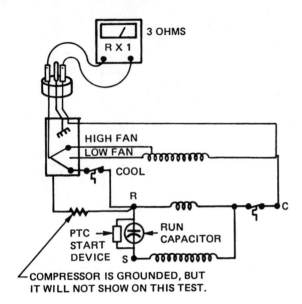

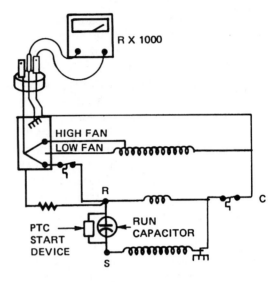

Figure 18-58. This unit has a grounded circuit, but it is not indicated by an ohmmeter test of the 230 V plug. An additional test is required.

Figure 18-59. The grounded circuit is in the compressor.

The technician moves the unit to the truck and takes it to the shop. Before taking the unit into the building, the technician recovers the refrigerant from the unit to an approved cylinder for contaminated refrigerants. Because the compressor failed, it is safe to assume that the refrigerant may be burned and contaminated.

The technician obtains a liquid-line drier, a suction-line drier, a ¼ inch tee fitting for the liquid line, a compressor, and a ¼ inch fitting for the process tube. When the refrigerant is removed from the unit, the technician cuts the compressor discharge line at the compressor using diagonal pliers because it is too close for the tubing cutters. Then the suction line is cut with a small tube cutter and the compressor is removed from the unit. A tee fitting is installed in the liquid line while it is still outside. A gage line is fastened to the tee fitting and nitrogen is purged through the unit (Figure 18-60). This will push much of the contamination from the unit. It is then taken into the shop to the workbench.

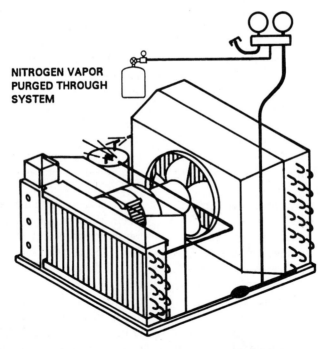

Figure 18-60. The system is purged using the test fitting installed in the liquid line.

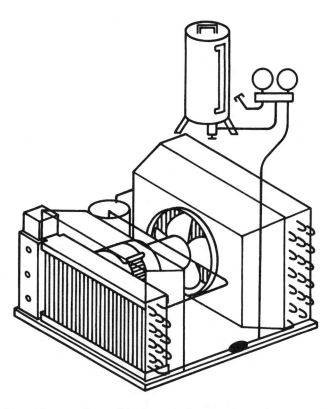

Figure 18-61. Liquid being charged into an air conditioning unit. Because liquid refrigerant is denser than vapor refrigerant, it takes less time to transfer it to the system.

The technician sets the new compressor in place and solders the discharge line to the new compressor. The suction line is cut to the correct configuration for the suction-line drier. The liquid line is now cut just before the strainer entering the capillary tube, and the new liquid-line drier is installed. The suction-line drier and the ¼ inch fitting for the process tube are soldered in place. The gages are fastened to the process port and the tee fitting in the liquid line. The unit is pressured with a small amount of R-22 (to 5 psig) and nitrogen is added to push the pressure up to 150 psig. The unit is then leak checked using an electronic leak detector. After the unit is leak checked, the trace refrigerant and nitrogen are purged from the system.

After the leak check, the vacuum pump is connected and started. While the unit is being evacuated, the capacitors and relay are checked. The run and start capacitors are both good. The relay has a measurable resistance in the coil and the contacts look good.

The charge for the unit is 18 ounces, plus the capacity of the small liquid-line drier, which is 1.9 ounces.

Note: This information should be packed in the drier directions but all manufacturers do not furnish it.

The total charge is 19.9 ounces. When a vacuum of 500 microns is obtained according to the micron gage, the charge is added to the system.

The charge is in a charging cylinder, such as explained in Unit 6. The high-pressure gage line is connected to the liquid line on the unit, and the charge is allowed to enter the system through the liquid line in the liquid state (Figure 18-61). When the complete charge is inside the unit, the liquid-line valve is pinched off using a special pinch-off tool and soldered shut.

The unit is started and allowed to run while the technician is monitoring the suction pressure and amperage. The unit is operating correctly and can be returned to the customer.

SUMMARY

- Room air conditioners may be designed for installation in windows or through the wall.
- These units may be designed for cooling only, both cooling and heating with electric strip heat, or cooling and heating with a heat pump cycle.
- The four major components in the cooling cycle for these units are the evaporator, compressor, condenser, and metering device.
- The metering device used in most room air conditioners is the capillary tube.
- Most units are designed to produce a heat exchange between the capillary tube and the suction line. This adds superheat to the suction line and subcools the refrigerant in the capillary tube.

- The heat pump cycle utilizes the reversing valve to redirect the refrigerant.
- The heat pump compressor is different from one found in a cooling-only unit. It must have enough pumping capacity for low-temperature operation.
- Window units may be installed in either double-hung or casement windows. Those installed in casement windows must be designed for that purpose.
- Before installing a unit, make sure that the electrical service is adequate.
- When installing these units, ensure that there is no possibility of direct recirculation of air through the unit.
- The selector switch is the primary control for room air conditioning units. It switches the unit from cooling to heating, between high and low cooling, between high and low heating, and between high and low fan operation. There may also be a control to switch between indoor air and outdoor air.
- The primary maintenance for these units involves keeping the filters and coils clean.
- Service gages should be installed only when absolutely necessary.
- All units are now required to be grounded. This involves a green wire that is run to every motor in the unit from the grounded neutral bus bar in the main control panel. Many older units may not have this ground wire
- Electrical service may involve the fan motor, thermostat, selector switch, compressor, and power cord.

REVIEW QUESTIONS

1. Describe the difference between a window air conditioning unit and a through-the-wall unit.
2. Describe the two types of window unit cases.
3. What are the two methods of providing heating with window or through-the-wall units?
4. What are the four major components used in the cooling cycle?
5. Briefly describe the cooling cycle in a window unit.
6. What materials are the evaporator tubing and fins typically made of?
7. What type of metering device is generally used in these units?
8. What is the component that controls the reverse cycle in the heat pump of a room air conditioner?
9. Briefly describe the heat pump heating cycle.
10. Why is the heat pump compressor different from a cooling-only compressor?
11. What does the selector switch normally control?
12. What are the two major maintenance requirements on room air conditioning units?
13. Describe the test that should performed before installing gages.
14. Describe the conditions at the evaporator if the charge is correct but the compressor is not pumping to capacity.
15. What substance may be used to increase the refrigerant pressure for the purpose of a leak check?

EVAPORATORS 19

OBJECTIVES

Upon completion of this unit, you should be able to

- Describe the purpose of an evaporator.
- Discuss the heat exchange from the air stream into the refrigerant in an evaporator
- Recognize a multi-circuit evaporator.
- Describe a simple performance test on a multi-circuit evaporator.

19.1 EVAPORATOR FUNCTIONS

Note: You should have a complete understanding of Unit 3 before attempting this unit.

The purpose of a refrigeration (cooling) system is to move heat from a place where it is not wanted to a place where it makes no difference. This heat must be absorbed into the refrigeration system, and then moved to the part of the system where it is rejected. The evaporator is the component that absorbs heat into the system. As explained in Unit 3, this heat is absorbed into the evaporator by passing air over a refrigerant coil. The refrigerant coil is at a temperature below that of the air passing over it. In a central air conditioning system, the refrigerant in the evaporator coil will typically have a temperature of 40°F while the air passing over the coil may be at 75°F (Figure 19-1). This is a 35°F temperature difference. You will recall that heat transfers naturally from a warmer substance to a cooler substance. In an evaporator, some of the heat from the air will transfer to the refrigerant in the coil. The refrigerant in the coil is maintained at a pressure that will cause it to change from a liquid into a vapor (Figure 19-2). The temperature of the air is reduced by about 20°F in the typical air conditioning process, so the leaving air is about 55°F (75°F - 20°F). This air is routed to the conditioned space where it is released and absorbs heat from the 75°F room air. In most applications, the cooling process also dehumidifies the conditioned air (discussed later).

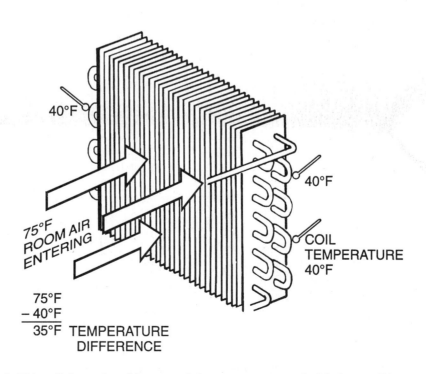

Figure 19-1. This coil shows the refrigerant and air temperatures at typical design conditions.

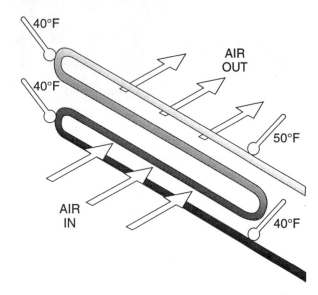

Figure 19-2. The refrigerant enters the evaporator as a liquid and is evaporated (boiled) into a vapor at a temperature of 40°F.

19.2 EVAPORATOR CONSTRUCTION AND OPERATION

Manufacturers strive to obtain the highest heat exchange possible from every square inch of coil surface so their equipment can be smaller, lighter, and more efficient. A typical evaporator coil uses copper tubes to circulate the refrigerant and aluminum fins on the coils to add heat exchange surface to the copper tubes (Figure 19-3). These aluminum fins extend the surface of the copper tubes because they are in very close contact with the tubes. This close contact is normally accomplished by inserting the copper tube in the fins and then expanding the tube to force it against the fins (Figure 19-4). This is necessary as the heat is transferred into the fins by the air, and must be able to transfer easily from the fins into the copper tube for transfer into the refrigerant. The aluminum-to-copper contact surface is called a *dissimilar metal junction*. It is subject to corrosion. When this occurs, the contact is diminished and heat transfer is reduced. Both age and use

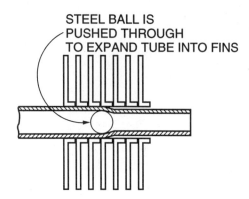

Figure 19-4. The tube is inserted into the hole in the fins, then expanded until it is in close contact with the fins.

of incorrect coil cleaning chemicals can accelerate this process.

Some manufacturers of residential and light commercial equipment use aluminum tubes and aluminum fins. This creates a much lighter and less expensive coil. Some applications use an all-copper coil, with copper fins and copper tubes. This is a very expensive coil and is only used for special applications.

Manufacturers use many different methods to improve the heat exchange between the air and the aluminum fins. For example, the greater the surface area of the fins, the greater the heat exchange. This can be accomplished by using pleated or corrugated fins (Figure 19-5). This pattern causes the air to scrub against the surface of the fins, which prevents still air from accumulating. The air actually wiggles through the fins as it works its way through the coil.

Coil tubes are also designed to achieve greater heat exchange by allowing the air to pass over the refrigerant several times. This is accomplished by arranging the

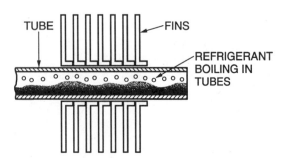

Figure 19-3. Fins on the tubes increase the tube surface area for greater efficiency.

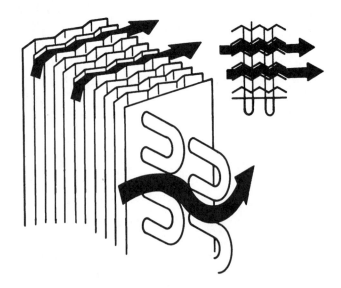

Figure 19-5. Pleated fins increase the surface area for maximum heat exchange.

EVAPORATOR CONSTRUCTION AND OPERATION 279

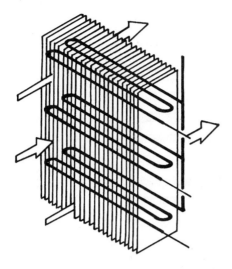

Figure 19-6. The refrigerant tubes are arranged in a serpentine fashion to increase the heat exchange.

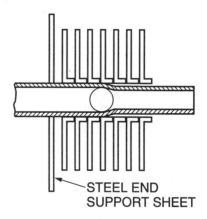

Figure 19-8. The refrigerant tubes and fins are held together by the coil end sheet.

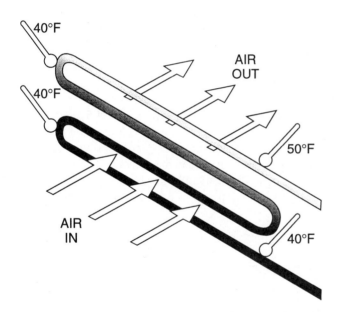

Figure 19-7. As the air passes through the coil, it becomes lower in temperature. The refrigerant temperature remains almost constant while in the liquid state.

Figure 19-9. The coil end sheet is fastened to the framework that holds the coil steady.

tubes in a serpentine fashion (Figure 19-6). Multiple passes are very common. As the air moves over the coil tubes, the air temperature is reduced, while the temperature of the liquid refrigerant remains about the same (Figure 19-7).

The coil tubes and fins are held together by the coil end sheet. The tubes are expanded into the end sheet, which is typically steel (Figure 19-8). This end sheet or casing may also be used to mount the coil in its installation by means of folded flanges on the ends with bolt holes for mounting. The end sheet may also be fastened to framework that passes under and over the coil to form a framed casing (Figure 19-9).

Large evaporator coils may incorporate more than one circuit so that the refrigerant does not have to make such a long trip through the evaporator. The longer the refrigerant circuit, the greater the pressure drop through the coil. It would be desirable to have no pressure drop, but this is not practical. Figure 19-10 illustrates what happens to the refrigerant as it passes through the ideal or perfect coil. Notice there is no pressure drop. The pressure drop under actual operating conditions is due to the friction. The pressure drop in the coil causes energy loss due to the expansion of the refrigerant vapor, which reduces the capacity of the compressor. The expansion is minimized by the additional circuits.

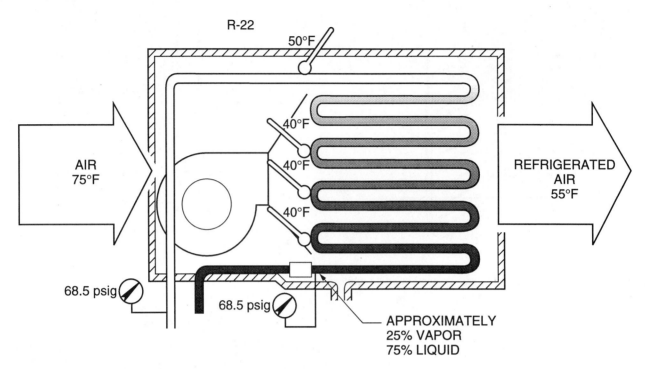

Figure 19-10. Conditions of the refrigerant as it passes through the evaporator coil.

Refrigerant pressure drop is related to temperature drop because of the pressure-temperature relationship of the refrigerant. You will recall that a refrigerant pressure drop will result in a lower boiling point. For example, suppose a refrigerant coil using R-22 has a pressure at the beginning of the coil of 76 psig, and the refrigerant is boiling at 45°F. If the coil has a pressure drop of 7.5 psig, the pressure at the end of the coil would be 68.5 psig, corresponding to a temperature of 40°F (Figure 19-11). This is an excessive pressure drop. For most systems, the pressure drop should not exceed the pressure that corresponds to 2°F. For R-22, this would be about 3 psig [42°F (71.4 psig) - 40°F (68.5 psig) = 71.4 - 68.5 = 2.9 psig pressure drop].

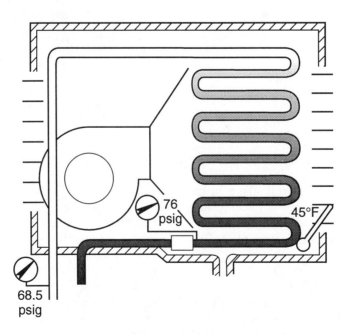

Figure 19-11. This coil has an excess pressure drop (7.5 psig).

EVAPORATOR CONSTRUCTION AND OPERATION

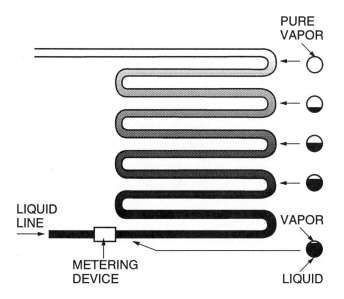

Figure 19-12. A cross-section of an evaporator. The technician should have a mental picture of what is happening at each point in the coil.

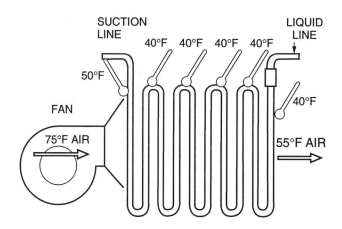

Figure 19-13. Airflow and refrigerant flow in an evaporator coil.

The other event that takes place inside the evaporator occurs when the refrigerant has all boiled into a vapor. This happens about ¾ of the way through the circuit. When the refrigerant enters the coil, it is about 25% vapor and 75% liquid (Figure 19-12). As this mixture moves through the coil circuit, more of the liquid is changed into a vapor. At the point where the refrigerant has completely changed into vapor, it is at the saturation temperature (typically the design temperature of 40°F) and is still capable of removing heat from 75°F air. Usually, about 10°F of superheat is added to the refrigerant, raising the vapor temperature to 50°F. The heat exchange between the vapor refrigerant and the liquid refrigerant is not as efficient as between the air and the liquid refrigerant, and is therefore kept to a minimum. However, the addition of some superheat is necessary to ensure that all liquid refrigerant is changed into a vapor.

The manufacturer also decides how the airflow must pass through the coil in relation to the direction of the refrigerant flow. Typically, the manufacturer moves the air through the coil in the opposite direction of the refrigerant flow Figure 19-13. The air entering the coil is at its warmest point, and is in contact with the warmest refrigerant, which is the vapor refrigerant at the end of the coil that has been superheated by 10°F. The 75°F entering air is in contact with the 50°F leaving refrigerant for a temperature difference at this point of 25°F. On the other side, the leaving air is put in contact with the coolest refrigerant for optimum heat exchange. The leaving air temperature may be 55°F with the refrigerant in the coil at 40°F.

Air Circulation Over the Evaporator Coil

Air passing through an evaporator coil also experiences a pressure drop. The greater the number of passes, the higher the pressure drop, and the more energy it takes to push the air through the coil. Air pressure drop through a coil must be kept within acceptable limits for the system to function properly. Pressure drop can be controlled to some extent by means of the fin spacing in the coils. The wider the space between the fins, the lower the pressure drop. The manufacturer must seek a balance between increased surface area (and therefore heat exchange) and minimal pressure drop.

Dehumidification

As mentioned earlier, moisture is removed from the air as it is moving over the cooling coils. This is where dehumidification takes place. Typically, about one pint of water per hour per ton of air conditioning may be removed from the air. This moisture forms on the fins of the evaporator coil (Figure 19-14). Moisture takes up space between the fins so the manufacturer usually provides separate pressure drop values for dry and wet coil applications.

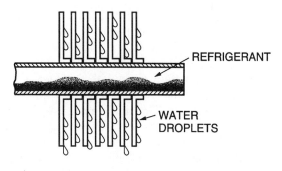

Figure 19-14. Moisture forms on the fins of the evaporator coil.

282 UNIT 19 EVAPORATORS

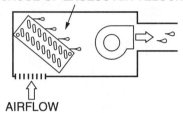

Figure 19-15. Excess velocity will wipe the water from the fins and push it into the air stream.

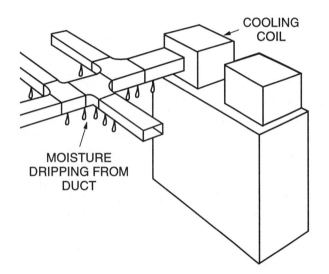

Figure 19-16. Water can be carried into the ductwork and drip out into the conditioned space.

The air passing over the coil may have enough velocity to wipe the condensing water off the fins and propel it into the air stream (Figure 19-15). This velocity must be kept within design standards (about 400 to 600 feet per minute). Above this value, airborne water can move into the duct work, where it will drip out into the conditioned space (Figure 19-16).

The Routing of the Evaporator Piping

The routing of the evaporator piping is important to controlling the pressure drop and returning any oil that is moving with the refrigerant back to the compressor. Typically, evaporators are piped with the liquid entering the top of the evaporator and exiting the bottom of the circuit (Figure 19-17). The liquid refrigerant will run down the evaporator like a waterfall and should be evaporated before it reaches the compressor suction line. The compressor suction line is typically piped to a level above the evaporator and then back downward to prevent gravity from allowing liquid refrigerant from entering the vapor line during the off cycle.

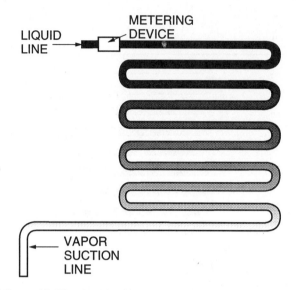

Figure 19-17. Liquid refrigerant should typically enter the top of the coil piping and flow downward, where gravity can help to move it through the coil.

Troubleshooting

The pressure-temperature conditions in a properly-operating evaporator are shown in Figure 19-18.

Poor evaporator operation can result from any of the following:

- Starved evaporator
- Flooded evaporator
- Dirty evaporator
- Reduced airflow through evaporator

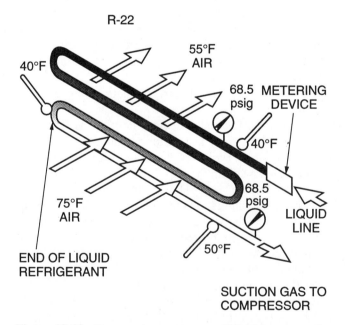

Figure 19-18. Pressure-temperature conditions in a properly-operating evaporator coil.

EVAPORATOR CONSTRUCTION AND OPERATION 283

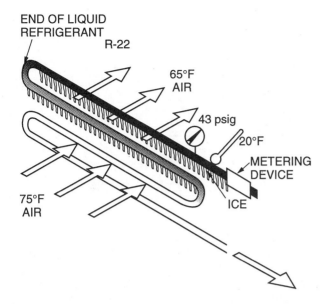

Figure 19-19. This coil has frozen due to a starved evaporator.

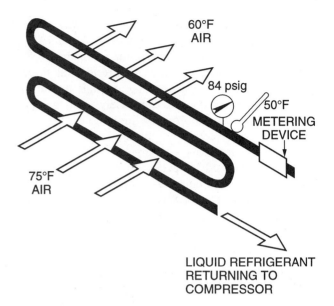

Figure 19-20. This is a flooded evaporator.

A starved evaporator is not receiving the necessary amount of refrigerant. The symptoms are coil sweating or frosting on the first portion of the coil. The reason for this is that the system compressor is still maintaining approximately the same amount of cooling and there is not enough refrigerant to fill the coil so it starts evaporating at a lower pressure. The pressure can become so low that the coil will freeze, even to the point that an ice buildup will occur (Figure 19-19). The starved evaporator can be caused by anything that prevents the correct amount of refrigerant from entering the evaporator, such as a pinched line, defective metering device, or low refrigerant charge.

A flooded evaporator is receiving too much refrigerant. The main symptom is reduced cooling capacity. Excess refrigerant causes the evaporator pressures and temperatures to rise, reducing the coil capacity. Another symptom is liquid refrigerant entering the vapor line and returning to the compressor (Figure 19-20). A flooded evaporator can be very harmful to the compressor.

A dirty evaporator is caused by improper filtration. When the air filters at the inlet to the evaporator allow dirt to escape, it will gradually build up on the evaporator coil. The evaporator coil is an excellent dust collector because the fins are close together and the coil is wet. Any dust particles that are not collected by the filter media will stick to the wet evaporator. Even filters that have been well maintained will allow a small amount of dust and dirt to escape and will eventually restrict the airflow to the coil. The symptoms of a dirty evaporator are frosting or freezing of the coil (Figure 19-21). The vapor line leaving the evaporator may even collect frost that may travel all the way to the compressor housing. An evaporator coil used for air conditioning should never show signs of frost or ice under normal operating conditions (remember, the design temperature is 40°F).

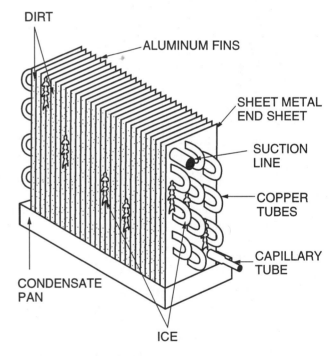

Figure 19-21. This coil is frozen because it is dirty.

A coil with restricted airflow will have the same symptoms as a coil with a dirty evaporator. Restricted airflow can result from several problems, such as: fan turning too slow (slipping fan belt, wrong pulley size, or wrong fan speed on multi-speed fan); fan running in reverse rotation; air vents shut off; dirty air filters; trash in the duct; or collapsed duct. The symptoms of restricted airflow are frosting or freezing of the coil. Again, there should never be signs of frost on a properly-operating air conditioning system.

Small evaporators perform in the same way as large evaporators. There should be between 8°F and 12°F of superheat at the end of the evaporator for the evaporator. A single-circuit evaporator in a small residential air conditioning system performs just like a single circuit in a large commercial evaporator. Conversely, the large multi-circuit evaporator is really just several single-circuit evaporators connected in parallel (Figure 19-22). To analyze the problems of a large evaporator, each circuit must be checked for performance. There may be as many as 20 circuits. Figure 19-23 shows a properly-operating multi-circuit evaporator with the correct refrigerant flow to each circuit. Figure 19-24 shows a multi-circuit

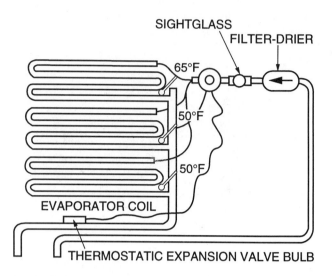

Figure 19-24. This multi-circuit evaporator is not operating correctly doe to a restriction in one of the circuits. The temperature is measured at the outlet of each circuit to determine which one is restricted.

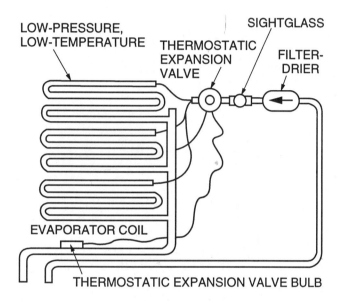

Figure 19-22. Several small evaporators can be combined to create one large evaporator.

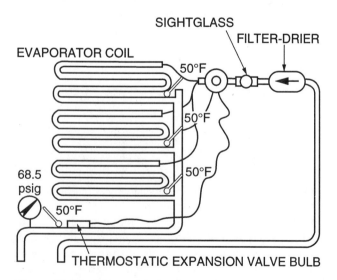

Figure 19-23. This multi-circuit evaporator is operating correctly with all circuits doing an equal amount of work.

evaporator that has one circuit that is restricted and starving. This symptom would be hard to detect without proper analysis. Proper analysis requires checking the pressure of the common suction header, converting the pressure to the coil evaporating temperature, and checking the temperature of the suction line in each circuit. A suction line with a higher-than-normal temperature is starved.

19.3 TYPES OF EVAPORATOR COILS

Evaporator coils are manufactured in several configurations: the "A" coil, slant coil, and slab coil are the most common types.

The "A" coil is used for smaller applications. It is constructed of two coils that are arranged in an "A" shape so that air can flow through either coil (Figure 19-25). The "A" coil can be installed in upflow, downflow, or horizontal applications. The refrigerant connections are on the end of the coil and may be piped at right angles for an exit at the side of the coil casing.

The slant coil is also commonly used in residential applications. It is called a slant coil, but it is really a small slab coil that is mounted at an angle (Figure 19-26). The refrigerant connections are on the end of the coil.

The slab coil is usually a large coil that is mounted on its side with all circuits entering the end of the coil (Figure 19-27). These coils have many circuits, and are used for large evaporators. Some of them are as large as the wall of a room in a house.

Evaporators are also classified according to the type of airflow. There are two distinct types of airflow through evaporator coils: draw through and blow through.

TYPES OF EVAPORATOR COILS 285

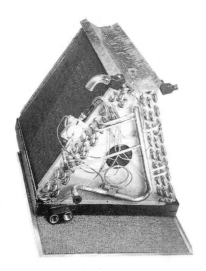

Figure 19-25. "A" coil. *Reproduced courtesy of Carrier Corporation.*

Figure 19-26. Slant-type evaporator coil. *Courtesy BDP Company.*

The draw-through coil has the fan located downstream of the coil and air is pulled through the coil (Figure 19-28). Since the fan is creating a vacuum at the coil outlet, this provides a nearly steady pull of air through each area of the coil since the vacuum is about the same at all

Figure 19-27. Large slab-type evaporator coils. *Courtesy BDP Company.*

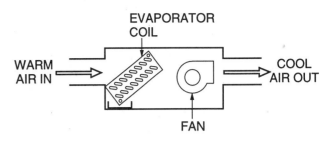

Figure 19-28. This coil has a draw-through airflow with even airflow over the coil.

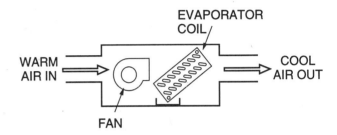

Figure 19-29. This coil has a blow-through airflow with much of the air flowing through the middle of the coil, and reduced airflow at the edges.

surfaces of the coil. This is usually the preferred method for coil location because of this even flow.

The blow-through coil has the fan located upstream of the coil and the coil is mounted in the fan discharge (Figure 19-29). The velocity of the fan may be such that it is greater at the top of the fan outlet, and certainly, the highest velocity is directly in front of the fan. In addition, the fan discharge area is much smaller than the coil surface area. To enhance coil performance, a static plate or spreader is placed in front of the fan to spread the air out as it passes through the coil (Figure 19-30). This helps to prevent high-velocity areas on the coil surface.

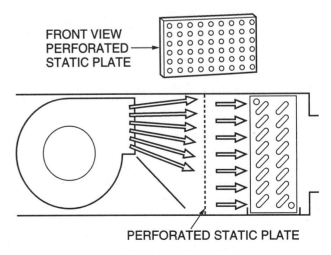

Figure 19-30. A static plate has been placed in front of the fan to distribute the air evenly over the coil.

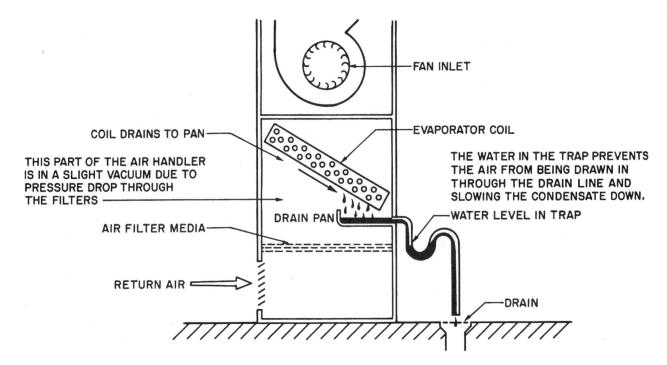

Figure 19-31. A slant coil and condensate drain.

19.4 CONDENSATE AND THE EVAPORATOR

The evaporator must be able to gather and drain the condensate. Most coils generate about one pint per hour per ton of air conditioning. For example, a ten-ton unit would create 240 pints of water per 24-hour period (10 pints/hour × 24 hours = 240 pints). This is 30 gallons of water per day, which is a considerable amount of water to handle. Water is collected on the fins, where it runs by gravity to the bottom of the coil and is collected in a condensate pan or drain.

When a coil is mounted in a slant position, the condensate must be able to follow the coil fins and travel to the condensate pan or drain (Figure 19-31). The air velocity across the coil must not be so high that it overcomes the force of gravity that is used to drain the condensate.

19.5 SENSIBLE AND LATENT HEAT REMOVAL

Evaporators are sized according to their ability to remove both sensible and latent heat from the air to be conditioned. You will recall that latent heat is the heat absorbed or lost during a change in state. The condensing of moisture out of the air represents latent heat removal. Sensible heat removal from the air to be conditioned can be recorded by monitoring the drop in temperature from the inlet to the outlet of the evaporator coil (Figure 19-32). The typical temperature drop is 20°F for most applications. These applications typically have an airflow across

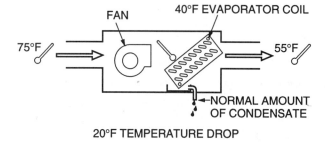

Figure 19-32. Sensible heat and the evaporator coil.

the evaporator of 400 cubic feet per minute (cfm) per ton of air conditioning. Therefore, a 10-ton system would circulate 4000 cfm (10 tons × 400 cfm/ton = 4000 cfm).

When there is a need for additional moisture removal, a higher temperature drop may be used, such as 22°F. This higher temperature drop may be accomplished by reducing the airflow across the coil, which will lower the evaporating temperature of the refrigerant. The lower evaporating temperature creates more of an attraction for the moisture in the air; therefore, more condensate is created and more humidity is removed (Figure 19-33). This lower airflow is typically used in applications where the humidity is high, such as in the tropics or locations close to the southern coastline. An airflow rate of 325 to 350 cfm/ton may be used in areas of high humidity.

Some locations typically have low humidity levels and it is desirable to remove only a small amount of the moisture from the air. The desert areas of the country

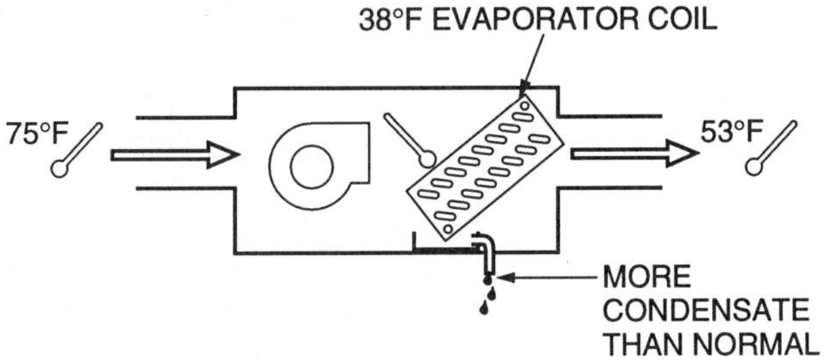

Figure 19-33. Latent heat and the evaporator coil.

would be in this category. When it is desirable to remove less humidity, more airflow is passed across the coil to raise the refrigerant evaporating point. This results in a lower temperature drop across the coil (typically 18°F). There will be less attraction to the moisture in the air, so less removal will take place. For these areas, 450 to 475 cfm/ton of air conditioning is the airflow used.

Checking Evaporator Airflow

Since the airflow determines the temperature drop of the air across an evaporator, a formula may be used to determine the desired temperature drop. When the designer knows the sensible heat capacity of a coil and the volume of air passing across the coil (CFM) is known, the temperature drop may be calculated using the sensible heat formula:

$$H_s = 1.08 \times cfm \times TD$$
H_s = sensible heat
1.08 = a constant
CFM = cubic feet of air per minute
TD = temperature difference

For example, suppose you know how much sensible heat capacity a coil has at a particular cfm from the manufacturer's data and you want to know the correct inlet/output temperatures to use in order to remove the correct amount of moisture for this application. The problem may call for 21,000 cfm of air to be moved across an evaporator coil that has a sensible heat removal capacity of 453,600 Btuh.

First, rearrange the formula to solve for TD:

$$TD = \frac{H_s}{1.08 \times cfm}$$

Next, insert the known values:

$$TD = \frac{453,600}{1.08 \times 21,000}$$

$$TD = 20°F$$

You can now measure the temperature in the return air and the supply air and set the airflow by the 20°F temperature difference.

SUMMARY

- The refrigeration (cooling) system moves heat from a place where it is not wanted to a place where it makes no difference. The evaporator is the system component that absorbs unwanted heat for transfer to a different location.
- Evaporators are typically made of copper coils with aluminum fins; however, there are also coils that use aluminum coils and aluminum fins or copper coils and copper fins.
- The fin-to-tube connection is very important and is accomplished by inserting the tube into the fin and expanding the tube inside the fin.
- For better heat exchange, the fins may be formed into a pattern that causes the air to move from side to side as it passes through the fins.
- Larger coils will have multiple circuits to reduce the pressure drop in the coil.
- The refrigerant pressure drop through the coil should be held to about the equivalent of 2°F from one end of the coil to the other.
- The temperature of a refrigerant coil is controlled by controlling the pressure in the coil.
- The typical design temperatures for an evaporator coil are about 40°F for the refrigerant evaporating temperature with 75°F air entering the coil.
- Manufacturers are working to make coils smaller, lighter, and more efficient.
- Both sensible and latent heat are removed from the air as it passes over the evaporator coil. Sensible heat removal lowers the air temperature. Latent heat removal results from a change in state, such as removing moisture from the air in the form of condensate.
- Typical evaporator problems include starved evaporators, flooded evaporators, dirty evaporators, and restricted airflow.
- Coils are manufactured in "A", slant, and slab configurations.
- Airflow through coils may be either draw through or blow through.
- Evaporators are sized to remove both sensible and latent heat. Latent heat removal can be controlled by the coil temperature, which can be controlled by the cfm across the coil. Most normal-humidity applications use an airflow of 400 cfm/ton with a temperature drop of about 20°F.
- Locations which require additional moisture removal may use airflows of 325 to 350 cfm/ton with a temperature drop of about 22°F.
- Locations which require less moisture removal may use airflows of 450 to 475 cfm/ton with a temperature drop of about 18°F.

UNIT 19 EVAPORATORS

REVIEW QUESTIONS

1. What is the purpose of the evaporator coil in an air conditioning system?.
2. What are the typical materials that evaporators are constructed of?
3. What is the purpose of the fins on a coil?
4. When is it best to add more circuits to a refrigerant coil?
5. What is the recommended refrigerant pressure drop for an evaporator coil?
6. What is the result of excess pressure drop through an evaporator coil?
7. What is the typical airflow through an evaporator coil under normal humidity removal conditions?
8. What is the typical airflow through an evaporator coil where high humidity is involved?
9. What is the typical airflow through an evaporator coil where low humidity is involved?
10. An evaporator has a sensible heat capacity of 320,000 Btuh and a temperature drop of 20°F. Calculate the airflow.

COMPRESSORS 20

OBJECTIVES

Upon completion of this unit, you should be able to

- Describe the purpose of the compressor in an air conditioning system.
- State the different types of compressors used for air conditioning in residential and commercial applications.
- Describe the refrigerant flow and action from the inlet to the outlet in a reciprocating compressor.
- List the main components of a reciprocating compressor.
- Describe the typical lubrication system in a reciprocating compressor.
- Discuss the differences between a hermetic compressor and a semi-hermetic compressor.
- Describe capacity control for a compressor and refrigeration system.
- State the two types of rotary compressors.
- Describe the pumping action of a scroll compressor.
- List three types of motor overload devices.
- Describe the design conditions for an air conditioning compressor.
- Troubleshoot basic compressor problems.

20.1 COMPRESSOR FUNCTIONS

The compressor is often thought of as the "heart" of the refrigeration system. It is the pump that moves the refrigerant in the system. All of the work energy applied to the refrigerant is applied in the compressor. The simplest description of a compressor is that it is a *vapor pump*.

In the refrigeration cycle, the compressor is the component in the line between the evaporator outlet and the condenser inlet. It maintains a low pressure in the evaporator by removing vapor refrigerant at the correct rate to maintain a specific evaporating temperature in the evaporator. The vapor pressure is raised to a pressure corresponding to the condensing temperature in the condenser while being pumped to the condenser (Figure 20-1). The common compressors used for air conditioning are the reciprocating, rotary, and scroll compressors. Other types of compressors are used for larger chilled water systems and are discussed in a later unit.

20.2 RECIPROCATING COMPRESSORS

The process that occurs in the compressor is called *vapor compression* and can be accomplished using several different methods. It is the process of reducing the volume

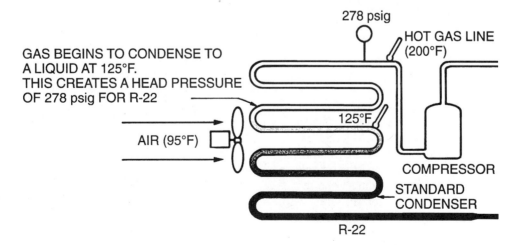

Figure 20-1. Compressor and condenser.

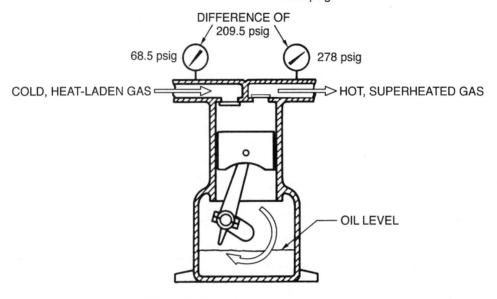

Figure 20-2. Reciprocating compressor.

of a vapor. The most common method uses a piston (much like the one in a small engine) attached to a crankshaft. This type of compressor is called a *reciprocating compressor* (Figure 20-2).

The term *reciprocating* refers to the back and forth or up and down motion of many compressors and engines. In an internal combustion engine, an explosion at the top of the stroke causes the piston to move away from the top (Figure 20-3). This action is transferred to a crankshaft and used to power something, such as a lawn mower or automobile. This type of reciprocating action is used to turn fuel into work energy. In the reciprocating compressor, the opposite occurs—work energy is applied to the crankshaft to turn it and compression occurs at the top of the piston (Figure 20-4). As the piston rises in the cylinder, pressure is created, causing the piston to be forced to the top of the stroke (Figure 20-5).

The refrigerant flow in the compressor is as follows:

1. Compression and control of refrigerant flow is accomplished using valves at the top of the stroke. The valve known as the *suction valve* allows refrigerant gas to flow from the evaporator outlet into the cylinder during the down stroke. As the piston starts down, a low-pressure area is created in the cylinder and refrigerant vapor flows into the cylinder (Figure 20-6). Refrigerant vapor will continue to flow until the cylinder is at the same pressure as the inlet vapor or the piston reaches bottom dead center and starts back up (Figure 20-7).

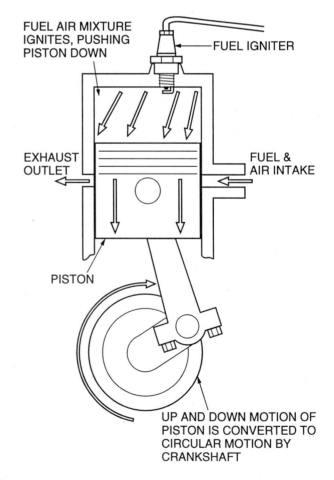

Figure 20-3. Reciprocating engine action.

RECIPROCATING COMPRESSORS 291

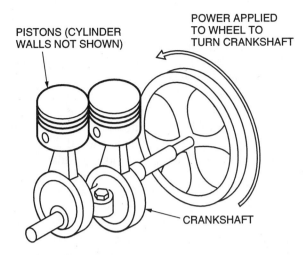

Figure 20-4. Energy applied to a crankshaft.

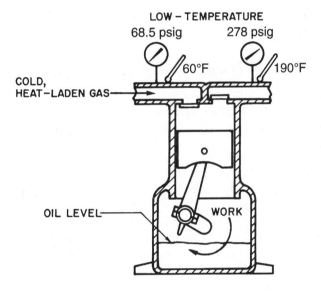

Figure 20-5. Piston rising in a cylinder creates pressure.

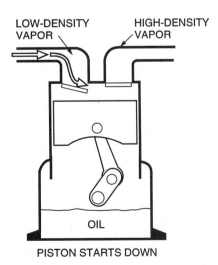

Figure 20-6. Refrigerant fills the cylinder on the down stroke.

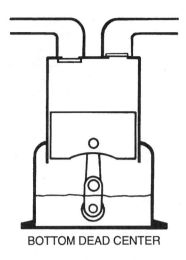

Figure 20-7. Piston at bottom dead center.

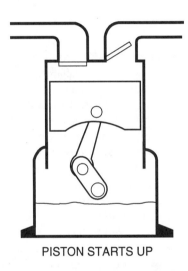

Figure 20-8. As the piston rises in the cylinder, the pressure increases.

2. When the piston reaches bottom dead center and starts back up, the refrigerant in the cylinder is being compressed and the pressure starts to rise. The suction valve will close due to differential pressure. As the piston approaches top dead center, the pressure becomes much greater at the top of the cylinder and the discharge valve begins to open (Figure 20-8).

3. When the piston reaches top dead center, the discharge valve opens completely and allows the refrigerant above the piston to escape into the compressor discharge line (Figure 20-9). When the piston starts back down, the pressure in the discharge line will close the discharge valve.

The valve action in a reciprocating compressor is accomplished using differential pressure and, in some cases, small springs to positively close the valves.

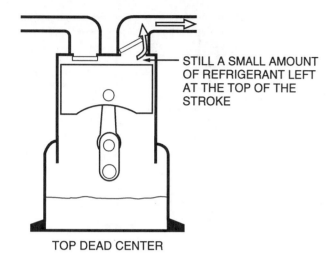

Figure 20-9. The discharge valve opens as the piston approaches the top of the cylinder and the refrigerant moves to the discharge line.

This is a very simple process compared to an internal combustion engine where a camshaft opens and closes the cylinder valves at a particular time.

Since the pumping action of the piston is up and down, and the direction of the gas flow is determined by the valves, the rotation of the compressor makes no difference—the vapor will move only in one direction. Some compressors are subject to reverse rotation, but they still pump the same at either rotation.

The reciprocating compressor is a positive displacement compressor, meaning when the piston rises in the cylinder, it is being forced to the top and if there is interference, damage will result. Remember, the compressor is a vapor pump. It is not designed to pump liquid. If liquid enters the cylinder in a reciprocating compressor, damage will likely occur because liquid cannot be compressed. In fact, anything else may happen (e.g., the piston, rod, or crankshaft may break or the motor may stall) but the liquid will not be compressed.

During compression, heat builds up in the top of the compression stroke. The refrigerant gas entering the compressor is typically superheated from the evaporator to about 50°F. Additional superheat is added to the suction gas as it travels down the suction line towards the compressor, so the gas entering the compressor housing may be 60°F (Figure 20-10). In a suction-cooled compressor, additional heat is added to the refrigerant in the compressor as the cold suction gas passes over the hot motor windings. Also, some of the mechanical energy of the device used to drive the compressor is converted to heat energy. All of this heat is concentrated and is contained in the compressor discharge gas. A typical air conditioning compressor using R-22 may have an entering refrigerant temperature of 60°F and a leaving refrigerant temperature of 200°F on an air-cooled system on a hot day.

Components of a Reciprocating Compressor

The reciprocating compressor contains several components: the piston and rings, rod and bearings, crankshaft, lubricating system, valve plate and valves and the compressor head.

The piston and rings accept much of the load because compression takes place at the top of the piston (Figure 20-11). The piston reaches top dead center and reverses rotation and returns to bottom dead center to be reversed again for every rotation the compressor makes. Many compressors turn at 3450 revolutions per minute (RPM). The piston is in constant motion whenever the compressor operates.

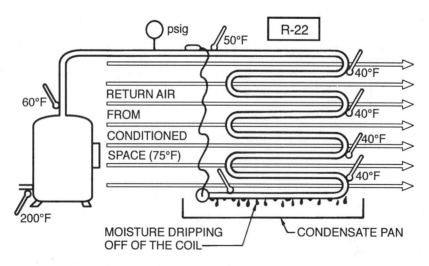

Figure 20-10. Superheat in the suction line raises the suction gas temperature to about 60°F.

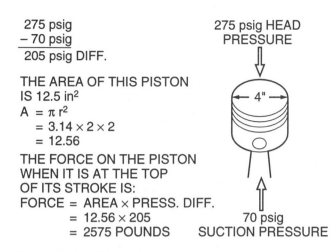

275 psig
− 70 psig
205 psig DIFF.

THE AREA OF THIS PISTON
IS 12.5 in²
A = π r²
  = 3.14 × 2 × 2
  = 12.56

THE FORCE ON THE PISTON
WHEN IT IS AT THE TOP
OF ITS STROKE IS:
FORCE = AREA × PRESS. DIFF.
      = 12.56 × 205
      = 2575 POUNDS

275 psig HEAD PRESSURE
70 psig SUCTION PRESSURE

Figure 20-11. The piston and rings are under a tremendous load.

The piston is round and machined to a very close tolerance so that it fits the cylinder correctly and fits the top of the cylinder as closely as possible. The closer the piston is to the top of the cylinder, the more refrigerant it will push out of the cylinder at the top of the stroke.

Pistons are typically made from aluminum or cast iron with grooves for the sealing rings normally located close to the top of the piston (Figure 20-12). The top two sealing rings are typically the compression rings responsible for maintaining the seal at the cylinder walls (Figure 20-13).

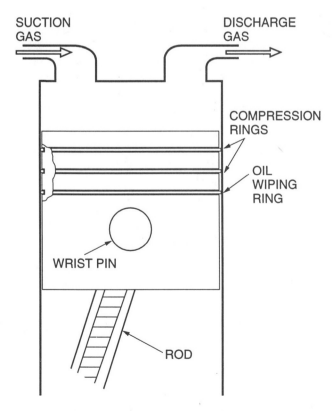

Figure 20-12. Cutaway of a piston and the ring grooves.

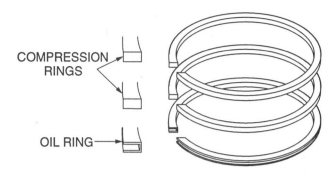

Figure 20-13. The oil ring has a different shape than the compression rings.

The bottom ring is an oil wiping ring. It has a special shape to return as much lubricating oil to the crankcase as possible. The rings move up and down in the cylinder, which is made of cast iron. These rings are made of very hard steel and should operate reliably for many years.

The compressor rod and bearings connect the piston to the crankshaft. The rod is often called a *connecting rod*. It typically has a small hole at the top where it is fastened to the piston by means of a wrist pin. A larger hole is at the bottom of the rod where it is fastened to the crankshaft, usually by means of a removable cap (Figure 20-14). The rod must withstand a large load because it transmits all of the power from the crankshaft to the piston. When the compressor is in the compression stroke, pressure is pushing the piston towards the crankshaft with the rod between the two. When the piston is on the down stroke, the rod is pulling the piston downward (Figure 20-15). These are two distinct types of load and the rod must be designed to accept both and last for years.

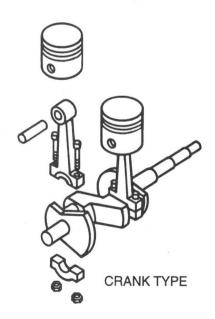

Figure 20-14. The rod has a large end that typically has a removable cap for disassembly.

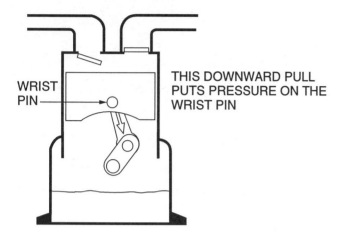

Figure 20-15. On the down stroke, the rod pulls the piston down.

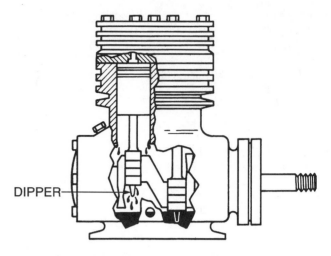

Figure 20-17. The crankshaft is an off-center device.

There is a bearing surface in each end of the rod that requires lubrication. The upper bearing surface at the wrist pin is often lubricated by means of oil that is splashed up into the inside of the piston and captured in a small reservoir at the top of the rod. The oil then lubricates the wrist pin bearing by means of gravity (Figure 20-16). The lower rod bearing is lubricated via pressure lubrication by means of holes drilled in the crankshaft.

The crankshaft is typically made of cast iron. It is the off-center device that changes the rotating motion of the drive mechanism to an up and down reciprocating motion (Figure 20-17). The off-center portion of the crankshaft where the rod fastens is known as the *throw*. In smaller compressors, the crankshaft stroke may be short and the compressor may have only one or two cylinders. Larger compressors may have cylinder strokes up to about four inches long with up to 12 cylinders. As the crankshaft becomes longer, more support must be added in the way of bearings in the middle of the shaft.

The crankshaft is a precision device that must be in correct balance for the compressor to operate without vibration. It has oil passages drilled from one end to the other to route oil to the various passages that lubricate the rod and bearings.

Another type of reciprocating compressor crankshaft device is called a *scotch yoke*. This device is used on small compressors up to about five tons. It also converts rotating action to reciprocating action.

The lubrication system is very important because every moving part must be kept lubricated for the compressor to continue running. The purpose of lubrication is to form a thin, slick barrier between all moving parts. This barrier prevents any metal-to-metal contact which would cause wear. Compressor wear is minimal when proper lubrication is available. Most compressor wear occurs during the first few seconds of operation due to the initial lack of lubrication before oil pressure is established. Once the compressor is turning and oil pressure is established, wear

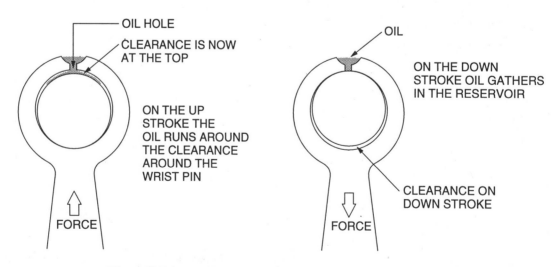

Figure 20-16. Lubrication of the wrist pin using gravity.

is reduced to a minimum. Because of this condition, is desirable to limit the number of compressor starts.

Most compressors operate with a positive pressure oil system using an oil pump to build the oil pressure. The oil pump is driven by the compressor motor. It is mounted on the end of the compressor crankshaft and turns at the same speed as the compressor. On larger compressors, the shaft is horizontal and the oil pump is at the end of the shaft under a domed housing with a keyway-type drive (Figure 20-18). On smaller compressors, the oil pump is still at the end of the shaft, but it is on the bottom submerged in the oil (Figure 20-19).

The oil pump on the larger compressor is positioned above the oil level in the compressor oil sump, so a suction pickup tube is used to move the oil from the oil sump to the oil pump. The oil pump picks up oil from the oil sump, which is at compressor suction pressure and raises the pressure to a value higher than suction pressure. Typically, the oil pressure is 20 to 50 psig higher than the suction pressure. This is called the *net oil pressure* and is found by subtracting the suction pressure from the oil pump outlet pressure. For a system operating correctly at 40°F evaporating temperature, the suction pressure should be 68.5 psig. The oil pressure should be from 88.5 psig (68.5 + 20 = 88.5) to 118.5 psig (68.5 + 50 = 118.5).

The oil pumps for larger compressors are typically bidirectional, meaning they will pump in either direction of rotation. This maintains oil pressure in case of reverse rotation.

As mentioned earlier, part of the system may be splash lubricated. Splash lubrication requires oil to be splashed around in the crankcase so that it can be gathered for gravity distribution to any component using gravity lubrication.

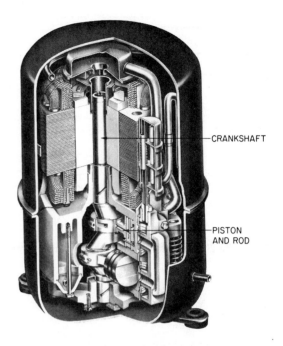

Figure 20-19. On this smaller compressor, the oil pump is at the bottom of the crankshaft. *Courtesy Tecumseh Products Company.*

Most residential and commercial refrigeration systems will continue to use R-22 until replacement refrigerants become available. R-22 has an attraction for the lubricating oil used in systems and will actually migrate to the oil. Figure 20-20 shows a container of R-22 and a container of refrigerant oil connected by a tube.

Figure 20-18. This crankshaft is in a horizontal position with the oil pump on the end of the shaft. *Courtesy Copeland Corporation.*

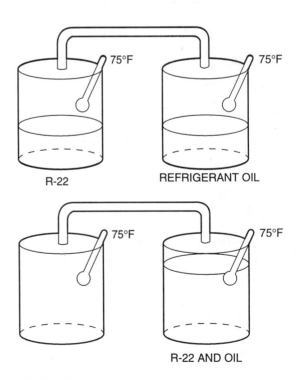

Figure 20-20. Over time, the refrigerant in the container on the left will migrate into the oil in the container on the right.

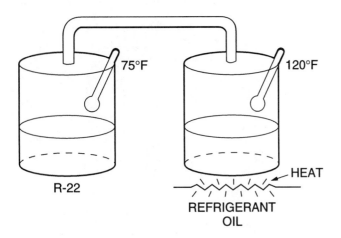

Figure 20-21. Heat will prevent the refrigerant from migrating to the oil.

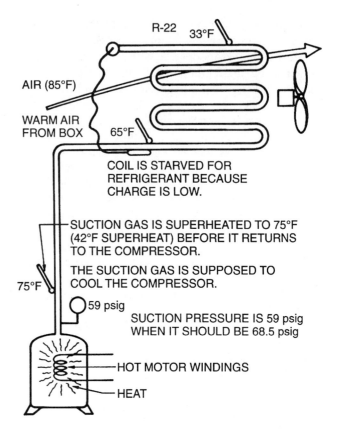

Figure 20-22. This compressor is overheating because the suction gas is too warm to cool the motor.

Over a period of several hours at room temperature, the R-22 migrates to the oil. If the proper amount of heat is added to the oil, this migration does not occur (Figure 20-21). Most compressors have a crankcase heater to warm the oil, which minimizes its attraction for R-22.

Correct lubrication also involves the correct oil consistency. The oil must have the right lubricity (lubrication quality) to lubricate correctly. Two conditions that may disturb the quality of lubrication in a compressor, are overheating and liquid dilution of the oil.

Overheating of the oil occurs when the compressor overheats. The oil transfers much of the heat in the compressor components, so when the compressor is too hot, the oil is too hot. Usually, a compressor becomes too hot when the suction cooling is inadequate due to a low charge or starved evaporator (Figure 20-22). The compressor can also become too hot when the head pressure is too high, such as occurs with a dirty condenser. This will be discussed in the next unit.

The lubricating oil will be diluted if liquid refrigerant enters the compressor crankcase and mixes with the oil. This is similar to operating an auto engine with oil that is too thin. It does not maintain a barrier between the moving parts and the metal starts to rub. You can tell when oil dilution by liquid occurs because the compressor crankcase will feel cool when it should be warm or hot. It is not unusual for the crankcase to become cool just after start-up because of liquid refrigerant boiling out of the oil, but if this condition continues, bearing damage will occur.

If you discover a lubrication problem at start-up, (i.e., the crankcase is too cool) the crankcase heater should be suspected if the compressor has one.

Actually, it is best not to start the compressor unless the crankcase is up to temperature. The manufacturer's instructions will tell you how long the crankcase heater must be energized before the compressor is started.

Note: Some compressors do not have crankcase heaters because of the small amount of refrigerant charge in the system.

The valve plate and valves are mounted at the top of the cylinder and are typically clamped between the cylinder head and cylinder wall (Figure 20-23). The valve plate is constructed of steel and is used to hold the valves in place. The valves may be either flapper-type valves or ring valves. Flapper-type valves are made of very high quality steel and are held stationary by pins or screws at the back. The discharge valves will be on the top of the valve plate and the suction valves on the bottom of the plate (Figure 20-24). These valves are removable and can be changed if one breaks; however, the complete valve plate assembly is usually replaced unless there is reason to believe there is no wear on the valve seat. The discharge valve will have a retainer to prevent the valve from bending backward when flow pushes the valve open (Figure 20-25).

The ring valve is also made of very high quality steel. It is typically used on larger compressors and has large ports to provide additional refrigerant flow (Figure 20-26).

The compressor head holds the valve plate in place. It is also the passage for the compressed gas to leave the

RECIPROCATING COMPRESSORS 297

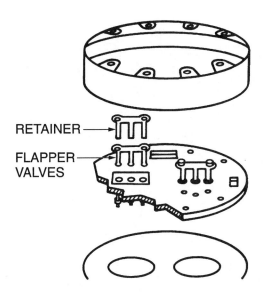

Figure 20-23. The valve plate is clamped between the cylinder walls and the compressor head and held tight by the head bolts.

Figure 20-24. The discharge valve will be on top of the valve plate, and is typically held down by pins or bolts. *Reproduced courtesy of Carrier Corporation.*

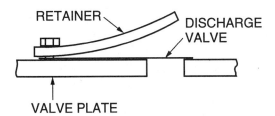

Figure 20-25. This retainer prevents the discharge valve from bending backwards should liquid refrigerant enter the cylinder.

compressor. It routes the hot compressed gas to the discharge line leaving the compressor. With hermetic compressors, the head is internal to the compressor shell. In semi-hermetic compressors, the head can be removed by bolts from the outside (Figure 20-27).

The compressor housing holds the compressor components together and keeps the system leak free.

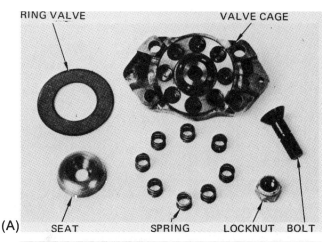
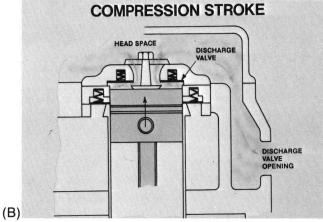
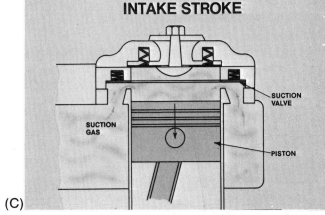
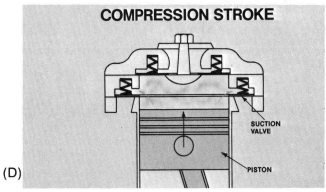

Figure 20-26. A ring-type refrigerant valve.

Figure 20-27. The compressor head on a semi-hermetic compressor. *Courtesy Copeland Corporation.*

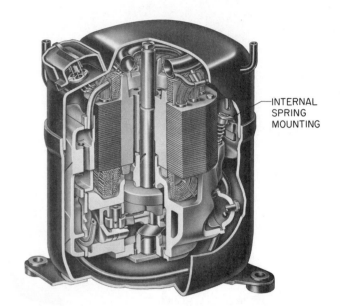

Figure 20-28. This compressor is suspended by springs inside the compressor shell. *Courtesy Tecumseh Products Company.*

There are three distinct types of compressor housing: hermetic, semi-hermetic, and open-drive.

20.3 HERMETIC COMPRESSORS

The hermetic compressor has a welded steel shell with the complete compressor and drive motor mounted inside. These compressors range in size from fractional tonnage units to about 40 tons. One advantage of this compressor is that it is factory sealed and welded shut so it is less likely to leak. Another advantage is that the motor is in a refrigerant atmosphere where it will remain clean and have a controlled cooling process for the motor windings. One disadvantage is that it must be cut open to be serviced, so all service must be accomplished at a compressor rebuild center. Another disadvantage is that if the the motor burns out, it will pollute the refrigerant in the system. Following a burnout, the refrigerant must be cleaned using proper filtration.

The hermetic compressor may be mounted inside the shell using one of two different methods: the compressor and motor housing may be pressed into the shell, as in some older compressors, or the compressor may be suspended from springs from inside the shell. The spring-mounted compressor is the most common type of reciprocating hermetic compressor. The springs help to isolate the compressor from the rest of the system framework to prevent vibration. The compressor housing is mounted on rubber-type neoprene mounts to the unit frame, which also helps to reduce vibration and noise (Figure 20-28).

The reciprocating hermetic compressor motor is cooled by means of cold refrigerant vapor from the evaporator. All refrigerant leaving the evaporator passes through the motor housing and then to the compressor cylinders. The superheat of the refrigerant leaving the evaporator must be controlled so that there is enough cooling capacity in the returning refrigerant gas.

20.4 SEMI-HERMETIC COMPRESSORS

The semi-hermetic compressor is bolted together to allow field disassembly for service or rebuild purposes. Access to all components is provided through bolted flanges. The motor is inside the housing and is suction cooled just like the hermetic compressor. These compressors range in size from fractional hp to about 150 hp. The fractional hp units are used for refrigeration and a few special air conditioning applications.

One advantage of a semi-hermetic compressor is that it can be disassembled in the field for repair. For example, if a compressor fails in a critical application, it may save time to rebuild it on-site or simply replace some of the internal components rather than removing and replacing the entire compressor. In many cases, it may also be less expensive.

Like a hermetic compressor, the semi-hermetic compressor must be isolated from the system to prevent vibration and noise. It is typically mounted on springs or isolation pads.

Note: The hold-down bolts on a semi-hermetic compressor are usually tightened down during shipment to prevent the compressor from moving around (Figure 20-29). Remember to remove or loosen the hold-down bolts at start-up or excess vibration will occur, which may cause a refrigerant line to fail. The compressor should have a certain amount of freedom during normal operation.

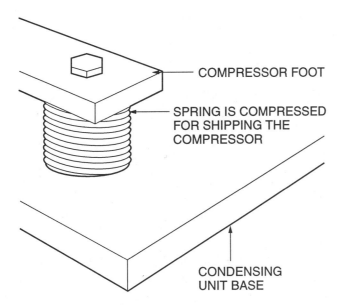

Figure 20-29. This compressor foot is tightened down for shipment and must be loosened before operation.

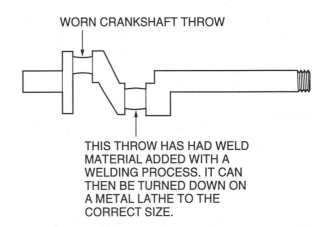

Figure 20-31. A worn crankshaft can be repaired by welding material to the worn surface. The surface is then machined down to the original size.

Semi-hermetic compressors are manufactured with replaceable components to be rebuilt many times. The main casting is the largest component. It receives all of the other components and is very durable. The cylinders are the components that receive the most wear and are typically replaceable with sleeves when the compressor is rebuilt.

The motor is a very expensive component in the compressor. It is removable and can often be rebuilt by rewinding it with new windings (Figure 20-30).

Another expensive component is the crankshaft. This device is under a tremendous load and subject to much abuse should lack of lubrication occur. The crankshaft can often be resurfaced by adding material to the worn surface and grinding it back to the correct size. This can be accomplished by welding the surface to add material (Figure 20-31). However, the high temperatures used when welding can damage the crankshaft. Another method known as *flame coating* adds small amounts of material without having to overheat the crankshaft. The shaft is then machined back down to the correct size.

20.5 OPEN-DRIVE COMPRESSORS

The open-drive compressor is bolted together just like the semi-hermetic compressor. The difference is that the open-drive compressor has the motor on the outside (Figure 20-32). The advantage of this is that the motor can be serviced from outside the refrigerant atmosphere.

Figure 20-30. This motor can be rebuilt by rewinding it if it burns out. *Courtesy Copeland Corporation.*

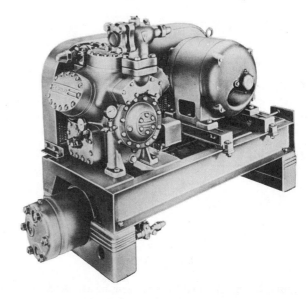

Figure 20-32. Belt-drive compressor. *Reproduced courtesy of Carrier Corporation.*

If the motor burns out, it will not pollute the refrigerant in the system. The heat from the motor is released outside the refrigerant circuit and is not handled by the refrigerant. The motor must be cooled by the ambient air, so adequate ventilation must be provided. A guard must be furnished to prevent personal injury. The guard may be removed for service.

An open-drive compressor can be either direct driven or belt driven.

Direct-Drive Compressors

Direct-drive compressors are installed in a special manner. When the compressor is direct driven, the motor shaft is placed at the end of the compressor shaft and a flexible coupling is installed between them. The flexible coupling helps to prevent minor misalignment from causing vibration to the compressor. There are several different styles of coupling; some have rubber inserts while others use a spring steel between the coupling halves.

The compressor and motor shafts must be aligned to the tolerance recommended by the manufacturer. There are two different alignments to be concerned with: angular and parallel. Angular misalignment occurs when the two shafts are not at the same angle (Figure 20-33). Parallel misalignment occurs when the two shafts are not parallel (Figure 20-34). Both types of alignment are checked by using a dial indicator gage and rotating the motor shaft through 360° of rotation (Figure 20-35). This is a time consuming, technical process and should only be attempted by an experienced technician.

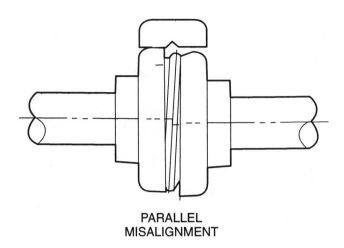

Figure 20-34. Parallel misalignment occurs when the motor shaft and compressor shaft are not parallel. *Courtesy Amtrol, Inc.*

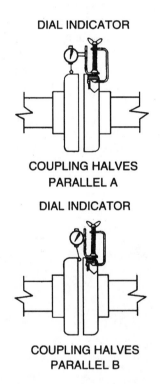

Figure 20-35. Alignment is accomplished by rotating the motor crankshaft and using a dial indicator to check the alignment. *Courtesy Amtrol, Inc.*

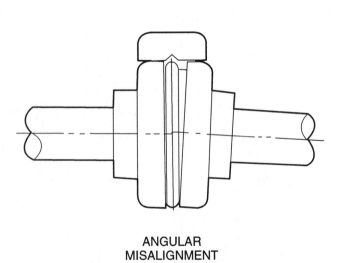

Figure 20-33. Angular misalignment occurs when the motor shaft and compressor shaft are at different angles. *Courtesy Amtrol, Inc.*

As the misalignment is found, shim stock (very thin sheets of steel) are placed under the motor or compressor feet to align the components (Figure 20-36). Once both

OPEN-DRIVE COMPRESSORS   301

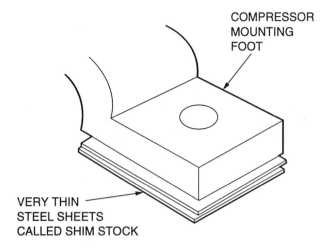

Figure 20-36. Thin pieces of steel called *shim stock* are used to raise the compressor or motor base as needed. The motor or compressor is tightened down on the shim stock.

Belt-Drive Compressors

Belt-drive compressors are installed differently from direct-drive compressors because the motor is installed and mounted beside the compressor (Figure 20-38). Belts are used to transfer power to the compressor shaft. The belts accept any vibration from the motor and protect the compressor. The compressor and motor pulleys must be in alignment for the belts to run correctly, but this is a much easier service procedure and can be accomplished using a straightedge rather than precision instruments (Figure 20-39).

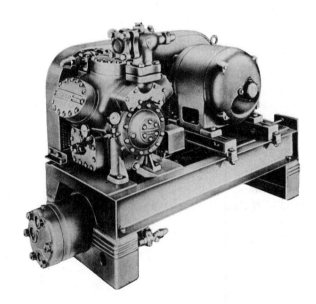

Figure 20-38. In a belt-drive compressor, the motor is mounted beside the compressor. *Reproduced courtesy of Carrier Corporation.*

components are tightened down and the alignment is accomplished, the unit is started up and allowed to run long enough to get the motor and compressor up to operating temperature. The alignment is then checked again as it may have changed with the temperature rise. When it is established that the units are in line, both the compressor and the motor base are drilled and a tapered pin is driven into the base to hold it absolutely steady (Figure 20-37). Concrete is then poured into the cavity of the base to further strengthen it. It should be apparent that this is a rather difficult procedure. To make things easier, many manufacturers furnish the motor compressor as a factory-mounted assembly.

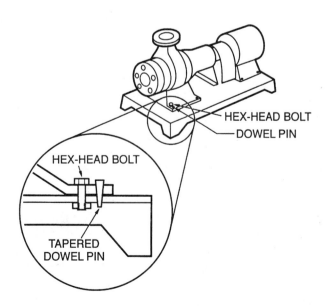

Figure 20-37. The motor and compressor are then fastened permanently by drilling the base with a tapered bit. A tapered pin is driven into the hole.

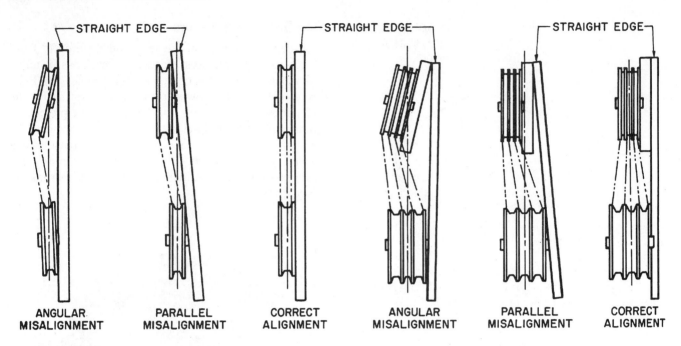

Figure 20-39. A straightedge can be used to align the compressor and motor pulleys in a belt-drive application.

The belts must have the correct tension to prevent slippage or to prevent them from being adjusted to an over-tight condition. Proper tension can only be accomplished using a belt tension gage (Figure 20-40).

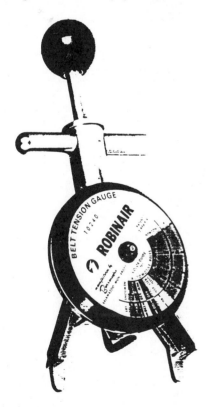

Figure 20-40. Belt tension gage. *Courtesy Robinair Division, SPX Corporation.*

The belts must also be of the correct length. Many technicians think that if they buy three belts of the same length (e.g., 50 inches) that they are all 50-inch belts. Manufacturing tolerances may cause one belt to be $50\frac{1}{4}$ inches while another belt may be $49\frac{3}{4}$ inches. This is not much, but is enough for the belts to pull unevenly. In order to acquire the correct belts for a compressor drive that uses three belts, a set of matched 50-inch belts must be purchased. They will be marked "matched" and shipped tied together to keep them from becoming separated.

The belt-drive installation must have a belt guard to prevent personnel from catching their hands or clothing in the moving belt. This belt guard is typically designed so that you cannot even reach over the back of the belt drive and become tangled in the belts (Figure 20-41). Belt guards are required by insurance companies and government safety agencies such as the Occupational Safety and Health Authority (OSHA).

Refrigerant Shaft Seal in Open-Drive Compressors

A refrigerant shaft seal prevents refrigerant from leaking out of the compressor housing while the shaft is turning or standing (Figure 20-42). Since the open-drive motor is on the outside of the casting and the compressor crankshaft is on the inside, a rotating shaft seal must hold the refrigerant in the casting. This seal is located where the compressor crankshaft protrudes from the compressor housing. It is typically very reliable unless it encounters undue stress, such as from a compressor that is out of

RECIPROCATING COMPRESSOR EFFICIENCY 303

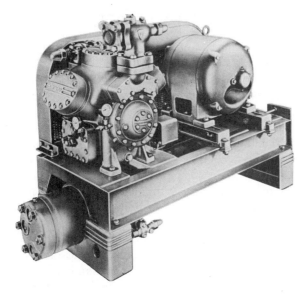

Figure 20-41. The belt guard is necessary to prevent possible injury. *Reproduced courtesy of Carrier Corporation.*

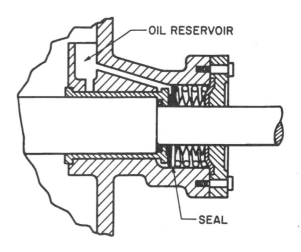

Figure 20-42. Crankshaft seal to prevent refrigerant from escaping while the shaft is turning.

balance and vibrating or from compressor-to-shaft misalignment with direct-drive compressors.

The seal usually has a spring that pushes the rotating sealing surface against the stationary sealing surface. Typically, ceramic or carbon is used as the rotating surface and is pushed against a stationary steel surface that is lubricated with oil from the system.

20.6 RECIPROCATING COMPRESSOR EFFICIENCY

The pumping efficiency of a reciprocating compressor largely depends on the amount of refrigerant that empties from the cylinder compared to the amount that enters the cylinder. The clearance that is left at the top of the piston stroke is the volume of refrigerant left. The piston cannot be allowed to completely rise to the top or it will strike the head of the compressor. There must be some clearance at the top of the cylinder. This clearance is called the *clearance volume* of the compressor (Figure 20-43). The more clearance volume, the more refrigerant there will be to re-expand on the down stroke after the discharge valve closes. Compressor manufacturers have worked to reduce this clearance volume to the absolute minimum. The valve plate for the discharge valve is one place where the clearance volume has been reduced by some manufacturers. This has been accomplished by making the discharge valve part of the valve plate rather than a separate valve above the plate (Figure 20-44).

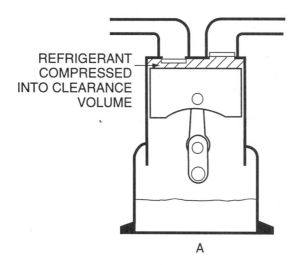
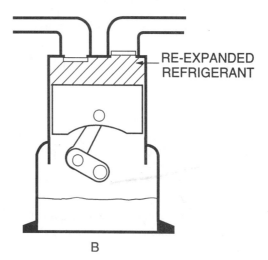

Figure 20-43. The clearance volume in a compressor cylinder contains high-pressure refrigerant that will re-expand when the piston starts down, taking up space that could be used for suction gas from the low-pressure side of the system.

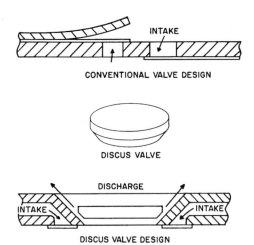

Figure 20-44. Clearance volume is reduced by making the discharge valve part of the valve plate. *Courtesy Copeland Corporation.*

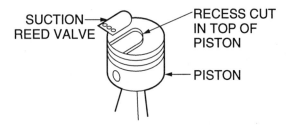

Figure 20-45. This suction valve is recessed in the piston to reduce clearance volume.

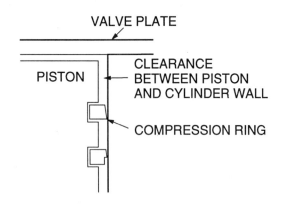

Figure 20-46. There is also clearance volume between the compression ring and the top of the piston.

The suction valve often has a recess machined in the piston so the clearance volume can be reduced to a minimum (Figure 20-45).

The clearance between the compression ring and the top of the piston also provides some volume for high-pressure refrigerant to be trapped (Figure 20-46).

20.7 COMPRESSOR CAPACITY CONTROL

Compressor capacity control is a method of varying the operating capacity of a compressor to meet a changing load. The air conditioning load on a commercial building varies depending upon the time of day, day of week, and time of year. It is desirable to have a system that can operate at various levels of capacity. For example, an office building that requires only 10 tons of cooling in the morning may require 40 tons during the afternoon when the sun is up and shining on the building. There may also be more people in the building and they may require more lighting and operate more equipment. It would be best to have a system that can operate at 10 tons in the morning and 40 tons in the afternoon. This can be accomplished by a system that has four compressors with 10 tons of for capacity each. This is a common method of capacity control. However, as compressor motors become larger, they also become more efficient, so it is desirable to use the largest motor possible.

When a compressor has more than one cylinder, the capacity of the compressor can be varied by preventing some of the cylinders from pumping. For example, suppose that a 40 ton compressor has eight cylinders. Each cylinder has a capacity of five tons (8 × 5 = 40). If the compressor has four heads with two cylinders under each head and the cylinders can be unloaded in pairs down to the last two cylinders, the compressor could have a capacity rating of 10, 20, 30, or 40 tons, depending on how many cylinders were pumping (Figure 20-47). The unloading of a compressor is typically expressed in terms of percent of load. The compressor above would be loaded at 100% at 40 tons, 75% at 30 tons, 50% at 20 tons and 25% at 10 tons. This compressor could be started in the morning and may not shut off for the entire day. It would just change its capacity to meet the needs of the building. This would greatly increase the life of the compressor, as start-up is when most of the wear occurs. Without capacity control, the 40 ton compressor would be stopped and started many times in a typical day.

Capacity control is used in many commercial operations on compressors starting at about 10 tons. Compressor capacity control is accomplished in several ways, depending on the design of the compressor. Common methods include: cylinder unloading, blocked suction unloading, and diversion gas unloading.

Cylinder Unloading

Cylinder unloading is accomplished by holding the suction valve open when reduced capacity is required. When the suction valve is held open, refrigerant gas and circulating oil are drawn into the cylinder on the down stroke as usual. However, on the up stroke, the refrigerant gas and oil are pushed back out into the suction line so the refrigerant is not moved into the high pressure side of the system (Figure 20-48). In other words, the piston is merely moving up and down in the cylinder. Lubrication

COMPRESSOR CAPACITY CONTROL

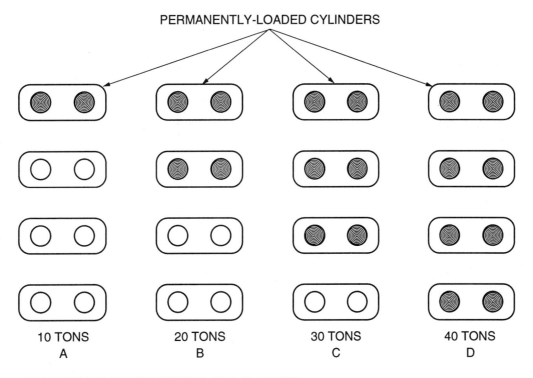

Figure 20-47. This compressor can operate at several different capacities.

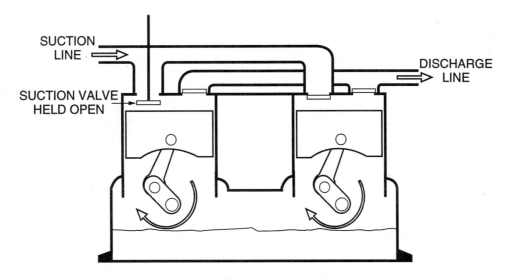

Figure 20-48. Cylinder unloading.

of the cylinder is accomplished, but no compression takes place and very little work is done, so there is a reduction in power consumption.

When an ammeter is applied to a compressor during this operation, the amperage pulled by the compressor motor will be proportioned very closely to the load on

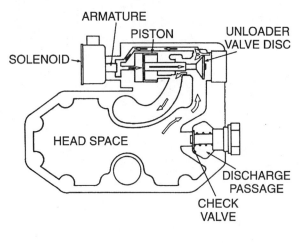

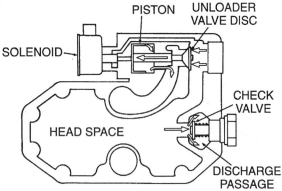

Figure 20-49. Gas diversion compression unloading. *Courtesy Trane Company.*

the compressor. For example, if the 40 ton compressor mentioned earlier had an amperage draw of 120 amps at 100% load, it would have approximately 90 amps at 75% (120 × .75 = 90), 60 amps at 50%, and 30 amps at 25%. This information can be valuable for troubleshooting purposes. All that needs to be done to determine the approximate load level is to measure the amperage.

Blocked Suction Unloading

Blocked suction unloading reduces power consumption by placing an obstacle in the suction line to prevent suction gas from entering the cylinder. When the piston goes down, no refrigerant is pulled into the cylinder, so there is none to pump out on the up stroke.

Gas Diversion Unloading

Gas diversion unloading reduces power consumption by using a diversion chamber to hold the refrigerant gas so that it is not compressed. This is accomplished by means of a solenoid valve that controls the entrance to the diversion chamber (Figure 20-49).

20.8 ROTARY COMPRESSORS

The rotary hermetic compressor is used for certain residential air conditioning applications under five tons.

One difference between the rotary hermetic compressor and the reciprocating hermetic compressor is that the rotary compressor is cooled by hot discharge gas. The compressor housing will be much hotter than that of a reciprocating compressor. This is usually indicated by a warning on the compressor housing.

Like the reciprocating compressor, the rotary compressor is also a positive displacement compressor; however, it is more tolerant of liquid refrigerant. Small amounts of liquid refrigerant can usually move through a rotary compressor without damage.

The rotary compressor is more efficient than the reciprocating compressor because it does not have the clearance volume at the top of the cylinder. For all practical purposes, the cylinder empties completely. There are two types of rotary compressors in use for air conditioning: the rotary vane and the stationary vane.

Rotary Vane Compressors

The rotary vane compressor has a rotating, drum-shaped cylinder with vanes mounted in the outer perimeter that are free to move in and out as the cylinder turns. The cylinder is mounted on a shaft in the center of the drum so that it turns in a circle. The cylinder is then mounted in a housing so that it is slightly off center. As the cylinder turns, the vanes move outward and rub against the outside of the housing, sealing refrigerant between the vanes. As the vanes pass the inlet port, refrigerant is pulled into the cavity between the vanes. As it makes the circle, the passage becomes smaller, compressing the refrigerant gas. As the refrigerant gas passes the discharge port, it escapes into the discharge line past a flapper-type check valve (Figure 20-50). Since there is no back stroke, there is no re-expansion of refrigerant to fill the cylinder as in a reciprocating compressor.

Rotary vane compressors are very efficient and are often used in high-quality vacuum pumps. They can reduce the pressure in a vessel to almost 29.92 in. Hg, whereas reciprocating compressors can only reduce the vacuum in a container to about 28 in. Hg.

The cylinder in a rotary vane compressor only turns in one direction, so it is not subject to the back-and-forth vibration found in a reciprocating compressor. This compressor is very quiet and steady in its operation.

Stationary Vane Compressors

The stationary vane compressor has a stationary vane that is mounted in the outer housing of the compressor. The cylinder is mounted inside the housing using an off-center shaft. The cylinder turns inside the housing much like a lobe on a cam. As the cylinder turns, it creates a low-pressure area that pulls in refrigerant gas from the suction line. The refrigerant is trapped between the rotating cylinder and the stationary vane. As the vane turns, the pressure increases until the gas is pushed out into the discharge line (Figure 20-51).

ROTARY COMPRESSORS

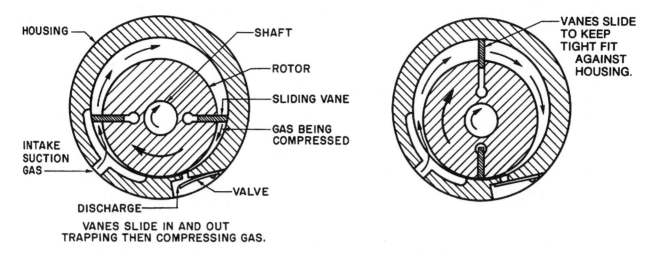

Figure 20-50. Gas being compressed into the discharge line on a rotary compressor.

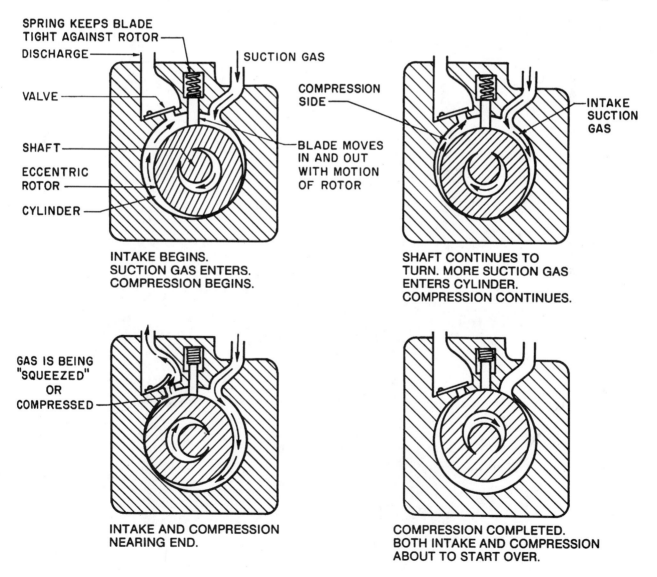

Figure 20-51. Rotary vane compressor operation.

20.9 SCROLL COMPRESSORS

The scroll compressor represents the latest in compressor technology (Figure 20-52). It was actually patented many years ago, but the technology was not available until recently to complete the intricate machine work required in its manufacture. The scroll compressor is now being used in many residential applications up to about five tons and is also being used in the 10 to 15 ton range for commercial applications. It is a hermetically sealed compressor.

The scroll compressor is a positive displacement compressor, but will allow liquid refrigerant to pass through the scrolls. If liquid refrigerant enters the compressor, the scrolls will separate and allow the liquid to pass through without damage to the compressor.

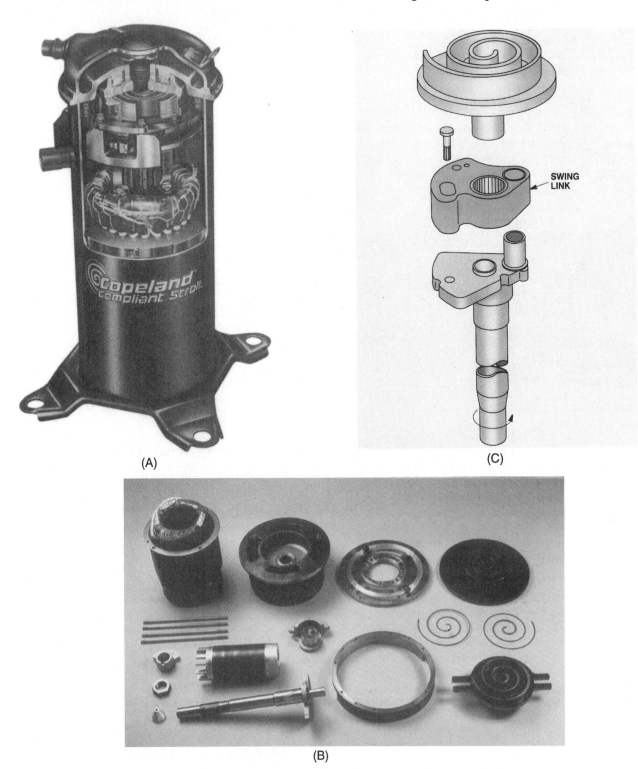

Figure 20-52. Scroll compressor. *Courtesy of: (A) Copeland Corporation, (B and C) Trane Company.*

The compression takes place between two spiral-shaped forms. One spiral is stationary and the other orbits within it (Figure 20-53). This orbiting movement draws gas into a pocket between the two spirals. As this action continues, the gas opening is sealed off and the gas is forced into a smaller pocket at the center (Figure 20-54).

Note: Figure 20-54 illustrates only one pocket of gas; actually, most of the scroll or spiral is filled with gas and compression or squeezing takes place in all the pockets at the same time.

The contact between the spirals sealing off the pockets is achieved using centrifugal force. This minimizes gas leakage. Because of its design, the scroll tips remain in contact without seals. Therefore, there is little wear while operating.

Figure 20-53. The compression process in the scrolls of a scroll compressor. *Courtesy Copeland Corporation.*

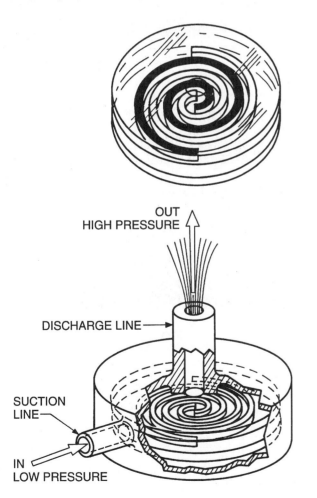

Figure 20-54. The discharge gas is released from the center of the scrolls.

There are several reasons why the scroll compressor is so efficient:

- The compressor has no valves so there is no clearance volume as in the reciprocating compressor. It is much like the rotary compressor; it turns in one direction and has no reverse stroke.
- The suction and discharge chambers are separated by refrigerant at lower temperature differences. This reduces heat transfer in the compressor.
- There are no flapper valves so there is no loss of pressure due to valve pressure drop.
- There are several steps of compression in the scrolls so there is very little vibration. This compressor is very quiet.

The scroll compressor has a check valve in the discharge line leaving the compressor. It prevents the pressures from equalizing back through the compressor during the off cycle. For example, if the head pressure at shutdown is 250 psig and the suction pressure is 70 psig, these pressures will try to equalize backward through the compressor during the off cycle and may successfully turn the compressor backwards for some time without lubrication. The check valve minimizes this reverse rotation.

The motor in the scroll compressor is suction cooled like the reciprocating compressor.

When the scroll compressor is shut down, the refrigerant gas trapped between the check valve and the compressor turns the compressor backwards for just an instant, producing a distinct sound. You should be aware that during short power interruptions, the compressor may start up backwards and try to operate. The compressor will not pump refrigerant so no cooling will take place. Since the compressor is refrigerant cooled, it will soon become hot and shut down because of the motor over-temperature condition. A time delay is often included as part of the control package to allow the compressor to completely stop before another start is allowed. When the scroll compressor is used for applications over five tons, a three-phase motor is used. Since the scroll compressor must turn in only one direction, various control arrangements are often applied to the system to ensure proper rotation. This is called *phase protection*. If the power phases are switched, the compressor will not start up backwards because the control will detect the reverse rotation.

20.9 MOTORS FOR HERMETIC AND SEMI-HERMETIC COMPRESSORS

Ordinary motors are not used for hermetic or semi-hermetic compressors because they are incompatible with the refrigerant and oil that circulate in the system. The motor is made up of the wire in the windings, the varnish or coating on the wire, the ties that hold parts of the windings together, the insulators in the slots and the motor leads that transfer power to the motor windings.

310 UNIT 20 COMPRESSORS

All of these materials must be carefully selected to accomplish their jobs and be compatible with the oil and refrigerant in the system.

The motor in smaller hermetic compressors is typically mounted in a vertical position (Figure 20-55). Often, the refrigerant gas enters the compressor in such a manner as to flow over the windings for the purpose of cooling them (Figure 20-56). Refrigerant gas makes an excellent cooling agent for the windings. The motor is a special motor with the rotor mounted on the crankshaft. The oil pump is at the bottom of the shaft.

Motors for larger semi-hermetic compressors are typically mounted in a horizontal fashion. Again, the crankshaft is part of the rotor assembly, with the oil pump on the end of the shaft (Figure 20-57).

(C)

Figure 20-57. Motor and crankshaft assembly for a semi-hermetic compressor. The oil pump is located on the end of the shaft. *Courtesy Copeland Corporation.*

Motor Protection

As mentioned earlier, all hermetic and semi-hermetic compressor motors (except rotary compressors) are suction gas cooled. The suction gas returning to the compressor must be cool enough to maintain the correct motor temperature or the motor will overheat. Typically, the motors run with 60°F return refrigerant gas that has been superheated about 20°F (Figure 20-58). When the refrigerant exceeds this temperature, the motor will overheat.

Internal overload protection devices in hermetic motors protect the motor from overheating. These devices are embedded in the windings and are wired in two different ways. One style breaks the line circuit inside the compressor. Because it is internal and carries

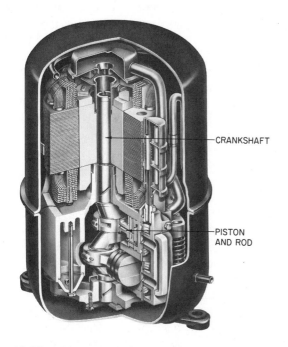

Figure 20-55. This motor is in a vertical position. *Courtesy Tecumseh Products Company.*

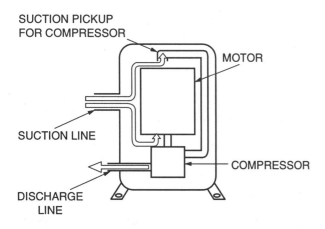

Figure 20-56. The cool suction gas flows over the motor windings to cool them.

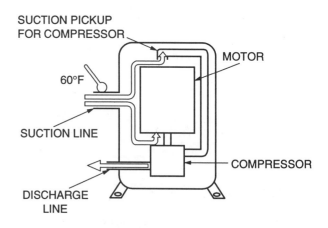

Figure 20-58. The suction refrigerant gas is about 60°F as it enters the motor cavity to cool the motor windings.

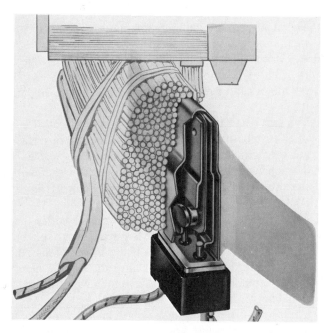

Figure 20-59. This motor winding thermostat is imbedded in the motor windings. *Courtesy Tecumseh Products Company.*

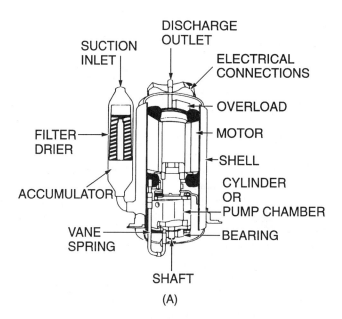

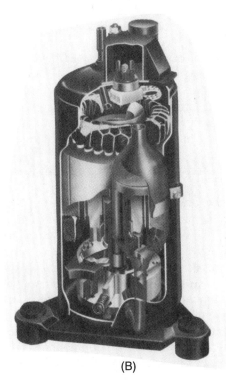

Figure 20-60. This motor is pressed into the compressor housing. *Courtesy Motors and Armatures, Inc.*

the line current, it is limited to smaller compressors. It has to be enclosed to prevent the electrical arc from affecting the refrigerant (Figure 20-59). If contact of this line type remained open, the compressor cannot be restarted. The compressor would have to be replaced.

Another type of motor overload protection device breaks the control circuit. This is wired to the outside of the compressor to the control circuit. If the pilot-duty type wired to the outside of the compressor were to remain open, an external overload device could be substituted.

The internal winding thermostat can protect the motor from several heat-related problems. For example, a low charge or starved coil both raise the superheat or reduce the ability of the suction gas to properly cool the motor, so the motor will gradually become hotter and trip on the internal winding thermostat. High motor amperage due to overload will also cause the motor to become hot and trip the motor winding thermostat. If a motor becomes overloaded very quickly, the control may not react immediately; however, it will react in time to preserve the compressor motor.

When a motor becomes too hot and the internal motor overload trips, it will be very slow to reset because the mass of the motor is hot. Often, it may take an hour or more to cool the compressor to the point where the thermostat resets itself. When the compressor is a welded hermetic compressor, the motor windings may be suspended in the vapor space inside the shell and will take longer to cool using ambient air. If the compressor motor is pressed into the shell, the shell can be cooled quickly using a fan (Figure 20-60).

SUMMARY

- The compressor is a vapor pump that maintains the correct low pressure in the evaporator and raises the pressure to the correct condenser pressure. This process is called *vapor compression*. It reduces the volume of a vapor.

- Common compressors for residential and commercial air conditioning are reciprocating, rotary, and scroll compressors.
- The reciprocating compressor uses mechanical energy to drive a piston to compress gas in a cylinder. A crankshaft is used to convert rotary motion to back and forth motion.
- Valves are used to control the direction of flow in a reciprocating compressor.
- When the piston starts down, a low-pressure area is created in the cylinder and suction gas enters the cylinder. When the piston passes bottom dead center, compression starts and the suction valve closes due to the difference in pressure. When the piston approaches top dead center, the discharge valve opens due to the difference in pressure and the refrigerant flows into the discharge line. As the piston passes top dead center, the process is repeated.
- The reciprocating compressor is known as a positive displacement compressor, meaning when a cylinder is full, it must be emptied on the up stroke or the compressor is likely to be damaged.
- Liquid will not compress. Remember, the compressor is a vapor pump and is intended to compress only vapor.
- One of the determining factors for efficiency in a reciprocating compressor is the clearance volume at the top of the stroke. The lower the clearance volume, the higher the efficiency.
- The lubrication system is responsible for lubricating every moving part, either by pressure or oil splashing to the correct places.
- The oil pump picks up oil in the oil sump at the same pressure as the suction pressure and pressurizes it to 20 to 50 psig above suction pressure. This is called the *net oil pressure*.
- The compressor housing holds the compressor and related components together. Housing types include hermetic, semi-hermetic, and open-drive.
- The hermetic compressor housing is made of steel and is welded together with all components mounted inside. This compressor is manufactured in sizes up to about 40 tons.
- The hermetic compressor is mounted inside the shell by either pressing the motor housing to the shell or suspending it from springs inside the shell.
- The hermetic compressor must be isolated from the system to prevent vibration and noise. In some applications, this is accomplished using springs and/or neoprene mounting methods.
- Hermetic compressors are cooled by passing the cool suction gas over the motor windings before it reaches the actual compressor.
- One disadvantage of the hermetic compressors is that if the motor burns out, it is in the refrigerant atmosphere and will pollute the refrigerant.
- The semi-hermetic compressor may be rebuilt in the field, if necessary.
- Access to all parts is furnished through bolted flanges.
- In a semi-hermetic or open-drive compressor, the motor can be rebuilt by removing it and rewinding it.
- The crankshaft is also an expensive component. It can often be rebuilt by adding material to the worn surface and then grinding it back to the correct size.
- The open-drive compressor is bolted together with the motor on the outside of the housing. It is air cooled.
- One advantage of the open-drive compressor is that the motor is outside the refrigerant atmosphere, cannot pollute the refrigerant if the motor burns out. Also, the refrigerant does not have to cool the motor and the compressor does not have to handle the additional heat from the motor windings.
- The direct-drive compressor must have correct shaft alignment. Achieving both angular and parallel alignment is a technical and time consuming process. Belt-driven compressors are easier to align.
- The crankshaft seal keeps the refrigerant from escaping.
- Compressor capacity control may be achieved using any of the following methods: cylinder unloading, blocked suction unloading, and gas diversion. All three methods reduce the power consumption of the compressor.
- Rotary compressors are more efficient than reciprocating compressors because they have no clearance volume.
- The scroll compressor is made up of a stationary spiral and an orbiting spiral. The scroll compressor is very efficient.
- The scroll compressor must turn in the correct rotation to pump, so if it is used in a three-phase system, phase protection is often applied to the control circuit to ensure correct phase connection.
- The scroll compressor has a check valve in the discharge line to prevent reverse rotation during the off cycle as the refrigerant on the high side tries to equalize through the compressor.
- Motors for hermetic and semi-hermetic compressors are manufactured using special materials that are compatible with the refrigerant and oil circulating in the system.
- Motors in smaller hermetic compressors are mounted vertically with the oil pump on the bottom, while in semi-hermetic compressors, the motor is mounted horizontally with the oil pump at the end of the crankshaft.
- Most motors in hermetic compressors are cooled by suction gas, except the motors in rotary compressors which are cooled by discharge gas and operate at higher temperatures.

REVIEW QUESTIONS

1. Provide a two-word description of a compressor.
2. Name the three common compressors used for residential and light commercial air conditioning systems.
3. What would happen to a reciprocating compressor if it tried to compress liquid refrigerant?
4. What is the function of the crankshaft in a compressor?
5. Where is the valve plate located in a compressor and what fastens it in place?
6. Name two methods by which hermetic compressors are cooled.
7. Name two advantages of a hermetic compressor.
8. Name two disadvantages of a hermetic compressor.
9. Name two advantages of an open-drive compressor.
10. Name two disadvantages of an open-drive compressor.
11. State three methods used for capacity control in large compressors.
12. List two advantages of capacity control.
13. Where is the oil pump located in a large semihermetic compressor?
14. What prevents a compressor motor from becoming too hot and burning out?

21 CONDENSERS

OBJECTIVES

Upon completion of this unit, you should be able to

- State the three purposes of a condenser.
- Describe refrigerant subcooling.
- Discuss refrigerant-to-earth heat exchangers.
- Describe the differences between pipe-in-pipe, shell-and-coil, and shell-and-tube condensers.
- Discuss the differences between a wastewater system and a cooling tower system.
- Describe fouling as it relates to water-cooled condenser heat exchange surfaces.
- Define ozination.
- Describe the bleed water process as it relates to a water-cooled condenser.
- State the differences between water-cooled condensers and air-cooled condensers.

21.1 CONDENSER FUNCTIONS

You will recall that the purpose of refrigeration is to remove heat from a place where it is not wanted and release it in a place where it makes no difference. The condenser is the device that releases the heat from a refrigeration system. The heat travels from the compressor to the condenser in the form of a highly superheated vapor. This heat must be rejected from the condenser into some medium (earth, air, or water) for release into the atmosphere.

The condenser has three functions:

1. To desuperheat the hot refrigerant gas down to the condensing temperature.
2. To condense the hot vapor refrigerant into a liquid.
3. To subcool the liquid refrigerant to a temperature below the condensing temperature. (All condensers do not have this feature.)

In addition to performing these three functions, the condenser must also provide the correct liquid pressure for entry into the expansion device. If the medium used to condense the refrigerant is too cold, the condensing temperature will be too low and the head pressure will be insufficient. Various methods can be used to maintain the correct condensing temperature and therefore, the correct head pressure. Some of these methods will be discussed in this unit.

To desuperheat the refrigerant is to cool the refrigerant down to the condensing temperature. When the refrigerant leaves the compressor, it may be at 170°F to 200°F when the condensing temperature may need to be 105°F to 125°F. For example, if the refrigerant leaves the compressor at 200°F and the condensing temperature is 125°F, the refrigerant contains 75°F of superheat (200°F - 125°F = 75°F). The first portion of the condenser removes the superheat (Figure 21-1).

To condense the refrigerant is to change it from a hot vapor into a hot liquid. For example, if the condensing

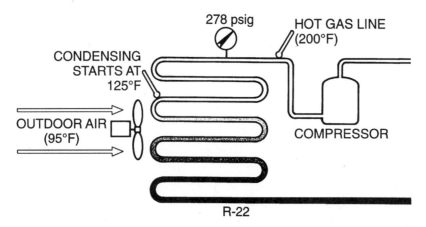

Figure 21-1. The first portion of the condenser is used to remove the superheat.

temperature is 125°F, the change of state will take place at 125°F.

Note: A condensing temperature of 125°F is used for air-cooled condensers and a condensing temperature of 105°F is used for water-cooled condensers.

As the refrigerant condenses inside the coil, drops of liquid refrigerant form and move to the bottom of the condenser tube (Figure 21-2). The liquid is then moved along the condenser tube by gravity or refrigerant velocity.

To subcool the refrigerant is to cool the liquid refrigerant to a temperature below the condensing temperature. The liquid refrigerant cannot be subcooled while it is in the vicinity of the condensing process because if more heat is removed, the refrigerant would just condense faster. The liquid must be moved to a place where removing more heat will not condense more liquid, but will cool the liquid that is already condensed. This is accomplished by collecting the liquid refrigerant and then passing the liquid through a medium that is cooler than the condensed refrigerant. Figure 21-3 shows an air-cooled condenser with a small receiver that collects the liquid, then passes it through the condenser again. This lowers the temperature by another 15°F. It is common for condensers to subcool the refrigerant from 10°F to 20°F below the condensing temperature.

There are three common mediums used to absorb the heat rejected from the condenser: earth, water, and air.

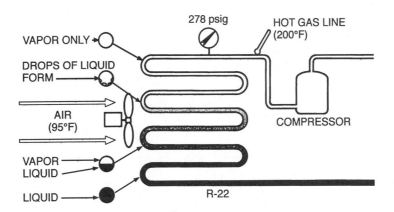

Figure 21-2. Drops of liquid refrigerant form as the refrigerant condenses from a vapor into a liquid.

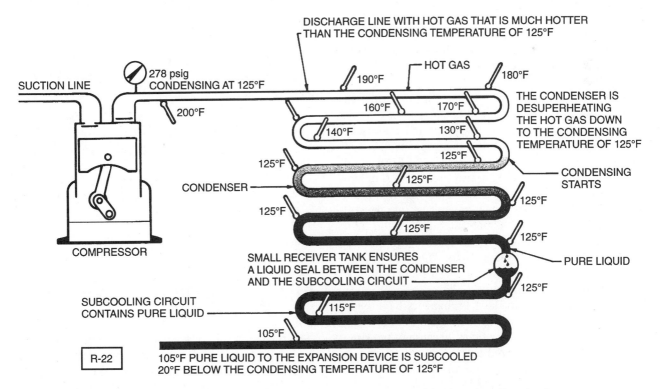

Figure 21-3. An air-cooled condenser with a small receiver used for the subcooling circuit.

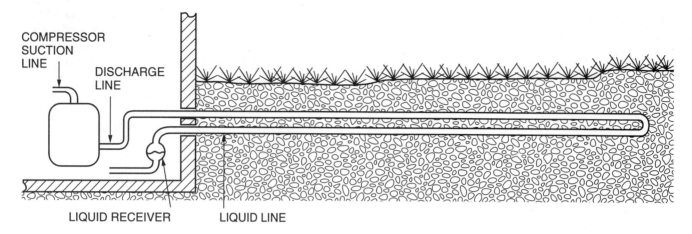

Figure 21-4. Ground-type heat exchanger.

21.2 REFRIGERANT-TO-EARTH HEAT EXCHANGERS

Exchanging heat with the earth or ground is not very common because the refrigerant must be piped to a ground coil, requiring the use of long piping circuits and large amounts of refrigerant. When this type of heat exchanger is used, the refrigerant gas is piped from the compressor to a large, underground refrigerant coil (Figure 21-4). The ground coil must be at least four feet below the surface so that common digging will not cause damage. Ideally, there should be no connections underground where leaks may occur, so a continuous coil is preferred for a ground coil.

The ground coil is made of a special piping material called *cupro-nickel*. This copper/nickel combination is more resistant to the corrosive soils found in some areas.

The type of soil has an effect on the rate of heat exchange. For example, if the soil contains a lot of clay, heat may eventually cause it to dry out and pull away from the coil (Figure 21-5). When this happens, the efficiency of the heat exchange suffers and high head pressure will occur. Sandy soil also presents a problem because sand is basically silicone or glass and it is not a good conductor. The best place to reject heat into the soil is where the soil is wet. Water surrounding the heat exchange coil will cause the heat to be absorbed into the soil through the water (Figure 21-6). This will help to make sandy soil a better conductor and prevent clay-type soil from heating and shrinking.

Ground-type refrigerant heat exchangers have been used to some extent in the past but it is not likely that they will be used widely in the future because of the large amount of refrigerant required and the inconvenience of burying the refrigerant coil. In addition, underground coils are difficult to service and may contaminate the ground water should leaks occur.

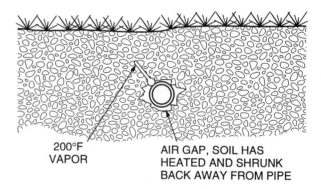

Figure 21-5. Heat may cause the soil to pull away from the heat exchanger.

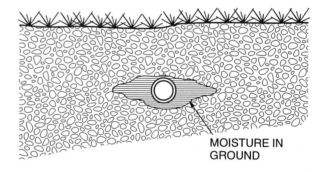

Figure 21-6. The moisture surrounding this coil helps to improve the heat exchange with the soil.

21.3 REFRIGERANT-TO-WATER HEAT EXCHANGERS

The refrigerant-to-water heat exchanger, usually called a *water-cooled condenser*, is very popular for certain types of equipment, particularly large systems. These condensers offer an excellent rate of heat exchange because refrigerant easily gives up heat to water.

REFRIGERANT-TO-WATER HEAT EXCHANGERS

Water-cooled systems typically operate at much lower condensing temperatures than ground or air-cooled systems. There are several types of refrigerant-to-water heat exchangers, including: pipe-in-pipe, shell-and-coil and shell-and-tube.

Pipe-in-Pipe Condensers

The pipe-in-pipe condenser is the simplest type of water-cooled condenser and is used in small applications up to about 10 tons. Figure 21-7 shows a condenser using a pipe-in-pipe heat exchanger. The hot refrigerant flows through the inner pipe while water flows through the outer pipe. As the refrigerant proceeds down the pipe, heat is removed. Let's assume that the refrigerant condenses at 105°F, the entering water is 85°F, and the refrigerant entering the condenser is 170°F. Any heat above the condensing temperature (105°F) is considered superheat and must be removed before condensing can occur (Figure 21-8). As the refrigerant proceeds down the inner pipe, the temperature drops because the refrigerant is giving up heat to the water. When the vapor refrigerant drops to the condensing temperature, liquid refrigerant begins to form.

As the refrigerant proceeds through the inner pipe, more and more refrigerant turns into a liquid until it is all liquid. At this point, the liquid is called *saturated* because

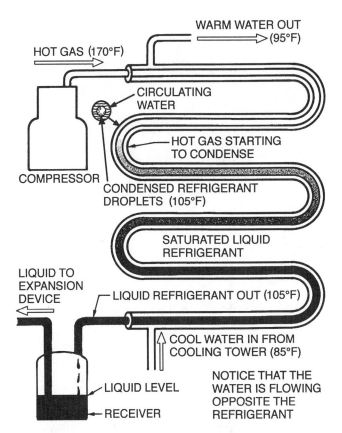

Figure 21-8. Superheat must be removed before condensing takes place.

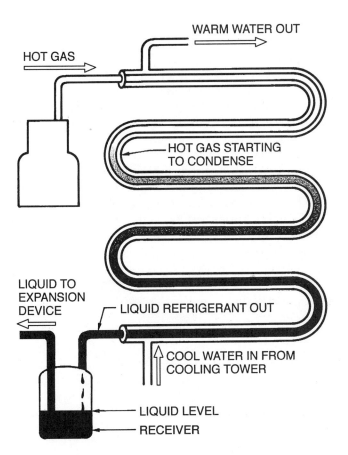

Figure 21-7. Pipe-in-pipe condenser.

if any heat were added to it, it would start changing back into a vapor. The liquid is saturated with heat at 105°F.

The entering water temperature is 85°F and gradually warms up as it proceeds through the condenser. Notice that the entering water is at the opposite end of the condenser. This provides a more efficient heat exchange because the entering water will have a lower temperature than the leaving refrigerant and more heat can be extracted from the refrigerant.

When the refrigerant has completely changed into a liquid, more heat can still be removed from it because it is at 105°F while the entering water is at 85°F. When additional heat is removed from the liquid, it is known as *subcooling*. This process drops the liquid temperature to a point below the saturation temperature. Remember, the cooler the liquid leaving the condenser, the more efficient the system.

The pipe-in-pipe condenser assembly would take up too much space if manufactured in a straight line. To make it more compact, it is often wound in a circular or oval shape (Figure 21-9).

Shell-and-Coil Condensers

The shell-and-coil condenser is another popular condenser because of its low first cost. Like the pipe-in-pipe condenser, this condenser is also used in small applications (up to about 10 tons). It consists of a coil of pipe

318 UNIT 21 CONDENSERS

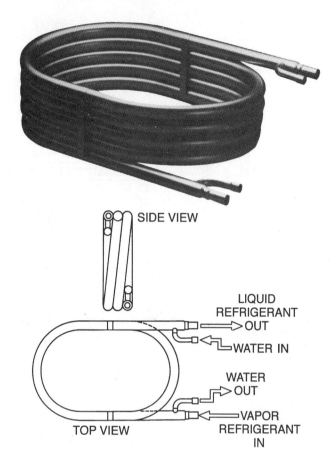

Figure 21-9. The pipe-in-pipe condenser is wound in a circular or oval shape to make it more compact. *Courtesy of: (A) Photo by Bill Johnson, (B) Noranda Metal Industries, Inc.*

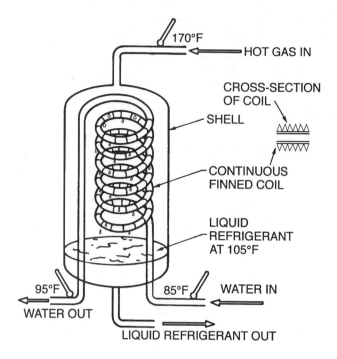

Figure 21-10. Shell-and-coil condenser.

that carries the water wound inside a shell that houses the refrigerant (Figure 21-10). The shell of the condenser is also used as a refrigerant receiver.

Refrigerant enters the top of the shell as a superheated vapor. Cool water enters the coil in the shell and as it moves though the coil, it absorbs heat and leaves the condenser at a much warmer temperature. The refrigerant condenses on the coils and drops down to the bottom of the shell where it is collected.

Shell-and-Tube Condensers

The shell-and-tube condenser is the most expensive of the three water-cooled condensers. This type of condenser is normally used in systems from about 10 tons up to several thousand tons. It is manufactured by fastening the water tubes to an end sheet on each end of the condenser that allows water to pass from end sheet to end sheet through the tubes (Figure 21-11). The refrigerant is contained in a shell that is fastened to the end sheets (Figure 21-12). The water enters the condenser at one end through a divided water box cover. When the water leaves the tubes on the other end, it is turned back through the tubes for another pass through the condenser (Figure 21-13). Condensers can be manufactured to make one, two, three, and four passes using dividers in the water box covers that cause the water to turn and pass back through the shell

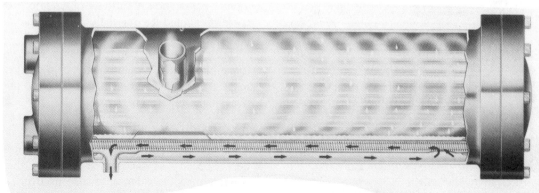

Figure 21-11. Shell-and-tube condenser. *Courtesy Trane Company.*

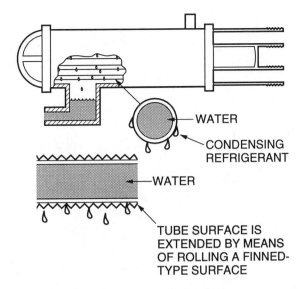

Figure 21-12. The refrigerant is contained inside the shell and condenses on the outer surface of the tubes.

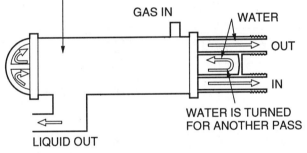

Figure 21-13. The water boxes contain the water and direct it through the condenser circuit.

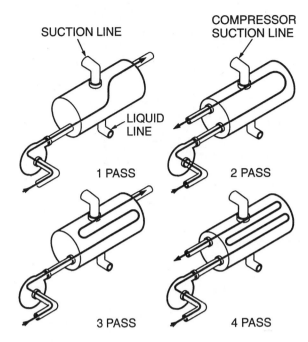

Figure 21-14. Condensers can be manufactured to make one, two, three, or four passes, depending on the water box construction.

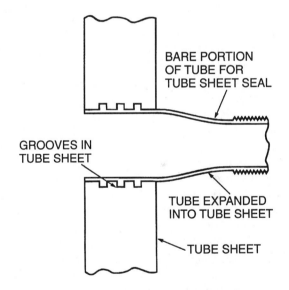

Figure 21-15. The holes in the end (tube) sheets have grooves cut around the inside to seal the copper tubes.

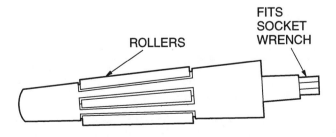

Figure 21-16. This rolling tool gradually expands the tube into the grooves in the end sheet.

(Figure 21-14). The longer the water is in contact with the refrigerant, the more heat it can remove.

The tubes are fastened to the end sheets using one of two methods: brazing or rolling. The connection must be very tight or refrigerant will leak into the water circuit, which is not only expensive, but will also affect system operation. Rolling is the most common method used to fasten the tubes because when the tubes are brazed, it is difficult to remove them for repair.

Rolling requires an end sheet with grooves cut around the inside of the holes where the tubes will fit (Figure 21-15). The tube is positioned inside the hole and a tapered rolling tool is inserted into the end of the tube. As the rolling tool is turned, it will gradually expand the copper tube into the grooves cut into the end sheet (Figure 21-16). This is a delicate process as too much pressure applied to the roller will distort the end sheet, causing leaks to the adjacent tube connections.

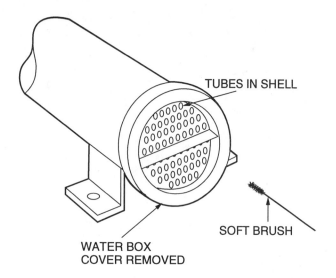

Figure 21-17. Tubes may be cleaned using soft brushes.

Fouling

All water contains some minerals that will eventually begin to deposit onto the tubes. These deposits (usually referred to as *fouling*) act as a barrier to the heat exchange process. Fouling must be kept to a minimum to ensure optimum system performance. Most manufacturers rate their equipment with a fouling factor of 0.0005" of deposit on the tubes. Any level of deposit beyond this rating will have an effect on efficiency.

When minerals are deposited on the water side of the tube surface in a shell-and-tube condenser, the water box ends can be removed from the condenser and the tubes can be physically cleaned using soft brushes (Figure 21-17). This is the primary advantage of a shell-and-tube condenser. The most common type of brush has nylon bristles and is fastened to a rod that is pushed through the tubes one at a time. The rod can be fastened to an electric drill motor that can be used to turn the brush as it passes through the tube. Special equipment is also available that flushes water through the tube while the brush is turned (Figure 21-18).

When pipe-in-pipe or shell-and-coil condensers become dirty or fouled, they cannot be cleaned physically using brushes. They must be cleaned using special chemicals that remove the film from the condenser (discussed later).

Water Supply Systems

All water-cooled condensers use either wastewater or cooling tower supply systems.

Wastewater Systems The wastewater system uses water from a large source, typically a lake or well, and circulates it only once. The water is then discharged back to the source or the soil.

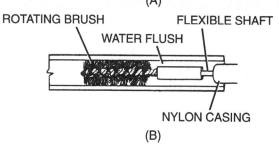

Figure 21-18. Special tube cleaning equipment turns the brush and injects water to wash away the dirt from the fouled tubes. *Courtesy Goodway Tools Corporation.*

If the source is a well, its temperature will vary from one part of the country to another. For example, in southern climates, the water temperature is typically 50°F to 60°F when taken from deep wells. In northern climates, the water temperature from a deep well may be significantly lower.

Lake water temperature ranges can be quite broad. Some lakes are in warm climates while others are in cold climates. A deep lake in a cold climate may have a water temperature of 50°F while a shallow lake in a warm climate may have water at 85°F. It would of course take much more 85°F water to remove the same amount of heat as 50°F water.

Wastewater from a manufacturing process may also be used for wastewater air conditioning. The water may first be taken from a well and used to cool a machine in a manufacturing plant, and still have enough capacity to cool the condenser in an air conditioner.

REFRIGERANT-TO-WATER HEAT EXCHANGERS

A typical water-cooled system may use 1.5 gallons per minute per ton of air conditioning. This means a 10 ton system would use 15 gallons per minute (10 × 1.5 = 15) which is 900 gallons per hour (15 × 60 = 900). This is a sizable amount of water so the water supply must be either inexpensive or free to be practical. For this reason, it is unlikely that city water would be used to cool an air conditioning condenser because of the expense of buying the water and wasting it. However, it has been used in the past for limited applications.

Wastewater systems are often used in small remote applications for spot cooling where an air-cooled system is not practical.

The quantity of water used by a wastewater system can be reduced by incorporating a water regulating valve. This valve is used to adjust the water supply based on the current need. For example, for an R-22 system with a condensing temperature of 105°F, the water can be regulated to maintain a head pressure corresponding to 105°F. For R-22 this would be 210.8 psig. A water regulating valve can be installed in the water entering the condenser. This valve has a sensing line that monitors the head pressure. As the head pressure rises, the valve increases the water flow. As the head pressure drops, the valve decreases the water flow. Figure 21-19 shows a water regulating valve installation. Water temperatures vary from season to season in some areas where city water is used, so a water regulating valve can be used to stabilize the head pressure.

Cooling Tower Systems A cooling tower system uses the same water over and over, removing the heat from the water and transferring it to the atmosphere. The heat travels from the refrigerant to the water in the condenser, and from the water to the atmosphere in the cooling tower. The cooling tower is either on the rooftop or behind the building. Air is passed over the water in the cooling tower where evaporation takes place, cooling the remaining water to a lower temperature. Typically, a cooling tower can cool the water to within 7°F of the entering wet bulb temperature of the air. For example, in a southern climate, the entering wet bulb temperature may be 78°F so the entering condenser water may be cooled to 85°F (78°F + 7°F = 85°F).

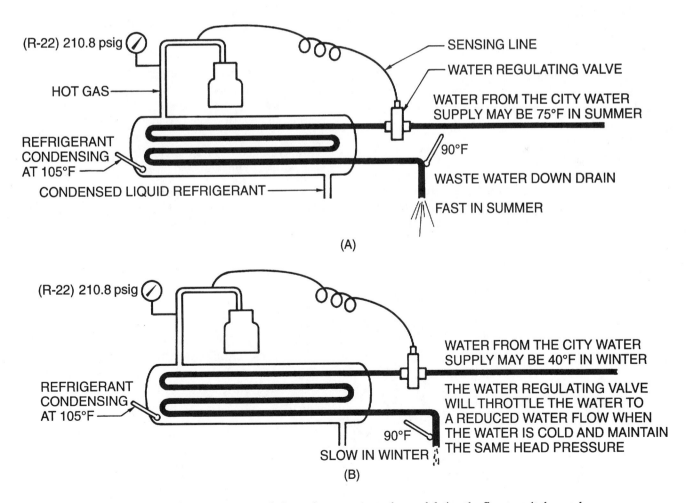

Figure 21-19. This water regulating valve saves water by modulating the flow to suit the need.

322 UNIT 21 CONDENSERS

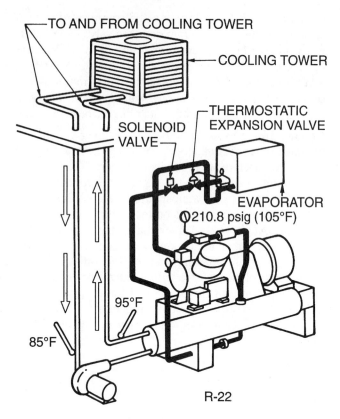

Figure 21-20. Cooling tower for a water-cooled condenser is located at a remote area outside the building.

Cooling tower systems usually have a 10°F temperature rise across the condenser, which is a 10°F drop across the cooling tower. For example, if the water entering the condenser is 85°F, the water leaving the condenser would be 95°F (Figure 21-20).

The typical design conditions must be known in order to troubleshoot condenser problems successfully. The water is always at a lower temperature than the refrigerant in the condenser. Remember, heat moves from a warmer substance to a cooler substance. In water-cooled condensers, there is a relationship between the condensing temperature and the leaving water temperature. The refrigerant condensing temperature is usually 7°F to 10°F warmer than the leaving condenser water (Figure 21-21). If the water flow is reduced due to a low water level in the tower, clogged filter, or other restriction, the water temperature *drop* across the condenser will increase. For example, when the water flow across a condenser is inadequate, the temperature rise of the water will increase because the same amount of heat is being added to less water (Figure 21-22). The condensing temperature of the refrigerant will rise and the head pressure will also rise.

Water Treatment

Water-cooled condensers are subject to two major types of buildup: algae and minerals. As mentioned earlier, the heat exchanger surface will typically have a fouling factor of 0.0005". Water-cooled condensers must be maintained with a minimal degree of fouling or efficiency will suffer. Algae is a low form of plant life. You have probably noticed it as the green slime that grows in stagnant water. The water in a cooling tower is moving very slowly and is very warm, both of which promote the growth of algae.

Most water contains a significant amount of minerals in suspension. As water is evaporated from a cooling tower, the remaining water becomes more and more concentrated with the minerals that are left behind. You may have noticed mineral buildup on a glass coffee pot. This is the same thing—the minerals are left behind when the water is boiled. These minerals cannot be allowed to concentrate in a water-cooled condenser or they will deposit on the warmest surface, the tubes leaving the condenser.

Minerals and algae are controlled by means of chemical additives, ozination, and bleed water.

Chemical Additives Chemicals may be added to the water to keep the minerals in suspension. Note that these chemicals do not remove the minerals, they simply keep

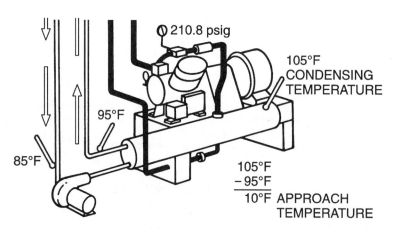

Figure 21-21. Condenser-to-leaving water temperature relationship for a water-cooled condenser.

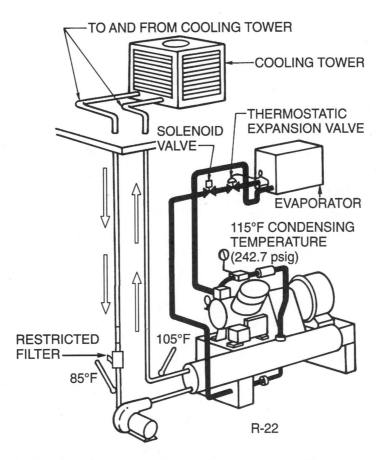

Figure 21-22. When the water flow through a condenser is reduced, the temperature difference increases because the same amount of heat is added to less water. This will also produce a rise in the head pressure.

them from depositing on the tubes. These chemicals are added to the water either by the attendant or automatically by chemical feed systems that monitor the water condition and add chemicals as needed. At the same time chemicals are added to suspend the minerals, other chemicals are added to kill the algae. These chemicals are expensive and can be a hazard to the ecology if not managed correctly.

Ozination Ozination is a water treatment method that adds ozone to the water. Ozone (O_3) is a concentrated form of oxygen. The ozone kills the algae and is used to reduce the chemical content of the water.

Bleed Water Bleed water is a method of mineral concentration control that allows a specific amount of water to escape (bleed) from the system. When water is bled out of the system, fresh makeup water is automatically added back into the system and dilutes the mineral content. This system must be monitored carefully to prevent excessive water loss which would affect system operation. In addition, bleed water is used in conjunction with chemical treatment, so when bleed water escapes, expensive chemicals also escape (Figure 21-23).

Cooling towers are discussed in more detail in Unit 30.

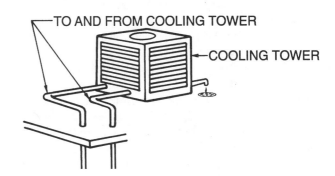

Figure 21-23. The small amount of bleed water released from this cooling tower is replaced by the makeup system. This also dilutes the mineral content of the tower water.

21.4 REFRIGERANT-TO-AIR HEAT EXCHANGERS

Condensers that release heat to the air are called *air-cooled condensers*. They are the most common type of condenser. They are simple and easy to maintain. The first air-cooled condensers were bare pipe condensers with a fan to move the air across the condenser tubes.

Figure 21-24. Aluminum fins provide additional surface area. *Reproduced courtesy of Carrier Corporation.*

Later, fins were added to the tubes to increase the area available for heat exchange (Figure 21-24). This provided additional condenser capacity. A typical air-cooled condenser has copper or aluminum tubes with aluminum fins. (Some special-application condensers use copper fins.)

Air-cooled condensers are constructed by pushing the tubes through the fins, then expanding the tubes into the fins for better contact and heat transfer. The condenser tube turns are then brazed (or aluminum welded in the case of aluminum tubes) to provide a leak-free connection (Figure 21-25). Steel framework is typically used to hold the condenser together and mount it to the condenser housing.

Air-cooled condensers have fans to move the air across the condenser. This also improves the efficiency of the condenser. Most air-cooled condensers use propeller-type fans. These fans move more air using a smaller space than other types, such as centrifugal fans. Propeller-type fans are noisier, but they are typically located where noise is not a problem, such as on a rooftop or behind a building. As the air-cooled condenser becomes larger, multiple fans are often used instead of one large fan (Figure 21-26). Partitions in the condenser can be used to position each fan in its own chamber. This allows the fans to be cycled off and on to provide head pressure control in mild weather. For example, if a condenser has six fans, they can be stopped one at a time until the last fan is operating. The last fan may also be speed controlled for fine tuning the head pressure. These fans can be controlled either thermostatically or by using pressure controls that cycle the fans when the head pressure becomes too low.

Air-cooled condensers range in size from fractional tonnage up to about 150 tons of air conditioning capacity.

Refrigerant typically enters the top of the air-cooled condenser and as it condenses, it flows via gravity to the bottom where it exits as a subcooled liquid.

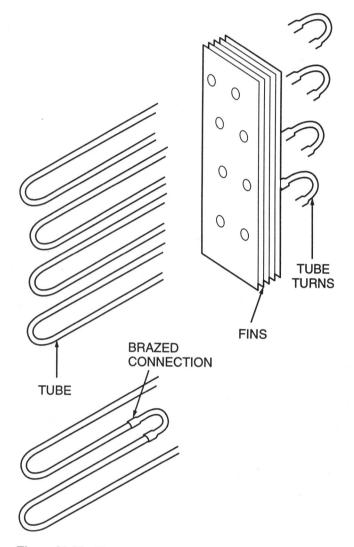

Figure 21-25. The tube turns are brazed (or aluminum welded in the case of aluminum tubes) for a leak-free connection.

Air-cooled condenser coils may be constructed in either a vertical or horizontal configuration, depending on the design engineer.

When the coil is mounted in a vertical position, the prevailing winds may affect condenser operation. Often, the air conditioner must be operated in mild or even cold weather to maintain the desired temperature in the conditioned space. A computer room in the middle of an office building would be one example; it may produce enough heat to require cooling even in the winter. If the air-cooled condenser is located where the prevailing wind blows through the coils, the head pressure may be too low. In this case, a wall or other wind barrier may need to be constructed.

When the condenser is mounted horizontally, the prevailing winds will not affect system operation. However, horizontal condensers are subject to blockage from leaves and other debris.

Figure 21-26. Condenser with multiple propeller-type fans for better head pressure control. *Reproduced courtesy of Carrier Corporation.*

SUMMARY

- The condenser has three primary functions: it desuperheats the refrigerant from the compressor, condenses the vapor refrigerant into a liquid, and subcools the liquid below the condensing temperature. It also maintains the correct pressure at the condenser outlet.
- Vapor leaving the compressor may be from 170°F to 200°F and must be desuperheated to the condensing temperature. For example, if the condensing temperature is 125°F and the hot gas is 200°F, 75°F of superheat must be removed before the refrigerant will start to condense.
- Refrigerant that is cooled below the condensing temperature is called a *subcooled liquid*. When liquid refrigerant is condensed at 125°F, it may be subcooled by another 10°F to 15°F before it leaves the condenser.
- Heat is rejected from the refrigeration system into one of three common mediums: the soil, water, or air. Ground-cooled condensers reject heat into the soil, water-cooled condensers reject heat into water, and air-cooled condensers reject heat into the atmosphere. Air-cooled condensers are the most common.
- Types of water-cooled condensers include: pipe-in-pipe, shell-and-coil, or shell-and-tube.
- Pipe-in-pipe and shell-and-coil condensers are the least expensive and are used on systems up to about 10 tons of capacity.
- The shell-and-tube condenser is the most expensive, but its construction allows the technician to physically clean the water tubes using brushes.
- There are two types of water supplies for water-cooled systems: wastewater and recirculated water using a cooling tower. Wastewater systems commonly use well or lake water and waste it to a drain.
- Cooling tower systems recirculate the water and use it until it evaporates. These systems are popular because they use less water. However, cooling tower systems require water treatment to prevent fouling of the condenser tubes.
- The water side of a condenser may become fouled due to algae or mineral deposits. Water treatment options include chemical additives, ozination, and bleed water.
- Air-cooled condensers are very popular because they do not use water and require less maintenance. These condensers consist of copper or aluminum tubes routed through aluminum fins that provide additional surface area for heat exchange. Air is usually forced across the condenser using propeller-type fans.

REVIEW QUESTIONS

1. Name the three functions of a condenser.
2. How is refrigerant subcooled?
3. What is the advantage of pipe-in-pipe and shell-and-coil condensers?
4. What is the advantage of a shell-and-tube condenser?

5. Where do minerals come from in water-cooled condensing systems?
6. Why must the minerals content be maintained at a certain level in a water-cooled system?
7. What is the fouling factor for most water-cooled condensers?
8. How may mineral deposits be removed from a pipe-in-pipe condenser?
9. How may mineral deposits be removed from a shell-and-tube condenser?
10. What are the tubes in an air-cooled condenser usually made of?
11. What are the fins in an air-cooled condenser usually made of?
12. What is the purpose of the fins in an air-cooled condenser?
13. How are the fins fastened to the tubes in an air-cooled condenser?
14. How is the head pressure controlled in air-cooled condensers?

EXPANSION DEVICES 22

OBJECTIVES

Upon completion of this unit, you should be able to

- Explain the purpose of an expansion device.
- Discuss how a thermostatic expansion valve functions.
- Identify each of the components in a thermostatic expansion valve.
- Explain the purpose of internal and external equalizers as they apply to thermostatic expansion valves.
- Describe how a thermostatic expansion valve responds to load changes.
- Describe the operation of capillary tube and orifice metering devices.
- Discuss pressure equalization during the off cycle with respect to capillary tube and orifice metering devices.

22.1 EXPANSION DEVICE FUNCTIONS

The expansion device, often called the *metering device*, is the fourth component necessary for the compression refrigeration cycle to function. It can either be a valve or a fixed-bore device. The expansion device is usually concealed inside the unit cabinet and may not be as easy to recognize as the evaporator, condenser, or compressor. In order to effectively troubleshoot air conditioning systems, it is important that you be able to recognize and understand the operation of various metering devices.

As shown in Figure 22-1, the expansion device is one of the dividing lines between the high-pressure side of the system and the low-pressure side of the system (the compressor is the other). This device is responsible for metering the correct amount of refrigerant to the evaporator. The evaporator performs best when it is as full of liquid refrigerant as possible without any running over into the suction line. Any liquid refrigerant that enters the suction line may reach the compressor because the suction line normally does not add enough heat to boil the liquid into a vapor.

The expansion device is usually installed in the liquid line between the condenser and the evaporator. On a hot day, the liquid line will be warm to the touch and can be followed to the expansion device, where there is a pressure drop and an accompanying temperature drop. For example, if the liquid line entering the expansion device is at 110°F, the pressure on the evaporator side will be about 68.5 psig, which corresponds to 40°F (Figure 22-2). This is a dramatic temperature drop and can be easily detected. This change can occur in a very short space—less than an inch on a valve. (It will occur more gradually on some fixed-bore devices.)

Expansion devices for air conditioning come in two different types: thermostatic expansion valves and fixed-bore devices such as the capillary tubes (Figure 22-3).

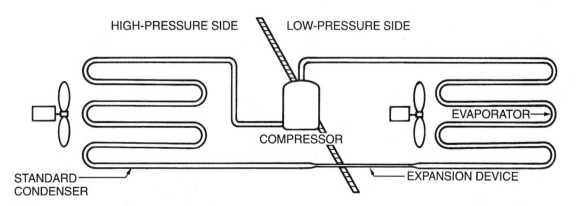

Figure 22-1. The expansion device is one of the dividing lines between the high-pressure and low-pressure sides of the system.

328 UNIT 22 EXPANSION DEVICES

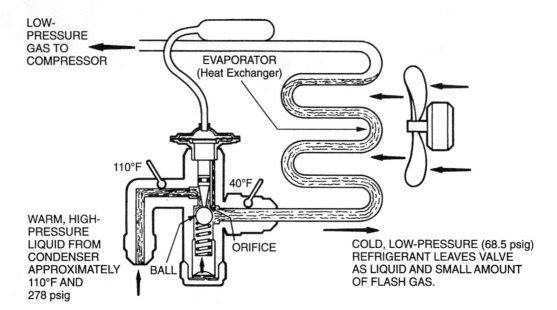

Figure 22-2. The expansion device has a dramatic temperature change from one side to the other. *Courtesy Parker Hannifin Corporation.*

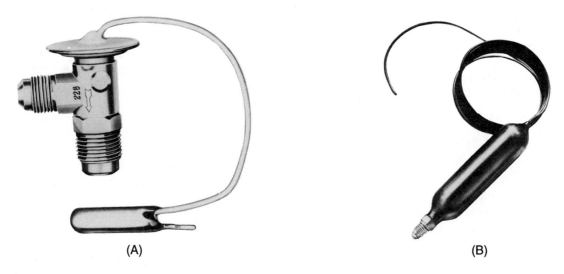

Figure 22-3. (A) Thermostatic expansion valve. (B) Fixed-bore metering device. *Courtesy of: (A) Photo by Bill Johnson, (B) Parker Hannifin Corporation.*

22.2 THERMOSTATIC EXPANSION VALVES

The thermostatic expansion valve (TXV) meters the refrigerant to the evaporator using a thermal sensing element to monitor the superheat. This valve changes dimension in the seat area in response to a thermal element. The thermostatic expansion valve maintains a constant superheat in the evaporator. When there is superheat, there is no liquid refrigerant in the suction line. Excess superheat is not desirable, but a small amount is necessary to ensure that no liquid refrigerant leaves the evaporator.

The thermostatic expansion valve consists of the valve body, diaphragm, needle and seat, spring, adjustment and packing gland, and sensing bulb and transmission tube (Figure 22-4).

Valve Body

The valve body is an accurately machined piece of solid brass or stainless steel that holds the rest of the components and fastens the valve to the refrigerant piping circuit (Figure 22-5). TXVs have several different configurations. Some of them are one piece and cannot be disassembled, and others are made so that they can be taken apart for service.

THERMOSTATIC EXPANSION VALVES 329

These valves may be fastened to the system using flare, flange, or solder connections. Flare or flange connections are normally used in systems where ease of service is desirable. They are easy to remove (Figure 22-6).

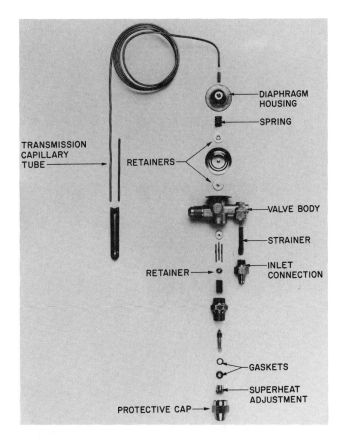

Figure 22-4. Exploded view of a thermostatic expansion valve. All components are visible in the order in which they go together in the valve. *Courtesy Singer Controls Division, Schiller Park, Illinois.*

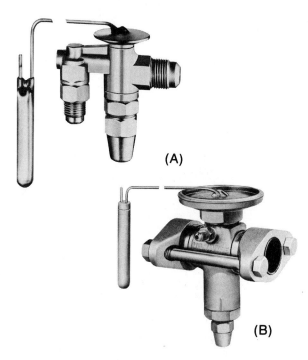

Figure 22-6. (A) Flare-type valve. (B) Flange-type valve. These valves can be replace easily when they are installed in an accessible location. *Courtesy Singer Controls Division, Schiller Park, Illinois.*

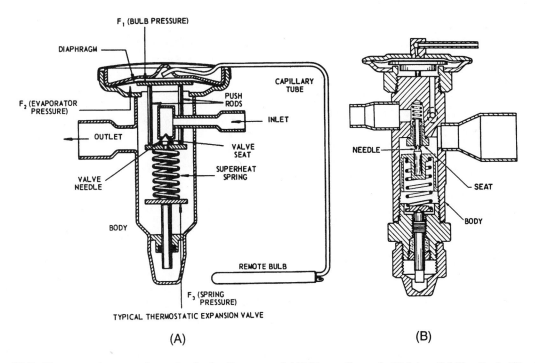

Figure 22-5. Thermostatic expansion valve body. *Courtesy of: (A) Singer Controls Division, Schiller Park, Illinois, (B) Sporlan Valve Company.*

UNIT 22 EXPANSION DEVICES

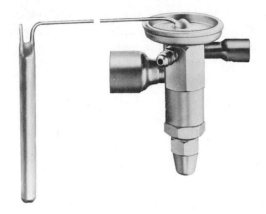

Figure 22-7. Solder-type valve that can be disassembled and rebuilt without taking it out of the system. *Courtesy Singer Controls Division, Schiller Park, Illinois.*

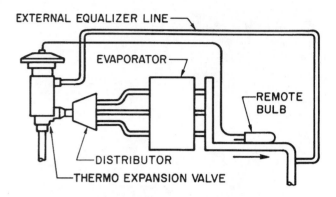

Figure 22-9. Evaporator with multiple circuits and an external equalizer line to keep pressure drops to a minimum.

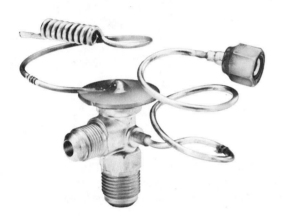

Figure 22-8. The third connection on this expansion valve is the external equalizer. *Courtesy Singer Controls Division, Schiller Park, Illinois.*

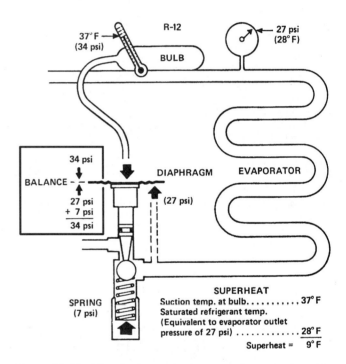

Figure 22-10. The diaphragm in the expansion valve is a thin metal membrane that responds to the pressure differences caused by changes in the system load. *Courtesy Parker Hannifin Corporation.*

If a soldered connection is used, a valve that can be disassembled and rebuilt in place is desirable (Figure 22-7).

Most TXVs have a screen at the inlet of the valve body. It has a very fine mesh to strain out any small particles that may interfere with valve operation.

Some valves have a special connection called an *external equalizer*. This connection is normally a ¼-inch flare or ¼-inch solder and is on the side of the valve close to the diaphragm (Figure 22-8). Some evaporators have several circuits and a method of distributing the refrigerant that will cause a pressure drop between the expansion valve outlet and the evaporator inlet. This installation must have an external equalizer for the expansion valve to maintain correct control of the refrigerant (Figure 22-9). Equalizers are discussed in more detail later in this unit.

Diaphragm

The diaphragm is a thin metal membrane located inside the valve body (Figure 22-10). It moves the needle in and out of the seat in response to system load changes. The diaphragm is commonly made of a hard metal such as stainless steel, but is thin enough to have a certain amount of flexibility.

Needle and Seat

The needle and seat control the flow of refrigerant through the valve. They are normally made of a hard metal, such as stainless steel, to prevent the refrigerant passing through from eroding the seat. The needle and seat are carefully machined to provide close control of the refrigerant (Figure 22-11). Some valve manufacturers have needle and seat mechanisms that can be adjusted to meet different capacities or to correct a problem.

THERMOSTATIC EXPANSION VALVES

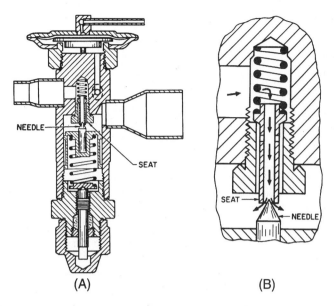

Figure 22-11. Needle and seat devices used in expansion valves. *Courtesy Sporlan Valve Company.*

The size of the needle and seat determines how much liquid refrigerant will pass through the valve with a specific pressure drop. For example, suppose the head pressure is 225 psig on the high-pressure side of the valve and 70 psig on the low-pressure side of the valve. This represents a pressure drop of 155 psig (225 - 70 = 155). At these conditions, a measured and predictable amount of liquid refrigerant will pass through the valve. If the same valve were used at pressures of 150 psig and 60 psig (a 90 psig drop), the valve would not be able to pass as much refrigerant. The conditions under which the valve will operate must be considered when selecting the valve.

Thermostatic expansion valves are rated in tons of refrigeration at a particular pressure drop condition. The standard pressure drop for air conditioning applications using R-22 is 100 psig across the valve orifice. The capacity and the working conditions of the system must be known to make the valve selection. For example, using the manufacturer's data in Figure 22-12, we see that a Model TRAE 20HW valve using R-22 has a capacity of 19.2 tons at 40°F and a pressure drop across the valve orifice of 100 psig. The model number of the valve, 20HW, means a nominal 20 ton valve. The word "nominal" means the size closest to standard tonnage ratings of equipment (10, 15, and 20 tons are typical).

When using these tables, you should be aware that the pressure difference from one side of the valve to the other is not necessarily the difference between the discharge pressure and the suction pressure. Any nonstandard pressure drop in the condenser, refrigerant distributor (if used), and interconnecting piping may cause a problem if it is not considered in valve selection. Figure 22-13 shows the actual operating conditions in a system using a distributor.

Note that the discharge and suction pressures are not the same as the pressure drop across the expansion valve. For this reason, selecting a valve for a 20 ton evaporator involves using the actual pressure drop across the valve and it may exceed 100 psig, so a smaller valve may need to be considered. A 20 ton evaporator will have more than one refrigerant circuit. For example, it may have 10 circuits, each with two tons of capacity. The TXV will be responsible for feeding the total quantity of refrigerant to the evaporator through a device called a *distributor* that will be used to distribute the refrigerant equally to each circuit (Figure 22-14). The distributor is a highly machined piece of brass that ensures that each circuit in the evaporator receives the correct amount of refrigerant. A nozzle placed between the valve and the distributor inlet creates a pressure drop and sprays the refrigerant into the distributor. This is necessary because the refrigerant is a partial liquid/partial vapor mixture. As mentioned earlier, approximately 25% of the liquid is flashed to a vapor, so the remaining 75% of the mixture

TRAE	R-22																							
CATALOG NUMBER	+50						+40						+20						0				
	Pressure Drop Across Valve —PSI—																							
	75	100	125	150	175	200	75	100	125	150	175	200	75	100	125	150	175	200	100	125	150	175	200	225
TRAE 10HW	10.3	11.9	13.3	14.6	15.8	16.9	10.2	11.8	13.2	14.5	15.6	16.7	10.0	11.5	12.9	14.8	15.2	16.3	10.4	11.6	12.7	13.7	14.7	15.6
TRAE 15HW	15.1	17.4	19.5	21.3	23.0	24.6	14.9	17.2	19.2	21.1	22.8	24.3	14.5	16.8	18.7	20.5	22.2	24.3	15.1	16.9	18.5	20.0	21.4	22.7
TRAE 20HW	16.8	19.4	21.7	23.8	25.7	27.5	16.6	19.2	21.5	23.5	25.4	27.2	16.2	18.7	20.9	22.9	24.8	26.5	16.9	18.9	20.7	22.4	23.9	25.4
TRAE 30HW	25.5	29.4	32.9	36.1	38.9	41.6	25.2	29.1	32.5	35.6	38.5	41.2	24.6	28.4	31.7	34.7	37.5	40.1	25.6	28.6	31.4	33.9	36.2	38.4
TRAE 40HW	35.0	40.5	45.2	49.6	53.5	57.2	34.6	40.0	44.7	49.0	52.9	56.6	33.8	39.0	43.6	47.8	51.6	55.2	35.2	39.4	43.1	46.6	49.8	52.8
TRAE 50HW	47.3	54.6	61.1	66.9	72.3	77.3	46.8	54.0	60.4	66.1	71.4	76.4	45.6	52.6	58.8	64.4	69.6	74.4	47.5	53.1	58.2	62.9	67.2	71.3
TRAE 60HW	51.6	59.6	66.7	73.0	78.8	84.3	51.0	58.9	65.9	72.1	77.9	83.3	49.7	57.4	64.2	70.3	75.9	81.2	51.8	58.0	63.5	68.6	73.3	77.8
TRAE 70HW	61.4	70.9	51.0	86.9	93.8	100.3	60.7	70.1	78.4	85.9	92.7	99.1	59.2	68.3	76.4	83.7	90.4	96.6	61.7	69.0	75.6	81.6	87.3	92.6

Figure 22-12. Part of a manufacturer's table showing the capacity of several valves at different pressure drops. Notice that the same valve has different capacities at different pressure drops (the greater the pressure drop, the greater the valve capacity). *Courtesy ALCO Controls Division, Emerson Electric Company.*

332 UNIT 22 EXPANSION DEVICES

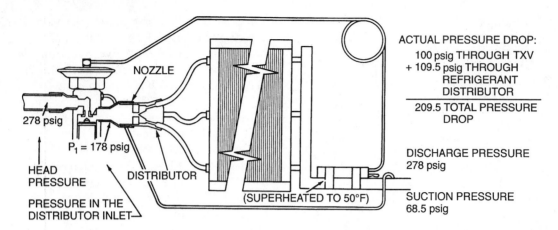

Figure 22-13. Actual operating pressures in a typical system. Note that the pressure drop across the expansion valve is not necessarily the difference between the head pressure and the suction pressure. The pressure drop through the refrigerant distributor is one of the big pressure drops that must be considered in valve selection. The refrigerant distributor is located between the expansion valve and the evaporator coil. *Courtesy Sporlan Valve Company.*

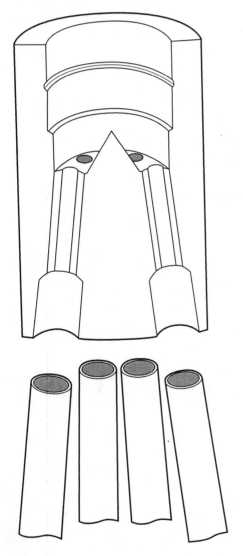

Figure 22-14. Refrigerant distributor for dispensing the correct amount of refrigerant to each circuit in a multi-circuit evaporator.

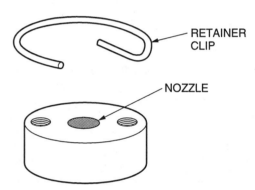

Figure 22-15. The nozzle creates a pressure drop just before the distributor to spray the refrigerant evenly into the distributor. This keeps the liquid from falling to the bottom and stratifying.

is liquid. Without the nozzle (Figure 22-15), the liquid would fall to the bottom and feed vapor to the top of the distributor and an imbalance of liquid and vapor would occur.

The manufacturer or design engineer usually chooses the valve, nozzle, and distributor. However, you should be familiar with the process in order to troubleshoot systems and to make modifications. One of the modifications that will be made in the future will occur when R-22 is replaced with an HFC refrigerant. The TXV and nozzle must be matched. When a refrigerant changeout occurs, the technician should contact the TXV manufacturer for the correct replacement nozzle. If only the valve is changed, there may be a mismatch and problems with refrigerant feeding the coil may occur.

Spring

The spring is one of the three forces that act on the diaphragm (the other two forces are the bulb pressure and the evaporator pressure). The spring raises the diaphragm and closes the valve by pushing the needle into the seat.

THERMOSTATIC EXPANSION VALVES   333

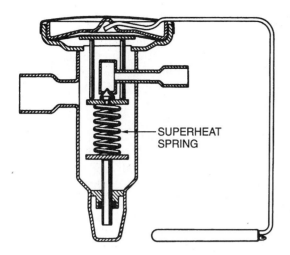

Figure 22-16. The spring tension in a TXV is set to maintain 8°F to 12°F of superheat in the evaporator. *Courtesy Singer Controls Division, Schiller Park, Illinois.*

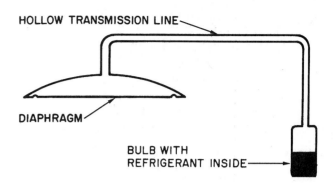

Figure 22-18. The transmission line is a small-diameter tube that transmits pressure between the sensing bulb and the diaphragm.

When a valve is adjustable, the adjustment applies more or less pressure to the spring to change the tension for different superheat settings. The spring tension is factory set for a predetermined superheat of 8°F to 12°F (Figure 22-16).

Adjustment and Packing Gland

The adjustment part of this valve can either be a screw-slot or square-head shaft, and is normally covered with a cap to prevent water, ice, or other foreign matter from collecting on the stem. The cap also serves as backup leak prevention. Most adjustment stems on expansion valves have a gasket or packing gland that can be tightened to prevent refrigerant leaks (Figure 22-17). A cap covers the stem and gland. Normally, one complete turn of the stem changes the superheat reading from 0.5°F to 1°F of superheat, depending on the manufacturer.

Sensing Bulb and Transmission Tube

The sensing bulb and transmission tube are extensions of the valve diaphragm. The bulb detects the temperature at the end of the evaporator on the suction line and transmits this temperature, converted to pressure, to the top of the diaphragm. The bulb contains a fluid, such as refrigerant, that responds to a pressure/temperature chart like R-22. When the suction line temperature goes up, this temperature change takes place inside the bulb. When there is a pressure change, the transmission line (which is nothing more than a small-diameter tube) allows the pressure between the bulb and the diaphragm to equalize back and forth (Figure 22-18).

The seat is stationary in the valve body, and the needle is moved by the diaphragm. One side of the diaphragm gets its pressure from the bulb, and the other side gets its pressure from the evaporator. As mentioned earlier, the diaphragm moves up and down in response to three different pressures: bulb pressure, evaporator pressure, and spring pressure. These three forces work together to position the valve needle at the correct position for the load conditions at any particular time (Figure 22-19). The bulb pressure acts to open the valve and the evaporator and spring pressures act to close the valve.

Figure 22-17. This expansion valve has a packing gland around the adjusting stem to prevent leaks when the stem is adjusted. *Photo by Bill Johnson.*

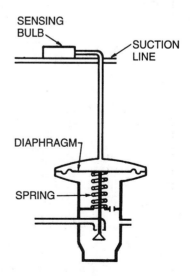

Figure 23-19. The sensing bulb charge, evaporator pressure, and spring all work as a team to position the needle in the seat to regulate the correct mount of refrigerant flow.

22.3 BULB CHARGES

The fluid inside the expansion valve bulb is known as the *charge* for the valve. There are four charges that can be obtained with a thermostatic expansion valve: liquid charge, cross liquid charge, vapor charge, and cross vapor charge. Liquid charge bulbs are the most common type of bulb used in air conditioning applications.

Liquid Charge Bulb

The liquid charge bulb contains a fluid with the same characteristics as the system refrigerant. The diaphragm and bulb are not actually full of liquid, but contain enough liquid so that they always have some liquid inside them. In other words, the liquid never completely vaporizes. On a graph, the pressure/temperature relationship for this type of bulb is almost a straight line. When the temperature goes up a degree, the pressure also goes up a specific amount and can be followed on a pressure/temperature chart.

Caution: As you will recall, when high temperatures are encountered, high pressures will exist. Be careful not to overheat the bulb while soldering nearby connections. The pressure inside the bulb can become great enough to rupture the diaphragm.

Cross Liquid Charge Bulb

The cross liquid charge bulb also contains a fluid but it does not follow the pressure/temperature relationship of the system refrigerant (Figure 22-20). It has a flatter curve and will close the valve faster on a rise in evaporator pressure. This closes the valve during the off cycle when the compressor shuts off and the evaporator pressure rises. This helps to prevent liquid refrigerant from flooding into the compressor at start-up.

Vapor Charge Bulb

The vapor charge bulb contains only a small amount of liquid refrigerant (Figure 22-21). It is sometimes called a *critical charge bulb*. When the bulb temperature rises, more and more of the liquid will boil into a vapor until there is no more liquid. When this point is reached, an increase in temperature will no longer produce a corresponding increase in pressure. At this point, the pressure curve will become horizontal (Figure 22-22).

When this type of bulb is used, care must be taken that the valve body does not get colder than the bulb, or the liquid in the bulb will condense above the diaphragm. When this happens, the bulb control at the end of the suction line will be lost to the valve diaphragm area and the valve will be controlled by the temperature of the liquid at the diaphragm. Small heaters may be installed at the valve body to keep this from happening in some specialty applications.

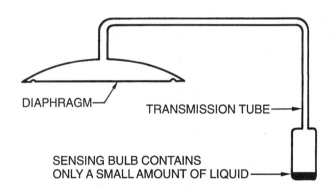

Figure 22-21. A vapor charge bulb has only a small amount of liquid in the sensing bulb.

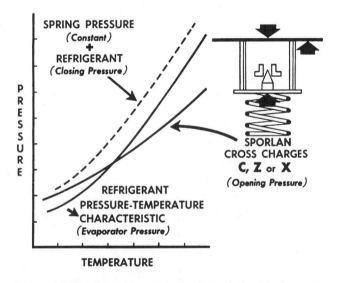

Figure 22-20. The pressure/temperature relationship for a cross liquid charge bulb. It does not follow the same pressure/temperature relationship as the refrigerant in the system. *Courtesy Sporlan Valve Company.*

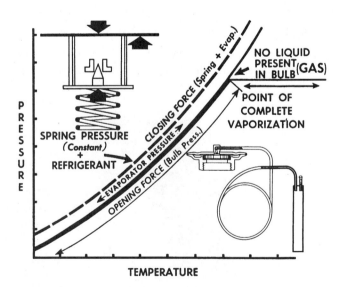

Figure 22-22. This graph shows the pressure/temperature relationship for a vapor charge bulb. *Courtesy Sporlan Valve Company.*

Cross Vapor Charge Bulb

The cross vapor charge bulb is similar to the vapor charge bulb, but it has a fluid with different characteristics than the refrigerant in the system. This produces a different pressure/temperature relationship. These bulbs are only used under special circumstances. Consult the manufacturer or your supplier for specific application details.

22.4 TXV OPERATION

The TXV responds to a change in load in the following manner. When the load increases (e.g., when the lights in an office building are turned on) the TXV opens, allowing more refrigerant into the coil. The evaporator needs more refrigerant at this time because the increased load is evaporating the refrigerant at a faster rate. The suction pressure also increases. When there is a load decrease (e.g., when the lights are turned off), the liquid in the evaporator evaporates slower and the suction pressure decreases. At this time, the TXV will throttle back by slightly closing the needle and seat to maintain the correct superheat.

All TXVs require some type of equalizer to sense the true suction pressure at the suction line. Air conditioning systems can use either internal or external equalizers, depending on the degree of pressure drop.

TXV Operation Using an Internal Equalizer

In systems with little or no pressure drop across the evaporator, an internal equalizer is used. A TXV of this type will operate as described below.

Note: This example uses a liquid-filled bulb.

1. Low-load conditions. The valve is operating in a stable condition known as *equilibrium* (Figure 22-23). The suction pressure is 68.5 psig using R-22, and the refrigerant is boiling in the evaporator at 40°F. The expansion valve is maintaining 10°F of superheat, so the suction line temperature is 50°F at the bulb location. The bulb has been on the line long enough to be at the same temperature as the line (50°F). For now, assume that the liquid in the bulb is at 84 psig, which corresponds to 50°F. The spring is exerting a pressure equal to the difference between the bulb pressure and the evaporator pressure. This holds the needle at the correct position in the seat to maintain this condition. The spring pressure in this example is 15.5 psi.

2. During TXV operation, the internal equalizer serves to register the true suction pressure by feeding evaporator inlet pressure to the bottom of the diaphragm through a small internal passage. The load changes as people enter and lights are turned on. (See Figure 22-24). When people enter the room and the lights are turned on, heat is added to the room, changing the load on the evaporator. The air entering the return air duct becomes warmer, and the load is added to the refrigeration coil by the air. This warmer air passing over the coil causes the liquid refrigerant inside the coil to boil faster. The suction pressure will also rise. The net effect of this condition will be that the last point of liquid in the coil will be farther from the end of the coil than when the coil was under the last equilibrium state of low-load conditions. The coil will start to starve for

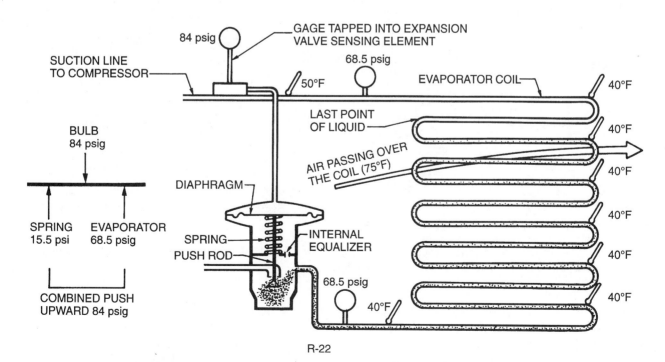

Figure 22-23. Thermostatic expansion valve under low-load conditions. The valve is said to be in *equilibrium* and the needle is stationary.

336 UNIT 22 EXPANSION DEVICES

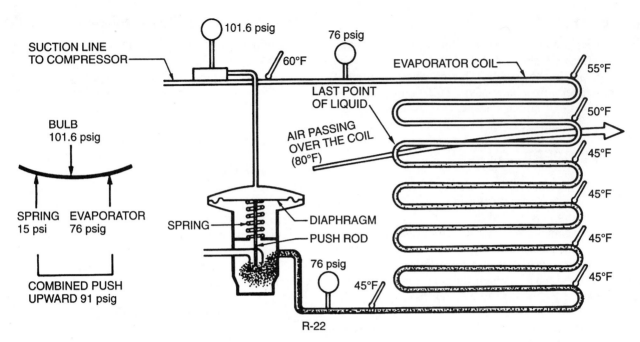

Figure 22-24. When heat is added to the conditioned space such as when people enter the building and lights are turned on, the load on the coil increases. The extra heat added to the air in the room is transferred into the coil, which causes a temperature increase in the suction line. This causes the valve to open and feed more refrigerant. If this condition is prolonged, the valve will reach a new equilibrium point.

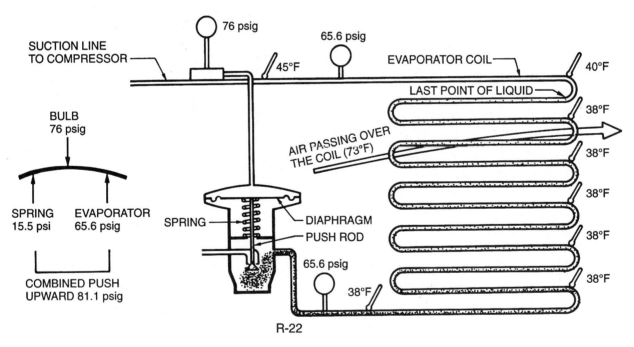

Figure 22-25. When there is a load decrease, such as when people leave the building and turn the lights off, the load is reduced. The valve will reach a new point of equilibrium at a low-load condition.

liquid refrigerant. The thermostatic expansion valve will feed more refrigerant to compensate for this shortage. When this condition has gone on for an extended period of time, the TXV will stabilize and reach a new point of equilibrium, where no further adjustment takes place.

3. The load changes at the end of the day when people leave and turn off the lights. (See Figure 22-25). When people leave and the lights are turned off, the load on the evaporator coil will decrease. There will no longer be enough of a load to boil the amount of refrigerant that the expansion valve is feeding into the coil, so the

expansion valve will start to flood the coil with liquid refrigerant. In response, the TXV throttles the refrigerant flow. When the TXV has operated at this condition for a certain period of time, it will stabilize and reach the original (low-load) point of equilibrium.

Note: The equilibrium point is the superheat setting of 10°F. The valve should always return to this setpoint with prolonged steady-state operation.

Eventually, the thermostat will stop the compressor completely because the air temperature in the conditioned space has reached the thermostat cut-out point.

TXV Operation Using an External Equalizer

Some evaporators have an excessive pressure drop from the inlet to the outlet. This pressure drop may be caused by a long piping circuit or a distributor located after the expansion valve. Excess pressure drop in a coil where a thermostatic expansion valve is used will cause the valve to starve the coil of refrigerant (Figure 22-26). Remember, the evaporator is most efficient when it has the maximum amount of refrigerant without flooding back to the compressor. If for any reason there is a pressure drop in excess of about 3 psig, a thermostatic expansion valve with an external equalizer should be used (Figure 22-27).

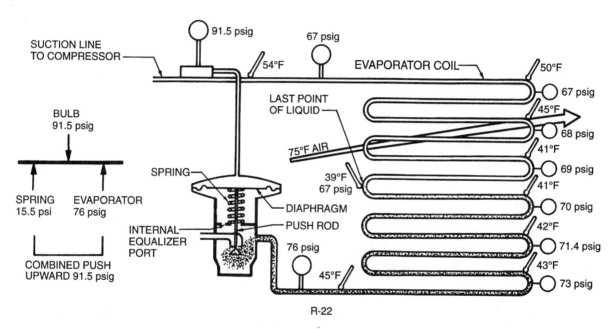

Figure 22-26. This evaporator has a pressure drop of 9 psig, which is causing the valve to starve the coil. Note that the refrigerant is not following the expected pressure/temperature relationship.

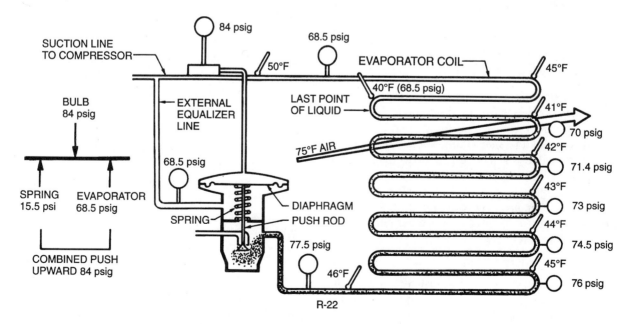

Figure 22-27. When a TXV with an external equalizer is added. the coil receives the correct amount of refrigerant for best efficiency.

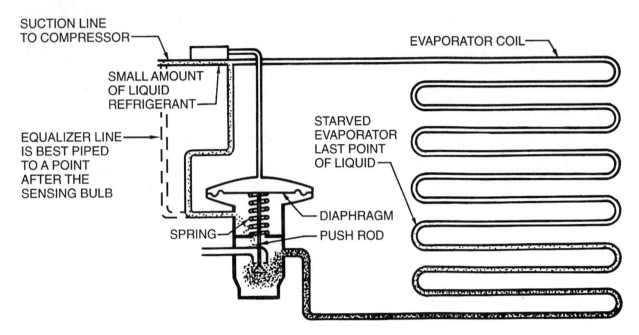

Figure 22-28. When the equalizer line is connected before the sensing bulb, any internal leak in the valve will simulate a flooded evaporator. For this reason, the equalizer line should always be piped to a point on the suction line after the sensing bulb.

The external equalizer line should always be piped to a point on the suction line after the expansion valve sensing bulb to prevent superheat problems caused by internal leaks in the valve. If a TXV with an external equalizer line has an internal leak, very small amounts of liquid will vent to the suction line through the external equalizer line. If the sensing bulb for the valve comes in contact with this liquid, it will throttle toward the closed position because this condition simulates a flooded coil (Figure 22-28). Sometimes, this condition can be sensed by feeling the equalizer line leading to the suction line. It should not become cold as it leaves the expansion valve.

22.5 TXV INSTALLATION AND SERVICE CONSIDERATIONS

When any thermostatic expansion valve is chosen, care should be taken that the valve is serviceable and will perform correctly. There are several things that should be considered: type of fastener (flare, solder, or flange), location of the valve for service and performance, and expansion valve bulb location. This valve has moving parts that are subject to wear. When a valve has to be replaced, an exact replacement is usually best. When this is not possible, your supplier can furnish the information needed to select another valve.

Sensing Element Installation

Particular care should be taken when installing the expansion valve sensing element. Each manufacturer has a recommended method for this installation, but they are all similar. The valve sensing bulb must be mounted at the end of the evaporator on the suction line. The best location is on a horizontal run where the bulb can be mounted flat and will not be raised by a fitting (Figure 22-29). On small suction lines, the bulb is mounted on top of the line. On large suction lines, the bulb is mounted on the side of the line toward the bottom. The bulb should not be located directly on the bottom of the line because the oil returning to the compressor will act as an insulator and may affect the sensing element. The object of the sensing element is to sense the temperature of the suction line. To do this, the line should be very clean, and the bulb must be fastened securely. To ensure an accurate reading, the bulb is insulated from the ambient temperature if it is much warmer than the suction line temperature. For example, if the ambient temperature for an air conditioning coil is 75°F and the suction line temperature is 50°F, insulation should be applied to the bulb.

Adjustments

Many technicians (and owners) will attempt to adjust a TXV. This may cause problems as it will change the pressures and temperatures in the coil. If the valve sensing element is mounted securely and in the correct location, the valve should operate properly as shipped from the factory. These valves are very reliable and normally do not require any adjustment. If there are signs that the valve has been adjusted, look for other problems, as the valve probably did not need adjustment to begin with. To prevent improper adjustments, many manufacturers are now producing non-adjustable valves (Figure 22-30).

SOLID-STATE CONTROLLED EXPANSION VALVES   339

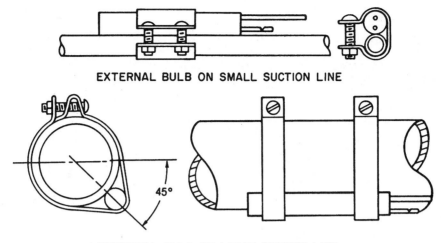

Figure 22-29. Best positions for mounting the expansion valve sensing bulb. *Courtesy ALCO Controls Division, Emerson Electric Company.*

Figure 22-30. This TXV is not adjustable. *Photo by Bill Johnson.*

22.7 SOLID-STATE CONTROLLED EXPANSION VALVES

The solid-state controlled expansion valve uses a thermistor as a sensing element (Figure 22-31). This thermistor varies the voltage to a bimetal element wrapped with a resistance heater. This valve normally uses 24 V as the control voltage.

When voltage is applied to the coil in the valve, the valve opens. Modulation is accomplished by varying the voltage (Figure 22-32). This valve is very versatile and can be used to accomplish different functions in the system (Figure 22-33). For example, if the voltage is cut off at the end of the cycle, the valve will shut off and the

22.6 BALANCED PORT EXPANSION VALVES

Low ambient temperatures have led manufacturers to develop TXV valves that will feed at a consistent rate even when the ambient temperature is low. These valves do not reduce refrigerant flow when the head pressure is low in mild weather. This provides the evaporator with the correct amount of refrigerant and allows it to operate at design conditions at lower outdoor temperatures.

Note: You cannot tell a balanced port expansion valve from a regular expansion valve by its appearance. However, it will have a different model number and you can look it up in the manufacturers' catalog to determine the exact type of valve.

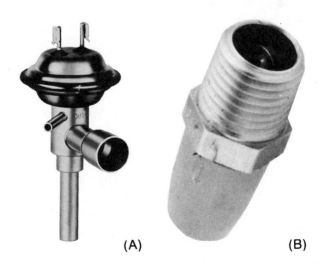

Figure 22-31. Expansion valve controlled with a thermistor and heat motor. *Courtesy Singer Controls Division, Schiller Park, Illinois.*

340 UNIT 22 EXPANSION DEVICES

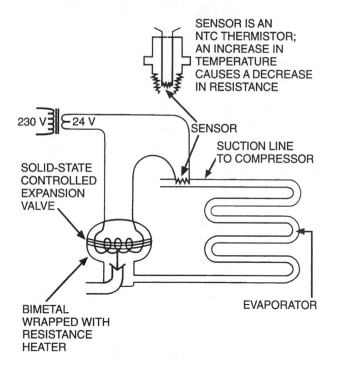

Figure 22-32. The temperature of the sensor controls the current flow to the bimetal element in the valve. More current flow causes the bimetal to warp and open the valve. Less current flow cools the bimetal and throttles the valve closed.

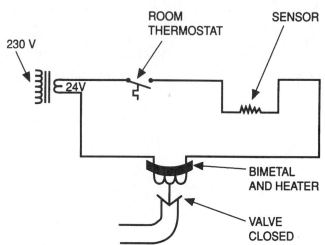

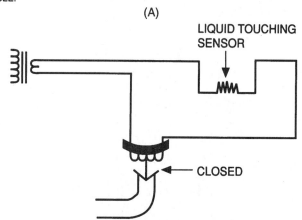

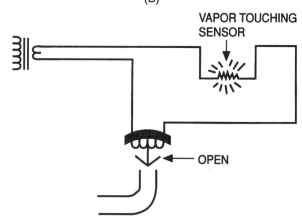

Figure 22-33. Different applications for the solid-state (thermistor) controlled valve.

system can be pumped down. The liquid refrigerant in the evaporator can be pumped to the condenser until the thermostat calls for cooling again. This protects the compressor from liquid flooding at start-up. On the other hand, if the voltage is allowed to remain on the element, the valve will remain open during the off cycle, and the pressures will equalize. This is advantageous if the compressor has a low starting torque, such as a single-phase compressor using a permanent split capacitor (PSC) motor.

To install this valve, the thermistor is inserted into the vapor stream at the end of the evaporator. It is very small and will respond quickly to temperature changes.

The solid-state controlled expansion valve responds to temperature changes in the sensing element just like a typical TXV except that it does not have a spring. When the thermistor is suspended in dry vapor, it is heated by the current passing through it. This creates a faster response than merely measuring the vapor temperature. When the valve opens and saturated vapor reaches the element, the valve begins to close slightly. This valve controls to a very low superheat, which allows the evaporator to utilize maximum surface area.

The solid-state controlled expansion valve is unique because in some models, refrigerant can flow in either direction through the valve body. For this reason, it is suitable for heat pump applications.

FIXED-BORE METERING DEVICES

Figure 22-34. Refrigerant receiver. When the load increases, more refrigerant is needed and moves from the receiver into the system. *Courtesy Refrigeration Research.*

Figure 22-35. A capillary tube metering device is a copper tube with a very small inside diameter. *Courtesy Parker Hannifin Corporation.*

22.8 RECEIVERS

A TXV allows more or less refrigerant flow, depending on the load. It needs a storage device (receiver) for refrigerant when it is not needed during load fluctuations (Figure 22-34). A receiver is a small tank located between the condenser and the expansion device. Normally, the condenser is close to the receiver. It has a king valve that functions as a service valve. This valve stops the refrigerant from leaving the receiver when the low side of the system is serviced. A receiver can serve both as a storage tank for different load conditions and as a tank into which the refrigerant can be pumped when servicing the system.

22.9 FIXED-BORE METERING DEVICES

The capillary tube and orifice metering device are known as *fixed-bore metering devices*. They have no moving parts and consist simply of a fixed hole in a housing.

Capillary Tubes

The capillary tube controls refrigerant flow by pressure drop. It is a copper tube with a very small calibrated inside diameter (Figure 22-35). The diameter and length of the tube determine how much liquid will pass through the tube at any given pressure drop (Figure 22-36). This relationship is used to arrive at the correct pressure drop which will allow the proper amount of refrigerant to pass through the capillary tube to the evaporator. The capillary tube can be quite long on some installations and may be wound in a coil to conserve space and prevent it from being damaged.

The capillary tube does not control superheat or pressure. It is a fixed-bore device with no moving parts. Since this device cannot adjust to a load change, it is normally used where the load is relatively constant.

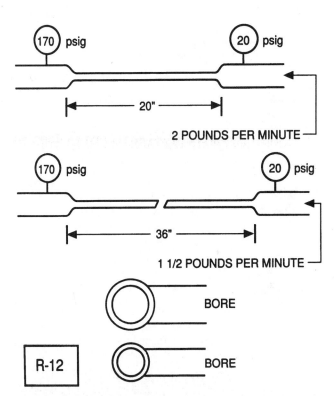

Figure 22-36. The length and bore (inside diameter) of the capillary tube determine the flow rate of the refrigerant. Refrigerant flow is measured in pounds per minute.

The capillary tube is a very inexpensive control device and is often used in small equipment. This device does not have a valve and will not stop the liquid from moving to the low side of the system during shutdown, so the pressures will equalize during the off cycle. This reduces the motor starting torque requirements for the compressor. Figure 22-37 illustrates a capillary tube in place at the evaporator inlet.

342 UNIT 22 EXPANSION DEVICES

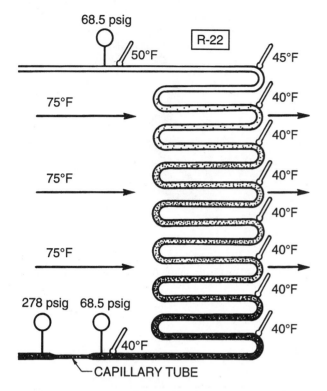

Figure 22-37. Typical operating conditions for an air conditioning system with a capillary tube metering device.

Because a capillary tube has no moving parts, it will not wear out. About the only problem it may have would be small particles that may fully or partially block the tube. The bore is so tiny that a small piece of flux, carbon, or solder will cause a problem if it reaches the tube inlet. To prevent this from happening, manufacturers always place a strainer or strainer-drier just before the capillary tube (Figure 22-38).

Capillary tubes are sized by the manufacturer and do not need to be resized when making system adjustments. If a new capillary tube is needed, the size of the tube and the length must be known in order to select a proper replacement.

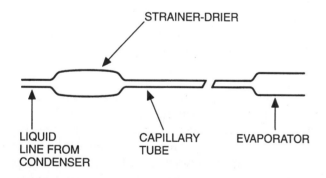

Figure 22-38. A strainer-drier protects the capillary tube from circulating particles.

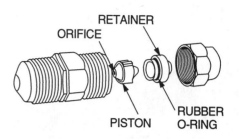

Figure 22-39. Orifice metering device.

Orifice Metering Devices

One popular metering device uses a removable brass piston that contains a fixed orifice (Figure 22-39). It has the advantage of being much smaller and more rugged that the capillary tube, and therefore less prone to damage. Like the capillary tube, the orifice is selected at the factory. This is a much shorter metering device than a capillary tube so the hole must be sized precisely. Each orifice size is assigned a number so that you can choose the correct replacement when necessary.

The orifice metering device is used in residential and light commercial air conditioning equipment up to about five tons of capacity.

Since the orifice metering device is basically a calibrated hole in a piece of brass, the pressures between the high and low sides will equalize during the off cycle. As with the capillary tube, this can be an advantage because PSC motors can be used. They are both economical and simple.

22.10 OPERATING CHARGE FOR FIXED-BORE DEVICES

Systems that use a capillary tube or orifice require only a small amount of refrigerant because they do not modulate the refrigerant to meet to the load. When the refrigerant charge is analyzed, we notice that when the unit is operating at the design conditions, there is a specific amount of refrigerant in the evaporator and a specific amount of refrigerant in the condenser. This is the amount of refrigerant required for a proper balance of charge in the system. Any other refrigerant that is in the system is in the pipes for circulating purposes only.

The amount of refrigerant in the system is very critical when using fixed-bore devices. It is easy to overcharge the system if you are not careful. In most capillary tube and orifice systems, the charge is printed on the nameplate of the equipment. The manufacturer always recommends measuring the refrigerant into these systems using either scales or liquid charging cylinders from a deep vacuum. A properly-charged system will maintain about 10°F of superheat at the end of the coil during design operating conditions (Figure 22-40).

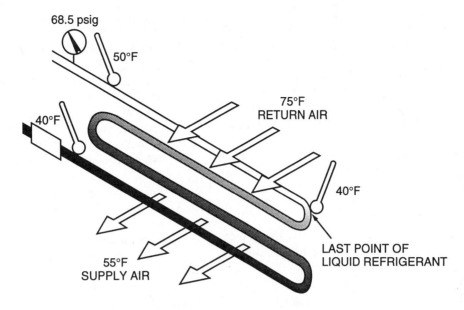

Figure 22-40. At design operating conditions, a properly-charged orifice metering device system will have approximately 10°F of superheat at the end of the coil.

In many capillary tube systems, the capillary tube is fastened to the suction line between the condenser and the evaporator to exchange heat between the capillary tube and the suction line. This design is used on all window air conditioners. It is called a *heat exchanger* and is discussed in more detail in the unit on window air conditioners. To troubleshoot a possible capillary tube problem, the correct superheat reading must be taken. The capillary tube should hold the refrigerant to about 10°F of superheat at the end of the evaporator.

Note: To obtain the correct superheat reading, it must be taken before the heat exchanger.

The capillary tube is very slow to respond to load changes or charge modifications. For example, if you add a small amount of refrigerant to a capillary tube system, it will take at least 15 minutes for the charge to adjust. The reason for this is that it takes time for the refrigerant to move from one side of the system to the other through the small bore. When you add refrigerant to the low side of the system, it moves into the compressor and is pumped into the condenser. It must now move to the evaporator before the charge will balance. Many technicians rush this process and overcharge this type of system. Manufacturers do not recommend adding refrigerant to top off the charge; instead, they recommend starting over from a deep vacuum and measuring a complete charge into the system.

The capillary tube metering device is used primarily on small air conditioning systems up to about five tons of capacity. These systems are hermetically sealed and have no bolted gasket connections. This ensures a leak-free system. These systems are factory assembled in a very clean environment and should provide years of trouble-free operation.

The refrigerant charge in an orifice metering device system is also critical. For a 3 ton system, some manufacturers indicate the charge is critical to 1/2 ounce of refrigerant. This is about one 3 foot gage line full of liquid refrigerant. Orifice systems should also be charged using accurate scales or liquid measuring devices. These systems should be charged from a deep vacuum to ensure the correct charge.

If a charge is added to an orifice metering device system while it is running (e.g., to top off the charge after a leak is found and repaired) the charge should be added very slowly in the vapor state. The orifice metering device will respond much faster to adding a charge of refrigerant than a capillary tube device because the orifice is much shorter than the capillary tube. The liquid refrigerant does not have to flow through the long capillary tube to reach the low-pressure side of the system.

SUMMARY

- The expansion device is one of the dividing lines between the high and low sides of the system (the compressor is the other).
- The thermostatic expansion valve maintains a constant superheat in the evaporator.
- The TXV consists of the body, diaphragm, needle and seat, spring, adjustment, and bulb and transmission tube.
- The TXV has three forces acting on it: bulb pressure, evaporator pressure, and spring pressure.
- The forces inside the expansion valve all work together to hold the needle and seat in the correct position so that the evaporator will have the correct amount of refrigerant under all load conditions.

- The TXV bulb contains one of four different charges: liquid charge, cross liquid charge, vapor charge, or cross vapor charge.
- The TXV bulb has to be mounted securely to the suction line to accurately sense the suction line temperature.
- The TXV responds to a load increase by feeding more refrigerant into the evaporator.
- An external equalizer prevents excessive pressure drop in the evaporator from causing the TXV to starve the evaporator.
- Balanced port expansion valves are used where low ambient temperatures exist.
- The solid-state controlled expansion valve uses a thermistor to monitor the suction line temperature to control refrigerant flow to the evaporator.
- The capillary tube metering device is a fixed-bore metering device that is usually made of copper. It has a very small inside diameter and no moving parts. Because it does not modulate the refrigerant flow to meet the load, the charge is critical.
- The capillary tube is commonly used in small systems.
- The orifice metering device consists of a calibrated hole in a piece of brass. It is very short compared to a capillary tube and also operates with a critical charge.

REVIEW QUESTIONS

1. What are the three forces acting on the TXV diaphragm?
2. What are the needle and seat normally made of?
3. The TXV maintains what condition in the evaporator?
4. How does a thermostatic expansion valve respond to an increased load in the evaporator?
5. Where is the bulb of the TXV mounted?
6. When do some TXVs require an external equalizer?
7. Draw a diagram showing how an external equalizer line is connected into a system.
8. How does a TXV respond to an decreased load in the evaporator?
9. What determines the amount of refrigerant that flows through a capillary tube metering device?
10. Why were balanced port expansion valves developed?
11. What is used to store the excess refrigerant in a TXV system when the load changes?
12. Name one advantage of an electronic expansion valve.
13. Name one advantage of a fixed-bore metering device.
14. Why is the charge critical in capillary tube and orifice metering device systems?
15. What is the recommended charging procedure for capillary tube and orifice metering device systems?

HEAT GAIN CALCULATIONS 23

OBJECTIVES

Upon completion of this unit, you should be able to

- **Explain the benefits of an accurate load estimate.**
- **Use the ASHRAE tables to find the outside design temperature for cooling in any major city in the U.S.**
- **State the accepted inside design temperature for cooling and explain why it may vary from person to person.**
- **Use the ASHRAE tables and the heat transfer formula to find the heat transfer rate through a wall, floor, door, window, roof, or ceiling.**
- **Explain the effects of solar heat gain.**
- **Discuss heat gain due to infiltration or ventilation.**
- **Explain why ductwork may need to be insulated if it runs through an unconditioned space.**
- **List several sources of internal heat gains.**
- **Explain the benefits of a room-by-room load calculation.**

23.1 SYSTEM DESIGN

Careful system design is important to successful operation of an air conditioning system. A properly-designed system will maintain the air temperature and humidity within the comfort range and will not produce drafts, either hot or cold, in the conditioned space. This design process may appear complicated, but is fairly straightforward when a systematic approach is followed.

The design sequence begins with a load calculation of the structure to find the total heat gain (sometimes referred to as the *cooling load*). This process determines how much heat is leaking into the structure. An ideal cooling system will remove the heat from a structure as fast as it is leaking in. Some tolerances may be allowed and will be discussed in this unit. Air distribution will be discussed in a later unit.

In addition to occupant comfort, the load calculation is also important from an economic standpoint—many lending institutions will not loan money on new construction unless they are assured that the cooling system is adequate. This is because they do not want to assume any problems if the loan defaults and they have to sell the structure.

This text is meant to provide an overview of the load calculation process. It is not meant to be used for the purpose of running a load calculation. There are various methods that designers/estimators use for the actual load calculation. The most common nationally-recognized method for residential use is *ACCA Manual J. ACCA Manual N* may be used for commercial buildings. These manuals contain detailed instructions for performing load calculations. Some designers/estimators prefer to use manufacturers' forms for load calculations. For example, if a company installs only one brand of equipment, they are likely to use an abbreviated form for calculating the heat gain. It is always recommended that a nationally-recognized method be used. This may be important if a dispute occurs. It is not uncommon for a customer to bring a contractor to court over the choice of equipment. Using a nationally-recognized procedure will provide an advantage.

There are times when you will need to know why the designer/estimator made certain decisions in choosing a particular system. You may also be caught in the middle of a dispute over why a system does not work correctly and will need to understand the basics of the design process in order to troubleshoot the system problem. For example, suppose you are called out to work on a cooling system that will not cool a structure. Naturally, the first step is to check the cooling system to make sure that it is performing up to capacity. If it is performing as designed and will not cool the structure, the heat gain calculation may be the problem. Either the designer/estimator did not account for something or some other problem has occurred. For example, the addition of a new room or even a set of sliding glass doors may have an impact on the original load estimate. Also, the homeowner may have inadvertently added to the load by other means. It is not uncommon for a homeowner to leave the vent covers off of the crawl space in the summer to allow warm air to move under the house. This keeps it from becoming too damp. However, the designer/estimator would normally assume that these covers would remain on. Many load problems may be uncovered by the observant technician and corrected without calling the designer/estimator. The designer/estimator may not even be available when the job is out of warranty because the owner may change service companies. Your job is to know the complete system and all problems that may impact system operation, including those that affect the original load estimate.

Equipment Sizing

The size of a cooling system must be selected carefully in order to ensure occupant comfort and optimum efficiency. In the summer, heat transfers from the outside of a building to the inside. This is why the load calculation is often called a *heat gain calculation*. Remember, heat transfers from hot to cold. The heat gain in a building can be compared to a water leak in a boat. Water is continuously leaking into the boat and a bilge pump is used to pump the water back into the ocean. If the bilge pump does not keep up, the boat will begin to sink. The water must to be removed as fast as it is leaking in. An air conditioner works much the same way—as heat leaks into a structure, the air conditioner pumps it back outside.

The temperature in a structure must be maintained by removing heat as fast as it leaks in on the highest demand day, which is the hottest part of the summer. The greater the temperature difference between the inside and the outside of the structure, the faster the heat transfer. For example, heat will transfer much faster on a day when the outdoor temperature is 100°F than on a day when the outdoor temperature is 80°F. The difference in temperature is the driving force that causes heat transfer.

Choosing the correct size cooling system is like choosing the bilge pump for a boat. If the pump is not large enough, the water will rise in the boat. If the pump is too large, it will cost more and will not run as efficiently. Similarly, if a cooling system is undersized, it will fall behind and the temperature will begin to rise. If the system is grossly undersized, the temperature will become hot enough to cause occupant discomfort. Conversely, if a cooling system is oversized, it will be more expensive to purchase and operate, and may not provide close control at part-load conditions.

In the past, many technicians have chosen the air conditioner size by merely looking at a structure and guessing. This could be called the "big house, big unit" method of equipment selection, and it frequently results in loss of business and dissatisfied customers. It is poor practice to select a big system because you know it will cool the structure on the hottest day with reserve capacity. If you adopt this method of equipment selection, your competitor will win the job by proposing a smaller, less expensive, and more efficient system. Another value-added service is suggesting ways to improve the home's efficiency, such as adding insulation or shading sunny windows.

23.2 DESIGN CONDITIONS

There are two conditions that must be known for accurate cooling system selection: the inside design temperature and the outside design temperature. For most residential and light commercial applications, the inside design temperature is typically 75°F with 50% relative humidity. This temperature may vary somewhat from person to person, depending on age, weight, activity level, and metabolism. Whenever possible, it is best to consult the occupants before selecting the inside design temperature.

The outside design temperature refers to the highest outdoor temperature that is normally encountered for any extended period of time. This tells you the design temperature difference. For example, suppose that the outside design temperature is 95°F and the inside design temperature is 75°F. The design temperature difference is 20°F. Remember, the temperature difference is the driving force for heat exchange.

ASHRAE, the American Society of Heating, Refrigerating, and Air Conditioning Engineers, Inc., provides a listing of the weather data for cities around the world. This design data includes the outside design temperature information needed to determine the temperature difference for any major city. This design data is calculated to $1/10$th of a °F and is rounded off to the nearest degree for practical use.

We will limit our discussion to cities in the U.S., but the same procedure can be applied to any country where weather data is available. Figure 23-1 is a page of the ASHRAE table showing several states: Ohio, Oklahoma, and Oregon (the entire table is reproduced in the back of this book). These states have very different climates, with Oklahoma being very hot and dry, Ohio being in a cool northern climate, and Oregon in a temperate coastal zone. The table lists data gathered at first-order weather stations (U.S. Air Force Bases and airports) around the country and includes both winter and summer data. We are only interested in the summer data contained in column 6. Figure 23-2 contains the entire U.S. data for column 6. This column lists the weather data for three conditions: 1%, 2.5%, and 5%. This data is gathered during the summer months of June, July, August, and September for a total of 2880 hours (24 hours × 120 days = 2880 hours). When you use the 1% value, it means that the temperature will not rise above the design value for more than 1% of the time during a typical summer. This amounts to 29 hours in an average summer (2880 × .01 = 28.8 or 29). The number of hours will be higher during unusually hot summers. Keep in mind that these hours are likely to be spread out over the entire summer. Normally, they will occur for one or two hours in the middle of the afternoon, usually between 2 pm and 4 pm on clear days. This heat gain is not instantaneous and may not have an effect on the space temperature for two to four hours, which would be between 4 pm and 6 pm. By this time, the sun is going down and its heat is diminishing. If the cooling system was not able to keep up during the midafternoon hours, it would not affect the space temperature until the end of the day.

The 2.5% value means that the temperature will be below the design value for 97.5% of the time during an average summer and above this for 2.5% of the time.

Climatic Conditions for the United States (*Continued*)

Col. 1	Col. 2		Col. 3		Col. 4	Winter,[b] °F Col. 5		Summer,[c] °F Col. 6			Col. 7	Col. 8			Prevailing Wind Col. 9				Temp., °F Col. 10	
State and Station[a]	Lat.		Long.		Elev.	Design Dry-Bulb		Design Dry-Bulb and Mean Coincident Wet-Bulb			Mean Daily Range	Design Wet-Bulb			Winter		Summer		Median of Annual Extr.	
	°	'N	°	'W	Feet	99%	97.5%	1%	2.5%	5%		1%	2.5%	5%	Knots[d]				Max.	Min.
OHIO																				
Akron, Canton AP	40	55	81	26	1208	1	6	89/72	86/71	84/70	21	75	73	72	SW	9	SW		94.4	−4.6
Ashtabula	41	51	80	48	690	4	9	88/73	85/72	83/71	18	75	74	72						
Athens	39	20	82	06	700	0	6	95/75	92/74	90/73	22	78	76	74						
Bowling Green	41	23	83	38	675	−2	2	92/73	89/73	86/71	23	76	75	73					96.7	−7.3
Cambridge	40	04	81	35	807	1	7	93/75	90/74	87/73	23	78	76	75						
Chillicothe	39	21	83	00	640	0	6	95/75	92/74	90/73	22	78	76	74	W	8	WSW		98.2	−2.1
Cincinnati Co	39	09	84	31	758	1	6	92/73	90/72	88/72	21	77	75	74	W	9	SW		97.2	−0.2
Cleveland AP	41	24	81	51	777	1	5	91/73	88/72	86/71	22	76	74	73	SW	12	N		94.7	−3.1
Columbus AP	40	00	82	53	812	0	5	92/73	90/74	87/72	24	77	75	74	W	8	SSW		96.0	−3.4
Dayton AP	39	54	84	13	1002	−1	4	91/73	89/72	86/71	20	76	75	73	WNW	11	SW		96.6	−4.5
Defiance	41	17	84	23	700	−1	4	94/74	91/73	88/72	24	77	76	74						
Findlay AP	41	01	83	40	804	2	3	92/74	90/73	87/72	24	77	76	74					97.4	−7.4
Fremont	41	20	83	07	600	−3	1	90/73	88/73	85/71	24	76	75	73						
Hamilton	39	24	84	35	650	0	5	92/73	90/72	87/71	22	76	75	73					98.2	−2.8
Lancaster	39	44	82	38	860	0	5	93/74	91/73	88/72	23	77	75	74						
Lima	40	42	84	02	975	−1	4	94/74	91/73	88/72	24	77	76	74	WNW	11	SW		96.0	−6.5
Mansfield AP	40	49	82	31	1295	0	5	90/73	87/72	85/72	22	76	74	73	W	8	SW		93.8	−10.7
Marion	40	36	83	10	920	0	5	93/74	91/73	88/72	23	77	76	74						
Middletown	39	31	84	25	635	0	5	92/73	90/72	87/71	22	76	75	73						
Newark	40	01	82	28	880	−1	5	94/74	92/73	89/72	23	77	75	74	W	8	SSW		95.8	−6.8
Norwalk	41	16	82	37	670	−3	1	90/73	88/73	85/71	22	76	75	73					97.3	−8.3
Portsmouth	38	45	82	55	540	5	10	95/76	92/74	89/73	22	78	77	75	W	8	SW		97.9	1.0
Sandusky Co	41	27	82	43	606	1	6	93/73	91/72	88/71	21	76	74	73					96.7	−1.9
Springfield	39	50	83	50	1052	−1	3	91/74	89/73	87/72	21	77	76	74	W	7	W			
Steubenville	40	23	80	38	992	1	5	89/72	86/71	84/70	22	74	73	72						
Toledo AP	41	36	83	48	669	−3	1	90/73	88/73	85/71	25	76	75	73	WSW	8	SW		95.4	−5.2
Warren	41	20	80	51	928	0	5	89/71	87/71	85/70	23	74	73	71						
Wooster	40	47	81	55	1020	1	6	89/72	86/71	84/70	22	75	73	72					94.0	−7.7
Youngstown AP	41	16	80	40	1178	−1	4	88/71	86/71	84/70	23	74	73	71	SW	10	SW			
Zanesville AP	39	57	81	54	900	1	7	93/75	90/74	87/73	23	78	76	75	W	6	WSW			
OKLAHOMA																				
Ada	34	47	96	41	1015	10	14	100/74	97/74	95/74	23	77	76	75						
Altus AFB	34	39	99	16	1378	11	16	102/73	100/73	98/73	25	77	76	75	N	10	S			
Ardmore	34	18	97	01	771	13	17	100/74	98/74	95/74	23	77	77	76						
Bartlesville	36	45	96	00	715	6	10	101/74	98/74	95/74	23	77	77	76						
Chickasha	35	03	97	55	1085	10	14	101/74	98/74	95/74	24	78	77	76						
Enid, Vance AFB	36	21	97	55	1307	9	13	103/74	100/74	97/74	24	79	77	76						
Lawton AP	34	34	98	25	1096	12	16	101/74	99/74	96/74	24	78	77	76						
McAlester	34	50	95	55	776	14	19	99/74	96/74	93/74	23	77	76	75	N	10	S			
Muskogee AP	35	40	95	22	610	10	15	101/74	98/75	95/75	23	79	78	77						
Norman	35	15	97	29	1181	9	13	99/74	96/74	94/74	24	77	76	75	N	10	S			
Oklahoma City AP	35	24	97	36	1285	9	13	100/74	97/74	95/73	23	78	77	76	N	14	SSW			
Ponca City	36	44	97	06	997	5	9	100/74	97/74	94/74	24	77	76	76						
Seminole	35	14	96	40	865	11	15	99/74	96/74	94/74	23	77	76	75						
Stillwater	36	10	97	05	984	8	13	100/74	96/74	93/74	24	77	76	75	N	12	SSW		103.7	1.6
Tulsa AP	36	12	95	54	650	8	13	101/74	98/75	95/75	22	79	78	77	N	11	SSW			
Woodward	36	36	99	31	2165	6	10	100/73	97/73	94/73	26	78	76	75					107.1	−1.3
OREGON																				
Albany	44	38	123	07	230	18	22	92/67	89/66	86/65	31	69	67	66					97.5	16.6
Astoria AP	46	09	123	53	8	25	29	75/65	71/62	68/61	16	65	63	62	ESE	7	NNW			
Baker AP	44	50	117	49	3372	−1	6	92/63	89/61	86/60	30	65	63	61					97.5	−6.8
Bend	44	04	121	19	3595	−3	4	90/62	87/60	84/59	33	64	62	60					96.4	−5.8
Corvallis	44	30	123	17	246	18	22	92/67	89/66	86/65	31	69	67	66	N	6	N		98.5	17.1
Eugene AP	44	07	123	13	359	17	22	92/67	89/66	86/65	31	69	67	66	N	7	N			
Grants Pass	42	26	123	19	925	20	24	99/69	96/68	93/67	33	71	69	68	N	5	N		103.6	16.4
Klamath Falls AP	42	09	121	44	4092	4	9	90/61	87/60	84/59	36	63	61	60	N	4	W		96.3	0.9
Medford AP	42	22	122	52	1298	19	23	98/68	94/67	91/66	35	70	68	67	S	4	WMW		103.8	15.0
Pendleton AP	45	41	118	51	1482	−2	5	97/65	93/64	90/62	29	66	65	63	NNW	6	WNW			
Portland AP	45	36	122	36	21	17	23	89/68	85/67	81/65	23	69	67	66	ESE	12	NW		96.6	18.3
Portland Co	45	32	122	40	75	18	24	90/68	86/67	82/65	21	69	67	66					97.6	20.5
Roseburg AP	43	14	123	22	525	18	23	93/67	90/66	87/65	30	69	67	66					99.6	19.5
Salem AP	44	55	123	01	196	18	23	92/68	88/66	84/65	31	69	68	66	N	6	N		98.9	15.9
The Dalles	45	36	121	12	100	13	19	93/69	89/68	85/66	28	70	68	67					105.1	7.9

Figure 23-1. A portion of the ASHRAE weather data table. *Reprinted by permission of American Society of Heating, Refrigerating and Air-Conditioning Engineers.*

Climatic Conditions for the United States

State and Station[a]	Summer,[b] °F Design Dry-Bulb and Mean Coincident Wet-Bulb			State and Station[a]	Summer,[b] °F Design Dry-Bulb and Mean Coincident Wet-Bulb		
	1%	2.5%	5%		1%	2.5%	5%
ALABAMA				Los Angeles AP	83/68	80/68	77/67
Alexander City	96/77	93/76	91/76	Los Angeles Co	93/70	89/70	86/69
Anniston AP	97/77	94/76	92/76	Merced, Castle AFB	102/70	99/69	96/68
Auburn	96/77	93/76	91/76	Modesto	101/69	98/68	95/67
Birmingham AP	96/74	94/75	92/74	Monterey	75/63	71/61	68/61
Decatur	95/75	93/74	91/74	Napa	100/69	96/68	92/67
Dothan AP	94/76	92/76	91/76	Needles AP	112/71	110/71	108/70
Florence AP	97/74	94/74	92/74	Oakland AP	85/64	80/63	75/62
Gadsden	96/75	94/75	92/74	Oceanside	83/68	80/68	77/67
Huntsville AP	95/75	93/74	91/74	Ontario	102/70	99/69	96/67
Mobile AP	95/77	93/77	91/76	Oxnard	83/66	80/64	77/63
Mobile Co	95/77	93/77	91/76	Palmdale AP	103/65	101/65	98/64
Montgomery AP	96/76	95/76	93/76	Palm Springs	112/71	110/70	108/70
Selma, Craig AFB	97/78	95/77	93/77	Pasadena	98/69	95/68	92/67
Talladega	97/77	94/76	92/76	Petaluma	94/68	90/66	87/65
Tuscaloosa AP	98/75	96/76	94/76	Pomona Co	102/70	99/69	95/68
				Redding AP	105/68	102/67	100/66
ALASKA				Redlands	102/70	99/69	96/68
Anchorage AP	71/59	68/58	66/56	Richmond	85/64	80/63	75/62
Barrow	57/53	53/50	49/47	Riverside, March AFB	100/68	98/68	95/67
Fairbanks AP	82/62	78/60	75/59	Sacramento AP	101/70	98/70	94/69
Juneau AP	74/60	70/58	67/57	Salinas AP	74/61	70/60	67/59
Kodiak	69/58	65/56	62/55	San Bernardino, Norton AFB	102/70	99/69	96/68
Nome AP	66/57	62/55	59/54	San Diego AP	83/69	80/69	78/68
				San Fernando	95/68	91/68	88/67
ARIZONA				San Francisco AP	82/64	77/63	73/62
				San Francisco Co	74/63	71/62	69/61
Douglas AP	98/63	95/63	93/63	San Jose AP	85/66	81/65	77/64
Flagstaff AP	84/55	82/55	80/54	San Luis Obispo	92/69	88/70	84/69
Fort Huachuca AP	95/62	92/62	90/62	Santa Ana AP	89/69	85/68	82/68
Kingman AP	103/65	100/64	97/64	Santa Barbara AP	81/67	77/66	75/65
Nogales	99/64	96/64	94/64	Santa Cruz	75/63	71/61	68/61
Phoenix AP	109/71	107/71	105/71	Santa Maria AP	81/64	76/63	73/62
Prescott AP	96/61	94/60	92/60	Santa Monica Co	83/68	80/68	77/67
Tucson AP	104/66	102/66	100/66	Santa Paula	90/68	86/67	84/66
Winslow AP	97/61	95/60	93/60	Santa Rosa	99/68	95/67	91/66
Yuma AP	111/72	109/72	107/71	Stockton AP	100/69	97/68	94/67
				Ukiah	99/69	95/68	91/67
ARKANSAS				Visalia	102/70	100/69	97/68
Blytheville AFB	96/78	94/77	91/76	Yreka	95/65	92/64	89/63
Camden	98/76	96/76	94/76	Yuba City	104/68	101/67	99/66
El Dorado AP	98/76	96/76	94/76	**COLORADO**			
Fayetteville AP	97/72	94/73	92/73	Alamosa AP	84/57	82/57	80/57
Fort Smith AP	101/75	98/76	95/76	Boulder	93/59	91/59	89/59
Hot Springs	101/77	97/77	94/77	Colorado Springs AP	91/58	88/57	86/57
Jonesboro	96/78	94/77	91/76	Denver AP	93/59	91/59	89/59
Little Rock AP	99/76	96/77	94/77	Durango	89/59	87/59	85/59
Pine Bluff AP	100/78	97/77	95/78	Fort Collins	93/59	91/59	89/59
Texarkana AP	98/76	96/77	93/76	Grand Junction AP	96/59	94/59	92/59
				Greeley	96/60	94/60	92/60
CALIFORNIA				Lajunta AP	100/68	98/68	95/67
Bakersfield AP	104/70	101/69	98/68	Leadville	84/52	81/51	78/50
Barstow AP	106/68	104/68	102/67	Pueblo AP	97/61	95/61	92/61
Blythe AP	112/71	110/71	108/70	Sterling	95/62	93/62	90/62
Burbank AP	95/68	91/68	88/67	Trinidad AP	93/61	91/61	89/61
Chico	103/69	101/68	98/67				
Concord	100/69	97/68	94/67	**CONNECTICUT**			
Covina	98/69	95/68	92/67	Bridgeport AP	86/73	84/71	81/70
Crescent City AP	68/60	65/59	63/58	Hartford, Brainard Field	91/74	88/73	85/72
Downey	93/70	89/70	86/69	New Haven AP	88/75	84/73	82/72
El Cajon	83/69	80/69	78/68	New London	88/73	85/72	83/71
El Centro AP	112/74	110/74	108/74	Norwalk	86/73	84/71	81/70
Escondido	89/68	85/68	82/68	Norwich	89/75	86/73	83/72
Eureka, Arcata AP	68/60	65/59	63/58	Waterbury	88/83	85/71	82/70
Fairfield, Travis AFB	99/68	95/67	91/66	Windsor Locks, Bradley Fld	91/74	88/72	85/71
Fresno AP	102/70	100/69	97/68	**DELAWARE**			
Hamilton AFB	89/68	84/66	80/65	Dover AFB	92/75	90/75	87/74
Laguna Beach	83/68	80/68	77/67	Wilmington AP	92/74	89/74	87/73
Livermore	100/69	97/68	93/67	**DISTRICT OF COLUMBIA**			
Lompoc, Vandenberg AFB	75/61	70/61	67/60	Andrews AFB	92/75	90/74	87/73
Long Beach AP	83/68	80/68	77/67	Washington, National AP	93/75	91/74	89/74

[a] AP, AFB, following the station name designates airport or military airbase temperature observations. Co designates office locations within an urban area that are affected by the surrounding area. Undersigned stations are semirural and may be compared to airport data.

[b] Summer design data are based on the 4-month period, June through September.

Figure 23-2. Weather data for 1%, 2.5%, and 5% columns. *Reprinted by permission of American Society of Heating, Refrigerating and Air-Conditioning Engineers.*

Climatic Conditions for the United States (Continued)

State and Station[a]	Summer,[b] °F Design Dry-Bulb and Mean Coincident Wet-Bulb			State and Station[a]	Summer,[b] °F Design Dry-Bulb and Mean Coincident Wet-Bulb		
	1%	2.5%	5%		1%	2.5%	5%
FLORIDA				Greenville	94/76	92/75	89/74
Belle Glade	92/76	91/76	89/76	Joliet	93/75	90/74	88/73
Cape Kennedy AP	90/78	88/78	87/78	Kankakee	93/75	90/74	88/73
Daytona Beach AP	92/78	90/77	88/77	La Salle/Peru	93/75	91/75	88/74
E Fort Lauderdale	92/78	91/78	90/78	Macomb	95/76	92/76	89/75
Fort Myers AP	93/78	92/78	91/77	Moline AP	93/75	91/75	88/74
Fort Pierce	91/78	90/78	89/78	Mt Vernon	95/76	92/75	89/74
Gainesville AP	95/77	93/77	92/77	Peoria AP	91/75	89/74	87/73
Jacksonville AP	96/77	94/77	92/76	Quincy AP	96/76	93/76	90/76
Key West AP	90/78	90/78	89/78	Rantoul, Chanute AFB	94/75	91/74	89/73
Lakeland Co	93/76	91/76	89/76	Rockford	91/74	89/73	87/72
Miami AP	91/77	90/77	89/77	Springfield AP	94/75	92/74	89/74
Miami Beach Co	90/77	89/77	88/77	Waukegan	92/76	89/74	87/73
Ocala	95/77	93/77	92/76				
Orlando AP	94/76	93/76	91/76	**INDIANA**			
Panama City, Tyndall AFB	92/78	90/77	89/77	Anderson	95/76	92/75	89/74
Pensacola Co	94/77	93/77	91/77	Bedford	95/76	92/75	89/74
St. Augustine	92/78	89/78	87/78	Bloomington	95/76	92/75	89/74
St. Petersburg	92/77	91/77	90/76	Columbus, Bakalar AFB	95/76	92/75	90/74
Sanford	94/76	93/76	91/76	Crawfordsville	94/75	91/74	88/73
Sarasota	93/77	92/77	90/76	Evansville AP	95/76	93/75	91/75
Tallahassee AP	94/77	92/76	90/76	Fort Wayne AP	92/73	89/72	87/72
Tampa AP	92/77	91/77	90/76	Goshen AP	91/73	89/73	86/72
West Palm Beach AP	92/78	91/78	90/78	Hobart	91/73	88/73	85/72
				Huntington	92/73	89/72	87/72
GEORGIA				Indianapolis AP	92/74	90/74	87/73
Albany, Turner AFB	97/77	95/76	93/76	Jeffersonville	95/74	93/74	90/74
Americus	97/77	94/76	92/75	Kokomo	91/74	90/74	88/73
Athens	94/74	92/74	90/74	Lafayette	94/74	91/73	88/73
Atlanta AP	94/74	92/74	90/73	La Porte	93/74	90/74	87/73
Augusta AP	97/77	95/76	93/76	Marion	91/74	90/73	88/73
Brunswick	92/78	89/78	87/78	Muncie	92/74	90/74	87/73
Columbus, Lawson AFB	95/76	93/76	91/75	Peru, Grissom AFB	90/74	88/73	86/72
Dalton	94/76	93/76	91/76	Richmond AP	92/74	90/74	87/73
Dublin	96/77	93/76	91/75	Shelbyville	93/74	91/74	88/73
Gainesville	96/77	93/77	91/77	South Bend AP	91/73	89/73	86/72
Griffin	93/76	90/75	88/74	Terre Haute AP	95/75	92/74	89/73
LaGrange	94/76	91/75	89/74	Valparaiso	93/74	90/74	87/73
Macon AP	96/77	93/76	91/75	Vincennes	95/75	92/74	90/73
Marietta, Dobbins AFB	94/74	92/74	90/74				
Savannah	96/77	93/77	91/77	**IOWA**			
Valdosta-Moody AFB	96/77	94/77	92/76	Ames	93/75	90/74	87/73
Waycross	96/77	94/77	91/76	Burlington AP	94/74	91/75	88/73
				Cedar Rapids AP	91/76	88/75	86/74
HAWAII				Clinton	92/75	90/75	87/74
Hilo AP	84/73	83/72	82/72	Council Bluffs	94/76	91/75	88/74
Honolulu AP	87/73	86/73	85/72	Des Moines AP	94/75	91/74	88/73
Kaneohe Bay MCAS	85/75	84/74	83/74	Dubuque	90/74	88/73	86/72
Wahiawa	86/73	85/72	84/72	Fort Dodge	91/74	88/74	86/72
				Iowa City	92/76	89/76	87/74
IDAHO				Keokuk	95/75	92/75	89/74
Boise AP	96/65	94/64	91/64	Marshalltown	92/75	90/75	88/74
Burley	99/62	95/61	92/66	Mason City AP	90/74	88/74	85/72
Coeur D'Alene AP	89/62	86/61	83/60	Newton	94/75	91/74	88/73
Idaho Falls AP	89/61	87/61	84/59	Ottumwa AP	94/75	91/74	88/73
Lewiston AP	96/65	93/64	90/63	Sioux City AP	95/74	92/74	89/73
Moscow	90/63	87/62	84/61	Waterloo	91/76	89/75	86/74
Mountain Home AFB	99/64	97/63	94/62				
Pocatello AP	94/61	91/60	89/59	**KANSAS**			
Twin Falls AP	99/62	95/61	92/60	Atchison	96/77	93/76	91/76
				Chanute AP	100/74	97/74	94/74
ILLINOIS				Dodge City AP	100/69	97/69	95/69
Aurora	93/76	91/76	88/75	El Dorado	101/72	98/73	96/73
Belleville, Scott AFB	94/76	92/76	89/75	Emporia	100/74	97/74	94/73
Bloomington	92/75	90/74	88/73	Garden City AP	99/69	96/69	94/69
Carbondale	95/77	93/77	90/76	Goodland AP	99/66	96/65	93/66
Champaign/Urbana	95/75	92/74	90/73	Great Bend	101/73	98/73	95/73
Chicago, Midway AP	94/74	91/73	88/72	Hutchinson AP	102/72	99/72	97/72
Chicago, O'Hare AP	91/74	89/74	86/72	Liberal	99/68	96/68	94/68
Chicago Co	94/75	91/74	88/73	Manhattan, Ft Riley	99/75	95/75	92/74
Danville	93/75	90/74	88/73	Parsons	100/74	97/74	94/74
Decatur	94/75	91/74	88/73	Russell AP	101/73	98/73	95/73
Dixon	93/75	90/74	88/73	Salina	103/74	100/74	97/73
Elgin	91/75	88/74	86/73	Topeka AP	99/75	96/75	93/74
Freeport	91/74	89/73	87/72	Wichita AP	101/72	98/73	96/73
Galesburg	93/75	91/75	88/74				

Figure 23-2. Weather data for 1%, 2.5%, and 5% columns (continued).

Climatic Conditions for the United States (Continued)

State and Station[a]	Summer,[b] °F Design Dry-Bulb and Mean Coincident Wet-Bulb			State and Station[a]	Summer,[b] °F Design Dry-Bulb and Mean Coincident Wet-Bulb		
	1%	2.5%	5%		1%	2.5%	5%
KENTUCKY				Port Huron	90/73	87/72	83/71
Ashland	94/76	91/74	89/73	Saginaw AP	91/73	87/72	84/71
Bowling Green AP	94/77	92/75	89/74	Sault Ste. Marie AP	84/70	81/69	77/66
Corbin AP	94/73	92/73	89/72	Traverse City AP	89/72	86/71	83/69
Covington AP	92/73	90/72	88/72	Ypsilanti	92/72	89/71	86/70
Hopkinsville, Ft Campbell	94/77	92/75	89/74				
Lexington AP	93/73	91/73	88/72	**MINNESOTA**			
Louisville AP	95/74	93/74	90/74	Albert Lea	90/74	87/72	84/71
Madisonville	96/76	93/75	90/75	Alexandria AP	91/72	88/72	85/70
Owensboro	97/76	94/75	91/75	Bemidji AP	88/69	85/69	81/67
Paducah AP	98/76	95/75	92/75	Brainerd	90/73	87/71	84/69
				Duluth AP	85/70	82/68	79/66
LOUISIANA				Fairbault	91/74	88/72	85/71
Alexandria AP	95/77	94/77	92/77	Fergus Falls	91/72	88/72	85/70
Baton Rouge AP	95/77	93/77	92/77	International Falls AP	85/68	83/68	80/66
Bogalusa	95/77	93/77	92/77	Mankato	91/72	88/72	85/70
Houma	95/78	93/78	92/77	Minneapolis/St. Paul AP	92/75	89/73	86/71
Lafayette AP	95/78	94/78	92/78	Rochester AP	90/74	87/72	84/71
Lake Charles AP	95/77	93/77	92/77	St. Cloud AP	91/74	88/72	85/70
Minden	99/77	96/76	94/76	Virginia	85/69	83/68	80/66
Monroe AP	99/77	96/76	94/76	Willmar	91/74	88/72	85/71
Natchitoches	97/77	95/77	93/77	Winona	91/75	88/73	85/72
New Orleans AP	93/78	92/78	90/77				
Shreveport AP	99/77	96/76	94/76	**MISSISSIPPI**			
				Biloxi, Keesler AFB	94/79	92/79	90/78
MAINE				Clarksdale	96/77	94/77	92/76
Augusta AP	88/73	85/70	82/68	Columbus AFB	95/77	93/77	91/76
Bangor, Dow AFB	86/70	83/68	80/67	Greenville AFB	95/77	93/77	91/76
Caribou AP	84/69	81/67	78/66	Greenwood	95/77	93/77	91/76
Lewiston	88/73	85/70	82/68	Hattiesburg	96/77	94/77	92/77
Millinocket AP	87/69	83/68	80/66	Jackson AP	97/76	95/76	93/76
Portland	87/72	84/71	81/69	Laurel	96/78	94/77	92/77
Waterville	87/72	84/69	81/68	McComb AP	96/77	94/76	92/76
				Meridian AP	97/77	95/76	93/76
MARYLAND				Natchez	96/78	94/78	92/77
Baltimore AP	94/75	91/75	89/74	Tupelo	96/77	94/77	92/76
Baltimore Co	92/77	89/76	87/75	Vicksburg Co	97/78	95/78	93/77
Cumberland	92/75	89/74	87/74				
Frederick AP	94/76	91/75	88/74	**MISSOURI**			
Hagerstown	94/75	91/74	89/74	Cape Girardeau	98/76	95/75	92/75
Salisbury	93/75	91/75	88/74	Columbia AP	97/74	94/74	91/73
				Farmington AP	96/76	93/75	90/74
MASSACHUSETTS				Hannibal	96/76	93/76	90/76
Boston AP	91/73	88/71	85/70	Jefferson City	98/75	95/74	92/74
Clinton	90/72	87/71	84/69	Joplin AP	100/73	97/73	94/73
Fall River	87/72	84/71	81/69	Kansas City AP	99/75	96/74	93/74
Framingham	89/72	86/71	83/69	Kirksville AP	96/74	93/74	90/73
Gloucester	89/73	86/71	83/70	Mexico	97/74	94/74	91/73
Greenfield	88/72	85/71	82/69	Moberly	97/74	94/74	91/73
Lawrence	90/73	87/72	84/70	Poplar Bluff	98/78	95/76	92/76
Lowell	91/73	88/72	85/70	Rolla	94/77	91/75	89/74
New Bedford	85/72	82/71	80/69	St. Joseph AP	96/77	93/76	91/76
Pittsfield AP	87/71	84/70	81/68	St. Louis AP	97/75	94/75	91/74
Springfield, Westover AFB	90/72	87/71	84/69	St. Louis Co	98/75	94/75	91/74
Taunton	89/73	86/72	83/70	Sikeston	98/77	95/76	92/75
Worcester AP	87/71	84/70	81/68	Sedalia, Whiteman AFB	95/76	92/76	90/75
				Sikeston	98/77	95/76	92/75
MICHIGAN				Springfield AP	96/73	93/74	91/74
Adrian	91/73	88/72	85/71				
Alpena AP	89/70	85/70	83/69	**MONTANA**			
Battle Creek AP	92/74	88/72	85/70	Billings AP	94/64	91/64	88/63
Benton Harbor AP	91/72	88/72	85/70	Bozeman	90/61	87/60	84/59
Detroit	91/73	88/72	86/71	Butte AP	86/58	83/56	80/56
Escanaba	87/70	83/69	80/68	Cut Bank AP	88/61	85/61	82/60
Flint AP	90/73	87/72	85/71	Glasgow AP	92/64	89/63	85/62
Grand Rapids AP	91/72	88/72	85/70	Glendive	95/66	92/64	89/62
Holland	88/72	86/71	83/70	Great Falls AP	91/60	88/60	85/59
Jackson AP	92/74	88/72	85/70	Havre	94/65	90/64	87/63
Kalamazoo	92/74	88/72	85/70	Helena AP	91/60	88/60	85/59
Lansing AP	90/73	87/72	84/70	Kalispell AP	91/62	87/61	84/60
Marquette Co	84/70	81/69	77/66	Lewistown AP	90/62	87/61	83/60
Mt Pleasant	91/73	87/72	84/71	Livingstown AP	90/61	87/60	84/59
Muskegon AP	86/72	84/70	82/70	Miles City AP	98/66	95/66	92/65
Pontiac	90/73	87/72	85/71	Missoula AP	92/62	88/61	85/60

Figure 23-2. Weather data for 1%, 2.5%, and 5% columns (continued).

Climatic Conditions for the United States (*Continued*)

State and Station[a]	Summer,[b] °F Design Dry-Bulb and Mean Coincident Wet-Bulb			State and Station[a]	Summer,[b] °F Design Dry-Bulb and Mean Coincident Wet-Bulb		
	1%	2.5%	5%		1%	2.5%	5%
NEBRASKA				**NEW YORK**			
Beatrice	99/75	95/74	92/74	Albany AP	91/73	88/72	85/70
Chadron AP	97/66	94/65	91/65	Albany Co	91/73	88/72	85/70
Columbus	98/74	95/73	92/73	Auburn	90/73	87/71	84/70
Fremont	98/75	95/74	92/74	Batavia	90/72	87/71	84/70
Grand Island AP	97/72	94/71	91/71	Binghamton AP	86/71	83/69	81/68
Hastings	97/72	94/71	91/71	Buffalo AP	88/71	85/70	83/69
Kearney	96/71	93/70	90/70	Cortland	88/71	85/71	82/70
Lincoln Co	99/75	95/74	92/74	Dunkirk	88/73	85/72	83/71
McCook	98/69	95/69	91/69	Elmira AP	89/71	86/71	83/70
Norfolk	97/74	93/74	90/73	Geneva	90/73	87/71	84/70
North Platte AP	97/69	94/69	90/69	Glens Falls	88/72	85/71	82/69
Omaha AP	94/76	91/75	88/74	Gloversville	89/72	86/71	83/69
Scottsbluff AP	95/65	92/65	90/64	Hornell	88/71	85/70	82/69
Sidney AP	95/65	92/65	90/64	Ithaca	88/71	85/71	82/70
				Jamestown	88/70	86/70	83/69
				Kingston	91/73	88/72	85/70
				Lockport	89/74	86/72	84/71
NEVADA				Massena AP	86/70	83/69	80/68
Carson City	94/60	91/59	89/58	Newburgh, Stewart AFB	90/73	88/72	85/70
Elko AP	94/59	92/59	90/58	NYC-Central Park	92/74	89/73	87/72
Ely AP	89/57	87/56	85/55				
Las Vegas AP	108/66	106/65	104/65	NYC-Kennedy AP	90/73	87/72	84/71
Lovelock AP	98/63	96/63	93/62	NYC-La Guardia AP	92/74	89/73	87/72
Reno AP	95/61	92/60	90/59	Niagara Falls AP	89/74	86/72	84/71
Reno Co	96/61	93/60	91/59	Olean	87/71	84/71	81/70
Tonopah AP	94/60	92/59	90/58	Oneonta	86/71	83/69	80/68
Winnemucca AP	96/60	94/60	92/60	Oswego Co	86/73	83/71	80/70
				Plattsburg AFB	86/70	83/69	80/68
				Poughkeepsie	92/74	89/74	86/72
				Rochester AP	91/73	88/71	85/70
NEW HAMPSHIRE				Rome, Griffiss AFB	88/71	85/70	83/69
Berlin	87/71	84/69	81/68				
Claremont	89/72	86/70	83/69	Schenectady	90/73	87/72	84/70
Concord AP	90/72	87/70	84/69	Suffolk County AFB	86/72	83/71	80/70
Keene	90/72	87/70	83/69	Syracuse AP	90/73	87/71	84/70
Laconia	89/72	86/70	83/69	Utica	88/73	85/71	82/70
Manchester, Grenier AFB	91/72	88/71	85/70	Watertown	86/73	83/71	81/70
Portsmouth, Pease AFB	89/73	85/71	83/70				
				NORTH CAROLINA			
				Asheville AP	89/73	87/72	85/71
NEW JERSEY				Charlotte AP	95/74	93/74	91/74
Atlantic City Co	92/74	89/74	86/72	Durham	94/75	92/75	90/75
Long Branch	93/74	90/73	87/72	Elizabeth City AP	93/78	91/77	89/76
Newark AP	94/74	91/73	88/72	Fayetteville, Pope AFB	95/76	92/76	90/75
New Brunswick	92/74	89/73	86/72	Goldsboro	94/77	91/76	89/75
Paterson	94/74	91/73	88/72	Greensboro AP	93/74	91/73	89/73
Phillipsburg	92/73	89/72	86/71	Greenville	93/77	91/76	89/75
Trenton Co	91/75	88/74	85/73	Henderson	95/77	92/76	90/76
Vineland	91/75	89/74	86/73	Hickory	92/73	90/72	88/72
				Jacksonville	92/78	90/78	88/77
				Lumberton	95/76	92/76	90/75
NEW MEXICO				New Bern AP	92/78	90/78	88/77
Alamagordo, Holloman AFB	98/64	96/64	94/64	Raleigh/Durham AP	94/75	92/75	90/75
Albuquerque AP	96/61	94/61	92/61	Rocky Mount	94/77	91/76	89/75
Artesia	103/67	100/67	97/67				
Carlsbad AP	103/67	100/67	97/67	Wilmington AP	93/79	91/78	89/77
Clovis AP	95/65	93/65	91/65	Winston-Salem AP	94/74	91/73	89/73
Farmington AP	95/63	93/62	91/61				
Gallup	90/59	89/58	86/58				
Grants	89/59	88/58	85/57	**NORTH DAKOTA**			
Hobbs AP	101/66	99/66	97/66	Bismarck AP	95/68	91/68	88/67
Las Cruces	99/64	96/64	94/64	Devils Lake	91/69	88/68	85/66
Los Alamos	89/60	87/60	85/60	Dickinson AP	94/68	90/66	87/65
Raton AP	91/60	89/60	87/60	Fargo AP	92/73	89/71	85/69
Roswell, Walker AFB	100/66	98/66	96/66	Grand Forks AP	91/70	87/70	84/68
Santa Fe Co	90/61	88/61	86/61				
Silver City AP	95/61	94/60	91/60	Jamestown AP	94/70	90/69	87/68
Socorro AP	97/62	95/62	93/62	Minot AP	92/68	89/67	86/65
Tucumcari AP	99/66	97/66	95/65	Williston	91/68	88/67	85/65

Figure 23-2. Weather data for 1%, 2.5%, and 5% columns (continued).

Climatic Conditions for the United States (*Continued*)

State and Station[a]	Summer,[b] °F Design Dry-Bulb and Mean Coincident Wet-Bulb			State and Station[a]	Summer,[b] °F Design Dry-Bulb and Mean Coincident Wet-Bulb		
	1%	2.5%	5%		1%	2.5%	5%
OHIO				**PENNSYLVANIA**			
Akron, Canton AP	89/72	86/71	84/70	Allentown AP	92/73	88/72	86/72
Ashtabula	88/73	85/72	83/71	Altoona Co	90/72	87/71	84/70
Athens	95/75	92/74	90/73	Butler	90/73	87/72	85/71
Bowling Green	92/73	89/73	86/71	Chambersburg	93/75	90/74	87/73
Cambridge	93/75	90/74	87/73	Erie AP	88/73	85/72	83/71
				Harrisburg AP	94/75	91/74	88/73
Chillicothe	95/75	92/74	90/73				
Cincinnati Co	92/73	90/72	88/72	Johnstown	86/70	83/70	80/68
Cleveland AP	91/73	88/72	86/71	Lancaster	93/75	90/74	87/73
Columbus AP	92/73	90/73	87/72	Meadville	88/71	85/70	83/69
Dayton AP	91/73	89/72	86/71	New Castle	91/73	88/72	86/71
Defiance	94/74	91/73	88/72	Philadelphia AP	93/75	90/74	87/72
				Pittsburgh AP	89/72	86/71	84/70
Findlay AP	92/74	90/73	87/72				
Fremont	90/73	88/73	85/71	Pittsburgh Co	91/72	88/71	86/70
Hamilton	92/73	90/72	87/71	Reading Co	92/73	89/72	86/72
Lancaster	93/74	91/73	88/72	Scranton/Wilkes-Barre	90/72	87/71	84/70
Lima	94/74	91/73	88/72	State College	90/72	87/71	84/70
Mansfield AP	90/73	87/72	85/72	Sunbury	92/73	89/72	86/70
				Uniontown	91/74	88/73	85/72
Marion	93/74	91/73	88/72				
Middletown	92/73	90/72	87/71	Warren	89/71	86/71	83/70
Newark	94/73	92/73	89/72	West Chester	92/75	89/74	86/72
Norwalk	90/73	88/73	85/71	Williamsport AP	92/73	89/72	86/70
Portsmouth	95/76	92/74	89/73	York	94/75	91/74	88/73
Sandusky Co	93/73	91/72	88/71				
Springfield	91/74	89/73	87/72	**RHODE ISLAND**			
Steubenville	89/72	86/71	84/70	Newport	88/73	85/72	82/70
Toledo AP	90/73	88/73	85/71	Providence AP	89/73	86/72	83/70
Warren	89/71	87/71	85/70				
Wooster	89/72	86/71	84/70	**SOUTH CAROLINA**			
Youngstown AP	88/71	86/71	84/70	Anderson	94/74	92/74	90/74
Zanesville AP	93/75	90/74	87/73	Charleston AFB	93/78	91/78	89/77
				Charleston Co	94/78	92/78	90/77
				Columbia AP	97/76	95/75	93/75
OKLAHOMA				Florence AP	94/77	92/77	90/76
Ada	100/74	97/74	95/74	Georgetown	92/79	90/78	88/77
Altus AFB	102/73	100/73	98/73				
Ardmore	100/74	98/74	95/74	Greenville AP	93/74	91/74	89/74
Bartlesville	101/73	98/74	95/74	Greenwood	95/75	93/74	91/74
Chickasha	101/74	98/74	95/74	Orangeburg	97/76	95/75	93/75
				Rock Hill	96/75	94/74	92/74
Enid, Vance AFB	103/74	100/74	97/74	Spartanburg AP	93/74	91/74	89/74
Lawton AP	101/74	99/74	96/74	Sumter, Shaw AFB	95/77	92/76	90/75
McAlester	99/74	96/74	93/74				
Muskogee AP	101/74	98/75	95/75	**SOUTH DAKOTA**			
Norman	99/74	96/74	94/74	Aberdeen AP	94/73	91/72	88/70
Oklahoma City AP	100/74	97/74	95/73	Brookings	95/73	92/72	89/71
				Huron AP	96/73	93/72	90/71
Ponca City	100/74	97/74	94/74	Mitchell	96/72	93/71	90/70
Seminole	99/74	96/74	94/73	Pierre AP	99/71	95/71	92/69
Stillwater	100/74	96/74	93/74	Rapid City AP	95/66	92/65	89/65
Tulsa AP	101/74	98/75	95/75				
Woodward	100/73	97/73	94/73	Sioux Falls AP	94/73	91/72	88/71
				Watertown AP	94/73	91/72	88/71
				Yankton	94/73	91/72	88/71
OREGON							
Albany	92/67	89/66	86/65				
Astoria AP	75/65	71/62	68/61	**TENNESSEE**			
Baker AP	92/63	89/61	86/60	Athens	95/74	92/73	90/73
Bend	90/62	87/60	84/59	Bristol-Tri City AP	91/72	89/72	87/71
Corvallis	92/67	89/66	86/65	Chattanooga AP	96/75	93/74	91/74
Eugene AP	92/67	89/66	86/65	Clarksville	95/76	93/74	90/74
Grants Pass	99/69	96/68	93/67	Columbia	97/75	94/74	91/74
Klamath Falls AP	90/61	87/60	84/59	Dyersburg	96/78	94/77	91/76
Medford AP	98/68	94/67	91/66				
Pendleton AP	97/65	93/64	90/62	Greeneville	92/73	90/72	88/72
Portland AP	89/68	85/67	81/65	Jackson AP	98/76	95/75	92/75
Portland Co	90/68	86/67	82/65	Knoxville AP	94/74	92/73	90/73
				Memphis AP	98/77	95/76	93/76
Roseburg AP	93/67	90/66	87/65	Murfreesboro	97/75	94/74	91/74
Salem AP	92/68	88/66	84/65	Nashville AP	97/75	94/74	91/74
The Dalles	93/69	89/68	85/66	Tullahoma	96/74	93/73	91/73

Figure 23-2. Weather data for 1%, 2.5%, and 5% columns (continued).

Climatic Conditions for the United States (*Continued*)

State and Station[a]	Summer,[b] °F Design Dry-Bulb and Mean Coincident Wet-Bulb			State and Station[a]	Summer,[b] °F Design Dry-Bulb and Mean Coincident Wet-Bulb		
	1%	2.5%	5%		1%	2.5%	5%
TEXAS				**VIRGINIA**			
Abilene AP	101/71	99/71	97/71	Charlottesville	94/74	91/74	88/73
Alice AP	100/78	98/77	95/77	Danville AP	94/74	92/73	90/73
Amarillo AP	98/67	95/67	93/67	Fredericksburg	96/76	93/75	90/74
Austin AP	100/74	98/74	97/74	Harrisonburg	93/72	91/72	88/71
Bay City	96/77	94/77	92/77	Lynchburg AP	93/74	90/74	88/73
Beaumont	95/79	93/78	91/78	Norfolk AP	93/77	91/76	89/76
				Petersburg	95/76	92/76	90/75
Beeville	99/78	97/77	95/77	Richmond AP	95/76	92/76	90/75
Big Spring AP	100/69	97/69	95/69	Roanoke AP	93/72	91/72	88/71
Brownsville AP	94/77	93/77	92/77	Staunton	93/72	91/72	88/71
Brownwood	101/73	99/73	96/73	Winchester	93/75	90/74	88/74
Bryan AP	98/76	96/76	94/76				
Corpus Christi AP	95/78	94/78	92/78	**WASHINGTON**			
				Aberdeen	80/65	77/62	73/61
Corsicana	100/75	98/75	96/75	Bellingham AP	81/67	77/65	74/63
Dallas AP	102/75	100/75	97/75	Bremerton	82/65	78/64	75/62
Del Rio, Laughlin AFB	100/73	98/73	97/73	Ellensburg AP	94/65	91/64	87/62
Denton	101/74	99/74	97/74	Everett, Paine AFB	80/65	76/64	73/62
Eagle Pass	101/73	99/73	98/73	Kennewick	99/68	96/67	92/66
El Paso AP	100/64	98/64	96/64	Longview	88/68	85/67	81/65
				Moses Lake, Larson AFB	97/66	94/65	90/63
Fort Worth AP	101/74	99/74	97/74	Olympia AP	87/66	83/65	79/64
Galveston AP	90/79	89/79	88/78	Port Angeles	72/62	69/61	67/60
Greenville	101/74	99/74	97/74	Seattle-Boeing Field	84/68	81/66	77/65
Harlingen	96/77	94/77	93/77	Seattle Co	85/68	82/66	78/65
Houston AP	96/77	94/77	92/77	Seattle-Tacoma AP	84/65	80/64	76/62
Houston Co	97/77	95/77	93/77	Spokane AP	93/64	90/63	87/62
				Tacoma, McChord AFB	86/66	82/65	79/63
Huntsville	100/75	98/75	96/75	Walla Walla AP	97/67	94/66	90/65
Killeen, Robert Gray AAF	99/73	97/73	95/73	Wenatchee	99/67	96/66	92/64
Lamesa	99/69	96/69	94/69	Yakima AP	96/65	93/65	89/63
Laredo AFB	102/73	101/73	99/74				
Longview	99/76	97/76	95/76	**WEST VIRGINIA**			
Lubbock AP	98/69	96/69	94/69	Beckley	83/71	81/69	79/69
				Bluefield AP	83/71	81/69	79/69
Lufkin AP	99/76	97/76	94/76	Charleston AP	92/74	90/73	87/72
McAllen	97/77	95/77	94/77	Clarksburg	92/74	90/73	87/72
Midland AP	100/69	98/69	96/69	Elkins AP	86/72	84/70	82/70
Mineral Wells AP	101/74	99/74	97/74	Huntington Co	94/76	91/74	89/73
Palestine Co	100/76	98/76	96/76	Martinsburg AP	93/75	90/74	88/74
Pampa	99/67	96/67	94/67	Morgantown AP	90/74	87/73	85/73
				Parkersburg Co	93/75	90/74	88/73
Pecos	100/69	98/69	96/69	Wheeling	89/72	86/71	84/70
Plainview	98/68	96/68	94/68				
Port Arthur AP	95/79	93/78	91/78	**WISCONSIN**			
San Angelo, Goodfellow AFB	101/71	99/71	97/70	Appleton	89/74	86/72	83/71
San Antonio AP	99/72	97/73	96/73	Ashland	85/70	82/68	79/66
Sherman, Perrin AFB	100/75	98/75	95/74	Beloit	92/75	90/75	88/74
				Eau Claire AP	92/75	89/73	86/71
Snyder	100/70	98/70	96/70	Fond Du Lac	89/74	86/72	84/71
Temple	100/74	99/74	97/74	Green Bay AP	88/74	85/72	83/71
Tyler AP	99/76	97/76	95/76	La Crosse AP	91/75	88/73	85/72
Vernon	102/73	100/73	97/73	Madison AP	91/74	88/73	85/71
Victoria AP	98/78	96/77	94/77	Manitowoc	89/74	86/72	83/71
Waco AP	101/75	99/75	97/75	Marinette	87/73	84/71	82/70
Wichita Falls AP	103/73	101/73	98/73	Milwaukee AP	90/74	87/73	84/71
				Racine	91/75	88/73	85/72
UTAH				Sheboygan	89/75	86/73	83/72
Cedar City AP	93/60	91/60	89/59	Stevens Point	92/75	89/73	86/71
Logan	93/62	91/61	88/60	Waukesha	90/74	87/73	84/71
Moab	100/60	98/60	96/60	Wausau AP	91/74	88/72	85/70
Ogden AP	93/63	91/61	88/61				
Price	93/60	91/60	89/59	**WYOMING**			
Provo	98/62	96/62	94/61	Casper AP	92/58	90/57	87/57
				Cheyenne	89/58	86/58	84/57
Richfield	93/60	91/60	89/59	Cody AP	89/60	86/60	83/59
St George Co	103/65	101/65	99/64	Evanston	86/55	84/55	82/54
Salt Lake City AP	97/62	95/62	92/61	Lander AP	91/61	88/61	85/60
Vernal AP	91/61	89/60	86/59	Laramie AP	84/56	81/56	79/55
				Newcastle	91/64	87/63	84/63
				Rawlins	86/57	83/57	81/56
VERMONT				Rock Springs AP	86/55	84/55	82/54
Barre	84/71	81/69	78/68	Sheridan AP	94/62	91/62	88/61
Burlington AP	88/72	85/70	82/69	Torrington	94/62	91/62	88/61
Rutland	87/72	84/70	81/69				

Figure 23-2. Weather data for 1%, 2.5%, and 5% columns (continued).

354 UNIT 23 HEAT GAIN CALCULATIONS

This represents only 72 hours when the outside temperature is above the design temperature (2880 hours × .025 = 72). This is an acceptable value for many designers, particularly in the southern U.S. For example, the design temperature using the 2.5% column for Cincinnati, Ohio is 90°F. If a designer uses this as the outside design temperature, it will normally only occur for a few hours in the early afternoon. The customer may be a retail store and may not even need the extra cooling in the midafternoon because of light customer traffic. By the time customers start to arrive in the early evening, the store temperature will have returned to normal. Similarly, many residential customers do not remain at home during the day and may not require the extra capacity provided by the 1% design value. If this is the case, it may benefit them to tolerate occasional temperature swings rather than oversize the system.

The only drawbacks of using the 2.5% value are poor control of space temperature during uncommonly warm summers and the fact that the system has very little reserve capacity for large gatherings, such as parties or times when the system is temporarily shut down for repairs or by power outages. For example, if the system is shut off for any reason on a day that is close to or above the design value, it will take a long time to pull the indoor temperature back down to the design temperature. For this reason, many designers prefer to use the 1% value.

When designing a system, it is unlikely that the available equipment size will have exactly the same value as the actual heat gain of the structure. For example, a cooling load may be 33,000 Btuh. A cooling system with a capacity of 36,000 Btuh would likely be chosen as the closest size. This is an oversize of 3000 Btuh. The exact design conditions and equipment choice are regional. It is easy to write specifications and descriptions as to how to design systems, but much more practical to use the experience of successful local contractors.

Different types of systems are popular in different parts of the country. For example, evaporative cooling systems are seldom used in the southeastern states because of the high humidity in that part of the country. Evaporative coolers (sometimes known as *swamp coolers*) require low humidity for use as comfort cooling systems so they are popular in the southwestern states.

Calculating the heat gain for a structure involves looking at all the places in the structure where heat can leak in. This includes any wall where a difference in temperature occurs. It also involves any opening where warm air may enter the structure and have to be cooled. Typical sources of heat gain include the following:

- Through outside walls.
- Through partitions to unconditioned spaces, such as garages or storage rooms.
- Through floors above crawl spaces or unconditioned spaces.
- Through windows via conduction.
- Around windows and doors via infiltration (air that leaks in due to pressure differences between the inside and outside of the structure), or ventilation (planned infiltration).
- Through ductwork located in unconditioned spaces.
- Through internal sources, such as people, lights, and equipment.
- Solar heat gain.

23.3 HEAT TRANSFER THROUGH BUILDING MATERIALS

The building materials used in a structure have a large impact on the amount of heat that leaks into the building. As you will recall, some materials are conductors and others are insulators. For example, steel is a conductor and styrofoam is an insulator. A steel door would therefore conduct a lot of heat. The heat transferred through a steel door may be reduced by using a sandwich method of construction with steel on the two outer surfaces and styrofoam in between.

The designer must know how much heat will transfer through various building materials in order to calculate the total heat gain of a structure. The amount of heat transferred through a building material is known as the *heat of transmission*. Actually, this is heat transfer via conduction. The heat of transmission is calculated using a heat transfer coefficient called the *U factor*. The U factor is expressed in Btu's per hour per °F temperature difference per square foot (Btuh/°F/sq. ft.). It may be used in the following formula for calculating the heat transfer through a building material:

$$Q = A \times U \times TD$$

where: Q = heat transfer in Btuh

A = area of the building material in square feet

U = overall heat transfer coefficient

TD = temperature difference from one side to the other in °F

For example, if a building material has a 1°F temperature difference from one side to the other, a U factor of .5, and an area of 1 square foot, it will transfer .5 Btu's of heat per hour (Btuh). (See Figure 23-3.)

$$Q = A \times U \times TD$$

where: Q = 1 sq. ft. × .5 Btuh × 1°F TD

Q = .5 Btuh

Finding the area of the building material is easy; it can be measured for an existing structure or taken from the blueprints for a structure that has not yet been built.

The temperature difference is found by subtracting the outside design temperature from the inside design temperature. For example, suppose the inside design temperature is 75°F and the outside design temperature is 95°F. The temperature difference is 20°F (95°F - 75°F = 20°F).

HEAT TRANSFER THROUGH BUILDING MATERIALS

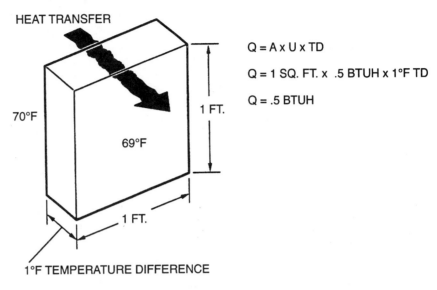

Figure 23-3. Overall heat transfer coefficient (U factor).

The U factor is found by testing materials and determining their rate of heat transfer. The U factors for most construction materials are published in tabular form. However, it is often necessary to construct or build a U factor when one is not published for a particular composite wall. A composite wall is one constructed of several different materials, such as a wall in a house that may have brick on the outside, followed by building board, then an insulated stud wall with sheetrock as the indoor wall surface (Figure 23-4). There are several materials in this wall, each with a different rate of heat transfer. The overall U factor for this composite wall can be calculated using a special formula (discussed later).

Heat Gain through Walls

The walls, ceilings, and floors make up the greatest portion of surface area in a structure and represent a large source of heat gain. As you proceed through the following explanations, remember that the walls contain the doors and windows. When only the wall is being

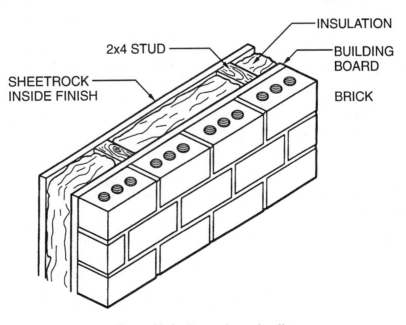

Figure 23-4. Composite stud wall.

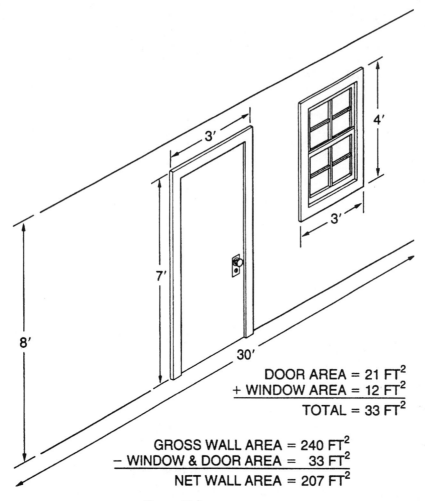

Figure 23-5. Net wall area calculation.

calculated, the area of the windows and doors must be subtracted from the wall area. This is called the *net wall area* (Figure 23-5).

The walls of a structure may be either above grade or below grade. Below-grade walls or basements are not included in heat gain calculations because the temperature of the ground is lower than the space temperature.

When a wall is made of several materials, the U factor (overall coefficient of heat transfer) may be found by determining the resistance to heat flow (R value) of each individual material in the wall, then adding them together to find the total resistance to heat transfer or R_t (R total). The R_t may then be converted to a U factor by dividing it into 1 (this is called the *reciprocal of R*). Stated algebraically:

$$U = \frac{1}{R_t}$$

The R_t for a wall is fairly easy to calculate using the table in Figure 23-6. This table provides information for many types of building materials. For example, bricks may seem like a simple building material but when we look at the resistances listed for brick, we find there are many varieties to consider, each with a different density. The density of brick runs from 70 to 150 pounds per cubic foot. Different bricks come from different parts of the country and are used for different applications. For the example in Figure 23-7, we will use a resistance value for brick of .20 per inch of thickness. This was found using the highest resistance for brick with a density of 110 pounds per cubic foot.

Concrete blocks also come in a wide variety of compositions and may be somewhat difficult to calculate. In this text, we will use an average resistance of 1.04 for concrete block. This was found by averaging the values listed for concrete (1.11 + .97 = 2.08 ÷ 2 = 1.04).

Before we calculate the total resistance to heat flow for a wall, there are two other factors that must be considered. There is an air film on each side of the wall that offers resistance to heat flow. On the inside of the house, this film is very still because the air in the room is not moving very fast. Heat does not readily transfer through still air.

HEAT TRANSFER THROUGH BUILDING MATERIALS

Typical Thermal Properties of Common Building and Insulating Materials—Design Values[a]

Description	Density, lb/ft³	Conductivity[b] (k), Btu·in / h·ft²·°F	Conductance (C), Btu / h·ft²·°F	Resistance[c] (R) Per Inch Thickness (1/k), °F·ft²·h / Btu·in	Resistance[c] (R) For Thickness Listed (1/C), °F·ft²·h / Btu	Specific Heat, Btu / lb·°F
BUILDING BOARD						
Asbestos-cement board	120	4.0	—	0.25	—	0.24
Asbestos-cement board .... 0.125 in.	120	—	33.00	—	0.03	
Asbestos-cement board .... 0.25 in.	120	—	16.50	—	0.06	
Gypsum or plaster board ...... 0.375 in.	50	—	3.10	—	0.32	0.26
Gypsum or plaster board ...... 0.5 in.	50	—	2.22	—	0.45	
Gypsum or plaster board ...... 0.625 in.	50	—	1.78	—	0.56	
Plywood (Douglas Fir)[d]	34	0.80	—	1.25	—	0.29
Plywood (Douglas Fir) .... 0.25 in.	34	—	3.20	—	0.31	
Plywood (Douglas Fir) .... 0.375 in.	34	—	2.13	—	0.47	
Plywood (Douglas Fir) .... 0.5 in.	34	—	1.60	—	0.62	
Plywood (Douglas Fir) .... 0.625 in.	34	—	1.29	—	0.77	
Plywood or wood panels .... 0.75 in.	34	—	1.07	—	0.93	0.29
Vegetable fiber board						
Sheathing, regular density[e] .... 0.5 in.	18	—	0.76	—	1.32	0.31
.... 0.78125 in.	18	—	0.49	—	2.06	
Sheathing intermediate density[e] .. 0.5 in.	22	—	0.92	—	1.09	0.31
Nail-base sheathing[e] .... 0.5 in.	25	—	0.94	—	1.06	0.31
Shingle backer .... 0.375 in.	18	—	1.06	—	0.94	0.31
Shingle backer .... 0.3125 in.	18	—	1.28	—	0.78	
Sound deadening board .... 0.5 in.	15	—	0.74	—	1.35	0.30
Tile and lay-in panels, plain or acoustic	18	0.40	—	2.50	—	0.14
.... 0.5 in.	18	—	0.80	—	1.25	
.... 0.75 in.	18	—	0.53	—	1.89	
Laminated paperboard	30	0.50	—	2.00	—	0.33
Homogeneous board from repulped paper	30	0.50	—	2.00	—	0.28
Hardboard[e]						
Medium density	50	0.73	—	1.37	—	0.31
High density, service-tempered grade and service grade	55	0.82	—	1.22	—	0.32
High density, standard-tempered grade	63	1.00	—	1.00	—	0.32
Particleboard[e]						
Low density	37	0.71	—	1.41	—	0.31
Medium density	50	0.94	—	1.06	—	0.31
High density	62.5	1.18	—	0.85	—	0.31
Underlayment .... 0.625 in.	40	—	1.22	—	0.82	0.29
Waferboard	37	0.63	—	1.59	—	—
Wood subfloor .... 0.75 in.	—	—	1.06	—	0.94	0.33
BUILDING MEMBRANE						
Vapor—permeable felt	—	—	16.70	—	0.06	
Vapor—seal, 2 layers of mopped 15-lb felt	—	—	8.35	—	0.12	
Vapor—seal, plastic film	—	—	—	—	Negl.	
FINISH FLOORING MATERIALS						
Carpet and fibrous pad	—	—	0.48	—	2.08	0.34
Carpet and rubber pad	—	—	0.81	—	1.23	0.33
Cork tile .... 0.125 in.	—	—	3.60	—	0.28	0.48
Terrazzo .... 1 in.	—	—	12.50	—	0.08	0.19
Tile—asphalt, linoleum, vinyl, rubber	—	—	20.00	—	0.05	0.30
vinyl asbestos						0.24
ceramic						0.19
Wood, hardwood finish .... 0.75 in.	—	—	1.47	—	0.68	
INSULATING MATERIALS						
Blanket and Batt[f,g]						
Mineral fiber, fibrous form processed from rock, slag, or glass						
approx. 3–4 in.	0.4–2.0	—	0.091	—	11	
approx. 3.5 in.	0.4–2.0	—	0.077	—	13	
approx. 3.5 in.	1.2–1.6	—	0.067	—	15	
approx. 5.5–6.5 in.	0.4–2.0	—	0.053	—	19	
approx. 5.5 in.	0.6–1.0	—	0.048	—	21	
approx. 6–7.5 in.	0.4–2.0	—	0.045	—	22	
approx. 8.25–10 in.	0.4–2.0	—	0.033	—	30	
approx. 10–13 in.	0.4–2.0	—	0.026	—	38	
Board and Slabs						
Cellular glass	8.0	0.33	—	3.03	—	0.18
Glass fiber, organic bonded	4.0–9.0	0.25	—	4.00	—	0.23
Expanded perlite, organic bonded	1.0	0.36	—	2.78	—	0.30
Expanded rubber (rigid)	4.5	0.22	—	4.55	—	0.40
Expanded polystyrene, extruded (smooth skin surface) (CFC-12 exp.)	1.8–3.5	0.20	—	5.00	—	0.29
Expanded polystyrene, extruded (smooth skin surface) (HCFC-142b exp.)[h]	1.8–3.5	0.20	—	5.00	—	0.29

Figure 23-6. This table shows the R values for many common building materials. *Reprinted by permission of American Society of Heating, Refrigerating and Air-Conditioning Engineers.*

Typical Thermal Properties of Common Building and Insulating Materials—Design Values[a] (Continued)

Description	Density, lb/ft³	Conductivity[b] (k), Btu·in / h·ft²·°F	Conductance (C), Btu / h·ft²·°F	Resistance[c] (R) Per Inch Thickness (1/k), °F·ft²·h / Btu·in	Resistance[c] (R) For Thickness Listed (1/C), °F·ft²·h / Btu	Specific Heat, Btu / lb·°F
Expanded polystyrene, molded beads	1.0	0.26	—	3.85	—	—
	1.25	0.25	—	4.00	—	—
	1.5	0.24	—	4.17	—	—
	1.75	0.24	—	4.17	—	—
	2.0	0.23	—	4.35	—	—
Cellular polyurethane/polyisocyanurate[i]						
(CFC-11 exp.) (unfaced)	1.5	0.16–0.18	—	6.25–5.56	—	0.38
Cellular polyisocyanurate[i]						
(CFC-11 exp.)(gas-permeable facers)	1.5–2.5	0.16–0.18	—	6.25–5.56	—	0.22
Cellular polyisocyanurate[i]						
(CFC-11 exp.)(gas-impermeable facers)	2.0	0.14	—	7.04	—	0.22
Cellular phenolic (closed cell)(CFC-11, CFC-113 exp.)	3.0	0.12	—	8.20	—	—
Cellular phenolic (open cell)	1.8–2.2	0.23	—	4.40	—	—
Mineral fiber with resin binder	15.0	0.29	—	3.45	—	0.17
Mineral fiberboard, wet felted						
Core or roof insulation	16–17	0.34	—	2.94	—	—
Acoustical tile	18.0	0.35	—	2.86	—	0.19
Acoustical tile	21.0	0.37	—	2.70	—	—
Mineral fiberboard, wet molded						
Acoustical tile[k]	23.0	0.42	—	2.38	—	0.14
Wood or cane fiberboard						
Acoustical tile,[k] ....0.5 in.	—	—	0.80	—	1.25	0.31
Acoustical tile[k] ....0.75 in.	—	—	0.53	—	1.89	—
Interior finish (plank, tile)	15.0	0.35	—	2.86	—	0.32
Cement fiber slabs (shredded wood with Portland cement binder)	25–27.0	0.50–0.53	—	2.0–1.89	—	—
Cement fiber slabs (shredded wood with magnesia oxysulfide binder)	22.0	0.57	—	1.75	—	0.31
Loose Fill						
Cellulosic insulation (milled paper or wood pulp)	2.3–3.2	0.27–0.32	—	3.70–3.13	—	0.33
Perlite, expanded	2.0–4.1	0.27–0.31	—	3.7–3.3	—	0.26
	4.1–7.4	0.31–0.36	—	3.3–2.8	—	—
	7.4–11.0	0.36–0.42	—	2.8–2.4	—	—
Mineral fiber (rock, slag, or glass)[g]						
approx. 3.75–5 in.	0.6–2.0	—	—	—	11.0	0.17
approx. 6.5–8.75 in.	0.6–2.0	—	—	—	19.0	—
approx. 7.5–10 in.	0.6–2.0	—	—	—	22.0	—
approx. 10.25–13.75 in.	0.6–2.0	—	—	—	30.0	—
Mineral fiber (rock, slag, or glass)[g]						
approx. 3.5 in. (closed sidewall application)	2.0–3.5	—	—	—	12.0–14.0	—
Vermiculite, exfoliated	7.0–8.2	0.47	—	2.13	—	0.32
	4.0–6.0	0.44	—	2.27	—	—
Spray Applied						
Polyurethane foam	1.5–2.5	0.16–0.18	—	6.25–5.56	—	—
Ureaformaldehyde foam	0.7–1.6	0.22–0.28	—	4.55–3.57	—	—
Cellulosic fiber	3.5–6.0	0.29–0.34	—	3.45–2.94	—	—
Glass fiber	3.5–4.5	0.26–0.27	—	3.85–3.70	—	—
METALS (See Chapter 36, Table 3)						
ROOFING						
Asbestos-cement shingles	120	—	4.76	—	0.21	0.24
Asphalt roll roofing	70	—	6.50	—	0.15	0.36
Asphalt shingles	70	—	2.27	—	0.44	0.30
Built-up roofing ....0.375 in.	70	—	3.00	—	0.33	0.35
Slate ....0.5 in.	—	—	20.00	—	0.05	0.30
Wood shingles, plain and plastic film faced	—	—	1.06	—	0.94	0.31
PLASTERING MATERIALS						
Cement plaster, sand aggregate	116	5.0	—	0.20	—	0.20
Sand aggregate ....0.375 in.	—	—	13.3	—	0.08	0.20
Sand aggregate ....0.75 in.	—	—	6.66	—	0.15	0.20
Gypsum plaster:						
Lightweight aggregate ....0.5 in.	45	—	3.12	—	0.32	—
Lightweight aggregate ....0.625 in.	45	—	2.67	—	0.39	—
Lightweight aggregate on metal lath ....0.75 in.	—	—	2.13	—	0.47	—
Perlite aggregate	45	1.5	—	0.67	—	0.32
Sand aggregate	105	5.6	—	0.18	—	0.20
Sand aggregate ....0.5 in.	105	—	11.10	—	0.09	—
Sand aggregate ....0.625 in.	105	—	9.10	—	0.11	—
Sand aggregate on metal lath ....0.75 in.	—	—	7.70	—	0.13	—
Vermiculite aggregate	45	1.7	—	0.59	—	—
MASONRY MATERIALS						
Masonry Units						
Brick, fired clay	150	8.4–10.2	—	0.12–0.10	—	—
	140	7.4–9.0	—	0.14–0.11	—	—
	130	6.4–7.8	—	0.16–0.12	—	—
	120	5.6–6.8	—	0.18–0.15	—	—
	110	4.9–5.9	—	0.20–0.17	—	0.19

Figure 23-6. This table shows the R values for many common building materials (continued).

Typical Thermal Properties of Common Building and Insulating Materials—Design Values[a] (Continued)

Description	Density, lb/ft³	Conductivity[b] (k), Btu·in / h·ft²·°F	Conductance (C), Btu / h·ft²·°F	Resistance [c] (R) Per Inch Thickness (1/k), °F·ft²·h / Btu·in	Resistance [c] (R) For Thickness Listed (1/C), °F·ft²·h / Btu	Specific Heat, Btu / lb·°F
Brick, fired clay continued	100	4.2–5.1	—	0.24–0.20	—	—
	90	3.6–4.3	—	0.28–0.24	—	—
	80	3.0–3.7	—	0.33–0.27	—	—
	70	2.5–3.1	—	0.40–0.33	—	—
Clay tile, hollow						
1 cell deep .... 3 in.	—	—	1.25	—	0.80	0.21
1 cell deep .... 4 in.	—	—	0.90	—	1.11	—
2 cells deep .... 6 in.	—	—	0.66	—	1.52	—
2 cells deep .... 8 in.	—	—	0.54	—	1.85	—
2 cells deep .... 10 in.	—	—	0.45	—	2.22	—
3 cells deep .... 12 in.	—	—	0.40	—	2.50	—
Concrete blocks[i]						
Limestone aggregate						
8 in., 36 lb, 138 lb/ft³ concrete, 2 cores	—	—	—	—	—	—
Same with perlite filled cores	—	—	0.48	—	2.1	—
12 in., 55 lb, 138 lb/ft³ concrete, 2 cores	—	—	—	—	—	—
Same with perlite filled cores	—	—	0.27	—	3.7	—
Normal weight aggregate (sand and gravel)						
8 in., 33–36 lb, 126–136 lb/ft³ concrete, 2 or 3 cores	—	—	0.90–1.03	—	1.11–0.97	0.22
Same with perlite filled cores	—	—	0.50	—	2.0	—
Same with verm. filled cores	—	—	0.52–0.73	—	1.92–1.37	—
12 in., 50 lb, 125 lb/ft³ concrete, 2 cores	—	—	0.81	—	1.23	0.22
Medium weight aggregate (combinations of normal weight and lightweight aggregate)						
8 in., 26–29 lb, 97–112 lb/ft³ concrete, 2 or 3 cores	—	—	0.58–0.78	—	1.71–1.28	—
Same with perlite filled cores	—	—	0.27–0.44	—	3.7–2.3	—
Same with verm. filled cores	—	—	0.30	—	3.3	—
Same with molded EPS (beads) filled cores	—	—	0.32	—	3.2	—
Same with molded EPS inserts in cores	—	—	0.37	—	2.7	—
Lightweight aggregate (expanded shale, clay, slate or slag, pumice)						
6 in., 16–17 lb 85–87 lb/ft³ concrete, 2 or 3 cores	—	—	0.52–0.61	—	1.93–1.65	—
Same with perlite filled cores	—	—	0.24	—	4.2	—
Same with verm. filled cores	—	—	0.33	—	3.0	—
8 in., 19–22 lb, 72–86 lb/ft³ concrete,	—	—	0.32–0.54	—	3.2–1.90	0.21
Same with perlite filled cores	—	—	0.15–0.23	—	6.8–4.4	—
Same with verm. filled cores	—	—	0.19–0.26	—	5.3–3.9	—
Same with molded EPS (beads) filled cores	—	—	0.21	—	4.8	—
Same with UF foam filled cores	—	—	0.22	—	4.5	—
Same with molded EPS inserts in cores	—	—	0.29	—	3.5	—
12 in., 32–36 lb, 80–90 lb/ft³ concrete, 2 or 3 cores	—	—	0.38–0.44	—	2.6–2.3	—
Same with perlite filled cores	—	—	0.11–0.16	—	9.2–6.3	—
Same with verm. filled cores	—	—	0.17	—	5.8	—
Stone, lime, or sand						
Quartzitic and sandstone	180	72	—	0.01	—	—
	160	43	—	0.02	—	—
	140	24	—	0.04	—	—
	120	13	—	0.08	—	0.19
Calcitic, dolomitic, limestone, marble, and granite	180	30	—	0.03	—	—
	160	22	—	0.05	—	—
	140	16	—	0.06	—	—
	120	11	—	0.09	—	0.19
	100	8	—	0.13	—	—
Gypsum partition tile						
3 by 12 by 30 in., solid	—	—	0.79	—	1.26	0.19
3 by 12 by 30 in., 4 cells	—	—	0.74	—	1.35	—
4 by 12 by 30 in., 3 cells	—	—	0.60	—	1.67	—
Concretes						
Sand and gravel or stone aggregate concretes (concretes with more than 50% quartz or quartzite sand have conductivities in the higher end of the range)	150	10.0–20.0	—	0.10–0.05	—	—
	140	9.0–18.0	—	0.11–0.06	—	0.19–0.24
	130	7.0–13.0	—	0.14–0.08	—	—
Limestone concretes	140	11.1	—	0.09	—	—
	120	7.9	—	0.13	—	—
	100	5.5	—	0.18	—	—
Gypsum-fiber concrete (87.5% gypsum, 12.5% wood chips)	51	1.66	—	0.60	—	0.21
Cement/lime, mortar, and stucco	120	9.7	—	0.10	—	—
	100	6.7	—	0.15	—	—
	80	4.5	—	0.22	—	—
Lightweight aggregate concretes						
Expanded shale, clay, or slate; expanded slags; cinders; pumice (with density up to 100 lb/ft³); and scoria (sanded concretes have conductivities in the higher end of the range)	120	6.4–9.1	—	0.16–0.11	—	—
	100	4.7–6.2	—	0.21–0.16	—	0.20
	80	3.3–4.1	—	0.30–0.24	—	0.20
	60	2.1–2.5	—	0.48–0.40	—	—
	40	1.3	—	0.78	—	—

Figure 23-6. This table shows the R values for many common building materials (continued).

Typical Thermal Properties of Common Building and Insulating Materials—Design Values[a] *(Concluded)*

Description	Density, lb/ft³	Conductivity[b] (k), Btu·in / h·ft²·°F	Conductance (C), Btu / h·ft²·°F	Resistance[c] (R) Per Inch Thickness (1/k), °F·ft²·h / Btu·in	Resistance[c] (R) For Thickness Listed (1/C), °F·ft²·h / Btu	Specific Heat, Btu / lb·°F
Perlite, vermiculite, and polystyrene beads	50	1.8–1.9	—	0.55–0.53	—	—
	40	1.4–1.5	—	0.71–0.67	—	0.15–0.23
	30	1.1	—	0.91	—	—
	20	0.8	—	1.25	—	—
Foam concretes	120	5.4	—	0.19	—	—
	100	4.1	—	0.24	—	—
	80	3.0	—	0.33	—	—
	70	2.5	—	0.40	—	—
Foam concretes and cellular concretes	60	2.1	—	0.48	—	—
	40	1.4	—	0.71	—	—
	20	0.8	—	1.25	—	—
SIDING MATERIALS (on flat surface)						
Shingles						
Asbestos-cement	120	—	4.75	—	0.21	—
Wood, 16 in., 7.5 exposure	—	—	1.15	—	0.87	0.31
Wood, double, 16-in., 12-in. exposure	—	—	0.84	—	1.19	0.28
Wood, plus insul. backer board, 0.3125 in.	—	—	0.71	—	1.40	0.31
Siding						
Asbestos-cement, 0.25 in., lapped	—	—	4.76	—	0.21	0.24
Asphalt roll siding	—	—	6.50	—	0.15	0.35
Asphalt insulating siding (0.5 in. bed.)	—	—	0.69	—	1.46	0.35
Hardboard siding, 0.4375 in.	—	—	1.49	—	0.67	0.28
Wood, drop, 1 by 8 in.	—	—	1.27	—	0.79	0.28
Wood, bevel, 0.5 by 8 in., lapped	—	—	1.23	—	0.81	0.28
Wood, bevel, 0.75 by 10 in., lapped	—	—	0.95	—	1.05	0.28
Wood, plywood, 0.375 in., lapped	—	—	1.59	—	0.59	0.29
Aluminum or Steel[m], over sheathing						
Hollow-backed	—	—	1.61	—	0.61	0.29
Insulating-board backed nominal 0.375 in.	—	—	0.55	—	1.82	0.32
Insulating-board backed nominal 0.375 in., foil backed	—	—	0.34	—	2.96	—
Architectural (soda-lime float) glass	158	6.9	—	—	—	0.21
WOODS (12% moisture content)[e,n]						
Hardwoods						0.39[o]
Oak	41.2–46.8	1.12–1.25	—	0.89–0.80	—	—
Birch	42.6–45.4	1.16–1.22	—	0.87–0.82	—	—
Maple	39.8–44.0	1.09–1.19	—	0.92–0.84	—	—
Ash	38.4–41.9	1.06–1.14	—	0.94–0.88	—	—
Softwoods						0.39[o]
Southern Pine	35.6–41.2	1.00–1.12	—	1.00–0.89	—	—
Douglas Fir-Larch	33.5–36.3	0.95–1.01	—	1.06–0.99	—	—
Southern Cypress	31.4–32.1	0.90–0.92	—	1.11–1.09	—	—
Hem-Fir, Spruce-Pine-Fir	24.5–31.4	0.74–0.90	—	1.35–1.11	—	—
West Coast Woods, Cedars	21.7–31.4	0.68–0.90	—	1.48–1.11	—	—
California Redwood	24.5–28.0	0.74–0.82	—	1.35–1.22	—	—

Figure 23-6. This table shows the R values for many common building materials (continued).

The resistance to heat flow from the film on the outside is dependent on the wind speed and is normally lower than the inside film factor. These two film factors can be expressed in the form of resistances:

- The inside film resistance is between .68 and 1.47 (average of 1.1).
- The outside film resistance for a wind speed of 7.5 mph is .25.

These two resistances are included when the total resistance of a wall is calculated. An example of how to calculate the R_t for a wall can be seen in Figure 23-7. This is a typical masonry wall using brick, polystyrene insulation, and concrete block with plaster on the inside. The U factor for this wall is 0.107 or 0.11. If the outside design temperature is 95°F and the inside design temperature is 75°F, the temperature difference through this wall would be 20°F. If the wall has no windows, is 30 feet long, and has an 8 foot ceiling, it will have a total area of 240 square feet (30 × 8 = 240). This can all be inserted into the formula for total heat gain:

$$Q = A \times U \times TD$$
$$Q = 240 \times 0.11 \times 20$$
$$Q = 528 \text{ Btuh}$$

In new construction, the R value may also be used to try various building materials in a composite wall to determine the best insulation for the money. For example, suppose a customer is building a house and would like to know the benefits of using two inches of polystyrene

HEAT TRANSFER THROUGH BUILDING MATERIALS

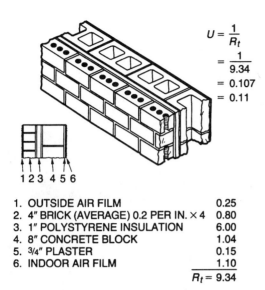

Figure 23-7. Calculating the R_t for a masonry wall.

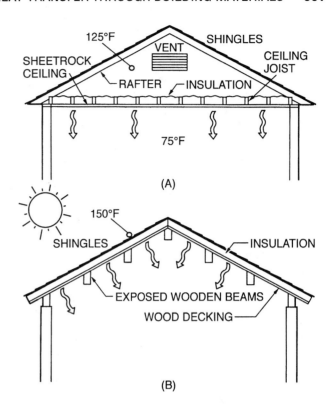

Figure 23-8. Heat transfer through a ceiling from an attic (A) or directly through a composite roof/ceiling combination (B).

insulation versus using only one inch. Polystyrene offers far more resistance than any of the other building materials in the structure, and adding another inch will greatly improve the efficiency. The R_t with the added insulation would be 15.34 (9.34 + 6 = 15.34). When this is converted to a U factor, it becomes .065. Notice that the larger the R_t, the smaller the U factor and the smaller the U factor, the less heat loss through the wall. This makes it easier for the public to understand the significance of the R value for insulation—bigger is better.

The wall will now transfer less heat:

$$Q = A \times U \times TD$$
$$Q = 240 \times .065 \times 20$$
$$Q = 312 \text{ Btuh}$$

For the addition of one inch of polystyrene, the wall transfers only 59% of the original heat (312 ÷ 528 = .59, the decimal equivalent of 59%). This represents a savings of 41% on the cooling bill as applied to this wall (100% − 59% = 41%). The whole house can be evaluated in this manner by taking it one wall at a time.

If the above wall was located between an enclosed garage and the conditioned space of the house, it would be treated as a partition. The same temperature difference would not be used because the inside of the garage would not be as hot as the outside air. For practical purposes, the inside temperature of the garage can be calculated as the average temperature between the inside and the outside, or 10°F (20 ÷ 2 = 10).

Heat Gain through Roofs and Ceilings

Heat is also gained by conduction through the top of the structure. The ceiling of the structure may be connected to a vented attic or it may terminate on the roof itself (Figure 23-8). Neither type will provide much resistance to heat transfer without insulation, which is the key to an efficient ceiling or roof. It is common to use a large amount of insulation to produce a very high R_t (30 to 50) because the heat above the ceiling creates a great deal of heat inside the structure. The roof is like a solar collector, storing heat in the air at the top of the house. It is important to keep the ceiling as cool as practical. A high R_t for a roof or ceiling is also more efficient for the heating season.

The ceiling heat gain is calculated just like the wall gain. Suppose the house has a floor area of 30 feet × 40 feet, and a ceiling with an R_t of 30, and a design temperature difference of 50°F. The U factor would be .033 (U = 1 ÷ R_t). The heat transfer through the ceiling is calculated as follows:

$$Q = A \times U \times TD$$
$$Q = 1200 \times .033 \times 50$$
$$Q = 1980 \text{ Btuh}$$

If the R_t for the ceiling was only 15, it would transfer much more heat to the conditioned space. Insulating a ceiling is something the homeowner can usually do as a home improvement job.

Heat Gain through Floors

Many houses are located over crawl spaces or basements. If the crawl space is not vented (or the vents are closed in summer), it is treated just like a basement.

No heat is transferred. If the vents are opened in summer, outside air will circulate and the temperature difference will be the same as the design temperature difference. For example, if the outdoor-to-indoor temperature difference is 20°F, this would be used for the floor. Sometimes, the crawl space vents are opened to keep the underside of the structure dry. If the homeowner intends to leave these vents open, the extra load should be accounted for in the heat gain calculation.

When a house is located on a slab with the slab on top of the ground, no heat is gained through the floor of the structure.

Heat Gain through Windows and Doors

The heat gain through windows and doors is one of the major load factors in a structure. These openings are connected directly to the outside air and are often on hinges or tracks. They can be readily opened and it is hard to seal the heat out of the structure.

The basic window is a single glaze (commonly called a *pane*). There is not much resistance to heat transfer using a single-glaze window. The U factor for plain glass in a typical single-glaze window is 1.1. Double-glaze windows are windows with two pieces of glass separated by a dead vapor space (Figure 23-9). Triple-glaze windows use three panes of glass and two vapor spaces. The vapor in the space may be dry nitrogen or some other inert gas, such as argon. Triple-glaze windows are the most efficient, but they are also the most expensive. If a structure already has single-glaze windows, it will be very expensive to replace them, so many homeowners add storm windows to improve the efficiency (Figure 23-10). The addition of a storm window will provide roughly the same resistance as a double-glaze window.

When the heat gain of a window is calculated, the U factor is based on the rough window opening, including the framing (Figure 23-11). Since there are different materials in the framing, each type of window has a different U factor. Figure 23-12 shows part of the ASHRAE table for windows. A triple-glaze window in the R (residential) column (C is for commercial windows) with ¼ inch between the glazes and a wood or vinyl frame has a U factor of .35. The same window with a common aluminum frame has a U factor of .72. If the aluminum frame window has a thermal break (insulation in the frame), the U factor is .50.

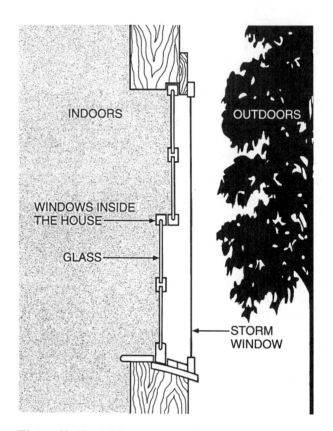

Figure 23-10. Adding a storm window will improve the efficiency.

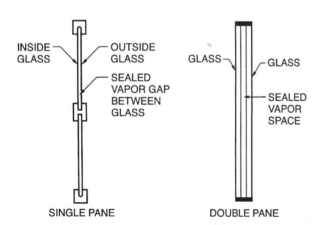

Figure 23-9. A double-glaze window uses two panes of glass.

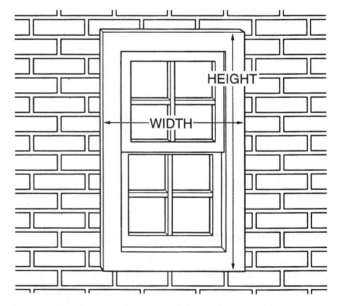

Figure 23-11. To measure a window, use the window frame opening.

Overall Coefficients of Heat Transmission of Various Fenestration Products (Concluded)

Part A: U-Values for Vertical Installation[a], Btu/h·ft²·°F

Glazing Type[b]	Glass Only		Aluminum Frame no thermal break ($U_f = 1.9$)		Aluminum Frame thermal break ($U_f = 1.0$)		Wood or Vinyl Frame ($U_f = 0.4$)	
	Center of Glass	Edge[c] of Glass	Product[d] Type R	C	Product[d] Type R	C	Product[d] Type R	C
Triple glass or double glass with polyester film suspended in between, $\varepsilon = 0.15$ on surface 2, 3, 4, or 5								
1/4 in. argon spaces	(0.27)	(0.46)	0.72	0.54	0.50	0.41	0.35	0.32
3/8 in. argon spaces	(0.22)	(0.45)	0.69	0.51	0.47	0.37	0.33	0.29
1/2 in. and greater argon spaces	(0.20)	(0.44)	0.68	0.50	0.46	0.36	0.31	0.28
Triple glass or double glass with polyester film suspended in between, $\varepsilon = 0.15$ on surfaces 2 or 3 and 4 or 5								
1/4 in. argon spaces	(0.22)	(0.45)	0.69	0.51	0.47	0.37	0.32	0.29
3/8 in. argon spaces	(0.17)	(0.43)	0.66	0.47	0.44	0.34	0.30	0.25
1/2 in. and greater argon spaces	(0.15)	(0.43)	0.65	0.46	0.43	0.32	0.29	0.24

Part B: U-Value Conversion Table for Sloped and Horizontal Glazing for Upward Heat Flow

Slope	U-Value, Btu/h·ft²·°F												
90° (vertical)	0.10	0.20	0.30	0.40	0.50	0.60	0.70	0.80	0.90	1.00	1.10	1.20	1.30
45°	0.14	0.25	0.36	0.47	0.57	0.68	0.79	0.90	1.00	1.11	1.22	1.33	1.44
0 (horiz.)	0.19	0.29	0.40	0.51	0.61	0.72	0.82	0.93	1.04	1.14	1.25	1.35	1.46

[a]All U-values are based on standard ASHRAE winter conditions of 70°F indoor and 0°F outdoor air temperature with 15 mph outdoor air velocity and zero solar flux. The outside surface coefficient at these conditions is approximately 5.1 Btu/hr·ft²·°F, depending on the glass surface temperature. With the exception of single glazing, small changes in the interior and exterior temperatures do not significantly affect overall U-values.

[b]Glazing layer surfaces are numbered from the outside to the inside. Double and triple refer to the number of glazing lites. All data are based on 1/8 in. glass unless otherwise noted. Thermal conductivities are: 0.53 Btu/h·ft·°F for glass, and 0.11 Btu/h·ft·°F for acrylic and polycarbonate.

[c]Based on aluminum spacers data. Edge of glass effect assumed to extend over the 2.5 in. band around perimeter of each glazing unit.

[d]Product types described elsewhere.

Figure 23-12. U factors for triple-glaze windows. *Reprinted by permission of American Society of Heating, Refrigerating and Air-Conditioning Engineers.*

You can see from this comparison that better windows provide significant energy savings. This is especially true in very hot climates, where the temperature difference is much higher. For example, suppose a room has two wood-framed, triple-glaze windows that are 5 feet high and 2.5 feet wide and the design temperature difference is 20°F. The area is 5 × 2.5 × 2 = 25 square feet. The TD is 20°F.

$Q = A \times U \times TD$

$Q = 25 \times .35 \times 20$

$Q = 175$ Btuh

This does not seem like much heat until you compare an equal wall area. Remember, the wall with two inches of polystyrene had a U factor of .065. The heat transfer rate for the same area of wall is as follows:

$Q = A \times U \times TD$

$Q = 25 \times .065 \times 20$

$Q = 32.5$ Btuh

This example shows the thermal difference between a wall and a window. The window transfers 5.4 times as much heat as the same area of wall (175 ÷ 32.5 = 5.4). When you consider that the window area of many homes is equal to 15% of the total wall area, it is easy to see why windows represent such a large portion of the total heat gain.

The area of a door is calculated in the same way as a window, using the outside dimensions of the framed opening. Outside doors may be constructed of wood, either solid or with a core of insulation, hollow metal or metal with a core of insulation, or glass. Many doors are combinations of two doors with the outer door called a *storm door*. A storm door serves much the same purpose as a storm window.

23.4 SOLAR HEAT GAIN

Up to this point, our discussion has dealt with conduction heat gains only, or heat transfer between the outside air and the inside air. Radiant heat gain due to solar heat is added to the conduction heat gain for a structure. When the sun shines on a building material, it absorbs heat. The rate at which it absorbs heat depends on several factors: the color and texture of the material and the angle of the sun.

The color of the building material has a significant impact on how much heat is absorbed into the structure. Dark materials absorb heat much faster than light materials. For this reason, many homes in northern climates use dark shingles for maximum heat gain in the winter months, while those in southern climates are often made of a lighter material to minimize heat gain in the summer.

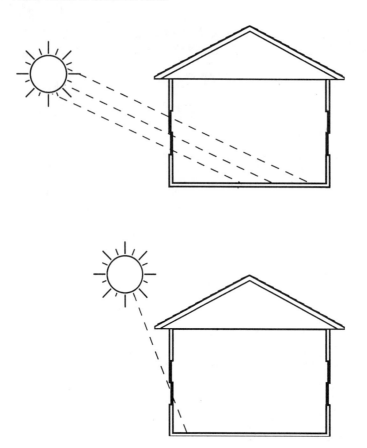

Figure 23-13. The sun's angle determines how much heat will be absorbed into the window.

The texture of the building material can also make a difference in the amount of solar heat absorbed. Rough materials absorb more heat than smooth materials.

When the sun is at a 90° angle to the material, it is shining straight down on it and the greatest amount of heat will be absorbed (Figure 23-13). For outside walls, this occurs in the mornings on eastern exposures and in the afternoons on western exposures. Fortunately, the sun is the farthest from the wall at these times and there is a chance that shade from another structure will help to minimize the heat gain. As the sun rises, it shines on the roof of the structure from midmorning to midafternoon when the sun is closest to the structure. If the structure has an attic, it can be ventilated using an exhaust fan. This will help to lower the attic temperature. If the structure has a ceiling/roof combination, good insulation is the only barrier.

Solar tables can be used to determine how the solar load affects the space temperature. These tables cover the time of day, latitude, and which direction the wall faces for a complete picture of how much solar heat gain can be expected. The complete solar table is quite extensive, but one page is shown in Figure 23-14. This table includes the following information:

- Direction the wall faces.
- Color of the wall D (dark) or L (light).
- Latitude (40° north or south).
- Shade, such as awnings or no shade. A north wall is always considered shaded.
- Type of building material.
- Temperature difference to be used for each different condition.
- Time of day.
- Day of year (the example page covers May 21 and July 23).

For example, suppose that a frame wall is dark in color, faces west, and is located at a northern latitude of 40°. The wall is 10 feet high, 40 feet long, and has a U factor of .08. What would be the highest solar heat gain for this wall on May 21?

In the table, we look across the frame wall section of west-facing walls and find that at 6 pm, this wall has a temperature difference of 50°F. When we apply this to the formula, we find that the wall has a heat gain of 1600 Btuh at this time of day.

$$Q = A \times U \times TD$$
$$Q = 400 \times .08 \times 50$$
$$Q = 1600 \text{ Btuh}$$

Note that this only applies to this particular exposure at this specific time of day. For example, the eastern exposure of this building has only a 16°F TD at 6 pm.

For commercial buildings, the designer must choose which time of day will have the largest impact on the expected building occupancy patterns. For example, a restaurant that will have dinner customers but is not open for breakfast should be designed to meet the solar heat gain that occurs in the late afternoon.

Windows are another source of solar heat gain. There are two types of solar radiation that affect windows: direct and indirect. Direct solar radiation is when the sun shines directly into the window and heats the conditioned space almost immediately as it shines on the interior. Indirect solar radiation is the light that bounces off of

Total Equivalent Temperature Differentials for Walls for May 21 and July 23*

NORTH LATITUDE		A.M. 8		10		12		P.M. 2		4		6		8		10		12		SOUTH LATITUDE
		D	L	D	L	D	L	D	L	D	L	D	L	D	L	D	L	D	L	
FRAME
	NE	24	12	26	14	16	12	14	12	16	16	16	16	12	12	8	6	4	4	SE
	E	32	16	38	20	34	18	14	14	16	16	16	16	12	12	8	8	4	4	E
	SE	15	8	28	18	30	20	26	18	18	16	16	16	12	12	8	6	4	4	NE
	S	-2	-2	6	2	24	14	32	22	28	22	18	16	12	12	8	8	4	4	N
	SW	-2	-2	2	0	8	6	28	24	42	30	44	30	26	22	8	6	4	4	NW
	W	-2	-2	2	2	8	8	22	14	42	30	(50)	36	24	24	10	10	4	4	W
	NW	-2	-2	2	0	8	8	14	12	26	22	42	28	36	26	8	6	4	4	SW
(SHADE)	N	-2	-2	0	0	6	6	12	12	16	16	14	14	10	10	6	6	2	2	S (SHADE)
4" BRICK OR STONE VENEER + FRAME
	NE	0	-2	26	14	22	12	12	8	14	12	16	14	14	14	12	12	8	6	SE
	E	4	2	32	16	33	19	16	16	14	14	16	16	14	14	12	10	8	8	E
	SE	4	0	22	12	30	18	28	18	20	16	16	16	14	14	12	10	8	8	NE
	S	-2	-2	0	0	14	8	26	18	28	20	22	18	14	14	10	10	6	6	N
	SW	2	0	2	0	4	4	14	10	34	24	38	28	36	26	12	10	8	8	NW
	W	2	0	2	2	6	4	12	10	28	20	42	30	44	30	18	16	8	8	W
	NW	-2	-2	0	0	4	4	10	8	14	14	32	24	36	26	14	12	8	8	SW
(SHADE)	N	-2	-2	0	0	2	2	8	8	12	12	14	14	10	10	6	6	4	4	S (SHADE)
8" HOLLOW TILE OR 8" CINDER BLOCK
	NE	2	2	2	2	22	12	18	12	12	8	14	12	14	12	14	12	10	10	SE
	E	6	4	14	6	26	14	28	16	22	14	16	14	16	12	12	10	10	10	E
	SE	4	2	4	2	18	10	22	14	22	16	16	14	16	14	14	12	10	8	NE
	S	2	2	2	2	4	2	14	8	26	16	28	18	22	16	14	12	10	8	N
	SW	4	2	4	2	4	2	6	6	14	12	28	20	32	22	28	20	10	10	NW
	W	6	4	6	4	6	4	8	6	12	10	20	16	32	24	34	24	20	16	W
	NW	2	2	2	2	4	2	6	4	10	8	14	12	24	20	32	24	12	10	SW
(SHADE)	N	0	0	0	0	0	0	2	2	8	8	12	12	12	12	12	12	8	8	S (SHADE)
8" BRICK OR 12" HOLLOW TILE OR 12" CINDER BLOCK
	NE	4	4	4	4	12	4	18	10	16	10	12	8	12	10	12	12	12	10	SE
	E	10	8	10	8	16	10	20	12	20	12	16	10	16	12	16	12	14	12	E
	SE	10	6	8	6	8	6	16	12	20	14	16	14	16	14	14	12	14	12	NE
	S	6	4	6	4	6	4	6	4	12	8	18	12	18	14	14	12	12	10	N
	SW	10	6	8	6	8	6	10	6	12	8	14	10	22	14	26	18	22	16	NW
	W	10	6	8	6	8	6	8	6	12	8	16	10	22	18	26	18	26	18	W
	NW	4	4	4	4	4	4	6	4	8	6	10	8	12	10	18	16	20	16	SW
(SHADE)	N	2	2	2	2	2	2	2	2	4	4	8	8	10	10	10	10	8	8	S (SHADE)
12" BRICK
	NE	10	8	10	8	10	6	10	6	12	6	14	8	14	8	12	8	12	8	SE
	E	14	10	14	10	14	10	12	8	14	10	16	12	16	12	16	10	16	10	E
	SE	12	8	12	8	12	8	12	8	12	8	14	10	16	12	16	12	14	12	N
	S	10	8	10	8	8	6	8	6	8	6	10	6	12	8	14	10	14	10	NE
	SW	12	8	12	8	12	8	12	8	12	8	12	8	14	10	16	12	16	12	NW
	W	14	10	14	10	14	10	12	8	12	8	12	8	12	8	14	10	18	12	W
	NW	10	8	10	8	10	8	10	6	10	6	10	6	10	8	12	8	12	8	SW
(SHADE)	N	6	6	4	4	4	4	4	4	4	4	4	4	4	4	6	6	8	8	S (SHADE)
8" CONCRETE OR STONE OR 6" OR 8" CONCRETE BLOCK
	NE	8	6	6	4	6	4	6	4	12	8	16	12	18	14	16	12	12	10	SE
	E	8	6	16	10	26	14	26	14	20	12	16	12	16	12	14	12	12	10	E
	SE	8	4	8	6	18	12	20	14	20	14	16	14	14	12	14	12	12	10	N
	S	4	3	4	3	6	3	14	8	18	14	20	14	16	14	12	10	10	8	NE
	SW	8	4	6	4	6	4	10	6	16	12	24	16	26	18	24	18	16	10	NW
	W	8	6	8	6	8	6	10	6	14	10	22	16	30	20	28	20	16	12	W
	NW	6	4	6	2	6	4	6	6	8	8	14	12	22	16	24	18	10	8	SW
(SHADE)	N	2	2	2	2	2	2	2	2	4	4	6	6	8	8	10	10	48	8	S (SHADE)
12" CONCRETE OR STONE
	NE	8	6	8	6	8	4	16	10	16	10	12	10	12	10	14	12	12	10	SE
	E	12	8	10	8	12	8	20	12	20	14	18	12	18	12	16	12	16	12	E
	SE	10	6	10	6	10	6	16	10	18	12	16	12	14	12	14	12	14	12	N
	S	8	6	6	6	6	4	6	4	12	8	16	12	18	14	16	12	12	10	NE
	SW	10	6	10	6	8	6	8	6	10	8	12	10	20	16	22	16	20	14	NW
	W	12	8	10	8	10	8	10	6	8	6	14	10	18	12	18	16	16	12	W
	NW	8	6	8	4	8	4	6	6	6	6	10	8	12	10	20	14	22	16	SW
(SHADE)	N	2	2	2	2	2	2	2	2	4	4	6	6	8	8	10	10	8	8	S (SHADE)

*Table is for 40 degrees North latitude. It may be used for 40 degrees South latitude for January 21 and November 21.

Figure 23-14. Page from a solar heat gain table. *Courtesy Trane Company.*

366 UNIT 23 HEAT GAIN CALCULATIONS

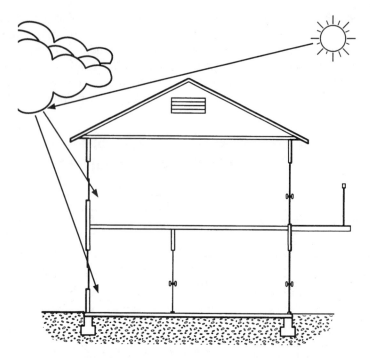

Figure 23-15. Indirect solar radiation.

clouds and other light-colored objects and then shines in the windows (Figure 23-15). The north side of a structure receives only indirect solar radiation because the sun never shines directly on the northern exposure of a building in the northern hemisphere.

Since windows represent such a large source of the heat gain, there has been much research into various ways to offset the solar gain through windows. Standard windows (even with double or triple panes) do not offer much resistance to the sun shining through unless they are treated with special coatings that help to reflect the sun's light and heat. Some windows with multiple panes have a coating on more than one surface and still offer good visibility.

Figure 23-16 shows a solar table for May 21 listing direct and diffuse solar radiation values for plain windows. Note that the western exposure has its greatest heat transfer at 4 pm, and it is expressed in Btuh rather than TD. The heat transfer for this window is 176.5 Btuh per square foot of gross window area. If this window was located in the above wall and was 4 feet × 10 feet or 40 square feet in area, the total heat gain would be 7060

Direct Solar and Diffuse Sky Radiation for Single Common Window Glass for May 21 and July 23*
(Btuh per sq ft of masonry opening)

TIME OF YEAR	SUN TIME	DIRECTIONS FOR NORTH LATITUDE (READ DOWN)									TIME OF YEAR
		N	NE	E	SE	S	SW	W	NW	HORIZ	
MAY 21 & JULY 23 NORTH LATITUDE	6 AM	26.3	109.2	125.0	60.6	7.2	6.4	5.5	5.2	24.6	NOV. 21 & JAN. 21 SOUTH LATITUDE
	7 AM	15.2	130.7	170.7	103.4	12.4	11.0	9.7	8.8	74.0	
	8 AM	10.1	110.5	176.5	129.4	17.8	14.4	13.0	11.4	122.4	
	9 AM	11.0	73.8	153.9	133.7	33.9	16.9	15.5	13.3	169.2	
	10 AM	11.4	32.4	114.2	121.6	54.3	18.2	16.7	14.2	202.3	
	11 AM	11.6	15.3	52.3	94.1	73.5	21.7	17.4	14.5	224.4	
	12 NOON	11.6	14.5	17.4	52.1	78.9	52.1	17.4	14.5	232.0	
	1 PM	11.6	14.5	17.4	21.7	73.5	94.1	52.3	15.3	224.4	
	2 PM	11.4	14.2	16.7	18.2	54.3	121.6	114.2	32.4	202.3	
	3 PM	11.0	13.3	15.5	16.9	33.9	133.7	153.9	73.8	169.2	
	4 PM	10.1	11.4	13.0	14.4	17.8	129.4	(176.5)	110.5	122.4	
	5 PM	15.2	8.8	9.7	11.0	12.4	103.4	170.7	130.7	74.0	
	6 PM	26.3	5.2	5.5	6.4	7.2	60.6	125.0	109.2	24.6	
		S	SE	E	NE	N	NW	W	SW	HORIZ	TIME OF YEAR
	DIRECTIONS FOR SOUTH LATITUDE (READ UP)										

*Table is for 40 degrees North latitude. It may be used for 40 degrees South latitude on January 21 and November 21 by reading up from the bottom.

Figure 23-16. At 4 pm, the western exposure has the greatest heat gain. *Courtesy Trane Company.*

Btuh (176.5 × 40 square feet = 7060). Remember, the complete wall without the window only had a heat gain of 1600 Btuh. This one window has 4.4 times the heat gain of a much larger wall (7060 ÷ 1600 = 4.4). It is easy to see why it would be beneficial to do something about this window; by itself, it represents more than half a ton of cooling. There are several methods for reducing the heat gain through a window, including adding an awning, tinted glass, a shade tree, drapes, or venetian blinds. If any of these are added to the window, the heat gain will be reduced significantly.

23.5 HEAT GAIN DUE TO INFILTRATION AND VENTILATION

We have talked about heat gains due to conduction and solar heat (radiation). Heat can also be gained through infiltration or ventilation.

Infiltration is air that enters the structure due to prevailing winds or breezes. When the air impinges on one side of a structure, it creates a slight positive pressure on that side and when it passes over the structure, it creates a slight negative pressure on the other side of the structure (Figure 23-17). This difference in pressure causes air to pass through the structure. It enters any available opening or crack in the structure and exits the same way. Actually, air enters one side and exits the other. This may be called *infiltration* on one side and *exfiltration* on the other.

There are many places where air can enter and escape a structure. Some of these include the following:

1. *Around windows and doors.* Some older windows and doors have very large cracks around them. You can tell an old window if it has weights to counterbalance the weight of the window.

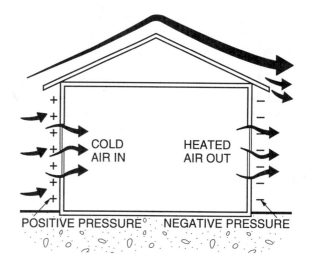

Figure 23-17. How air infiltrates into a structure.

2. *Where plumbing fixtures enter the structure if above grade.* This will only be seen in warmer climates. In colder climates, the plumbing must be well below the frost line or it will freeze. If the plumbing runs in the outside walls, there is a space around the pipes or vents for air to travel by the pipes. It is uncommon to find plumbing in the outside walls in colder climates; however, the vent pipes penetrate the ceiling into the attic and air can travel via this route if it is not sealed (Figure 23-18).

3. *Where electrical wires penetrate the structure.* For example, a porch light may have an inside switch. Air can enter the light fixture and exit via the switch or even a plug-in receptacle on an outside wall.

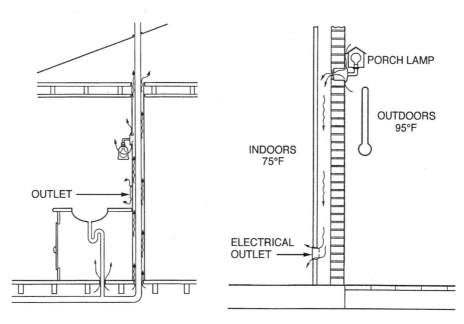

Figure 23-18. Infiltration around plumbing and electrical fixtures.

It should be remembered that a structure should not be too tight or internal pollution will occur. This has created a great concern for indoor air quality, often called *IAQ*. The materials used in building the structure will release chemicals for a long time after the construction is complete. Carpet, drapes, upholstered furniture, paints, and wall coverings all release particles into the air. In addition, personal hygiene products, cleaning solvents, and other sources, such as cooking, are among some of the common indoor pollutants. This release of pollutants is called *off-gassing*.

In a structure that is too tight, the oxygen level in the structure is reduced and replaced with the CO_2 expelled when people exhale. CO_2 buildup occurs in buildings with high occupancies for long periods of time, such as office buildings. Internal pollution is one of today's big issues. In some urban areas, many buildings have been found to contain indoor air with more pollutants than the outdoor air.

Another indoor air pollutant that has received much attention recently is radon gas that is released from the ground below the structure. This gas is known to cause cancer and is present in the soil under many houses (Figure 23-19). Many homes located in known radon gas localities must be designed to remove the gas. This is accomplished using a separate ventilation system under the house that does not affect the cooling system (Figure 23-20).

Residential Infiltration and Ventilation

Calculating the infiltration load on a residential structure is probably the most difficult part of a load calculation because there is no way of telling how much air will move through the structure without performing an actual

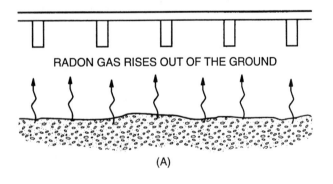

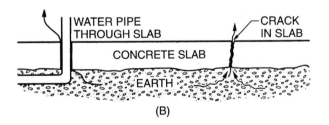

Figure 23-19. Radon gas pollution.

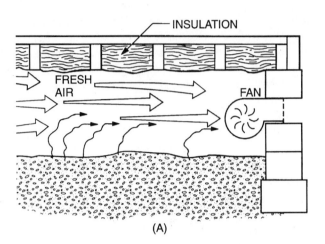

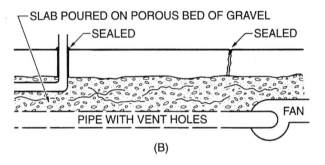

Figure 23-20. Venting a house to prevent radon gas pollution.

test. Some companies are using a test device that pushes air into the structure using a blower and then measures the entering air. The chimney or any flues will have to be plugged to prevent air from escaping at these points. The structure is pressurized to about .1 in. w.c. and the entering air is measured. A residence with good construction and all precautions taken to seal the structure will have about .3 to .5 air changes per hour. A structure constructed using common building practices may leak from .5 to 1 air change per hour. Poorly constructed or older homes will normally leak at a rate of more than 1 air change per hour. An air change means the total volume of air in the house is changed in one hour. For example, a house that is 30 feet × 40 feet has 1200 square feet. If the ceiling height is 8 feet, one air change is 9600 cubic feet per hour (30 × 40 × 8 = 9600). This is a leak rate of 160 cubic feet per minute (9600 ÷ 60 minutes/hour = 160 cfm). A very well designed house would leak at the rate of .5 air changes/hour (or less) which would be 4800 cubic feet/hour or 80 cfm.

Another method of estimating infiltration is to use the known passage of air through the cracks around the windows and doors. Modern windows and doors are tested using a standard wind velocity. The typical test calls for a wind velocity of 25 mph across the window or door and rates it as to how much air will leak through per foot of crack. A good window will typically leak .25 cfm per foot of crack. Poor windows may leak 1 cfm per foot of crack.

HEAT GAIN DUE TO INFILTRATION OR VENTILATION 369

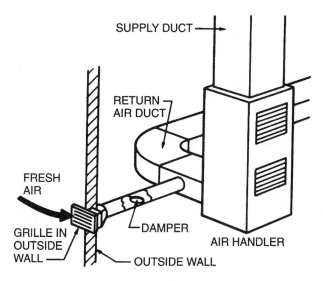

Figure 23-21. Ventilation air for a structure.

When the leak rate per foot of crack is known, the total feet of crack for the windows in the structure may be measured and divided by two (remember, air leaks in one side and out the other) to determine the total cfm of air leaking in for the structure.

Ventilation is the planned introduction of outside air to create a slight positive pressure in the structure. When it is determined that a house is too tight and not enough infiltration occurs, ventilation is often used to dilute the indoor air with the proper amount of outdoor air. This air typically enters the structure through the air conditioning system and is cooled before it enters the conditioned space (Figure 23-21). When there is more air induced into the house than will leak in, the air will leak out the cracks where infiltration would occur. When the ventilation rate is known, there is no infiltration calculation.

Commercial Building Ventilation

All large buildings require that a certain amount of fresh air be forced into the building for the purpose of preventing indoor air pollution. Some buildings have a suggested ventilation rate of 20 cfm per person. In the case of sports arenas, there are large crowds and large amounts of fresh air are required. Allowing smoking in buildings increases the fresh air requirements so many building owners do not allow it. This provides energy savings because it costs less to condition recirculated air than it does to condition outside ventilation air. ASHRAE studies the effects of indoor air pollution and provides periodic updates and recommendations. These guidelines are often adopted by local authorities.

Once the infiltration or ventilation rate is known, it must be determined how much heat must be removed by the cooling system to offset this heat gain.

There are two types of heat entering the structure by means of infiltration or ventilation: sensible heat and latent heat. As you will recall, sensible heat is heat that records directly on a thermometer as it causes a rise in space temperature. Latent heat is heat in the form of moisture or humidity that produces a change in state.

Sensible heat calculations are relatively easy using the sensible heat formula. For example, suppose the temperature difference for a structure is 20°F, and 160 cfm of air is estimated to leak into the structure at a rate of one air change/hour. The sensible heat gain (Q_s) may be calculated as follows:

$$Q_s = 1.08 \text{ (a constant)} \times \text{cfm} \times \text{TD}$$
$$Q_s = 1.08 \times 160 \times 20$$
$$Q_s = 3456 \text{ Btuh}$$

This represents the largest single heat gain we have encountered thus far. It can be seen why a tighter structure will be much less expensive to cool. If the house had only .5 air change/hour, the load would have been 1728 Btuh ($1.08 \times 80 \times 20 = 1728$ Btuh).

The latent heat gain (Q_l) for the structure is determined using the following formula:

$$Q_l = .68 \times \text{cfm} \times \text{grains difference}$$

where: .68 = a constant

grains difference = a value calculated from the psychrometric chart

This calculation is discussed in more detail in Unit 17, which should be reviewed if there are any questions. The above structure may be used as an example of how latent heat calculations are performed. Suppose that this structure is located in a city where the grains difference on a design day is 26 grains per pound of air. The latent heat calculation is as follows:

$$Q_l = .68 \times \text{cfm} \times \text{grains difference}$$
$$Q_l = .68 \times 160 \times 26$$
$$Q_l = 2828.8 \text{ Btuh}$$

The amount of latent versus sensible heat must be known by the system designer for the purpose of equipment sizing. This is called the *sensible heat ratio* (SHR). The SHR may be found using the following formula:

$$\text{SHR} = \text{sensible heat} \div \text{total heat}$$

For example, suppose a building has a heat gain of 26,000 Btuh sensible heat and 8,000 Btuh latent heat. This is a total heat gain of 34,000 Btuh. Using the SHR formula, we find:

$$\text{SHR} = \text{sensible heat} \div \text{total heat}$$
$$\text{SHR} = 26,000 \div 34,000$$
$$\text{SHR} = .76$$

This tells the designer/estimator that the air conditioning load on this structure is 76% sensible heat gain and 24% latent heat gain. The equipment selected must maintain the correct temperature and humidity for this space.

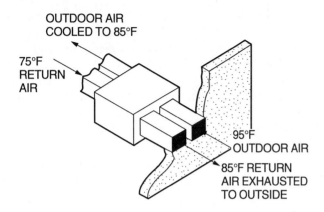

Figure 23-22. Air-to-air heat exchanger for exchanging heat between the inside and outside air.

Fresh air for ventilation is more expensive to condition than recirculated air, so other methods are being developed to minimize this extra load for both residential and commercial installations. A heat exchanger can be used to exchange heat between the inside air that is exhausted and the outside air that is brought in for ventilation. This can be accomplished by means of an air-to-air heat exchanger (Figure 23-22). These systems will become more commonplace as energy costs increase and more indoor air pollution is encountered.

23.6 HEAT GAIN THROUGH DUCTWORK

Another source of heat gain is ductwork routed through an unconditioned space. A cooling system will typically operate with a supply air temperature of 55°F. If the supply duct is routed through the attic of the structure, the ambient temperature may easily be 125°F, producing a 70°F temperature difference from the inside to the outside of the duct. If the duct is not insulated, it will gain a lot of heat under these conditions (Figure 23-23).

Typically, when the duct is routed through an unconditioned space, most designers use a 10% duct gain as a rule of thumb. It is a reasonable figure and may be used to size the system. For example, suppose the total heat gain for the structure is 73,000 Btuh. Adding the duct gain will push the unit size up to 80,300 Btuh (73,000 × 110% or 1.1 = 80,300).

23.7 INTERNAL HEAT GAINS

Heat is also generated from internal sources including people, lights, and equipment.

People Loads

The human body gives off heat as a part of the life process. Seated at rest, a typical adult male gives off approximately 390 Btuh of total heat. If the same adult male is doing light bench work in a factory, 800 Btuh of total heat is given off, and if involved in athletics, 2000 Btuh of heat is given off. These figures were taken from the table in Figure 23-24. Note the different columns and the footnotes at the bottom. The figures are adjusted for the estimated number of women and children that would be involved in the various activities.

To find the internal heat gain, the designer/estimator examines the application and estimates the number of occupants and their activity level at peak load hours. For example, suppose a dancing/supper club was being designed to accommodate 500 people. The designer may assume that half the people will be seated while the other half are doing moderate dancing. The designer may use the restaurant adjusted figure of 275 Btuh sensible heat and 105 Btuh latent heat for the 250 people that are estimated to be seated at dinner. Notice that the footnote allows for the heat given off by the food being served per person. The heat gain for the seated customers would be:

$$250 \times 275 = 68,750 \text{ Btuh sensible heat}$$

$$250 \times 105 = 26,250 \text{ Btuh latent heat}$$

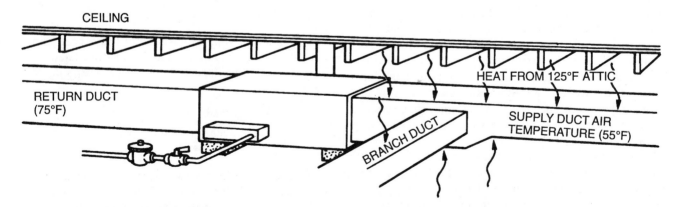

Figure 23-23. Heat gain between an unconditioned attic and conditioned air in a duct.

INTERNAL HEAT GAINS

Rates of Heat Gain from Occupants of Conditioned Spaces

Degree of Activity		Total Heat, Btu/h Adult Male	Adjusted, M/F[a]	Sensible Heat, Btu/h	Latent Heat, Btu/h	% Sensible Heat that is Radiant[b]	
						Low V	High V
Seated at theater	Theater, matinee	390	330	225	105		
Seated at theater, night	Theater, night	390	350	245	105	60	27
Seated, very light work	Offices, hotels, apartments	450	400	245	155		
Moderately active office work	Offices, hotels, apartments	475	450	250	200		
Standing, light work; walking	Department store; retail store	550	450	250	200	58	38
Walking, standing	Drug store, bank	550	500	250	250		
Sedentary work	Restaurant[c]	490	550	275	275		
Light bench work	Factory	800	750	275	475		
Moderate dancing	Dance hall	900	850	305	545	49	35
Walking 3 mph; light machine work	Factory	1000	1000	375	625		
Bowling[d]	Bowling alley	1500	1450	580	870		
Heavy work	Factory	1500	1450	580	870	54	19
Heavy machine work; lifting	Factory	1600	1600	635	965		
Athletics	Gymnasium	2000	1800	710	1090		

Notes:
1. Tabulated values are based on 75 °F room dry-bulb temperature. For 80 °F room dry bulb, the total heat remains the same, but the sensible heat values should be decreased by approximately 20%, and the latent heat values increased accordingly.
2. Also refer to Table 4, Chapter 8, for additional rates of metabolic heat generation.
3. All values are rounded to nearest 5 Btu/h.

[a] Adjusted heat gain is based on normal percentage of men, women, and children for the application listed, with the postulate that the gain from an adult female is 85% of that for an adult male, and that the gain from a child is 75% of that for an adult male.
[b] Values approximated from data in Table 6, Chapter 8, where V is air velocity with limits shown in that table.
[c] Adjusted heat gain includes 60 Btu/h for food per individual (30 Btu/h sensible and 30 Btu/h latent).
[d] Figure one person per alley actually bowling, and all others as sitting (400 Btu/h) or standing or walking slowly (550 Btu/h).

Figure 23-24. Heat gain table for human occupancy. *Reprinted by permission of American Society of Heating, Refrigerating and Air-Conditioning Engineers.*

The heat gain for the dancing customers would be:

$250 \times 305 = 76{,}250$ Btuh sensible heat

$250 \times 105 = 26{,}250$ Btuh latent heat

All of these figures would be added to find the total internal heat gain due to people for the structure (197,500 Btuh).

Lighting Loads

Heat gain from lights is all sensible heat and can represent a significant load where much lighting is required, such as in a department store. It is very important for the customers to be able to see the merchandise. There are two basic types of lighting, incandescent and fluorescent, and each has a very different load.

Incandescent Lighting Incandescent light is pure resistance heat converted into light. It is produced when electrical energy is applied to a resistance element, causing it to glow. A lightbulb is about 5% efficient for light energy. This means that only about 5% of the energy that goes into the lightbulb is given off as light and the other 95% is given off as heat. The wattage of an incandescent lightbulb can be directly converted to heat using the following formula:

$Q_s = \text{watts} \times 3.413$ Btu's per watt

For example, suppose that a retail store uses 100 incandescent lightbulbs of 150 watts each. The electric load for the lighting would be $100 \times 150 = 15{,}000$ watts. The total sensible heat load is calculated as follows:

$Q_s = \text{watts} \times 3.413$

$Q_s = 15{,}000 \times 3.413$

$Q_s = 51{,}195$ Btuh

Fluorescent Lighting Fluorescent lighting is much more efficient than incandescent lighting. With fluorescent lighting, about 60% of the electrical input is given off as light. Therefore, there is more illumination per watt input. This is the main reason that fluorescent lighting is used so widely in commercial applications. To some extent, it is also used in residential applications, such as kitchens and workrooms.

Fluorescent lighting works by igniting a gas in a tube. Once the gas inside the tube is ignited, a ballast is energized to reduce the voltage to the tube. This ballast increases the actual wattage by 20% so a factor of 1.2 is applied to the rated wattage of the fluorescent tube. The formula for sensible heat gain in fluorescent lighting is as follows:

$Q_s = \text{watts} \times 3.413 \times 1.2$

For example, suppose a retail store has 10,000 watts of fluorescent lighting. The heat gain would be:

$Q_s = \text{watts} \times 3.413 \times 1.2$

$Q_s = 10{,}000 \times 3.413 \times 1.2$

$Q_s = 40{,}956$ Btuh

Remember, there will be much more illumination per watt using fluorescent lighting.

372 UNIT 23 HEAT GAIN CALCULATIONS

Many estimators use 10 watts per foot as a rule of thumb. For example, a fixture that is 4 feet long with 4 tubes is equal to 16 feet of lighting. This equals 160 watts of heat.

Fluorescent lighting has one more consideration that must be taken into account. The fixture may be entirely in the conditioned space or it may be recessed into the ceiling. If it is recessed into the ceiling above the conditioned space and the area above the ceiling is not conditioned, some of the heat will be applied to the space above the ceiling and not to the room load. The manufacturer of the light fixture determines how much heat will be applied to the conditioned space (usually about 70% of the heat from the fixture).

Equipment Loads

The internal heat gain from equipment is another large heat gain for some buildings. The heat gain from equipment can be both sensible and latent. If the equipment is purely electrical, such as an electric motor, it will be sensible heat only. If the heat is from electric energy that is heating water in an open container, such as an electric steam table in a restaurant, the heat gain will be both sensible and latent.

There are a variety of tables that provide the sensible and latent heat gains for different types of equipment. Figure 23-25 shows the heat gains for different electric motors. Notice that the table designates whether the motor is inside the driven equipment or outside the equipment, such as an air handler. The air handler may be located inside the conditioned space or outside of it. The location of the motor determines how much of the work energy that is converted to heat will be transferred into the conditioned space or the load on the air conditioning equipment.

The table in Figure 23-26 shows part of a complete table of appliances and the heat given off by these appliances. Note that the table applies different heat gains depending on whether or not an exhaust hood is used to remove some of the heat from the conditioned space.

Heat Gain from Typical Electric Motors

Motor Nameplate or Rated Horsepower	Motor Type	Nominal rpm	Full Load Motor Efficiency %	Location of Motor and Driven Equipment with Respect to Conditioned Space or Airstream		
				A Motor in, Driven Equipment in, Btu/h	B Motor out, Driven Equipment in, Btu/h	C Motor in, Driven Equipment out, Btu/h
0.05	Shaded pole	1500	35	360	130	240
0.08	Shaded pole	1500	35	580	200	380
0.125	Shaded pole	1500	35	900	320	590
0.16	Shaded pole	1500	35	1160	400	760
0.25	Split phase	1750	54	1180	640	540
0.33	Split phase	1750	56	1500	840	660
0.50	Split phase	1750	60	2120	1270	850
0.75	3-Phase	1750	72	2650	1900	740
1	3-Phase	1750	75	3390	2550	850
1.5	3-Phase	1750	77	4960	3820	1140
2	3-Phase	1750	79	6440	5090	1350
3	3-Phase	1750	81	9430	7640	1790
5	3-Phase	1750	82	15,500	12,700	2790
7.5	3-Phase	1750	84	22,700	19,100	3640
10	3-Phase	1750	85	29,900	24,500	4490
15	3-Phase	1750	86	44,400	38,200	6210
20	3-Phase	1750	87	58,500	50,900	7610
25	3-Phase	1750	88	72,300	63,600	8680
30	3-Phase	1750	89	85,700	76,300	9440
40	3-Phase	1750	89	114,000	102,000	12,600
50	3-Phase	1750	89	143,000	127,000	15,700
60	3-Phase	1750	89	172,000	153,000	18,900
75	3-Phase	1750	90	212,000	191,000	21,200
100	3-Phase	1750	90	283,000	255,000	28,300
125	3-Phase	1750	90	353,000	318,000	35,300
150	3-Phase	1750	91	420,000	382,000	37,800
200	3-Phase	1750	91	569,000	509,000	50,300
250	3-Phase	1750	91	699,000	636,000	62,900

Figure 23-25. Heat gains for electric motors. *Reprinted by permission of American Society of Heating, Refrigerating and Air-Conditioning Engineers.*

Recommended Rate of Heat Gain from Restaurant Equipment Located in Air-Conditioned Area (*Concluded*)

Appliance	Size	Input Rating, Btu/h Maximum	Standby	Recommended Rate of Heat Gain,[a] Btu/h Without Hood Sensible	Latent	Total	With Hood Sensible
Toaster (bun toasts on one side only)	1400 buns/h	5120	—	2730	2420	5150	1640
Toaster (large conveyor)	720 slices/h	10920	—	2900	2560	5460	1740
Toaster (small conveyor)	360 slices/h	7170	—	1910	1670	3580	1160
Toaster (large pop-up)	10 slice	18080	—	9590	8500	18080	5800
Toaster (small pop-up)	4 slice	8430	—	4470	3960	8430	2700
Waffle iron	75 in^2	5600	—	2390	3210	5600	1770
Electric, Exhaust Hood Required							
Broiler (conveyor infrared), per square foot of cooking area/minute	2 to 102 ft^2	19230	—	—	—	—	3840
Broiler (single deck infrared), per square foot of broiling area	2.6 to 9.8 ft^2	10870	—	—	—	—	2150
Charbroiler, per square foot of cooking surface	1.5 to 4.6 ft^2	7320	—	—	—	—	307
Fryer (deep fat), per pound of fat capacity	15 to 70 lb	1270	—	—	—	—	14
Fryer (pressurized), per pound of fat capacity	13 to 33 lb	1565	—	—	—	—	59
Oven (large convection), per cubic foot of oven space	7 to 19 ft^3	4450	—	—	—	—	181
Oven (large deck baking with 537 ft^3 decks), per cubic foot of oven space	15 to 46 ft^3	1670	—	—	—	—	69
Oven (roasting), per cubic foot of oven space	7.8 to 23 ft^3	27350	—	—	—	—	113
Oven (small convection), per cubic foot of oven space	1.4 to 5.3 ft^3	10340	—	—	—	—	147
Oven (small deck baking with 272 ft^3 decks), per cubic foot of oven space	7.8 to 23 ft^3	2760	—	—	—	—	113
Range (burners), per 2 burner section	2 to 10 brnrs	7170	—	—	—	—	2660
Range (hot top/fry top), per square foot of cooking surface	4 to 8 ft^2	7260	—	—	—	—	2690
Range (oven section), per cubic foot of oven space	4.2 to 11.3 ft^3	3940	—	—	—	—	160
Range (stockpot)	1 burner	18770	—	—	—	—	6990
Gas, No Hood Required							
Broiler, per square foot of broiling area	2.7 ft^2	14800	660[b]	5310	2860	8170	1220
Cheese melter, per square foot of cooking surface	2.5 to 5.1 ft^2	10300	660[b]	3690	1980	5670	850
Dishwasher (hood type, chemical sanitizing), per 100 dishes/h	950 to 2000 dishes/h	1740	660[b]	510	200	710	230
Dishwasher (hood type, water sanitizing), per 100 dishes/h	950 to 2000 dishes/h	1740	660[b]	570	220	790	250
Dishwasher (conveyor type, chemical sanitizing), per 100 dishes/h	5000 to 9000 dishes/h	1370	660[b]	330	70	400	130
Dishwasher (conveyor type, water sanitizing), per 100 dishes/h	5000 to 9000 dishes/h	1370	660[b]	370	80	450	140
Griddle/grill (large), per square foot of cooking surface	4.6 to 11.8 ft^2	17000	330	1140	610	1750	460
Griddle/grill (small), per square foot of cooking surface	2.5 to 4.5 ft^2	14400	330	970	510	1480	400
Hot plate	2 burners	19200	1325[b]	11700	3470	15200	3410
Oven (pizza), per square foot of hearth	6.4 to 12.9 ft^2	4740	660[b]	623	220	843	85
Gas, Exhaust Hood Required							
Braising pan, per quart of capacity	105 to 140 qt	9840	660[b]	—	—	—	2430
Broiler, per square foot of broiling area	3.7 to 3.9 ft^2	21800	530	—	—	—	1800
Broiler (large conveyor, infrared), per square foot of cooking area/minute	2 to 102 ft^2	51300	1990	—	—	—	5340
Broiler (standard infrared), per square foot of broiling area	2.4 to 9.4 ft^2	1940	530	—	—	—	1600
Charbroiler (large), per square foot of cooking area	4.6 to 11.8 ft^2	16500	510	—	—	—	790
Charbroiler (small), per square foot of cooking area	1.3 to 4.5 ft^2	19700	510	—	—	—	950
Fryer (deep fat), per pound of fat capacity	11 to 70 lb	2270	660[b]	—	—	—	160
Oven (bake deck), per cubic foot of oven space	5.3 to 16.2 ft^3	7670	660[b]	—	—	—	140
Oven (convection), per cubic metre of oven space	7.4 to 19.4 ft^3	8670	660[b]	—	—	—	250
Oven (pizza), per square foot of oven hearth	9.3 to 25.8 ft^2	7240	660[b]	—	—	—	130
Oven (roasting), per cubic foot of oven space	9 to 28 ft^3	4300	660[b]	—	—	—	77
Oven (twin bake deck), per cubic foot of oven space	11 to 22 ft^3	4390	660[b]	—	—	—	78
Range (burners), per 2 burner section	2 to 10 brnrs	33600	1325	—	—	—	6590
Range (hot top or fry top), per square foot of cooking surface	3 to 8 ft^2	11800	330	—	—	—	3390
Range (large stock pot)	3 burners	100000	1990	—	—	—	19600
Range (small stock pot)	2 burners	40000	1330	—	—	—	7830
Steam							
Compartment steamer, per pound of food capacity/h	46 to 450 lb	280	—	22	14	36	11
Dishwasher (hood type, chemical sanitizing), per 100 dishes/h	950 to 2000 dishes/h	3150	—	880	380	1260	410
Dishwasher (hood type, water sanitizing), per 100 dishes/h	950 to 2000 dishes/h	3150	—	980	420	1400	450
Dishwasher (conveyor, chemical sanitizing), per 100 dishes/h	5000 to 9000 dishes/h	1180	—	140	330	470	150
Dishwasher (conveyor, water sanitizing), per 100 dishes/h	5000 to 9000 dishes/h	1180	—	150	370	520	170
Steam kettle, per quart of capacity	13 to 32 qt	500	—	39	25	64	19

[a] In some cases, heat gain data are given per unit of capacity. In those cases, the heat gain is calculated by: q = (recommended heat gain per unit of capacity) * (capacity)

[b] Standby input rating is given for entire appliance regardless of size.

Figure 23-26. Sensible and latent heat gains for appliances. *Reprinted by permission of American Society of Heating, Refrigerating and Air-Conditioning Engineers.*

For example, suppose a restaurant has an electric broiler with a hood to cook steaks where the customer can see the process. The heat given off will be 307 Btuh per square foot of surface area. If the broiler is 3 feet × 3 feet this would be 9 square feet times 307 Btuh/square foot or 2763 Btuh for this appliance.

Room-by-Room Load Calculations

A room-by-room load calculation is used to calculate how much heat gain is applied to each room in a structure. This allows the designer/estimator to furnish the correct air condition to each room for maintaining the desired conditions.

Room-by-room heat gains are calculated in the same way as the complete structure except that each room is treated as a standalone area. There is a lot of detail in a room-by-room calculation, so many contractors will take short cuts. One short cut that is often used is to perform a whole structure calculation and then divide each room area by the total. This prorates the load to each room by using a percentage. For example, if a house has 1200 square feet and the living room has 200 square feet, it is 16.7% of the total area of the house (200 ÷ 1200 = .1666 or 16.7%). If the total heat gain for the house is 30,300 Btuh, the heat added to this room would be 5060 (30,300 × .167 = 5060).

The above method of dividing the heat gain into the respective rooms is not as accurate as a true room-by-room calculation, but is often used. An experienced estimator will adjust each room load to account for extra windows and additional sun load.

Many designers/estimators use computer programs to estimate the heat gain for a structure. This is helpful because the types of construction for a particular part of the country may be very similar. Once the designer/estimator has entered several structures into the computer, the common features can be used in all other similar structures. This is typically done by entering a construction number for a typical type of building. The next time a structure with the same construction is encountered, most of the data is already there, and only the room areas need to be entered. Even the temperature difference is already in the computer.

The computer enables the designer/estimator to change the building materials for a particular structure to experiment with the most efficient structure for the least amount of money.

SUMMARY

- A successful cooling system is one that maintains the correct comfort range and air movement in the conditioned space.
- The design sequence begins with the load calculation, and the load calculation starts with knowing the inside and outside design temperatures. Typically, the inside design temperature is 75°F. The outside design weather conditions are available from the ASHRAE Fundamentals Handbook.
- Sizing a system for the 2.5% weather data column is fairly common because the temperature should not rise above the design temperature for more than 72 hours during a normal summer, and, depending on occupancy schedules, these hours may not affect the conditions in the structure.
- The cooling system must be chosen to closely match the heat gain. Usually, a unit size cannot be found that exactly fits the cooling load, so the next size up is chosen, producing a slightly oversized system.
- The building materials used will affect the rate of heat gain. Some materials are conductors, such as steel and others are insulators, such as styrofoam.
- The amount of heat gain may be calculated using the formula $Q = A \times U \times TD$, where Q = heat gain, A = area, U = overall heat transfer coefficient, and TD = temperature difference. All of the variables in the formula are easy to arrive at except the U factor. It may be calculated using the R value (resistance to heat transfer) of the various building materials. The total R value is known as R_t. The U factor is equal to the reciprocal of the R value $(1 \div R_t)$.
- The wall area may be found using a blueprint or by measuring the actual areas of the walls.
- The temperature difference can be found by subtracting the inside design temperature from the outside design temperature.
- The walls of a structure usually make up the greatest exposure or area of the structure. The walls may either be above ground level (above grade) or below ground level (below grade). Each type of wall is calculated differently.
- Windows are either single, double, or triple glaze. Glaze refers to the number of panes of glass.
- The windows in a structure gain much more heat than the materials in the walls.
- Doors are calculated much like the windows.
- Infiltration is air that leaks into a structure due to the prevailing winds. This creates a pressure difference between the inside and the outside of the structure, causing airflow through the structure. This air must be cooled when it enters. Ventilation is planned infiltration. A slight pressure is maintained on the structure to prevent infiltration.
- Infiltration is typically estimated by the number of air changes per hour to the structure. A tight home will have between .3 and .5 air changes per hour, while a loose home may have 1 or more air changes per hour.
- Infiltration may also be estimated by the size of the cracks around windows and doors. Good windows are tested for their leak rate, typically with a 25 mph wind. A leak rate of .25 cubic feet per minute per foot of crack is considered a good window.
- Heat gain through infiltration or ventilation is one of the big heat gains in any structure and is also one of the most difficult to measure.

- A structure must have either adequate infiltration or ventilation to prevent indoor air pollution.
- The air duct system is a source of heat gain if the duct is routed through unconditioned spaces.
- Room-by-room load calculations are important for some types of cooling systems because the air must be distributed correctly.
- Computer estimating programs are available and help in performing both total load calculations and room-by-room load calculations.

REVIEW QUESTIONS

1. What are the effects of an oversized cooling system?
2. What is the 1% outdoor design temperature for Oklahoma City, Oklahoma?
3. What is the typical indoor design temperature for cooling?
4. If a cooling system is designed for the 2.5% weather data column, how many hours will the temperature be above the design temperature?
5. Name various ways by which heat can enter through windows.
6. Name four places in a structure where heat can leak in.
7. A wall located in Oklahoma City, Oklahoma has a U factor of .046 and an area of 400 square feet. What is the estimated heat gain for this wall using the 2.5% value and an indoor design temperature of 75°F?
8. The R_t for a wall is 21. What is the U factor?
9. A wall has a U factor of .51 and the owner wants a U factor of .09. How many inches of polystyrene insulation will be required to achieve this U factor?
10. Name one way to simplify a room-by-room load calculation.

24 APPLICATION OF AIR CONDITIONING EQUIPMENT

OBJECTIVES

Upon completion of this unit, you should be able to

- **State the differences between water-cooled systems and air-cooled systems.**
- **Explain the advantages of rooftop package equipment.**
- **Choose the correct pipe size for a basic refrigeration system.**
- **Describe some basic piping practices.**
- **Describe three types of special-application air conditioning systems.**

24.1 APPLICATION OF EQUIPMENT

Proper application of air conditioning equipment involves choosing reliable equipment that meets the customer's budget and will satisfy the comfort requirements of the structure without using excess space or energy. For each application, there may be a number of acceptable solutions. Each building must be carefully evaluated to determine the best type of cooling system for the application. For example, strip shopping malls are now very popular across the country. For years, most designers chose to use a large central system to serve the entire mall. The major drawback of this approach was that if the system stopped operating, the entire shopping center was without air conditioning. A more common approach today is to apply package air conditioning equipment to each store using rooftop units (Figure 24-1). This limits any system problems to the inoperative unit and allows each tenant to have individual control of space temperature. It also allows for individual tenant metering, so each store can be charged for their own utility consumption instead of having to divide a central utility bill. One disadvantage of this method is that a single large compressor, if sized correctly, would be more

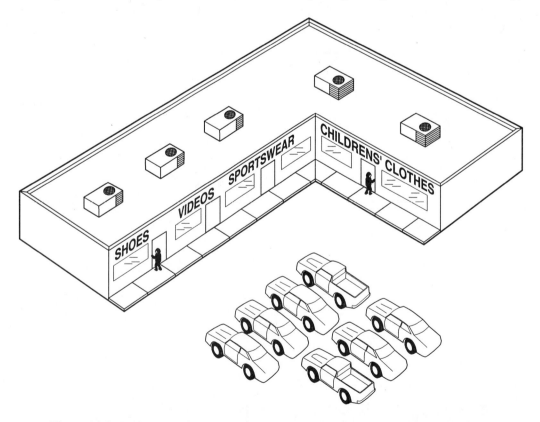

Figure 24-1. Packaged rooftop air conditioning units used in a shopping center application.

efficient than several smaller compressors. Generally, the larger the motor, the more efficient the system. However, most designers are willing to trade slightly diminished efficiency for the closer control and backup capabilities provided by several individual units.

The above example illustrates some of the decisions the designer must make to provide successful air conditioning installations. There are many other factors that enter into the equipment application process. In order to effectively service and troubleshoot air conditioning systems, you must know enough about both the system and the designer's intentions to recognize whether a system has a design problem, an application problem, or an equipment problem.

24.2 TYPES OF EQUIPMENT

The equipment used on residential and commercial systems is typically either package equipment or split systems. The actual equipment choice is often determined by the type of building. A multi-story building will often use a single compressor located in the basement or on the roof and multiple evaporator coils (Figure 24-2). This works well, but requires a considerable amount of refrigerant and increases the possibility of loss of charge due to all of the piping. The reason this system is popular in some areas is that it is less expensive and easier to route refrigerant piping than it is to locate a single coil and fan and run the ductwork from one floor to another (Figure 24-3).

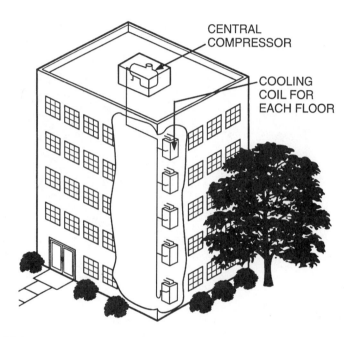

Figure 24-2. This system has a single compressor and multiple evaporators.

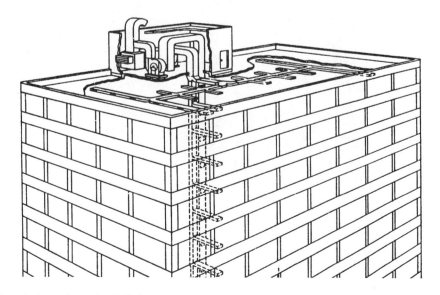

Figure 24-3. Routing air ducts through a building is expensive and takes up a lot of valuable space. *Courtesy Trane Company.*

The ductwork takes up much more floor space when routed from floor to floor, and floor space is money.

Package air conditioning systems are divided into two categories: air cooled and water cooled.

Air-Cooled Package Systems

Air-cooled package systems are popular because they are factory assembled and checked out before shipping to the job. All that has to be done to complete installation is to connect power to the unit and follow the manufacturer's instructions for start-up. This is appealing to many companies.

The ductwork must be connected to the unit from outside the structure and the duct must then pass through the side or top of the building. These systems can range in size from about 1½ tons to several hundred tons of capacity (Figure 24-4). As the systems become larger, they use multiple compressors and a heating system is often built into the unit.

There are two basic duct configurations for these package systems. The duct may be either fastened to the end of the unit for mounting on the ground next to the building or to the bottom of the unit for roof mounting (Figure 24-5).

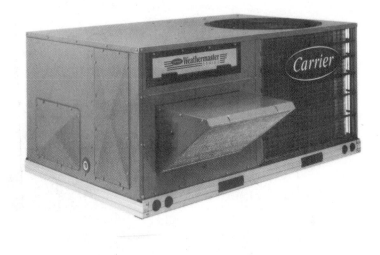

(B)

Figure 24-4. (A) Small package unit (B) Large package unit. *Reproduced courtesy of Carrier Corporation.*

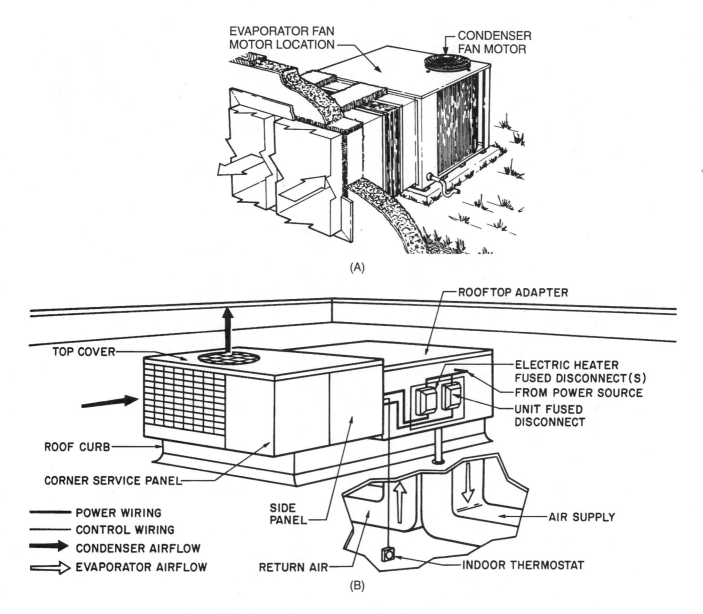

Figure 24-5. End and bottom duct configurations for package systems.

End-mount duct systems can have three different types of connections: round, over-and-under rectangular, and side-by-side rectangular (Figure 24-6). All of these duct connections must be applied in such a manner that the duct connection to the package unit is airtight. The round configuration is most often used for connections to mobile homes. The over-and-under and the side-by-side rectangular configurations are used to fasten the duct to the unit and run the duct in the crawl space under a structure (Figure 24-7). If the duct must be run through the roof, an adapter can be purchased or field-fabricated to turn the duct down to the roof. This installation must be waterproof and well insulated to prevent the weather from affecting the system through rain and/or heat exchange. A factory-made adapter is often the best choice.

Larger systems have the duct fastened to the bottom of the unit and are mounted on roof curbs (Figure 24-8). This prevents weather or water from affecting the efficiency of the system. These systems work best when installed on the roof of a single-story building. An opening is cut in the roof and the roof curb is fastened to the roof with the opening in the center of the roof curb. The unit is then placed on the roof curb. For best results, the roof curb should be installed by a professional roofing contractor.

The opening in the roof allows the duct to be routed to the unit from below. A supply and return duct system may be attached to the unit so the building has a conventional duct system except that it is routed through the roof.

380 UNIT 24 APPLICATION OF AIR CONDITIONING EQUIPMENT

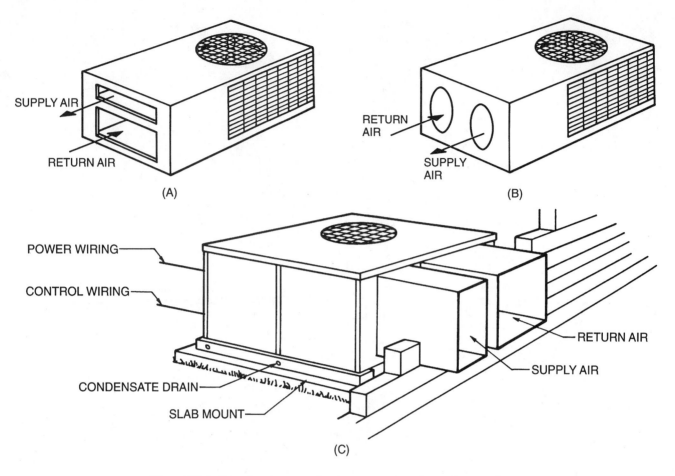

Figure 24-6. Round and rectangular duct connections for package equipment.

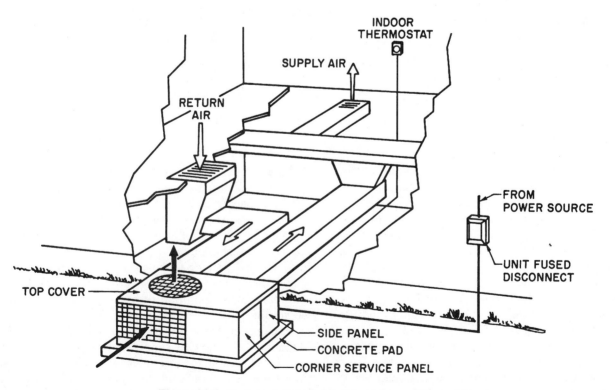

Figure 24-7. Package unit ducted to the crawl space of a residence.

TYPES OF EQUIPMENT   381

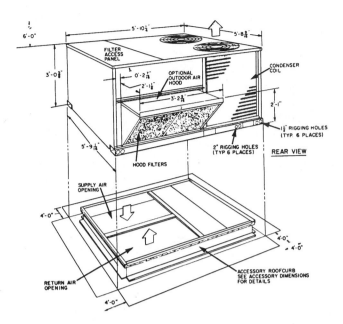

Figure 24-8. Roof curb installation. *Courtesy Heil Heating and Cooling Products.*

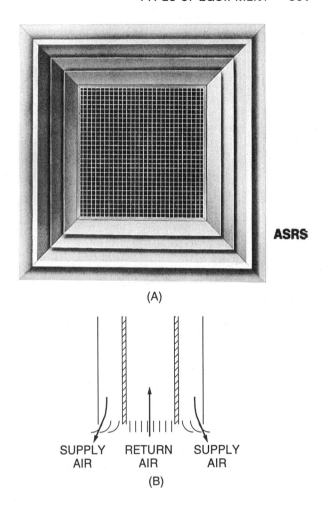

Figure 24-9. Supply and return air grille as one assembly. *Courtesy (A) Hart and Cooley.*

Another type of duct system used with roof-mounted package units is called a *concentric duct system*. It uses one large grille for the complete system. The grille is mounted on the ceiling below the unit and has the return air portion of the grille in the middle and the supply portion on the sides (Figure 24-9).

Package systems are relatively easy to install and maintain. The entire system is on the rooftop or behind the building so that all service can be accomplished without disturbing the occupants (Figure 24-10). Even the filters can be changed on the outside of the building when the filter rack is built into the unit.

Major maintenance, such as a compressor changeout or rebuild, must be handled from the rooftop. This can be accomplished using cranes, dollies, and hoists.

The cranes are used to position the equipment on the rooftop. Dollies are used to move the components to the unit, which may be located in the middle of the roof. Hoists may be used to lift the compressor up into the unit when the unit is outside the reach of a crane.

Figure 24-10. With a package unit, all service can be accomplished outside of the building. *Courtesy Trane Company.*

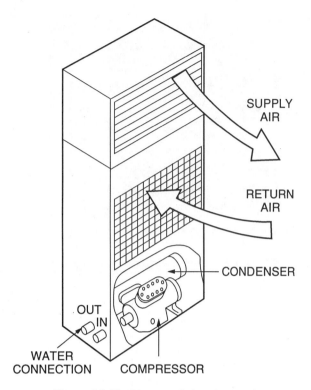

Figure 24-11. Water-cooled package unit.

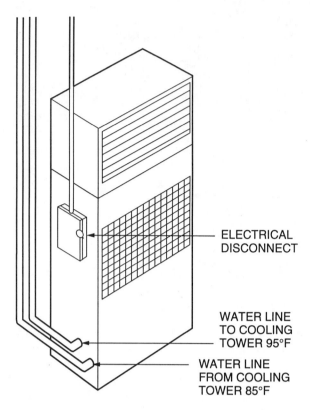

Figure 24-12. Water and power hookups are all that is required to install the water-cooled package unit.

Building security is another consideration in applying air-cooled package equipment because in many systems, the return air duct is large enough for a person to enter the building by climbing inside the unit. Security is often provided by means of alarm systems and/or steel bars placed in the return air duct.

Water-Cooled Package Systems

This type of system is often located in the conditioned space. It is also factory-assembled and has many of the advantages of air-cooled equipment, but is applied in a different manner and for different reasons. A typical water-cooled air conditioner may be a free-standing unit with the compressor located in the bottom of the unit (Figure 24-11). To complete the installation of this unit, all the contractor must do is connect power and water cooling lines to the unit (Figure 24-12). The air side of the unit may have ductwork fastened to the air discharge to route air to remote places or the unit may have a plenum on top of the unit.

One of the advantages of water-cooled equipment is that water lines can be routed to the unit even though the unit may be several floors below the cooling tower. Water lines are economical to install and easier to maintain than refrigerant lines.

The maintenance of water-cooled equipment is more involved than that for air-cooled equipment because of the water circuit. The water-cooled condenser is located in the bottom of the unit next to the compressor and will require attention from time to time. You will recall that water contains minerals and these minerals form deposits on the condenser surface. Most water-cooled package units up to about 7½ tons of capacity use shell-and-coil condensers that may be cleaned using chemicals. The larger units may use shell-and-tube condensers that can be cleaned mechanically (Figure 24-13). The shell-and-tube condenser installation must have room at the end of the condenser to rod the tubes (Figure 24-14). The cooling tower must also be maintained. This is discussed in a separate unit on cooling towers.

Some buildings may have several water-cooled package units operating from one cooling tower (Figure 24-15). This is an easy installation because the piping can all be manifolded to a common supply and return pipe to the cooling tower, which may be located on the roof or behind the building.

Split Systems

Split system air conditioning equipment describes any system in which the compressor, condenser, and evaporator are in different locations and are interconnected with refrigerant lines. This equipment typically falls into one of two categories: small systems where the interconnecting piping is factory supplied or large built-in-place systems where the contractor supplies the piping. The small equipment using factory-furnished interconnecting piping is sized from about 1½ tons to 5 tons of capacity and is applied to residential and small commercial applications.

TYPES OF EQUIPMENT 383

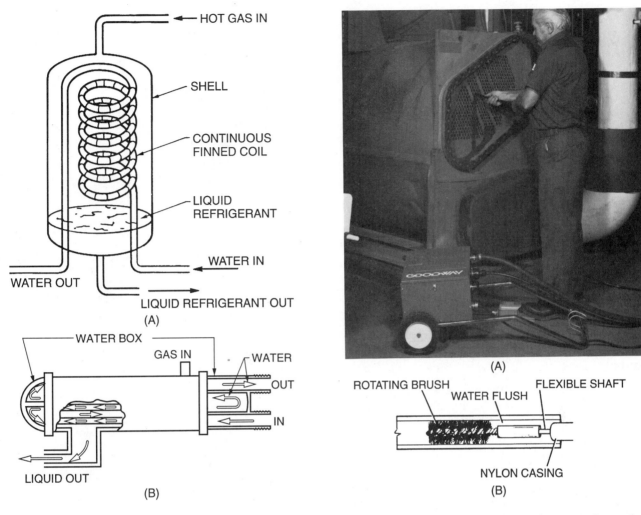

Figure 24-13. (A) Shell-and-coil condenser and (B) Shell-and-tube condenser.

Figure 24-14. The shell-and-tube condenser can be mechanically cleaned using brushes if enough room is provided at the end of the condenser. *Courtesy Goodway Tools Corporation.*

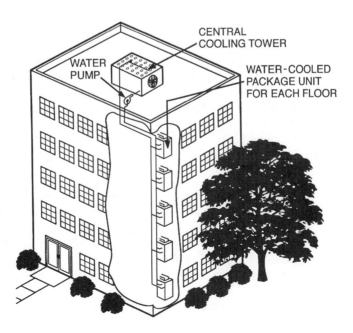

Figure 24-15. Multiple water-cooled units using one cooling tower.

Figure 24-16. Precharged line set with quick-connect fittings. *Photo by Bill Johnson.*

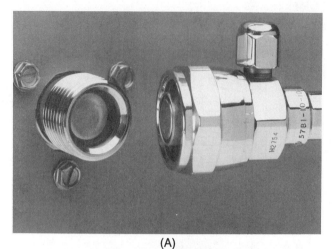

(A)

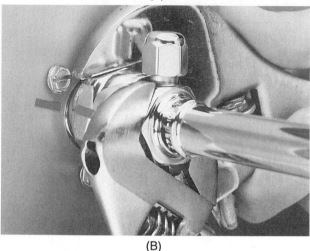

(B)

Figure 24-17. When the connections are made, full refrigerant flow is established. *Courtesy Aeroquip Corporation.*

There are two categories of interconnecting piping systems: the precharged line set and the non-precharged line set. Each has a different method of installation and start-up.

Precharged Line Sets The precharged line set is normally available in lengths of 10, 20, 30, 40, and 50 feet. The line set includes both the suction and liquid lines and has the correct amount of refrigerant precharged into the line set for the length of line furnished. The evaporator portion of the system contains the correct operating charge for the evaporator and the condenser portion of the system contains the correct operating charge for the condenser. The refrigerant is contained in the lines by means of special fittings applied to the ends of the lines. These fittings are designed in such a manner that when they are fastened to the condenser and evaporator, the refrigerant is released to the system and the pipe has a complete refrigerant charge. Connections are accomplished using a set of cutters located inside the fittings that cut away diaphragms inside the fittings on the unit and on the line set (Figure 24-16). When the technician makes the connection (e.g., the suction line to the evaporator) the line becomes a full flow refrigerant line with very little restriction (Figure 24-17). These line sets are very popular because they offer a completely factory-sealed system that is precharged with clean refrigerant for the five line lengths mentioned.

The fittings on the ends of the line set must be compatible with the fittings on the unit in order to make the connection.

If the line set must be altered (either lengthened or shortened) the affected line set should be treated as an independent vessel and must be recharged with the correct amount of refrigerant (Figure 24-18). This defeats the purpose of a precharged line set, but may be necessary to achieve the desired length. For example, the installer may receive a 20 foot line set when a 30 foot length is is required and rather than wait for the new line set to be ordered, the existing line set is altered.

It is recommended that a line set not be altered when it is too long. The tubing can be coiled in a circle to accomplish the correct length. When coiling the tubing in a circle, always lay the coil down—never coil it with the coil standing up or the coils will trap oil. With the coil laying down, the refrigerant should flow from the top to the bottom of the coil to move the oil along with the refrigerant (Figure 24-19). Refrigerant and oil flow are discussed in more detail later in this unit.

Non-Precharged Line Sets The non-precharged line set contains special adapters on the ends of the lines that connect to the evaporator and condenser. The suction line is insulated. This line set is furnished with dry nitrogen charged into the lines and neoprene plugs inserted into the ends of the tubing. When the plugs are removed, the dry nitrogen escapes (Figure 24-20). When you hear the nitrogen escape, you know that the line set is leak free because it did not lose its nitrogen charge until you released it.

TYPES OF EQUIPMENT   385

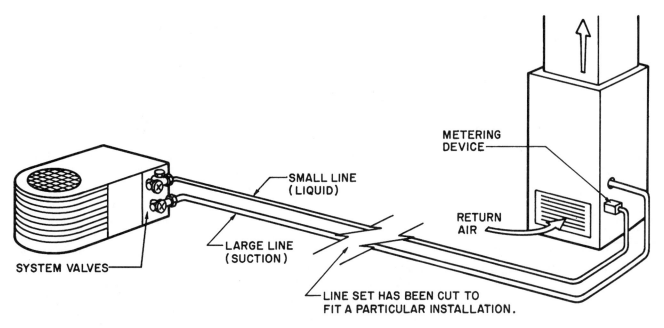

Figure 24-18. Altering the length of a line set.

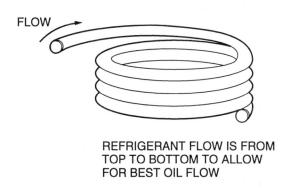

Figure 24-19. If a line set must be coiled, it should be done as shown here.

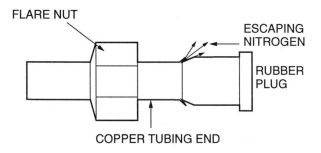

Figure 24-20. If you hear the nitrogen escape when the tube plugs are removed, you know that the tube is leak free.

The evaporator also contains dry nitrogen with neoprene plugs that when removed will vent the nitrogen from the evaporator coil. Again, hearing the factory-charged nitrogen escape assures you that the coil is leak free.

The installer typically fastens the lines starting with the suction line on the evaporator, then the liquid line on the evaporator, and finally, the liquid line on the condenser. These connections are tightened to the recommended torque. The the suction line on the condenser is then fastened finger tight and the valve on the condenser liquid line is slightly opened, allowing refrigerant to push the nitrogen through the liquid line, evaporator, and suction line where it escapes at the suction line connection until the nitrogen has been completely replaced with refrigerant. The system is now ready to start from a refrigerant charge standpoint if the line set is the correct length for the charge stored in the condensing unit.

The connections at the end of the line set are typically compression fittings (Figure 24-21). The evaporator and condensing unit must have a matching set of fittings for the line set to be compatible with the unit. These fittings are often manufactured in such a manner that the installing technician can solder the end of the tubing into the compatible fitting on the unit (Figure 24-22). When this is accomplished, the system is now hermetically sealed except for the service access ports.

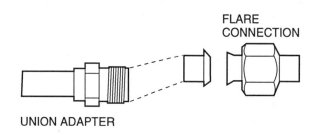

Figure 24-21. Compression fittings for line sets.

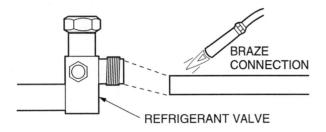

Figure 24-22. These fittings may also be soldered or silver brazed at the unit.

Figure 24-24. Insulation must be field-installed when using hard-drawn tubing. *Photo by Bill Johnson.*

Some technicians feel that soldering the connections provides an advantage over flare connections.

The refrigerant charge for this system is typically stored in the condensing unit for a line length of 30 feet and is released to the system by opening the service isolation valves after installation (Figure 24-23). Since the line set may be other than 30 feet, the refrigerant charge must be altered at start-up if the line set is not the correct length. When you start the system, the 30 foot line set charge may be recovered or adjusted by weight to attain the correct charge. All installation practices are discussed in the unit on installation.

For split systems below 5 tons of capacity that do not use factory line sets, the contractor must furnish the tubing and pipe the system with the correct size and type of pipe for the installation. This tubing is furnished in soft copper rolls that are uncoiled as needed. Air conditioning systems that are above 5 tons use tubing sizes that are not typically furnished in rolls. Tubing sizes of 1 1/8 inch outside diameter (OD) and larger are usually furnished in straight lengths. This tubing is not pliable and is called *hard-drawn tubing*. Since this tubing is furnished in straight lengths, fittings must be soldered or brazed to accomplish any turns. This adds considerable work to the installation and increases the risk of refrigerant leaks. It requires a special piping crew to ensure a leak-free installation. When the tube connections are fastened together with high-temperature brazing alloys, dry nitrogen should be passed through the connections while they are being made and until the point that they become cool to prevent oxidation inside the piping. Review the piping unit for these procedures.

The suction line for this tubing must be field insulated by the installing contractor as part of the job (Figure 24-24). All of the above tasks must be understood by the applications designer before the installation can be bid or completed.

24.3 PIPING PRACTICES

Piping practices are the skills required to correctly pipe a refrigeration system at an acceptable cost. All properly-operating refrigeration systems have only oil and refrigerant circulating in the interconnecting piping. Some oil is pumped out of the compressor in the normal pumping process. (Some compressors pump more oil than others, but all of them pump oil.) Once oil leaves the discharge port of the compressor, there are two ways it can be returned to the compressor crankcase, either by using an oil separator or through good piping practices.

Oil Separators

An oil separator is a component that can be placed in the discharge line just after the compressor and is used to return the oil to the compressor crankcase (Figure 24-25). It should be understood that oil separators do not ensure 100% oil return because between 1 and 2% of the oil pumped will still escape the oil separator in the form of a hot vapor. This oil will begin to accumulate in the discharge line and will not be returned until the compressor has been running for long enough at a high enough velocity to move the oil up the discharge line to the condenser and through the system. Since many compressors provide capacity control down to 25% of total capacity, it may be hard for the system to return this oil to the compressor. Long periods of running at 25% load will cause problems if the oil separator alone is relied upon for oil return.

Oil separators must be located where they will always be at a higher temperature than the condenser or they must have heat applied to keep them warm. If not, liquid refrigerant will accumulate in the separator during the off cycle. When the compressor starts up, the separator float

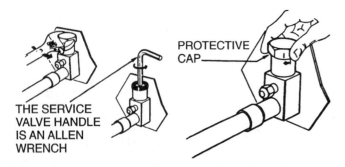

Figure 24-23. The charge from the condensing unit is released to the entire system by opening the service isolation valves. *Courtesy Aeroquip Corporation.*

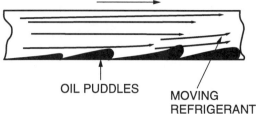

Figure 24-27. The refrigerant velocity is the force that returns oil to the compressor in a refrigerant line.

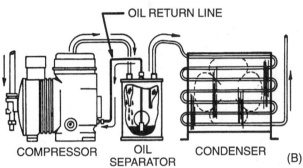

Figure 24-25. Oil separator application. *Courtesy AC&R Components, Inc.*

will allow liquid refrigerant to return to the compressor crankcase just like oil because it is only a liquid float-operated device (Figure 24-26).

Oil Return and Piping Practices

The method the technician uses to route and size the pipe is the best line of defense for proper oil return. The velocity of the refrigerant is the force that moves the oil in the refrigerant line, whether it be horizontal or vertical (Figure 24-27).

There are three different conditions that affect oil movement in the system:

1. Oil travels very easily when moving with liquid refrigerant because it dissolves in the refrigerant, much as motor oil dissolves in gasoline.
2. Oil in the discharge line moves easily because the oil is hot. Some of it may be in the vapor state in a hot discharge line, which may operate at 200°F or higher.
3. Oil in the suction line moves slowly because the suction line leaving the evaporator may be at only 50°F. The oil is much thicker at this lower temperature.

It is generally assumed that in order for oil to move in a vapor refrigerant line, the velocity of the refrigerant must be kept within the following limits:

1. Minimum of 500 feet per minute (fpm) in horizontal gas lines, both suction and hot gas.
2. Minimum of 1000 fpm in vertical gas lines, both suction and hot gas.
3. Maximum of 4000 fpm for any gas line, either suction or hot gas.

Refrigerant line sizing will be discussed in detail later in this unit.

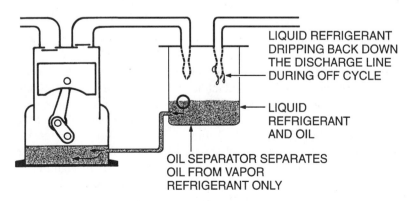

Figure 24-26. Liquid refrigerant will be returned by the oil separator just like liquid oil.

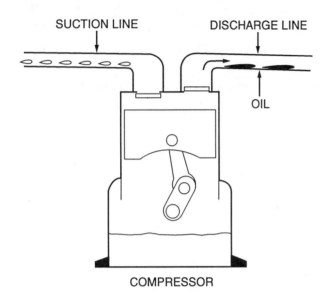

Figure 24-28. Both the suction and discharge lines are sloped in the direction of refrigerant flow. This helps to prevent oil from draining back to the compressor during the off cycle.

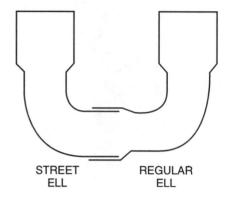

Figure 24-29. A street ell and a short radius ell are used to make a close coupled oil trap.

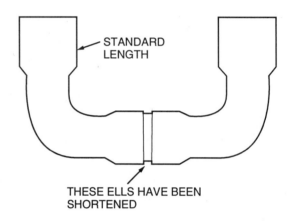

Figure 24-30. Two short radius ells are modified to make a close coupled oil trap.

Oil movement in refrigerant vapor lines is the most critical aspect of pipe system design. Special piping practices must be followed to ensure proper lubrication of the compressor. Since the oil is moving with the refrigerant, the rate of oil return will be influenced by both gravity, which helps the oil move through the system, and correct piping practices, which prevent excess oil from reaching places where it may do harm. As it turns out, one practice will help to achieve the best results from both of these influences and that is to slope the pipe slightly in the direction of the refrigerant flow. The pipe is sloped at a rate of 1/2 inch per 10 feet of run. When the pipe is a horizontal run of hot gas, it keeps the oil moving towards the condenser for return. It also keeps the oil from draining back to the compressor head during the off cycle (Figure 24-28). Oil standing on the compressor head could leak through the compressor valves and fill the cylinder with oil. This oil would cause damage at start-up as it is not compressible.

Oil in a vertical line would also drain back to the compressor during the off cycle so a special device called a *trap* is installed to collect the oil during the off cycle. The trap also gathers oil during the running cycle. This increases the velocity of the refrigerant at the trap and when it reaches a great enough velocity all of the oil in the trap is carried up the riser. Traps are made of two elbows that are fastened as close together as possible. Often, a street ell and a standard close radius ell are used (Figure 24-29). If these are not available, two standard short radius ells may be fastened together by trimming the fittings and using a very short stub of tubing (Figure 24-30).

Traps should be applied when the gas line rises more than 8 feet, and then a trap should be applied for every 25 feet of vertical rise in the gas line.

The suction line follows the same rules. It should be sloped downward in the direction of the refrigerant flow to keep the oil moving towards the compressor crankcase. If the piping is sloped backwards, oil will be trapped in low areas, starving the compressor of lubrication. During the off cycle, a considerable amount of oil can build up because the evaporator may also drain oil to the low spot. If the puddle becomes large enough, the returning oil may all reach the compressor cylinders at once and cause oil slugging of the compressor, which may result in serious damage.

24.4 REFRIGERANT LINE SIZING

The refrigerant lines are used to circulate both vapor and liquid refrigerant throughout the system. All reciprocating, rotary, and scroll compressors pump oil out the discharge line of the compressor. Once this oil leaves the compressor discharge valve, there is normally only one way back to the compressor. The oil must make the complete route through the condenser, liquid line, evaporator, and suction line before it again reaches the compressor sump (Figure 24-31). This is a long trip and depends not only on the effects of gravity and good piping practices, but also the refrigerant temperature and velocity.

REFRIGERANT LINE SIZING

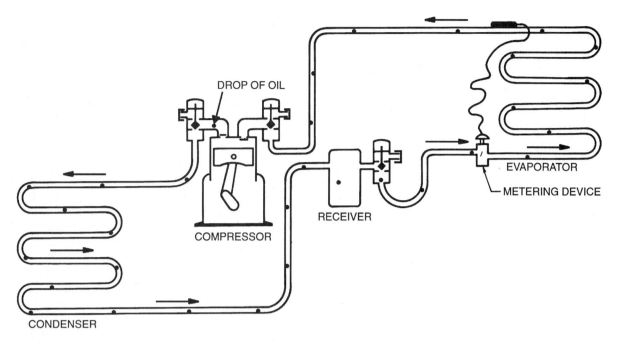

Figure 24-31. Oil leaving the compressor discharge must make the complete refrigerant circuit before returning to the compressor crankcase.

Oil traveling with hot refrigerant vapor moves fairly quickly because the oil is thinned by the heat, so the oil leaving the compressor moves easily toward the condenser. When the oil reaches the condenser, it will mix readily with the warm condensed refrigerant in the liquid state (Figure 24-32). This is much like mixing gasoline with motor oil or water with warm syrup. They mix easily and stay mixed. When the oil reaches the evaporator, it is still mixed with the liquid refrigerant but the refrigerant is now boiling into a vapor. As the oil moves through the evaporator, the refrigerant is separated out as a vapor and the oil is again the only liquid in the pipe. It must be pushed along by the velocity of the refrigerant vapor, only now it is at a temperature of about 40°F (Figure 24-33). This would be similar to moving motor oil on a painted concrete floor using an air hose except the refrigerant in the system is contained in a pipe. The oil travels along the bottom of the pipe.

The velocity of the refrigerant in a refrigerant line is controlled by the size of the line and the quantity of refrigerant moving in the line.

Figure 24-32. When the oil reaches the condenser, it is readily mixed with the liquid refrigerant.

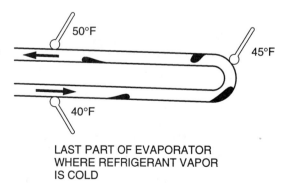

Figure 24-33. This oil is moving very slowly because the refrigerant vapor is cool.

There are many different methods used to size refrigerant lines. All manufacturers recommend using their own method for sizing piping for their equipment. We will use the *Trane* method, but the same pipe size would normally be obtained using any other comparable method.

When sizing pipe, several things must be considered. For example, the designer will want to select the smallest practical pipe size because of the economy of buying and working with smaller piping.

As we go through an actual pipe sizing exercise, you may ask yourself if all of this is necessary. If you intend to do a professional job, the answer is yes. To successfully troubleshoot air conditioning systems, you must be able to spot field problems with piping. You should also be able to check the pipe size to verify these problems. Some of the problems caused by poor pipe sizing are noise, low pumping capacity, and poor oil return to the compressor. When a system does not return the correct amount of oil to the compressor, the first thing to suspect is oversized piping or a lack of risers. When a system has low pumping capacity, undersized piping should be suspected.

Resistance and Equivalent Lengths

When sizing pipe, the first thing that must be known is how far the refrigerant must travel. When the refrigerant travels through the pipe, resistance is created by every foot of pipe. This is caused by friction as the refrigerant travels in the pipe. Resistance is also created by any fittings, valves, or other components in the piping system. For example, when the refrigerant travels around an elbow, centrifugal force will move the refrigerant to the far side of the elbow, causing a velocity change and eddy currents on the near side of the elbow (Figure 24-34). The friction of the refrigerant traveling through a fitting is typically expressed in *equivalent feet of pipe*. This provides a simple way of translating a fitting friction loss into the same velocity loss that would occur in a certain length of straight piping. For example, Figures 24-35 and 24-36 show that a 2¹/₈ inch OD long radius elbow has the same friction loss as 3.4 feet of straight pipe. It can also be seen that a 2¹/₈ inch OD long radius elbow is preferable to a short radius elbow because the short radius elbow has an equivalent length of 5.2 feet of pipe.

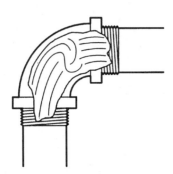

Figure 24-34. Eddy currents are formed as refrigerant moves around elbows. This causes a pressure drop.

EQUIVALENT LENGTHS OF NONFERROUS VALVES AND FITTINGS
Based on Type "L" copper tubing with flare, sweat or flanged fittings

LINE SIZE INCHES OD	GLOBE VALVE & SOL. VALVE	ANGLE VALVE	SHORT RADIUS ELL	LONG RADIUS ELL	TEE LINE FLOW & SIGHT GLASSES	TEE. BRANCH FLOW
1/2	70	24	4.7	3.2	1.7	6.6
5/8	72	25	5.7	3.9	2.3	8.2
3/4	75	25	6.5	4.5	2.9	9.7
7/8	78	28	7.8	5.3	3.7	12
1-1/8	87	29	2.7	1.9	2.5	8
1-3/8	102	33	3.2	2.2	2.7	10
1-5/8	115	34	3.8	2.6	3.0	12
2-1/8	141	39	(5.2)	(3.4)	3.8	16
2-5/8	159	44	6.5	4.2	4.6	20
3-1/8	185	53	8.0	5.1	5.4	25
3-5/8	216	66	10	6.3	6.6	30
4-1/8	248	76	12	7.3	7.3	35
5-1/8	292	96	14	8.8	7.9	42
6-1/8	346	119	17	10	9.3	50

Figure 24-35. This table shows the equivalent lengths of pipe for various fittings.

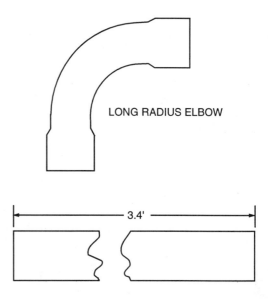

Figure 24-36. This 2¹/₈ inch long radius elbow causes the same amount of pressure drop as 3.4 feet of pipe.

Other typical accessory parts such as valves and sight glasses are listed in this table for the convenience of the application designer.

Sizing the Liquid Line

The liquid line will be the first line to be sized because it is the least complicated. Liquid refrigerant has considerable weight and can have quite a bit of power when in motion. When the heavy liquid is moving in a pipe and one of the valves is suddenly shut off, the liquid must come to a stop very quickly. This results in a lot of energy being released at the point where the valve closed. This is called *liquid hammer* and is comparable to an iron rod of the same weight suddenly meeting an obstacle such as a concrete wall. Needless to say, this causes a considerable amount of stress on the system. For this reason, the recommended maximum velocity for refrigerant traveling in a liquid line is 360 fpm. This is approximately 4 miles per hour.

We will size this piping for a 40 ton, air-cooled system using R-22 that is expected to operate at an evaporator temperature of 40°F and a condensing temperature of 125°F. This system will be built in place, with the compressor located on the second floor, the evaporator located on the first floor, and the condenser and liquid receiver located on the roof (Figure 24-37).

The compressor has a piping arrangement in which the suction and discharge lines are fastened to the compressor to prevent the compressor vibration from being transmitted into the piping. This arrangement shows the piping changing direction several times before it reaches the compressor. Many technicians use the practice of changing directions three times before entering the compressor and three times when leaving the compressor with the suction and discharge line piping (Figure 24-38).

The table for sizing the liquid line is shown in Figure 24-39. Note that this table is printed for water-cooled equipment because it is calculated for an evaporator temperature of 40°F and a condensing temperature of 105°F. A correction table for conditions other than these is shown in Figure 24-40. The correction table allows the applications designer to make corrections for different conditions. For example, the 40 ton system that we are using as an example will have a correction factor of 1.12 because of the 125°F condensing temperature. This factor

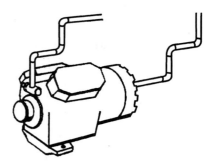

Figure 24-38. To avoid excess vibration, make three directional changes with the piping when entering or leaving the compressor.

is multiplied times the actual unit tonnage to arrive at a corrected tonnage of 44.8 tons (40 × 1.12 = 44.8).

Referring back to Figure 24-37, we see that the actual length of pipe for the system is 39 feet from the condenser outlet to the entrance to the expansion valve. We must estimate the equivalent length of pipe for purposes of finding a trial pipe size, so let's use 250 feet since there are a lot of fittings in this system. Referring to Figure 24-39 and locating 250 feet, we find that the estimated pipe size is 1 3/8 inches and the pressure drop is 4 psig. The maximum allowable pressure drop for R-22 is 6 psig, so this is acceptable. This 1 3/8 inch pipe will carry an actual load of 46 tons of capacity.

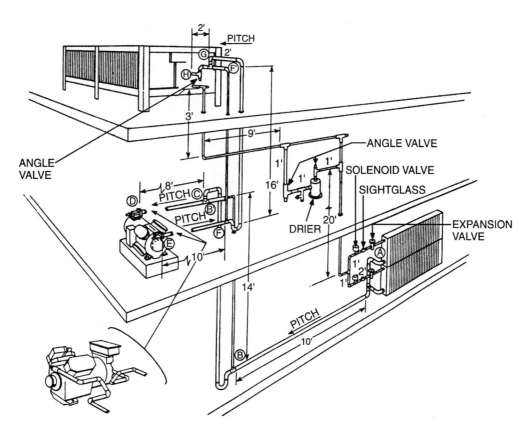

Figure 24-37. Built-in-place refrigeration system. *Courtesy Trane Company.*

TONNAGE CAPACITIES OF LINES DELIVERING LIQUID REFRIGERANT-22 FROM RECEIVER TO EVAPORATOR

Based on 40 F suction and 105 F condensing. For other conditions apply correction factor to design tons before entering this table.

EQUIV. LENGTH IN FEET	TOTAL PRESS. DROP PSI	COPPER TUBE SIZE—OD—TYPE L											
		3/8	1/2	5/8	3/4	7/8	1 1/8	1 3/8	1 5/8	2 1/8	2 5/8	3 1/8	3 5/8	4 1/8
20	0.5	1.5	3.4	6.5	11.0	17.6	35.5	61.0	99.0	204.0	350.0	565.0	890.0	1155.0
	1	2.2	5.1	9.6	16.1	26.0	52.0	90.0	146.0	298.0	520.0	835.0	1310.0	1698.0
	2	3.2	7.6	14.2	23.9	38.0	76.0	133.0	212.0	436.0	768.0	1225.0	1920.0	2490.0
	3	4.1	9.7	18.0	30.0	48.0	96.0	166.0	268.0	548.0	970.0	1540.0	2408.0	3125.0
	4	4.7	11.4	21.1	35.0	56.0	112.0	197.0	322.0	640.0	1145.0	1810.0	2840.0	3670.0
	5	5.4	13.0	24.0	40.0	64.0	128.0	220.0	354.0	730.0	1300.0	2070.0	3225.0	4180.0
	6	5.9	14.3	26.8	44.5	71.0	141.0	245.0	385.0	810.0	1425.0	2300.0	3550.0	4650.0
	7	6.5	15.8	29.0	48.6	77.5	153.0	265.0	430.0	885.0	1560.0	2505.0	3890.0	5000.0
30	0.5	1.1	2.5	4.9	8.2	13.2	26.4	46.0	74.0	152.0	267.0	420.0	660.0	860.0
	1	1.7	3.8	7.3	12.1	19.9	39.0	68.0	110.0	222.0	395.0	625.0	970.0	1285.0
	2	2.5	5.7	10.9	17.9	29.1	56.8	100.0	160.0	330.0	578.0	915.0	1440.0	1880.0
	3	3.1	7.2	13.7	22.4	36.0	71.0	127.0	201.0	415.0	736.0	1160.0	1800.0	2350.0
	4	3.6	8.5	16.0	26.3	42.5	84.0	148.0	236.0	487.0	860.0	1370.0	2115.0	2780.0
	5	4.1	9.6	18.0	30.0	48.2	94.0	168.0	268.0	550.0	985.0	1530.0	2400.0	3110.0
	6	4.5	10.8	20.1	33.0	53.5	104.0	186.0	297.0	610.0	1080.0	1700.0	2670.0	3470.0
	7	5.0	11.8	22.0	36.0	58.2	114.0	202.0	324.0	665.0	1180.0	1870.0	2900.0	3800.0
40	0.5	1.0	2.3	4.3	7.3	11.6	23.1	40.1	65.0	136.0	235.0	376.0	590.0	778.0
	1	1.5	3.4	6.4	10.9	17.1	34.0	59.8	95.0	200.0	345.0	558.0	877.0	1140.0
	2	2.2	5.1	9.6	16.0	25.3	50.0	88.5	140.0	291.0	510.0	820.0	1290.0	1685.0
	3	2.8	6.4	12.1	20.0	32.0	63.0	110.0	178.0	366.0	640.0	1025.0	1600.0	2100.0
	4	3.3	7.6	14.2	23.4	37.5	74.0	130.0	210.0	430.0	755.0	1200.0	1900.0	2500.0
	5	3.7	8.6	16.1	26.5	42.0	84.0	148.0	236.0	487.0	860.0	1380.0	2140.0	2810.0
	6	4.1	9.5	18.0	29.6	47.0	93.0	163.0	260.0	540.0	950.0	1505.0	2380.0	3105.0
	7	4.5	10.5	19.6	32.1	51.0	100.0	179.0	281.0	589.0	1020.0	1650.0	2595.0	3400.0
50	1	1.2	2.8	5.4	8.8	14.2	28.0	49.0	79.0	161.0	283.0	460.0	720.0	950.0
	2	1.8	4.2	7.9	13.0	21.0	41.0	73.0	116.0	240.0	420.0	678.0	1065.0	1200.0
	3	2.2	5.4	9.9	16.4	26.3	52.0	91.0	146.0	300.0	530.0	845.0	1320.0	1400.0
	4	2.7	6.3	11.8	19.3	31.0	61.0	108.0	170.0	350.0	625.0	995.0	1570.0	1760.0
	5	3.0	7.2	13.2	21.9	35.0	69.0	121.0	192.0	400.0	710.0	1130.0	1795.0	2080.0
	6	3.3	8.0	14.9	24.0	39.0	76.0	138.0	213.0	440.0	795.0	1270.0	1990.0	2350.0
	7	3.6	8.8	16.0	26.2	42.5	83.0	148.0	232.0	479.0	860.0	1380.0	2125.0	2820.0
75	1	1.0	2.4	4.4	7.4	12.0	23.4	41.0	66.0	133.0	233.0	370.0	587.0	770.0
	2	1.5	3.5	5.6	10.9	17.7	34.6	60.0	97.0	198.0	348.0	545.0	860.0	1125.0
	3	1.9	4.5	8.2	13.7	22.0	43.0	75.0	121.0	249.0	435.0	690.0	1100.0	1405.0
	4	2.3	5.3	9.7	16.0	25.8	51.0	88.0	142.0	289.0	510.0	810.0	1290.0	1670.0
	5	2.6	6.0	11.0	18.1	29.3	57.5	100.0	160.0	330.0	580.0	920.0	1450.0	1900.0
	6	2.8	6.6	12.1	20.0	32.4	63.8	111.0	179.0	360.0	642.0	1008.0	1600.0	2100.0
	7	3.1	7.3	13.2	21.9	35.2	69.0	121.0	195.0	394.0	700.0	1105.0	1730.0	2300.0
100	2	1.3	2.9	5.6	9.2	14.9	29.0	51.0	81.0	168.0	295.0	470.0	740.0	980.0
	3	1.6	3.7	7.0	11.6	18.7	36.5	64.0	102.0	210.0	365.0	590.0	925.0	1210.0
	4	1.9	4.4	8.2	13.5	22.0	43.0	75.0	121.0	250.0	435.0	695.0	1095.0	1420.0
	5	2.1	5.0	9.3	15.4	25.0	49.0	85.0	138.0	280.0	495.0	790.0	1220.0	1605.0
	6	2.4	5.6	10.3	17.0	27.4	54.0	94.0	151.0	310.0	542.0	865.0	1360.0	1800.0
	7	2.6	6.1	11.1	18.6	30.0	59.0	102.0	165.0	340.0	600.0	955.0	1499.0	1985.0
125	2	1.1	2.6	4.8	8.0	12.9	25.0	43.5	70.0	147.0	256.0	410.0	640.0	840.0
	3	1.4	3.2	6.0	10.0	16.0	31.4	55.0	88.0	182.0	320.0	510.0	800.0	1055.0
	4	1.6	3.8	7.2	11.9	19.0	37.0	64.0	103.0	213.0	375.0	600.0	940.0	1220.0
	5	1.8	4.4	8.1	12.4	21.2	42.0	74.0	118.0	241.0	430.0	680.0	1060.0	1400.0
	6	2.0	4.9	9.0	14.9	23.8	46.5	81.0	165.0	270.0	475.0	750.0	1185.0	1540.0
	7	2.2	5.3	9.8	16.1	25.9	50.0	89.0	141.0	290.0	520.0	820.0	1290.0	1690.0
150	2	1.0	2.4	4.4	7.4	11.9	23.0	41.0	65.0	133.0	233.0	375.0	588.0	780.0
	3	1.3	3.0	5.6	9.4	15.0	29.0	51.0	81.5	168.0	295.0	470.0	740.0	985.0
	4	1.5	3.5	6.6	11.0	17.4	34.0	60.0	95.0	197.0	346.0	558.0	867.0	1150.0
	5	1.7	4.1	7.5	12.6	20.0	38.8	68.5	110.0	222.0	395.0	630.0	980.0	1300.0
	6	2.0	4.5	8.3	13.9	21.9	43.0	76.0	120.0	247.0	435.0	700.0	1095.0	1445.0
	7	2.1	4.9	9.0	15.0	24.0	47.0	82.0	131.0	269.0	470.0	760.0	1185.0	1580.0
175	2	0.9	2.1	3.9	6.4	10.4	20.6	35.5	57.0	118.0	208.0	330.0	518.0	680.0
	3	1.1	2.6	4.9	8.2	13.0	25.8	45.0	72.0	149.0	260.0	415.0	642.0	850.0
	4	1.3	3.1	5.8	9.6	15.3	30.0	53.0	84.0	171.0	305.0	486.0	760.0	1000.0
	5	1.5	3.5	6.6	10.9	17.2	34.0	60.0	95.0	195.0	350.0	548.0	855.0	1120.0
	6	1.6	3.9	6.6	12.0	19.1	38.0	66.0	105.0	216.0	385.0	610.0	950.0	1250.0
	7	1.8	4.3	8.0	14.2	21.0	41.0	72.0	115.0	238.0	420.0	660.0	1030.0	1370.0
200	2	0.8	2.0	3.7	6.2	10.0	20.0	35.0	55.0	115.0	200.0	320.0	500.0	660.0
	3	1.1	2.5	4.7	7.8	12.8	25.0	43.0	69.0	142.0	250.0	400.0	620.0	820.0
	4	1.3	3.0	5.6	9.2	15.0	29.4	51.0	82.0	169.0	300.0	470.0	735.0	960.0
	5	1.4	3.4	6.3	10.5	17.0	33.0	58.0	93.0	190.0	336.0	540.0	830.0	1100.0
	6	1.6	3.8	7.0	11.6	18.9	37.0	64.0	101.0	210.0	370.0	600.0	925.0	1210.0
	7	1.7	4.2	7.6	12.7	20.4	40.0	70.0	111.0	230.0	405.0	640.0	1000.0	1325.0
→ 250	2	0.8	1.8	3.4	5.6	9.0	18.0	31.5	50.0	102.0	180.0	290.0	450.0	590.0
	3	0.9	2.3	4.2	7.0	11.3	22.5	39.0	63.0	129.0	228.0	360.0	560.0	740.0
	→ 4	1.1	2.7	5.0	8.3	13.4	26.0	(46.0)	74.0	151.0	267.0	425.0	660.0	860.0
	5	1.3	3.1	5.6	9.4	15.1	30.0	52.0	84.0	172.0	302.0	480.0	750.0	990.0
	6	1.4	3.4	6.2	10.4	16.9	33.0	58.0	93.0	190.0	335.0	540.0	840.0	1100.0
	7	1.5	3.8	6.8	11.3	18.2	35.7	63.0	101.0	209.0	365.0	580.0	915.0	1200.0
	8	1.6	4.1	7.4	12.2	19.9	38.5	68.5	110.0	225.0	396.0	630.0	988.0	1300.0
300	2	0.7	1.6	3.0	5.0	8.0	16.0	28.0	44.0	91.5	160.0	227.0	400.0	525.0
	3	0.9	2.0	3.8	6.3	10.0	20.0	35.0	55.0	115.0	200.0	320.0	500.0	660.0
	4	1.0	2.4	4.5	7.4	12.0	23.1	41.0	65.0	134.0	238.0	380.0	595.0	765.0
	5	1.2	2.7	5.1	8.4	13.4	26.0	46.0	74.0	151.0	268.0	430.0	654.0	885.0
	6	1.3	3.0	5.6	9.3	14.9	29.0	51.0	82.0	168.0	299.0	470.0	740.0	960.0
	7	1.4	3.4	6.1	10.0	16.1	32.0	56.0	88.0	180.0	320.0	510.0	800.0	1050.0
	8	1.5	3.6	6.6	11.0	17.3	34.0	60.0	96.0	197.0	348.0	558.0	867.0	1135.0
350	2	0.6	1.5	2.8	4.7	7.4	14.9	25.8	40.5	83.0	148.0	232.0	363.0	480.0
	3	0.8	1.9	3.5	5.9	9.4	18.3	32.0	51.0	105.0	185.0	295.0	460.0	600.0
	4	0.9	2.2	4.2	6.9	11.0	21.5	38.0	60.0	123.0	218.0	347.0	540.0	705.0
	5	1.1	2.5	4.8	7.8	12.4	24.3	42.0	67.8	140.0	247.0	390.0	610.0	802.0
	6	1.2	2.8	5.2	8.6	13.8	27.0	47.0	75.0	152.0	270.0	430.0	678.0	895.0
	7	1.3	3.1	5.6	9.4	15.0	29.2	51.0	81.0	169.0	298.0	470.0	740.0	990.0
	8	1.4	3.3	6.1	10.0	16.1	31.6	54.8	88.0	180.0	318.0	508.0	897.0	1045.0
400	2	0.6	1.4	2.5	4.2	6.9	13.5	23.3	38.0	77.0	136.0	219.0	340.0	450.0
	3	0.7	1.7	3.2	5.3	8.6	17.0	29.5	47.0	98.0	171.0	272.0	430.0	560.0
	4	0.9	2.0	3.8	6.3	10.1	20.0	34.6	53.0	115.0	202.0	320.0	500.0	658.0
	5	1.0	2.3	4.3	7.1	11.5	22.3	39.5	63.0	130.0	230.0	360.0	570.0	740.0
	6	1.1	2.6	4.8	7.9	12.7	25.0	43.0	70.0	144.0	252.0	400.0	635.0	820.0
	7	1.2	2.8	5.2	8.6	13.9	27.0	47.0	76.0	158.0	276.0	440.0	690.0	895.0
	8	1.3	3.1	5.6	9.2	15.0	29.2	51.0	82.0	170.0	299.0	472.0	747.0	962.0

NOTE: BOLD FACE FIGURES ARE MAXIMUM RECOMMENDED TOTAL PRESSURE DROPS. SHADED AREAS ARE FOR GENERAL INFORMATION ONLY.

Figure 24-39. Table for sizing the liquid line. *Courtesy Trane Company.*

REFRIGERANT LINE SIZING

LIQUID LINE TONNAGE CORRECTION FACTORS, R-22

COND. TEMP.	SUCTION TEMPERATURE																
	-30	-25	-20	-15	-10	-5	0	5	10	15	20	25	30	35	40	45	50
85	1.01	1.00	0.99	0.99	0.98	0.97	0.96	0.95	0.95	0.94	0.93	0.93	0.92	0.91	0.91	0.90	0.90
90	1.04	1.03	1.02	1.01	1.00	0.99	0.99	0.98	0.97	0.96	0.96	0.95	0.94	0.94	0.93	0.92	0.92
95	1.07	1.06	1.05	1.04	1.03	1.02	1.01	1.00	1.00	0.99	0.98	0.97	0.97	0.96	0.95	0.95	0.94
100	1.09	1.09	1.08	1.07	1.06	1.05	1.04	1.03	1.02	1.01	1.00	1.00	0.99	0.98	0.98	0.97	0.96
105	1.13	1.12	1.11	1.09	1.08	1.08	1.07	1.06	1.05	1.04	1.03	1.02	1.02	1.01	1.00	0.99	0.99
110	1.16	1.15	1.14	1.13	1.12	1.11	1.10	1.09	1.08	1.07	1.06	1.05	1.04	1.03	1.03	1.02	1.01
115	1.20	1.18	1.17	1.16	1.15	1.14	1.13	1.12	1.11	1.10	1.09	1.08	1.07	1.06	1.06	1.05	1.04
120	1.24	1.22	1.21	1.20	1.19	1.17	1.16	1.15	1.14	1.13	1.12	1.11	1.10	1.09	1.09	1.08	1.07
125	1.28	1.27	1.25	1.24	1.23	1.22	1.20	1.19	1.18	1.17	1.16	1.15	1.14	1.13	(1.12)	1.11	1.10
130	1.33	1.31	1.30	1.28	1.27	1.25	1.24	1.23	1.22	1.21	1.19	1.18	1.17	1.16	1.15	1.15	1.14
135	1.38	1.36	1.35	1.33	1.32	1.30	1.29	1.28	1.26	1.25	1.24	1.23	1.22	1.20	1.19	1.19	1.18
140	1.43	1.42	1.40	1.38	1.37	1.35	1.34	1.32	1.31	1.30	1.28	1.27	1.26	1.25	1.24	1.23	1.22
145	1.50	1.48	1.46	1.44	1.43	1.41	1.39	1.38	1.36	1.35	1.33	1.32	1.31	1.30	1.28	1.27	1.26

Figure 24-40. Tonnage correction factor table for Figure 24-39. *Courtesy Trane Company.*

As stated earlier, all fittings and valves cause a pressure drop as the refrigerant flows through them. This flow is calculated in engineering manuals as pressure drop in psig for different flow rates. The method used here is simpler. It shows the pressure drop in the fittings in equivalent feet of pipe. This simplifies the calculation because the equivalent fitting length can then be added to the pipe length for a total length of travel. Figure 24-35 listed the equivalent lengths of common fittings. The list of valves and fittings and their equivalent feet of pipe that the liquid refrigerant must flow through in this system is as follows:

Fittings (1⅜ inch)	Equivalent Feet of Pipe
1. Angle valve leaving condenser	33.0
2. 3 long radius elbows to the solenoid valve (3 × 2.2 ft.)	6.6
3. 3 branch flow tees (3 × 2.7 ft.)	8.1
4. 2 angle valves at the drier (2 × 33 ft.)	66.0
5. 1 line flow tee fitting	2.7
6. Sight glass	2.7
7. Solenoid valve	102.0
Total feet for fittings	221.1
Actual straight pipe length	39.0
Total equivalent feet	260.1

We actually have a liquid line capable of handling 46 tons of capacity of refrigerant flow, so the line has a little extra capacity since our corrected capacity was 44.8 tons. The total of 260 feet found here will use some of that extra capacity as we used only 250 feet of equivalent length to determine the pipe size.

The liquid line filter-drier is estimated to have a pressure drop of 2 psig. (This is listed in the drier manufacturer's literature.) When we add the estimated 4 psig drop of the piping to the 2 psig drop of the liquid line drier, the total pressure drop for the liquid line is 6 psig, right at the maximum recommended pressure drop of 6 psig for R-22. Since most condensers have between 10°F and 20°F of subcooling, this will prevent any flash gas from reaching the expansion valve.

Sizing the Gas Lines

The gas lines (both discharge and suction) are sized in a similar manner with some different considerations. The velocity of refrigerant in the gas lines is more critical than in the liquid line. It is harder to move oil uphill in vertical risers than in horizontal runs because the oil must be pushed up the pipe. As stated earlier, the minimum recommended velocity in a horizontal gas line is 500 fpm for proper oil return. The minimum velocity for a vertical gas line is 1000 fpm. The maximum velocity for any gas line is 4000 fpm. Velocities greater than 4000 fpm will cause noise and excess pressure drop. Excess pressure drop is more acceptable in a hot gas line than in a suction line because it is easier for the compressor to push gas than it is for it to pull it. Pressure drop in the suction line causes reduced suction pressure and less dense suction gas, which lowers compressor capacity.

Another factor that must be considered when sizing gas lines is whether or not the compressor has the capacity to unload. The compressor used in this example is a 40 ton compressor and is capable of unloading down to 10 tons of capacity. The gas lines must have the ability to maintain the correct minimum velocities at both 10 tons and 40 tons of capacity to ensure proper oil return. This is accomplished by means of double risers for both the gas lines. The discharge line is used to lift the hot gas to the condenser on the roof at both 10 tons and 40 tons of capacity (Figure 24-41). The suction line must also lift the suction gas from the coil in the basement to the compressor at both 10 tons and 40 tons of capacity.

A double gas riser has two pipes to carry the gas; one is rated at 10 tons and the other is rated at 30 tons for a total of 40 tons. The piping arrangement is such that when the compressor is unloaded to 10 tons, all of the vapor and oil will travel up the small riser due to an oil trap in the larger of the two lines (Figure 24-42). When the compressor capacity changes and more refrigerant gas is flowing, the oil trap is cleared by the moving refrigerant and gas flows in both risers (Figure 24-43).

The discharge line may be sized by following similar methods to the liquid line. There are two different

394 UNIT 24 APPLICATION OF AIR CONDITIONING EQUIPMENT

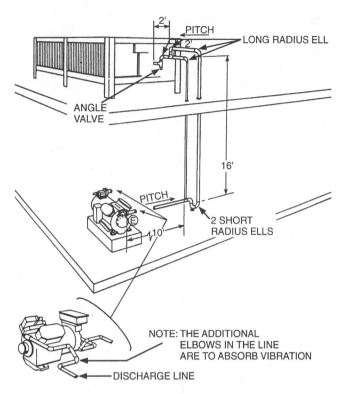

Figure 24-41. Discharge piping configuration. *Courtesy Trane Company.*

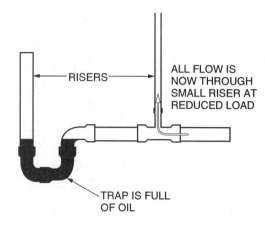

Figure 24-42. Refrigerant vapor and oil are moving up the small riser only because the compressor is running at reduced load.

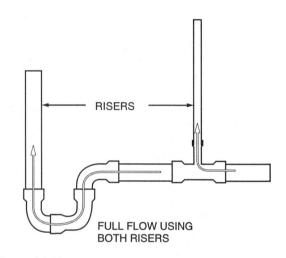

Figure 24-43. This oil trap is operating at full capacity with the refrigerant vapor moving in both lines and the oil moving in the large line because the compressor is running at full load.

discharge lines: the line that carries all of the refrigerant all of the time, which is the horizontal portion of the line, and the double riser. The first thing we need to know is the line length of the horizontal line from the compressor to the double riser and the equivalent length of the fittings. The line length in Figure 24-41 for the discharge line is 10 feet. Notice the method used to enter the compressor in the small drawing to the bottom left of the system. As stated earlier, the number of turns is used to prevent vibration transmission from the compressor to the refrigerant lines.

The correction factor for sizing the discharge line is found using the table in Figure 24-44. It is .96. This factor is multiplied times the compressor tonnage of 40 to arrive at a corrected tonnage of 38.4 for use in Figure 24-45. A trial pipe size of $1^5/_8$ inches is found to have a pressure drop of 4 psig.

DISCHARGE GAS LINE TONNAGE CORRECTION FACTORS, R-22

COND. TEMP.	SATURATED EVAPORATOR TEMPERATURE °F																
	-30	-25	-20	-15	-10	-5	0	5	10	15	20	25	30	35	40	45	50
85	1.21	1.20	1.19	1.18	1.17	1.16	1.15	1.14	1.13	1.12	1.11	1.11	1.10	1.09	1.08	1.08	1.07
90	1.18	1.17	1.16	1.15	1.14	1.13	1.12	1.11	1.10	1.10	1.09	1.08	1.07	1.07	1.06	1.05	1.05
95	1.16	1.15	1.14	1.13	1.12	1.11	1.10	1.09	1.08	1.07	1.07	1.06	1.05	1.04	1.04	1.03	1.02
100	1.14	1.13	1.12	1.11	1.10	1.09	1.08	1.07	1.06	1.05	1.05	1.04	1.03	1.02	1.02	1.01	1.00
105	1.13	1.12	1.11	1.09	1.08	1.08	1.07	1.06	1.05	1.04	1.03	1.02	1.02	1.01	1.00	0.99	0.99
110	1.12	1.10	1.09	1.08	1.07	1.06	1.05	1.04	1.03	1.03	1.02	1.01	1.00	0.99	0.99	0.98	0.97
115	1.11	1.09	1.08	1.07	1.06	1.05	1.04	1.03	1.02	1.01	1.01	1.00	0.99	0.98	0.98	0.97	0.96
120	1.10	1.09	1.08	1.06	1.05	1.04	1.03	1.02	1.01	1.01	1.00	0.99	0.98	0.97	0.97	0.96	0.95
125	1.10	1.08	1.07	1.06	1.05	1.04	1.03	1.02	1.01	1.00	0.99	0.98	0.97	0.97	0.96	0.95	0.94
130	1.09	1.08	1.07	1.06	1.05	1.04	1.02	1.01	1.00	1.00	0.99	0.98	0.97	0.96	0.95	0.95	0.94
135	1.10	1.08	1.07	1.06	1.05	1.04	1.03	1.02	1.01	0.99	0.99	0.97	0.96	0.95	0.95	0.94	0.94
140	1.10	1.09	1.08	1.06	1.05	1.04	1.03	1.02	1.01	1.00	0.99	0.98	0.97	0.96	0.95	0.94	0.93
145	1.12	1.10	1.09	1.07	1.06	1.05	1.04	1.02	1.01	1.00	0.99	0.98	0.97	0.96	0.96	0.95	0.94

Figure 24-44. Tonnage correction factor table for sizing the discharge line. *Courtesy Trane Company.*

TONNAGE CAPACITIES OF DISCHARGE LINES DELIVERING HOT REFRIGERANT-22 VAPOR FROM COMPRESSOR TO CONDENSER

Based on 40 F suction and 105 F condensing. For other conditions apply correction factor to design tons before entering this table.

EQUIV. LENGTH IN FEET	TOTAL PRESS. DROP PSI	COPPER TUBE SIZE — OD — TYPE L											
		½	⅝	¾	⅞	1⅛	1⅜	1⅝	2⅛	2⅝	3⅛	3⅝	4⅛	5⅛
20	1	1.1	2.1	3.6	5.5	11.2	19.9	31.0	64.2	116.0	181.0	266.0	370.0	675.0
	2	1.7	3.1	5.3	8.1	16.8	29.2	45.8	95.0	170.0	265.0	390.0	542.0	1000.0
	3	2.1	3.9	6.7	10.2	21.0	36.5	57.5	120.0	211.0	332.0	484.0	688.0	1270.0
	4	2.5	4.6	7.8	12.0	24.8	41.9	67.8	141.0	250.0	393.0	578.0	810.0	1485.0
	5	2.8	5.1	8.9	13.6	28.1	48.8	76.2	160.0	282.0	445.0	657.0	920.0	1680.0
	6	3.1	5.7	9.8	15.1	30.2	54.0	84.4	178.0	314.0	490.0	725.0	1010.0	1860.0
	7	3.4	6.2	10.7	16.6	33.8	58.9	92.0	193.0	340.0	538.0	798.0	1100.0	2010.0
30	1	0.93	1.7	2.9	4.5	9.2	16.0	25.0	53.0	94.0	148.0	217.0	302.0	557.0
	2	1.4	2.5	4.3	6.6	13.8	23.8	37.2	77.5	140.0	216.0	320.0	445.0	808.0
	3	1.7	3.1	5.2	8.3	17.0	29.9	46.2	97.5	173.0	270.0	402.0	565.0	1035.0
	4	2.0	3.7	6.3	9.7	20.1	35.0	54.0	115.0	205.0	318.0	468.0	659.0	1200.0
	5	2.3	4.2	7.2	11.1	22.9	39.7	61.5	130.0	229.0	360.0	536.0	742.0	1370.0
	6	2.5	4.6	7.9	12.3	25.0	44.1	69.0	144.5	251.0	400.0	592.0	829.0	1515.0
	7	2.7	5.0	8.6	13.2	27.4	47.8	74.0	158.0	278.0	435.0	641.0	902.0	1625.0
40	1	0.8	1.4	2.5	3.8	7.8	13.8	21.3	44.8	79.0	125.0	185.0	260.0	478.0
	2	1.2	2.1	3.7	5.6	11.6	20.0	31.1	65.6	118.0	183.0	270.0	380.0	696.0
	3	1.5	2.7	4.6	7.1	14.5	25.1	39.2	82.5	148.0	228.0	339.0	476.0	875.0
	4	1.7	3.1	5.4	8.3	17.0	29.5	46.0	97.0	174.0	268.0	400.0	556.0	1025.0
	5	1.9	3.5	6.2	9.4	19.2	33.5	52.3	110.0	196.0	305.0	451.0	635.0	1160.0
	6	2.1	4.0	6.8	10.3	21.3	37.2	58.1	128.0	218.0	338.0	502.0	702.0	1280.0
	7	2.3	4.3	7.4	11.3	23.0	40.1	62.4	134.5	237.0	363.0	545.0	760.0	1390.0
50	1	0.7	1.3	2.2	3.4	6.9	12.2	19.0	39.8	70.0	110.0	165.0	230.0	419.0
	2	1.0	1.9	3.2	5.0	10.2	18.1	28.0	58.0	114.0	162.0	240.0	336.0	608.0
	3	1.3	2.4	4.1	6.3	12.9	22.6	34.8	73.6	131.0	205.0	304.0	420.0	775.0
	4	1.5	2.8	4.8	7.4	15.1	26.5	(40.9)	86.0	154.0	241.0	358.0	495.0	905.0
	5	1.7	3.2	5.4	8.3	17.0	29.9	47.0	98.2	175.0	283.0	404.0	562.0	1020.0
	6	1.9	3.5	6.0	9.2	18.9	33.0	51.6	109.6	193.0	302.0	446.0	624.0	1140.0
	7	2.1	3.8	6.5	10.0	20.7	36.2	55.8	117.5	211.0	328.0	489.0	678.0	1245.0
75	1	0.5	1.0	1.8	2.7	5.6	9.6	15.0	31.8	56.8	88.0	130.0	180.0	333.0
	2	0.8	1.5	2.6	4.0	8.2	14.2	22.0	46.5	82.4	130.0	191.0	265.0	497.0
	3	1.0	1.9	3.3	5.0	10.3	17.9	28.0	58.6	104.0	162.0	241.0	332.0	629.0
	4	1.2	2.2	3.9	5.9	12.1	21.0	32.6	69.1	122.0	193.0	283.0	391.0	732.0
	5	1.4	2.5	4.4	6.7	13.8	23.9	37.3	78.2	139.0	218.0	320.0	448.0	829.0
	6	1.5	2.8	4.9	7.4	15.2	26.5	41.5	86.1	152.0	241.0	355.0	494.0	920.0
	7	1.7	3.1	5.3	8.0	16.8	29.9	44.5	96.0	167.0	262.0	388.0	539.0	998.0
100	1	0.5	0.9	1.5	2.3	4.8	8.4	13.0	27.2	48.0	75.0	112.0	158.0	288.0
	2	0.7	1.3	2.2	3.4	7.1	12.3	19.1	40.0	70.8	112.0	167.0	230.0	420.0
	3	0.9	1.6	2.8	4.3	8.8	15.4	24.0	50.7	89.0	140.0	209.0	291.0	531.0
	4	1.0	1.9	3.3	5.0	10.5	18.1	28.3	59.2	104.0	165.0	245.0	340.0	622.0
	5	1.2	2.2	3.8	5.8	11.8	20.8	32.0	66.8	118.0	187.0	277.0	384.0	710.0
	6	1.3	2.4	4.2	6.4	13.0	22.9	35.5	74.1	131.0	206.0	308.0	429.0	780.0
	7	1.4	2.6	4.5	6.9	14.1	24.8	39.1	80.3	142.0	224.0	332.0	464.0	842.0
125	3	0.8	1.4	2.5	3.8	7.8	13.7	20.9	44.1	78.0	122.0	182.0	251.0	468.0
	4	0.9	1.7	2.9	4.4	9.1	15.8	24.6	51.8	92.0	142.0	212.0	299.8	546.0
	5	1.1	1.9	3.3	5.0	10.4	17.9	27.8	58.6	104.0	162.0	232.0	335.0	620.0
	6	1.2	2.1	3.6	5.6	11.5	20.0	30.8	64.5	115.0	180.0	268.0	370.0	685.0
	7	1.3	2.3	3.9	6.0	12.5	21.3	33.2	70.2	126.0	194.0	291.0	402.0	748.0
150	3	0.7	1.3	2.2	3.4	7.0	12.2	19.1	40.0	70.0	110.0	165.0	228.0	420.0
	4	0.8	1.5	2.6	4.0	8.3	14.3	22.2	47.1	83.0	130.0	194.0	267.0	492.0
	5	0.9	1.7	3.0	4.5	9.4	16.2	25.1	53.1	94.0	149.0	220.0	302.0	559.0
	6	1.0	1.9	3.3	5.0	10.4	18.0	28.1	59.5	104.0	163.0	242.0	337.0	621.0
	7	1.1	2.1	3.6	5.4	11.2	19.7	30.2	64.0	113.0	178.0	264.0	368.0	676.0
175	3	0.6	1.2	2.0	3.1	6.4	11.2	17.2	36.1	63.9	100.0	150.0	209.0	388.0
	4	0.8	1.4	2.4	3.7	7.6	13.2	20.3	42.8	75.4	119.0	177.0	244.0	450.0
	5	0.9	1.6	2.7	4.2	8.6	15.0	23.0	48.8	85.2	135.0	200.0	278.0	516.0
	6	1.0	1.7	3.0	4.6	9.5	16.6	25.5	53.9	94.8	150.0	220.0	308.0	569.0
	7	1.0	1.9	3.3	5.0	10.2	18.0	27.9	58.1	103.0	162.0	240.0	332.0	622.0
200	3	0.6	1.1	2.0	3.0	6.2	10.8	16.8	35.0	63.0	96.8	146.0	201.0	366.0
	4	0.7	1.3	2.3	3.5	7.3	12.7	19.8	41.2	74.0	116.0	170.0	238.0	430.0
	5	0.8	1.5	2.6	4.0	8.3	14.3	22.2	47.0	84.0	130.0	193.0	270.0	490.0
	6	0.9	1.7	2.9	4.4	9.1	15.9	24.8	52.0	93.0	143.0	210.0	299.0	542.0
	7	1.0	1.8	3.2	4.8	9.9	16.2	26.8	56.0	100.0	157.0	231.0	323.0	591.0
250	3	0.5	1.0	1.6	2.5	5.2	9.1	14.1	29.5	52.0	82.0	122.0	170.0	315.0
	4	0.6	1.1	1.9	3.0	6.1	10.7	16.7	34.2	61.0	96.0	143.0	200.0	370.0
	5	0.7	1.3	2.2	3.4	6.9	12.2	18.9	39.4	69.4	110.0	162.0	226.0	425.0
	6	0.8	1.4	2.4	3.8	7.6	13.4	21.0	43.5	76.3	121.0	180.0	250.0	470.0
	7	0.9	1.5	2.7	4.1	8.3	14.6	23.9	47.3	83.6	133.0	198.0	272.0	516.0
300	3	0.5	0.9	1.5	2.4	4.9	8.5	13.1	27.9	49.5	76.0	114.0	158.0	286.0
	4	0.6	1.0	1.8	2.8	5.8	10.0	15.4	32.6	57.9	90.5	134.0	185.0	337.0
	5	0.7	1.2	2.1	3.2	6.5	11.2	17.7	37.0	65.4	101.0	152.0	211.0	385.0
	6	0.7	1.3	2.3	3.5	7.3	12.6	19.7	40.9	72.8	112.0	169.0	232.0	424.0
	7	0.8	1.4	2.5	3.8	7.9	13.8	21.1	44.9	79.5	123.0	183.0	254.0	460.0
350	3	0.4	0.8	1.4	2.1	4.3	7.6	11.8	24.6	43.5	68.0	100.0	141.0	266.0
	4	0.5	0.9	1.6	2.5	5.1	8.9	13.9	29.0	51.0	80.0	119.0	165.0	312.0
	5	0.6	1.1	1.8	2.8	5.8	10.1	15.8	33.1	58.1	90.0	134.0	189.0	355.0
	6	0.6	1.2	2.0	3.1	6.4	11.2	17.3	36.4	64.0	100.0	148.0	209.0	393.0
	7	0.7	1.3	2.2	3.4	7.0	12.1	18.9	39.6	70.0	109.0	161.0	227.0	428.0
400	3	0.4	0.8	1.4	2.1	4.3	7.6	11.7	24.4	43.2	67.0	98.0	135.0	252.0
	4	0.5	0.9	1.6	2.5	5.1	8.9	13.8	28.8	49.6	78.0	117.0	160.0	295.0
	5	0.6	1.1	1.8	2.8	5.8	10.1	15.7	32.9	57.9	88.0	132.0	175.0	335.0
	6	0.6	1.2	2.0	3.1	6.4	11.2	17.3	36.2	63.7	98.0	147.0	200.0	365.0
	7	0.7	1.3	2.2	3.4	7.0	12.1	18.8	39.4	69.7	107.0	159.0	210.0	401.0

NOTE: FIGURES IN BOLD FACE TYPE ARE MAXIMUM RECOMMENDED PRESSURE DROPS CALCULATED TO MINIMIZE HOT GAS LINE TEMPERATURE PENALTY. SHADED AREAS ARE FOR GENERAL INFORMATION ONLY.

Figure 24-45. Table for sizing the discharge line. *Courtesy Trane Company.*

The actual equivalent pipe length is as follows:

Fittings (1⅝ inch)	Equivalent Feet of Pipe
4 long radius elbows	10.4
(4 × 2.6 ft.)	
Actual straight pipe length	10.0
Total equivalent feet	20.4

The trial pipe size seems reasonable but needs to be checked against the velocity table in Figure 24-46. Notice that a new correction factor table is used for this velocity and is found using Figure 24-47. The correction factor is calculated by multiplying 40 tons by the correction factor of .88 to arrive at the corrected tonnage of 35.2 tons. When the velocity chart is checked, the velocity is approximately 2700 fpm. This is within the acceptable range of 500 fpm to 4000 fpm.

The corrected tonnage for the riser is the same as for the horizontal run (38.4 tons). Since the riser is divided into two pipes, with one carrying 25% of the load, the corrected tonnage for the small line will be 9.6 tons (.25 × 38.4 = 9.6). The large riser will be sized on the remaining corrected tonnage of 28.8 tons (38.4 − 9.6 = 28.8).

Now let's find the trial pipe size for the large riser carrying 28.8 tons of capacity by using 50 feet of estimated equivalent length of piping. From Figure 24-45, we find that we have two choices, 1⅜ inches at a pressure drop of 5 psig or 1⅝ inches at a pressure drop of 3 psig. Let's check both of these pipe sizes and determine which would be the best choice.

Fittings (1⅜ inch)	Equivalent Feet of Pipe
1. 1 line flow tee fitting	2.7
2. 3 short radius elbows	9.6
(3 × 3.2)	
3. 2 long radius elbows	4.4
(2 × 2.2)	
4. 1 branch flow tee	10.0
Total length of fittings	26.7
Actual straight pipe length	30.0
Total equivalent feet	56.7

Now let's check the velocity for the 1⅜ inch pipe size selection. Using Figure 24-47, we find the correction factor to be the same as before (.88). The corrected tonnage is 28.8. Using Figure 24-46, the velocity for 28.8 tons of capacity in a 1⅜ inch pipe is about 3200 fpm, well within the 4000 fpm maximum. We can use the same chart to check the velocity of the other possible line choice, 1⅝ inch pipe. It is about 2250 fpm. Either of these pipes will carry the load and be within the recommended velocity. Some conservative designers would choose the 1⅝ inch pipe. However, it will cost more for materials and installation to run the larger pipe, so many designers would choose the 1⅜ inch pipe. For best economy, we will use the smaller pipe.

The next pipe to be sized is the small riser. Let's use 50 feet of pipe again for the trial pipe size. Figure 24-45 shows that, at 50 feet and 9.6 tons of capacity, we find only one good choice, 1⅛ inch pipe. Let's total the pipe lengths.

Fittings (1⅛ inch)	Equivalent Feet of Pipe
1. 1 Branch flow tee	8.0
2. 1 Long radius elbow	1.9
3. 1 Line flow tee	2.5
Total length of fittings	12.4
Actual straight pipe length	30.0
Total equivalent feet	42.4

Now let's check the velocity of the refrigerant at 9.6 tons times the correction factor of .88. We find the corrected tonnage to be 8.45 tons. Going to Figure 24-46, we find the velocity of 8.45 tons of capacity in an 1⅛ inch pipe is about 1600 fpm. This is acceptable.

If we use the same chart to find whether the common discharge line leaving the compressor would have been suitable for a single riser, we can see that at 9.6 tons of capacity, the velocity would only be about 700 fpm with the 1⅝ inch common pipe. This velocity is unacceptable because it would not carry the oil up to the condenser, so it would not return to the compressor.

Sizing the suction line is much the same as the discharge line and is a more common calculation. It is easy to see why most designers will use a remote condensing unit with the compressor built in because the discharge line does not have to be sized; the only calculation is for the suction line.

Refer to Figure 24-48 for sizing the suction line. We will start with the common portion of the suction line between the compressor and the riser. The suction line is going to be a much larger pipe because the refrigerant is not as compressed in the suction line. The suction pressure is typically 68.5 psig while the discharge pressure is 278 psig for a condensing temperature of 125°F for R-22.

The first thing we will do is apply the correction factor to the suction line. The correction factor for 125°F condensing temperature and 40°F evaporator temperature is 1.12, as shown in Figure 24-49. The corrected tonnage is 40 × 1.12 = 44.8. Let's use 30 feet as the trial pipe length for the suction line. In Figure 24-50, we find that the first pipe size that will handle 44.8 tons is 2⅝ inch pipe.

Now we'll add the actual pipe fittings to determine whether this size is correct.

Fittings (2⅝ inch)	Equivalent Feet of Pipe
1. 3 long radius elbows	12.6
(3 × 4.2)	
2. 1 line flow tee fitting	4.6
Total length of fittings	17.2
Actual straight pipe length	10.0
Total equivalent feet	27.2

REFRIGERANT LINE SIZING 397

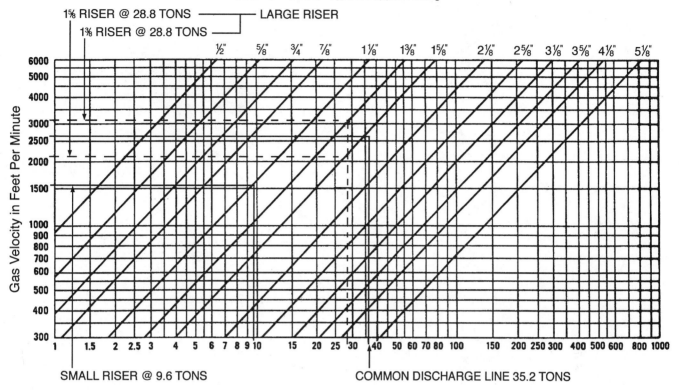

Figure 24-46. Discharge line refrigerant velocity chart. *Courtesy Trane Company.*

DISCHARGE GAS VELOCITY CORRECTION FACTORS, R-22

COND. TEMP.	SATURATED SUCTION TEMPERATURE																
	-30	-25	-20	-15	-10	-5	0	5	10	15	20	25	30	35	40	45	50
85	1.70	1.64	1.59	1.54	1.49	1.46	1.42	1.38	1.35	1.31	1.28	1.26	1.23	1.21	1.18	1.17	1.15
90	1.63	1.58	1.52	1.48	1.44	1.40	1.36	1.32	1.29	1.26	1.23	1.20	1.18	1.15	1.13	1.12	1.10
95	1.57	1.51	1.47	1.42	1.38	1.34	1.30	1.27	1.24	1.21	1.18	1.15	1.13	1.10	1.08	1.07	1.05
100	1.52	1.46	1.42	1.37	1.33	1.29	1.26	1.22	1.19	1.16	1.13	1.11	1.08	1.06	1.04	1.02	1.01
105	1.47	1.42	1.37	1.33	1.29	1.25	1.21	1.18	1.15	1.12	1.09	1.07	1.04	1.02	1.00	0.98	0.97
110		1.37	1.33	1.29	1.25	1.21	1.17	1.14	1.11	1.08	1.05	1.03	1.01	0.98	0.96	0.95	0.93
115			1.29	1.25	1.21	1.17	1.14	1.10	1.08	1.05	1.02	0.99	0.97	0.95	0.93	0.91	0.90
120			1.26	1.22	1.18	1.14	1.11	1.07	1.05	1.02	0.99	0.97	0.94	0.92	0.90	0.89	0.87
125			1.24	1.19	1.16	1.12	1.08	1.05	1.02	0.99	0.97	0.94	0.92	0.90	0.88	0.86	0.85
130				1.17	1.13	1.09	1.05	1.02	0.99	0.97	0.94	0.92	0.89	0.87	0.85	0.84	0.82
135				1.15	1.11	1.07	1.04	1.01	0.97	0.95	0.92	0.90	0.87	0.85	0.83	0.82	0.81
140					1.09	1.05	1.01	0.98	0.95	0.92	0.90	0.87	0.85	0.83	0.82	0.80	0.79
145						1.04	1.00	0.97	0.94	0.91	0.88	0.86	0.84	0.82	0.80	0.79	0.78

Figure 24-47. Tonnage correction table for Figure 24-46. *Courtesy Trane Company.*

398 UNIT 24 APPLICATION OF AIR CONDITIONING EQUIPMENT

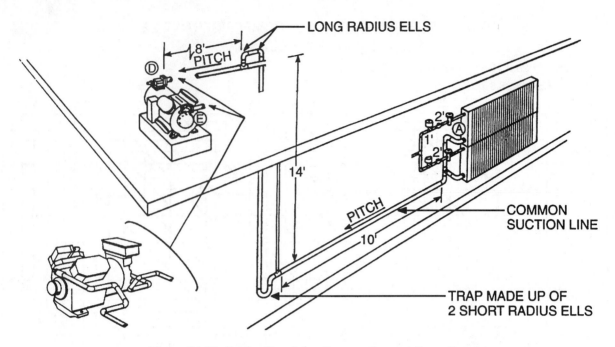

Figure 24-48. Suction line piping diagram. *Courtesy Trane Company.*

SUCTION LINE TONNAGE CORRECTION FACTORS, R-22

COND. TEMP.	SUCTION TEMPERATURE																
	-30	-25	-20	-15	-10	-5	0	5	10	15	20	25	30	35	40	45	50
85	2.13	2.00	1.87	1.75	1.65	1.56	1.45	1.37	1.28	1.20	1.13	1.07	1.02	0.96	0.91	0.87	0.84
90	2.19	2.05	1.92	1.80	1.69	1.60	1.49	1.41	1.31	1.23	1.16	1.09	1.04	0.99	0.93	0.89	0.86
95	2.25	2.11	1.97	1.84	1.73	1.64	1.53	1.44	1.35	1.26	1.19	1.12	1.06	1.01	0.95	0.91	0.88
100	2.31	2.17	2.02	1.89	1.78	1.68	1.56	1.48	1.38	1.30	1.22	1.15	1.09	1.04	0.98	0.94	0.90
105	2.38	2.22	2.08	1.94	1.83	1.73	1.61	1.52	1.42	1.33	1.25	1.18	1.12	1.06	1.00	0.96	0.92
110	2.45	2.29	2.14	2.00	1.88	1.78	1.65	1.56	1.46	1.37	1.29	1.21	1.15	1.09	1.03	0.98	0.94
115	2.52	2.36	2.21	2.06	1.94	1.83	1.70	1.61	1.50	1.41	1.32	1.25	1.18	1.12	1.06	1.01	0.97
120	2.61	2.44	2.28	2.13	2.00	1.88	1.75	1.66	1.55	1.45	1.36	1.28	1.22	1.15	1.09	1.04	1.00
125	2.70	2.53	2.36	2.20	2.07	1.95	1.81	1.71	1.60	1.50	1.41	1.33	1.26	1.19	(1.12)	1.07	1.03
130	2.79	2.61	2.44	2.28	2.14	2.01	1.87	1.77	1.65	1.54	1.45	1.37	1.29	1.23	1.15	1.11	1.06
135	2.91	2.72	2.54	2.37	2.22	2.09	1.94	1.83	1.71	1.60	1.50	1.41	1.34	1.27	1.19	1.15	1.09
140	3.02	2.83	2.63	2.46	2.31	2.17	2.02	1.90	1.77	1.66	1.56	1.46	1.39	1.31	1.24	1.18	1.13
145	3.16	2.95	2.75	2.56	2.40	2.26	2.10	1.98	1.84	1.73	1.62	1.52	1.44	1.37	1.28	1.23	1.17

Figure 24-49. Tonnage correction table for suction line. *Courtesy Trane Company.*

Now we will check the velocity of this pipe using Figure 24-51, for the correction factor and Figure 24-52, for the velocity. The corrected tonnage is found by multiplying the correction factor of 1.12 × 40 tons. We find that the correct tonnage is the same as that used for the pipe chart (44.8). This is a coincidence, as the correction factor must be used for each chart or table. The velocity is found to be approximately 2700 fpm.

Now we will size the large riser. The corrected tonnage for the large riser is 33.6 tons (44.8 × .75 = 33.6). We will use 50 feet for the trial pipe size at 33.6 tons.

We find that we only have one good choice, 2⅛ inch pipe. Let's total the fittings to make sure.

Fittings (2⅛ inch)	Equivalent Feet of Pipe
1. 1 line flow tee fitting	3.8
2. 3 short radius elbows (3 × 5.2)	15.6
3. 2 long radius elbows (2 × 3.4)	6.8
4. 1 branch flow tee fitting	16.0
Total length of fittings	42.2
Actual straight pipe length	14.0
Total equivalent feet	56.2

REFRIGERANT LINE SIZING 399

TONNAGE CAPACITIES OF SUCTION LINES DELIVERING REFRIGERANT-22 VAPOR FROM EVAPORATOR TO COMPRESSOR

SMALL & LARGE RISER
COMMON SUCTION LINE

Based on 40 F suction and 105 F condensing. For other conditions apply correction factor to design tons before entering this table.

EQUIV. LENGTH IN FEET	TOTAL PRESS. DROP PSI	COPPER TUBE SIZE — OD — TYPE L											
		½	⅝	¾	⅞	1⅛	1⅜	1⅝	2⅛	2⅝	3⅛	3⅝	4⅛	5⅛
10	0.5	0.8	1.4	2.5	3.9	7.7	13.6	21.3	44.5	78.5	125	186	260	475
	1.0	1.1	2.1	3.6	5.6	11.3	19.9	31.2	66.0	115.0	181	272	382	699
	2.0	1.6	3.0	5.3	8.1	16.4	28.7	44.8	94.0	166.0	263	395	551	1010
	3.0	2.1	3.8	6.6	10.1	20.4	35.9	56.0	117.0	208.0	326	495	688	1260
	4.0	2.4	4.4	7.7	11.9	24.1	42.2	66.0	137.0	243.0	385	578	807	1490
20	0.5	0.5	1.0	1.7	2.7	5.3	9.5	14.7	30.7	54.0	85.5	128	179	327
	1.0	0.8	1.4	2.5	3.9	7.7	13.6	21.3	44.5	78.5	125.0	186	260	475
	2.0	1.1	2.1	3.6	5.6	11.3	19.9	31.2	66.0	115.0	181.0	272	382	699
	3.0	1.4	2.6	4.5	7.0	14.1	24.6	38.9	80.5	142.0	226.0	336	472	867
	4.0	1.6	3.0	5.3	8.1	16.4	28.7	44.8	94.0	166.0	263.0	395	551	1010
30	0.5	0.4	0.8	1.4	2.1	4.3	7.6	11.9	24.8	43.5	69	103	144	263
	1.0	0.6	1.2	2.0	3.1	6.3	11.0	17.2	35.9	63.5	100	150	211	385
	2.0	0.9	1.7	2.9	4.6	9.2	16.0	25.2	53.0	92.5	147	219	306	565
	3.0	1.1	2.1	3.6	5.6	11.3	19.9	31.2	66.0	115.0	181	272	382	699
	4.0	1.3	2.5	4.2	6.6	13.3	23.3	36.2	76.0	134.0	213	318	440	821
40	0.5	0.4	0.7	1.2	1.8	3.7	6.5	10.2	21.3	37.2	59.0	87.5	123	225
	1.0	0.5	1.0	1.7	2.7	5.3	9.5	14.7	30.7	54.0	85.5	128.0	179	327
	2.0	0.8	1.4	2.5	3.9	7.7	13.6	21.3	44.5	78.5	125.0	186.0	260	475
	3.0	1.0	1.8	3.1	4.8	9.7	17.0	26.6	56.0	98.5	156.0	233.0	324	598
	4.0	1.1	2.1	3.6	5.6	11.3	19.9	31.2	66.0	115.0	181.0	272.0	382	699
50	0.5	0.3	0.6	1.1	1.6	3.3	5.7	9.0	18.9	33.0	52.5	77.5	110	200
	1.0	0.5	0.9	1.5	2.4	4.7	8.3	13.0	27.3	47.8	76.0	113.0	159	290
	2.0	0.7	1.3	2.2	3.4	6.9	12.1	19.0	39.5	70.0	110.0	165.0	232	425
	3.0	0.9	1.6	2.8	4.3	8.6	15.1	23.7	49.5	86.5	139.0	207.0	287	532
	4.0	1.0	1.8	3.2	4.9	10.0	17.6	27.8	57.5	102.0	162.0	241.0	335	621
75	0.5		0.5	0.9	1.3	2.7	4.7	7.2	15.1	26.5	42.2	63	87.5	160
	1.0	0.4	0.7	1.2	1.9	3.8	6.7	10.6	22.0	38.5	61.5	91	128.0	233
	2.0	0.6	1.0	1.8	2.8	5.5	9.7	15.6	31.8	56.0	89.0	132	185.0	336
	3.0	0.7	1.3	2.2	3.4	6.9	12.1	19.0	39.5	70.0	110.0	165	232.0	425
	4.0	0.8	1.5	2.6	4.0	8.0	14.2	22.1	46.5	81.5	129.0	193	270.0	499
100	0.5		0.4	0.7	1.1	2.3	4.0	6.2	12.9	22.5	35.9	53.5	75.5	137
	1.0	0.3	0.6	1.1	1.6	3.3	5.7	9.0	18.9	33.0	52.5	77.5	110.0	200
	2.0	0.5	0.9	1.5	2.4	4.7	8.3	13.0	27.3	47.8	76.0	113.0	159.0	290
	3.0	0.6	1.1	1.9	2.9	5.9	10.3	16.3	33.9	60.0	94.5	141.0	198.0	362
	4.0	0.7	1.3	2.2	3.4	6.9	12.1	19.0	39.5	70.0	110.0	165.0	232.0	425
125	0.5		0.3	0.7	1.0	2.0	3.5	5.4	11.5	20.0	31.8	47.2	67.0	122
	1.0		0.5	0.9	1.4	2.9	5.1	8.0	16.7	29.1	46.5	69.5	96.5	177
	2.0	0.4	0.8	1.4	2.1	4.3	7.4	11.6	24.3	42.5	67.0	101.0	141.0	258
	3.0	0.5	1.0	1.7	2.6	5.2	9.2	14.4	30.0	52.5	83.5	125.0	174.0	319
	4.0	0.6	1.1	2.0	3.0	6.1	10.7	16.8	34.9	62.0	98.5	146.0	205.0	374
150	0.5		0.3	0.6	0.9	1.8	3.2	5.0	10.4	18.0	28.7	43.2	60.0	110
	1.0		0.5	0.9	1.3	2.7	4.7	7.2	15.1	26.5	42.2	63.0	87.5	160
	2.0	0.4	0.7	1.2	1.9	3.8	6.7	10.6	22.0	38.5	61.5	91.0	128.0	233
	3.0	0.5	0.9	1.5	2.4	4.7	8.3	13.0	27.3	47.8	76.0	113.0	159.0	290
	4.0	0.6	1.0	1.8	2.7	5.5	9.7	15.5	31.8	56.0	88.5	132.0	185.0	335
175	0.5			0.5	0.8	1.7	2.9	4.6	9.6	16.6	26.6	39.5	59.0	101
	1.0		0.5	0.8	1.2	2.4	4.3	6.7	14.0	24.3	38.5	57.0	80.5	147
	2.0	0.3	0.7	1.1	1.7	3.5	6.2	9.7	20.2	35.6	57.0	83.5	117.0	214
	3.0	0.4	0.8	1.4	2.2	4.4	7.7	12.0	25.3	43.9	70.0	105.0	146.0	267
	4.0	0.5	0.9	1.6	2.5	5.0	9.0	14.0	29.2	51.0	81.5	122.0	170.0	311
200	0.5			0.5	0.8	1.6	2.7	4.3	8.9	15.5	24.7	36.9	51.5	93.5
	1.0		0.4	0.7	1.1	2.3	4.0	6.2	12.9	22.5	35.9	53.5	75.5	137.0
	2.0	0.3	0.6	1.1	1.6	3.3	5.7	9.0	18.9	33.0	52.5	77.5	110.0	200.0
	3.0	0.4	0.8	1.3	2.0	4.1	7.2	11.2	23.4	40.5	65.0	97.0	136.0	249.0
	4.0	0.5	0.9	1.5	2.4	4.7	8.3	13.0	27.3	47.8	76.0	113.0	159.0	290.0
250	0.5			0.4	0.7	1.4	2.4	3.7	7.9	13.7	21.8	29.2	45.9	83
	1.0		0.3	0.7	1.0	2.0	3.5	5.4	11.5	20.0	31.8	47.2	67.0	122
	2.0		0.5	0.9	1.4	2.9	5.1	8.0	16.7	29.1	46.5	69.5	96.5	177
	3.0	0.4	0.7	1.2	1.8	3.6	6.3	10.0	20.7	36.2	57.5	86.0	120.0	220
	4.0	0.4	0.8	1.4	2.1	4.2	7.4	11.6	24.2	42.0	67.0	101.0	140.0	257
300	0.5			0.4	0.7	1.3	2.2	3.4	7.1	12.5	19.9	29.6	41.2	76
	1.0		0.3	0.6	0.9	1.8	3.2	5.0	10.4	18.0	28.7	43.2	60.0	110
	2.0		0.5	0.9	1.3	2.7	4.7	7.2	15.1	26.5	42.2	63.0	87.5	160
	3.0	0.3	0.6	1.1	1.6	3.3	5.7	9.0	18.9	33.0	52.5	77.5	110.0	200
	4.0	0.4	0.7	1.2	1.9	3.8	6.7	10.6	22.0	38.5	61.5	91.0	128.0	233
350	0.5			0.4	0.6	1.2	2.0	3.2	6.6	11.4	18.3	27.3	38.2	70
	1.0			0.5	0.8	1.7	2.9	4.6	9.6	16.6	26.6	39.5	59.0	101
	2.0		0.5	0.8	1.2	2.4	4.3	6.7	14.0	24.3	38.5	57.0	80.5	147
	3.0		0.6	1.0	1.5	3.0	5.3	8.3	17.3	30.2	47.8	71.0	100.0	182
	4.0	0.3	0.7	1.1	1.7	3.5	6.2	9.7	20.2	35.0	57.0	83.5	117.0	214
400	0.5			0.3	0.5	1.1	1.9	3.0	6.1	10.6	16.9	25.3	35.6	64.0
	1.0			0.5	0.8	1.6	2.7	4.3	8.9	15.5	24.7	36.9	51.5	93.5
	2.0		0.4	0.7	1.1	2.3	4.0	6.2	12.9	22.5	35.9	53.5	75.5	137.0
	3.0		0.5	0.9	1.4	2.8	5.0	7.7	16.3	28.2	44.9	67.0	93.5	171.0
	4.0	0.3	0.6	1.1	1.6	3.3	5.7	9.0	18.9	33.0	52.5	77.5	110.0	200.0
450	0.5				0.5	1.0	1.8	2.8	5.7	10.0	16.0	23.8	33.2	60.5
	1.0			0.5	0.7	1.5	2.6	4.0	8.3	14.6	23.3	34.2	48.5	88.0
	2.0		0.4	0.7	1.1	2.1	3.7	5.7	12.1	21.2	33.6	50.0	70.0	129.0
	3.0		0.5	0.9	1.3	2.7	4.7	7.2	15.1	26.5	42.2	63.0	87.5	160.0
	4.0	0.3	0.6	1.0	1.5	3.1	5.4	8.5	17.8	31.0	49.2	73.5	103.0	187.0
500	0.5				0.5	1.0	1.7	2.6	5.4	9.4	15.1	22.4	31.4	57
	1.0			0.4	0.7	1.4	2.4	3.7	7.9	13.7	21.8	29.2	45.9	83
	2.0		0.3	0.7	1.0	2.0	3.5	5.4	11.5	20.0	31.8	47.2	67.0	122
	3.0		0.5	0.8	1.2	2.5	4.4	6.8	14.3	24.9	39.5	59.0	82.5	152
	4.0	0.3	0.5	0.9	1.4	2.9	5.1	8.0	16.7	29.1	46.5	69.5	96.5	177

NOTE: FIGURES IN BOLD FACE TYPE ARE MAXIMUM RECOMMENDED TONNAGES AT PRESSURE DROPS CALCULATED TO MINIMIZE SUCTION LINE TEMPERATURE PENALTY. SHADED AREAS ARE FOR GENERAL INFORMATION ONLY.

Figure 24-50. Suction line sizing table. *Courtesy Trane Company.*

400 UNIT 24 APPLICATION OF AIR CONDITIONING EQUIPMENT

SUCTION GAS VELOCITY CORRECTION FACTORS, R-22

COND. TEMP.	SUCTION TEMPERATURE																
	-30	-25	-20	-15	-10	-5	0	5	10	15	20	25	30	35	40	45	50
85	3.99	3.53	3.14	2.80	2.50	2.25	2.01	1.81	1.63	1.47	1.33	1.21	1.10	1.00	0.91	0.83	0.76
90	4.10	3.63	3.22	2.87	2.57	2.30	2.06	1.86	1.67	1.51	1.37	1.24	1.12	1.02	0.93	0.85	0.78
95	4.21	3.72	3.51	2.95	2.64	2.36	2.12	1.91	1.71	1.55	1.40	1.27	1.15	1.05	0.95	0.87	0.79
100	4.32	3.83	3.40	3.03	2.71	2.42	2.17	1.96	1.76	1.59	1.43	1.30	1.18	1.07	0.98	0.89	0.81
105	4.45	3.93	3.49	3.11	2.78	2.49	2.23	2.01	1.80	1.63	1.47	1.33	1.21	1.10	1.00	0.91	0.83
110	4.58	4.05	3.60	3.20	2.86	2.56	2.29	2.06	1.85	1.67	1.51	1.37	1.24	1.13	1.03	0.94	0.85
115	4.72	4.18	3.71	3.30	2.95	2.63	2.36	2.12	1.91	1.72	1.55	1.41	1.28	1.16	1.06	0.96	0.88
120	4.88	4.31	3.82	3.41	3.04	2.72	2.43	2.19	1.97	1.77	1.60	1.45	1.31	1.19	1.09	0.99	0.90
125	5.06	4.47	3.97	3.53	3.15	2.81	2.52	2.26	2.04	1.83	1.66	1.50	1.36	1.23	(1.12)	1.02	0.93
130	5.23	4.62	4.10	3.64	3.25	2.90	2.60	2.34	2.10	1.89	1.71	1.54	1.40	1.27	1.15	1.05	0.96
135	5.45	4.80	4.26	3.79	3.37	3.01	2.70	2.42	2.17	1.96	1.77	1.60	1.45	1.31	1.19	1.09	0.99
140	5.66	4.99	4.42	3.93	3.50	3.13	2.80	2.51	2.25	2.03	1.83	1.66	1.50	1.36	1.24	1.13	1.03
145	5.92	5.22	4.62	4.10	3.65	3.26	2.91	2.62	2.35	2.11	1.90	1.72	1.56	1.41	1.28	1.17	1.07

Figure 24-51. Tonnage correction factor for suction line refrigerant velocity. *Courtesy Trane Company.*

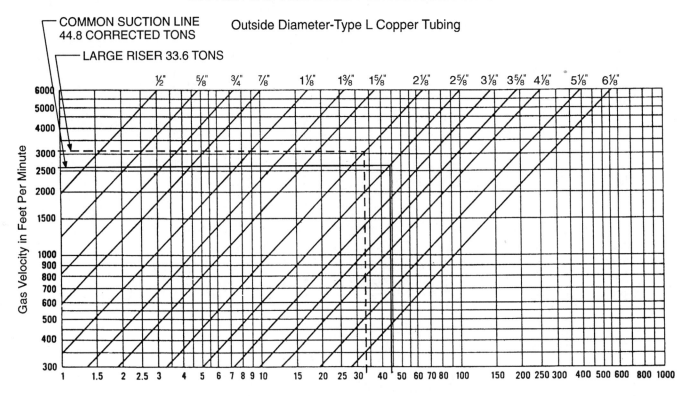

Figure 24-52. Suction line velocity chart. *Courtesy Trane Company.*

The 56.2 feet is close enough to the 50 feet used for the trial pipe size.

Now let's check the velocity. The corrected tonnage for the large riser is 33.6 tons from the previous calculation. The velocity chart (Figure 24-52), shows a velocity of about 3100 fpm for 2⅛ inch pipe at 33.6 tons. This is acceptable.

The small riser is sized by again using 50 feet as the trial pipe length and a corrected tonnage of 11.2 tons (44.8 × .25 = 11.2). The chart shows that 1⅜ inch pipe that is carrying a capacity of 11.2 tons will have a pressure drop of about 2 psig.

Let's total the actual pipe length and check the velocity.

Fittings (1³⁄₈ inch)	Equivalent Feet of Pipe
1. 1 branch flow tee fitting	10.0
2. 1 long radius elbow	2.2
3. 1 line flow tee fitting	2.7
Total length of fittings	14.9
Actual straight pipe length	14.0
Total equivalent feet	28.9

We could recalculate this using 1¹⁄₈ inch pipe, but before we do, let's look at the velocity using 1³⁄₈ inch pipe. Coincidentally, the corrected tonnage for the velocity table is the same (11.2 tons). The velocity chart shows about 2500 fpm for 1³⁄₈ inch pipe. Look up to 1¹⁄₈ inch pipe and you will find that it is above the 4000 fpm mark and is therefore inadequate.

As with the discharge line, we can easily see that a single suction riser cannot be used in this application. The corrected tonnage at the minimum load is 11.2 tons. When we go to the velocity chart at 11.2 tons for the single riser size of 2⁵⁄₈ inches, we see that the velocity is about 700 fpm, far below the minimum of 1000 fpm for vertical risers.

As seen from this exercise, this is quite a long process, but must be completed carefully to ensure correct system operation. This is a simplified version compiled and published in the *Trane Reciprocating Refrigeration Manual*.

Special piping applications would involve multiple compressors with common suction and discharge lines. This is used where multiple compressors are applied to provide the desired capacity control for a particular application. For example, in a building where an engineer wants to use reciprocating compressors and the building load is more than a single compressor can handle, two or more compressors can be manifolded together. Figure 23-53 shows two large compressors that can operate either independently or at the same time. Notice that the two compressor crankcases are manifolded together so that a common oil level is maintained in both compressors. This must be done so that if a single compressor is operating alone, it will not receive all of the return oil. Free movement of oil from compressor to compressor must be allowed. Also notice that the common vent line is both above and below the oil level (Figure 24-54). This prevents the vapor pressure from becoming greater in one compressor than the other, causing oil movement when it is not wanted.

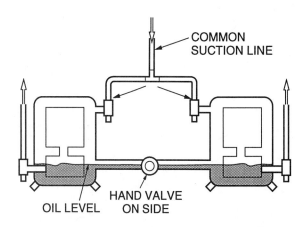

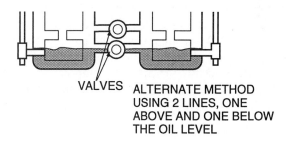

Figure 24-54. The vent lines are located both above and below the oil level to prevent unequal pressures from causing oil migration.

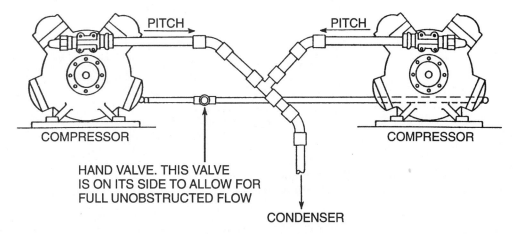

Figure 24-53. Two compressors manifolded together and the interconnection oil lines. *Courtesy Trane Company.*

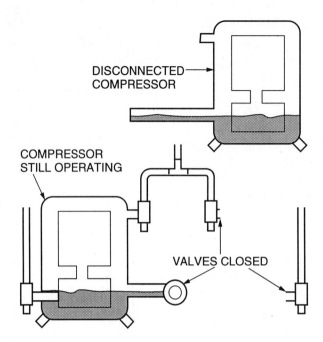

Figure 24-55. If one compressor burns out, it can be valved off and repaired or changed.

This practice has been used for years for open-drive compressors and it works well. However, it can cause some problems with hermetic compressors if there is a motor burnout. In this case, there will be some oil contamination of the good compressor if the other compressor burns out. This may be dealt with if discovered before the contamination harms the second compressor. If you are working with dual compressors and one burns out, valve it off and remove it from the system (Figure 24-55).

Then pump the refrigerant into the condenser and replace the filter-drier cores. Check the oil for acid content. Run the system as long as possible to circulate oil through the filters. Change the filters as many times as needed to clean the system. The inoperative compressor can then be changed and the new one started up. This can be accomplished with a minimum of problems if it is done in a timely manner. The filter-drier arrangement shown in Figure 24-56 allows the second compressor to run and

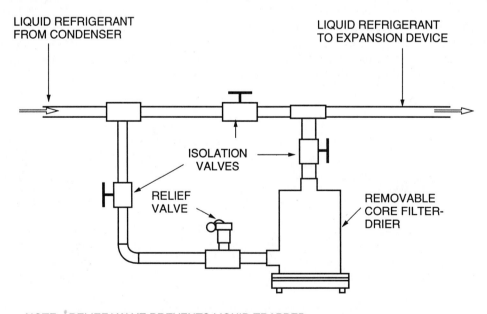

Figure 24-56. The filter-drier in this piping arrangement may be used to change drier cores without stopping the compressor. This is used in critical applications.

operate while the drier cores are changed. This arrangement would be used for critical applications where the compressor must continue to run, such as in large computer rooms.

24.5 SPECIAL-APPLICATION COOLING EQUIPMENT

Special-application equipment is any nonstandard equipment used to condition a space. There are three types of special-application equipment that will be discussed here: evaporative cooling, ductless air conditioning, and portable air conditioning.

Evaporative Cooling

Evaporative cooling is a method that uses the evaporation of water to lower the temperature of the air. It is most commonly used in warm, dry areas where more evaporation would occur, although it can be effective for some purposes in more humid areas. This type of cooling works on the principle that when water is evaporated, heat is absorbed. If heat is absorbed from the air or from a surface covered with the evaporating water, that air or surface will be cooled. For example, if you wet your arm and blow on it, your arm will feel cooler as the moisture evaporates. Figure 24-57 shows an example of an evaporative cooling unit. Keep in mind that this is not refrigerated air conditioning, but the cooling of air by the evaporation of water.

Figure 24-58. A psychrometer indicates both wet and dry bulb temperatures.

Using ordinary evaporative cooling, the dry bulb (DB) temperature of the air can be cooled close to the wet bulb (WB) temperature. As discussed previously, the DB temperature is the temperature taken with a standard thermometer. The WB temperature is the temperature recorded by a thermometer which has its bulb enclosed by a wet sock, which is placed in a moving airstream or the thermometer moved briskly through the air. A sling psychrometer (Figure 24-58) will indicate both the DB and WB temperatures. Evaporative cooling is limited by the fact that it cannot produce cooling below the WB temperature. When conditions can only be satisfied with temperatures below the WB temperature, refrigerated cooling is required.

Significant cooling can be achieved by evaporative cooling in those areas of the country that are very warm and dry. For example, the design conditions in Phoenix, AZ are 105°F DB and 71°F WB. Evaporative cooling equipment could produce cooling to approximately 75°F DB. This is much cooler than the outside temperature of 105°F. A typical DB temperature in Atlanta, GA may be 90°F, with a WB of 73°F. The inside DB temperature may be lowered to about 78°F through evaporative cooling. On a design temperature day, the WB temperature may be 78°F and the air may be cooled to 82°F, which may only be satisfactory for special applications, such as laundries or kitchens where the temperature would typically be much higher than about 82°F.

The cost to purchase, install, and the operate evaporative cooling equipment are generally much lower than those for refrigerated cooling, so this type of equipment is popular in areas and situations where it can be used effectively.

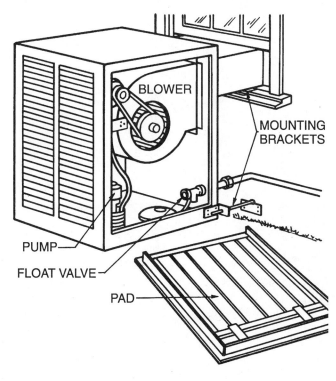

Figure 24-57. Evaporative cooling unit.

404 UNIT 24 APPLICATION OF AIR CONDITIONING EQUIPMENT

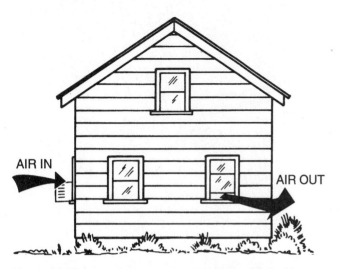

Figure 24-59. An exhaust system must be provided with evaporative cooling because of the volume of fresh air used.

Evaporative cooling relies on a large volume of air being ducted from the outside to the inside of the structure. This application uses 100% outside air and has a tendency to pressurize the structure. We will see later in this discussion that the air in a structure may be changed every 1 to 2 minutes using evaporative cooling. This is a lot of air, so provisions must be made to exhaust the air that enters the structure (Figure 24-59).

Operation Outside air is typically pulled through the evaporative cooler media where evaporation takes place and the air temperature is reduced. At the same time the air temperature is reduced, the moisture is transferred to the air so the leaving air is very humid. This works out to be very satisfactory for applications in the southwestern part of the U.S. because the humidity of the outside air tends to be very low. In the case of a laundry or kitchen in the more humid parts of the country, additional humidity is not a problem as the exhaust system removes it from the area.

Figure 24-60 shows a cutaway of an evaporative cooling unit. Some systems may have a prefilter through which the outside air is first drawn. This prefilter keeps insects out, reduces the formation of algae by keeping out the sunlight, and filters dust and dirt from the system. The air then passes through a wetted media (evaporative pad). This may be a standard fiber media or it may be made from various materials including special cellulose paper with anti-rot salts and rigidifying agents, or large glass fibers bound with inorganic, noncrystalline fillers.

(A) Celldek Media
(B) Optional Maxaire Prefilter
(C) Washer Cabinet and Water Pan
(D) Float Operated Valve
(E) Bleed-off Valve
(F) Water Distribution Manifold
(G) Blower
(H) Motor
(J) Blower Cabinet
(K) Pump

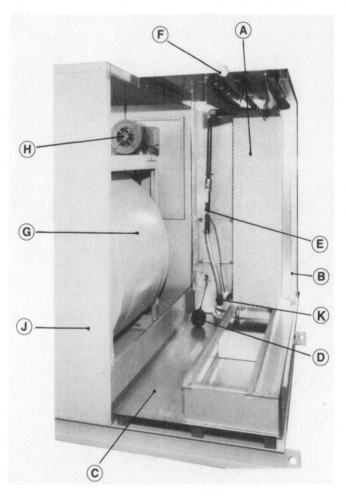

Figure 24-60. Evaporative cooling unit. *Courtesy Alton Company.*

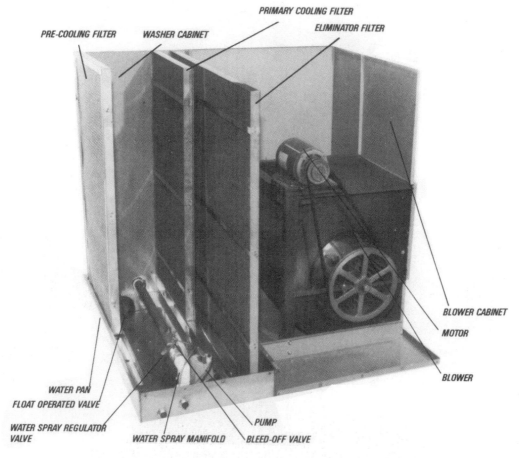

Figure 24-61. Water distribution manifold. *Courtesy Alton Company.*

These media may be designed to provide an air washing action to remove dust and dirt from the airstream and/or the media may be designed so that it creates turbulence to mix the moisture and air to achieve maximum heat transfer. The water may be distributed to the media through a spray manifold, as shown in Figure 24-61. Another manufacturer uses a slinger assembly to sling the water onto the media (Figure 24-62).

The air is then blown into the conditioned space either directly or through a duct system. Centrifugal, forward curved, or propeller fans may be used. As the moisture evaporates, latent heat is removed from the air. Water that runs through the media is collected in a pan or sump. A bleedoff valve is often used to allow a prescribed amount of water to be removed and replaced by fresh water. This helps to keep the buildup of minerals in the water under control, preventing clogging of parts of the system. Generally, a float valve is utilized to maintain the desired water level in the sump. In systems utilizing a manifold water distribution system to the media, a pump is included to pump the water from the sump to the manifold.

Another design utilizes a coil, and a secondary and intermediate airstream, as well as the primary airstream, to produce greater efficiency and cooler temperatures (Figure 24-63). Cool water from the sump is pumped through a finned-tube coil. The primary airstream flows through the fins of the coil, cooling the air. This primary air then flows through the evaporative media, which cools it further. The water takes on some heat as it proceeds through the coil to the top of the evaporative media. As it passes through the media it is further cooled by the secondary, intermediate, and primary airstreams. This water can be cooled as much as 5°F below the outside WB temperature.

Maintenance The evaporative cooler is an excellent air filter because all of the air passes through a bath of water. This is good because the outside air contains contaminants and they are collected in the water. This provides excellent air quality but creates a maintenance problem. These contaminants must be cleaned from the water basin in the unit.

Regular maintenance is required or the system will become contaminated with dirt and mold, which can be hazardous to health. Maintenance requires regular washing of the water basin and the evaporative media. The water pump also requires lubrication of the bearings.

406 UNIT 24 APPLICATION OF AIR CONDITIONING EQUIPMENT

SLINGER ASSEMBLY

Figure 24-62. Slinger assembly method of water distribution. *Courtesy of Bessam-Aire.*

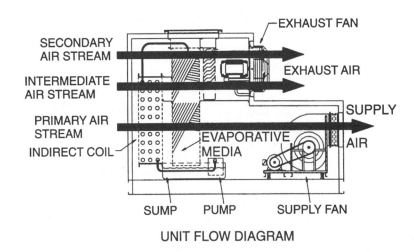

UNIT FLOW DIAGRAM

Figure 24-63. High-efficiency evaporative cooler. *Courtesy Aztech Sensible Cooling, Inc.*

Sizing Choosing the correct evaporative cooler is accomplished by using a manufacturer's catalog. Figure 24-64 shows a selection procedure used by one manufacturer where a map of the country is used to select a zone and a chart is used to select the correct number of air changes. The steps used for the selection are as follows:

Area Cooling:

Step 1. Select the zone from the "Cooling Zone Map".
Step 2. Decide if the interior heat load is "Normal" or "Excessive". "Normal" means no heat-producing equipment or large crowds. "Excessive" means an area with heat-generating equipment or a large number of people. Examples of heat-generating equipment are large motors, stoves, ovens, or industrial process equipment.
Step 3. Decide if the area to be cooled is protected from the heat of the sun. "Good" sun protection is when the building is insulated or the exterior walls are shaded. Windows do not face the sun during the heat of the day. "Poor" sun protection is when the building is uninsulated, sun beams down on the building, and windows face the sun during the afternoon hours.
Step 4. Refer to the "Air Change Table" and determine the number of minutes required per air change.
Step 5. Calculate the building size (length in feet × width in feet × height in feet = _____ cubic feet).

Note: If the building has a very high ceiling, use a height of 15 to 18 feet. The Turbospray unit should be installed so it discharges into the lower part of the room while an exhaust fan draws air from the upper level.

Step 6. Reduce the building size by the room taken up with machinery, inventory, etc. (building size in cubic feet - machinery, etc. in cubic feet = space to cool in cubic feet).
Step 7. Divide the space to cool in cubic feet from Step 6 by the number of minutes required per air change from Step 4. This gives the cfm rating of the Turbospray.
Step 8. Refer to the "Air Delivery Table" to select the Turbospray model number. If the required cfm rating falls between models, select the larger of the two.

SELECTING A TURBOSPRAY

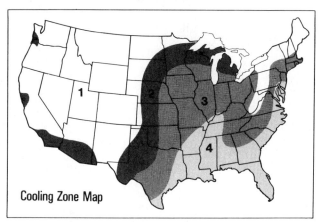

Cooling Zone Map

The Turbospray can be used to area cool or spot cool. Area cooling is used to cool an entire work area where workers constantly move about and cooling requirements do not vary greatly from one location to the next. Spot cooling is used when building volume is large and workers are located near machines that give off heat. Cool air from the Turbospray is distributed through ductwork and discharged directly on the workers.

AIR CHANGE TABLE

Interior Heat Load	Sun Protection	Minutes Per Air Change			
		Zone 1	Zone 2	Zone 3	Zone 4
Excessive	Poor	1½	1	¾	½
	Good	2	1½	1	¾
Normal	Poor	2	1½	1	¾
	Good	2	2	1½	1

Figure 24-64. Sizing an evaporative cooler. *Courtesy Alton Company.*

TURBOSPRAY SERIES — AIR DELIVERY TABLE

MODEL NO.	PUMP HP	CFM	ISP ("W.C.")	1/8" HP	1/8" RPM	1/4" HP	1/4" RPM	1/2" HP	1/2" RPM	3/4" HP	3/4" RPM	1" HP	1" RPM	1 1/4" HP	1 1/4" RPM
HEDM-550 (WDM-550) FFA-15.6φ 18" FC DWDI	1/6	3000	0.21	0.5	340	0.5	390	0.75	500	–	–	–	–	–	–
		4000	0.25	0.75	370	0.75	420	1.5	500	1.5	580	–	–	–	–
		5000	0.30	1	400	1.5	450	1.5	530	2	595	3	660	3	710
		5500	0.33	1.5	420	1.5	470	2	545	3	610	3	665	3	725
		6000	0.37	2	470	2	500	2	560	3	630	3	680	5	735
		6500	0.41	2	495	2	520	3	580	3	640	5	700	5	750
HEDM-750 (WDM-750) FFA-19.5φ 20" FC DWDI	1/6	5000	0.25	0.75	320	1	360	1.5	425	–	–	–	–	–	–
		6000	0.29	1	350	1.5	385	1.5	450	2	510	–	–	–	–
		7000	0.34	1.5	380	1.5	415	2	470	3	530	3	585	3	640
		8000	0.40	2	415	3	445	3	500	3	550	5	595	5	650
		9000	0.48	3	455	3	480	5	530	5	580	5	620	5	665
		10000	0.57	5	495	5	520	5	565	5	610	5	650	7.5	690
HEDM-1250 (WDM-1250) FFA-23.8φ 25" FC DWDI	1/6	8000	0.31	1.5	265	1.5	300	2	365	–	–	–	–	–	–
		9000	0.36	1.5	280	2	315	3	370	3	420	–	–	–	–
		10000	0.41	2	300	3	325	3	380	5	430	5	475	–	–
		11000	0.48	3	325	3	345	5	395	5	440	5	480	7.5	525
		12000	0.55	3	345	5	370	5	410	5	455	7.5	495	7.5	535
		13000	0.65	3	370	5	390	5	430	7.5	475	7.5	515	7.5	545
HEDM-1680 (WDM-1550) FFA-36.0φ 27" FC DWDI	1/6	14000	0.37	3	280	5	305	5	345	5	385	7.5	425	–	–
		15000	0.41	5	290	5	315	5	350	7.5	390	7.5	430	–	–
		16000	0.45	5	308	5	330	5	370	7.5	405	7.5	440	7.5	474
		17000	0.49	5	320	7.5	340	7.5	380	7.5	415	7.5	445	10	480
		18000	0.55	7.5	340	7.5	355	7.5	395	7.5	425	10	455	10	485
		19000	0.61	7.5	355	7.5	375	7.5	410	10	440	10	465	15	495
		20000	0.67	7.5	375	10	390	10	425	10	455	15	480	15	505
HEDM-2480 (2) (WDM-1250) FFA-47.6φ 27" FC DWDI	1/6 (Qty. 2)	18000	0.36	5	315	7.5	330	7.5	365	7.5	400	10	430	10	465
		19000	0.38	7.5	325	7.5	345	7.5	375	10	415	10	440	10	470
		20000	0.41	7.5	340	7.5	355	10	390	10	425	10	450	15	475
		21000	0.44	7.5	355	7.5	370	10	400	10	435	15	465	15	485
		22000	0.48	10	370	10	385	10	410	15	445	15	475	15	500
		23000	0.51	10	385	10	400	15	425	15	450	15	485	15	510
		24000	0.55	10	400	15	410	15	440	15	465	15	495	15	525
HEDM-2880 (2) (WDM-1550) FFA-72φ 30" FC DWDI	1/6 (Qty. 2)	21000	0.27	5	250	7.5	270	7.5	300	7.5	335	10	370	10	400
		22000	0.29	7.5	260	7.5	275	7.5	310	10	340	10	375	15	405
		23000	0.30	7.5	265	7.5	285	10	315	10	345	10	380	15	405
		24000	0.31	7.5	275	10	290	10	325	10	350	15	385	15	410
		25000	0.32	10	285	10	300	10	330	15	355	15	385	15	415
		26000	0.34	10	290	10	305	15	335	15	365	15	390	15	420
		27000	0.35	10	300	10	315	15	345	15	375	15	395	15	425
		28000	0.37	15	310	15	320	15	350	15	380	15	400	20	425

Figure 24-64. Sizing an evaporative cooler (continued).

Example:

Step 1. Small plastics factory located in Houston, Texas. See "Cooling Zone Map" to get Zone 4.

Step 2. Factory has mold machines that give off heat. Interior heat load is "Excessive".

Step 3. Factory is exposed to sun throughout the day. Sun protection is "Poor".

Step 4. From "Air Change Table" get .5 minute per air change.

Step 5. Calculate the size of the factory (50 ft. length × 30 ft. width × 10 ft. height = 15,000 cubic feet).

Step 6. Room occupied by machinery and inventory is 2200 cubic feet (15,000 cubic feet - 2200 cubic feet = 12,800 cubic feet).

Step 7. Cfm rating = $\dfrac{12{,}800 \text{ cubic feet}}{.5 \text{ min.}} = 25{,}600$ cfm

Step 8. From the "Air Delivery Table" select model HEDM-2880.

Ductless Air Conditioning

As the name implies, ductless air conditioning has no bulky ducts that must be run from the indoor air handling unit to the spaces to be cooled. Rather than ducts, there is an indoor unit housing a fan, evaporator, and controls for each area to be conditioned. In typical installations, there is normally one outdoor unit for every one or two indoor units. The indoor unit(s) are connected to the outdoor unit with piping for the suction and liquid lines.

Figure 24-65 shows a photo of a typical ductless unit. There are many styles and sizes available. Figure 24-66 shows a photo of a typical outdoor unit. Tubing and control wiring must be run between units and condensate drains must be provided for each indoor unit. Some units are manufactured to be controlled with a remote control (Figure 24-67).

Ductless indoor units can often be added in buildings with hot water or baseboard electric heating when a duct-type system would be difficult to install.

The indoor units are designed for many different types of installations such as on the floor, on the wall, on the ceiling, or in the ceiling.

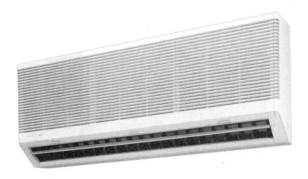

Figure 24-65. Ductless air conditioner evaporator section. *Courtesy Toshiba Corporation.*

Figure 24-66. Ductless air conditioner condenser section. *Courtesy Enviromaster International Corporation.*

Figure 24-67. Remote control for ductless air conditioning. *Courtesy Hitachi America, Ltd.*

Portable Air Conditioning

Portable air conditioning may be used when a backup system is needed, such as when the central system is shut down or for spot cooling in extra warm areas on very warm days. It may also be used for special event cooling such as in a tent. There are many other situations when portable or temporary cooling may be used. Figure 24-68 shows a photo of a portable air conditioner. Units such as these are available in many different sizes.

For portable air conditioning to be effective, the condenser intake and exhaust should be ducted away from the space to be cooled. The exhaust air should also be deflected away from the intake air duct. This can be accomplished by providing ducts to a common ceiling plenum. If a ceiling plenum is not available, the ducts can be run to an outside wall, to another room, or through a window or windows. There are many other possibilities. The important thing is to provide condenser intake and exhaust air from a different area than the conditioned space and not to allow the exhaust air directly into the intake duct.

The condensate may be collected in a container located within the cabinet or if convenient, may be piped by gravity or pumped to a drain.

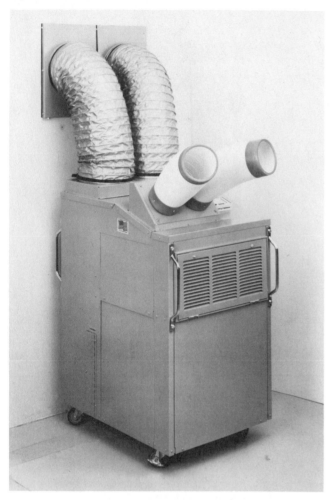

Figure 24-68. Portable air conditioner for spot cooling. *Photo by Air Rover.*

SUMMARY

- There are two types of piping arrangements for split system equipment up to 5 tons: precharged line sets with special fittings and non-precharged line sets. Either type is typically available in lengths of 10, 20, 30, 40, and 50 feet.
- Good piping practices and correct pipe sizing are used to ensure proper oil return to the compressor as all compressors pump some amount of oil out into the system.
- Oil easily mixes with liquid refrigerant and usually travels through the system without problems.
- The maximum recommended velocity for refrigerant traveling in a liquid line is 360 feet per minute (fpm). Excess velocity will cause an increased pressure drop.
- Refrigerant velocity is what moves the oil through the vapor lines in a refrigeration system. The refrigerant velocity must be a minimum of 500 fpm in a horizontal refrigerant vapor line and 1000 fpm in a vertical vapor line. The maximum recommended refrigerant velocity is 4000 fpm. Velocities above 4000 fpm will cause noise and excess pressure drop.
- Pipe sizing controls the velocity of the refrigerant in the piping, which ensures proper oil return.
- Evaporative cooling is an alternative method of cooling a structure by means of the evaporation of water to remove heat from air instead of mechanical refrigeration. It is much less expensive to manufacture and is also less expensive to operate. This process works best in hot, dry climates.
- Evaporative cooling uses 100% outside air.
- Ductless air conditioners are systems that do not use ductwork. These are typically small systems that are used for spot cooling.
- Portable air conditioning systems are air-cooled package systems and are used for temporary cooling. The condenser air is piped away from the general area to be cooled.

REVIEW QUESTIONS

1. Name two different types of equipment commonly used for central air conditioning.
2. What line set lengths are typically furnished by most manufacturers?
3. What is the big disadvantage of water-cooled equipment?
4. What is the big advantage of air-cooled equipment?
5. What moves oil through refrigerant vapor lines?
6. What are the minimum and maximum recommended velocities for refrigerant vapor lines?
7. What is the purpose of the trap at the bottom of a refrigerant vapor line riser?
8. What is the maximum recommended velocity for a refrigerant liquid line?
9. A system using R-22 has a capacity of 50 tons with an evaporator temperature of 40°F and a condensing temperature of 120°F. What is the corrected tonnage for sizing a suction line?
10. What is the recommended suction line size for a line with 150 equivalent feet of pipe?

AIR DISTRIBUTION AND BALANCE 25

OBJECTIVES

Upon completion of this unit, you should be able to

- Describe the prime mover of air in an air conditioning system.
- Describe characteristics of propeller and centrifugal blowers.
- Take basic air pressure measurements.
- Measure air quantities.
- List the different types of air measuring devices.
- Describe the common types of motors and drives.
- Describe duct systems.
- Explain what constitutes good airflow through a duct system.
- Describe a return air system.
- Plot airflow conditions on the air friction chart.

25.1 FORCED-AIR SYSTEMS

As indicated previously, in many situations the air must be conditioned in order to maintain occupant comfort. One way to condition air is to use a fan to move the air over the conditioning equipment. This equipment may consist of a cooling coil, a heating device, a humidifier or dehumidifier, or a device to clean the air. The forced air system uses the same room air over and over again. Air from the room enters the system, is conditioned, and then returns to the room. Fresh air enters the structure either via infiltration around the windows and doors or through ventilation from a fresh air inlet connected to the outside (Figure 25-1).

The forced air system is different from a natural draft system, where the air passes naturally over the conditioning equipment. Baseboard hot water heat is an example of natural draft heat. The warmer water in the pipe heats the air in the vicinity of the pipe. The warmer air expands and rises. Cooler air from the floor takes the place of the heated air (Figure 25-2). There is very little concern for the amount of air moving in a natural convection system.

Air Quantity

The object of a forced air system is to distribute the correct quantity of conditioned air to the occupied space. When this occurs, the air mixes with the room air and creates a comfortable atmosphere in that space. Different spaces

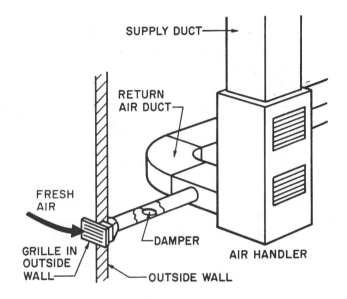

Figure 25-1. Ventilation.

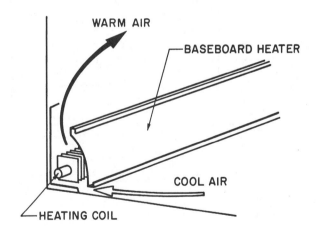

Figure 25-2. Heated air will expand and rise.

have different air quantity requirements, so the same structure may have a variety of cooling loads. For example, a house has rooms of various sizes with individual cooling loads and different occupancy schedules. A bedroom requires less heating and cooling than a large living room. Different amounts of air need to be delivered to each room to maintain comfort conditions (Figures 25-3 and 25-4).

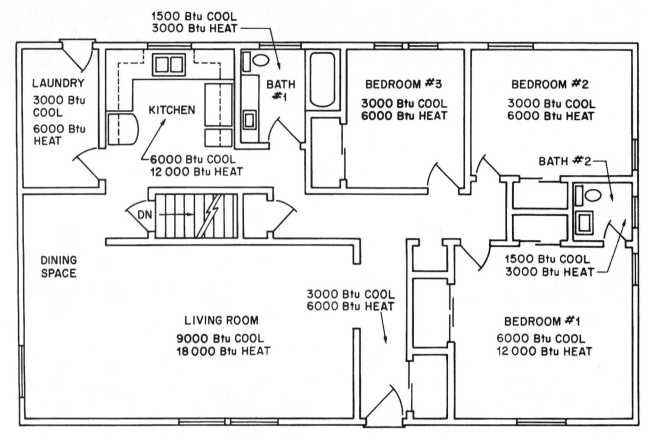

Figure 25-3. This floor plan shows the heating and cooling requirements for each room.

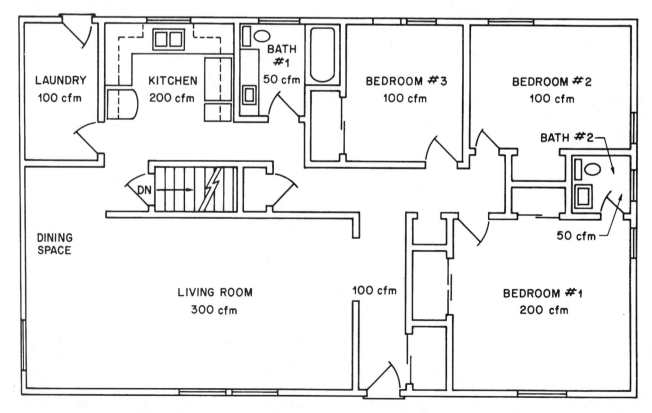

Figure 25-4. This floor plan shows the air quantities delivered to each room.

DUCT SYSTEM PRESSURES 413

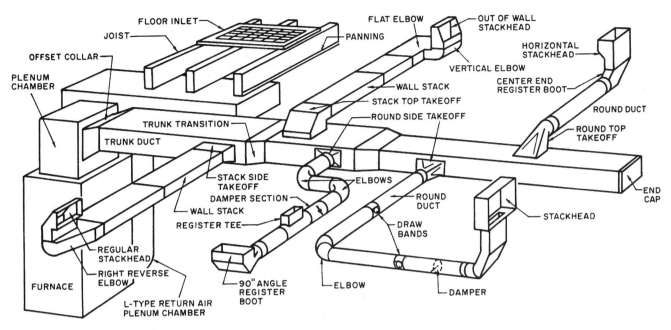

Figure 25-5. Duct fittings.

Another example is a small office building with a large cooling load in the lobby and a small load in the individual offices. The correct amount of air must be delivered to each part of the building so that no area is either overcooled or undercooled.

Components

The components that make up the forced air system include the blower (fan), the air supply system, the return air system, and the grilles and registers where the circulated air enters the room and returns to the conditioning equipment. See Figure 25-5 for an example of duct fittings. When these components are correctly chosen, they work together as a system to maintain the following air conditions:

1. No air movement should be felt in any conditioned space that is normally occupied (i.e., air movement in a conditioned but normally unoccupied space, such as a storage room, might be acceptable).
2. No air noise should be noticed in the conditioned space.
3. The occupants should not notice any temperature swings.
4. The occupants should not be aware that the system is either on or off unless it stops for a long time and the temperature changes.

Unobtrusive system operation is vital to occupant satisfaction.

25.2 DUCT SYSTEM PRESSURES

The blower or fan provides the pressure difference that forces the air into the duct system, through the grilles and registers, and into the room. Air has both weight and resistance to movement. This means that it takes energy to move the air through the duct system to the conditioned space. The fan may be required to push enough air through the evaporator and ductwork for 3 tons of air conditioning. Typically, 400 cubic feet of air must be moved per minute (cfm) per ton of air conditioning. In other words, this system would move 400 cfm × 3 tons or 1200 cubic feet of air per minute. Air has a weight of one pound per 13.35 cubic feet. This fan would be moving 90 pounds per minute (1200 cfm ÷ by 13.35 cubic feet per pound = 89.88) or 5400 pounds per hour (90 pounds per minute × 60 minutes per hour = 5400) or 129,600 pounds per day (5400 pounds per hour × 24 hours per day = 129,600). The blower motor consumes the energy required to move the air (Figure 25-6).

The pressure in a duct system for a residence or a small office building is too small to be measured in pounds per square inch (psi). It is measured in a unit of pressure that is still force per unit of area but in a smaller graduation. The pressure in ductwork is measured in inches of water column (in. w.c.). A pressure of 1 in. w.c. is the pressure necessary to raise a column of water one inch. Air pressure in a duct system is measured with a manometer, which uses colored oil that rises up a tube and is calibrated to read in in. w.c. Figure 25-7 shows an inclined manometer which provides accuracy at very low pressures. Figure 25-8 shows some other instruments that may be used to measure very low air pressures. Note that these devices are all graduated in inches of water column even though they may not contain water.

The atmosphere exerts a pressure of 14.696 psi at sea level at 70°F. Atmospheric pressure will support a column of water that is 34 feet high. This is equal to 14.696 psi.

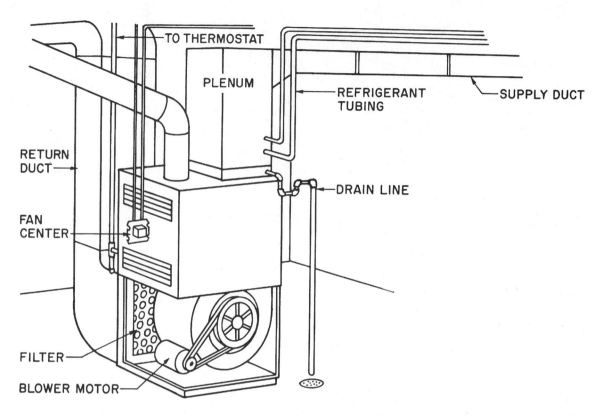

Figure 25-6. The blower motor consumes the energy required to move air through the ductwork.

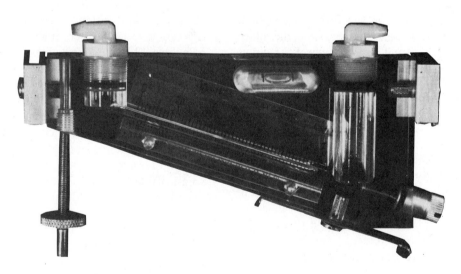

Figure 25-7. Water manometer. *Photo by Bill Johnson.*

One psi will raise a column of water 27.7 inches or 2.31 feet (Figure 25-9). The average duct system will not exceed a pressure of 1 in. w.c. A pressure of 0.05 psig will support a column of water 1.39 inches high (27.7 × 0.05 = 1.39 inches). As you can see, air pressure in ductwork is measured in some very low figures.

Identifying System Pressures

A duct system is pressurized by three forces—*static pressure, velocity pressure, and total pressure*. Static pressure is the same as the pressure on an enclosed vessel, such as a cylinder of refrigerant. This is similar to the pressure of the refrigerant in a cylinder that is pushing outward.

DUCT SYSTEM PRESSURES 415

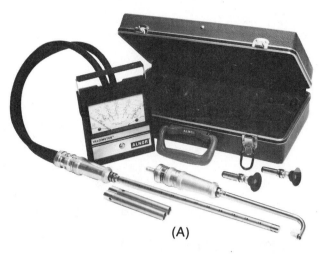

(A)

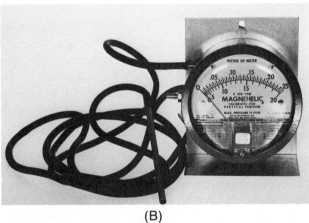

(B)

Figure 25-8. Other instruments used to measure air pressures. *Courtesy of: (A) Alnor Instrument Company, (B) Photo by Bill Johnson.*

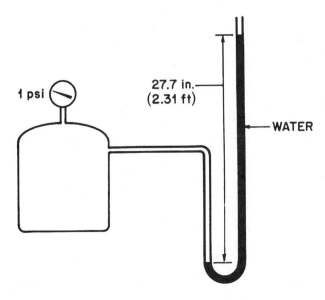

Figure 25-9. Vessel with a pressure of 1 psi inside. The water level in the manometer is at 27.7 inches (2.31 feet).

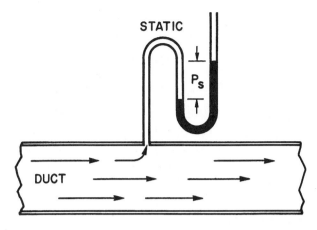

Figure 25-10. Manometer connected to measure the static pressure.

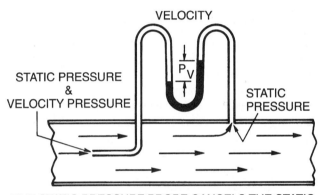

THE STATIC PRESSURE PROBE CANCELS THE STATIC PRESSURE AT THE VELOCITY PRESSURE END AND INDICATES TRUE VELOCITY PRESSURE

Figure 25-11. Manometer connected to measure the velocity pressure.

Figure 25-10 shows a manometer for measuring static pressure. Notice the position of the sensing tube. The probe has a very small hole in the end, so the air rushing by the probe opening will not cause incorrect readings.

The air in a duct system is moving along the duct and therefore has velocity. The velocity and weight of the air create velocity pressure. Figure 25-11 shows a manometer for measuring velocity pressure in an air duct. Notice the position of the sensing tube. The air velocity goes straight into the tube inlet which registers both velocity and static pressure. This probe/manometer arrangement reads the velocity pressure by canceling the static pressure with the second probe. The static pressures balance each other and the velocity pressure is the difference.

The total pressure of a duct can also be measured using a manometer, except that it is applied a little differently (Figure 25-12). Notice that the velocity component

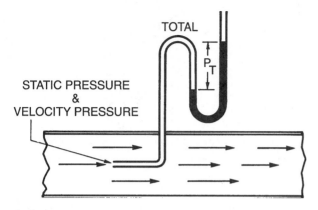

Figure 25-12. Manometer connected to measure the total air pressure.

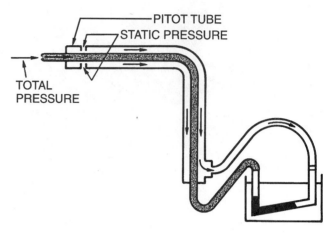

Figure 25-14. Pitot tube set up to measure velocity pressure.

or probe of the manometer is positioned such that the air is directed into the end of the tube. This will register the combined effects of the static and velocity pressures.

Air Measuring Instruments

The water manometer has been mentioned as an air pressure measuring instrument. The instrument used to measure the actual air velocity is the velometer. This instrument actually measures how fast the air is moving past a particular point in the system. Figure 25-13 shows two types of velometers. These instruments should be used as directed by the manufacturer.

(A)

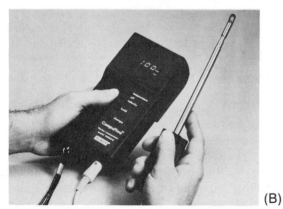

(B)

Figure 25-13. Two types of velometers. *Courtesy of: (A) Photo by Bill Johnson, (B) Alnor Instrument Company.*

A special device called a *pitot tube* was developed many years ago and is used with special manometers for checking duct air pressure at most pressure levels (Figure 25-14).

25.3 TYPES OF FANS

The blower or fan can be described as a device that produces airflow or movement. There are several different types of blowers, but all can be described as *nonpositive displacement air movers*. Remember, the compressor is a positive displacement pump. In other words, when the cylinder is full of refrigerant (or air for an air compressor), the compressor must empty that cylinder or it will break something. The fan is not a positive displacement device and therefore cannot build up the same amount of pressure as a compressor. However, the fan has other characteristics that have to be dealt with.

The two fans that we will discuss in this text are the propeller fan and the forward-curve centrifugal fan, also called the *squirrel cage fan wheel*.

Propeller Fans

The propeller fan is used for exhaust and condenser fan applications. It will handle large volumes of air at low pressure differentials. The propeller can be cast iron, aluminum, or stamped steel and is set into a housing called a *venturi* to encourage airflow in a straight line from one side of the fan to the other (Figure 25-15). The propeller fan makes more noise than a centrifugal fan so it is normally used where noise is not a factor.

Centrifugal Fans

The squirrel cage or centrifugal fan has several advantages. It builds more pressure from the inlet to the outlet and moves more air against more pressure. This fan has a forward-curve blade and a cutoff to shear the air spinning around the fan wheel. This air is then thrown by centrifugal action to the outer perimeter of the fan wheel.

TYPES OF FANS 417

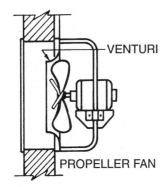

Figure 25-15. Propeller-type fan.

Some of it would keep going around with the fan wheel if it were not for the shear that cuts off the air and sends it out the fan outlet (Figure 25-16). The centrifugal fan is very quiet when properly applied. It meets all requirements of duct systems up to very large systems that are considered high-pressure systems. High-pressure systems have pressures that exceed 1 in. w.c. and use different types of fans, some of them similar to the forward-curve centrifugal fan.

One characteristic that makes troubleshooting the centrifugal fan easier than a propeller fan is the relationship between the volume of air and the power consumption. This fan uses energy at the rate at which it moves air through the ductwork. Therefore, the current draw of the fan motor is proportional to the pounds of air it moves or pumps. The pressure the fan pumps against has little to do with the amount of energy used when operating at close to design conditions. For example, a fan motor that pulls full load amperage at the rated fan capacity will pull less than full load amperage at any value less than the fan capacity. If the fan is supposed to pull 10 A at a maximum capacity of 3000 cfm, it will only pull 10 A while moving this amount of air. The weight of this volume of air can be calculated by dividing (3000 cfm ÷ 13.35 ft.^{3}/lb. = 224.7 lbs.). If the fan inlet is blocked, the suction side of the fan will be starved for air and the current will decrease. If the discharge side of the fan is blocked, the pressure will increase on the discharge side of the fan and the current will decrease because the fan is not handling as many pounds of air (Figure 25-17). The air is merely spinning around in the fan and housing and not being forced into the ductwork. This particular type of fan can be checked for airflow with an ammeter when making simple field measurements. Remember, if the current decreases, the airflow also decreases. If the airflow increases, the current also increases. For example, if the door on the blower compartment is opened, the fan current will go up because the fan will have access to the large opening of the fan compartment.

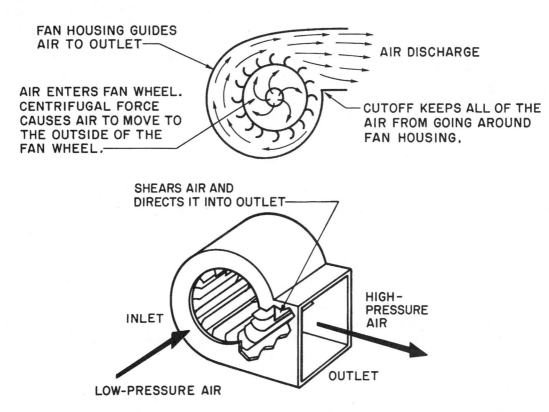

Figure 25-16. Centrifugal fan.

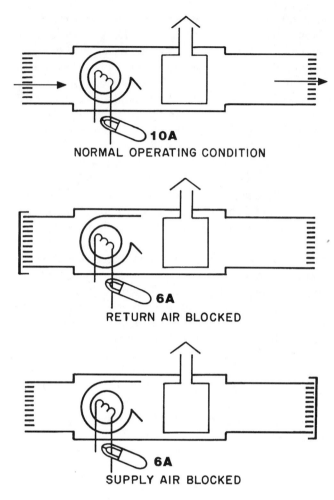

Figure 25-17. Various airflow situations for a typical fan.

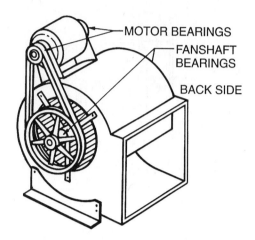

Figure 25-18. This belt-drive uses two pulleys to transfer the motor energy to the fan wheel. The speed of the fan can be modified by adjusting or replacing the pulleys.

Figure 25-19. Direct-drive fan motor. *Courtesy W.W. Grainger, Inc.*

Fan Drives

The centrifugal blower must be turned by a motor. Two drive mechanisms are used: belt drive and direct drive. The belt-drive blower was used exclusively for many years. Early belt-drive motors were usually rated at 1800 rpm and actually ran at 1750 rpm under load. They normally operated very quietly. The motor usually had a start capacitor and would go from a stopped position to 1750 rpm in about one second. The motor generally made more noise at start-up than when running. Today, manufacturers are designing equipment to be more compact and use smaller belt-drive blowers with 3600 rpm motors. These motors actually turn at 3450 rpm under load. They go from 0 rpm to 3450 rpm in about one second and can make quite a large noise on start-up. Sleeve bearings and resilient (rubber) mountings are used to keep bearing noise out of the blower section. Belt-drive blowers have two bearings on the fan shaft and two bearings on the motor. Sometimes, these bearings are permanently lubricated by the manufacturer, so you cannot oil them.

Maintenance points on a belt-drive motor include the drive pulley on the motor, the driven pulley on the fan shaft, and the belt. This type of motor has many uses because the pulleys can be adjusted or changed to change the fan speeds (Figure 25-18). If the pulleys are changed, the belts may also need to be changed.

Recently, most manufacturers have been using direct-drive blowers (Figure 25-19) for fractional horsepower applications. The motor is mounted on the blower housing, usually with rubber mounts, with the fan shaft extending into the fan wheel. The motor is a PSC motor that starts up very slowly, taking several seconds to get up to full speed. It is very quiet and does not have a belt or pulleys that might wear out or require adjustment. Shaded-pole motors are used on some direct-drive blowers; however, they are not as efficient as PSC motors.

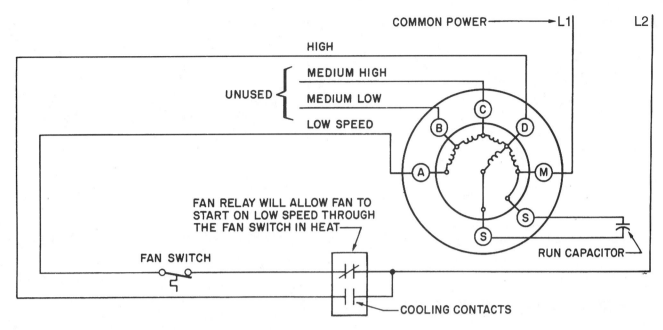

Figure 25-20. Wiring diagram of a multiple-speed motor.

With PSC motors, the fan wheel bearing is located in the motor, which reduces the bearing surfaces from four to two. The bearings may be permanently lubricated at the factory. The front bearing near the fan wheel may be hard to lubricate if a special oil port is not furnished. The fan turns at the same speed as the motor, so multispeed motors are common. The air volume may be adjusted by selecting a different fan motor speed. The motor may have up to four different speeds that can be changed by switching wires at the motor terminal box. Common speeds range from about 1500 rpm down to about 800 rpm. The motor can be operated at a faster speed in the summer to allow additional airflow for cooling (Figure 25-20).

25.4 SUPPLY DUCT SYSTEM

The supply duct system distributes air to the terminal units, registers, or diffusers in the conditioned space. Starting at the fan outlet, the duct can be fastened directly to the blower or blower housing or have a vibration eliminator between the blower and the ductwork. A vibration eliminator (Figure 25-21) is recommended on all installations but is not always used. If the blower is quiet, it may not be necessary.

The duct system must be designed to allow air moving toward the conditioned space to travel as freely as possible, but the duct must not be oversized. Oversized duct is not economical and can cause airflow problems. Duct systems can be designed in various configurations, including: plenum, extended plenum, reducing plenum, or perimeter loop (Figure 25-22). Each system has advantages and disadvantages.

Note: Because duct system design affects both heating and cooling, this text makes reference to both modes.

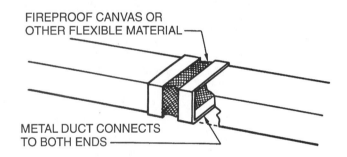

Figure 25-21. Vibration eliminator.

Plenum System

The plenum system is an individual supply system that makes it well suited for applications where the room outlets are all located close to the unit (Figure 25-23). This type of system is fairly easy to install and is very economical from a first-cost standpoint. The supply diffusers (where the air is diffused and blown into the room) are normally located on the outside walls and are used for heating systems that have very warm or hot air as the heating source. Plenum systems work better on fossil fuel systems (coal, oil, or gas) than with heat pumps because the leaving air temperatures are much warmer in fossil fuel systems. For example, the supply air temperature on a heat pump without strip heat is rarely more than 100°F, whereas on a fossil fuel system it could easily reach 130°F. Plenum systems are not limited to outside wall installations. They are often applied to inside walls for short runs (e.g., in apartment houses because of low installation costs).

420 UNIT 25 AIR DISTRIBUTION AND BALANCE

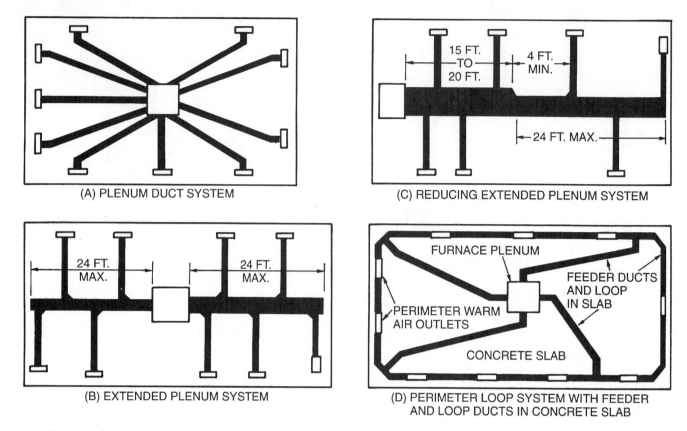

Figure 25-22. (A) Plenum system. (B) Extended plenum system. (C) Reducing plenum system. (D) Perimeter loop system.

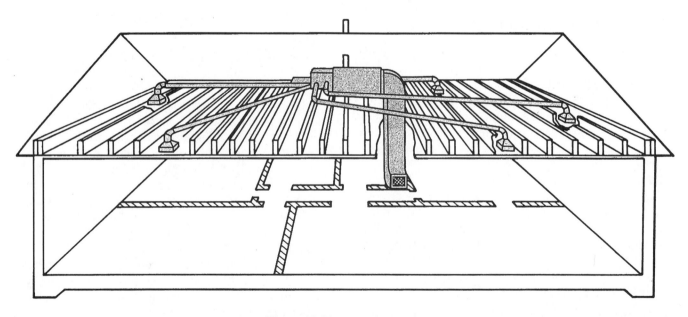

Figure 25-23. Plenum system.

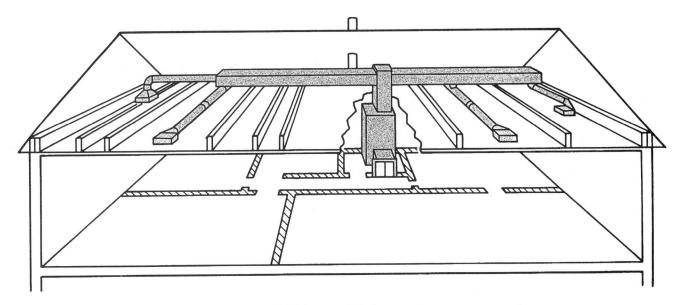

Figure 25-24. Extended plenum system.

The return air system can be a single return located at the air handler, which makes materials very economical. The single-return system will be discussed in more detail later in this book.

Extended Plenum System

The extended plenum system can be applied to a long structure such as a ranch-style house. This system uses a longer plenum to move the air closer to the farthest point (Figure 25-24). The center portion is called the *trunk duct* and can be round, square, or rectangular. The system uses small ducts called *branches* to complete the connection to the terminal units. These small ducts can also be round, square, or rectangular. In small sizes, they are usually round duct because it is less expensive to manufacture and assemble. An average home may have 6 inch round duct for the branches.

Reducing Plenum System

The reducing plenum system reduces the trunk duct size as branch ducts are added (Figure 25-25). This system has the advantage of saving materials and maintaining the same pressure from one end of the duct system to the other when properly sized. This ensures that each branch duct has approximately the same pressure pushing air into its takeoff from the trunk duct.

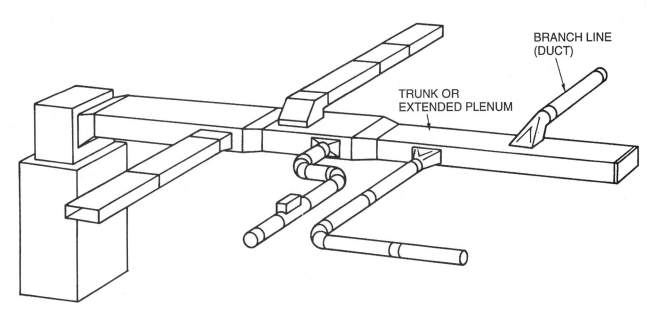

Figure 25-25. Reducing plenum system. *Courtesy Climate Control.*

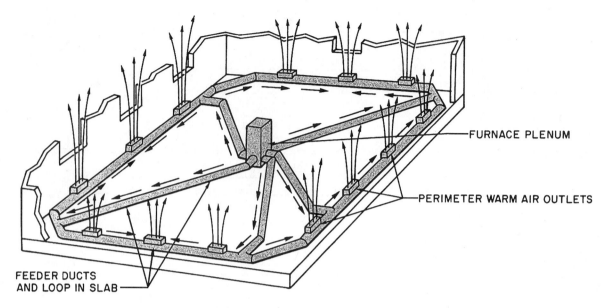

Figure 25-26. Perimeter loop system.

Perimeter Loop System

The perimeter loop system (Figure 25-26) is particularly well suited for installation in a concrete floor in a colder climate. The loop can be run under the slab close to the outside walls with the outlets next to the wall. In winter, there is warm air in the whole loop when the furnace fan is running, and this keeps the slab at a more even temperature. The loop has a constant pressure around the system and provides the same pressure to all outlets.

25.5 DUCT SYSTEM STANDARDS

All localities have state and/or local codes or standards that are applied to the installation of air conditioning systems. Many state and local standards incorporate other known standards, such as BOCA (Building Officials and Code Administrators International). You must become familiar with the code requirements for your locality. These requirements are not standard across the nation. The intention of the standards is to maintain a minimum level of materials, design concept, and craftsmanship.

25.6 DUCT MATERIALS

The ductwork that carries the air from the fan to the conditioned space can be made of several different materials. For many years, galvanized sheet metal was used exclusively, but it is expensive to manufacture and assemble at the job site. However, galvanized sheet metal is by far the most durable material and is still used in walls where easy access for servicing is not available. Aluminum, fiberglass ductboard, spiral metal duct, and flexible duct have all been used successfully.

Note: The duct material selected must meet any applicable codes for fire protection.

Installation

When installing any type of ductwork, make every effort to adhere to the following guidelines:

- Protect the duct during the construction of the job. Stray material can make its way into the ducts and block the air. When insulation is applied to the inside of the duct, it must be fastened correctly or it may come loose and fall into the airstream.
- Do not collapse the duct and then insulate over the collapsed portion. You must replace or repair the damaged duct.
- Ductwork that is run below a concrete slab can easily be damaged to the point that it will not pass the correct amount of air.

Galvanized Steel Duct

Galvanized steel duct comes in several different thicknesses, called the *gauge* of the metal. When a metal thickness is 28 gauge, it means that it takes about 28 layers of the metal to make a stack that is one inch high (i.e., the metal is about $1/28$ of an inch thick). The duct can be thinner when the dimensions of the ductwork are small. When the ductwork is larger, it must be more rigid or it will swell and make noises when the fan starts or stops. Often, the duct manufacturer will cross-break or make a slight bend from corner to corner on large fittings to make the duct more rigid. Figure 25-27 shows a table that can be used as a guideline for choosing duct metal thickness.

Metal duct is normally furnished in 4 foot lengths and can be round, square, or rectangular. Smaller round duct can be purchased in lengths up to 10 feet. Duct lengths can be fastened together using special fasteners, such as S fasteners and drive clips if the duct is square or

DUCT MATERIALS 423

GAGES OF METAL DUCTS AND PLENUMS USED FOR COMFORT HEATING OR COOLING FOR A SINGLE DWELLING UNIT

	COMFORT HEATING OR COOLING			Comfort Heating Only
	Galvanized Steel			
	Nominal Thickness (In Inches)	Equivalent Galvanized Sheet Gage No.	Approximate Aluminum B & S Gage	Minimum Weight Tin-Plate Pounds Per Base Box
Round Ducts and Enclosed Rectangular Ducts				
14″ or less	0.016	30	26	135
Over 14″	0.019	28	24	—
Exposed Rectangular Ducts				
14″ or less	0.019	28	24	—
Over 14″	0.022	26	23	—

Figure 25-27. Table of recommended metal thicknesses for different sizes of duct.

rectangular, or self-tapping sheet metal screws if the duct is round (Figure 25-28). These fasteners make a secure connection that is almost airtight at the low pressures at which the duct is normally operated. If there is any question of the air leaking out, special tape can be applied to ensure a tight connection.

Warning: To avoid the risk of electric shock, always use a grounded, double insulated, or cordless portable electric drill when installing or drilling into metal duct. Also, be aware that sheet metal can cause serious cuts. Wear gloves when installing or working with sheet metal.

Aluminum duct follows the same guidelines as galvanized duct. The cost of aluminum prevents it from being used in many applications.

Fiberglass Duct

Fiberglass duct is furnished in two styles: flat sheets for fabrication and round prefabricated duct. Fiberglass duct is normally one inch thick with an aluminum foil backing (Figure 25-29). The foil backing has a fiber reinforcement to make it strong. When the duct is fabricated, the fiberglass is cut to form the edges, and the reinforced foil backing is left intact to support the connection. These duct systems are easily transported and assembled in the field.

Fiberglass duct can be fabricated using several different methods. Special knives that cut the board in such a manner as to produce overlapping connections can be used in the field by placing the ductboard on any flat surface and using a straightedge to guide the knife. Special ductboard machines can also be used to fabricate the duct at the job site or in the shop. An operator has to be able to set up the machine for different sizes of duct and fittings. When the duct is made in the shop, it can be cut and left flat and stored in the original boxes. This makes transportation easy. The pieces can be marked for quick assembly at the job site.

The machines or knives used to cut away the fiberglass do so in such a way that the duct can be folded with the foil on the back side. When two pieces are fastened together, an overlap of foil is left so that one piece can be stapled to the other using a special type of stapler. Then, the connection is taped and the tape is pressed on using a special iron. One of the advantages of fiberglass duct is that the insulation is already on the duct when it is assembled.

Spiral Metal Duct

Spiral metal duct is commonly used in large systems. It is normally manufactured at the job site using a special machine. The duct comes in rolls of flat, narrow metal. The machine winds the metal off the spool and folds a seam in it. It is possible to fabricate very long duct runs using this method.

Flexible Duct

Flexible round duct comes in sizes up to about 24 inches in diameter. Some of it has a reinforced aluminum foil backing and is shipped in a short box with the duct material compressed. Without the insulation on it, the duct looks like a coil spring with a flexible foil backing.

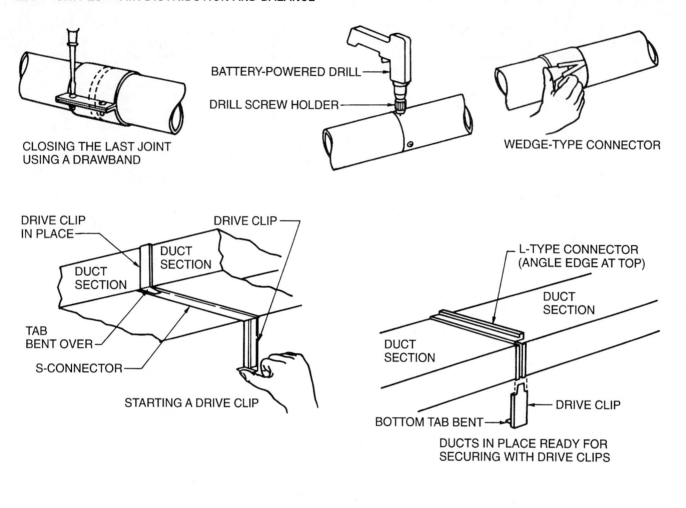
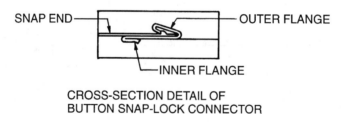
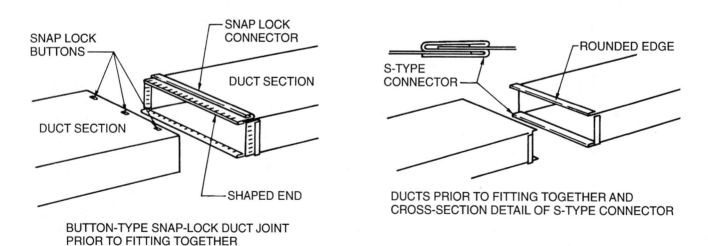

Figure 25-28. Fasteners for square and round duct for low-pressure systems only.

DUCT MATERIALS 425

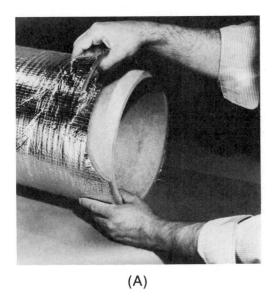
(A)

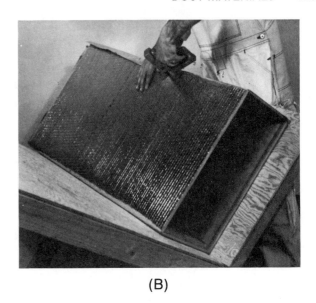
(B)

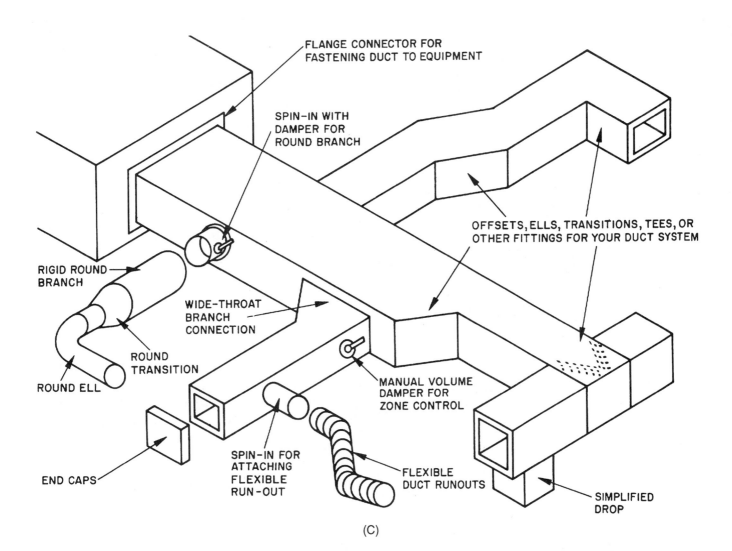
(C)

Figure 25-29. Ducts manufactured out of compressed fiberglass with a foil backing. *Courtesy Manville Corporation.*

426 UNIT 25 AIR DISTRIBUTION AND BALANCE

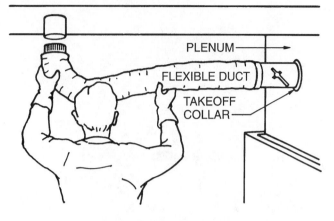

Figure 25-30. Flexible duct.

Some flexible duct comes with either vinyl or foil backing and is preinsulated (Figure 25-30). This type of duct is available in lengths of 25 feet to a box. It is compressed into the box and looks like a caterpillar when the box is opened and the duct is allowed to expand.

Flexible duct is easy to route around corners. When using any type of flexible duct, keep the duct as short as practical and avoid tight turns that may cause the duct to collapse. One disadvantage of this type of duct is that it creates a larger friction loss than metal duct. For best airflow, flexible duct should be stretched as tight as is practical. One advantage of flexible duct is that it serves as a sound attenuator to reduce blower noise down the duct.

25.7 COMBINATION DUCT SYSTEMS

Duct systems can be combined in various ways.

1. Square or rectangular metal trunk lines with square or rectangular metal branch (Figure 25-31).
2. Square or rectangular metal trunk lines with round metal branch ducts (Figure 25-32).
3. Square or rectangular metal trunk lines with round fiberglass branch ducts.
4. Square or rectangular metal trunk lines with flexible branch ducts.
5. Square or rectangular fiberglass ductboard trunk lines with round fiberglass branch ducts.

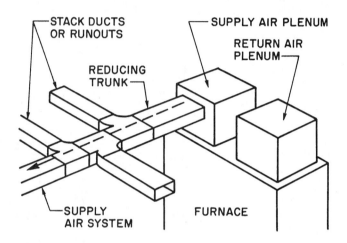

Figure 25-31. Square or rectangle duct.

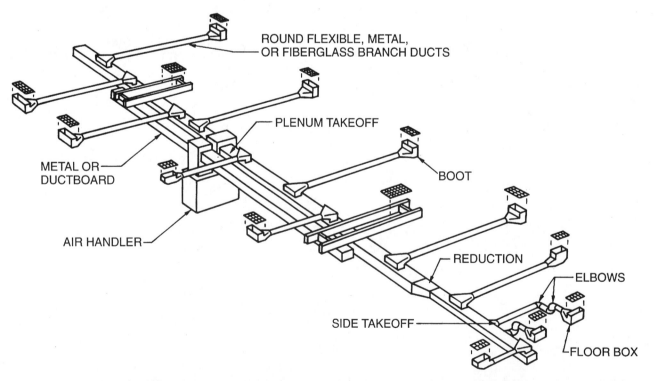

Figure 25-32. Rectangular metal trunk duct with round metal branch ducts.

6. Square or rectangular fiberglass ductboard trunk lines with round metal branch ducts.
7. Square or rectangular fiberglass ductboard trunk lines with flexible branch ducts.
8. All round metal duct with round metal branch ducts.
9. All round metal trunk lines with flexible branch ducts.

25.8 INSTALLATION CONSIDERATIONS

There are a variety of factors that contribute to the proper operation of an air distribution system. Some of these include the proper use of fittings, balancing dampers, and duct insulation.

Air Movement through Fittings

Special attention should be given to the point where the branch duct leaves the main trunk duct to achieve the correct amount of air into the branch duct. The branch duct must be fastened to the main trunk line with a takeoff fitting. A takeoff that has a larger throat area than the branch (runout) duct will allow the air to leave the trunk duct with a minimum of effort. This is called a *streamlined takeoff* (Figure 25-33). The takeoff encourages the air moving down the duct to enter the takeoff to the branch duct.

Air moving in a duct has inertia—in other words, it wants to continue moving in a straight line. If air has to turn a corner, the turn should be carefully designed because centrifugal force will create eddy currents that present a significant pressure drop (Figure 25-34). A square-throated elbow offers more resistance to airflow than a round-throated elbow. If the duct is rectangular or square, turning vanes will improve the airflow around a corner (Figure 25-35).

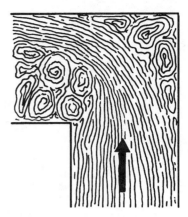

Figure 25-34. When air tries to go around a corner, it creates eddy currents that present a significant pressure loss.

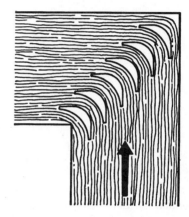

Figure 25-35. Square elbow with turning vanes.

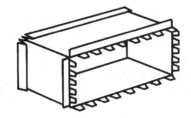

NOT THE BEST CHOICE FOR A TAKEOFF
(A)

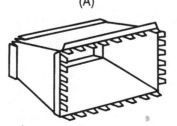

LARGE AREA FASTENS TO DUCT CREATES A LOW-PRESSURE AREA FOR AIR TO MOVE INTO
(B)

Figure 25-33. (A) Standard takeoff fitting. (B) Streamlined takeoff fitting.

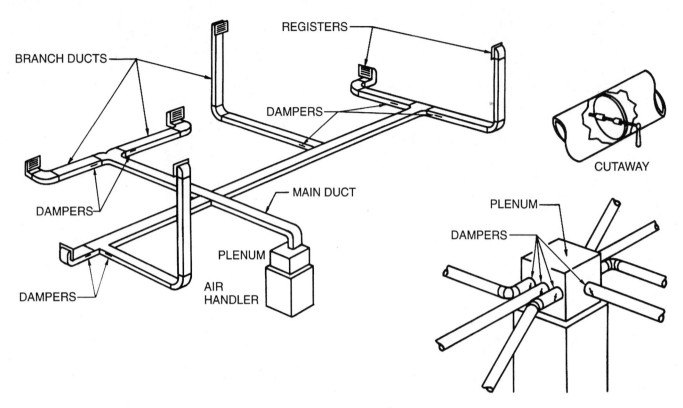

Figure 25-36. Balancing dampers.

Balancing Dampers

A well-designed system will incorporate balancing dampers in the branch ducts to balance the air in the various parts of the system. The dampers should be located as close as practical to the trunk line, with the damper handles uncovered if the duct is insulated. The place to balance the air is near the trunk, so if there is any air velocity noise it will be absorbed in the branch duct before it enters the room. A damper consists of a piece of metal shaped like the inside of the duct with a handle protruding through the side of the duct to the outside. The handle allows the damper to be turned at an angle to the airstream to reduce the airflow (Figure 25-36).

Duct Insulation

When ductwork passes through an unconditioned space, heat transfer may take place between the air in the duct and the air in the unconditioned space. If the heat exchange adds or removes very much heat from the conditioned air, insulation should be applied to the ductwork. A 15°F temperature difference from inside the duct to outside the duct is considered the maximum difference allowed before insulation is necessary.

In a fiberglass duct, the insulation is provided by the manufacturer. Metal duct can be insulated in one of two ways: on the outside or on the inside. When applied to the outside, the insulation is usually fiberglass with a foil or vinyl backing. It comes in several thicknesses, with two inches being the most common. The backing creates a moisture/vapor barrier. This is important where the duct may operate below the dew-point temperature of the surrounding air and moisture would condense on the duct. The insulation is joined by overlapping it and stapling it. It is then taped to prevent moisture from entering the seams. External insulation can be added after the duct has been installed if the duct has enough clearance all around.

When applied to the inside of the duct, the insulation is either glued or fastened to tabs mounted on the duct using spot welds. This insulation must be applied when the duct is being manufactured.

25.9 BLENDING THE CONDITIONED AIR WITH ROOM AIR

When the air reaches the conditioned space, it must be properly distributed so that every room will be comfortable without the occupants being aware that an air conditioning system is operating. This means the air system must distribute the air to the proper area of the conditioned space for optimum air blending. The following guidelines can be used for room air distribution.

1. Whenever possible, air should be directed on the walls because they normally present the largest load. For example, in the winter air can be directed on the

BLENDING THE CONDITIONED AIR WITH ROOM AIR   429

outside walls and windows to cancel the load from the cold wall and keep the wall warmer. This will help to prevent the wall from absorbing heat from the room air. The same distribution will also work in the summer; in this case, it will keep the wall cool and help to prevent the room air from absorbing heat from the wall. A diffuser spreads the air to blanket the outside wall (Figure 25-37).

2. Warm air should be distributed from the floor because it tends to rise (Figure 25-38).

3. Cool air should be distributed from the ceiling because it tends to fall (Figure 25-39).
4. The amount of throw (how far the diffuser will blow air into the room) depends on the air pressure behind the diffuser and the style of the diffuser blades. The air pressure in the duct behind the diffuser creates the velocity of the air leaving the diffuser.

Figure 25-40 shows several air registers and diffusers, including low sidewall, high sidewall, floor, ceiling, and baseboard types.

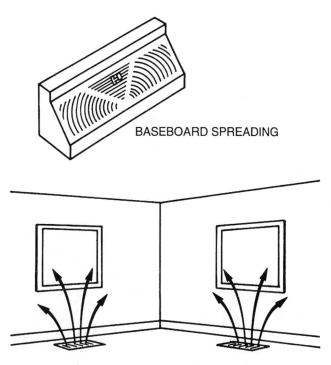

Figure 25-37. Baseboard and floor diffusers.

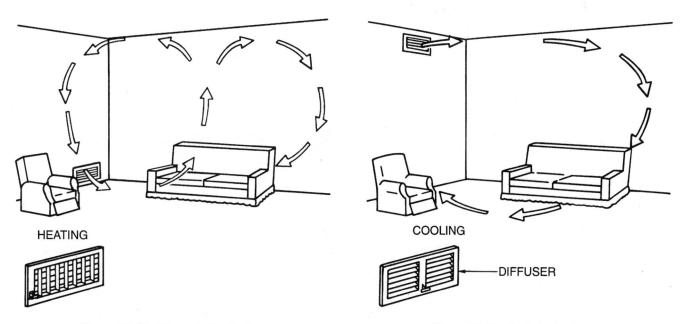

Figure 25-38. Warm air distribution.

Figure 25-39. Cool air distribution.

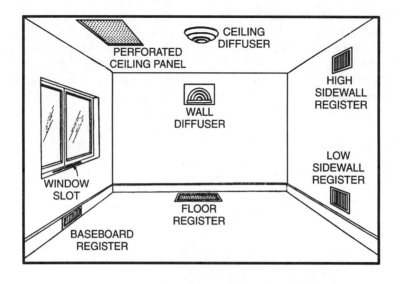

	FLOOR	BASEBOARD	LOW SIDEWALL	HIGH SIDEWALL	CEILING
COOLING PERFORMANCE*	Excellent	Excellent if used with perimeter systems	Excellent if designed to discharge upward	Good	Good
HEATING PERFORMANCE	Excellent	Excellent if used with perimeter systems	Excellent if used with perimeter systems	Fair – should not be used to heat slab houses in northern climates	Fair – should not be used to heat slab houses in northern climates
INTERFERENCE WITH DECOR	Easily concealed because it fits flush with the floor and can be painted to match	Not quite so easy to conceal because it projects from the baseboard	Hard to conceal because it is usually in a flat wall	Impossible to conceal because it is above furniture and in a flat wall	Impossible to conceal but special decorative types are available
INTERFERENCE WITH FURNITURE PLACEMENT	No interference – located at outside wall under a window	No interference – located at outside wall under a window	Can interfere because air discharge is not vertical	No interference	No interference
INTERFERENCE WITH FULL-LENGTH DRAPES	No interference – located 6 or 7 inches from the wall	When drapes are closed, they will cover the outlet	When located under the window, drapes will close over it	No interference	No interference
INTERFERENCE WITH WALL-TO-WALL CARPET	Carpeting must be cut	Carpeting must be notched	No interference	No interference	No interference
OUTLET COST	Low	Medium	Low to medium, depending on the type selected	Low	Low to high – wide variety of types are available
INSTALLATION COST	Low because the sill need not be cut	Low when fed from below – sill need not be cut	Medium – requires wall stack and cutting of plates	Low on furred ceiling system; high when using under-floor system	High because attic ducts require insulation

Figure 25-40. Air registers and diffusers. From Lang, *Principles of Air Conditioning,* © 1979 Delmar Publishers, Inc.

25.10 RETURN AIR DUCT SYSTEM

The return air duct is constructed in much the same manner as the supply duct except that some installations are built with central returns instead of individual room returns. Individual return air systems have a return air grille in each room that has a supply diffuser (with the exception of restrooms and kitchens). The individual return system works better than a central return system, but it is also more expensive. The return air duct is normally sized at least slightly larger than the supply duct, so there is less resistance to airflow in the return system than in the supply system. (This is discussed in more detail later on when we review duct sizing.) Figure 25-41 shows an example of a system with individual room returns.

The central return system is usually satisfactory for a single-story residence. Larger return air grilles are located so that air from common rooms can easily move back to the common returns. For air to return to central returns, there must be a pathway, such as doors with grille work, open doorways, and undercut doors in common hallways. These open areas transmit noise from room to room and may be undesirable in some areas of the home where privacy is a concern.

In a multi-story structure, a central return is installed on each level. Remember, cold air naturally moves downward and warm air moves upward. This results in some degree of stratification. Figure 25-42 illustrates air stratification in a two-story home.

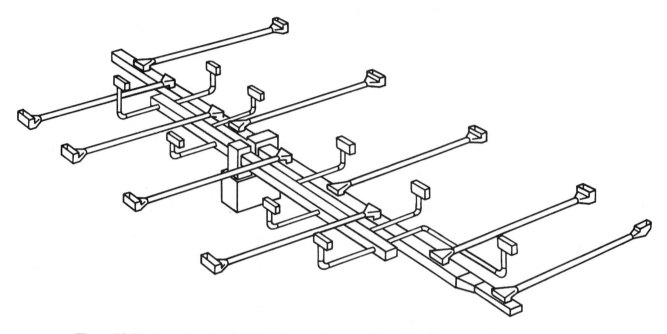

Figure 25-41. Duct plan of an individual return air system. *Reproduced courtesy of Carrier Corporation.*

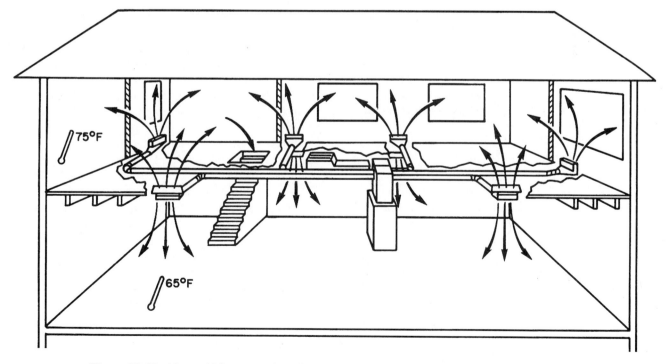

Figure 25-42. Air stratifying even when distributed evenly because warm air rises and cold air falls.

Figure 25-43. Return air grilles.

A properly-constructed central return air system helps to eliminate fan noise in the conditioned space. The return air plenum should not be located on the air handler itself because the fan noise will be noticeable even at a distance of several feet. Instead, the return air grille should be located around an elbow from the furnace. If this cannot be done, the return air plenum can be insulated on the inside to help deaden the fan noise.

Return air grilles are normally large and meant to be decorative. They do not have any other function unless they house a filter. They are usually made of stamped metal or have a metal frame with grille work (Figure 25-43).

25.11 FRICTION LOSS

To move air takes energy because of the following: air has weight, the air tumbles down the duct, rubbing against itself and the ductwork, and fittings create resistance to airflow.

Friction loss in ductwork is due to the actual rubbing action of the air against the sides of the duct and the turbulence of the air rubbing against itself while moving down the duct.

Friction due to rubbing cannot be eliminated but can be minimized with good design practices. Proper duct sizing helps to achieve best air system performance. Friction also depends upon the surface of the ductwork (i.e., the smoother the duct surface, the less friction there is). In addition, the slower the air is moving, the less friction there will be. It is beyond the scope of this text to go into details of air system design. However, the following information can be used as a basic guideline for typical residential and light commercial installations.

Each foot of duct offers a known resistance to airflow. This is called *friction loss*. It can be determined from tables and special slide calculators designed for this purpose. The following example is used to explain friction loss in a typical air system (Figure 25-44).

1. This example uses a ranch-style home requiring 3 tons of cooling.
2. Cooling is provided by a 3 ton cooling coil in the ductwork.
3. The heat and fan are provided by a furnace with an input of 100,000 Btuh and an output of 80,000 Btuh.
4. The fan has a capacity of 1360 cfm of air while operating against 0.40 in. w.c. static pressure with the system fan operating at medium to high speed. The system only needs 1200 cfm of air in the cooling mode. The system fan will easily be able to achieve this with a small amount of reserve capacity using a 1/2-hp motor (Figure 25-45). The cooling mode usually requires more air than the heating mode. Cooling normally requires 400 cfm of air per ton (3 tons × 400 = 1200 cfm).
5. The system has 11 outlets that each require 100 cfm and two outlets that each require 50 cfm, for a total of 13 outlets. Most of these outlets are on the exterior walls of the house and distribute the conditioned air on the outside walls. The two smaller outlets are located in the bathrooms.
6. The return air is taken into the system from a common hallway, with one return at each end of the hall.
7. While reviewing this example, think of the entire house as the air system. The supply air must leave the supply registers and sweep the walls. It then makes its way across each room to the door adjacent to the hall. The air is at room temperature at this time and goes under the door to make its way to the return air grille. Each door must have about 1 inch of space under it.
8. The return air grille is where the duct system starts. It has a slight negative pressure in relation to the room pressure, which gives the air incentive to enter the system. The filters are located in the return air grilles. The pressure on the fan side of the filter is -0.03 in. w.c., which is less than the pressure in the room, so the room pressure pushes the air through the filter into the return duct.

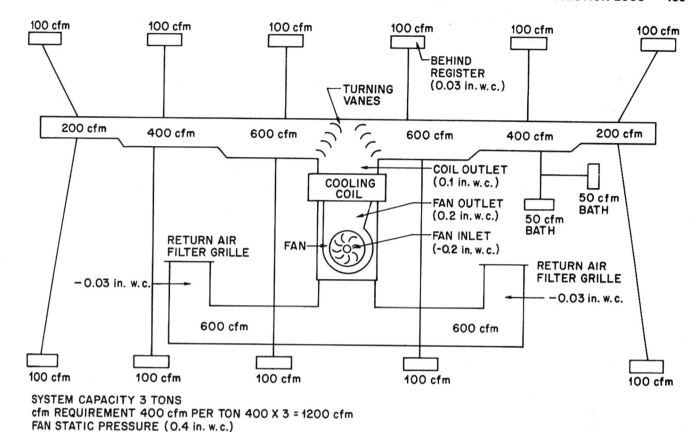

Figure 25-44. Typical residential air system.

SIZE	Blower Motor HP	Speed	External Static Pressure in. w.c.							
			0.1	0.2	0.3	0.4	0.5	0.6	0.7	0.8
048100	1/2 PSC	High	1750	1750	1720	1685	1610	1530	1430	—
		Med-High	1360	1370	1370	1360	1340	1315	—	—
		Med-Low	1090	1120	1140	1130	1100	—	—	—
		Low	930	960	980	980	965	945	—	—

Figure 25-45. Manufacturer's airflow table. *Courtesy BDP Company.*

9. As the air proceeds down the duct toward the fan, the pressure continues to decrease. The lowest pressure in the system is in the fan inlet, which is -0.20 in. w.c., significantly below the room pressure.
10. When the air is forced through the fan, the pressure increases. The greatest pressure in the system is at the fan outlet (0.20 in. w.c. above the room pressure). The pressure difference between the inlet and the outlet of the fan is 0.40 in. w.c.
11. The air then moves through the cooling coil where it enters the supply duct system at a pressure of 0.10 in. w.c.
12. The air undergoes a slight pressure drop as it travels around the corner of the tee that splits the duct into two reducing plenums, one for each end of the house. This tee contains turning vanes to help reduce the pressure drop as the air goes around the corner.
13. The first section of each reducing trunk has to handle an equal amount of air (600 cfm each). Two branch ducts are supplied in the first trunk run, each with an air quantity of 100 cfm. This reduces the capacity of the trunk to 400 cfm on each side. A smaller trunk can be used at this point, and materials can be saved.

14. The duct is reduced to a smaller size to handle 400 cfm on each side. Because another 200 cfm of air is distributed to the conditioned space, another reduction can be made.
15. The last part of the reducing trunk on each side of the system needs to handle only 200 cfm for each side of the system.

This system will distribute the air for this house with minimal noise and maximum comfort. The pressure in the duct will be about the same all along the duct because as air is distributed off the trunk line, the duct size is reduced to keep the pressure inside the duct at the prescribed value.

At each branch duct, dampers should be installed to balance the system air supply to each room. This system will furnish 100 cfm to each outlet except the bathrooms, but if a room is not occupied and does not require 100 cfm, the dampers can be adjusted to deliver the desired airflow.

The return air system is the same size on each side of the system. It returns 600 cfm per side with the filters located in the return air grilles in the halls. The furnace fan is located far enough from the grilles so that the blower operating noise will not disturb the occupants.

There are several books available on duct sizing. Manufacturer's representatives may also be able to help you with specific applications. Many manufacturers offer classes in duct sizing.

25.12 MEASURING AIR MOVEMENT FOR BALANCING

Air balancing is sometimes accomplished by measuring the air leaving each supply register. When one outlet has too much air, the damper in that run is throttled to reduce the airflow. Naturally, this redistributes air to the other outlets and will increase their airflow.

The air quantity delivered by an individual duct can be measured in the field with some degree of accuracy by using instruments to determine the velocity of the air in the duct. For instance, a velometer can be inserted into a cross-section of the duct, and an average of the readings taken to determine the velocity. This is called *traversing* the duct. For example, if the air in a 1 ft.^{2} duct (18 in. × 8 in. = 144 in.^{2}), was traveling at a velocity of one foot per minute (fpm), the volume of air passing any point in the duct would be 1 cfm. If the velocity were 100 fpm, the volume of air passing the same point would be 100 cfm. The cross-sectional area of the duct is multiplied by the average velocity of the air to determine the volume of the moving air (Figure 25-46).

Warning: Wear goggles whenever working around duct that has been opened with the fan running.

When you know the average velocity of the air in any duct, you can determine the cfm using the following formula:

$$\text{cfm} = \text{area (sq. ft.)} \times \text{velocity (fpm)}$$

For example, suppose a duct is 20 inches × 30 inches and the average velocity is 850 fpm. The cfm can be found by first finding the area in square feet.

$$\text{area} = \text{width} \times \text{height}$$

$$\text{area} = 20 \text{ in.} \times 30 \text{ in.}$$

$$\text{area} = \frac{600 \text{ in.}^2}{144 \text{ in.}^2 \text{ per ft.}^2}$$

$$\text{area} = 4.17 \text{ ft.}^2$$

$$\text{cfm} = \text{area} \times \text{velocity}$$

$$\text{cfm} = 4.17 \text{ ft.}^2 \times 850 \text{ fpm}$$

$$\text{cfm} = 3545 \text{ cfm}$$

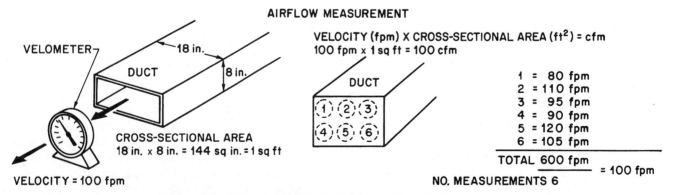

Figure 25-46. Measuring duct area and air velocity in a cross-section of duct.

Now suppose the duct were round and had a diameter of 12 inches with an average velocity of 900 fpm.

$$\text{area} = \text{pi} \times r^2$$

where: pi = a constant which we will round to 3.14
r² = the square of the radius (the radius is equal to half the diameter)

$$\text{area} = 3.14 \times (6 \times 6)$$

$$\text{area} = \frac{113 \text{ in.}^2}{144 \text{ in.}^2 \text{ per ft.}^2}$$

$$\text{area} = 0.78 \text{ ft.}^2$$

$$\text{cfm} = \text{area} \times \text{velocity}$$

$$\text{cfm} = 0.78 \text{ ft.}^2 \times 900 \text{ fpm}$$

$$\text{cfm} = 706 \text{ cfm}$$

Special techniques and good instrumentation must be used to find the correct average velocity in a duct. For example, readings should not be taken within 10 duct diameters of the nearest fitting. This means that long straight runs must be used for taking accurate readings. This is not always possible because of the many fittings and takeoffs in a typical system, particularly residential systems.

The proper duct traverse must be used and it is different for round duct versus rectangular duct. Figure 25-47 shows the actual traverse patterns for rectangular and round duct that should be used to achieve the most accurate average reading.

Air Friction Charts

The previous system can be plotted on the friction chart in Figure 25-48. This chart covers volumes of air up to 2000 cfm. Using the 400 cfm/ton cooling value mentioned earlier, we can use this chart for systems up to 5 tons of cooling (2000 cfm ÷ 400 cfm/ton = 5 tons). Figure 25-49 shows a chart used for larger systems (up to 100,000 cfm or 250 tons).

The friction charts have cfm in the left column and round duct sizes listed at an angle from left to right toward the top of the page. These round duct sizes can be converted to square or rectangular duct by using the table in Figure 25-50. The round duct sizes are for air with a density of 0.075 lb./ft.³ using galvanized pipe. Other charts are available for fiberglass ductboard and flexible duct. In Figures 25-48 and 25-49, the air velocity in the duct is shown on the diagonal lines that run from left to right toward the bottom of the page. The friction loss in inches of water column per 100 feet of duct is shown along the top and bottom of the chart. For example, a run of round duct that carries 100 cfm (on the left of the chart) can be plotted over the intersection of the 6 inch duct line. When this pipe is carrying 100 cfm of air, it has a velocity of just over 500 fpm and a pressure drop (friction loss) of 0.085 in. w.c. per 100 feet. A 50 foot run would have half the pressure drop or 0.0425 in. w.c. per 50 feet.

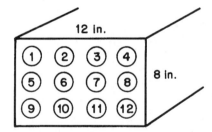

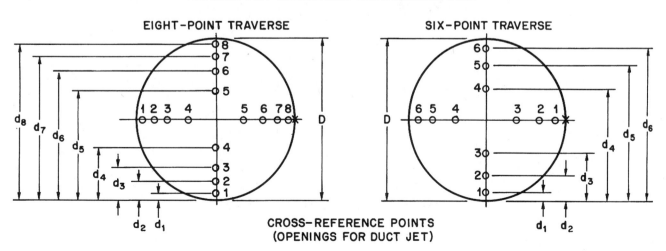

Figure 25-47. This figure shows the patterns used to traverse rectangular and round duct.

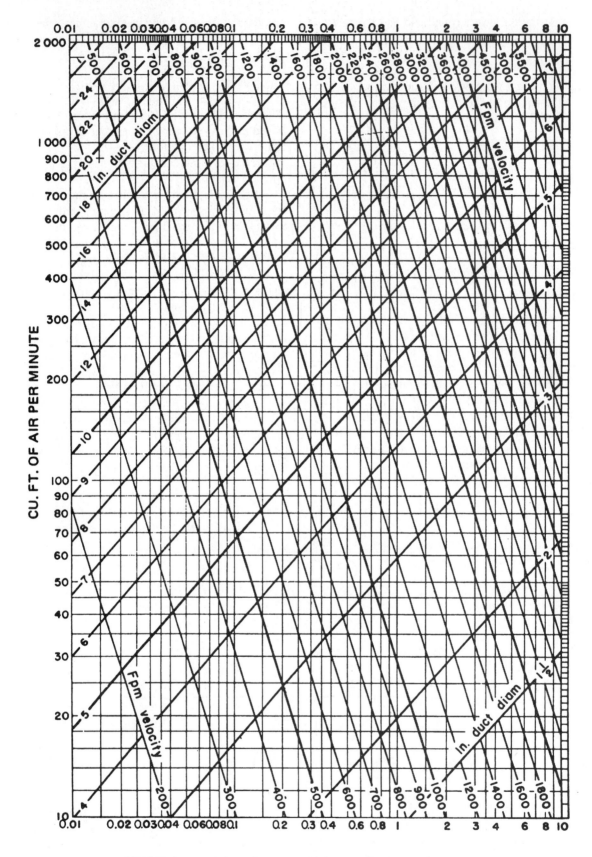

Figure 25-48. Air friction chart. *Reprinted by permission of American Society of Heating, Refrigerating and Air-Conditioning Engineers.*

MEASURING AIR MOVEMENT FOR BALANCING 437

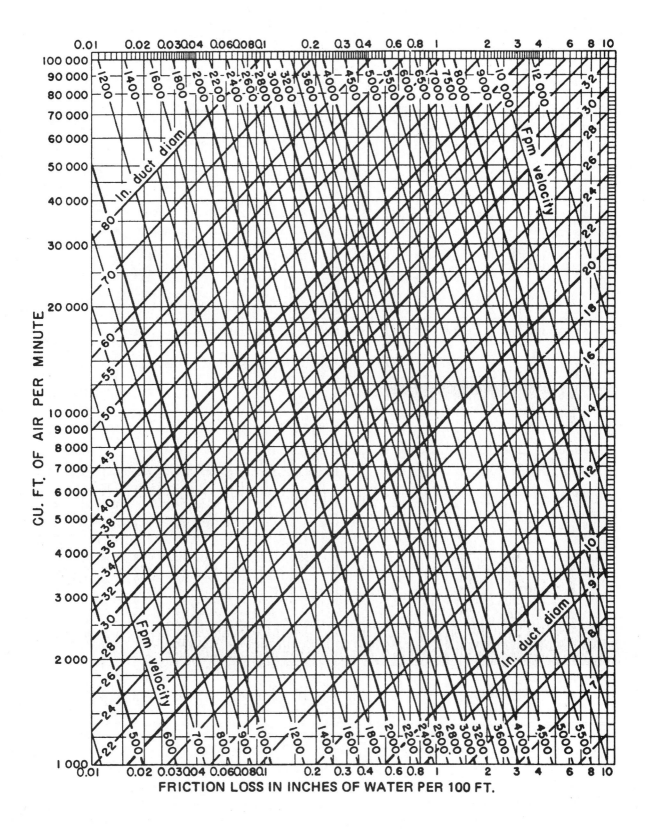

Figure 25-49. Friction chart for larger air systems. *Reprinted by permission of American Society of Heating, Refrigerating and Air-Conditioning Engineers.*

438 UNIT 25 AIR DISTRIBUTION AND BALANCE

Lgth Adj.[b]	\multicolumn{20}{c	}{Length of One Side of Rectangular Duct (a), in.}	Lgth Adj.[b]
	6	7	8	9	10	11	12	13	14	15	16	17	18	19	20	22	24	26	28	30	
6	6.6																				6
7	7.1	7.7																			7
8	7.6	8.2	8.7																		8
9	8.0	8.7	9.3	9.8																	9
10	8.4	9.1	9.8	10.4	10.9																10
11	8.8	9.5	10.2	10.9	11.5	12.0															11
12	9.1	9.9	10.7	11.3	12.0	12.6	13.1														12
13	9.5	10.3	11.1	11.8	12.4	13.1	13.7	14.2													13
14	9.8	10.7	11.5	12.2	12.9	13.5	14.2	14.7	15.3												14
15	10.1	11.0	11.8	12.6	13.3	14.0	14.6	15.3	15.8	16.4											15
16	10.4	11.3	12.2	13.0	13.7	14.4	15.1	15.7	16.4	16.9	17.5										16
17	10.7	11.6	12.5	13.4	14.1	14.9	15.6	16.2	16.8	17.4	18.0	18.6									17
18	11.0	11.9	12.9	13.7	14.5	15.3	16.0	16.7	17.3	17.9	18.5	19.1	19.7								18
19	11.2	12.2	13.2	14.1	14.9	15.7	16.4	17.1	17.8	18.4	19.0	19.6	20.2	20.8							19
20	11.5	12.5	13.5	14.4	15.2	16.0	16.8	17.5	18.2	18.9	19.5	20.1	20.7	21.3	21.9						20
22	12.0	13.0	14.1	15.0	15.9	16.8	17.6	18.3	19.1	19.8	20.4	21.1	21.7	22.3	22.9	24.0					22
24	12.4	13.5	14.6	15.6	16.5	17.4	18.3	19.1	19.9	20.6	21.3	22.0	22.7	23.3	23.9	25.1	26.2				24
26	12.8	14.0	15.1	16.2	17.1	18.1	19.0	19.8	20.6	21.4	22.1	22.9	23.5	24.2	24.9	26.1	27.3	28.4			26
28	13.2	14.5	15.6	16.7	17.7	18.7	19.6	20.5	21.3	22.1	22.9	23.7	24.4	25.1	25.8	27.1	28.3	29.5	30.6		28
30	13.6	14.9	16.1	17.2	18.3	19.3	20.2	21.1	22.0	22.9	23.7	24.4	25.2	25.9	26.6	28.0	29.3	30.5	31.7	32.8	30
32	14.0	15.3	16.5	17.7	18.8	19.8	20.8	21.8	22.7	23.5	24.4	25.2	26.0	26.7	27.5	28.9	30.2	31.5	32.7	33.9	32
34	14.4	15.7	17.0	18.2	19.3	20.4	21.4	22.4	23.3	24.2	25.1	25.9	26.7	27.5	28.3	29.7	31.0	32.4	33.7	34.9	34
36	14.7	16.1	17.4	18.6	19.8	20.9	21.9	22.9	23.9	24.8	25.7	26.6	27.4	28.2	29.0	30.5	32.0	33.3	34.6	35.9	36
38	15.0	16.5	17.8	19.0	20.2	21.4	22.4	23.5	24.5	25.4	26.4	27.2	28.1	28.9	29.8	31.3	32.8	34.2	35.6	36.8	38
40	15.3	16.8	18.2	19.5	20.7	21.8	22.9	24.0	25.0	26.0	27.0	27.9	28.8	29.6	30.5	32.1	33.6	35.1	36.4	37.8	40
42	15.6	17.1	18.5	19.9	21.1	22.3	23.4	24.5	25.6	26.6	27.6	28.5	29.4	30.3	31.2	32.8	34.4	35.9	37.3	38.7	42
44	15.9	17.5	18.9	20.3	21.5	22.7	23.9	25.0	26.1	27.1	28.1	29.1	30.0	30.9	31.8	33.5	35.1	36.7	38.1	39.5	44
46	16.2	17.8	19.3	20.6	21.9	23.2	24.4	25.5	26.6	27.7	28.7	29.7	30.6	31.6	32.5	34.2	35.9	37.4	38.9	40.4	46
48	16.5	18.1	19.6	21.0	22.3	23.6	24.8	26.0	27.1	28.2	29.2	30.2	31.2	32.2	33.1	34.9	36.6	38.2	39.7	41.2	48
50	16.8	18.4	19.9	21.4	22.7	24.0	25.2	26.4	27.6	28.7	29.8	30.8	31.8	32.8	33.7	35.5	37.2	38.9	40.5	42.0	50
52	17.1	18.7	20.2	21.7	23.1	24.4	25.7	26.9	28.0	29.2	30.3	31.3	32.3	33.3	34.3	36.2	37.9	39.6	41.2	42.8	52
54	17.3	19.0	20.6	22.0	23.5	24.8	26.1	27.3	28.5	29.7	30.8	31.8	32.9	33.9	34.9	36.8	38.6	40.3	41.9	43.5	54
56	17.6	19.3	20.9	22.4	23.8	25.2	26.5	27.7	28.9	30.1	31.2	32.3	33.4	34.4	35.4	37.4	39.2	41.0	42.7	44.3	56
58	17.8	19.5	21.2	22.7	24.2	25.5	26.9	28.2	29.4	30.6	31.7	32.8	33.9	35.0	36.0	38.0	39.8	41.6	43.3	45.0	58
60	18.1	19.8	21.5	23.0	24.5	25.9	27.3	28.6	29.8	31.0	32.2	33.3	34.4	35.5	36.5	38.5	40.4	42.3	44.0	45.7	60
62		20.1	21.7	23.3	24.8	26.3	27.6	28.9	30.2	31.5	32.6	33.8	34.9	36.0	37.1	39.1	41.0	42.9	44.7	46.4	62
64		20.3	22.0	23.6	25.1	26.6	28.0	29.3	30.6	31.9	33.1	34.3	35.4	36.5	37.6	39.6	41.6	43.5	45.3	47.1	64
66		20.6	22.3	23.9	25.5	26.9	28.4	29.7	31.0	32.3	33.5	34.7	35.9	37.0	38.1	40.2	42.2	44.1	46.0	47.7	66
68		20.8	22.6	24.2	25.8	27.3	28.7	30.1	31.4	32.7	33.9	35.2	36.3	37.5	38.6	40.7	42.8	44.7	46.6	48.4	68
70		21.1	22.8	24.5	26.1	27.6	29.1	30.4	31.8	33.1	34.4	35.6	36.8	37.9	39.1	41.2	43.3	45.3	47.2	49.0	70
72			23.1	24.8	26.4	27.9	29.4	30.8	32.2	33.5	34.8	36.0	37.2	38.4	39.5	41.7	43.8	45.8	47.8	49.6	72
74			23.3	25.1	26.7	28.2	29.7	31.2	32.5	33.9	35.2	36.4	37.7	38.8	40.0	42.2	44.4	46.4	48.4	50.3	74
76			23.6	25.3	27.0	28.5	30.0	31.5	32.9	34.3	35.6	36.8	38.1	39.3	40.5	42.7	44.9	47.0	48.9	50.9	76
78			23.8	25.6	27.3	28.8	30.4	31.8	33.3	34.6	36.0	37.2	38.5	39.7	40.9	43.2	45.4	47.5	49.5	51.4	78
80			24.1	25.8	27.5	29.1	30.7	32.2	33.6	35.0	36.3	37.6	38.9	40.2	41.4	43.7	45.9	48.0	50.1	52.0	80
82				26.1	27.8	29.4	31.0	32.5	34.0	35.4	36.7	38.0	39.3	40.6	41.8	44.1	46.4	48.5	50.6	52.6	82
84				26.4	28.1	29.7	31.3	32.8	34.3	35.7	37.1	38.4	39.7	41.0	42.2	44.6	46.9	49.0	51.1	53.2	84
86				26.6	28.3	30.0	31.6	33.1	34.6	36.1	37.4	38.8	40.1	41.4	42.6	45.0	47.3	49.6	51.7	53.7	86
88				26.9	28.6	30.3	31.9	33.4	34.9	36.4	37.8	39.2	40.5	41.8	43.1	45.5	47.8	50.0	52.2	54.3	88
90				27.1	28.9	30.6	32.2	33.8	35.3	36.7	38.2	39.5	40.9	42.2	43.5	45.9	48.3	50.5	52.7	54.8	90
92					29.1	30.8	32.5	34.1	35.6	37.1	38.5	39.9	41.3	42.6	43.9	46.4	48.7	51.0	53.2	55.3	92
96					29.6	31.4	33.0	34.7	36.2	37.7	39.2	40.6	42.0	43.3	44.7	47.2	49.6	52.0	54.2	56.4	96

Figure 25-50. Chart used for conversions between round and square or rectangular duct. *Reprinted by permission of American Society of Heating, Refrigerating and Air-Conditioning Engineers.*

The friction chart can be used by the designer to size the ductwork before bidding the job. This chart provides the sizes that will be used for figuring the duct materials. The duct should be sized using the charts in Figure 25-48 through 25-50, and the recommended velocities taken from the table in Figure 25-51. In the previous, example a 4 inch pipe could have been used, but the velocity would have been nearly 1200 fpm. This would be noisy, and the fan may not have enough capacity to push sufficient air through the duct.

High-velocity systems have been designed and used successfully in small applications. Such systems normally have a high air velocity in the trunk and branch ducts, and the velocity is then reduced at the register to avoid drafts.

Structure	Supply Outlet	Return Openings	Main Supply	Branch Supply	Main Return	Branch Return
Residential	500–750	500	1,000	600	800	600
Apartments, Hotel Bedrooms, Hospital Bedrooms	500–750	500	1,200	800	1,000	800
Private Offices, Churches, Libraries, Schools	500–1,000	600	1,500	1,200	1,200	1,000
General Offices, Deluxe Restaurants, Deluxe Stores, Banks	1,200–1,500	700	1,700	1,600	1,500	1,200
Average Stores, Cafeterias	1,500	800	2,000	1,600	1,500	1,200

Figure 25-51. Chart showing recommended velocities for different duct designs. From Lang, *Principles of Air Conditioning,* © 1987 by Delmar Publishers, Inc.

However, if the air velocity is not reduced at the register, the register has a streamlined effect and is normally located in the corners of the room where someone is not likely to walk under it.

The friction chart can also be used in the field to troubleshoot airflow problems. Airflow problems can result from poor system design, installation, or maintenance. They can also result when the homeowner covers or closes air registers or diffusers.

System design problems can result from a poor fan choice or incorrect duct sizes. Design problems may also result from not understanding the fact that the air friction chart is for 100 feet of duct only. You should have an understanding of how to size duct in order to eliminate poor air system design as a possible source of system malfunctions.

To size duct properly, you must know how to calculate the total pressure drop. The pressure drop is calculated and read on the friction chart in feet of duct, from the air handler to the terminal outlet. This is a straightforward measurement. The fittings are calculated in a different manner. The pressure drop across different types of fittings in in. w.c. have been determined by laboratory experiment. These pressure drops have been converted to equivalent feet of duct. You must know what types of fittings are used in the system and then a chart may be consulted for the equivalent feet of duct (Figure 25-52). The equivalent feet of duct for all of the fittings in a particular run must be added together and then combined with the actual duct length to determine the proper duct size for a particular run.

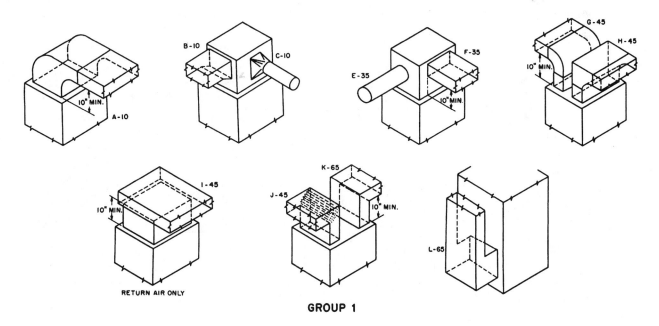

Figure 25-52. Various duct fittings and their equivalent lengths. *Courtesy Air Conditioning Contractors of America (ACCA).*

440 UNIT 25 AIR DISTRIBUTION AND BALANCE

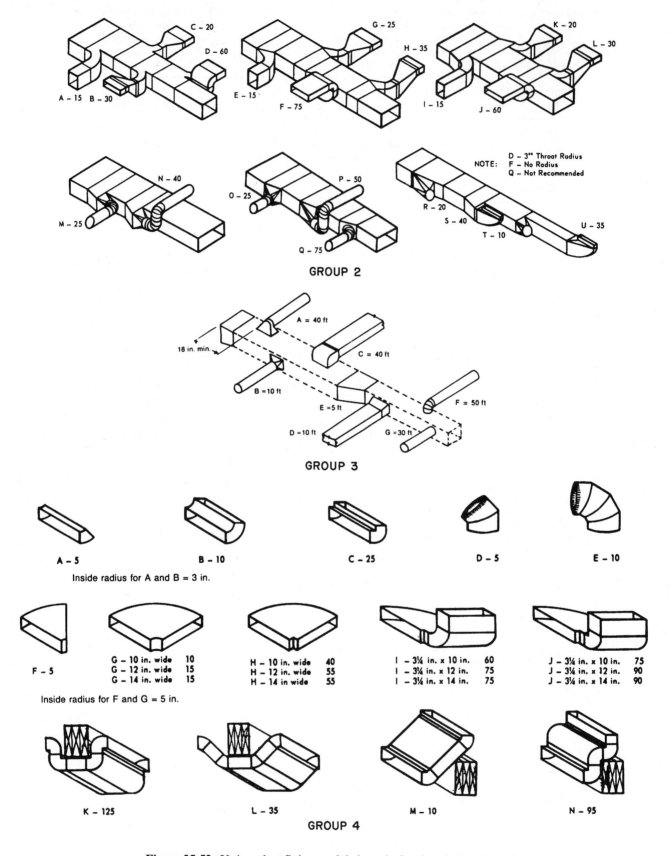

Figure 25-52. Various duct fittings and their equivalent lengths (continued).

MEASURING AIR MOVEMENT FOR BALANCING 441

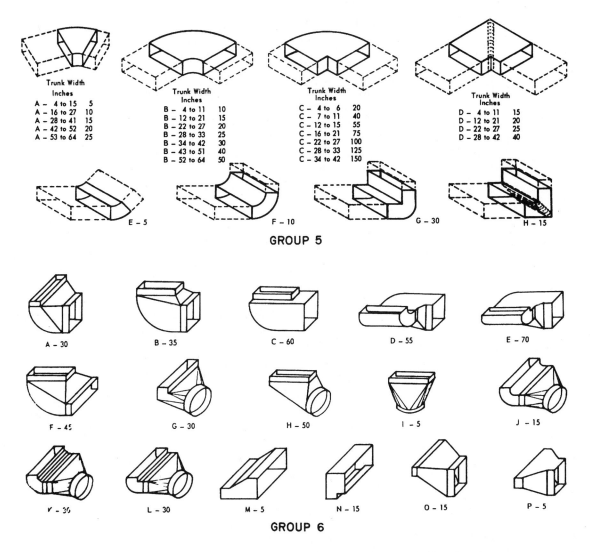

Figure 25-52. Various duct fittings and their equivalent lengths (continued).

When a duct run is under 100 feet, a corrected friction factor must be used or the duct will not have enough resistance, which will cause excessive airflow. The corrected friction factor may be determined by multiplying the design friction loss in the duct system times 100 (the friction loss used on the air friction chart) and dividing by the actual length of run. For example, a system may have a duct run handling 300 cfm of air to an inside room. This duct is very close to the main duct and air handler. The total equivalent length is 60 effective feet, including the friction loss in the fittings and the straight duct length (Figure 25-53).

The fan may be capable of moving the correct airflow in the supply duct with a friction loss of 0.36 in. w.c. The cooling coil may have a loss of 0.24 in. w.c. and the supply registers may have a loss of 0.03 in. w.c. When the coil and register losses are subtracted from the friction loss for the supply duct, the duct must be designed for a loss of 0.09 in. w.c. [0.36 in. w.c. - (0.24 in. w.c. + 0.03 in. w.c.) = 0.09 in. w.c.].

The following is the mistake the designer made in sizing this ductwork. The designer chose 9 inch duct from the air friction chart at the junction of 0.09 in. w.c. and 300 cfm without correcting for the fact that the duct is only 60 feet long. We must correct the friction loss for 60 feet of duct using the following formula:

$$\text{corrected friction factor} = \frac{\text{design static pressure} \times 100}{\text{total equivalent length of duct}}$$

$$\text{corrected friction factor} = \frac{0.09 \times 100}{60}$$

corrected friction factor = 0.15 in. w.c.

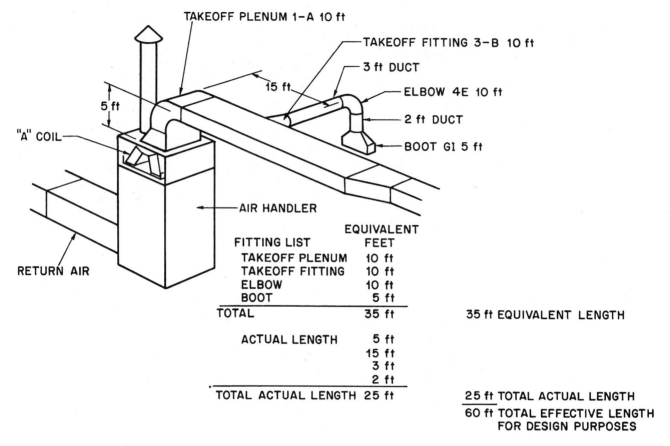

Figure 25-53. This ductrun has a total equivalent length of 60 feet. The designer must use a corrected friction factor to avoid oversizing.

Using this corrected friction factor, we then refer to the air friction chart and plot at 300 cfm, then move to the right to 0.15 in. w.c. The duct size now becomes 8 inches. This would not be a significant savings in materials, but would reduce the airflow by the correct amount. An alternative is to use a throttling damper where the branch line leaves the trunk duct. *The key to this problem is correcting the friction loss to an adjusted value.* This causes the chart to undersize the duct for the purpose of reducing the airflow to this run. Otherwise, the air will have so little resistance that too much air will flow from this run, starving other runs. One run that is oversized will not make too much difference, but a whole system sized using the same technique will have problems.

This example illustrates a run that is less than 100 feet. The same technique will apply to a run that is over 100 ft, except the run will be undersized and starved for air unless the friction loss is corrected. For example, suppose another run in the same system requires 300 cfm of air and has a total equivalent length of 170 feet (Figure 25-54). When the designer sized the duct for 300 cfm at a friction loss of 0.09 in. w.c., a 9 inch round duct was chosen. Let's suppose we find that the airflow is too low. We can check the duct size using the following formula:

$$\text{corrected friction factor} = \frac{\text{design static pressure} \times 100}{\text{total equivalent length of duct}}$$

$$\text{corrected friction factor} = \frac{0.09 \times 100}{170}$$

$$\text{corrected friction factor} = 0.053 \text{ in. w.c.}$$

When the duct is sized using the corrected friction factor, the adjustment causes the chart to oversize the duct and it moves the correct amount of air. In this case, the technician enters the chart at 300 cfm and moves to the right to 0.053 in. w.c. and finds the correct duct size to be 10 inches instead of 9 inches. This is why this system is starved for air. All duct is first sized in round, then converted to square or rectangular.

Note: All runs of duct are measured from the air handler to the end of the run.

As shown from the previous examples, when some duct is undersized and some oversized, the result will probably produce problems. You should pay close attention to any job with airflow problems and try to find the original duct sizing calculations. This will enable you to fully analyze a problem duct installation. Any equipment supplier can help you choose the correct fan and equipment. They may even help with the duct design.

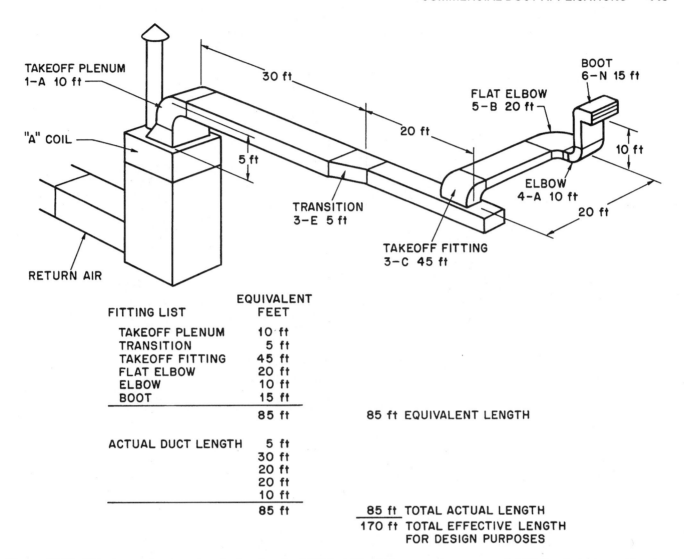

Figure 25-54. This ductrun has a total equivalent length of 170 feet. The designer must use a corrected friction factor to avoid undersizing.

25.13 COMMERCIAL DUCT APPLICATIONS

Commercial buildings are all air conditioned with some form of forced air. It is easy to imagine that as the space to be conditioned becomes larger, the air distribution system also becomes larger. Air distribution systems in large buildings must use different methods than those used in residential and light commercial applications. At the same time, commercial buildings require closer temperature control. To avoid tenant dissatisfaction, building owners and managers want each space to be comfortable.

Residential and light commercial systems normally use a single-duct system for both heating and cooling. This limits the system to either mode, which may be a disadvantage if internal loads vary widely. The forced air for heating is simply moved down the same duct as the cooling air (Figure 25-55). As buildings become larger, it becomes important to separate these two airstreams so a building can have cooling in one part of the building and heating in another. Large buildings have a heat gain in the core of the building even in the coldest of winter weather, so there may be a year-round need for cooling. At the same time, the same building may have a need for heating at the perimeter because it is in contact with the cold outdoors. Now we are considering either one or two complete duct systems for many applications. Often, a building duct system may be divided into zones to satisfy each of the different types of load.

Residential and light commercial systems are not as concerned with the physical size of the duct as larger building applications, so they typically use low-velocity, low-pressure duct systems. Large building applications are interested in saving space or building volume in any way possible. The duct system is one area that requires quite a bit of space or volume, particularly now that we are discussing two duct systems for each major zone or conditioned space. To reduce the size of the ductwork, these buildings use high-velocity, high-pressure duct

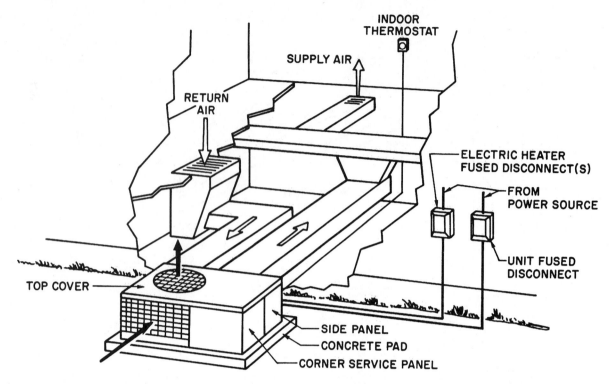

Figure 25-55. This duct system is used for both heating and cooling.

systems. This reduces the cross-sectional area of the duct (i.e., the size and volume). These systems may use velocities up to about 4500 fpm. For example, the system design chart shown earlier in Figure 25-51 limited the velocity to 2000 fpm in a main supply duct for average stores to prevent air noise in the duct. Other methods may be used to control the air noise in the duct so the air velocity may be increased and smaller duct can be used for the system. For example, suppose a designer had a choice of using 2000 fpm or 4000 fpm for a duct that was to handle 20,000 cfm of air. The round duct size would be 43 inches for 2000 fpm and 31 inches for 4000 fpm design velocities (Figure 25-56). This is a difference of 12 inches. When a building owner is considering multiple floors, space between floors can be saved as the duct system would typically be located between the floors and the duct is the largest equipment located in this space. When 12 inches can be saved per floor, 12 feet can be gained for every 12 floors. This is about the equivalent of an actual floor for every 12 floors of a building. At this point, the savings become significant.

The air moving in the ducts is moving much too fast to be utilized at a room register for distribution. This air must be slowed down before distribution. The velocity reduction is accomplished in pressure-reducing boxes called *sound attenuating boxes* that reduce the velocity and pressure of the air (Figure 25-57). The sound attenuating box does to the airstream what a muffler does to the exhaust explosions of an automobile; it reduces the sound level.

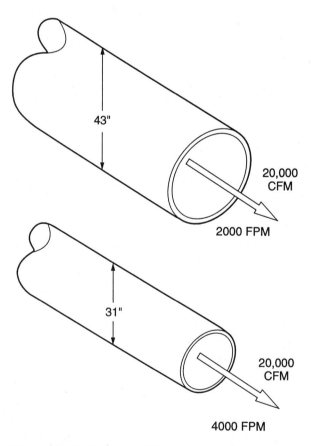

Figure 25-56. The same airflow requirements can be supplied by different air velocities.

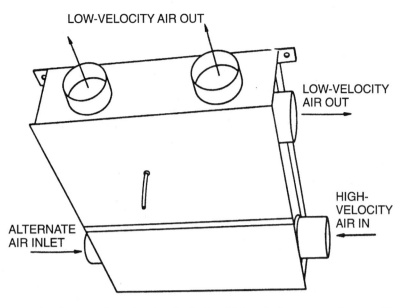

Figure 25-57. Sound attenuating box to reduce the air velocity in a high-velocity system. *Courtesy Trane Company.*

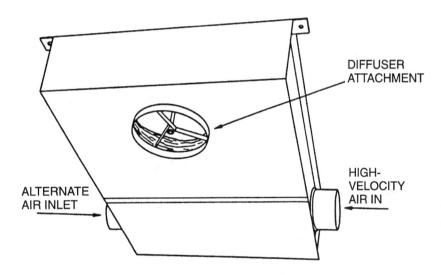

Figure 25-58. The sound attenuating box is located at the end of the run, just before the air distribution register. *Courtesy Trane Company.*

These boxes may be located at the end of the duct very close to the air distribution registers (Figure 25-58).

Single-Duct Applications

Single-duct applications are considered for buildings that may have only a single zone. For example, a multi-story building where the floor area is not large enough to have internal core heat from lights, computers, and people may use a single-duct system on each floor (Figure 25-59). Realize that each individual floor may have either heating or cooling operating, and that both cannot be operating at the same time on the same floor. However, it is possible to have cooling on one floor and heating on another so that different tenant applications can be addressed. For example, a computer company with a high internal heating load in the winter can be placed on a floor of their own. Remember, each floor has only one thermostat for a single-duct system unless some of the spaces have heat added to the leaving air to prevent the spaces from becoming too cool. This heat is called reheat and could be controlled by individual room thermostats (Figure 25-60). Reheat tempers the air by adding a small amount of heat, but is inefficient because the air is first cooled and then heated. This is not a common practice in modern systems because it wastes energy.

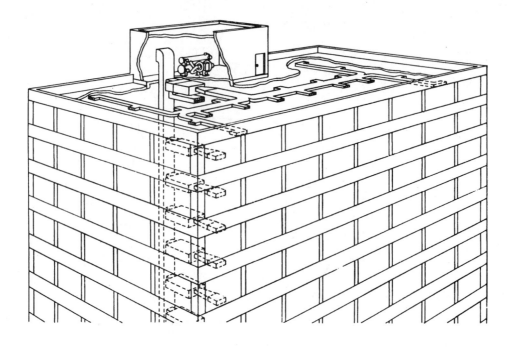

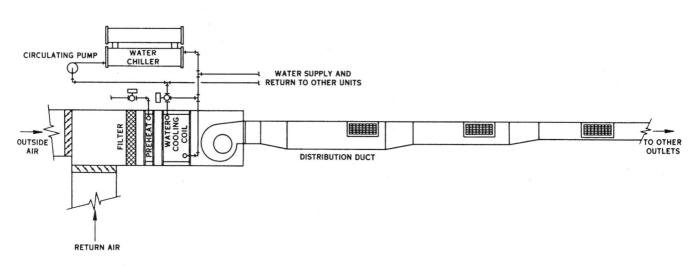

Figure 25-59. Single-duct application in a multi-story building. *Courtesy Trane Company.*

Multi-Zone Applications

A multi-zone application in its simplest form is a unit that uses one cooling coil, one heating method, and one fan with multiple duct connections. The industry terms used for the different cooling and heating methods are *hot deck* and *cold deck*, which refer to the heating and cooling sections of the unit, respectively. The cold deck can either be a direct expansion evaporator with circulating refrigerant or a chilled water coil with water circulating at about 45°F. The hot deck can either be electric strip heat, a steam coil, or a hot water coil (Figure 25-61). There is one duct for each zone which handles the heating and cooling. The unit has a set of dampers for each zone (one heating and one cooling) that are controlled by the room thermostat (Figure 25-62). These dampers modulate open and closed for the best control. The advantage of this system is that each zone has its own thermostatic control. *Modulation* means that the dampers can be moved very slowly from one position to another with any number of increments between open and closed. *Non-modulated dampers* would be either open or closed with no positions in between.

An example of how a multi-zone unit may maintain conditions in an office building is as follows. Consider Zone 1 on the multi-zone unit in Figure 25-63. It is the middle of the day and the weather is warm. The zone has a considerable internal heating load during this part of the day. The thermostat is calling for maximum cooling.

COMMERCIAL DUCT APPLICATIONS **447**

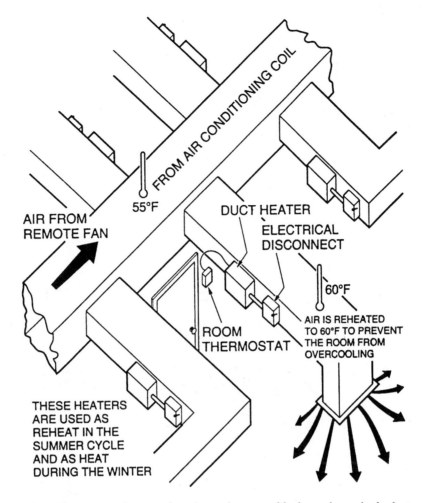

Figure 25-60. Reheat is used to prevent the room from becoming too cold when using a single-duct system.

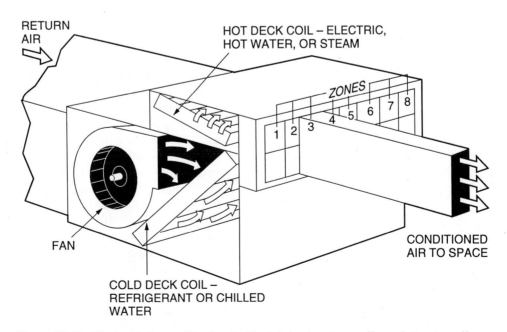

Figure 25-61. The hot deck can either be electric strip heat, a steam coil, or a hot water coil.

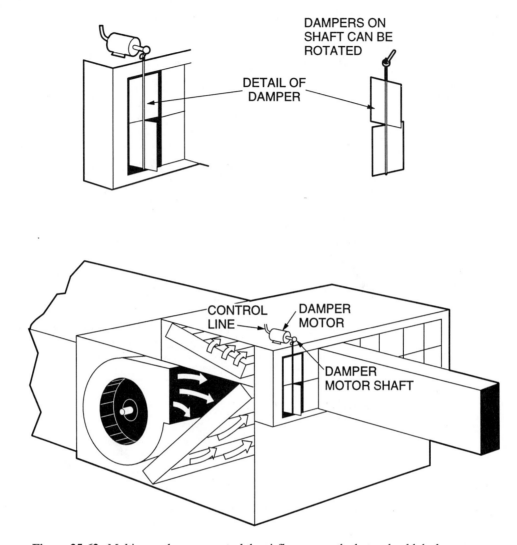

Figure 25-62. Multi-zone dampers control the airflow across the hot and cold decks.

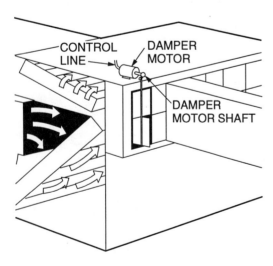

Figure 25-63. One zone of a multi-zone unit.

The cold deck damper is open, providing maximum cooling using 55°F air from the cold deck (Figure 25-64). This zone is used at night for a second shift of people who enter data into the computers. The cooling zone is satisfied in the early evening so the thermostat calls for the multi-zone damper to reposition for less cooling. The heating system is not operating (it is off because the outdoor temperature it is not cold enough to operate the boiler). The air from the hot deck is actually return air temperature of 75°F so the air from the cold deck is blended with the return air temperature for a mixed air temperature of 65°F, which is not enough to overcool the space. If the building space becomes cooler, the damper will be repositioned to furnish only recirculated return air of 75°F. As the seasons change towards cooler weather, the hot deck temperature will begin to rise and provide heat through the hot deck to the heating

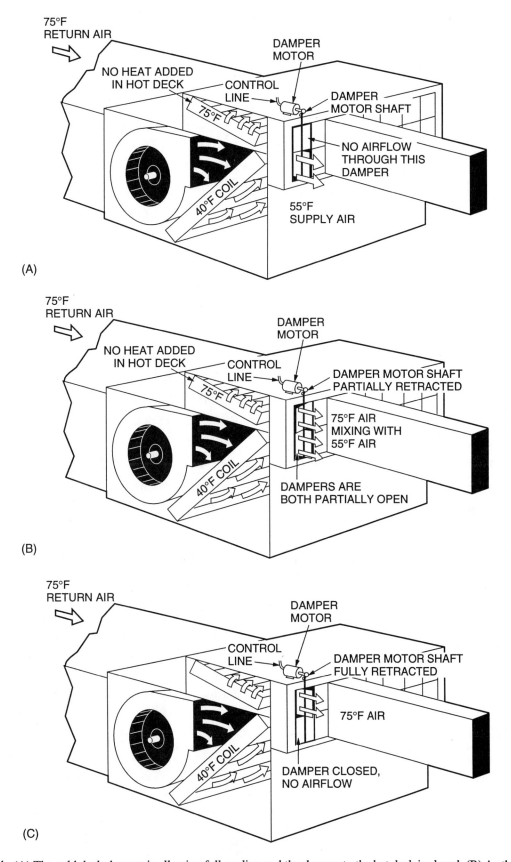

Figure 25-64. (A) The cold deck damper is allowing full cooling and the damper to the hot deck is closed. (B) As the room thermostat is satisfied, the dampers reposition to partially open the hot deck damper and allow the return air temperature of 75°F to mix with the 55°F cold deck air. (C) Later, the thermostat only allows the 75°F return air to enter the room.

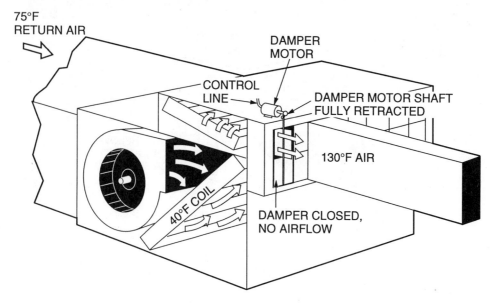

Figure 25-65. The hot deck is now providing all of the airflow and heat to the conditioned space.

duct (Figure 25-65). This system uses an outdoor thermostat—the colder the outdoors, the warmer the hot deck. When the outdoor temperature reaches a predetermined level depending on the system design, the cooling system shuts off and only heating is provided. This may occur at 60°F for some systems and at lower temperatures for others.

The disadvantage of this system is that during certain periods of mild weather, the system is both heating and cooling the air to this particular zone. This is not desirable but can be minimized by using outdoor thermostats and proper control settings. Another disadvantage is that while return air is mixing with conditioned supply air, proper humidity control may not be maintained in climates where the humidity is high.

Dual-Duct Applications

Dual-duct systems typically use a central station fan with a hot and cold deck to provide heating and cooling (Figure 25-66). This system requires two separate duct systems to each zone, one for heating and another for cooling (Figure 25-67). Since there are both heating and cooling ducts, there must be some method to distribute the air at the terminal location where conditioning is required. A mixing box may be used to furnish either hot air, cold air, or a mix of hot and cold air, much like the multi-zone system (Figure 25-68). The mixing box typically has three duct connections: hot, cold, and mixed air. The mixing box also reduces the velocity and pressure of the air so that it can be distributed to the air registers.

Variable Air Volume (VAV) Systems

All of the above systems are called *constant volume systems*. The air flows at a constant volume, furnished by the fan or fans. Single-duct systems using large fans typically use large amounts of power to drive the fan or fans. The power consumption is in direct proportion to the quantity of air moved through the system. If this air quantity can be reduced on demand for load changes, power consumption can also be reduced. This is accomplished by reducing the air volume to the whole system or to parts of the system. Reduced air volume to the conditioned space can be used to control space temperature. For example, the cooling requirement for a space can be controlled by controlling the quantity of air moving into the space. A cooling capacity of 2000 cfm of air is twice the cooling capacity of 1000 cfm of air. If the air volume in a duct serving a conditioned space can be reduced, the cooling capacity can also be reduced. This can be accomplished by using a VAV box (Figure 25-69). This will naturally increase the static pressure in the duct feeding the VAV box and also the static pressure at the fan outlet so having a building with many VAV boxes that may throttle at the same time creates a need to control the static pressure in the main duct system to reduce the fan power requirements and prevent duct damage due to high static pressure. There are several methods used to control the static pressure in the main duct, including fan outlet dampers, bypass dampers, variable fan inlet vanes, and variable fan speeds. They are all controlled by the duct static pressure.

COMMERCIAL DUCT APPLICATIONS 451

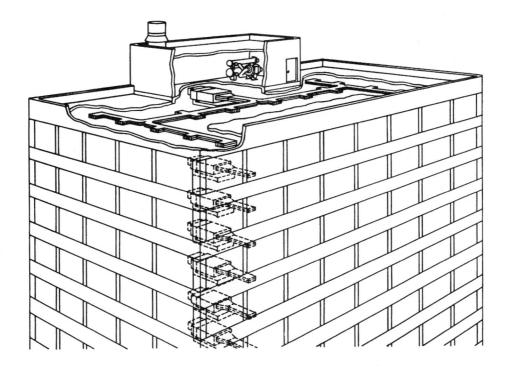

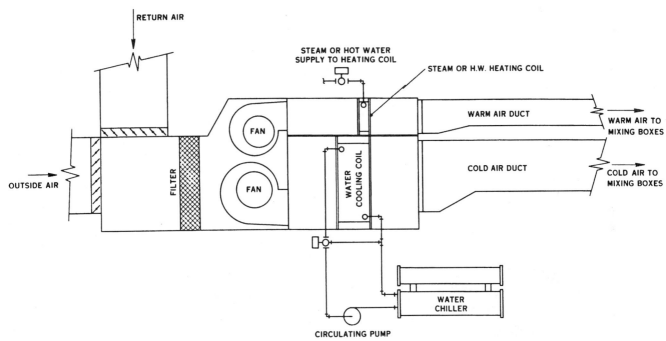

SYSTEM WITH CENTRAL FAN FOR HEATING, CENTRAL FAN FOR COOLING, WATER COOLING COIL AND A REFRIGERATION UNIT FOR CHILLING WATER, OR A REFRIGERATION COMPRESSOR AND DIRECT EXPANSION COOLING COIL FOR COOLING THE AIR, A STEAM OR HOT WATER COIL, FILTERS, DOUBLE-DUCT AIR DISTRIBUTION, MIXING BOXES FOR THE CONDITIONED SPACES AND CONTROLS.

Figure 25-66. Hot and cold deck coils for a dual-duct application. *Courtesy Trane Company.*

452 UNIT 25 AIR DISTRIBUTION AND BALANCE

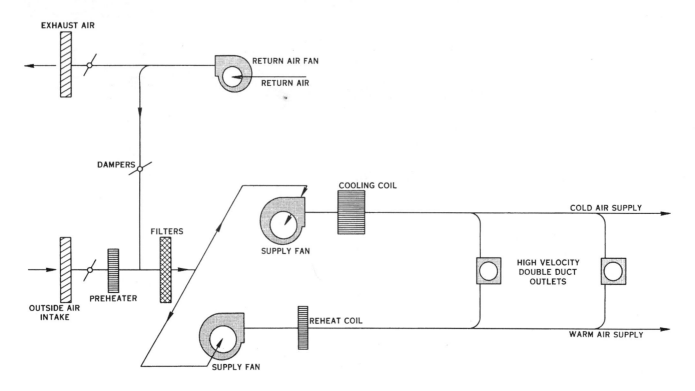

Figure 25-67. A separate duct system must be supplied to each zone. *Courtesy Trane Company.*

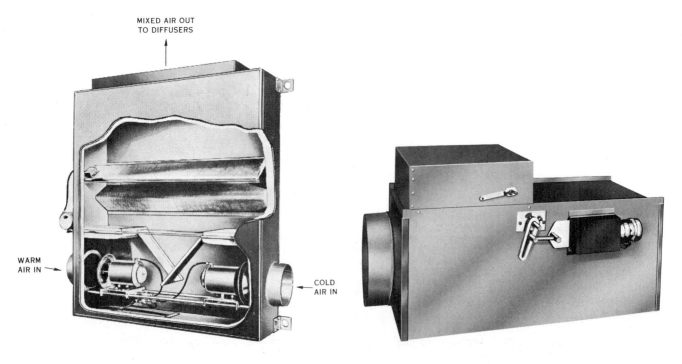

Figure 25-68. A mixing box at the terminal unit provides the correct air temperature to the room. *Courtesy Trane Company.*

Figure 25-69. A variable air volume (VAV) box varies the airflow to the conditioned space. *Courtesy York International.*

COMMERCIAL DUCT APPLICATIONS 453

Figure 25-70. A fan outlet damper is used to reduce the static pressure in the main duct. *Reproduced courtesy of Carrier Corporation.*

Figure 25-72. Variable inlet vanes in a centrifugal fan. *Reproduced courtesy of Carrier Corporation.*

Fan Outlet Dampers Some systems may use a fan outlet damper that modulates to merely throttle the air supply to the main duct system (Figure 25-70). This creates a high static pressure between the fan outlet and the damper and results in some energy savings, but not as much as that provided by other methods.

Bypass Dampers A modulating damper can be mounted in a bypass from the fan outlet to the fan inlet. As the static pressure in main duct system begins to increase because VAV boxes are throttling, this damper modulates open, maintaining a constant pressure in the main duct (Figure 25-71). This method provides no power savings because the air is still being moved, but it does allow temperature control.

Variable Fan Inlet Vanes Fans may be equipped with variable inlet vanes that reduce the air quantity that reaches the fan. These are typically triangular metal devices that are located in the fan inlet (Figure 25-72). They serve two purposes: to reduce the airflow to the fan by offering a restriction and pre-rotate the air as it enters the fan wheel. This device can be controlled by the static pressure in the main duct to throttle the air in a modulating manner. Large amounts of energy can be saved with this device.

Variable Fan Speeds The speed of the fan wheel can be controlled to some extent to maintain a constant static pressure in the main duct system as the VAV boxes throttle the airflow. This can be accomplished in two ways: either with variable-drive couplings or using electronic variable-drive motor speed control. Either of these methods will reduce the airflow and save power.

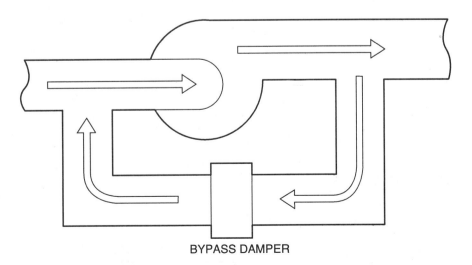

Figure 25-71. This bypass damper arrangement limits the static pressure in the main duct.

454 UNIT 25 AIR DISTRIBUTION AND BALANCE

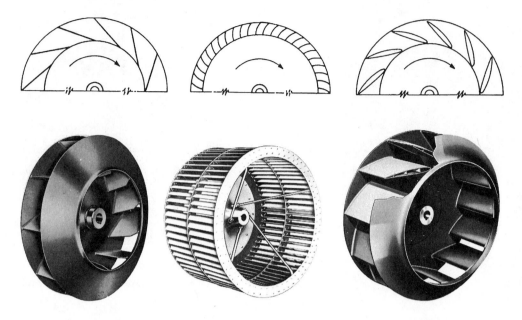

Figure 25-73. Various centrifugal fan blade designs. *Courtesy Trane Company.*

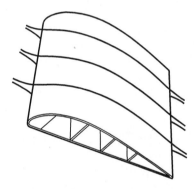

Figure 25-74. This airfoil centrifugal fan blade provides better efficiency than a straight blade.

Types of Fans Used in Commercial Systems

There are two basic types of fans used in large commercial systems: centrifugal and axial.

The centrifugal fan used for commercial applications is designed and built in several different configurations: forward curved, backward curved, radial, and backward inclined. Figure 25-73 shows various blade designs. Each design has a different set of operating conditions and is carefully chosen to suit the particular application. One variation to the above designs is the airfoil-type centrifugal blade (Figure 25-74). This blade design is much more intricate to manufacture than the straight blade design, and is therefore more expensive. However, the air foil design is also more efficient than a straight blade design.

Large centrifugal fans are designed as either single or double inlet fan wheels. These fans are typically installed in the middle of the floor in a fan room (Figure 25-75).

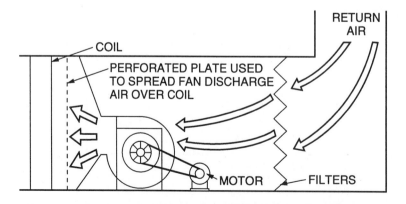

Figure 25-75. Centrifugal fan installed in a fan room.

COMMERCIAL DUCT APPLICATIONS 455

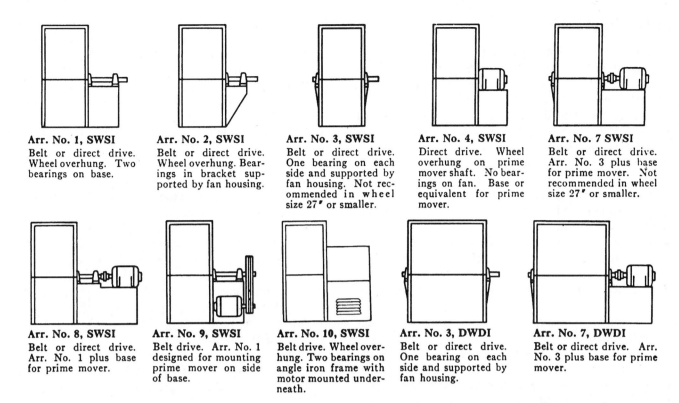

Figure 25-76. Motor and fan arrangements. *Reproduced courtesy of Carrier Corporation.*

This requires extra floor space that may present a problem in some buildings. There are many different types of motor mount configurations for these fans and the design engineer must choose the best configuration for the application (Figure 25-76).

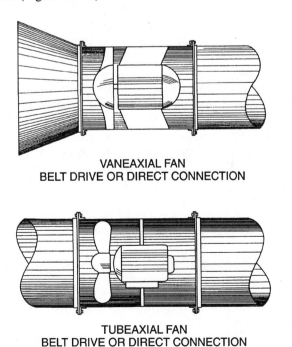

Figure 25-77. Axial fan that mounts in the ductwork. *Reproduced courtesy of Carrier Corporation.*

The axial fan is much like a propeller-type fan, and is mounted in a round housing (Figure 25-77). This fan is particularly useful because it can be mounted in a round portion of the duct off the floor, which takes up less floor space. The motor is typically mounted on the outside of the housing with a belt run through the housing (Figure 25-78).

As you can see from the above information, experience is a valuable asset when designing or troubleshooting any air system.

Caution: Manufacturers and distributors are both valuable sources of application data. In addition, you should not hesitate to consult with the system designer and/or the experienced technician if you have any questions about an ineffective or malfunctioning system.

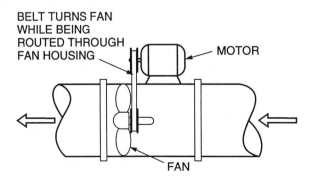

Figure 25-78. The motor on this axial fan is located outside the fan housing.

SUMMARY

- Infiltration is air leaking into a structure.
- Ventilation is air being induced into the HVAC equipment and heated or cooled before it enters the conditioned space.
- The duct system distributes air to the conditioned space. It consists of the blower or fan, the supply duct, and the return duct.
- The blower or fan uses energy to move the air.
- Propeller and centrifugal fans are commonly used in residential and small commercial systems.
- The propeller-type fan is used to move a lot of air against a small pressure drop.
- The centrifugal-type fan is used to move large amounts of air in ductwork, which offers a larger resistance to air movement.
- Fans are turned either by belt-drive or direct-drive motors.
- Unlike a compressor, a fan is not a positive displacement device.
- Small centrifugal fans use energy in proportion to the amount of air they move.
- Duct can be made from aluminum, galvanized steel, flexible tubes, and fiberglass ductboard.
- The pressure the fan creates in the duct is very small and is measured in inches of water column (in. w.c.).
- 1 in. w.c. is the amount of pressure needed to raise a column of water one inch.
- The atmospheric pressure of 14.696 psia will support a column of water that is 34 feet high.
- 1 psi will support a column of water that is 27.1 inches high, or 2.31 feet high.
- Moving air in a duct system creates static pressure, velocity pressure, and total pressure.
- Static pressure is the pressure pushing outward on the duct.
- Velocity pressure is the pressure created by the velocity of the air moving in the duct.
- Total pressure is the sum of the velocity pressure and the static pressure.
- A pitot tube is a probe device used to measure air pressures.
- The air velocity in a duct (in fpm) can be multiplied by the cross-sectional area of the duct (in sq. ft.) to obtain the amount of air passing that particular point in cubic feet per minute (cfm).
- Typical supply duct systems are plenum, extended plenum, reducing plenum, and perimeter loop.
- The plenum system is economical and can be easily installed.
- The extended plenum system takes the trunk duct closer to the farthest outlets but is more expensive.
- The reducing plenum system reduces the trunk duct when some of the air volume has been reduced.
- The perimeter loop system supplies an equal pressure to all of the outlets. It is commonly used in concrete slab floors.
- Branch ducts should always have balancing dampers to balance the air to each room in the conditioned space.
- When the air is distributed to the conditioned space, it is normally used to blanket the outside walls to cancel the heating or cooling load.
- Warm air distributes better from the floor because it rises.
- Cold air distributes better from the ceiling because it falls.
- The amount of throw tells how far the air from a diffuser will travel into the conditioned space.
- Return air systems include individual room returns and common or central returns.
- Each foot of duct (supply or return) has a friction loss that can be plotted on a friction chart for round duct. Round duct sizes can be converted to square or rectangular equivalents for sizing and friction readings.
- Residential and light commercial systems may use only one duct system for both heating and cooling.
- As commercial buildings become larger, different types of duct systems are used.
- Some large commercial buildings have a cooling load in the core of the building even in the winter, so they may have two duct systems, one for heating and the other for cooling.
- Larger air systems can be reduced in size by using high-velocity, high-pressure duct systems. This saves space.
- The air velocity and pressure must be reduced before the air is introduced into the room air distribution register. This is accomplished using sound attenuating boxes.
- With single-duct systems, it is not possible to have both heating and cooling at the same time without some sort of reheat of the air before it enters the conditioned space. This is not energy efficient.
- Multi-zone applications allow heating or cooling in different zones of the same building at the same time.
- Dual-duct applications are similar to multi-zone units except the air is mixed at a mixing box near the terminal air distribution unit.
- Variable air volume (VAV) systems use reduced air volume to control the conditioned space temperature.
- Reduced air volume can also reduce the energy consumption for the building through control of the fan power.
- There are two basic types of fans for moving air in a duct system: centrifugal and axial.

REVIEW QUESTIONS

1. Name the two ways that a structure obtains fresh air.
2. Name two types of blowers that move the air in a forced air system.
3. Which type of blower is used to move large amounts of air against low pressures?
4. Which type of blower is used to move large amounts of air in ductwork?

5. Name two reasons why ductwork resists airflow.
6. Name four types of duct distribution systems.
7. Name two types of blower drives.
8. What units are used to express pressure in ductwork?
9. What is a common instrument used to measure pressure in ductwork?
10. Why is pounds per square inch not used to measure the pressures in ductwork?
11. Name the three types of pressures created by the airflow in ductwork.
12. Name two types of return air systems.
13. What component distributes the air in the conditioned space?
14. Where is the best place to distribute warm air? Why?
15. Where is the best place to distribute cold air? Why?
16. What chart is used to size ductwork?
17. Name four materials used to manufacture duct.
18. What fitting leaves the trunk duct and directs the air into the branch duct?
19. The duct sizing chart is expressed in round duct. How can this be converted to square or rectangular duct?
20. What is used to temper the air in a single-duct air distribution system?
21. How can a duct system size be reduced in a large office building?
22. How can the velocity be reduced in a high-velocity duct system?
23. What is a disadvantage of a dual-duct system?
24. Name two types of circulating fans for a large duct system.

26 INSTALLATION

OBJECTIVES

Upon completion of this unit, you should be able to

- List three crafts involved in air conditioning installation.
- Identify various duct system installations.
- Describe the installation of metal duct.
- Describe the installation of ductboard systems.
- Describe the installation of flexible duct.
- Recognize good installation practices for package air conditioning equipment.
- Discuss different connections for package air conditioning equipment.
- Describe the installation of split-system air conditioning equipment.
- Recognize correct refrigerant piping practices.
- List the basic start-up procedures for air conditioning equipment.

26.1 EQUIPMENT INSTALLATION

Installing air conditioning equipment requires three crafts: ductwork fabrication and installation, electrical, and mechanical, which includes refrigeration. Some contractors use separate crews to carry out the different tasks. Others may perform the duties of two of the crafts within their own company and subcontract the third to a specialized contractor. Some small contractors do all three jobs with a few highly-skilled people. The three job disciplines are often licensed at local and state levels; they may be licensed by different agencies. The contractor should strive for all three licenses and therefore be able to work in any of the areas.

26.2 DUCT INSTALLATION

This section covers basic fabrication and installation procedures for square, rectangular, and round metal duct.

Warning: Local codes must be followed while performing all work.

Square and Rectangular Duct

Square and rectangular metal duct is fabricated in a sheet metal shop by qualified sheet metal layout and fabrication personnel. The duct is then moved to the job site and assembled. Since the all-metal duct system is rigid, all dimensions have to be precise or the system will not fit together. The duct must sometimes rise over objects or go beneath them and still measure out correctly so the takeoff will reach the correct branch location. The branch duct must be of the correct length for it to reach the terminal point (the boot for the room register). Figure 26-1 shows an example of a duct system layout.

Warning: Be careful when handling sharp tools, fasteners, and metal duct. Wear gloves whenever possible.

Square or rectangular metal duct is assembled using S-fasteners and drive clips, which make the duct connections nearly airtight. If further sealing is required, the connection can be caulked and taped. While the duct is being assembled, it has to be fastened to the structure for support. This can be accomplished in several ways. It may lay flat in an attic installation or be hung by hanger straps. The duct should be supported so that it will be steady when the fan starts and will not transmit noise to the structure. The rush of air (which has weight) down the duct will move the duct if it is not fastened. Vibration eliminators under the fan section will prevent the transmission of fan vibrations down the duct. Flexible duct connectors can be installed between the fan and the metal duct to minimize vibration. These are always recommended, but are not always used.

Pre-fabricated metal duct can be purchased from supply houses in several sizes for small and medium systems. This makes metal duct systems available to the small contractor who may not have a sheet metal shop. The assortment of standard duct sizes can be assembled using a variety of standard fittings (Figure 26-2) to build a system that appears to be custom-made for the job.

Round Metal Duct

Round metal duct systems are easy to install and are available from some supply houses in standard sizes for small and medium systems. These systems use reducing fittings from a main trunk line and may be assembled in the field. Self-tapping sheet metal screws are popular fasteners (Figure 26-3). The screws are held by a magnetized screwholder while an electric drill turns and starts the screw.

DUCT INSTALLATION 459

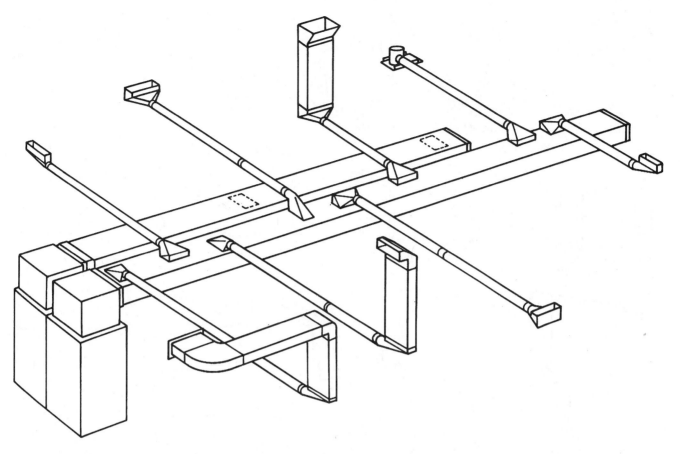

Figure 26-1. Duct system layout.

Warning: Use only grounded, double insulated, or cordless electrical tools when drilling into metal ductwork.

Each connection should have a minimum of three evenly-spaced screws to keep the duct steady. A good installer can fasten the connections as fast as the screws can be placed in the screwholder. A cordless, reversible variable-speed drill is the ideal tool to use.

Round metal duct requires more clearance space than square or rectangular duct. It must be supported and mounted at the correct intervals to keep it straight. Exposed round metal duct may look less attractive than square or rectangular duct, so it is often used in places that are out of sight.

Metal Duct Insulation

Insulation for metal duct can be applied to either the inside or the outside of the duct. This prevents heat transfer between the duct and any unconditioned space, and also prevents condensation damage. When insulation is applied to the inside, the job is usually done in the fabrication shop. The insulation can be fastened with tabs, glue, or both. The tabs are fastened to the inside of the duct and have a shaft that looks like a nail protruding from the duct wall (Figure 26-4). The liner and washer may also be fastened using a tab and shaft. The tab has a base that the shaft is fastened to, and this base is usually fastened to the inside of the duct using glue. Spot welding is the most permanent method, but is both difficult to do and expensive. An electrical spot welder must be used to weld the tab.

The liner can be glued to the duct, but it may come loose and block airflow, perhaps years after the installation date. This is difficult to find and repair, particularly if the duct is in a wall or framed in by the building structure. The glue must be applied correctly and even then it may not hold forever. Using tabs in combination with glue is a more permanent method.

The liner is normally fiberglass and is coated on the air side to keep the airflow from eroding the fibers. Many pounds of air pass through the duct each season. An average air conditioning system handles 400 cfm per ton. Therefore, a 3 ton system handles 1200 cfm or 72,000 cfh (1200 cfm ¥ 60 min./hr. = 72,000 cfh). This volume of 72,000 cfh is equal to 5393 pounds per hour (72,000 ÷ 13.35 ft.3/lb. = 5393 lb./hr.) of air. This is more than 2 1/2 tons of air per hour. Because of this high airflow, the duct liner takes a lot of abuse.

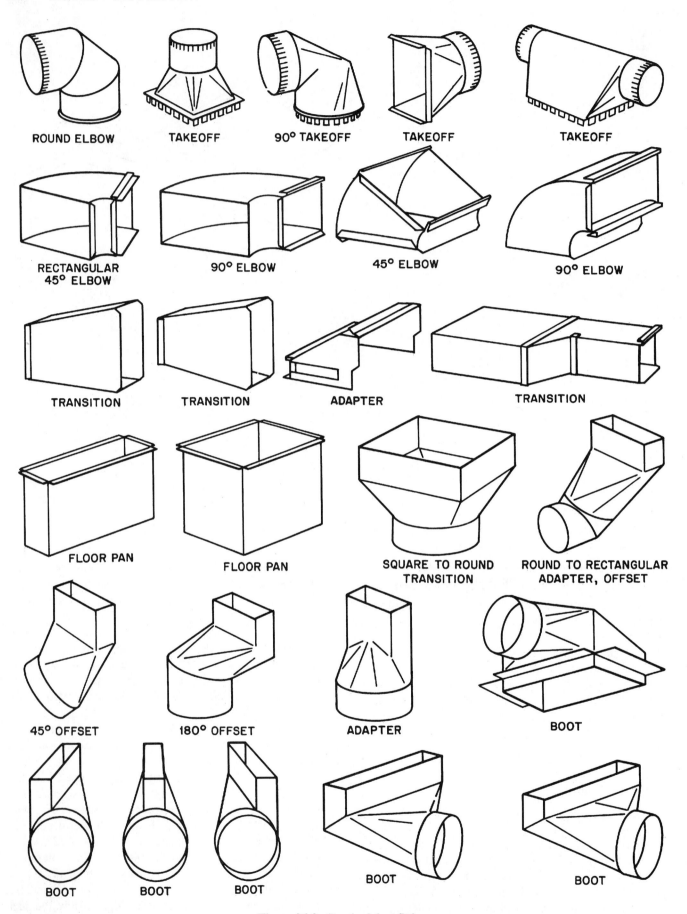

Figure 26-2. Standard duct fittings.

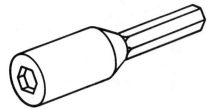

Figure 26-3. Self-tapping sheet metal screw and magnetic screw holder.

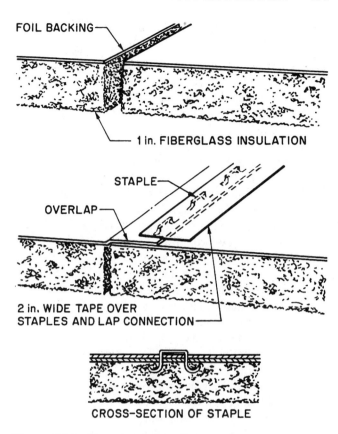

Figure 26-5. The outer layers of the two sections are overlapped and the backing of one piece is stapled to the backing of the other piece.

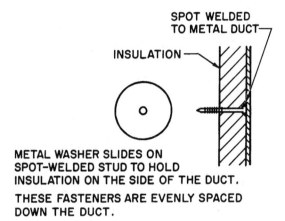

Figure 26-4. Tab to hold fiberglass duct liner to the inside of a duct.

Ductboard Systems

Fiberglass ductboard is very popular with many contractors because it requires little special training to construct a system. Special knives can be used to fabricate the duct in the field. When these knives are not available, simple cuts and joints can be made with a utility knife.

Warning: Be careful when using any sharp instrument. Always cut on a sound, level surface.

This duct has the insulation already attached to the outer layer. This layer is made of foil with a fiber running through it to give it strength. When assembling this duct, it is important to cut some of the insulation away so that the outer layer will overlap the meeting surface.

Caution: Fiberglass fibers can irritate your skin and eyes. Gloves, goggles, and protective clothing should be worn while handling or cutting fiberglass.

The two pieces are strengthened by stapling them together and then taped to make the connection airtight (Figure 26-5).

Fiberglass ductboard can be made into almost any configuration. It is lightweight and easy to transport because the duct can be cut, laid flat, and assembled at the job site. The original shipping boxes can be used to transport the ductboard and to keep it dry. Metal duct fittings, on the other hand, are large, awkward to handle, and take up a lot of space in a truck.

Round fiberglass duct is as easy to install as straight ductboard because it can also be cut with a knife.

Like metal duct, fiberglass duct must also be supported to keep it straight. The weight of the ductboard itself will cause it to sag over long spans. This type of system requires the use of a broad type of hanger that will not cut the ductboard cover.

Fiberglass ductboard deadens sound because the inside of the duct has a coating that helps to reduce any air or fan noise that may be transmitted into the duct. Unlike metal duct, fiberglass ductboard is not rigid and does not transmit sound.

Flexible Duct

Flexible duct has a flexible liner and may have a fiberglass outer jacket for insulation if needed (e.g., where heat exchange takes place between the air in the duct and any unconditioned space). The outer jacket is held by a moisture-resistant cover made of fiber-reinforced foil or vinyl. The duct may be used for the supply or return and should be run in a direct path whenever possible. Sharp bends can greatly reduce the airflow and should be avoided. In addition, runs of flexible duct should be avoided because they present excessive friction loss.

Flexible duct may be used to connect the main supply trunk to the boot at the room diffuser. The boot is the fitting that goes through the floor or ceiling. In this case, it has a round connection on one end for the flex duct and a rectangular connection on the other end where the register is attached. The duct flexibility makes it a popular connector for metal duct systems. The metal duct can be routed close to the boot and the flexible duct can be used to make the final connection. Flexible duct used at the end of a metal duct run will help to reduce any noise that may be traveling through the duct.

Flexible duct must be properly supported using support bands that are at least one inch wide. Proper support will prevent the duct from collapsing and reducing the inside dimension. Some flexible duct has built-in eyelet holes for hanging the duct. Flexible duct should be stretched to a comfortable length to keep the liner from closing and creating friction loss.

26.3 ELECTRICAL INSTALLATION

There are some guidelines that you should be familiar with regarding electrical installation to make sure that the unit has the correct power supply and that the power supply is safe for you, the owner, and the equipment. The control voltage for the space thermostat is often installed by the air conditioning contractor even if the line voltage power supply is installed by a licensed electrical technician.

The power supply must include the correct voltage and wiring practices, including the correct wire size for the current consumed. The law requires the manufacturer to provide a nameplate with each electrical device that gives the voltage requirements and the current draw (Figure 26-6). The applied voltage (the voltage that the unit will actually be using) should be within ±10% of the rated voltage of the unit. For example, if the rated voltage of a unit is 230 V, the maximum operating voltage of the unit would be 253 volts (230 × 1.10 = 253). The minimum voltage would be 207 volts (230 × 0.90 = 207).

Caution: If the unit is operated for a long periods at voltages beyond these limits, the motor and control will be damaged.

When package equipment is installed, there is one power supply for the unit. In a split system, there are two power supplies, one for the inside unit and one for the outside unit. Both power supplies will be connected to a main panel, but there will be a separate fuse or breaker for each device (Figure 26-7).

Note: There should be a disconnect or cutoff switch within 25 feet of each unit and within sight of the unit. If the disconnect is out of sight, such as around a corner, someone may inadvertently turn on the power while you are working on the unit.

Warning: Exercise extreme caution when working with any electrical circuit. All safety rules must be adhered to. Particular care should be taken when working with the primary power supply to any building because the fuse that protects that circuit may be on the power pole outside. For example, a screw driver touched across phases may throw off pieces of hot metal before the outside fuse responds.

Wire sizing tables specify the wire size required by each component. *The National Electrical Code®* (NEC) provides installation standards, including workmanship, wire sizes, methods of routing wires, and types of enclosures for wiring and disconnects. Use it for all electrical wiring installations unless a local code prevails.

The control voltage wiring in air conditioning equipment is the line voltage reduced through a step-down transformer. It is installed with color-coded or numbered wires so that the circuit can be followed through the various components. For example, the 230 V power supply is often at the air handler, which may be in a closet, attic, basement, or crawl space. The interconnecting wiring may have to leave the air handler and be routed to the room thermostat and the condensing unit at the back of the house. The air handler may be used as the junction for these connections (Figure 26-8).

The control wire is a light-duty wire because it carries a low voltage and current. The standard wire size is 18 gauge. A standard air conditioning cable has four wires, each of 18 gauge in the same plastic-coated sheath.

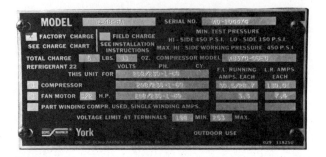

Figure 26-6. Air conditioning unit nameplate. *Photo by Bill Johnson.*

ELECTRICAL INSTALLATION 463

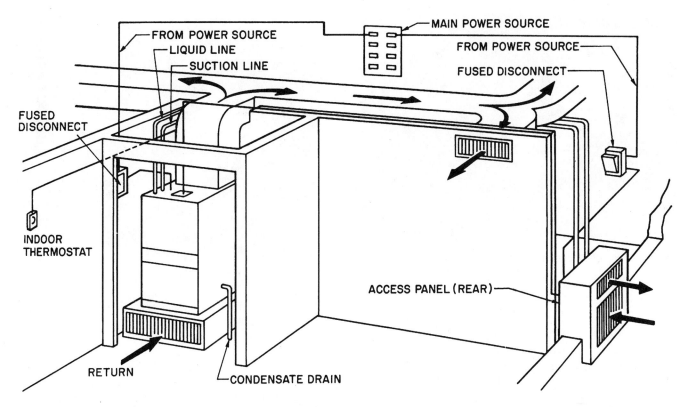

Figure 26-7. Wiring connecting the indoor and outdoor units to the main power source.

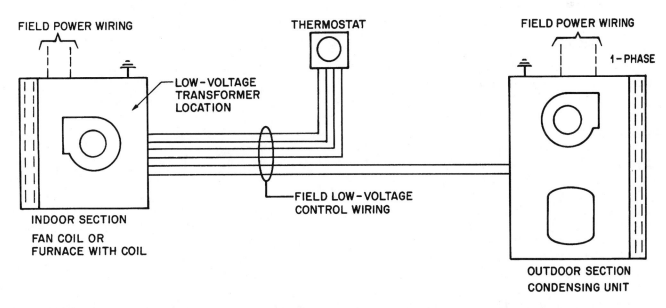

Figure 26-8. Pictorial diagram showing the wiring routed to the room thermostat, air handler, and condensing unit.

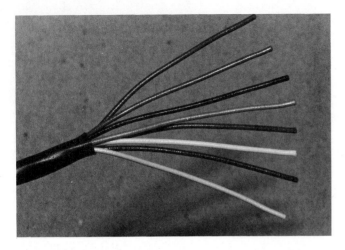

Figure 26-9. 18–8 thermostat wire. *Photo by Bill Johnson.*

Figure 26-10. Air-to-air package unit. *Courtesy Heil Heating and Cooling Products.*

This is called *18–4 wire* (18 gauge, 4 wires). Some wires may have eight conductors and are called *18–8 wire*, (Figure 26-9). There are red, white, yellow, and blue wires in the four-wire cable. The cable or sheath can be installed by the air conditioning contractor in most areas because it is low voltage. An electrical license may be required in some localities.

Caution: Although low-voltage circuits are generally harmless, you can receive a dangerous shock if you are wet and touch live wires. Exercise caution when working with any electrical circuit.

26.4 REFRIGERATION SYSTEM INSTALLATION

The mechanical or refrigeration part of the air conditioning system is included in either a package or split system.

Package Systems

Package or self-contained equipment is equipment in which all components are located in one cabinet or housing. The window air conditioner is a small package unit. Larger package systems may provide 100 tons of air conditioning capacity.

The package unit is available in several different configurations for different applications. Common configurations include the following:

1. An air-to-air application is similar to a window air conditioner except that it has two motors. This is the most common type of package unit. The term *air-to-air* is used because the refrigeration unit absorbs its heat from the air and rejects it into the air (Figure 26-10).
2. Figure 26-11 shows an air-to-water unit. This system absorbs heat out of the air in the conditioned space and rejects the heat into water. The water is wasted or passed through a cooling tower to reject the heat to the atmosphere. This system is sometimes called a *water-cooled package unit.*

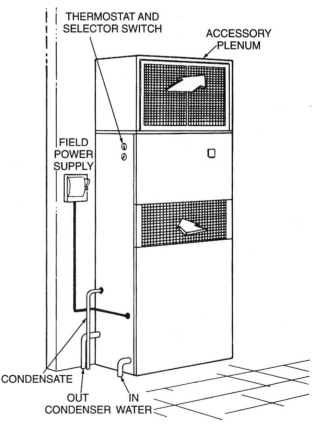

Figure 26-11. Air-to-water package unit. *Reproduced courtesy of Carrier Corporation.*

Figure 26-12. Water-to-water package unit. *Reproduced courtesy of Carrier Corporation.*

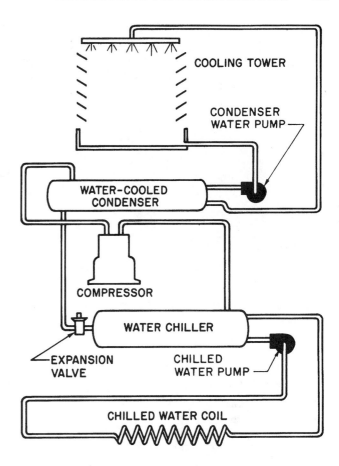

Figure 26-13. The total water-to-water system must have two pumps to move the water in addition to the fans required to cool the water.

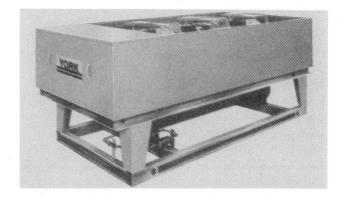

Figure 26-14. Water-to-air package unit. *Courtesy York International.*

3. Water-to-water equipment (Figure 26-12) has two water heat exchangers and is used in large commercial systems. The water is cooled and then circulated through the building to absorb heat (Figure 26-13). This system uses two pumps and two water circuits in addition to the fans to circulate the air in the conditioned space. To properly maintain this system, you need to understand how to service pumps and water circuits.
4. Water-to-air equipment (Figure 26-14) absorbs heat from the circulating water circuit and rejects the heat directly into the air. This equipment is used for large commercial systems.

The air-to-air system installation requires that the unit be positioned on a firm foundation. The unit may be furnished with a roof curb for rooftop installations. The roof curb will raise the unit off the roof and provide waterproof duct connections to the conditioned space below. When a unit is to be placed on a roof in new construction, the roof curb can be shipped separately and a roofer can install it. The air conditioning contractor can then position the package unit on the roof curb for a watertight installation. The foundation for another type of installation may be located outside on the ground next to the conditioned space. Outside pads may be constructed of a high-impact plastic, concrete, or metal.

Vibration Isolation The foundation of the unit should be installed in such a manner that the unit vibration is not transmitted into the building. The unit may require vibration isolation from the building structure. Common methods for preventing vibration include rubber and cork pads and spring isolators. Rubber and cork pads are the least expensive method for simple installations (Figure 26-15). The pads come in sheets that can be cut to the desired size. They are placed under the unit at the point where the unit rests on its foundation. Some units include raised areas on the bottom of the unit for attachment of the mounting pads.

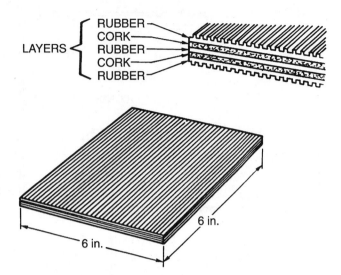

Figure 26-15. Rubber and cork pads may be placed under a unit to reduce the vibration.

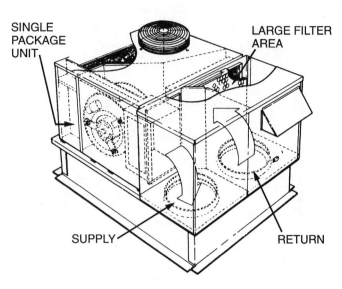

Figure 26-16. Duct connector for rooftop installation. *Courtesy Climate Control.*

Spring isolators may also be used. They need to be matched to the particular weight of the unit. If undersized, the springs will compress and lose their effectiveness under the excessive load. If oversized, they may be too stiff for the lighter load.

Duct Connections for Package Equipment The ductwork entering the structure may be furnished by the manufacturer or fabricated by the contractor. Manufacturers furnish a variety of connections for rooftop installations but do not normally have many factory-made fittings for going through a wall. See Figure 26-16 for a rooftop installation.

Package equipment comes in two different duct connection configurations: side-by-side and over-and-under. In the side-by-side configuration, the return duct connection is beside the supply connection (Figure 26-17). The connections are almost the same size. There will be a large difference in the connection sizes on an over-and-under unit. If the side-by-side unit is installed in a crawl space, the return duct must run on the correct side of the supply duct if the crawl space is low. The narrow confines of a typical crawl space make it difficult to cross a return duct over or under a supply duct. The side-by-side duct design makes it easier to connect the duct to the unit because of the sizes of the connections on the unit. They are closer to being square than the connections on an over-and-under duct design.

The over-and-under duct configuration is more difficult to install because of the oblong connections (Figure 26-18). They are wide because they extend from one side of the unit to the other, and shallow because one is on top of the other. A duct transition fitting is almost always necessary to connect this unit to the ductwork. The transition is normally from wide, shallow duct to nearly square duct.

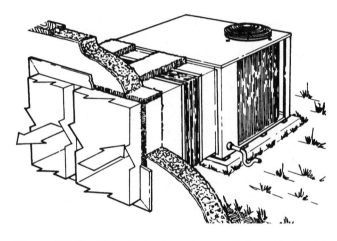

Figure 26-17. This air-to-air package unit has side-by-side duct connections. *Courtesy Climate Control.*

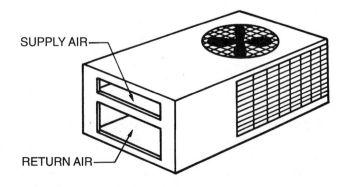

Figure 26-18. Over-and-under duct connections.

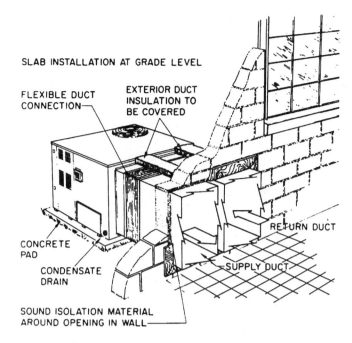

Figure 26-19. Air-to-air package unit installed through a wall. *Courtesy Climate Control.*

When over-and-under systems are used, the duct system may be designed to be shallow and wide to simplify the duct transition.

The duct connection must be watertight and insulated (Figure 26-19). Some contractors cover the duct with a weatherproof hood.

At some point the package equipment will need to be replaced. The duct system usually outlasts two or three units. Manufacturers may change their equipment style from over-and-under to side-by-side. This complicates the new choice of equipment because an over-and-under unit may not be as easy to find as a side-by-side unit, and the duct may already be in place for an over-and-under connection (Figure 26-20).

Package air conditioners have no field-installed refrigerant piping. The refrigerant piping is manufactured within the unit at the factory and the refrigerant charge is included in the price of the equipment. The equipment is ready to operate except for electrical and duct connections. However, one precaution must be observed—most of these systems use R-22 and the oil in the system has an affinity for R-22. The R-22 in the unit will all be in the compressor if it is not forced out by a crankcase heater.

Note: All manufacturers that use crankcase heaters on their compressors supply a warning to leave the crankcase heater energized for some time, perhaps as long as 12 hours, before starting the compressor.

The installing contractor must coordinate the electrical connection and start-up times closely, because the unit should not be started immediately when the power is connected.

REFRIGERATION SYSTEM INSTALLATION 467

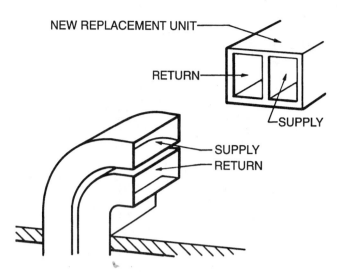

Figure 26-20. Air-cooled package unit with over-and-under duct connections. The manufacturer has changed design to side-by-side duct connections and the installing contractor has a problem.

Split Systems

In split-system air conditioning equipment installations, the condenser is in a different location from the evaporator. For this reason, it can be treated as two separate installation procedures.

Evaporator Section The evaporator is normally located close to the fan section regardless of whether the fan is in a furnace or in a special air handler or blower. The air handler (fan section) and coil must be located on a solid base or suspended from a strong support. Upflow and downflow equipment often has a fireproof, rigid base for the air handler to rest on. Some vertical-mount air handler installations have wall-mounted support for the unit.

When the unit is installed horizontally, it may rest on the ceiling joists in an attic or on a foundation of blocks or concrete in a crawl space (Figure 26-21). The air

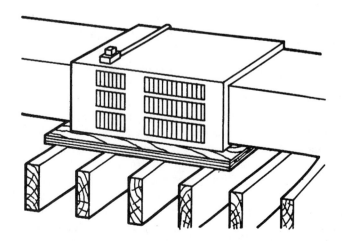

Figure 26-21. Air handler located in an attic crawl space.

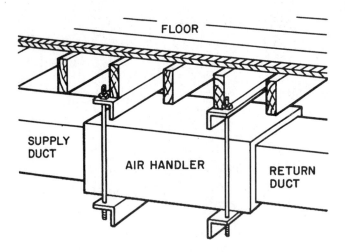

Figure 26-22. The air handler is hung from above and the ductwork is connected and hung at the same height.

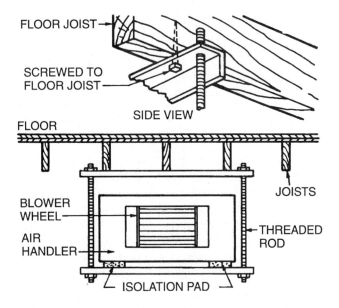

Figure 26-23. Trapeze hanger with vibration isolation pads.

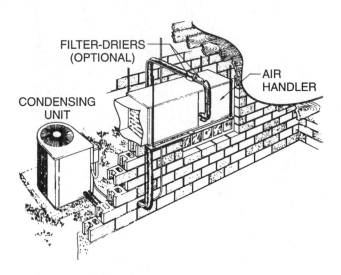

Figure 26-24. Electric furnace with air conditioning coil. *Courtesy Climate Control.*

When an air handler is located in a crawl space, side access is important, so the air handler should be located off the ground next to the floor joists (Figure 26-25). The ductwork will also be installed off the ground. If the air handler has top access panels, it will have to be located lower than the duct and the duct must make a transition downward to the air handler (Figure 26-26).

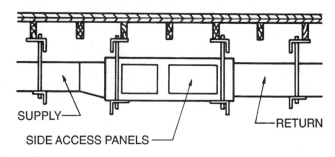

Figure 26-25. Slide access for an air handler in a crawl space.

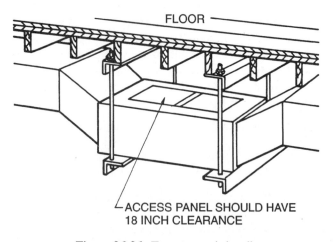

Figure 26-26. Top-access air handler.

handler may also be hung from the floor or ceiling joists (Figure 26-22). If the air handler is hung from above, vibration isolators are often placed underneath it to keep fan noise or vibration from being transmitted into the structure. Figure 26-23 shows a trapeze hanger using vibration isolation pads.

The air handler should always be installed so that it will be easily accessible for future service. The air handler contains the blower and sometimes the controls and heat exchanger.

Most manufacturers have designed their air handlers so that when installed vertically, they are totally accessible from the front. This works well for closet installations where there is insufficient room between the closet walls and the side walls of the air handler. Figure 26-24 shows an electric furnace used as an air conditioning air handler.

REFRIGERATION SYSTEM INSTALLATION 469

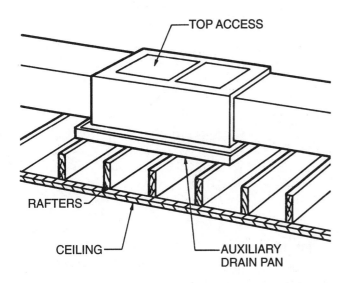

Figure 26-27. Evaporator (air handler) installation in an attic crawl space.

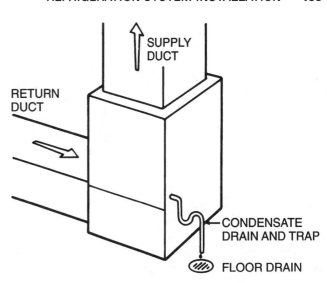

Figure 26-28. Condensate piped to a drain below the evaporator drain pan.

The best location for a top-access air handler is in an attic crawl space where the air handler can be positioned on the joists. Access to the unit is provided from above (Figure 26-27).

Condensate Drain Piping – When the evaporator is installed, provisions must be made for the condensate that will be collected in the air conditioning cycle. An air conditioner in a climate with average humidity will collect about three pints of condensate per hour of operation for each ton of air conditioning. Therefore, a 3-ton system will condense about nine pints per hour of operation. This is more than a gallon of condensate per hour or more than 24 gallons in a 24 hour operating period. This can add up to a great deal of water over a short period of time. If the unit is near a floor drain, simply pipe the condensate to the drain (Figure 26-28). A trap in the drain line will hold some water and keep air from coming into the unit from the termination point of the drain. The drain may terminate in an area where foreign particles may be pulled into the drain pan. The trap will prevent this (Figure 26-29). If there is no drain close to the unit, the condensate must be drained or pumped to another location (Figure 26-30).

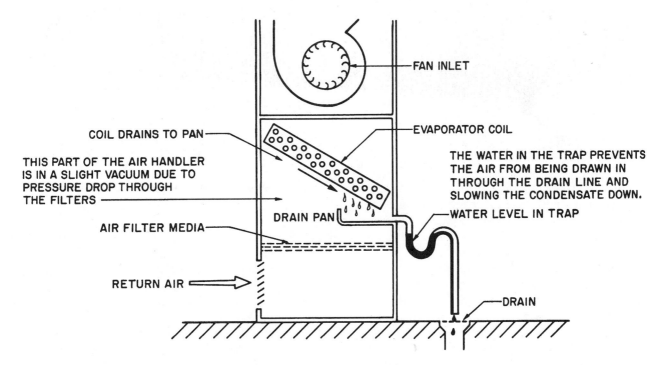

Figure 26-29. Cutaway of a trap.

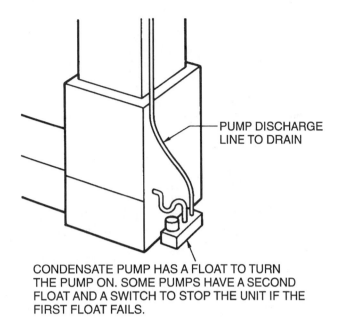

Figure 26-30. The available drain in this installation is above the drain connection on the evaporator, so the condensate must be pumped up to the higher level.

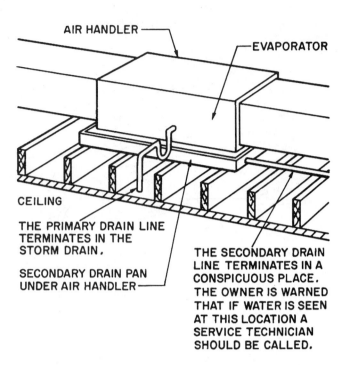

Figure 26-32. Auxiliary drain pan installation.

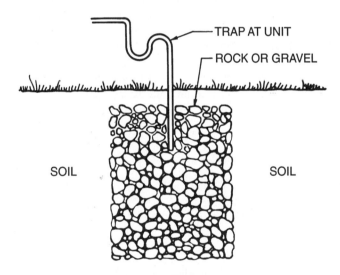

Figure 26-31. Dry well for condensate.

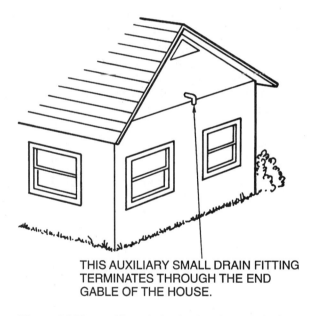

Figure 26-33. Auxiliary drain piped to the end of a house.

Some locations call for the condensate to be piped to a dry well. A dry well is a hole in the ground filled with stones and gravel. The condensate is drained into the well and absorbed into the ground (Figure 26-31). For this installation to be successful, the soil must be able to absorb the amount of water collected by the unit.

When the evaporator and drain are located above the conditioned space, an auxiliary drain pan under the unit is recommended (Figure 26-32). Airborne particles, such as dust and pollen, can get into the drain. Algae will also grow in the water in the lines, traps, and pans and may eventually plug the drain. If the drain system is plugged, the auxiliary drain pan will catch the overflow and keep the water from damaging adjacent surfaces. This auxiliary drain should be piped to a conspicuous place. The owner should be warned that if water ever comes from this drain line, a service call is necessary. Some contractors pipe this drain to the end of the house and out next to a driveway or patio so that if water ever drains from this point, it will be readily noticed (Figure 26-33).

REFRIGERATION SYSTEM INSTALLATION 471

Condensing Unit – The condensing unit location is remote from the evaporator. The following must be considered carefully when installing a condensing unit:

1. Proper air circulation
2. Electrical and piping considerations
3. Service accessibility
4. Proper drainage of rainwater and groundwater
5. Solar influence
6. Appearance

Proper Air Circulation – The unit must have adequate air circulation. The air discharge from the condenser may be from the side or from the top. Discharge air must not hit any object and circulate back through the condenser. The air leaving the condenser is warm or hot and will create high head pressure and poor operating efficiency if it recirculates through the condenser. Follow the manufacturer's instructions for minimum clearances.

Electrical and Piping Considerations – The refrigerant piping and electrical service must be connected to the condensing unit. Piping is discussed in more detail later in this unit, but for now, just remember that the piping will be routed between the evaporator and the condenser. When the condenser is located next to a house, the piping must be routed between the house and the unit, usually behind or beside the house with the piping routed next to the ground. If the unit is placed too far from the house, the piping and electrical service become natural obstructions between the house and the unit. Children may jump on the piping and electrical conduit. If the unit is placed too close to the house, it may be difficult to remove the service panels. Before positioning the unit, study the electrical and refrigeration line connections and make them as short as practical, leaving adequate room for service.

Service Accessibility – Unit placement can make the difference between good service and barely adequate service. For example, a unit is often positioned so that you can touch a particular component but are unable to see it, or, by shifting positions, you can see the component but are not able to touch it. Care should be taken to provide both visual and physical access to all system components.

Proper Drainage of Rainwater and Groundwater – The natural drainage of the groundwater and rainwater or roof water should be considered in unit placement.

Caution: The unit should not be located in a low place where groundwater will rise in the unit. If this happens, the controls may short to ground, and the wiring will be harmed.

All units should be placed on some type of base pad. Concrete and high-impact plastic are commonly used. Metal frames can also be used to raise the unit when needed.

The roof drainage for a structure may run off into gutters. The condensing unit should not be located where the gutter drain or roof drainage will pour down onto the unit. Condensing units are made to withstand rainwater but not large volumes of drainage. If the unit is a top-discharge unit, the drainage from above will fall directly into the fan motor, which is drip-proof, not flood-proof.

Solar Influence – If possible, position the condensing unit on the shady side of the house, because the sun shining on the panels and coils will lower the efficiency. However, the difference is not crucial, and it would not pay to pipe the refrigerant tubing and electrical lines long distances just to keep the sun from shining on the unit.

Shade helps to cool the unit, but it may also cause problems. Some trees have small leaves, sap, berries, or flowers that may fall into the unit or harm the finish of the unit. For example, pine needles may fall into the unit, and the pine pitch that falls on the unit's cabinet may harm the finish and outweigh the benefit of placing the unit in the shade of this type of tree.

Appearance – The condensing unit should be located in an unobtrusive location where it will not make objectionable noise. When located on the side of a house, the unit may be hidden from the street by a low shrub. If the unit is a side-discharge type, the fan discharge must be away from the shrub or the shrub will not live. If the unit has a top discharge, the shrub is unaffected but the unit's sound will rise. This noise may be objectionable in a bedroom located above the unit.

Locating the condensing unit at the back of the house usually places the unit closer to the evaporator and means shorter piping, but the back of the house is where patios and porches are often located, and the homeowner, while sitting outside, may not want to hear the unit. In such a case, a side location at the end of the house where there are no bedrooms may be the best choice.

Each location has its own considerations. Always consult with the owner about locating the various components. A floor plan of the structure may help. Some companies use large graph paper to draw a rough floor plan to scale when estimating the job. The equipment can be located on the rough floor plan to help solve these problems. The floor plan can be reviewed with the homeowner to help them understand location decisions.

Installing Refrigerant Piping The refrigerant piping is always a major consideration when installing a split-system air conditioner. The choice of the piping system may make a difference in the start-up time for the system. The piping should always be kept as short as practical. For an air conditioning installation there are three methods used to connect the evaporator and condensing unit on almost all equipment under 5 tons: contractor-furnished piping, flare or compression fittings with the manufacturer furnishing the tubing, and precharged tubing with sealed quick-connect fittings. In all of these systems, the operating charge for the system is shipped in the equipment.

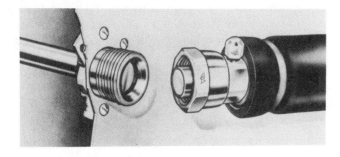

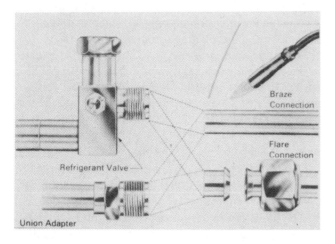

Figure 26-34. Two basic piping connections for split-system air conditioning. *Courtesy Aeroquip Corporation.*

Refrigerant Charge – Regardless of who furnishes the connecting tubing, the complete operating charge for the system is normally furnished by the manufacturer for units under 5 tons. The charge is shipped in the condensing unit. The manufacturer furnishes enough charge to operate the unit with a predetermined line length, typically 30 feet. The manufacturer holds and stores the refrigerant charge in the condensing unit with service valves if the tubing uses flare or compression fittings. When quick-connect line sets are used, the correct operating charge for the line set is included in the actual lines. Figure 26-34 shows both a service valve and a quick-connect fitting.

When service valves are used, the piping is fastened to the service valves using flare or compression fittings. Some manufacturers provide the option of soldering to the service valve connection.

The piping is always either hard-drawn or soft copper. In installations where the piping is exposed, straight pipe may look better. In this case, hard-drawn tubing may be used with factory elbows for the turns. When the piping is not exposed, soft copper is easily formed around corners where a long radius will be satisfactory.

Line Set – When the manufacturer furnishes the tubing, it is called a line set. The suction line is insulated, and the tubing may be charged with nitrogen contained by rubber plugs in the tube ends. When the rubber plugs are removed, the nitrogen rushes out with a loud hiss. This indicates that the tubing does not have any leaks.

When tubing is used, uncoil it from the end of the coil. Place one end of the coil on the ground and roll it while keeping your foot on the tubing end on the ground.

Caution: Be careful not to kink the tubing when going around corners, or it may collapse. Because of the insulation on the pipe, you may not even see the kink.

Tubing Leak Test and Evacuation – When the piping is routed and in place, the following procedure is commonly used to make the final connections:

1. The tubing is fastened at the evaporator end. When the tubing has flare nuts, a drop of oil applied to the back of the flare will help prevent the flare nut from turning the tubing when the flare connection is tightened. The suction line may be as large as 7/8 inch OD tubing. Large tubing will have to be made very tight. Use two adjustable wrenches to make the connection as shown in Figure 26-35). The liquid line will not be any smaller than 1/4 inch OD tubing, and it may be as large as 1/2 inch OD on larger systems.
2. The smaller (liquid) line is fastened to the condensing unit service valve.
3. The larger (suction) line is fastened hand tight.
4. The liquid-line service valve is opened slightly, and refrigerant is allowed to purge through the liquid line, the evaporator, and the suction line. This pushes the nitrogen out of the piping and the evaporator, and it escapes at the hand-tightened suction line flare nut (Figure 26-36).

Note: A small amount of refrigerant may be released with the nitrogen charge. This is allowable as it is considered a de minimus (minimal) amount of refrigerant.

Line sets come in standard lengths of 10, 20, 30, 40, or 50 feet. If lengths other than these are required, the manufacturer should be consulted. Some manufacturers have a maximum allowable line length, normally in the

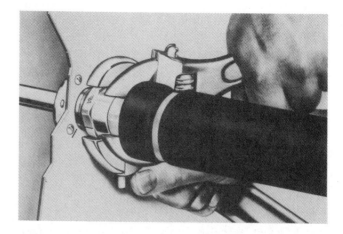

Figure 26-35. Proper method for tightening fitting. *Courtesy Aeroquip Corporation.*

REFRIGERATION SYSTEM INSTALLATION 473

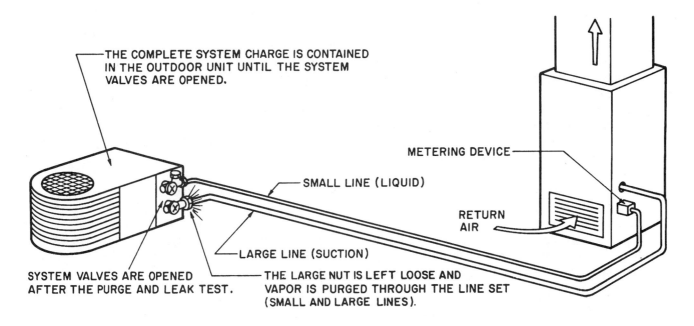

Figure 26-36. Purging the manufacturer's nitrogen charge.

vicinity of 50 feet. If the line length has to be changed from the standard length, the manufacturer will also recommend how to adjust the unit charge for the new line length. As mentioned earlier, most units are shipped with a charge for 30 feet of line. If the line is shortened, refrigerant must be removed. If the line is lengthened, refrigerant must be added.

Altering Line Set Lengths – When line sets must be altered, they should be treated as self-contained systems (Figure 26-37). The following procedure should be followed:

1. Alter the line sets as needed for the proper length. The nitrogen charge will escape during this alteration.

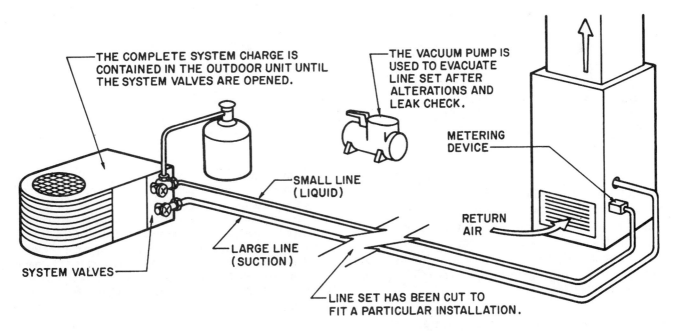

Figure 26-37. Altering the line set. After the line set has been altered, it is connected as shown in Figure 26-36 and purged. It is then checked for leaks. The line set and evaporator are evacuated. After evacuation, a vapor holding charge is added (about 20 psig of the refrigerant characteristic to the system).

To keep the line as clean as possible, do not remove the rubber plugs.

2. When the alterations are complete, the lines may be quickly pressurized to about 25 psig to leak test any connections you have made. If you wish to pressure test using higher pressures, you may safely pressurize up to 150 psig using refrigerant or nitrogen. If you use refrigerant only, it will have to be recovered when the line is purged. The line's service port is common to the line side of the system. If the valves are not opened at this time, the line set and evaporator may be thought of as a separate sealed system for the purpose of pressure testing and evacuation.
3. After the leak test has been completed, evacuate the line set (and evaporator if connected). Break the vacuum to about 10 psig and go through the line set purge just as though it were a factory installation. After starting the unit, you will have to alter the charge to meet the manufacturer's guidelines.

Precharged Line Sets (Quick-Connect Line Sets) — Precharged line sets with quick-connect fittings are shipped in most of the standard lengths. The difference between line sets and precharged line sets is that the correct refrigerant operating charge is shipped in the precharged line set. Refrigerant does not have to be added or taken out unless the line set is altered. The following procedures are recommended for connecting precharged line sets with quick-connect fittings:

1. Roll the tubing out straight.
2. Determine the routing of the tubing from the evaporator to the condenser and put it in place.
3. Remove the protective plastic caps from the evaporator fittings. Place a drop of refrigerant oil (this is sometimes furnished with the tubing set) on the neoprene O-rings of each line fitting. The O-ring is used to prevent refrigerant from leaking out while the fitting is being connected, and it serves no purpose after the connection is tightened. It may tear while making the connection if it is not lubricated.
4. Start the threaded fitting and tighten hand tight. Making sure that several threads can be tightened by hand will ensure that the fitting is not cross-threaded.
5. When the fitting is tightened hand tight, finish tightening the connection using a wrench. You may hear the hiss of a slight amount of escaping vapor while tightening the fitting; this purges the fitting of any air. The O-ring should be seated at this time. Once you have started tightening the fitting, do not stop until you are finished. If you stop in the middle, some of the system charge may be lost.
6. Tighten all fittings as indicated.
7. After all connections have been tightened, perform a leak test on all new connections.

Altering Precharged Line Sets — When the lengths of quick-connect line sets are altered, the charge will also have to be altered. The line set may be treated as a self-contained system for the purpose of alteration. The recommended procedure is as follows:

1. Before connecting the line set, recover the refrigerant from the line set. Do not just cut it because the liquid and suction lines contain some liquid refrigerant.

Warning: Failure to recover the refrigerant is illegal.

2. Cut the line set and alter the length as needed.
3. Pressure test the line set.
4. Evacuate the line set to a low vacuum.
5. Valve off the vacuum pump and pressurize to about 10 psig on just the line set and connect it to the system using the procedures already given.
6. Read the manufacturer's recommendations as to the amount of charge for the new line lengths and add this refrigerant to the system. If there are no recommendations, see Figure 25-38 for a table of liquid line capacities. The liquid line should contain the most refrigerant, and the proper charge should be added to the liquid line to make up for what was lost when the line was cut. No extra refrigerant charge needs to be added for the suction line if it is filled with vapor before assembly.

Piping Practices — The piping practices described here are typical of most manufacturer's recommendations. However, the specific manufacturer's recommendations should always be followed. Each manufacturer ships

LIQUID LINE DIAMETER INCHES	OUNCES OF R-22 PER FOOT OF LENGTH OF LIQUID LINE
$\frac{3}{8}$	0.58
$\frac{5}{16}$	0.36
$\frac{1}{4}$	0.21

If 3 feet of liquid line must be added to a ($\frac{3}{8}$ in.) 30-foot line, an additional 1.74 ounces must be added to the system upon starting the unit. (3 × 0.58 = 1.74)

If the unit is shipped with a precharged line set, the complete charge for that particular line set is contained in the lines. If this set is altered, the complete liquid line length must be used. For example, if a 50-foot set ($\frac{1}{4}$ in.) is cut to 25 feet of length, the charge is exhausted from the lines, and 5.25 ounces must be added to the charge contained in the condenser when started. (25 × 0.21 = 5.25)

Figure 26-38. Typical liquid line capacities.

installation and start-up literature inside the shipping crate. If it is not there, request it from the manufacturer and use it.

26.5 EQUIPMENT START-UP

The final step in the installation is starting the equipment. The manufacturer will furnish start-up instructions. After the system is in place, leak-checked, has the correct factory charge furnished by the manufacturer, and is wired for line and control voltage, follow the guidelines below for equipment start-up:

Warning: Before starting any system, examine all electrical connections and moving parts. The equipment may have faults and defects that can be harmful (e.g., a loose pulley may fly off when the motor starts to turn). Remember that vessels and hoses are under pressure. Always be mindful of potential electrical shock.

1. The line voltage must be connected to the unit and turned on. This should be done by a licensed electrician. Disconnect a low-voltage wire, such as the Y wire at the condensing unit, to prevent the compressor from starting. Turn on the line voltage. The line voltage allows the crankcase heater to heat the compressor crankcase. This applies to any unit with a crankcase heater. Heat must be applied to the compressor crankcase for the amount of time recommended by the manufacturer (usually no more than 12 hours). Heating the crankcase boils any refrigerant out of the crankcase before the compressor is started. If the compressor is started with liquid in the crankcase, some of the liquid will reach the compressor cylinders and may cause damage. The oil will also foam and provide only marginal lubrication until the system has run for some time.
2. It is normally a good idea to plan on energizing the crankcase heater in the afternoon and starting the unit the next day. Before you leave, make sure that the crankcase heater is hot.
3. If the system has service valves, open them. Some units have valves that do not have back seats (Figure 26-39). In this case, do not try to back-seat these valves or you may damage them. Open the valve until resistance is felt, then stop.
4. Check the line voltage at the installation site and make sure it is within the recommended limits.
5. Turn the power off, then check all electrical connections (including those made at the factory) to ensure that they are tight and secure.
6. Set the fan switch on the room thermostat to FAN ON and check the indoor fan for proper operation, rotation, and current draw. You should feel air at all registers, normally about two to three feet above a floor register. Make sure there are no air blockages at the supply and return openings.

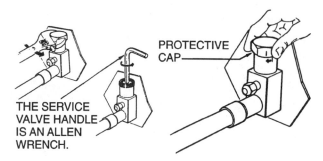

Figure 26-39. Service valves furnished with equipment may not have back seats. *Courtesy Aeroquip Corporation.*

7. Turn the fan switch to FAN AUTO and with the HEAT-OFF-COOL selector switch in the OFF position, replace the Y wire at the condensing unit. The power at the unit should be turned off while making this connection.
8. Place your ammeter on the common wire to the compressor, have someone move the HEAT-OFF-COOL selector switch to COOL, and slide the temperature setting to call for cooling. The compressor should start. Some manufacturer's literature will call for you to have a set of gages on the system at this time. Be careful of the line length on the gage you install on the high-pressure side of the system. If the system has a critical charge and a six foot gage line is installed, the line will fill up with liquid refrigerant and will alter the charge, possibly enough to affect performance. A short gage line is recommended (Figure 26-40). Leak test the gage port when the gages are removed. Replace the gage port cover.
9. If the manufacturer recommends that gages be installed for start-up, install them before you start the system. When gages are not installed, there are certain signs that indicate correct performance. The suction line coming back to the compressor should be cool, although the degree of coolness may vary. Two things will cause the suction-line temperature to vary and still be correct: the metering device and the ambient temperature. Most modern systems are fixed-bore metering devices (capillary tube or orifice). When the outside temperature is mild (e.g., 75°F to 80°F), the suction line will not be as cool as on a very hot day because the condenser becomes more efficient and liquid refrigerant is retained in the condenser, partially starving the evaporator. If the day is cool (e.g., 65°F to 70°F), some of the air to the condenser may be blocked, which causes the head pressure to rise and the suction line to become cooler. On a very hot day (85°F to 90°F) the ambient temperature in the conditioned space causes the evaporator to have a large load, which prevents the suction line from becoming very cold until the inside temperature is reduced to near the design temperature (about 75°F).

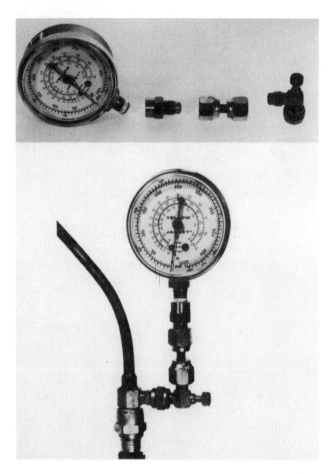

Figure 26-40. A short gage line for connecting to the high-pressure side of the system. *Photos by Bill Johnson.*

The amperage of the compressor is a good indicator of system performance. If the outside temperature is hot and there is a full load on the evaporator, the compressor will be pumping at near capacity. The motor current draw will be close to the nameplate amperage. It is also normal for the compressor amperage to be slightly below the nameplate rating. It is very rare for the compressor amperage to be more than the nameplate rating due to only the system load.

When you are certain that the unit is running satisfactorily, inspect the installation. Check the following:

1. All air registers are open.
2. There are no air restrictions.
3. The duct is hung correctly and all connections are taped.
4. All panels are in place with all screws intact.
5. All construction debris is picked up.
6. The customer knows how to operate the system.
7. All warranty information is filled out.
8. The customer has the operation manual.
9. The customer knows how to contact you.

SUMMARY

- The installation of air conditioning equipment normally involves three crafts: ductwork fabrication, electrical, and mechanical or refrigeration.
- Duct systems are normally constructed of square, rectangular, or round metal, ductboard, or flexible material.
- Square and rectangular metal duct systems are assembled using S-fasteners and drive clips.
- Round metal duct comes in many standard sizes with fittings.
- Round duct must be fastened with sheet metal screws at each connection.
- The first fitting in a duct system may be a vibration eliminator to keep any fan noise or vibration from being transmitted into the duct.
- Insulation may be applied to the inside or outside of any metal duct system that may exchange heat with an unconditioned space.
- When applied to the inside, insulation can be either glued or both glued and fastened with a tab.
- Fiberglass ductboard is compressed fiberglass with a reinforced foil backing that helps to support it and create a vapor barrier.
- Flexible duct has a flexible liner that may have a cover of fiberglass that is held in place with vinyl or reinforced foil.
- The electrical installation includes choosing the correct enclosures, wire sizes, and fuses or breakers.
- The electrical contractor will normally install the line voltage wiring, and the air conditioning contractor will usually install the low-voltage control wiring. The NEC and local codes should be consulted before any wiring is done.
- The low-voltage control wiring is normally color-coded or numbered.
- Air conditioning equipment is manufactured in package systems and split systems.
- Package equipment is completely assembled in one cabinet.
- Package equipment may use two types of duct connections: over-and-under and side-by-side.
- An air-to-air package equipment installation consists of placing the unit on a foundation, connecting the ductwork, and connecting the electrical service and control wiring.
- The duct connections in package equipment may be made through a roof or through a wall at the end of the structure.
- Roof installations have waterproof roof curbs and factory-made duct systems.
- Isolation pads or springs placed under the equipment prevent equipment noise from traveling into the structure.
- Outside units may be installed on pads or frames. These pads are normally made of high-impact plastic, concrete, or metal.

- In split-system air conditioning equipment, the evaporator is at an inside air handler (blower). The air handler may be an existing furnace or a separate blower package for cooling only. The compressor is located outside with the condenser.
- Two refrigerant lines connect the evaporator and the condensing unit; the large line is the insulated suction line and the small line is the liquid line.
- The air handler and condenser should be installed so that they are easily accessible for service.
- A condensate drain must be provided for the evaporator.
- A secondary drain pan should be provided if the evaporator is located above the conditioned space.
- The condensing unit should be located as close as practical to the evaporator and the electrical service. When located on a pad, it should not be positioned where it will be flooded by groundwater or where rainwater from the roof will drain directly into the top of the unit.
- The condensing unit should not be located where the operating noise will be disturb occupants.
- Line sets come in standard lengths of 10, 20, 30, 40, and 50 feet. The system charge is normally for 30 feet when the whole charge is stored in the condensing unit. The refrigerant charge must be adjusted if a line set length is altered.
- The start-up procedure for the equipment is included in the manufacturer's literature. Before start-up, check the electrical connections, fans, airflow, and refrigerant charge.

REVIEW QUESTIONS

1. Name the three crafts normally involved in an air conditioning installation.
2. Which craft normally installs the refrigerant lines and starts the system?
3. Which type of duct system is popular because it can be installed with little special training?
4. What fasteners are used to attach square or rectangle metal duct at the joints?
5. What is normally used to fasten round metal duct at the joints?
6. Name two methods for fastening insulation to the duct when it is insulated on the inside.
7. What happens if the insulation on the inside of a duct becomes loose?
8. What material is most duct insulation made from?
9. Why is a flexible connector installed in a metal duct system?
10. What happens if flexible duct is turned too sharply around a corner?
11. Why must flexible duct be pulled as tight as practical?
12. How should flexible duct be hung from above?
13. What is the difference between a package unit and a split system?
14. Name the two duct connection configurations for a package unit.
15. Name the two refrigerant lines connecting the condensing unit to the evaporator.

27 TYPICAL OPERATING CONDITIONS

OBJECTIVES

Upon completion of this unit, you should be able to

- Explain what conditions affect evaporator pressures and temperatures.
- Define how the various conditions in the evaporator and ambient air affect condenser performance.
- Explain the relationship between the evaporator and the rest of the system.
- Describe the relationship between condenser efficiency and total system performance.
- Compare high-efficiency equipment with standard equipment.
- Establish reference points when working on unfamiliar equipment in order to find the typical operating conditions.
- Describe how humidity affects equipment suction and discharge pressures.
- Explain three methods that manufacturers use to make air conditioning equipment more efficient.

27.1 MEASURING OPERATING CONDITIONS

Air conditioning technicians must be able to evaluate both mechanical and electrical systems. Mechanical operating conditions are determined or evaluated using gages and thermometers; electrical conditions are measured using electrical instruments.

Caution: It is often necessary to take electrical, pressure, and temperature readings while the equipment is in operation. Observe all electrical and mechanical safety precautions. Wear goggles and gloves when attaching gages to a system.

27.2 MECHANICAL OPERATING CONDITIONS

Air conditioning equipment is designed to operate at its rated capacity and efficiency for one set of design conditions. These conditions are generally considered to be at an outside temperature of 95°F and an inside temperature of 80°F, with a relative humidity of 50%. This rating is established by the Air Conditioning and Refrigeration Institute (ARI). Equipment must have a standard rating in order for buyers to have a common basis with which to compare one piece of equipment with another. All equipment in the ARI directory is rated under the same conditions. When an estimator or buyer finds that a piece of equipment is rated at 3 tons, he or she knows that it will provide around 36,000 Btuh of capacity under the stated conditions. When the conditions are different, the equipment will perform differently. For example, most homeowners will not be comfortable at 80°F with a relative humidity of 50%. They will normally operate their system at about 75°F at the same relative humidity. The equipment will have slightly less capacity at 75°F than it had at 80°F. If the designer wants the system to have a capacity of 3 tons at 75°F and 50% relative humidity, the manufacturer's literature may be consulted to make the proper choice.

Note: This 75°F, 50% relative humidity condition will be used for our design conditions because it represents common operating conditions in the field. ARI rates condensers with 95°F air passing over them. A new, standard efficiency condenser will condense refrigerant at about 125°F with 95°F air passing over it. As the condenser ages, dirt accumulates on the outdoor coil and the efficiency decreases. The refrigerant will then condense at a higher temperature. This higher temperature can easily approach 130°F. This is the value often found in the field. This text will use a condensing temperature of 125°F, assuming that the equipment is properly maintained.

The inside relative humidity also adds a significant load to the evaporator coil and has to be considered as part of the load. When conditions vary from the design conditions, the equipment will vary in capacity. The pressures and temperatures will also change.

It may not be easy to recognize small changes in load conditions on the gages and instruments used in the field. For example, an increase in humidity is not followed by a proportional rise in suction pressure and amperage. Perhaps the most noticeable aspect of an increase in humidity is an increase in the condensate accumulated in the condensate drain system.

System Component Relationships Under Load Changes

If the outside temperature increases (i.e., from 95°F to 100°F) the equipment will be operating at a higher head pressure and will not have as much capacity. The capacity also varies when the space temperature goes up or down or when the humidity changes. There is a measurable relationship between the various components in the system that will help you to trace the source of operating problems. The evaporator absorbs heat, so when anything happens to increase the amount of heat that is absorbed into the system, the system pressures will rise. Conversely, the condenser rejects heat, so if anything happens to prevent the condenser from rejecting heat from the system, the system pressures will also rise. The compressor pumps heat-laden vapor. Vapor at different pressure levels and saturation points (in reference to the amount of superheat) will hold different amounts of heat.

Evaporator Operating Conditions

The evaporator will normally have a boiling temperature of 40°F when operating at the 75°F, 50% relative humidity design conditions. This results in a typical suction pressure of 70 psig for R-22. (The actual pressure corresponding to 40°F is 68.5 psig; for our purposes, we will round this off to 70 psig.) This example is at design conditions and a steady-state load. At these conditions, the evaporator is boiling the refrigerant exactly as fast as the expansion device is metering it into the evaporator. In this system at these conditions, the evaporator has a return air temperature of 75°F and the air has a relative humidity of 50%. The liquid refrigerant goes nearly to the end of the coil, and the coil has a superheat of 10°F. This coil is operating as intended at design conditions (Figure 27-1).

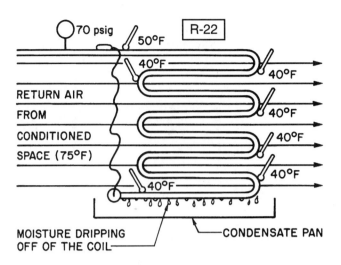

Figure 27-1. Evaporator operating at typical conditions. The refrigerant is boiling at 40°F in the coil. This corresponds to 68.5 psig (approximately 70 psig) for R-22.

Late in the day, after the sun has been shining on the house, the heat load inside the house becomes greater. A new condition is established. The example evaporator in Figure 27-2 has a fixed-bore metering device that will only feed a certain amount of liquid refrigerant. If the space temperature in the house climbs to 77°F, it will cause the liquid refrigerant in the evaporator to boil faster (Figure 27-3). This causes the suction pressure and superheat to rise slightly. The new suction pressure is 73 psig, and the new superheat is 13°F. This is well within the range of typical operating conditions for an evaporator.

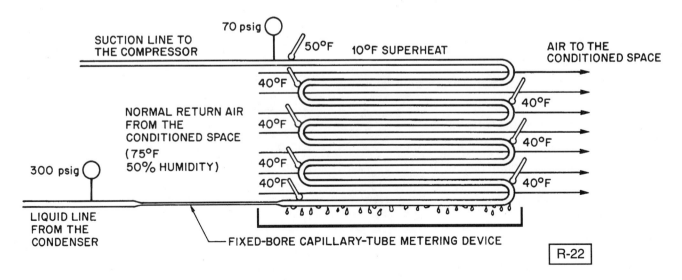

Figure 27-2. This system uses a fixed-bore metering device and is operating at an efficient state of 10°F superheat.

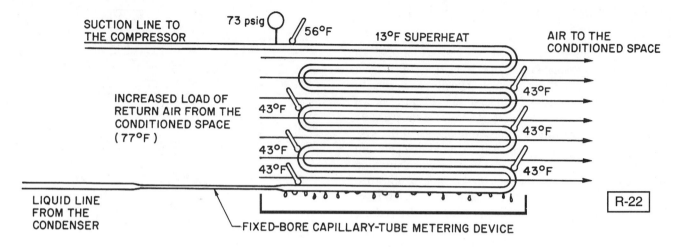

Figure 27-3. When a system uses a fixed-bore metering device, an increase in load will cause the suction pressure to rise.

The system actually has a little more capacity at this point if the outside temperature is not much above the design temperature. For example, if the head pressure goes up because the outside temperature is 100°F, the suction pressure will go even higher than 73 psig because it will be influenced by the increased head pressure (Figure 27-4).

From the previous discussion, it can be seen that there are many different conditions that will affect the operating pressures and the temperature in the conditioned space. There can actually be as many different pressures as there are different inside and outside temperature and humidity variations. This can be very confusing, particularly to the new service technician. There are some common conditions, however, that you can use when troubleshooting. This is necessary because there are very few times when you will be working on a piece of equipment when the conditions are perfect. Most of the time, when you are assigned the job, the system has been off for some time or not operating correctly for long enough that the conditioned space temperature and humidity are higher than normal (Figure 27-5). After all, a rise in space temperature is what often prompts a customer to call for service.

High Evaporator Load with Cool Condenser A temperature rise in the conditioned space is not the only thing that will cause the system to undergo pressure and capacity changes. The reverse can happen if the inside temperature is warm and the outside temperature is cooler than normal (Figure 27-6). For example, before going to work, the homeowners may turn the air conditioner off to save electricity. They may not get home until after dark to turn the air conditioner back on. By this time, it may be 75°F outside but still be 80°F inside.

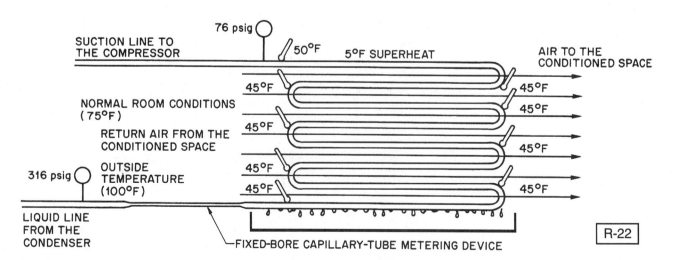

Figure 27-4. An increase in head pressure has increased the flow of refrigerant through the fixed-bore metering device and decreases the amount of superheat.

MECHANICAL OPERATING CONDITIONS 481

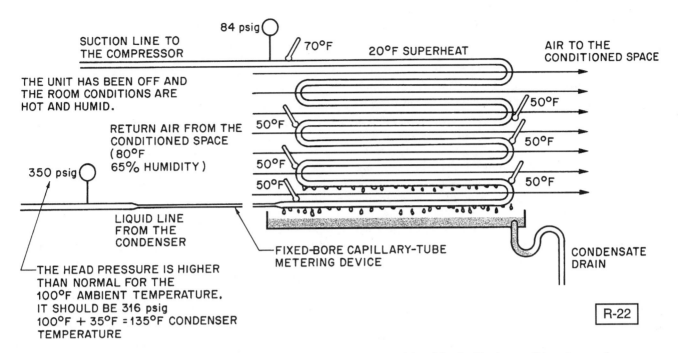

Figure 27-5. This system has been off long enough that the temperature and humidity inside the conditioned space have gone up. Notice the excess moisture forming on the coil and going down the drain.

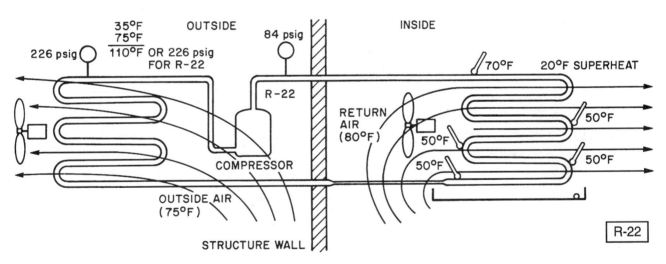

Figure 27-6. This system is operating with the outside ambient air cooler than the inside space temperature air. As the return air cools down, the suction pressure and refrigerant boiling temperature will both rise.

The air passing over the condenser is now cooler than the air passing over the evaporator. The evaporator may also have a large humidity load. The condenser becomes so efficient that it will hold some of the charge in the condenser. The reason is that the condenser starts to condense refrigerant in the first part of the condenser. Therefore, more of the refrigerant charge is in the condenser tubes, and this will slightly starve the evaporator. The system may not have enough capacity to cool the home for several hours. The condenser may hold back enough refrigerant to cause the evaporator to operate below freezing and possibly even freeze up before it can cool the house and satisfy the thermostat. This condition has been improved by some manufacturers by means of two-speed fan operation for the condenser fan. The fan is controlled by a single-pole, double-throw thermostat that operates the fan at high speed when the outdoor temperature is hot and low speed when the outdoor temperature is mild. Typically, the fan will run on high speed when the outdoor temperature is above 85°F and at low speed when it is below 85°F. When the fan is operating at low speed in mild weather, the head pressure is increased, reducing the risk of a frozen evaporator.

27.3 EQUIPMENT GRADES

Manufacturers continually adjust the design of air conditioning equipment to make it more efficient. There are normally three grades of equipment: economy, standard efficiency, and high efficiency. Some manufacturers produce all three grades of equipment and offer them to the supplier. Other manufacturers only offer one grade. Economy and standard grades are about equal in efficiency, but their appearance and the materials they are made of are different. High-efficiency equipment may be much more efficient and will not have the same operating characteristics. A condenser will normally condense the refrigerant at a temperature of about 30°F higher than the ambient temperature. For example, when the outside temperature is 95°F, the average R-22 system will condense the refrigerant at 125°F, with a corresponding head pressure of 278 psig. High-efficiency air conditioning equipment may have a much lower head pressure. The extra efficiency is gained by using a larger condenser surface and the same compressor, or buying a more efficient compressor, or both. The condensing temperature may be only 20°F more than the ambient temperature. This would bring the head pressure down to a temperature corresponding to 115°F or 243 psig (Figure 27-7). The compressor will not use as much power under these conditions. When large condensers are used, the head pressure is reduced. Lower power requirements are a result. More efficient condensers require some means of head pressure control, such as a two-speed fan, as mentioned above.

27.4 MANUFACTURER'S LITERATURE

In order to effectively service or troubleshoot an air conditioning system, you need to know the typical operating pressures at different conditions. Some manufacturers furnish a chart with the unit that lists the suction and discharge pressures at different conditions (Figure 27-8). Others publish a bulletin that lists all of their equipment along with the typical operating pressures and temperatures. Still others furnish this information with the unit in the installation and start-up instructions (Figure 27-9). If the homeowner has retained this booklet, it will be helpful when troubleshooting.

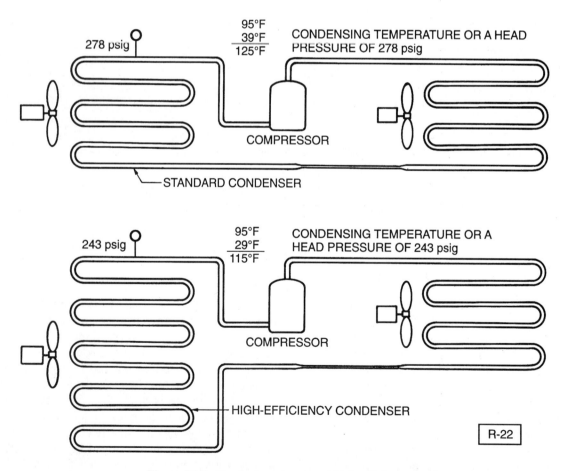

Figure 27-7. Standard condenser and high-efficiency condenser.

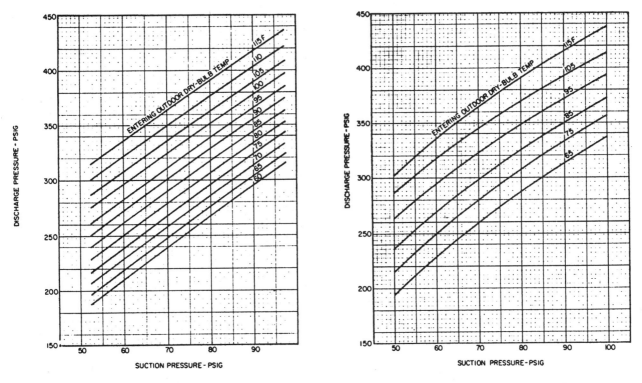

Figure 27-8. Charts furnished by some manufacturers to check unit performance. *Reproduced courtesy of Carrier Corporation.*

REFRIGERANT CHARGING

SAFETY PRECAUTION: To prevent personal injury, wear safety glasses and gloves when handling refrigerant. Do not overcharge system. This can cause compressor flooding.

1. Operate unit a minimum of 15 minutes before checking charge.
2. Measure suction pressure by attaching a gage to suction valve service port.
3. Measure suction line temperature by attaching a service thermometer to unit suction line near suction valve. Insulate thermometer for accurate readings.
4. Measure outdoor coil inlet air dry-bulb temperature with a second thermometer.
5. Measure indoor coil inlet air wet-bulb temperature with a sling psychrometer.
6. Refer to table. Find air temperature entering outdoor coil and wet-bulb temperature entering indoor coil. At this intersection note the superheat.
7. If unit has higher suction line temperature than charted temperature, add refrigerant until charted temperature is reached.
8. If unit has lower suction line temperature than charted temperature, bleed refrigerant until charted temperature is reached.
9. If air temperature entering outdoor coil or pressure at suction valve changes, charge to new suction line temperature indicated on chart.
10. This procedure is valid, independent of indoor air quantity.

SUPERHEAT CHARGING TABLE
(SUPERHEAT ENTERING SUCTION SERVICE VALVE)

Outdoor Temp (°F)	INDOOR COIL ENTERING AIR °F WB														
	50	52	54	56	58	60	62	64	66	68	70	72	74	76
55	9	12	14	17	20	23	26	29	32	35	37	40	42	45
60	7	10	12	15	18	21	24	27	30	33	35	38	40	43
65	—	6	10	13	16	19	21	24	27	30	33	36	38	41
70	—	—	7	10	13	16	19	21	24	27	30	33	36	39
75	—	—	—	6	9	12	15	18	21	24	28	31	34	37
80	—	—	—	—	5	8	12	15	18	21	25	28	31	35
85	—	—	—	—	—	—	8	11	15	19	22	26	30	33
90	—	—	—	—	—	—	5	9	13	16	20	24	27	31
95	—	—	—	—	—	—	—	6	10	14	18	22	25	29
100	—	—	—	—	—	—	—	—	8	12	15	20	23	27
105	—	—	—	—	—	—	—	—	—	5	9	13	17	22	26
110	—	—	—	—	—	—	—	—	—	—	6	11	15	20	25
115	—	—	—	—	—	—	—	—	—	—	—	—	8	14	18	23

Figure 27-9. Manufacturer's charging chart included in the installation and start-up literature for the equipment. *Reproduced courtesy of Carrier Corporation.*

Three factors must be taken into consideration when the manufacturer publishes typical operating conditions:

1. The load on the outdoor coil is influenced by the outdoor temperature.
2. The sensible heat load on the indoor coil is influenced by the indoor dry bulb temperature.
3. The latent heat load on the indoor coil is influenced by the humidity. The humidity is determined by measuring the wet bulb temperature of the indoor air.

To determine the performance of the equipment, you must record the required temperatures and plot them against the manufacturer's graph or table.

27.5 ESTABLISHING A REFERENCE POINT ON UNKNOWN EQUIPMENT

If you arrive at a job and find no manufacturer's literature and cannot obtain it, what should be done? The first thing is to try to establish some known condition as a reference point. For example, are you dealing with standard equipment or high efficiency equipment? This will help to establish a reference point for the suction and head pressures mentioned earlier. High-efficiency equipment is often larger than normal. The equipment is not always marked as high efficiency, so you may have to compare the size of the condenser to a standard condenser to determine the head pressure. It should be obvious that a larger or oversized condenser would have a lower head pressure. For example, a 3 ton compressor will have a full load amperage (FLA) rating of about 17 A at 230 V. The amperage rating of the compressor may help to determine the rating of the equipment (Figure 27-10). Although a 3 ton, high-efficiency piece of equipment will have a lower amperage rating than a standard piece of equipment, the ratings will be close enough to make comparisons for determining the capacity of the equipment. If the condenser is very large for a 3 ton amperage rating, it is probably high-efficiency equipment, and the head pressure will not be as high as in a standard piece of equipment.

High-efficiency air conditioning equipment often uses a thermostatic expansion valve (TXV) rather than a fixed-bore metering device because it provides certain amount of additional efficiency. The evaporator may also be larger than normal. An oversized evaporator will add to the efficiency of the system. It is more difficult to make a determination regarding the evaporator size because it is enclosed in a casing or the ductwork, and it is not as easy to see.

APPROXIMATE FULL LOAD AMPERAGE VALUES FOR ALTERNATING CURRENT MOTORS

Motor	Single Phase		3-Phase-Squirrel Cage Induction		
HP	115 V	230 V	230 V	460 V	575 V
$\frac{1}{6}$	4.4	2.2			
$\frac{1}{4}$	5.8	2.9			
$\frac{1}{3}$	7.2	3.6			
$\frac{1}{2}$	9.8	4.9	2	1.0	0.8
$\frac{3}{4}$	13.8	6.9	2.8	1.4	1.1
1	16	8	3.6	1.8	1.4
$1\frac{1}{2}$	20	10	5.2	2.6	2.1
2	24	12	6.8	3.4	2.7
3	34	17	9.6	4.8	3.9
5	56	28	15.2	7.6	6.1
$7\frac{1}{2}$			22	11.0	9.0
10			28	14.0	11.0

Does not include shaded pole.

Figure 27-10. Table of current ratings for different motor sizes at various voltages. *Courtesy BDP Company.*

Establishing a Reference Point on Unknown Equipment

The operating conditions of a system can vary so much that all possibilities cannot be covered; however, a few general statements committed to memory may help.

Operating Conditions Near Design Conditions for Standard Equipment

1. The boiling temperature of the refrigerant is 40°F (70 psig for R-22), which is 35°F less than the return air at 75°F (Figure 27-11).
2. The head pressure should correspond to a temperature that is about 30°F above the outside ambient temperature. For example, when the outside temperature is 95°F, the condensing temperature should be 125°F, as shown in Figure 27-12. The head pressure would be 278 psig for R-22.

Space Temperature Higher than Normal for Standard Equipment

1. The suction pressure will be higher than normal. Normally, the refrigerant boiling temperature is about 35°F cooler than the entering air temperature. Therefore, a normal entering air condition of 75°F would produce a boiling temperature of 40°F. This is true when the humidity is at 50%. When the space temperature is higher than normal because the equipment has been off for a long time, the return air temperature may go up to 85°F and the humidity may also be high. The boiling temperature may then go up to 50°F, with a corresponding suction pressure of 84 psig for R-22 (Figure 27-13). This higher-than-normal suction pressure may also cause the discharge pressure to rise.

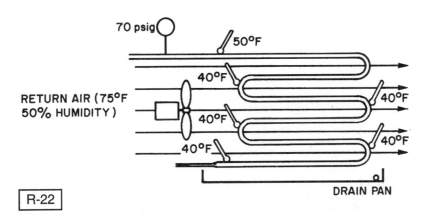

Figure 27-11. Evaporator operating at close to design conditions.

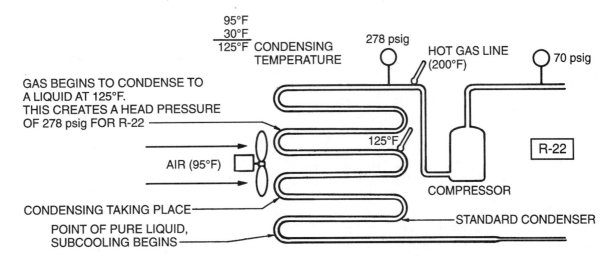

Figure 27-12. Normal conditions of a standard condenser operating on a 95°F day.

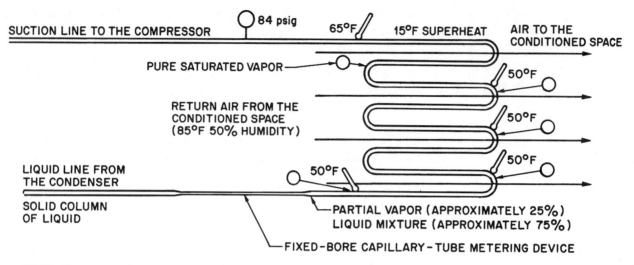

Figure 27-13. Pressures and temperatures as they may occur in an evaporator when the space temperature and humidity are above design conditions.

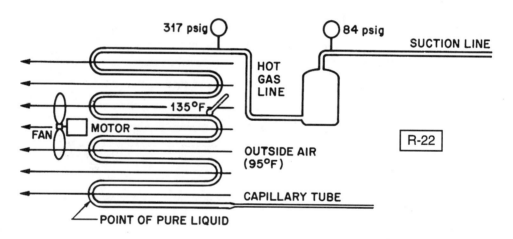

Figure 27-14. This condenser is operating at design conditions as far as the outdoor ambient air is concerned, but the pressure is high because the evaporator is under a load that is above design conditions.

2. The discharge pressure is influenced by the outside temperature and the suction pressure. For example, the discharge pressure should correspond to a temperature of no more than 30°F above the ambient temperature under normal conditions. Therefore, if the suction pressure rises to 84 psig, the discharge pressure is also going to rise. It may move up to a new condensing temperature that is 10°F higher than normal. This would mean that the condensing temperature for R-22 could be 135°F (95°F + 30°F + 10°F = 135°F) which is equal to 317 psig (Figure 27-14). When the unit begins to reduce the space temperature and humidity, the evaporator pressure will begin to go down because the load on the evaporator is reduced. The head pressure will also come down.

Operating Conditions Near Design Conditions for High-Efficiency Equipment

1. In some cases, the evaporator may operate at slightly higher pressures and temperatures on high-efficiency equipment because the evaporator is larger. The refrigerant boiling temperature may be 45°F with the larger evaporator at design conditions, which would correspond to a suction pressure of 76 psig for R-22. The high-efficiency evaporator operating at 45°F would have a temperature difference of 30°F compared to the return air (75°F - 45°F = 30°F).

2. The refrigerant may condense at a temperature as low as only 20°F above the outside ambient temperature.

ESTABLISHING A REFERENCE POINT ON UNKNOWN EQUIPMENT 487

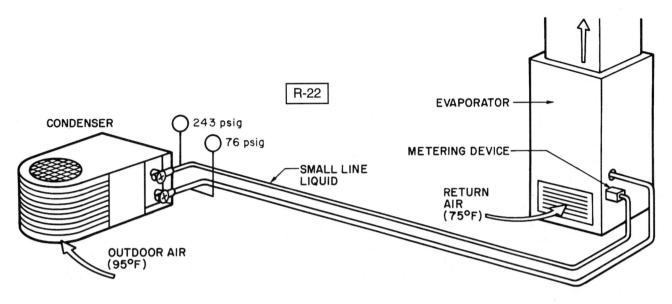

Figure 27-15. High-efficiency system operating conditions (45°F evaporator temperature and 115°F condensing temperature).

For a 95°F day with R-22, the head pressure may be as low as 243 psig (Figure 27-15). If the condensing temperature is as high as 30°F above the ambient temperature, you should suspect a problem. For example, the head pressure should not exceed 277 psig on a 95°F day for R-22. This will be covered in more detail later in this text.

Other-than-Design Conditions for High-Efficiency Equipment

1. When the unit has been off long enough for the load to build up, the space temperature and humidity will be above design conditions and the high-efficiency system pressures will be higher than normal, just as with the standard system. With standard equipment, when the return air temperature is 75°F and the humidity is approximately 50%, the refrigerant boils at about 40°F. This is a temperature difference of 35°F. A high-efficiency evaporator is larger, and the refrigerant may boil at a temperature difference of only 30°F, or around 45°F when the space temperature is 75°F. A higher boiling point will indicate a system problem (Figure 27-16). The exact boiling temperature relationship depends on the manufacturer and the coil surface area.
2. Like the evaporator, the high-efficiency condenser will also operate at a higher pressure when the load is

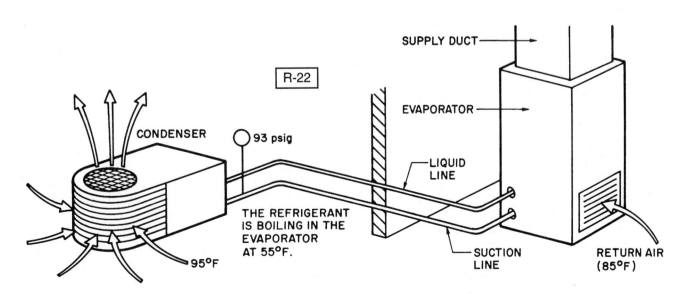

Figure 27-16. Evaporator in a high-efficiency system with a load above design conditions.

increased. The head pressure will not be as high as it would be with a standard condenser because the condenser has additional surface area.

The capacity of a high-efficiency system will not reach the rated capacity when the outdoor temperature is much below design. The earlier condition of the homeowners who shut off the air conditioning system before going to work and then turned it back on when they came home from work will be much worse with a high-efficiency system. The fact that the condenser became too efficient at night when the air was cooler with a standard condenser will also be much more evident with a larger, high-efficiency condenser. Most manufacturers produce two-speed condenser fans so that a lower fan speed can be used in mild weather to help compensate for this temperature difference. If you attempt to analyze a component of high-efficiency equipment on a mild day, you will find that the head pressure is low. This will cause the suction pressure to also be low. Using the coil-to-air relationships for the condenser and evaporator will help determine the correct pressures and temperatures.

Remember these two statements:

1. The evaporator absorbs heat, which is related to its operating pressures and temperatures.
2. The condenser rejects heat and has a predictable relationship with both the load and the ambient temperature.

27.6 EQUIPMENT EFFICIENCY RATINGS

Manufacturers have a method of rating equipment so the designer and owner can easily make comparisons between various systems. This rating was originally called the *Energy Efficiency Ratio* or *EER* and is actually the output in Btuh divided by the input in watts of power used to produce the output. For example, a system may have an output of 36,000 Btuh with an input of 4000 watts.

$$\frac{36,000 \text{ Btuh}}{4000 \text{ watts}} = \text{an EER of } 9$$

The larger the EER rating, the more efficient the equipment. For example, suppose the 36,000 Btuh air conditioner only required an input of 3600 watts.

$$\frac{36,000 \text{ Btuh}}{3600 \text{ Btuh}} = \text{an EER of } 10$$

This equipment provides the same capacity using less power and is therefore more efficient.

The EER rating is a steady-state rating and does not account for the time the unit operates before reaching peak efficiency. This operating time has an unknown efficiency. It also does not account for shutting the system down at the end of the cycle (when the thermostat is satisfied) leaving a cold coil in the duct. The cold coil

MODEL	CAPACITY (Btuh)	SEER
A	24,000	9.00
B	24,000	10.00
C	24,000	10.50
D	24,000	11.00
E	24,000	11.50
F	24,000	12.00

Figure 27-17. Examples of SEER ratings.

continues to absorb heat from the ambient air, not the conditioned space. Refrigerant equalizes to the cold coil and must be pumped out at the beginning of the next cycle. This accounts for some of the inefficiency at the beginning of the cycle.

The picture is not complete using the EER rating system, so a rating of seasonal efficiency has been developed, called the *Seasonal Energy Efficiency Ratio* or *SEER*. This rating is tested and verified by ARI and accounts for the start-up and shutdown cycles. Ratings of all available equipment are published by ARI, and manufacturers list the ratings in their catalogs. A typical rating table is shown in Figure 27-17.

27.7 TYPICAL ELECTRICAL OPERATING CONDITIONS

Electrical operating conditions are measured using a volt-ohmmeter and an ammeter. Three major power-consuming devices may have to be analyzed from time to time: the indoor fan motor, the outdoor fan motor, and the compressor. The control circuit is considered a separate function.

The starting point for considering electrical operating conditions is to find the system supply voltage. For residential units, 230 V is the most common system voltage. Light commercial equipment will nearly always use 208 V or 230 V single-phase or three-phase power. Both single-phase and three-phase power may be obtained from a three-phase power supply. The equipment rating may be 208/230 V. The reason for the two different ratings is that 208 V is the supply voltage that some power companies provide, and 230 V is the supply voltage provided by other power companies. Some light commercial equipment may use three-phase, 460 V power if the equipment is located at a large commercial installation. For example, an office may have a 3 ton air conditioning unit that operates separately from the main central system. If the supply voltage is 460 V, three phase, the small unit may operate from the same power supply. When 208/230 V equipment is used at a commercial installation, the compressor may be three phase, and the fan motor may be single phase. The number of phases that the power company furnishes makes a difference in the method of starting the compressor. Single-phase compressors may have a start-assist device, such as a PTC device or a start relay and start capacitor. Three-phase compressors do not use start-assist devices.

Matching the Unit to the Correct Power Supply

The typical operating voltages for any air conditioning system must be within the manufacturer's specifications. This is ±10% of the rated voltage. For the 208/230 V motor, the minimum allowable operating voltage would be 208 × 0.90 = 187.2 V. The maximum allowable operating voltage would be 230 × 1.10 = 253 V. Notice that the calculations used 208 V as the base for figuring the low voltage and 230 V for calculating the highest voltage. This is because this application is 208/230 V. If the motor is rated at 208 V or 230 V only, that value (208 V or 230 V) is used for evaluating the voltage.

Equipment may sometimes be started under conditions that exceed the rated limits. For example, if 180 V is measured, the motor should not be started, because the voltage will drop further with the current draw of the motor. If the voltage reads 260 V, the motor may be started, because the voltage will drop slightly when the motor is started. If the voltage drops to within the allowable limits, the motor is allowed to run.

Starting the Equipment

When the correct rated voltage is known and the minimum and maximum voltages are determined, the equipment may be started if the voltages are within the limits. The three motors (indoor fan, outdoor fan, and compressor) can be checked for the correct current draw.

The indoor fan builds the air pressure to move the air through the ductwork, filters, and grilles to the conditioned space. By law, the voltage characteristics must be printed on the motor in such a manner that they will not come off. This information may be printed on the motor, but the motor might be mounted so that they cannot be easily seen. In some cases, the motor may be inside the squirrel cage blower. If so, removing the motor is the only way of determining the fan current. When the supplier can be easily contacted, you may be able to obtain this information without removing the motor. If the motor electrical characteristics cannot be obtained from the motor nameplate, they might be listed on the unit nameplate. However, be aware that the fan motor is often changed to a larger motor to provide additional fan capacity.

Finding a Point of Reference for an Unknown Motor Rating

When a motor is mounted so that the electrical characteristics cannot be determined, you must improvise. The following paragraphs describe some of the ways in which you can determine the amperage for unknown equipment.

Determining the Fan Motor Amperage We know that air conditioning systems normally move about 400 cfm of air per ton. This can help you determine the amperage of the indoor fan motor by comparing the fan amperage of an unknown system to the amperage of a known system. All you need is the approximate system capacity. As discussed earlier, you can find this by comparing the compressor amperages of the unit in question with that of a known unit. For example, the compressor amperage of a 3 ton system is about 17 A. If you notice that the amperage of the compressor on the system you're checking is 17 A, you can assume that the system is close to 3 tons. The fan motor for a 3 ton system should be about $1/3$ hp for a typical duct system. The fan motor amperage for a $1/3$ hp PSC motor operating at 230 V is 3.6 A at 230 V. If the fan in question was pulling 5 A, suspect a problem.

The fan motor may have originally been shipped with the furnace and replaced later when air conditioning was added. In this case, the furnace nameplate will not provide the correct fan motor data. The condensing unit will have a nameplate for the condenser fan motor. This motor should be sized fairly close to its actual load and should pull an amperage that is close to the nameplate value.

Determining the Compressor Running Amperage
The compressor current draw may not be as easy to determine as the fan motor current draw because not all compressor manufacturers stamp the compressor full load amperage on the compressor nameplate. There are so many different compressor sizes that it is hard to state the correct full load amperage. For example, motors normally come in the following increments: 1, $1\frac{1}{2}$, 2, 3, and 5 hp. A unit that is rated at only 34,000 Btuh may be called a 3 ton unit, although it actually takes 36,000 Btuh to be a 3 ton unit. Ratings that are rounded to the closest standard rating are known as *nominal ratings*. For example, a typical 3 ton unit would have a 3 hp motor. A unit with a rating of 34,000 Btuh does not need a full 3 hp compressor motor, but is supplied with one because there is no standard horsepower motor to meet its needs. If the motor amperage for a 3 hp motor were stamped on the compressor, it could cause confusion because the motor may never operate at that amperage. The unit nameplate lists nominal electrical information.

Note: The manufacturer may stamp the compressor FLA (also listed as rated load amperage or RLA) on the unit nameplate. This amperage should not be exceeded.

It is rare for a compressor motor to operate at its FLA rating. If design or nearly-design conditions are in effect, the compressor will operate at close to full load. When the unit is operating at conditions that are significantly higher than design, such as when the unit has been off for some time in very hot weather, it might appear that the compressor is operating at an amperage that is higher than the FLA. However, there are usually other conditions that keep the compressor from drawing too much current (discussed below). The compressor is pumping vapor and vapor is very light. It takes a substantial increase in pressure difference to create a significantly greater work load.

High Voltage, the Compressor, and Current Draw – A motor operating at a voltage that is higher than the voltage rating of the motor is one condition that will prevent the motor from drawing too much current. A motor rated at 208/230 V has an amperage rating at some value between 208 V and 230 V. Therefore, if the voltage is 230 V, the amperage may be lower than the nameplate amperage even during overload. The compressor motor may be larger than needed and may not reach its rated horsepower until it gets to the maximum rating for the system. It may be designed to operate at 105°F or 115°F outdoor ambient for very hot regions, but when the unit is rated at 3 tons, it would be rated down at the higher temperatures. The unit nameplate may contain the compressor amperage at the highest operating condition for which the unit is rated (115°F ambient temperature).

Current Draw and the Two-Speed Compressor – Some air conditioning manufacturers use two-speed compressors to achieve better seasonal efficiencies. These compressors may use a motor capable of operating as a either two-pole or a four-pole motor. A four-pole motor runs at 1800 rpm, and a two-pole motor runs at 3600 rpm. A motor control circuit will slow the motor to half-speed in mild weather.

Variable speed may also be obtained using electronic circuits. The system efficiency is greatly improved because there is no need to stop the compressor and restart it at the beginning of the next cycle; the speed will automatically be adjusted for the capacity needed. However, this means more factors to consider when troubleshooting. The manufacturer's literature must be consulted to establish typical running conditions under other-than-design conditions. The equipment should perform just as typical high-efficiency equipment would when operating at design conditions.

SUMMARY

- To effectively service or troubleshoot an air conditioning system, you must know the typical mechanical and electrical operating conditions.
- Mechanical conditions deal with pressures and temperatures.
- The inside conditions that affect the load on the system are the space temperature and the humidity.
- When the discharge pressure rises, the system capacity decreases.
- When the suction pressure rises, the head pressure also rises.
- When the head pressure rises, the suction pressure also rises.
- High-efficiency systems often use larger or oversized evaporators, and the refrigerant will boil at a temperature of about 45°F at typical operating conditions. This is a temperature difference of 30°F compared to the return air.
- High-efficiency equipment has a lower operating head pressure than standard equipment partly because the condenser is larger. It has a different relationship to the outdoor temperature, normally +20°F to 25°F above the ambient temperature.
- The electrical instruments used to check working conditions are the ammeter and the volt-ohmmeter.
- The first thing you need to know for electrical troubleshooting is what the operating voltage for the unit is supposed to be.
- Typical voltages are 230 V single phase for residential and 208/230 V single or three phase or 460 V three phase for commercial installations.
- Equipment manufacturers require that operating voltages be maintained within ±10% of the rated voltage of the equipment.
- The unit nameplate may now include the compressor FLA. It should not be exceeded.
- Some manufacturers are now using two-speed motors and variable speed motors to provide variable capacity.

REVIEW QUESTIONS

1. What are the standard conditions at which air conditioning equipment is designed to operate?
2. What is the typical temperature relationship between a standard air-cooled condenser and the ambient temperature?
3. What is the temperature relationship between a high-efficiency condenser and the ambient temperature?
4. How is high-efficiency obtained with a condenser?
5. How does the temperature relationship between the condenser and the ambient temperature change as the ambient temperature increases?
6. What is the temperature relationship between the boiling refrigerant temperature and the entering air temperature with a standard evaporator?
7. What happens to the head pressure when the suction pressure rises?
8. What can cause the suction pressure to rise?
9. What happens to the suction pressure when the head pressure rises?
10. What can cause the head pressure to rise?
11. What two instruments are used to troubleshoot the system when an electrical problem is suspected?
12. What is the common voltage supplied to a residence?
13. How can an unknown current be found for a fan motor?
14. What are the typical motor horsepower sizes?
15. Name two methods that manufacturers use to vary the capacity of residential and small commercial compressors.

TROUBLESHOOTING 28

OBJECTIVES

Upon completion of this unit, you should be able to

- Select the correct instruments for checking an air conditioning unit with a mechanical problem.
- Determine the typical operating suction pressures for both standard and high-efficiency equipment at design conditions.
- Calculate the correct operating suction pressures for both standard and high-efficiency air conditioning equipment at other-than-design conditions.
- Calculate the standard operating discharge pressures at various ambient conditions.
- Select the correct instruments to troubleshoot electrical problems in an air conditioning system.
- Check the line voltage and control voltage power supplies.
- Troubleshoot basic electrical problems in an air conditioning system.
- Use an ohmmeter to check various components in an electrical system.

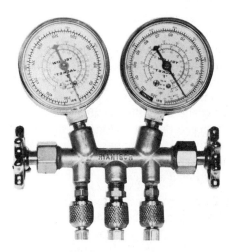

Figure 28-1. Gage manifold. The low-side gage is on the left and the high-side gage is on the right. *Courtesy Robinair Division, Sealed Power Corporation.*

28.1 TROUBLESHOOTING A/C SYSTEMS

Troubleshooting air conditioning equipment involves evaluating both mechanical and electrical systems, and they may have symptoms that overlap. For example, if an evaporator fan motor capacitor fails, the motor will slow down and begin to overheat. It may even get hot enough for the internal motor overload protection to trip. While it is running slowly, the suction pressure will go down and produce symptoms of a restriction or low charge. If you diagnose the problem based on suction pressure readings alone, you will fail to restore the system to correct operation.

Figure 28-2. Gages have refrigerant temperature relationships printed on them for R-12, R-22, and R-502. *Photo by Bill Johnson.*

28.2 MECHANICAL TROUBLESHOOTING

Gages and temperature-testing equipment are used when performing mechanical troubleshooting. The gages used are those on the gage manifold, as shown in Figure 28-1. The suction or low-side gage is on the left side of the manifold and the discharge or high-side gage is on the right side. The most common refrigerant used in air conditioning is R-22. An R-22 temperature chart is printed on each gage for determining the saturation temperature for the low-side and high-side pressures in the system. Since these same gages are used for refrigeration, the gages also include temperature scales for R-12 and R-502 (Figure 28-2). The R-12 scale will not be needed for residential air conditioning unless the equipment is old. Older automobiles also used R-12. Today, the common refrigerant used in automobile air conditioning is R-134a.

491

28.3 GAGE MANIFOLD USAGE

The gage manifold displays the low-side and high-side pressures while the unit is operating. These pressures can be converted to the saturation temperatures for the evaporating (boiling) refrigerant and condensing refrigerant by using the pressure/temperature relationship chart. The low-side pressure can be converted to the boiling temperature. For example, if the boiling temperature for a system is about 40°F, it can be converted to 70 psig for R-22. The superheat at the evaporator should be close to 10°F at this time. It is difficult to read the suction pressure at the evaporator because it normally has no gage port. Therefore, you should take the pressure and temperature readings at the condensing unit suction line to best evaluate the system performance. Guidelines for checking the superheat at the condensing unit will be discussed later in this unit. If the suction pressure were 48 psig, the refrigerant would be boiling at about 24°F, which is cold enough to freeze the condensate on the evaporator coil and is too low for continuous operation (Figure 28-3). The probable causes of low boiling temperatures are low charge or restricted airflow.

Warning: Be extremely careful when installing or removing manifold gages. High-pressure refrigerant will injure your skin and eyes. Always wear goggles and gloves. The danger from attaching the high-pressure gages can be reduced by shutting off the unit and allowing the pressures to equalize to a lower pressure. Liquid R-22 can cause serious frostbite since it boils at -44°F at atmospheric pressure.

The high-side gage may be used to convert pressures to condensing temperatures. For example, if the high-side gage reads 278 psig and the outside ambient temperature is 80°F, the head pressure may seem too high. The gage manifold chart shows that the condenser is condensing at 125°F. However, a condenser should not condense at a temperature of more than 30°F higher than the ambient temperature. The ambient temperature is 80°F + 30°F = 110°F, so the condensing temperature is actually 15°F too high. Probable causes are a dirty condenser or an overcharge of refrigerant.

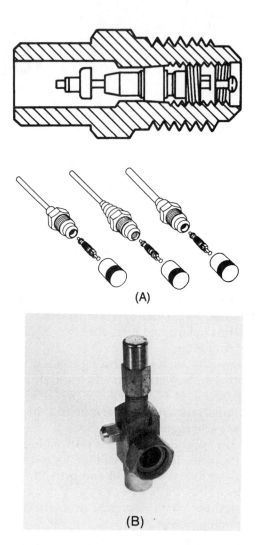

Figure 28-4. Two valves commonly used with taking pressure readings on modern air conditioning equipment. *Courtesy of: (A) J/B Industries, (B) Photo by Bill Johnson.*

The gage manifold is used to determine system pressures. Two types of pressure connections are used with air conditioning equipment: the Schrader valve and the service valve (Figure 28-4). The Schrader valve is a pressure connection only. The service valve can be used to

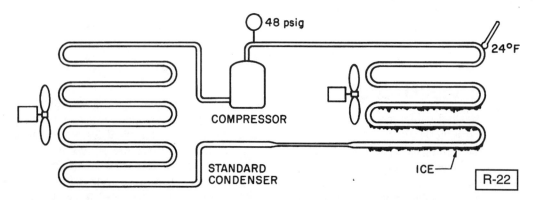

Figure 28-3. This coil was operated below freezing until the condensate on the coil froze.

GAGE MANIFOLD USAGE 493

Figure 28-5. This service valve has a Schrader port. *Photo by Bill Johnson.*

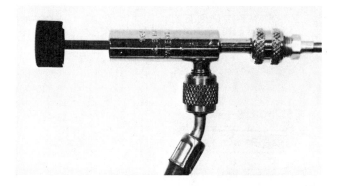

Figure 28-7. Hand valve used for depressing the Schrader valve stem to control the pressure. *Photo by Bill Johnson.*

isolate the system for service. It may have a both Schrader port for the gage connection and a service valve for isolation (Figure 28-5).

When to Connect the Gages

When servicing small systems, a gage manifold should not be connected every time the system is serviced because a small amount of refrigerant escapes each time the gage is connected. Some residential and small commercial systems have a critical refrigerant charge. When the high-side line is connected, high-pressure refrigerant will condense in the gage line. The refrigerant will escape when the gage line is disconnected from the Schrader port. If a gage line full of liquid refrigerant is lost while checking pressures, it may be enough to affect the system charge. A short gage line connector for the high side will help to prevent refrigerant loss (Figure 28-6).

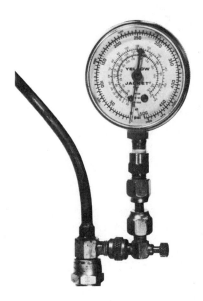

Figure 28-6. This short service connection is used on the high-side gage port to prevent refrigerant loss from the gage line. *Photo by Bill Johnson.*

Note: This connector can only be used to check pressures; you cannot use it to transfer refrigerant out of the system because it is not a manifold.

Another method of connecting the gage manifold to the Schrader valve service port is with a small hand valve which is used for depressing the stem in the Schrader valve (Figure 28-7). The hand valve can be used to keep the refrigerant in the system when disconnecting the gages. To use the valve, follow these steps:

1. Turn the stem on the valve out; this allows the Schrader valve to close.
2. Make sure that a plug is in the center line of the gage manifold.
3. Open the gage manifold handles. This will cause the gage lines to equalize to the low side of the system. The liquid refrigerant will move from the high-pressure gage line to the low side of the system. The only refrigerant that will be lost is an insignificant amount of vapor in the three gage lines. This vapor is at the suction pressure while the system is running and should be about 70 psig.

Caution: This procedure should only be performed using a clean and purged gage manifold (Figure 28-8).

Low-Side Gage Readings

When using the gage manifold on the low side of the system, you can compare the actual evaporating pressure to the normal evaporating pressure. This verifies that the refrigerant is boiling at the correct temperature for the low side of the system at a particular load condition. It has been indicated previously that there are both standard and high-efficiency systems, and that high-efficiency systems often have oversized evaporators. This makes the suction pressure slightly higher than normal. A standard system usually has a refrigerant boiling temperature of about 35°F cooler than the entering air temperature at the standard operating condition of 75°F return air with 50% relative humidity.

494   UNIT 28   TROUBLESHOOTING

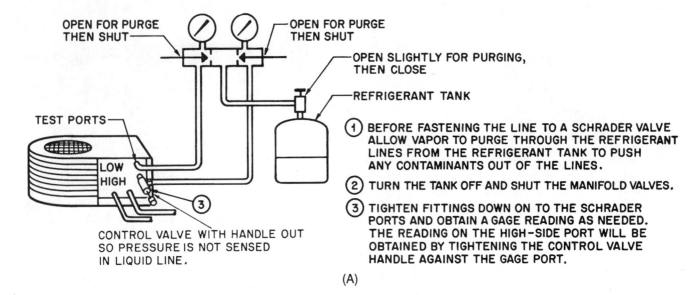

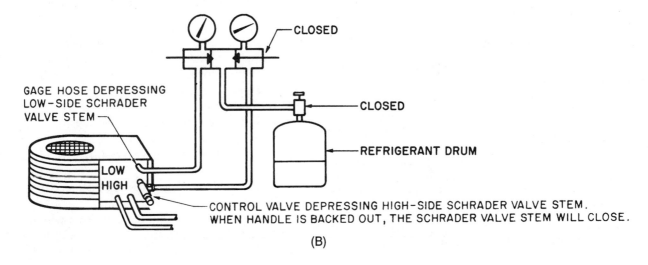

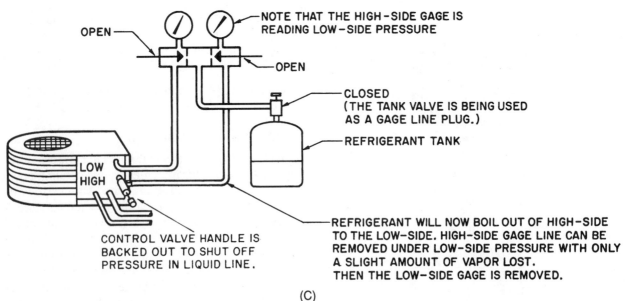

Figure 28-8. Method of pumping refrigerant condensed in the high-pressure line over into the suction line. (A) Purging manifold. (B) Taking pressure readings. (C) Removing liquid refrigerant from the high-side gage line.

If the space temperature is 85°F and the humidity is 70%, the evaporator has an oversized load. It is absorbing an extra heat load, both sensible heat and latent heat, from the moisture in the air. You need to wait a sufficient time for the system to reduce the load before you can determine if the equipment is functioning correctly. At this time, gage readings will not reveal the kind of information that will verify the system performance unless there is a manufacturer's performance chart available.

High-Side Gage Readings

Gage readings obtained on the high-pressure side of the system are used to check the relationship of the condensing refrigerant to the ambient air temperature. Standard air conditioning equipment condenses the refrigerant at no more than 30°F higher than the ambient temperature. For a 95°F entering air temperature, the head pressure should correspond to 95°F + 30°F = 125°F for a corresponding head pressure of 278 psig for R-22. If the head pressure shows that the condensing temperature is higher than this, something is wrong.

When checking the condenser entering air temperature to calculate the head pressure, be sure to check the actual temperature at the location. Never simply accept the weather report as the ambient temperature. For example, air conditioning equipment located on a black roof has tremendous solar influence from the air being pulled across the roof (Figure 28-9). In this case, the temperature can vary greatly from that recorded on the ground. Also, if the condenser is located close to any obstacles, such as below a sundeck, air may be circulated back through the condenser and cause the head pressure to be higher than normal. (Temperature measurements are discussed in more detail in the next section.)

High-efficiency condensers perform in the same way as standard condensers except they operate at lower pressures and condensing temperatures. High-efficiency condensers normally condense the refrigerant at a temperature as low as 20°F above than the ambient temperature. (Remember, standard condensers condense the refrigerant at a temperature that is 30°F above the ambient temperature.) On a 95°F day, the head pressure corresponds to a temperature of 95°F + 20°F = 115°F, which is a pressure of 243 psig for R-22.

28.4 TEMPERATURE READINGS

Temperature readings are vital to proper system maintenance and troubleshooting. The four-lead electronic temperature tester performs very well (Figure 28-10).

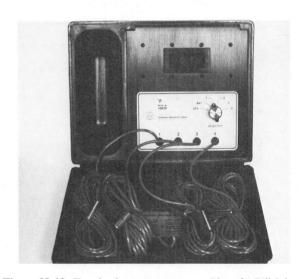

Figure 28-10. Four-lead temperature tester. *Photo by Bill Johnson.*

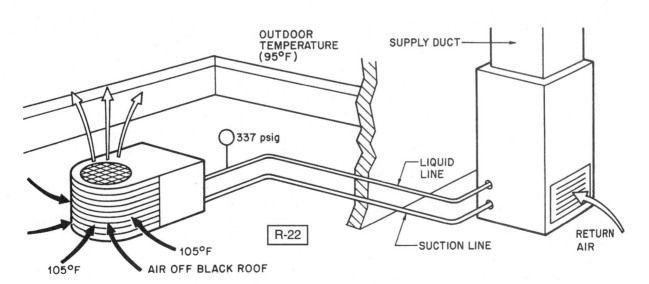

Figure 28-9. Condenser located low on a roof. Hot air from the roof enters it. A better installation would have the condenser mounted up about 20 inches so that the air near the roof surface could mix with the ambient air. The ambient air is 95°F but air off the roof is 105°F.

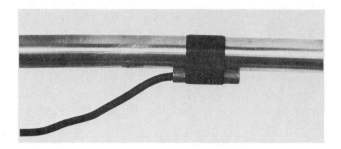

Figure 28-11. Temperature lead attached to a refrigerant line in the correct manner. It must be insulated. *Photo by Bill Johnson.*

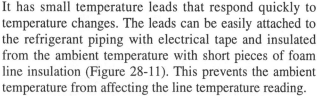

Figure 28-12. Glass thermometers strapped to a refrigerant line. *Photo by Bill Johnson.*

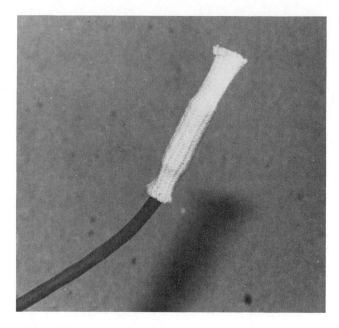

Figure 28-13. This electronic thermometer temperature lead has a damp cotton sock wrapped around it to convert it to a wet bulb lead. *Photo by Bill Johnson.*

It has small temperature leads that respond quickly to temperature changes. The leads can be easily attached to the refrigerant piping with electrical tape and insulated from the ambient temperature with short pieces of foam line insulation (Figure 28-11). This prevents the ambient temperature from affecting the line temperature reading.

Electronic temperature testers are more accurate and easier to use than the glass thermometers used in the past. It is nearly impossible to get a true line reading by strapping a glass thermometer to a copper line (Figure 28-12).

Temperatures vary from system to system, and you must be prepared to accurately record these temperatures in order to evaluate the various types of equipment. Some technicians record temperature readings of various equipment under different conditions for future reference. The common temperatures used would be the inlet air wet bulb (WB) and dry bulb (DB), the outdoor air dry bulb, and the suction line temperature. Sometimes, the compressor discharge temperature also needs to be measured. A thermometer with a range from -50°F to +250°F may be used for all of these tests.

Inlet Air Temperatures

It may be necessary to know the inlet air temperature of the evaporator for a complete system analysis. A wet bulb reading for determining the humidity may be required. Such a reading can be obtained by covering a temperature lead with a cotton sock that has been saturated with distilled water (Figure 28-13). Both wet bulb and dry bulb readings may be obtained by placing a dry bulb temperature lead next to a wet bulb temperature lead in the return airstream. The velocity of the return air will be enough to accomplish the evaporation required for the wet bulb reading.

Evaporator Outlet Temperatures

The evaporator outlet air temperature is seldom important. If needed, it may be obtained in the same manner as the inlet air temperature. The temperature drop across an evaporator coil is about 20°F at typical operating conditions of 75°F and 50% relative humidity. Therefore, the outlet air DB temperature will normally be about 20°F less than the inlet air DB temperature. If the conditioned space temperature is high with a high humidity, the temperature drop across the same coil will be much lower because of the latent heat load of the moisture in the air.

If a WB reading is taken, there will be approximately a 10°F WB drop from the inlet to the outlet at standard operating conditions. The outlet humidity will be almost 90%. It is very humid because the air has contracted or shrunk in the cooling process. This air will mix with the room air and will soon drop in humidity because it will expand as it reaches the room air temperature.

Caution: The temperature lead must not be allowed to touch the moving fan while taking air temperature readings.

Suction Line Temperatures

The temperature of the suction line returning to the compressor and the suction pressure can help you to understand the characteristics of the suction gas. The

TEMPERATURE READINGS **497**

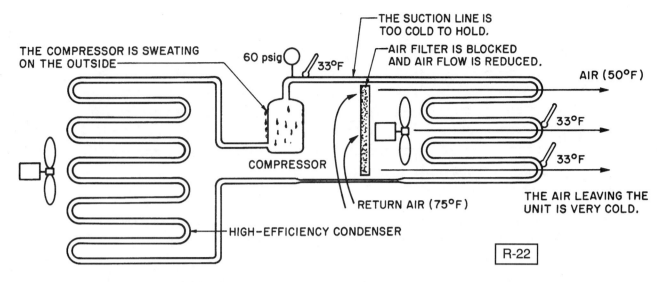

Figure 28-14. This evaporator is flooded with refrigerant.

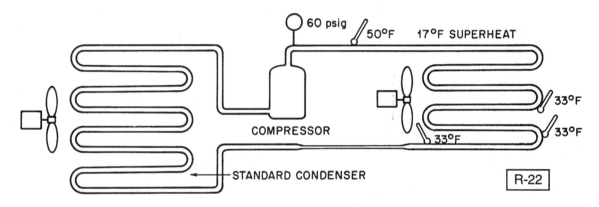

Figure 28-15. This evaporator is starved because of a low refrigerant charge.

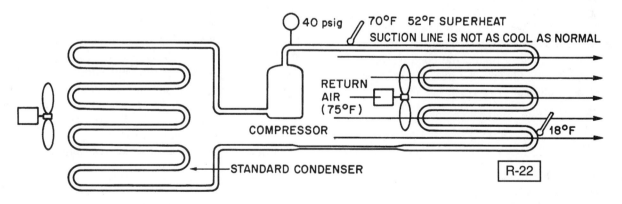

Figure 28-16. If the suction temperature is too low, and the suction line is warmer than normal, the evaporator is starved for refrigerant. The unit may have a low refrigerant charge.

suction gas may be part liquid if the filters are clogged or restricted or if the evaporator coil is dirty (Figure 28-14). The suction gas may have a high superheat if the unit has a low charge or if there is a refrigerant restriction (Figure 28-15). The combination of suction line temperature and pressure readings will help you determine whether the system has a low charge or a restricted air filter in the air handler. For example, if the suction pressure is too low and the suction line is warm, the system has a starved evaporator (Figure 28-16). If the suction line temperature

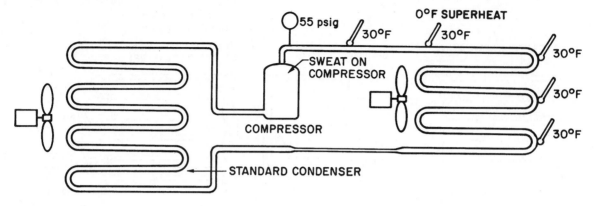

Figure 28-17. When the suction pressure is too low and the superheat is also low, the unit is not boiling the refrigerant in the evaporator. The coil is flooded with liquid refrigerant.

is cold and the pressure indicates that the refrigerant is boiling at a low temperature, the coil is not absorbing heat as it should. The coil may be dirty, or there may not be enough airflow (Figure 28-17). The cold suction line indicates that the unit has enough charge, because the evaporator must be full for liquid refrigerant to be returning to the suction line.

Discharge Line Temperatures

The temperature of the discharge line may tell you that something is wrong inside the compressor. If there is an internal refrigerant leak from the high-pressure side to the low-pressure side, the discharge gas temperature will increase. Normally, the discharge line temperature at the compressor would not exceed 220°F for an air conditioning application even in very hot weather (115°F). When a high discharge line temperature is discovered, the probable cause is an internal leak. You can prove this by building up the head pressure as high as 300 psig and then shutting off the unit. If there is an internal leak, this pressure difference between the high and the low sides can often be heard as a whistle as it equalizes through the compressor. If the suction line at the compressor shell starts to warm up immediately, the heat is coming from the compressor discharge.

Warning: The discharge line of a compressor may be as hot as 220°F under normal conditions, so use caution while attaching a temperature lead to this line.

Liquid Line Temperatures

The liquid line temperature may be used to check the subcooling efficiency of a condenser. Most condensers will subcool the refrigerant to between 10°F and 20°F below the condensing temperature of the refrigerant. If the condensing temperature is 125°F on a 95°F day, the liquid line leaving the condenser may be 105°F to 115°F when the system is operating normally. If there is a slight undercharge, there may be slightly less subcooling and the system efficiency will be reduced. The condenser performs three functions:

1. It removes the superheat from the discharge gas.
2. It condenses the refrigerant into a liquid.
3. It subcools the liquid refrigerant below the condensing temperature.

All three of these functions must be successfully accomplished for the condenser to operate at its rated capacity.

It's a good idea to take the time to completely check out a working system operating at the correct pressures and temperatures. Apply the temperature probes and gages to all points to actually verify the readings. This will provide reference points to remember when troubleshooting.

28.5 CHARGING PROCEDURES

While establishing field charging procedures, keep the following in mind. The charge consists of the correct amount of refrigerant in the evaporator, condenser, liquid line, discharge line between the compressor and condenser, and suction line. The discharge and suction lines do not hold as much refrigerant as the liquid line because the refrigerant is much denser in the liquid state. Actually, the liquid line is the only interconnecting line that contains a large amount of refrigerant. When a system is operating correctly under design conditions, there should be a prescribed amount of refrigerant in the condenser, evaporator, and liquid line.

Understanding the following statements will be very helpful when checking or adjusting a refrigerant charge:

1. The amount of refrigerant in the evaporator can be measured using the super heat method.
2. The amount of refrigerant in the condenser can be measured using the subcooling method.
3. The amount of refrigerant in the liquid line may be determined by measuring the length and calculating the refrigerant charge. However, for field service work, if the evaporator is performing correctly, you can assume that the liquid line has the correct charge.

A field charging procedure may be used to check the charge of some common systems. You will sometimes need typical reference points to add small amounts of refrigerant for adjusting the charge in equipment that has no charging directions. This can occur due to an overcharge or undercharge from the factory or from a previous technician's work. It can also occur due to system leaks.

Caution: System leaks should always be repaired to prevent further loss of refrigerant to the atmosphere.

You must establish charging procedures to use in the field for all types of equipment. These procedures will help you to get the system back on line under emergency situations. The following paragraphs describe various methods used for different types of equipment.

Field Charging Systems with Fixed-Bore Metering Devices

Fixed-bore metering devices such as the capillary tube or fixed orifice do not throttle the refrigerant to meet the load as with a thermostatic expansion valve. These devices allow refrigerant flow based on the difference between the inlet and outlet pressures. The one time when the system can be checked for the correct charge and everything will read normal is at the typical operating conditions of 75°F, 50% humidity return air and 95°F outside ambient air. If other conditions exist, different pressures and superheat readings will occur. The item that most affects the readings is the outside ambient temperature. When it is lower than normal, the condenser will become more efficient and will condense the refrigerant sooner in the coil. This will have the effect of partially starving the evaporator of refrigerant.

When you need to check or adjust a system charge, the best method is to follow the manufacturer's instructions. If they are not available, typical operating conditions may be simulated by reducing the airflow across the condenser to cause the head pressure to rise. On a 95°F day, the highest condenser head pressure is usually 278 psig for R-22 (95°F ambient + 30°F added condensing temperature = 125°F condensing temperature or 278 psig). Since the high pressure pushes the refrigerant through the metering device, when the head pressure is up to the high (normal) end of the operating conditions, there is no refrigerant held back in the condenser. Refrigerant that is in the condenser when it should be in the evaporator starves the evaporator.

When the condenser is pushing the refrigerant through the metering device at the correct rate, the remainder of the charge must be in the evaporator. A superheat check at the evaporator is not always easy with a split-system air conditioner, so a superheat check at the condensing unit can be made. The suction line from the evaporator to the condensing unit may be long or short. Let's use two different lengths for a test comparison: up to 30 feet, and from 30 to 50 feet. When the system is correctly charged, the superheat should be 10° to 15°F at the condensing unit with a line length of up to 30 feet. The superheat should be 15° to 18°F when the line is 30 to 50 feet long. Both of these conditions are with a head pressure of 278 psig ±10 psig. At these conditions, the actual superheat at the evaporator will be close to the correct superheat of 10°F. When using this method, be sure that you allow enough time for the system to adjust after adding refrigerant before you draw any conclusions (Figure 28-18).

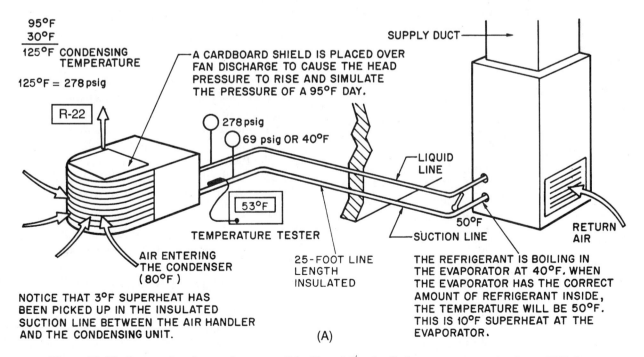

Figure 28-18. System charging can be accomplished by raising the discharge pressure to simulate a 95°F day.

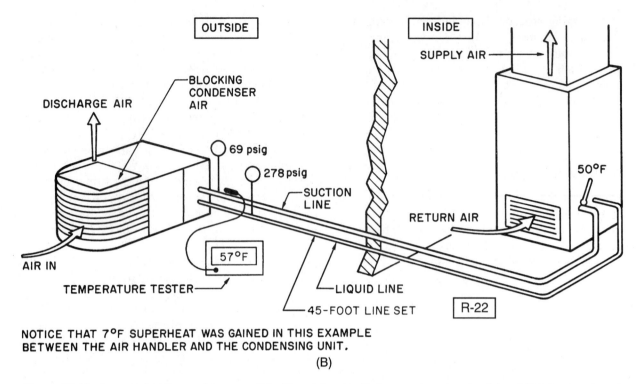

NOTICE THAT 7°F SUPERHEAT WAS GAINED IN THIS EXAMPLE BETWEEN THE AIR HANDLER AND THE CONDENSING UNIT.
(B)

Figure 28-18. System charging can be accomplished by raising the discharge pressure to simulate a 95°F day (continued).

Orifice-type metering devices have a tendency to "hunt" while reaching steady-state operation. This hunting can be observed by watching the suction pressure rise and fall accompanied by the superheat as the suction line temperature changes. When this occurs, you will have to use averaging to arrive at the proper superheat for the coil. For example, if it is varying between 6°F and 14°F, an average value of 10°F superheat may be used.

Field Charging Systems with Thermostatic Expansion Valves

The thermostatic expansion valve (TXV) system can be charged in much the same way as the fixed-bore system, with some modifications. The condenser on a TXV system will also hold refrigerant back in mild ambient conditions. This system uses a refrigerant reservoir or receiver to store refrigerant, which helps to compensate for lower ambient conditions. To check the charge, restrict the airflow across the condenser until the head pressure simulates a 95°F ambient day, which equates to a 278 psig head pressure for R-22. Using the superheat method will not work for TXVs because if there is an overcharge, the superheat will remain the same. Superheats of 15°F to 18°F are not unusual when measured at the condensing unit for TXV systems. If the sightglass is full, the unit has at least enough refrigerant, but it may have an overcharge. If the unit does not have a sightglass, a measure of the condenser subcooling may tell you what you want to know. For example, a typical subcooling circuit will subcool the liquid refrigerant from 10°F to 20°F cooler than the condensing temperature. A temperature lead attached to the liquid line should read 115°F to 105°F, or 10°F to 20°F cooler than the condensing temperature of 125°F (Figure 28-19). If the subcooling temperature is 20°F to 25°F cooler than the condensing temperature, the unit has an overcharge of refrigerant and the bottom of the condenser is acting as a large subcooling surface.

The charging procedures just described will also work for high-efficiency equipment. In this case, the head pressure does not need to be quite as high. A head pressure of 243 psig will be sufficient for an R-22 system when charging.

Warning: Refrigerant in cylinders and in the system is under high pressures. Use proper safety precautions when transferring refrigerant, and be careful not to overfill tanks or cylinders when recovering refrigerant. Do not refill disposable cylinders. Use only DOT-approved tanks or cylinders. The high-side pressure may be reduced for attaching and removing gage lines by shutting off the unit and allowing the unit pressures to equalize.

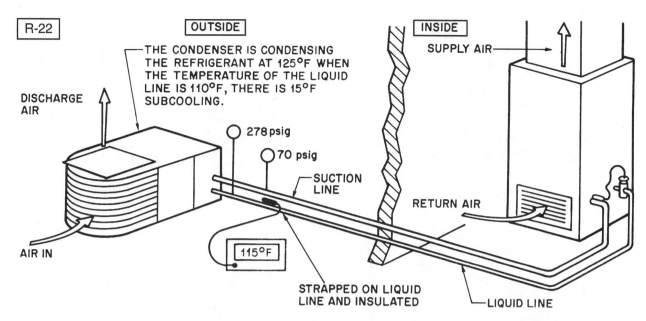

Figure 28-19. A unit with a thermostatic expansion valve cannot be charged using the superheat method. The head pressure is raised to simulate a 95°F day and the temperature of the liquid line is checked for the subcooling level. A typical system will have 10°F to 20°F of subcooling when the condenser contains the correct charge.

28.6 ELECTRICAL TROUBLESHOOTING

Electrical troubleshooting is often required at the same time as mechanical troubleshooting. The VOM (volt-ohm-milliammeter) and clamp-on ammeter are the primary instruments used.

You need to know what the readings should be in order to know whether the existing readings on a particular unit are correct. This may not be easy to determine because the desired readings may not be available. Some manufacturers furnish tables that list typical horsepower-to-amperage ratings for various motors. This can be very helpful when determining the correct amperage for a particular motor.

For a residence or small commercial building, one main power panel will normally serve the entire building. This panel is divided into many circuits. For a split system, there are usually separate breakers (or fuses) in the main panel for the inside air handler or furnace and the outdoor unit. For a package or self-contained system, there is usually one breaker (or fuse) to serve the unit. The power supply voltage is stepped down by the control transformer to the control voltage of 24 V.

You should begin any electrical troubleshooting process by verifying that the power supply is energized and the supply voltage is correct. One way to do this is to go to the room thermostat and see if the indoor fan will start using the FAN ON switch. See Figure 28-20 for a wiring diagram of a typical split-system air conditioner.

The air handler or furnace, where the low-voltage transformer is located, is frequently under the house or in the attic. This quick check with the FAN ON switch can save you a trip under the house or to the attic. If the fan will start with the fan relay, several things are apparent:

1. The indoor fan works.
2. There is control voltage.
3. There is line voltage to the unit because the fan will run.

When taking a service call over the phone, ask the homeowner if the indoor fan will run. If it won't, bring a transformer with you to the job. This could save a trip to the supply house or the shop.

Warning: All safety practices must be observed when troubleshooting electrical systems. Many times the system must be inspected while the power is on. Only the insulated meter leads should ever touch the hot terminals. Special care should be taken while troubleshooting the main power supply because the fuses are designed to allow great amounts of current to flow before they blow (e.g., when a screwdriver slips in the panel and shorts across hot terminals). For this reason, you should never use a screwdriver in an energized panel.

If the power supply voltages are correct, move on to the various components. The path to the load may be the next item to check. If you're trying to get the compressor to run, remember that the compressor motor is operated by the compressor contactor. Is the contactor energized? Are the contacts closed? See Figure 28-21 for a wiring diagram.

502 UNIT 28 Troubleshooting

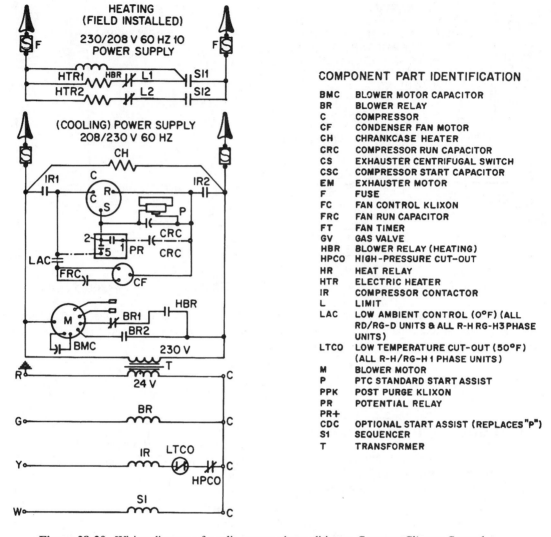

Figure 28-20. Wiring diagram of a split-system air conditioner. *Courtesy Climate Control.*

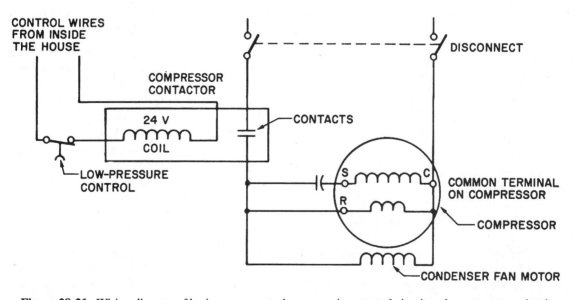

Figure 28-21. Wiring diagram of basic components that appear in a control circuit and a compressor circuit.

Note that the only thing that will keep the contactor coil from being energized is the thermostat, the path, or the low-pressure control. If the outdoor fan is operating and the compressor is not, the contactor is energized because it also starts the fan. If the fan is running and the compressor is not, either the path (wiring or terminals) are not making good contact or the compressor internal overload protector is open.

Compressor Overload Problems

When the compressor overload is open, touch the motor housing to see if it is hot. If you cannot hold your hand on the compressor shell, the motor is too hot. Ask yourself these questions: Can the charge be low (is this compressor suction-gas cooled)? Or, is the start-assist circuit not working so the compressor is not starting?

Allow the compressor to cool before restarting it. It is best that the unit be fixed so that it will not come back on for several hours. The best way to do this is to remove a low-voltage wire. If you pull the disconnect switch and come back the next day, the refrigerant charge may have migrated to the crankcase because there was no crankcase heat. This is not good practice. If you want to start the unit within the hour, pull the disconnect switch and run a small amount of water through a hose over the compressor.

Caution: Standing water poses a potential electrical hazard. Ensure that all electrical components are protected with plastic or other waterproof covering. Don't come in contact with the water or electrical current when working around live electricity.

The compressor will take about 30 minutes to cool. Have the gages on the unit and a cylinder of refrigerant connected because when the compressor is started up by closing the disconnect, it may need refrigerant. If the system has a low charge and you have to get set up to charge after starting the system, the compressor may cut off again from overheating before you have a chance to get the gages connected.

Compressor Electrical Checkup

You may need to perform an electrical check of the compressor if the compressor will not start or if a circuit protector has opened. For example, suppose that the compressor can be heard trying to start. It will make a humming noise but will not turn. Check the compressor with an ohmmeter to see if all of the windings are correct. Remember, a load must have the correct resistance. Let's say a compressor specification calls for the run winding to have a resistance of 4 Ω and the start winding a resistance of 15 Ω. If the ohmmeter indicates that the start winding has only 10 Ω, then the winding has a short circuit (sometimes called a *shunt*). This will change the winding characteristics, and the compressor will not start. It is defective and must be changed. See Figure 28-22. It is important to use quality instruments when making these measurements.

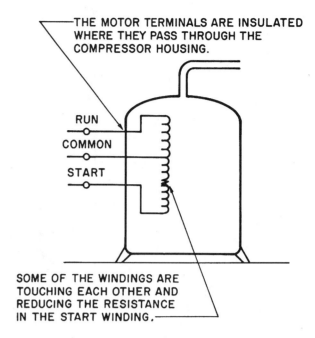

Figure 28-22. Compressor with shorted winding.

The ohmmeter check may show that the compressor has an open circuit in the start or run windings. Suppose that the same symptom of a hot compressor is discovered. The compressor is allowed to cool and it still will not start. An ohm check shows that the start or run winding is open. This compressor is also defective and must be replaced (Figure 28-23).

When the continuity check indicates that the common circuit in the compressor is open, it could be that the internal overload protector is open because the motor may not have cooled enough. The motor is suspended in a vapor space inside the shell, and it takes time to cool. Wait another 15 minutes and try again.

Caution: Never use an ohmmeter to check a live circuit because the current may destroy the meter movement.

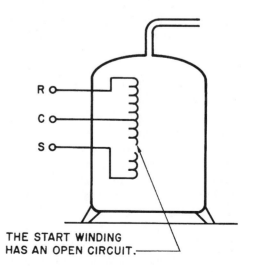

Figure 28-23. Compressor with open start winding.

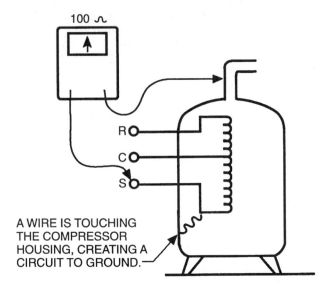

Figure 28-24. Compressor with winding shorted to the casing.

Troubleshooting Fuses and Circuit Breakers

One service call that must be treated cautiously is when a circuit protector such as a fuse or breaker opens the circuit. The compressor and fan motors have internal protection that will normally guard them from minor problems. The breaker or fuse is for large current surges in the circuit. If a protective device disconnects a circuit, don't simply reset it. Perform a resistance check of the compressor section, including the fan motor.

Caution: The compressor might be grounded (i.e., it has a circuit to the case of the compressor), and it will be harmful to try to start it. Be sure to isolate the compressor circuit before condemning the compressor. Take the motor leads off the compressor to check for a ground circuit in the compressor (Figure 28-24).

28.7 PREVENTIVE MAINTENANCE

Air conditioning equipment preventive maintenance involves the indoor airside, the outdoor airside for air-cooled units and the water circuit for water-cooled units, and the electrical system.

The indoor airside maintenance involves coil, motor, and filter maintenance.

The evaporator coil operates below the dew point temperature of the air and is wet. It is an excellent collector of any dust particles that may pass through the filter or leak in around loose panel compartment doors. The coil must be cleaned at the first sign of dirt buildup. Most air handlers have draw-through coil/fan combinations. The air passes through the coil before the fan. When the fan blades become loaded with dirt, it is a sure sign that the coil is dirty. The dirt had to pass through the coil first, and it is wet much of the time. When the fan is dirty, the coil must be cleaned.

The coil may be cleaned in place in the unit using two methods. One is the use of a special detergent manufactured to work while the coil is wet. This detergent is sprayed on the coil and into the core of the coil with a hand pump-type sprayer, similar to a garden sprayer. When the unit is started, the condensate will carry the dirt down the coil and down the condensate drain line. This type of cleaner is for light-duty cleaning. Care must be taken that the condensate drain line does not become clogged with the dirt from the coil and pan.

The coil may also be cleaned by shutting the unit down and applying a more powerful detergent to the coil, forcing it into the coil core. After the detergent has had time to work, the coil is then sprayed with a water source, such as a water hose. Care must be taken not to force too much water into the coil, or the drain pan will overflow. The water must not be applied faster than the drain can accept it. Special pressure cleaners that have 500 to 1000 psig nozzle pressures may be used. These units have a low water flow, about 5 gallons per minute, and will not overflow an adequate drain system. The high nozzle pressure allows the water to clean the core of the coil.

It is always best to back-wash a coil when cleaning it with water. The water is forced through the coil in the opposite direction to the airflow for back-washing. The reason for this is most of the dirt will accumulate on the inlet of the coil and the dirt buildup will be progressively less as the air moves through the coil. If the coil is not back-washed, you may only drive dirt accumulation to the center of the coil.

Warning: Never use hot water or a steam cleaner on refrigeration equipment if there is refrigerant in the unit as pressures will rise high enough to burst the unit at the weakest point. This may be the compressor shell. A refrigerant system must be open to the atmosphere before hot water or steam is used for cleaning.

In some cases, the coil cannot be cleaned in the unit and must be removed for cleaning. If the unit has service valves, the refrigerant may be pumped into the condenser/receiver and the coil removed from the unit. The coil will not have any refrigerant in it and may be cleaned with approved detergent and hot water or with a steam cleaner. Approved cleaners may be purchased at any air conditioning supply house.

Caution: Follow the directions. Make sure no water can enter the coil through the piping connections.

The outdoor unit may be either air-cooled or water-cooled. Air-cooled units have fan motors that must be lubricated. Other motors only require lubrication after several years of operation. Then a recommended amount of approved oil is added to the oil cup. Some motors require more frequent lubrication.

The fan blades should be checked to make sure they are secure on the shaft. At the same time, the rain shield on top of the fan motor should be checked, if applicable.

It should not allow water to enter the motor on upflow units where the motor is exposed to the elements.

Condenser coils may be easier to clean because they are on the outside. Coils may become dirty from pulling dust through the coils. It is hard to look at any coil and tell how much dirt is in the core. The coil should be cleaned at the first sign of dirt buildup.

Warning: Before cleaning, make sure that all power is turned off and locked so that it will not accidentally be turned on while you are servicing the equipment. The fan motor must be covered and care should be taken that water does not enter the controls. They may not all be in the control cabinet.

Apply an approved detergent to the coil using a hand pump sprayer. Soak the coil to the middle. Let the detergent soak for ¼ to ½ hour, then back-wash the coil using a garden hose or spray cleaner.

Water-cooled equipment must be maintained like air-cooled equipment. Two things are done as part of the maintenance program to minimize the mineral concentration in the water. One is to make sure the water tower has an adequate water bleed system. This is a measured amount of water that goes down the drain. This water is added back to the system with the supply water makeup line. Owners will often shut the bleed system off because all they see is water going down the drain. They don't know that the purpose of this is to dilute the minerals in the water that collect due to evaporation.

The other maintenance procedure is to use the correct water treatment. A water treatment specialist should supervise the setup of the treatment program. This will ensure the best quality water in the system.

Electrical preventive maintenance consists of examining any contactors and relays for frayed wire and pitted contacts. When these occur, they should be taken care of before leaving the job. There can be no prediction as to how long they will last. Single-phase compressors will just stop running if a wire burns in two or a contact burns away. A three-phase compressor will try to run on two phases and this is very hard on the motor. Wires should be examined to make sure they are not rubbing on the frame and wearing the insulation off the wire. Wires resting on copper lines may rub a hole in the line, resulting in a refrigerant leak.

28.8 TECHNICIAN SERVICE CALLS

Troubleshooting can take many forms and cover many situations. The following troubleshooting examples will help you understand what the service technician does while solving actual problems.

Service Call 1

A customer calls to report that the central air conditioner at their residence is not cooling enough and runs continuously. The problem is a low refrigerant charge.

The technician arrives and finds the unit running. The temperature indicator on the room thermostat shows the thermostat is set at 72°F, and the thermometer on the thermostat indicates that the space temperature is 80°F. The air feels very humid. The technician notices that the indoor fan motor is running. The velocity of the air coming out of the registers seems adequate, so the filters are not restricted.

The technician goes to the condensing unit and hears the fan running. The air coming out of the fan is not warm. This indicates that the compressor is not running. The door to the compressor compartment is removed and it is noticed that the compressor is hot to the touch. This is an indication that the compressor has been trying to run. The gages are installed on the service ports, and they both read the same because the system has been off. The gages read 144 psig, which corresponds to 80°F. The residence is 80°F inside, and the ambient is 85°F, so it can be assumed that the unit has some liquid refrigerant in the system because the pressure corresponds so closely to the chart. If the system pressure were 100 psig, it would be obvious that very little liquid refrigerant is left in the system. A large leak should be suspected.

The technician decides that the unit must have a low charge and that the compressor is off because of the internal overload protector. The technician pulls the electrical disconnect and takes a resistance reading across the compressor terminals with the motor leads removed. The meter shows an open circuit from common to run and common to start. It shows a measurable resistance between run and start. This indicates that the motor winding thermostat must be open. This indication is verified by the hot compressor.

All electrical components and terminals are covered with plastic. The electrical disconnect is checked to ensure that it is still pulled. A water hose is connected, and a small amount of water is allowed to run over the top of the compressor shell to cool it. A cylinder of refrigerant is connected to the gage manifold so that when it is time to start the compressor, the technician will be ready to add gas and keep the compressor running. The gage lines are purged of any air that may be in them. While the compressor is cooling, the technician changes the air filters and lubricates the condenser and indoor fan motors. After about 30 minutes, the compressor seems cool. The water hose is removed, and the water around the unit is allowed a few minutes to run off. When the disconnect switch is closed, the compressor starts. The suction pressure drops to 40 psig. The normal suction pressure is 70 psig for the system because the refrigerant is R-22. Refrigerant is added to the system to bring the charge up to normal. The space temperature is 80°F, so the suction pressure will be higher than normal until the space temperature becomes normal. The technician must have a reference point to get the correct charge. The following reference points are used in this situation because there is no factory chart.

1. The outside temperature is 90°F; the normal operating head pressure should correspond to a temperature of 90°F + 30°F = 120°F or 260 psig. It should not exceed this when the space temperature is down to a normal 75°F. This is a standard unit. The technician restricts the airflow to the condenser and causes the head pressure to rise to 278 psig as refrigerant is added.
2. The suction pressure should correspond to a temperature of about 35°F cooler than the space temperature of 80°F, or 45°F, which corresponds to a pressure of 76 psig for R-22.
3. The system has a capillary tube metering device, so some conclusions can be drawn from the temperature of the refrigerant coming back to the compressor. A thermometer lead is attached to the suction line at the condensing unit. As refrigerant is added, the technician notices that the refrigerant returning from the evaporator is getting cooler. The evaporator is about 30 feet from the condensing unit, and some heat will be absorbed into the suction line returning to the condensing unit. The technician uses a guideline of 15°F of superheat with a suction line of this length. This assumes that the refrigerant leaving the evaporator has about 10°F of superheat and that another 5°F of superheat is absorbed along the line. When these conditions are reached, the charge is very close to correct and no more refrigerant is added.

A leak check is performed and a flare nut is found to be leaking. It is tightened and the leak is stopped. The technician loads the truck and leaves. A call later in the day shows that the system is working correctly.

Service Call 2

A residential customer calls. Their air conditioning unit has been cooling correctly until this afternoon, when it quit cooling. This is a unit that has been in operation for several years. The problem is a complete loss of charge.

The technician arrives at the residence and finds the thermostat to be set at 75°F and the space temperature to be 80°F. The air coming out of the registers feels the same as the return air temperature. There is plenty of air velocity at the registers, so the filters appear to be clean.

The technician goes to the back of the house and finds that the fan and compressor in the condensing unit are not running. The breaker at the electrical box is in the ON position, so power must be available. A voltage check shows that the voltage is 235 V. A look at the wiring diagram shows that there should be 24 V between the C and Y terminals to energize the contactor. The voltage actually reads 25 V, slightly above normal, but so is the line voltage of 235 V. The conclusion is that the thermostat is calling for cooling, but the contactor is not energized. The only safety control in the contactor coil circuit is the no-charge protector (low-pressure control), so the unit must be out of refrigerant.

Gages are fastened to the gage ports, and the pressure is 0 psig. The technician connects a cylinder of refrigerant and starts adding refrigerant for the purpose of leak checking. When the pressure is up to about 2 psig, the refrigerant is stopped. Nitrogen is added to push the pressure up to 50 psig and gas can be heard leaking from the vicinity of the compressor suction line. A hole is found in the suction line where the cabinet had been rubbing against it. This accounts for the fact that the unit worked well up to a point and quit working almost immediately.

The trace refrigerant and nitrogen is allowed to escape from the system, and the hole is patched with silver solder. A liquid line filter-drier is installed in the liquid line, and a triple evacuation is performed to remove any contaminants that may have been pulled into the system. The system is charged and started. The technician follows the manufacturer's charging chart to ensure the correct charge.

Service Call 3

A commercial customer calls to report that the air conditioning unit at a small office building was not cooling. The unit was operating and cooling yesterday afternoon when the office closed. The problem is that someone turned the thermostat down to 55°F late yesterday afternoon, and the condensate on the evaporator froze solid overnight while trying to pull the space temperature down to 55°F.

The technician arrives at the site and notices that the thermostat is set to 55°F. An inquiry shows that one of the employees was too warm in an office at the end of the building and turned down the thermostat to cool that office. The technician notices there is no air coming out of the registers. The air handler is located in a closet at the front of the building. Examination shows that the fan is running, and the suction line is frozen solid at the air handler.

The technician stops the compressor and leaves the evaporator fan operating by turning the HEAT-OFF-COOL selector switch to OFF and the fan switch to ON. It is going to take a long time to thaw the evaporator (about an hour). The technician leaves the following directions with the office manager:

1. Let the fan run until air comes out of the registers.
2. When air is felt at the registers, wait 30 minutes and turn the thermostat back to COOL to start the compressor.

The technician then looks in the ceiling to check the air damper on the duct run serving the back office of the employee who turned down the thermostat. The damper is nearly closed; it probably was brushed against by a telephone technician who had been working in the ceiling. The damper is reopened.

A call later in the day indicates that the system is working again.

Service Call 4

A residential customer reports that the air conditioning unit is not cooling correctly. The problem is a dirty condenser. The unit is in a residence and is next to the side yard. The homeowner mows the grass, and the lawnmower throws grass on the condenser. This is the first hot day of summer, and the unit had been cooling the house until the weather became hot.

The technician arrives at the job and notices that the thermostat is set for 75°F but the space temperature is 80°F. There is plenty of slightly-cool air coming out of the registers. The technician goes to the side of the house where the unit is located. The suction line feels cool, but the liquid line is very hot. An examination of the condenser coil shows that the coil is dirty and is clogged with grass.

The technician shuts off the unit power using the breaker at the unit and takes enough panels off to spray coil cleaner on the coil. A high detergent coil cleaner is applied to the coil and allowed to stand for about 15 minutes. While this is taking place, the motors are oiled and the filters are changed. To keep the condenser fan motor from getting wet when the coil cleaner is washed off, it is covered with a plastic bag.

A water hose with a nozzle to concentrate the water stream is used to wash the coil in the opposite direction of the unit airflow. One application is not enough, so the coil cleaner is sprayed on the coil again and allowed to stand for another 15 minutes. The coil is washed again and is now clean.

The unit is assembled and started. The suction line feels cool to the touch and the liquid line feels warm. The technician decides not to put gages on the system and leaves. A call back later in the day indicates that the system is operating correctly.

Service Call 5

A homeowner indicates that their air conditioning unit is not cooling. This is a residential high-efficiency unit. The problem is the thermostatic expansion valve is defective.

The service technician arrives and finds the house warm. The thermostat is set to 74°F, and the house temperature is 82°F. The indoor fan is running, and plenty of air is coming out of the registers.

The technician goes to the condensing unit and finds that the fan and compressor are running, but they quickly stop. The suction line is not cool. Gages are attached to the gage ports. The low-side pressure is 25 psig, and the head pressure is 170 psig. This corresponds very closely to the ambient air temperature. The unit is off because of the low-pressure control. When the pressure rises, the compressor restarts but stops in about 15 seconds. The liquid line sightglass is full of refrigerant. The technician concludes that there is a restriction on the low side of the system. The expansion valve is a good place to start when there is nearly a complete blockage. The system does not have service valves, so the charge has to be removed and recovered.

The expansion valve is soldered into the system. It takes an hour to complete the expansion valve change. When the valve is changed, the system is leak tested and evacuated three times. A charge is measured into the system, and the system is started. It is evident from the beginning that the valve change has repaired the unit; all pressures are normal.

Service Call 6

An office manager calls to report that the air conditioning unit in a small office building is not cooling. The unit was cooling correctly yesterday afternoon when the office closed. The problem is that an electrical storm during the night tripped a breaker on the air conditioner air handler. The power supply is at the air handler.

The technician arrives and goes to the space thermostat. It is set at 73°F, and the temperature is 78°F. There is no air coming out of the registers. The fan switch is turned to ON, and the fan does not start. It is decided to check the low-voltage power supply, which is in the attic at the air handler. The technician finds the tripped breaker. Before resetting it, the technician decides to check the unit electrically.

A resistance check of the fan circuit and the low-voltage control transformer proves there is a measurable resistance. The circuit seems to be safe. The circuit breaker is reset and stays set. The thermostat is set to COOL, and the system starts. A call later in the day shows that the system is operating correctly.

Service Call 7

A residential customer calls because their air conditioning unit is not running. The homeowner found the breaker at the condensing unit tripped and reset it several times, but it did not stay set. The problem is the compressor motor winding is grounded to the compressor shell and tripping the breaker. The homeowner is warned that it is dangerous to reset a breaker more than once.

The technician arrives at the job and goes straight to the condensing unit with electrical test equipment. The breaker is in the tripped position and is moved to the OFF position. The voltage is checked at the load side of the breaker to ensure that there is no voltage. The breaker has been reset several times and is not to be trusted. When it is determined that the power is definitely off, the ohmmeter is connected to the load side of the breaker. The ohmmeter reads infinity, meaning there is no circuit. The compressor contactor is not energized, so the fan and compressor are not included in the reading. The technician pushes in the armature of the compressor contactor to make the contacts close. The ohmmeter now reads zero resistance, indicating that a short exists. It cannot be determined whether the fan or the compressor is the problem, so the compressor wires are disconnected from the bottom of the contactor. The short does not exist when the compressor is disconnected. This verifies that the short is in the compressor circuit, possibly the wiring. The meter is moved to the compressor terminal box and the wiring is disconnected. The ohmmeter is attached to the motor terminals at the compressor. The short is still there, so the compressor motor is condemned.

The technician must return the next day to change the compressor. The technician disconnects the control wiring and insulates the disconnected compressor wiring so that power can be restored. This must be done to keep crankcase heat on the unit until the following day. If it is not done, most of the refrigerant charge will be in the compressor crankcase and too much oil will be removed with the refrigerant when it is recovered.

The technician performs one more task before leaving the job. The Schrader valve fitting at the compressor is slightly depressed to allow a very small amount of refrigerant to escape. The refrigerant has a strong acid odor. A suction line filter-drier with a high acid removal core needs to be installed along with the normal liquid line filter-drier.

The technician returns on the following day, recovers the refrigerant from the system, and then changes the compressor, adding the suction and liquid line filter-driers. The unit is leak tested, evacuated, and the charge measured into the system with a set of accurate scales. The system is started, and the technician asks the customer to leave the air conditioner running (even though the weather is mild) to keep refrigerant circulating through the filter-driers in case any acid is left in the system.

The technician returns on the fourth day and measures the pressure drop across the suction line filter-drier to make sure that it is not stopped up with acid from the burned-out compressor. It is well within the manufacturer's specifications and is left in the line.

Service Call 8

A residential customer calls. The customer hears the indoor fan motor running for a short time, then stopping. This happens repeatedly. The problem is that the indoor fan motor is cycling on and off on its internal thermal overload protector because the fan capacitor is defective. The fan will start and run slowly, then stop.

The technician knows from the work order to go straight to the indoor fan section. It is in the crawl space under the house, so electrical instruments are carried on the first trip. The fan has been off long enough to cause the suction pressure to go so low that the evaporator coil is frozen. The breaker is turned off during the check. The technician suspects the fan capacitor or the bearings, so a fan capacitor is brought to the service call. After bleeding the charge from the fan capacitor, it is checked with an ohmmeter to see if it will charge and discharge. It is important to ensure that there is no electrical charge in the capacitor before checking it with an ohmmeter. A 20,000 ohm resistor is used between the two terminals to bleed off any charge. The capacitor will not charge and discharge, so it is changed.

The technician oils the fan motor and starts it from under the house with the breaker. The fan motor is drawing the correct amperage. The coil is still frozen, and the technician must set the space thermostat to operate only the fan until the homeowner feels air coming out of the registers. Then the fan should be operated for an additional $1/2$ hour to melt the rest of the ice from the coil before the compressor is started.

Note: The following service calls do not have the solutions provided. The solutions are included in the Instructor's Guide.

Service Call 9

A commercial customer reports that an air conditioning unit in a small office building is not cooling on the first warm day of summer.

The technician first goes to the thermostat and finds that it is set to 75°F. The space temperature is 82°F. The air feels very humid. There is no air coming from the registers. The air handler is in the attic. An examination of the condensing unit shows that the suction line has ice all the way back to the compressor. The compressor is still running.

What is the likely problem and the recommended solution?

Service Call 10

A residential customer calls. The customer reports that the unit in their residence was serviced in the early spring under a regular service contract. The earlier work order shows that refrigerant was added by a new technician.

The technician finds that the thermostat is set to 73°F, and the house is 77°F. The unit is running, and cold air is coming out of the registers. Everything seems normal until the condensing unit is examined. The suction line is cold and sweating, but the liquid line is only warm.

Gages are fastened to the gage ports, and the suction pressure is 85 psig, far from the correct pressure of 74 psig (77°F - 35°F = 42°F, which corresponds to about 74 psig). The head pressure is supposed to be 260 psig, which corresponds to a condensing temperature of 120°F (85°F + 35°F = 120°F or 260 psig). The head pressure is 350 psig. The unit shuts off after about 10 minutes running time.

What is the likely problem and the recommended solution?

Service Call 11

A residential customer calls. The unit is not cooling enough to reduce the space temperature to the thermostat setting. It is running continuously on this first hot day of summer.

The technician arrives and finds the thermostat is set to 73°F; the house temperature is 78°F. Air is coming out of the registers, so the filters must be clean. The air is cool but is not as cold as it should be. The air temperature should be about 55°F, and it is 63°F.

At the condensing unit, the technician finds that the suction line is cool but is not cold. The liquid line seems extra cool. It is 90°F outside, and the condensing unit should be condensing at about 120°F. If the unit had 15°F of subcooling, the liquid line should be warm to the touch, yet it is not.

Gages are fastened to the service ports to check the suction pressure. The suction pressure is 95 psig, and the discharge pressure is 225 psig. The airflow is restricted to the condenser, and the head pressure gradually climbs to 250 psig; the suction pressure goes up to 110 psig. The compressor should have a current draw of 27 A, but it only draws 15 A.

What is the likely problem and the recommended solution?

Service Call 12

A commercial customer calls to indicate that the air conditioning unit in a small office building was running but suddenly shut off.

The technician arrives, goes to the space thermostat, and finds that it is set to 74°F. The space temperature is 77°F. The fan is not running. The fan switch is moved to ON, but the fan motor does not start. It is decided that the control voltage power supply should be checked first. The power supply is in the roof condensing unit on this particular installation. The ladder is placed against the building away from the power line entrance. Electrical test instruments are taken to the roof along with tools to remove the panels. The breakers are checked and seem to be in the correct ON position. The panel is removed where the low voltage terminal block is mounted. A voltmeter check shows there is line voltage but no control voltage.

What is the likely problem and the recommended solution?

Caution: Until you gain experience, all troubleshooting should be performed under the supervision of a qualified person. If you do not know whether a situation is safe, consider it unsafe and take all necessary precautions.

HVAC Golden Rules

- Try to park your vehicle so that you do not block the customer's driveway.
- Always carry the proper tools and avoid extra trips through the customer's house.
- Make and keep firm appointments. Call ahead if you are delayed. The customer's time is valuable.
- Ask the customer about the problem. The customer can often provide details about system operation that will help you to solve the problem and save time.

Added Value to the Customer Here are some simple, inexpensive procedures that may be included in the basic service call:

- Touch test the suction and liquid lines to determine possible system overcharges or undercharges.
- Clean or change filters as needed.
- Lubricate all bearings as needed.
- Replace all panels with the correct fasteners.
- Inspect all contactors and wiring. This may lead to additional service.
- Make sure the condensate line is draining properly. Clean if needed. Algaecide tablets may be necessary.
- Check evaporator and condenser coils for blockage or dirt.
- Advise the customer about proper operation, such as avoiding recirculation of conditioned air into the return air.
- If removal of the unit is necessary, have the proper help and use drop cloths to prevent dirt from the unit from falling on the floor.

SUMMARY

- Troubleshooting air conditioning equipment involves both mechanical and electrical problems.
- Mechanical troubleshooting involves the use of gages and thermometers.
- The gage manifold is used along with the pressure/temperature chart to determine the boiling temperature for the system low side and the condensing temperature for the system high side.
- The superheat for an operating coil is used to prove coil performance.
- The typical customer operates the equipment at 75°F return air with a humidity of 50%. There is a relationship between the evaporator temperature and the return air at these typical conditions. For a standard unit, it is 35°F. The evaporator normally boils at 35°F lower than the 75°F return air or 40°F.
- A high-efficiency evaporator normally has a boiling temperature of 45°F.
- Gages fastened to the high side of the system are used to check the head pressure. The head pressure is related to the refrigerant condensing temperature.
- The condensing refrigerant temperature has a relationship to the medium to which it is giving up heat.
- A standard unit normally condenses the refrigerant at no more than 30°F higher than the air to which the heat is rejected.
- A high-efficiency condensing unit normally condenses the refrigerant at a temperature as low as 20°F warmer than the air used as a condensing medium.
- High efficiency is obtained with additional condenser surface area and more efficient compressors.
- Superheat is normally checked at the condensing unit using an electronic temperature tester.
- The temperature tester may be used to check both WB temperatures (using a wet wick on one bulb) and DB temperatures.

- The condenser has three functions: to remove the superheat from the discharge gas, to condense the hot gas into a liquid, and to subcool the refrigerant.
- A typical condenser may subcool the refrigerant 10°F to 20°F lower than the condensing temperature.
- Two types of metering devices are normally used on air conditioning equipment: the fixed bore (orifice or capillary tube) and the thermostatic expansion valve.
- The fixed-bore metering device uses the pressure difference between the inlet and outlet of the device for refrigerant flow. It does not vary in size.
- The thermostatic expansion valve modulates or throttles the refrigerant to maintain a constant superheat.
- To correctly charge an air conditioning unit, the manufacturer's recommendations must be followed.
- The thermostatic expansion valve system normally has a sightglass in the liquid line to aid in charging. A subcooling temperature check may be used when there is no sightglass.
- The tools for electrical troubleshooting are the ammeter and the volt-ohmmeter.
- Before checking a unit electrically; the proper voltage and current draw of the unit should be known.
- The main power panel may be divided into many circuits. The air conditioning system is normally on two separate circuits for a split system and one circuit for a package system.
- When a compressor is hot, such as when it has been running on an undercharge, the compressor internal overload protector may stop the compressor.
- When a hot compressor is started, assume that the system is low in refrigerant. A cylinder of refrigerant should be connected, so refrigerant may be added before the compressor shuts off again.
- A compressor can be checked electrically using an ohmmeter. The run and start windings should have a known resistance, and there should be no circuit to ground.

REVIEW QUESTIONS

1. What are the two main troubleshooting areas in the air conditioning service field?
2. What information is provided by the low-side pressure reading?
3. What information is provided by the high-side pressure reading?
4. What is the correct operating superheat for an air conditioning coil?
5. What are the typical indoor air conditions at which most homeowners operate an air conditioning system?
6. What is the temperature at which the typical air conditioning evaporator coil operates?
7. What is the typical temperature difference between the entering air and the boiling refrigerant temperatures on a standard air conditioning evaporator?
8. What is the best method of charging an air conditioning system that has a refrigerant undercharge?
9. When the outside ambient air temperature is 90°F, at what temperature should the refrigerant in the condenser be condensing, and what is the head pressure for an R-22 system?
10. How can a service technician measure the superheat at the condenser for a split system?
11. When the ambient air temperature is 95°F and the unit is a high-efficiency unit, what is the lowest condensing temperature at which the refrigerant would normally operate, and what would the head pressure be for R-22?
12. How is high efficiency accomplished in an air conditioning condensing unit?
13. Name two methods for fastening gage manifolds to an air conditioning system.
14. What instrument is used to measure the current draw of a compressor?
15. What instrument is used to measure continuity in a compressor winding?
16. What normally happens when a compressor that is electrically grounded tries to start?
17. What should a technician do if the circuit breaker is tripped?

CHILLED-WATER AIR CONDITIONING SYSTEMS 29

OBJECTIVES

Upon completion of this unit, you should be able to

- List different types of chilled-water air conditioning systems.
- Describe how chilled-water air conditioning systems operate.
- State the types of compressors often used in high-pressure chillers.
- Describe the operation of a centrifugal compressor in a high-pressure chiller.
- Explain the difference between direct expansion and flooded chiller evaporators.
- Explain what is meant by the term *approach temperature* as it relates to a water-cooled condenser.
- State two types of condensers used in chilled-water systems.
- Explain subcooling.
- List the types of metering devices used in high-pressure chillers.
- List the refrigerants typically used in low-pressure chillers.
- State the types of compressors used in low-pressure chiller systems.
- Describe the metering devices used in low-pressure chillers.
- Explain the purge system used in a low-pressure chiller condenser.
- Describe the absorption cooling system process.
- State the refrigerant generally used in large absorption chillers.
- State the compound normally used in salt solutions in large absorption chillers.
- State the types of electric motors typically used in chiller air conditioning systems.
- Discuss the various start mechanisms for these motors.
- Describe a load-limiting device used in chiller motors.
- Discuss various motor overload protection devices and systems.

29.1 CHILLED-WATER APPLICATIONS

Chilled-water systems are used in larger central air conditioning applications because of the ease with which chilled water can be circulated in the system. If refrigerant were piped to all floors of a multi-story building, there would be too many areas where leaks could occur, in addition to the expense of the large amount of refrigerant required to charge the system.

The design temperature for boiling R-22 in a cooling coil 40°F. When water is cooled to a similarly low temperature, it can also be used to condition air, even in single-room applications (Figure 29-1). This is the logic behind the use of circulating chilled-water systems. A refrigeration system is used to cool the water to about 45°F and the water is circulated throughout the building to heat exchange coils that absorb heat from the building air.

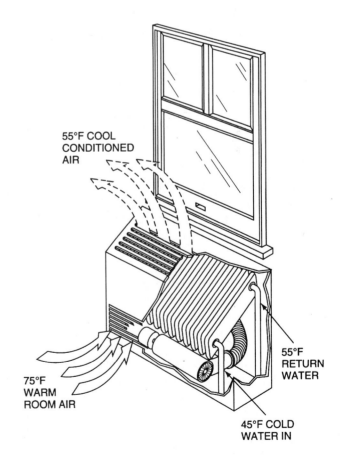

Figure 29-1. Water in a fan coil unit for cooling a room.

511

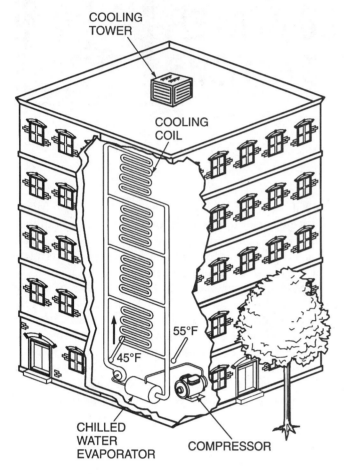

Figure 29-2. Water used as a secondary refrigerant.

When water is used for circulation in a building, it is called a *secondary refrigerant* (Figure 29-2). It is much less expensive than the refrigerant used to cool the water, which is known as the *primary refrigerant*.

Chillers

A chiller refrigerates the circulating water in a chilled-water system. As the water passes through the evaporator section of the unit, the temperature of the water is lowered. It is then circulated throughout the building where it picks up heat. The typical design temperatures for a circulating chilled-water system are 45°F water furnished to the building and 55°F water returned to the chiller from the building. The heat from the building adds 10°F to the water which returns to the chiller. Here the heat is removed and the water is recirculated.

There are two basic categories of chillers: the compression cycle chiller and the absorption chiller. The compression cycle chiller uses a compressor to provide the pressure differences inside the chiller in order to boil and condense refrigerant. Compression cycle chillers may be classified as either high-pressure or low-pressure systems. The absorption chiller uses salt and water rather than refrigerant to provide pressure differences required for cooling. These chillers are very different and will be discussed separately.

29.2 HIGH-PRESSURE COMPRESSION CYCLE CHILLERS

A high-pressure compression cycle chiller typically operates the evaporator at pressures above atmospheric pressure. The compression cycle chiller has the same four basic components as a typical air conditioner: a compressor, evaporator, condenser, and metering device. However, these components are generally larger to handle more refrigerant and may use different refrigerants rather than R-22.

Compressors Used in High-Pressure Chillers

The heart of the compression cycle refrigeration system is the compressor. As mentioned previously, there are several types of compressors. The compressors common to high-pressure chillers are the reciprocating, scroll, screw, and centrifugal. The compressor can be described as a *vapor pump*. You should think of the compressor as the component in the line that lowers the evaporator pressure to the desired boiling point of the refrigerant. Typically, this is about 38°F for a chiller. It then builds the pressure in the condenser to the point where the vapor will condense to a liquid for reuse in the evaporator. The typical condensing temperature is 105°F. You can use these temperatures to determine if a typical chiller is operating within the design parameters. The compressor must be of a design that will pump and compress the vapor to meet the needs of a particular installation.

Reciprocating Compressors Large reciprocating compressors used for water chillers operate the same as those for any other reciprocating compressor application, with a few exceptions. A review of the unit on compressors will help you to understand how compressors function. These compressors range in size from about $\frac{1}{2}$ hp to approximately 150 hp, depending on the application. Most manufacturers have stopped using one large compressor for a large reciprocating chiller and have started using multiple small compressors. These are positive displacement compressors and cannot pump liquid refrigerant without the risk of damage to the compressor.

Several refrigerants have been used for reciprocating compressor chillers; R-500, R-502, R-12, R-134a, and R-22 are the most common. R-22 is by far the most popular.

The large reciprocating compressor will have many cylinders to produce the pumping capacity needed to move large amounts of refrigerant. Some of these compressors have as many as 12 cylinders. This is a unit with a lot of moving parts and high internal friction. If one cylinder of the compressor fails, the whole system is offline. With multiple compressors, if one compressor fails, the others can carry the load. Multiple compressors provide some degree of protection against total system failure. Because of this feature and the ability to provide capacity control, many manufacturers prefer to use multiple compressors.

All large chillers must have some means of capacity control or the compressor will cycle on and off. This is not satisfactory as most compressor wear occurs during start-up before oil pressure is established. A better design approach is to keep the compressor online and operate it at reduced capacity. Reduced-capacity operation also smooths out temperature fluctuations that occur from shutting off the compressor and waiting for the water to warm up to bring the compressor back on.

Cylinder Unloading – Reduced capacity for a reciprocating compressor is accomplished via cylinder unloading. For example, suppose a 100 ton compressor with eight cylinders is used for the chiller for a large office building and the chiller has 12.5 tons of capacity per cylinder. When all eight cylinders are pumping, the compressor has a capacity of 100 tons (8 × 12.5 = 100). As the cylinders are unloaded, the capacity is reduced. For example, the cylinders may unload in pairs which would be 25 tons per unloading step. The compressor may have three steps of unloading so the compressor actually has four different capacities: 100 tons or full capacity (eight cylinders pumping), 75 tons (six cylinders pumping), 50 tons (four cylinders pumping), and 25 tons (two cylinders pumping). In the morning when the system first starts, the building may only need 25 tons of cooling. As the outdoor temperature rises, the chiller may need more capacity and the compressor will automatically load two more cylinders to provide 50 tons of cooling. As the temperature continues to rise, the compressor can load up to 100% capacity or 100 tons. If the building stays open at night, such as a hotel, the compressor will start to unload as the outside temperature cools. It will unload down to 25 tons, then if this is still too much capacity, the chiller will shut off. When the chiller is restarted, it will start at the reduced capacity, which lowers the starting current. A compressor cannot be unloaded to zero pumping capacity or it would not move any refrigerant through the system to return the oil that is in the system. Usually, compressors will unload down to 25% or 50% of their full load pumping capacity.

Another big advantage of cylinder unloading is that the power to operate the compressor is reduced as the capacity is reduced, much like the operation of an automobile. As you reduce the load on the engine, it takes less gas to operate it. The reduction in power consumption is not in direct proportion with compressor capacity, but the power consumption is greatly reduced at part load.

Cylinder unloading is accomplished in several ways; blocked suction unloading and lifting the suction valve are the most common methods. Blocked suction unloading is accomplished with a solenoid valve in the suction passage to the cylinder to be unloaded (Figure 29-3). If the refrigerant gas cannot reach the cylinder, no gas is

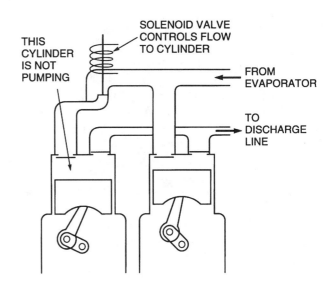

Figure 29-3. Blocked suction unloader for a reciprocating compressor.

pumped. If a compressor has four cylinders and the suction gas is blocked to one of the cylinders, the capacity of the compressor is reduced by 25% and the compressor then pumps at 75% capacity. The power consumption also goes down approximately 25%. Power consumption is related to the amperage draw of the compressor. Amperage is typically measured in the field using a clamp-on ammeter. When a compressor is running at 50% capacity, the amperage will be about 50% of the full load amperage.

Note: Although the power consumption of the compressor is actually measured in watts, using amperage for a measure of compressor capacity is close enough for field troubleshooting.

Suction valve unloading is accomplished in a different manner. It operates on the principle that if the suction valve is lifted off the seat of a cylinder while the compressor is pumping, the cylinder will stop pumping. Gas that enters the cylinder will be pushed back out into the suction side of the system on the up stroke. There is no resistance to pumping the refrigerant back into the suction side of the system, so it requires no energy. The power consumption will be reduced with blocked suction unloading. One of the advantages of lifting the suction valve is that the gas that enters the cylinder will contain oil and good cylinder lubrication will occur even while the cylinder is not pumping.

Compressor unloading could be accomplished by allowing hot gas back into the cylinder but this is not practical because power reduction will not occur. When the gas has been pumped from the low-pressure side of the system to the high-pressure side, the work has been accomplished. Also, the cylinder would become overheated by having to recompress the hot gas.

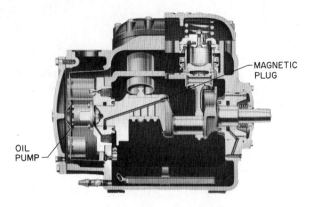

Figure 29-4. Pressure lubrication system for a reciprocating compressor using an oil pump. *Courtesy Trane Company.*

Except for cylinder unloading, the reciprocating chiller compressor is the same as smaller reciprocating compressors. All compressors over 5 hp have pressure-lubricating systems. The pressure for lubricating the compressor is provided by an oil pump that is typically mounted on the end of the compressor shaft and is driven by the shaft (Figure 29-4). The oil pump picks up oil in the sump at evaporator pressure and delivers the oil to the bearings at a pressure of about 30 to 60 psig greater than the suction pressure. This is called the *net oil pressure* (Figure 29-5). These compressors also have an oil safety shutdown in case of oil pressure failure. This device has a time delay of about 90 seconds to allow the compressor time to start and establish oil pressure before it shuts the compressor off.

Scroll Compressors The scroll compressor is a positive displacement compressor. The scroll compressors applied to chillers are larger than those described earlier in this text. Scroll compressors used in chillers are in the 10 to 15 ton range but operate the same as the smaller compressors. They are welded hermetic compressors. When these compressors are used in chillers, the capacity control of the chiller is maintained by cycling the compressors off and on in increments of 10 and 15 tons. For example, a 60 ton chiller may have four 15 ton compressors and be able to operate at 100% (60 tons), 75% (45 tons), 50% (30 tons), and 25% (15 tons).

Some of the advantages of the scroll compressor include the following:

- Efficiency
- Quiet
- Fewer moving parts
- Size and weight
- Ability to pump liquid refrigerant without compressor damage

The scroll compressor offers little resistance to refrigerant flow from the high side of the system to the low side during the off cycle. A check valve is provided to prevent backward flow when the system is shut down, (Figure 29-6).

Lubrication for the scroll compressor is provided by an oil pump at the bottom of the crankshaft. The oil pump picks up oil and lubricates all moving parts on the shaft.

Rotary (Screw) Compressors Most all major manufacturers are building rotary (screw) compressors for larger capacity chillers. This equipment uses high-pressure refrigerants. The screw compressor is capable of handling large volumes of refrigerant using few moving parts (Figure 20-7). Like the scroll compressor, this type of compressor is a positive displacement compressor that is able to handle some liquid refrigerant without incurring compressor damage. This is not true of a reciprocating compressor, which cannot handle liquid refrigerant. Screw compressors are manufactured in sizes from about 50 tons up to 700 tons of capacity. These compressors are both reliable and trouble-free.

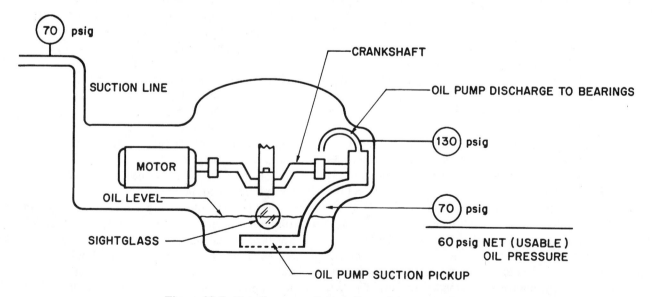

Figure 29-5. Net oil pressure for a reciprocating compressor.

HIGH-PRESSURE COMPRESSION CYCLE CHILLERS 515

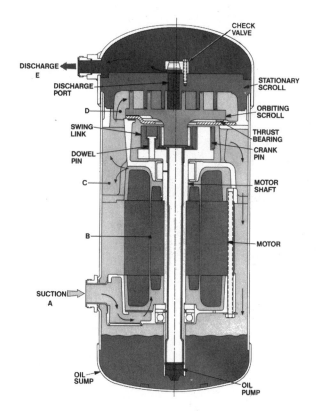

Figure 29-6. Check valve in a scroll compressor. *Courtesy Trane Company.*

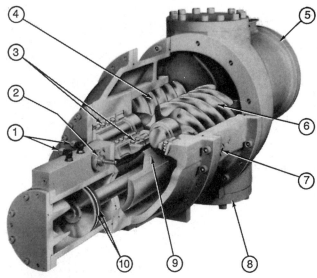

1 - Control Oil Lines
2 - Capacity Control Solenoid Valve
3 - Discharge Bearingh Assemblies
4 - Male Rotor
5 - Semi-Hermetic Motor
6 - Female Rotor
7 - Rotor Oil Injection Port
8 - Suction Inlet Flange
9 - Capacity Control Slide Valve
10 - Slide Piston Seals

Figure 29-7. A screw compressor has few moving parts. *Reproduced courtesy of Carrier Corporation.*

Manufacturers offer both semi-hermetic compressors and open-drive screw compressors. The open-drive models have direct-drive motors and must include a shaft seal to contain the refrigerant where the shaft penetrates the compressor shell.

Capacity control is accomplished in a screw compressor by means of a slide valve that blocks the suction gas before it enters the rotary screws in the compressor. This slide valve is typically operated by the oil pressure in the system. The slide valve may be moved to the completely unloaded position before the compressor shuts down so that on start-up, the compressor is unloaded, reducing the inrush current. Most of these compressors can function from about 10% of capacity up to 100% of capacity using sliding graduations because of the nature of the slide valve unloader. This is different than the steps of capacity control provided in a reciprocating compressor, which unloads one or two cylinders at a time.

The nature of the screw compressor is to pump a great deal of oil while compressing refrigerant, so these compressors typically have an oil separator to return as much oil to the compressor reservoir as possible (Figure 29-8). Instead of an oil pump, the oil is moved to the rotating parts of the compressor by means of the pressure differential within the compressor. Oil is also accumulated in the oil reservoir and moved to the components that need lubrication by means of gravity. The rotating screws are very close together but do not touch. The gap between the screws is sealed by oil that is pumped into the rotary screws as they turn. This oil is separated from the hot gas in the discharge line and returned to the oil pump inlet through an oil cooler.

This oil separation is necessary because if too much oil reaches the system, it will result in poor heat exchange in the evaporator and loss of capacity. The natural place to separate the oil from the refrigerant is in the discharge line. Different manufacturers use different methods but all of them have some means of oil separation.

Centrifugal Compressors The centrifugal compressor uses only the centrifugal force applied to the refrigerant to move the refrigerant from the low-pressure side of the system to the high-pressure side (Figure 29-9). It is like a large fan that creates a pressure difference from one side of the compressor to the other. It does not have a great deal of force, but some companies manufacture centrifugal compressors for use in high pressure systems. Centrifugal compressors can handle a large volume of refrigerant. In order to provide the pressure difference from the evaporator to the condenser for high-pressure systems the compressor is turned very fast by means of a gear box and/or multiple stages of compression. Typically,

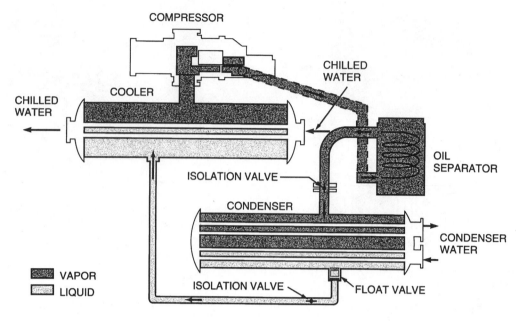

Figure 29-8. Oil separator for a rotary screw compressor. *Reproduced courtesy of Carrier Corporation.*

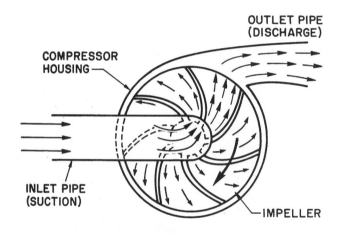

THE TURNING IMPELLER IMPARTS CENTRIFUGAL FORCE ON THE REFRIGERANT FORCING THE REFRIGERANT TO THE OUTSIDE OF THE IMPELLER. THE COMPRESSOR HOUSING TRAPS THE REFRIGERANT AND FORCES IT TO EXIT INTO THE DISCHARGE LINE. THE REFRIGERANT MOVING TO THE OUTSIDE CREATES A LOW PRESSURE IN THE CENTER OF THE IMPELLER WHERE THE INLET IS CONNECTED.

Figure 29-9. Centrifugal action used to compress refrigerant.

3600 rpm is used for direct-drive reciprocating, scroll, and screw compressors. When faster compressor speeds are needed for centrifugal systems, a gear box is used to increase the speed of the compressor. Speeds as high as 30,000 rpm are used for some single-stage centrifugal compressors.

One distinct characteristic of a centrifugal compressor is that if the head pressure becomes too high or the evaporator pressure is too low, the compressor cannot overcome the pressure differential and it will stop pumping. The motor and compressor still turn but refrigerant stops moving from the low-side of the system to the high-side. The compressor may make a very loud whistling sound. This is called a *surge*. Even though this is a very loud noise, it will normally not cause damage to the compressor or motor unless it is allowed to continue for a long period of time. If damage does occur due to a prolonged surge, it is likely to be to the thrust surface of the bearings or to the high-speed gear box.

The gear box used to obtain the high speeds in a centrifugal compressor adds some friction to the system and causes a slight loss of efficiency due to the horsepower needed to turn the gears. The gear box typically has two gears; the drive gear is the large gear and the driven gear is the small gear. The small gear turns faster than the larger drive gear. Figure 29-10 is an illustration of a gear box.

The clearances in the impeller of a centrifugal compressor are very critical and are measured in thousandths of an inch. The parts must be very close together or refrigerant will bypass from the high-pressure side of the compressor to the low-pressure side.

Lubrication is accomplished in a centrifugal compressor using a separate motor and oil pump. This motor is a three-phase fractional horsepower motor (usually $1/4$ hp). A three-phase motor is used because there is no internal motor start winding as with a single-phase motor. This separate oil pump motor allows the manufacturer to start the oil pump before the centrifugal motor and establish lubrication before the compressor and gear box start to turn. It also allows the oil pump to run as the centrifugal impeller and gear box wind down at the end of a cycle when the unit is shut off.

HIGH-PRESSURE COMPRESSION CYCLE CHILLERS 517

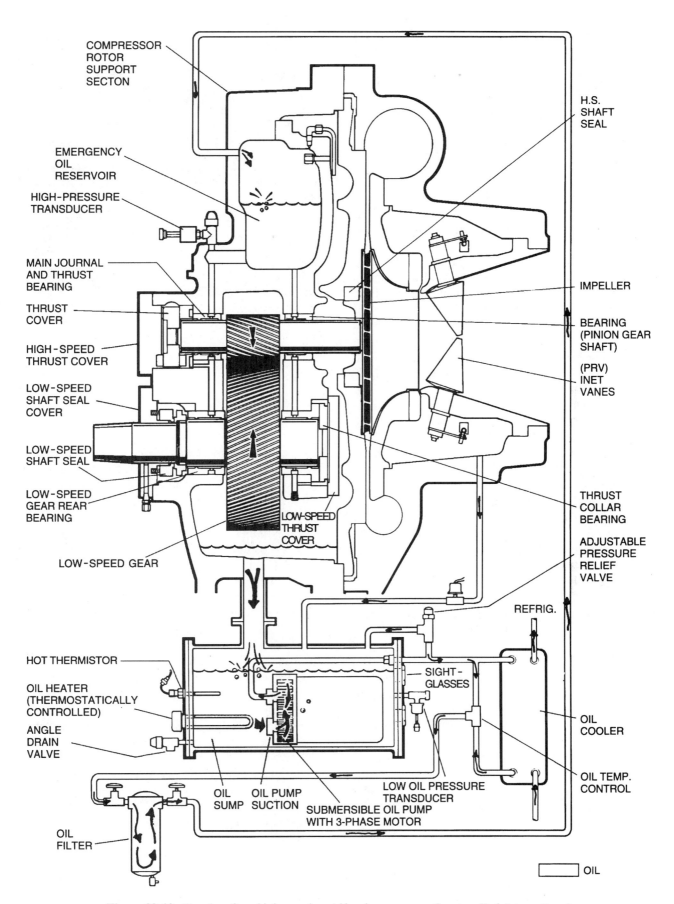

Figure 29-10. Gear box for a high-speed centrifugal compressor. *Courtesy York International.*

There is enough oil in the reservoir to prevent bearing and gear failure in the event of a power failure. The oil pump is usually a gear-type positive displacement oil pump. It will normally furnish a net oil pressure of about 15 psig to the bearings, and with some compressors, to an upper sump or emergency oil reservoir where oil is fed via gravity to the bearings and gear box in the event of a power failure (refer back to Figure 29-10). There are many acceptable methods of routing the oil. The only requirement is that the oil be able to reach the furthest bearings with enough pressure to circulate the oil around the bearings.

The lubrication system has a heater in the oil sump to prevent liquid refrigerant from migrating to the oil sump when the system is off. The heater typically maintains the oil sump at about 140°F. Without this heater, the oil would soon become saturated with liquid refrigerant. Even with the heater operating, some liquid will migrate to the oil sump. When the compressor is started, the oil may have a tendency to foam for the first few minutes of operation. You should keep an eye on the oil sump and run the compressor at reduced load until any foaming in the oil is reduced. Foaming oil contains liquid refrigerant and does not provide good lubrication until the refrigerant is boiled away.

When a chiller is operating, the oil is heated as it lubricates the moving parts. It also absorbs heat from the bearings in order to cool them. The oil will become overheated if it is not cooled, so an oil cooler is used in the circuit during operation. The oil is typically pumped from the oil sump through a filter and an oil cooler heat exchanger, with either water or refrigerant used to remove some of the heat from the oil. This oil cooler cools the oil from a sump temperature of between 140°F and 160°F down to about 120°F. It will then pick up heat again during lubrication.

The lubrication system for a centrifugal chiller is a sealed system and oil is not intended to be mixed with the refrigerant and separated for recovery as in the reciprocating, scroll, or screw systems. If the oil gets into the refrigerant, it is very difficult to remove because only vapor should leave the evaporator. When refrigerant in the evaporator becomes oil-logged or saturated, the heat exchange process becomes less efficient. Oil-logged refrigerant must be distilled to remove the oil. This is a time-consuming process. The intent of the manufacturer is to keep the two separated.

Capacity control in a centrifugal chiller is accomplished by means of guide vanes at the entrance to the impeller eye. These guide vanes are usually triangular devices located in a circle that can rotate to allow full flow or reduced flow down to about 15% or 20%, depending on the clearances in the vane mechanism (Figure 29-11). The guide vanes also serve another purpose in that they help the refrigerant enter the impeller eye by starting the refrigerant in a rotation pattern that matches the rotation of the impeller. The guide vanes are

Figure 29-11. Prerotation guide vanes are used to provide capacity control in a centrifugal compressor. *Courtesy York International.*

often called the *prerotation guide vanes*. They are controlled using electronic or pneumatic (air-driven) motors to vary the capacity of the centrifugal compressor. The controller sends either a pneumatic air signal or an electronic signal to the inlet guide vane operating motor, which is either a pneumatic or electronic type. As the building load is satisfied, the controller signals the vane motor to start closing the vanes. The capacity of the motor is controlled by either the entering or leaving chilled water, depending on the design. If the guide vanes are closed, the compressor only pumps at 15% to 20% of its rated capacity. When open all the way, the compressor pumps at 100% capacity. A 1000 ton chiller can operate from about 150 to 200 tons of capacity up to 1000 tons of capacity in any number of steps.

The guide vanes can also be used for two other purposes: to prevent motor overload and to start the motor at reduced capacity to reduce the inrush current on start-up.

When a chiller is operating at a condition where the chilled water is above the design temperature (e.g., when a building is hot on Monday morning start-up) the compressor motor would run at an overloaded condition. The return chilled water may be 75°F instead of the design temperature of 55°F. The chiller control panel will have a control known as a *load limiter* that will sense the motor amperage and partially close the guide vanes to limit the compressor amperage to the full load value. When the compressor is started, the prerotation guide vanes are closed and do not open until the motor is up to speed.

HIGH-PRESSURE COMPRESSION CYCLE CHILLERS 519

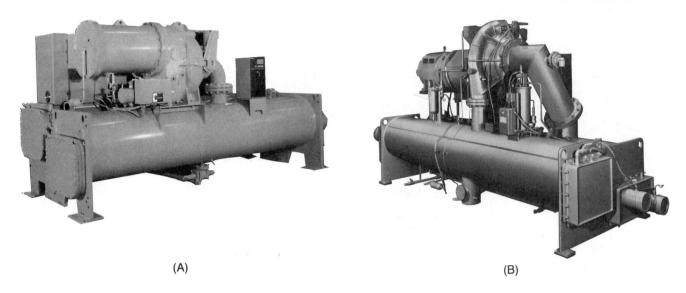

(A) (B)

Figure 29-12. (A) Hermetically-sealed centrifugal chiller. (B) Open-drive centrifugal chiller. *(A) Reproduced courtesy of Carrier Corporation, (B) Courtesy York International.*

Therefore, the compressor starts up unloaded. This reduces the power required to start the compressor.

The gear box and impeller have bearings to support the weight of the turning shaft. These bearings are typically made of a very soft material known as *babbitt*. The babbitt is backed and supported by a steel sleeve. With a single-stage compressor, there must also be a thrust bearing to counter the thrust of the refrigerant entering the compressor impeller. This thrust bearing is on the opposite end of the impeller shaft from the impeller.

The motor rotation for many centrifugal chillers is critical. If the motor is started in the wrong direction, it can damage some chillers. Reverse rotation will be discussed later in this unit.

Some centrifugal compressors are hermetically sealed while others have open drives (Figure 29-12). Hermetic compressors have refrigerant-cooled motors. Figure 29-13 shows a refrigerant-cooled motor. Liquid refrigerant is allowed to flow around the motor housing. Systems with open-drive motors use the ambient air in the equipment room to cool the motor. The heat from a large motor is considerable and must be exhausted from the equipment room.

Evaporators Used in High-Pressure Chillers

The evaporator is the component that absorbs heat into the refrigeration system. In a chiller, liquid refrigerant is boiled into a vapor by the circulating water.

The heat exchange surface is typically copper for water chillers. Other materials may be used if corrosive fluids are circulated for manufacturing processes. In typical smaller air conditioning systems, air is on one side of the heat exchange process and liquid or vapor refrigerant is on the other. The rate of heat exchange between air and vapor refrigerant is fair. The heat exchange between the air and liquid refrigerant is somewhat better and the best heat exchange is between water and liquid refrigerant. This is where the water chiller acquires part of its versatility. The heat exchange surface can be small and still produce the desired results.

The evaporators used in high-pressure chillers are either direct expansion evaporators or flooded evaporators.

Direct Expansion Evaporators Direct expansion evaporators are also known as *dry-type evaporators*, meaning they have an established superheat at the evaporator outlet. They normally use thermostatic expansion valves for metering the refrigerant. Direct expansion evaporators are used for smaller chillers, up to approximately 150 tons on older types and to about 50 tons for more modern equipment. Direct expansion chillers introduce the refrigerant into the end of the chiller barrel with the water being introduced into the side of the shell (Figure 29-14). The water is on the outside of the tubes and baffles cause the water to be in contact with as many tubes as possible for good heat exchange. The problem with this arrangement is that if the water side of the circuit ever gets fouled or dirty, it cannot be cleaned using brushes. The only way it can be cleaned is with the use of chemicals. These chillers are different from flooded chillers, where the water circulates in the tubes and the refrigerant circulates around the tubes.

520 UNIT 29 CHILLED-WATER AIR CONDITIONING SYSTEMS

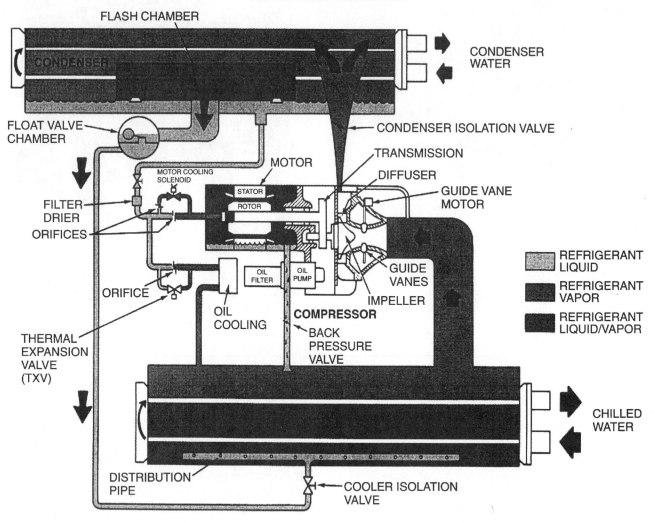

Figure 29-13. Refrigerant-cooled hermetic centrifugal compressor. *Reproduced courtesy of Carrier Corporation.*

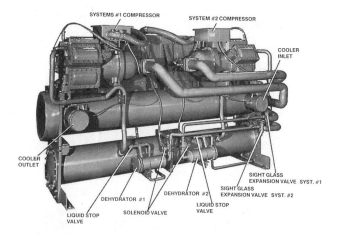

Figure 29-14. Direct expansion chiller with expansion device on the end of the chiller. *Courtesy York International.*

Flooded Evaporators Flooded chillers introduce the refrigerant at the bottom of the chiller barrel and the water circulates through the tubes. There are some advantages in this in that the tubes may be totally submerged under the refrigerant for the optimum heat exchange (Figure 29-15). The tubes may also be physically cleaned using several different methods, such as brushes. The water side of a chiller should not become dirty unless the chiller is applied to an open process, such as a manufacturing operation. Most chiller systems use a closed water circuit that should remain clean unless there are many leaks and water must be continually added. Flooded chillers use a much greater refrigerant charge than direct expansion chillers, so leak monitoring must be part of regular maintenance with high-pressure systems.

HIGH-PRESSURE COMPRESSION CYCLE CHILLERS 521

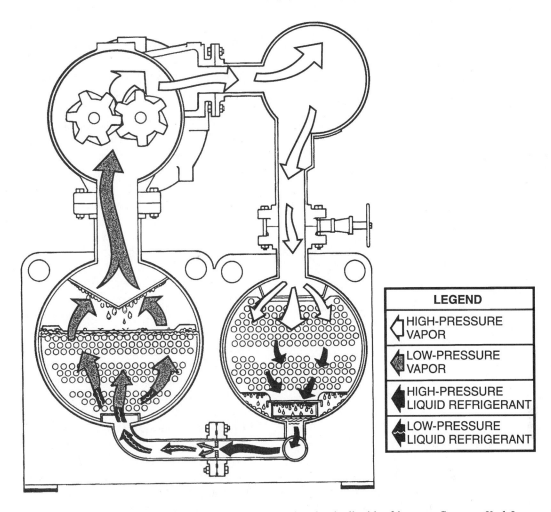

Figure 29-15. Flooded evaporators have the water tubes submerged under the liquid refrigerant. *Courtesy York International.*

When the water is introduced into the end of the chiller, the water is contained in water boxes. When these water boxes have removable covers, they are known as *marine water boxes*. When the piping is attached to the cover, the piping must be removed to remove the water box cover. These are known as *standard covers* (Figure 29-16). The water box is also used to direct the water for different applications. For example, the water may be passed through the chiller one, two, three, or four times for different applications (Figure 29-17). The water box directs the water by means of partitions. This enables the manufacturer to use one chiller with various combinations of water boxes and partitions for different applications.

When water passes through a one-pass chiller, it is only in contact with the refrigerant for a short period of time. Using two, three, or four passes will keep the water in contact longer, but also creates a greater pressure drop and requires more pumping horsepower.

Figure 29-16. Standard water boxes with piping connections. *Courtesy York International.*

522 UNIT 29 CHILLED-WATER AIR CONDITIONING SYSTEMS

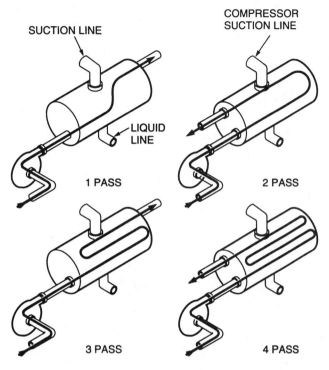

Figure 29-17. Various chiller piping arrangements.

The design water temperatures are typically 55°F inlet water and 45°F outlet water for a two-pass chiller. The refrigerant is absorbing heat from the water so it is typically about 7°F cooler than the leaving water. This is called the *approach temperature* (Figure 29-18).

The approach temperature is very important when troubleshooting chiller performance. If the chiller tubes become dirty, the approach temperature rises because the heat exchange is poor. The approach temperature is different for chillers with a different number of passes. Direct expansion chillers have only one pass of water (with baffles to sweep the tubes) and the approach temperature may be about 8°F. Chillers with three-pass evaporators will have an approach temperature of about 5°F and chillers with four passes may have an approach temperature of 3°F or 4°F. The longer the refrigerant is in contact with the water, the lower the approach temperature will be.

When a chiller is first started, a record should be kept of all operating data. This is known as an *operating-performance log*. This step is often omitted with small chillers because the installing contractor chooses to eliminate the expense. However, it is very important to maintain a log of machine performance at full load conditions. Manufacturers (rather than contractors) often choose to start up larger chillers because of the cost of warranty. They will always make entries in an operating log and keep it on file for future reference. Figure 29-19 shows an example of a manufacturer's log sheet.

Flooded chillers generally have a means of measuring the actual refrigerant temperature in the evaporator, either by using a thermometer well or by providing a direct readout on the control panel. Direct expansion chillers typically have a pressure gage on the low-pressure side of the system and the pressure must be converted using a pressure/temperature chart to obtain the evaporating refrigerant temperature.

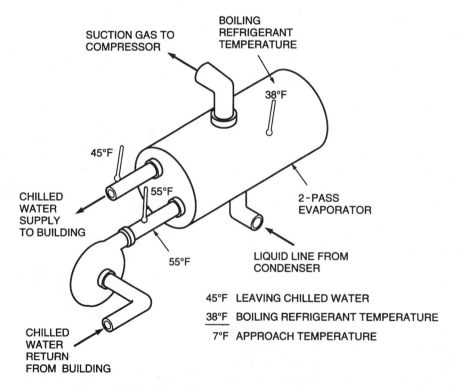

Figure 29-18. Approach temperature for a two-pass chiller.

HIGH-PRESSURE COMPRESSION CYCLE CHILLERS 523

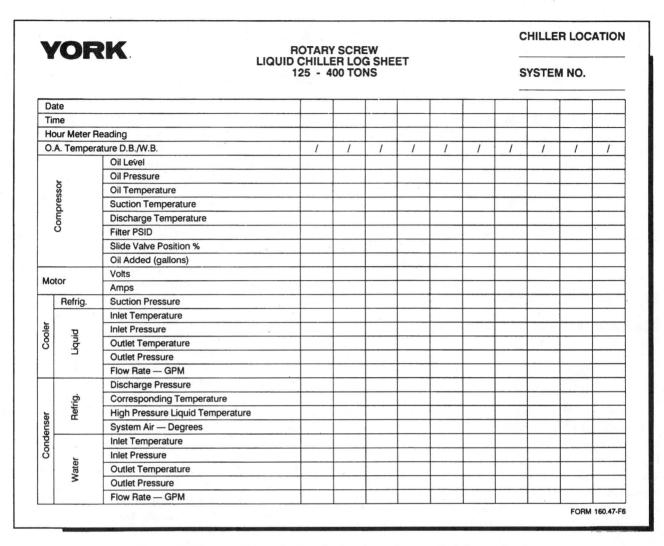

Figure 29-19. Manufacturer's operating log sheet. *Courtesy York International.*

When water is boiled in an open pot and the heat is turned up to the point where the water is boiling vigorously, water will splash out of the pot. Refrigerant in a flooded chiller acts in much the same way; the compressor is removing vapor from the chiller cavity, which will accelerate the process due to vapor velocity. Liquid eliminators are often placed above the tubes to prevent liquid from carrying over into the compressor suction intake at full load. The suction gas is often saturated as it enters the suction intake of a compressor on a flooded chiller. This is different from a direct expansion chiller, which will always have some superheat leaving the evaporator.

The heat exchange surface of the tubes in both direct expansion and flooded evaporators may be improved by machining fins on the outside of the tubes. Direct expansion evaporators may also have inserts that cause the refrigerant to sweep the sides of the tubes or the inside of the tubes may be machined like the bore of a rifle (Figure 29-20). These processes cause slight pressure drops but this is outweighed by the enhanced heat exchange.

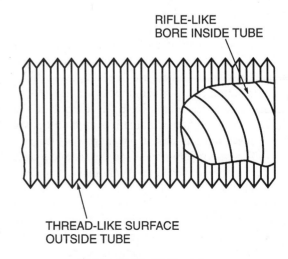

Figure 29-20. Chiller tube.

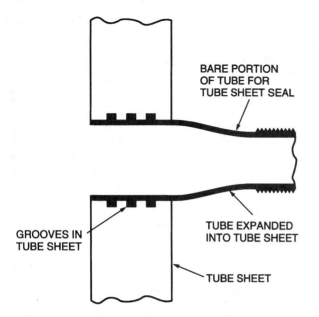

Figure 29-21. How leaks are prevented in chiller tube connections.

Evaporators are constructed in a shell with end sheets to hold the tubes. Typically, the tubes are attached to the tube sheets with a process called *rolling*. The tube sheet has grooves cut in the holes where the tubes fit into the sheet. After the tube is inserted in the tube sheet hole, it is expanded with a roller into the grooves for a leak-free connection (Figure 29-21). Some chiller evaporators may have silver-soldered tubes in the tube sheet. These tubes are very hard to replace.

The evaporator has typical working pressures for both the refrigerant circuit and the water circuit. High-pressure chillers must be able to accommodate the refrigerant used, generally R-22 for newer models. The water-side of the chiller may have one of two working pressures, 150 or 300 psig. The 300 psig chillers are used where the chiller is located on the bottom floor of a tall building where the standing column of water creates a higher pressure.

The evaporator is normally manufactured, then leak tested. After it has proven to be leak free, it is then insulated. Many larger chillers are insulated using rubber-type foam insulation that is glued directly to the shell. Smaller chillers may be insulated by being placed in a skin of sheet metal, with foam insulation applied between the chiller shell and the skin. If any repair is necessary, the skin will have to be removed and the foam cut away. The skin can later be replaced and foam insulation added back to insulate the chiller barrel.

Some evaporators may be located outside in the case of an air-cooled package chiller, so freeze protection for the water in the shell is necessary for these chillers. This can be accomplished by adding resistance heaters to the chiller barrel under the insulation. These heaters may be wired into the chiller control circuit and energized by a thermostat if the ambient temperature approaches freezing and water is not being circulated.

Condensers Used in High-Pressure Chillers

The condenser is the component in the system that transfers heat out of the system. High-pressure chillers use either water-cooled or air-cooled condensers. Usually, the heat is transferred to the atmosphere. In some manufacturing processes, the heat may be recovered and used for other applications.

Water-Cooled Condensers Water-cooled condensers used for high-pressure chillers are of the shell-and-tube type with the water circulating in the tubes and the refrigerant flowing around the tubes. The shell must have an adequate working pressure to accommodate the refrigerant used and a water-side working pressure of between 150 and 300 psig, like the evaporators mentioned earlier. The hot discharge gas is normally discharged into the top of the condenser (Figure 29-22). The refrigerant is condensed into a liquid and drips down to the bottom of the condenser, where it gathers and drains into the liquid line. As with evaporators, the heat exchange can be improved through the use of extended surfaces on the refrigerant-side of the condenser tube. The inside of the condenser tubes may also be grooved as with evaporator tubes to provide improved water-side heat exchange.

Condensers must also have a method for the water to be piped into the shell. There are two basic types of connections for the water box that is attached to the condenser shell. In a standard water box, the piping is attached to the removable water box and some of the piping must be removed before the inspection cover can be removed on the piping end. A marine water box has a removable cover for easy access to the piping end of the machine. In most installations, condenser tube access is more important in the condenser than in the evaporator because the open cooling tower provides potential for dirt and other debris to enter the tubes.

Water-Cooled Condenser Subcooling – The condensing temperature for a water-cooled condenser is around 105°F on a day when 85°F water is supplied to the condenser and the chiller is operating at full load. The saturated liquid refrigerant would be 105°F. Many condensers have subcooling circuits that reduce the liquid temperature to below 105°F. If the entering water at 85°F is allowed to exchange heat with the saturated refrigerant, the liquid line temperature can be reduced. If it can be reduced to 95°F, a considerable amount of capacity can be gained by the evaporator. This subcooling can increase the unit capacity about 1% per degree of subcooling. For a 300 ton chiller, 10°F of subcooling would increase the capacity about 30 tons. The only expense is the extra circuit in the condenser. To subcool refrigerant, it must be isolated from the condensing process using a separate circuit.

HIGH-PRESSURE COMPRESSION CYCLE CHILLERS 525

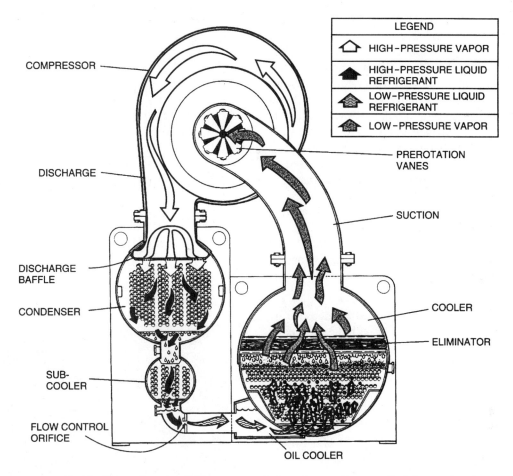

Figure 29-22. Subcooling circuit in a water-cooled condenser. *Courtesy York International.*

Like the evaporator, the condenser also has a relationship between the refrigerant condensing temperature and the leaving water temperature. This is known as the *condenser approach temperature*. Most water-cooled condensers are two-pass condensers and are designed for 85°F entering water and 95°F leaving water with a condensing temperature of about 105°F (Figure 29-23). This is an approach temperature of 10°F. There are other condenser configurations that have one, three, or four passes, each with different approach temperatures. As with evaporators, the longer the refrigerant is in contact with the water, the closer the approach temperature. The operating log will reveal the original approach temperature. It should be the same for similar conditions at a later date, provided the tubes are clean.

Since the condenser rejects heat from the system, its capacity to reject heat determines the head pressure for the refrigeration process. There must be enough pressure in the condenser to push the refrigerant through the expansion device. The head pressure must be controlled. There are two times when the head pressure may be too low and problems may occur. These times are at start-up and during very cold weather when there is still a call for air conditioning due to large internal heat gains.

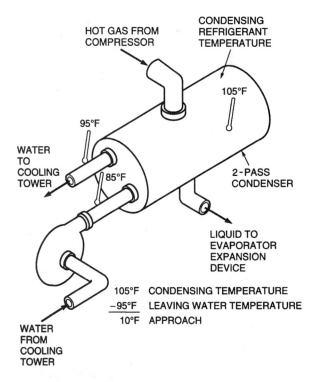

Figure 29-23. Condenser approach temperature.

526 UNIT 29 CHILLED-WATER AIR CONDITIONING SYSTEMS

When there is a call to start a water-cooled chiller and the water in the cooling tower is cold, it may cause the head pressure to become so low that the suction pressure will be low enough to trip either the low-pressure control (direct expansion chillers) or the evaporator freeze control (flooded chillers). In many systems, continued operation will also cause oil migration. A bypass valve is typically located in the condenser circuit to bypass water during start-up to prevent nuisance shutdowns due to cold condenser water. This same bypass may be used to bypass water during normal operation and is discussed in more detail in Unit 30.

When water-cooled condensers are used, the heat from the refrigerant is transferred into the water circulating in the condenser circuit. The heat must now be transferred to the atmosphere in a typical installation. This is done by means of a cooling tower (Figure 29-24). Cooling towers are discussed in more detail in Unit 30.

Air-Cooled Condensers Air-cooled condensers are generally constructed of copper tubes with the refrigerant circulating inside and aluminum fins on the outside to provide a larger heat exchange surface (Figure 29-25).

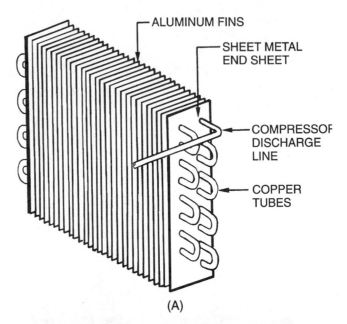

(B)

Figure 29-25. Air-cooled condenser.

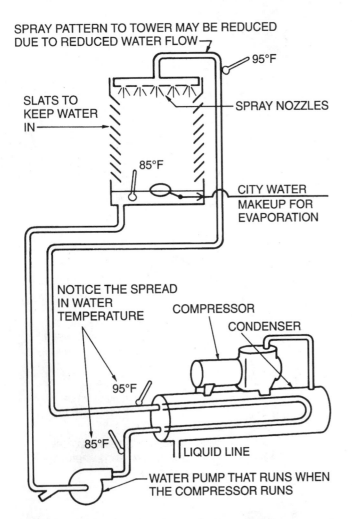

Figure 29-24. Heat moves from the refrigerant to the water and from the water to the atmosphere in the cooling tower.

These air-cooled heat exchange surfaces have been used for many years with great success. Some manufacturers furnish copper finned tubes or special coatings for the aluminum fins for locations where salt air may corrode the aluminum.

These coils are positioned differently by each manufacturer for their own design purposes. Some coils are positioned horizontally and others are positioned vertically.

Multiple fans are used for many of these condensers. The fans may be cycled on and off to provide head pressure control.

Air-cooled condensers are available in sizes from very small units to those that provide several hundred tons of capacity. Air-cooled condensers eliminate the need for water towers and the problems that occur when using water. Systems that use air-cooled condensers do not

HIGH-PRESSURE COMPRESSION CYCLE CHILLERS 527

operate at the low condensing temperatures and head pressures found in water-cooled condensers. Water-cooled condensers typically condense at about 105°F. Air-cooled condensers condense at temperatures between 20°F and 30°F above the entering air temperature, depending on the condenser surface area per ton of capacity. For example, an R-22 system would have a head pressure of about 211 psig at a condensing temperature of 105°F as indicated in the pressure/temperature chart for R-22. An air-cooled condenser on a 95°F day would have a head pressure of between 243 psig (95°F + 20°F = 115°F) and 278 psig (95°F + 30°F = 125°F). This is a considerable difference in head pressure and will increase the operating cost of the chiller. However, air-cooled chillers are popular because they require less maintenance.

Air-Cooled Condenser Subcooling – Subcooling in an air-cooled condenser is just as important as in a water-cooled unit. It provides about 1% of additional capacity per degree of subcooling. Subcooling is accomplished in an air-cooled condenser by means of a small reservoir in the condenser to separate the subcooling circuit from the main condenser (Figure 29-26). A condenser operating on a 95°F day can expect to achieve 10°F to 15°F of subcooling.

Metering Devices Used in High-Pressure Chillers

The metering device releases the liquid refrigerant into the evaporator at the correct rate. Four types of metering devices may be used in large chillers: thermostatic expansion valves, orifice metering devices, high-side and low-side floats, and electronic expansion valves.

Thermostatic Expansion Valves The thermostatic expansion valve (TXV) was discussed in detail in Unit 22. The TXVs used with chillers are of the same type, only larger. The TXV maintains a constant superheat at the end of the evaporator. Usually this is 8°F to 12°F of superheat. The more superheat, the more vapor there will be in the evaporator and the lower the rate of heat exchange. The more liquid in the evaporator, the better the heat exchange, so TXVs are not used except on smaller chillers (up to about 150 tons).

Orifice Metering Devices The orifice metering device is a fixed-bore metering device that merely creates a restriction in the liquid line between the condenser and the evaporator (Figure 29-27). Orifice devices are trouble-free because they have no moving parts.

The flow of refrigerant through an orifice is constant for a given pressure drop. When there is a greater pressure difference, additional flow will occur. The flow during increased load conditions is driven by the higher head pressures at greater loads. When the head pressure is allowed to rise, more flow will take place. Any chiller with an orifice for a metering device has a critical charge. The unit must not be overcharged or liquid refrigerant will enter the compressor inlet. As mentioned previously, this could be detrimental to a reciprocating compressor, so orifices may not be used with these compressors. Small amounts of liquid refrigerant, often in the form of a wet vapor, will not harm scroll, screw, or most centrifugal compressors.

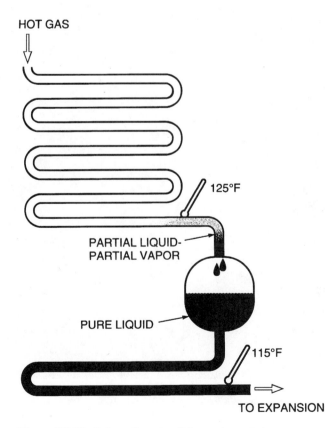

Figure 29-26. Subcooling circuit in an air-cooled compressor.

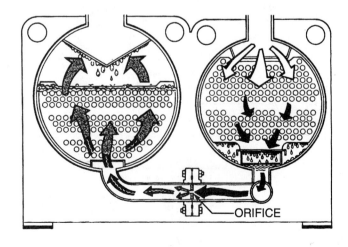

Figure 29-27. Orifice metering device. *Courtesy York International.*

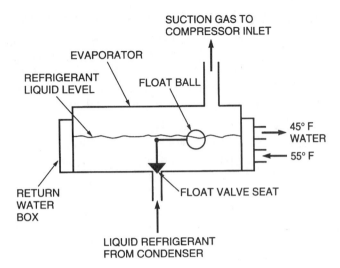

Figure 29-28. Low-side float.

Float-Type Metering Devices There are two kinds of float-type metering devices used in chilled-water systems: the low-side float and the high-side float. The low side float rises and throttles back the refrigerant flow when the liquid refrigerant level rises in the low side of the chiller (Figure 29-28). This float is located at the correct level to produce the level of refrigerant required for the chiller.

The high-side float is located in the liquid line entering the evaporator. When the level of refrigerant is greater in the liquid line, it is because the evaporator needs liquid refrigerant, so the float rises, allowing liquid refrigerant to enter the evaporator (Figure 29-29).

Floats can be a problem if they rub the side of the float chamber, creating additional resistance or perhaps even causing a hole to develop in the float. If a hole develops in the float, it will sink. In the case of a low-side float, it will open and allow full refrigerant flow to the evaporator and no refrigerant will be held back in the liquid line and in the bottom of the condenser. If a hole develops in the high-side float, it will sink and block liquid to the evaporator.

The refrigerant charge is critical for both low-side and high-side floats. An overcharge with a low-side float will back refrigerant up into the liquid line and condenser. An overcharge with a high-side float will flood the evaporator.

Electronic Expansion Valves The electronic expansion valve for large chillers is similar to the one explained in Unit 22. These valves operate with a thermistor to monitor the refrigerant temperature. The liquid does not reach the sensing element; instead, an electronic circuit is used to check the actual temperatures. This is much like a thermometer except it does not read out in temperature. A pressure transducer may be used to give the electronic circuit a pressure reading that can then be converted to an equivalent temperature. With the pressure reading from the evaporator converted to a temperature and the actual suction line temperature reading, the superheat can be determined and very close control can be achieved.

The electronic expansion valve looks similar to a solenoid valve in the line (Figure 29-30). The thermistor sensor can be inserted in the evaporator piping.

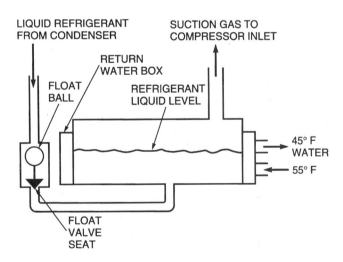

Figure 29-29. High-side float.

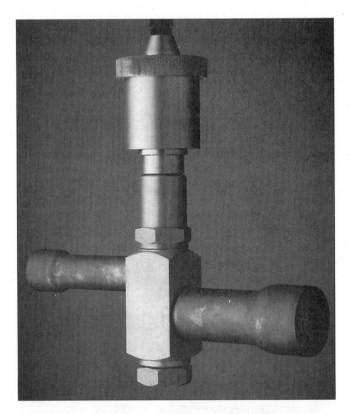

Figure 29-30. Electronic expansion valve. *Courtesy Trane Company.*

LOW-PRESSURE COMPRESSION CYCLE CHILLERS 529

An advantage of the electronic expansion valve is the versatility of its electronic control circuit. The valve can be used for a wider variation in loads than a typical thermostatic expansion valve. It can also be used to allow more refrigerant to flow during low head pressure conditions at low ambient temperatures. This is particularly helpful for air-cooled units when the weather is cold outside. The large liquid-handling capacity of the electronic expansion valve can allow maximum liquid refrigerant flow. This provides the condenser with much more capacity to pass on to the rest of the system. For example, if it is 40°F outside and the building is calling for cooling, the condenser can condense the refrigerant at about 60°F to 70°F. This is a head pressure of 102 to 121 psig for R-22. R-12 and R-500 would also have lower head pressures corresponding to these temperatures. This is not enough head pressure for a typical thermostatic expansion valve, but an electronic expansion valve (because of its flow capability) can use this low pressure to feed the evaporator. If the refrigerant is then subcooled by an additional 10°F, the evaporator can be furnished with liquid refrigerant at 50°F to 60°F. The compressor is only required to pump from an evaporator pressure of approximately 65.6 psig (corresponding to a 38°F evaporator boiling temperature) up to a 121 psig maximum head pressure. This is a big difference from a typical summer condition where the evaporator suction pressure may be 65.6 psig and the head pressure may be 278 psig (corresponding to a condensing temperature of 125°F for R-22).

Warning: Exercise care when working around high-pressure systems. Do not attempt to loosen fittings or connections when the system is pressurized. Follow recommended procedures when using nitrogen.

29.3 LOW-PRESSURE COMPRESSION CYCLE CHILLERS

Low-pressure chillers typically use R-11, R-113, or R-123. In the past, most small low-pressure chillers up to approximately 150 tons used R-113 and chillers above this range used R-11. Many of the newer chillers are using R-123. Some use R-114 for special applications. About 1990, the trend started moving away from all CFC refrigerants. All CFC refrigerants have now been phased out of production. This includes R-11, R-113, and R-114, which are all low-pressure refrigerants. Figure 29-31 shows a pressure/temperature chart for the above refrigerants. As replacement refrigerants are developed for CFC refrigerants, you will need to acquire and become familiar with their pressure/temperature relationship charts.

These chillers have the same components as high-pressure chillers: a compressor, evaporator, condenser, and metering device.

VAPOR PRESSURES

TEMP °F	113	11	114	134a
−150.0				
−140.0				29.6
−130.0				29.4
−120.0				29.1
−110.0			29.7	28.7
−100.0			29.5	28.0
−90.0		29.7	29.3	27.1
−80.0		29.6	29.0	25.7
−70.0		29.4	28.6	24.0
−60.0		29.2	28.0	21.6
−50.0	29.6	28.9	27.1	18.6
−40.0	29.5	28.4	26.1	14.7
−35.0	29.4	28.1	25.4	12.3
−30.0	29.3	27.8	24.7	9.7
−25.0	29.2	27.4	23.8	6.8
−20.0	29.0	27.0	22.9	3.6
−15.0	28.8	26.6	21.8	0.0
−10.0	28.7	26.0	20.6	2.0
−5.0	28.4	25.4	19.3	4.1
0.0	28.2	24.7	17.8	6.5
5.0	27.9	23.9	16.2	9.1
10.0	27.5	23.1	14.4	12.0
15.0	27.2	22.1	12.4	15.1
20.0	26.7	21.1	10.2	18.4
25.0	26.3	19.9	7.8	22.1
30.0	25.7	18.6	5.1	26.1
35.0	25.1	17.1	2.2	30.4
40.0	24.4	15.6	0.4	35.0
45.0	23.7	13.8	2.1	40.0
50.0	22.9	12.0	3.9	45.4
55.0	21.9	9.9	5.9	51.2
60.0	20.9	7.7	8.0	57.4
65.0	19.8	5.2	10.3	64.0
70.0	18.6	2.6	12.7	71.1
75.0	17.3	0.1	15.3	78.6
80.0	15.8	1.6	18.2	86.7
85.0	14.2	3.3	21.2	95.2
90.0	12.5	5.0	24.4	104.3
95.0	10.6	6.9	27.8	113.9
100.0	8.6	8.9	31.4	124.1
105.0	6.4	11.1	35.3	134.8
110.0	4.0	13.4	39.4	146.3
115.0	1.4	15.9	43.8	158.4
120.0	0.7	18.5	48.4	171.1
125.0	2.1	21.3	53.3	184.5
130.0	3.7	24.3	58.4	196.7
135.0	5.3	27.4	63.9	213.5
140.0	7.1	30.8	69.6	229.2
145.0	9.0	34.3	75.6	245.6
150.0	11.1	38.1	82.0	262.8

Figure 29-31. Pressure/temperature chart including low-pressure refrigerants.

Compressors Used in Low-Pressure Chillers

All low-pressure chillers (as well as some high-pressure chillers) use centrifugal compressors. These low-pressure centrifugal compressors also turn from the motor speed for direct-drive compressors or at very high speeds (about 30,000 rpm) for gear-drive units.

The centrifugal compressor has an impeller that turns and creates the low-pressure area in the center where the suction line is fastened to the housing (Figure 29-32). The compressed refrigerant gas is trapped in the outer shell, which is known as the *volute* and is then guided down the discharge line to the condenser (Figure 29-33).

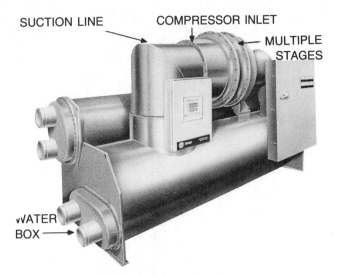

Figure 29-32. Multiple stages of compression. *Courtesy Trane Company.*

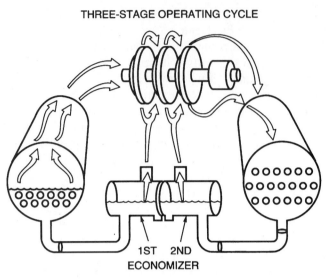

Figure 29-34. Three-stage centrifugal chiller using economizers to provide additional capacity. *Courtesy Trane Company.*

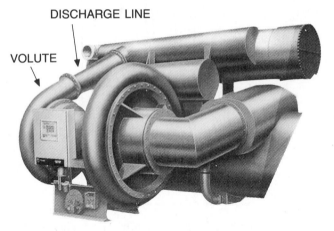

Figure 29-33. Hot gas entering the condenser. *Courtesy Trane Company.*

Centrifugal compressors can be manufactured to produce more pressure by using two more or compressors and operating them in series, called *multi-stage* or by operating one stage of compression at high speed. With multi-stage compressors, the discharge from one compressor becomes the suction for the next compressor. It is common for compressors in air conditioning applications to have up to three stages of compression (Figure 29-34). One of the advantages of multi-stage operation is that the compressor can be operated at slow speeds, typically the speed of the motor (about 3600 rpm).

Refrigerant vapor can also be drawn off from the refrigerant leaving the condenser to subcool the liquid entering the metering device. The refrigerant vapor is drawn off using the intermediate compression stages so the liquid temperature corresponds to the compressor intermediate suction pressure, which is less than the pressure corresponding to the condensing temperature. This process is known as an *economizer cycle*, and it provides the evaporator with extra capacity for its respective size at no additional pumping cost.

Single-stage compression can be provided by using a single compressor that turns at high speed (up to about 30,000 rpm). This is accomplished using a 3600 rpm motor and a gear box to step up the speed of the compressor. The advantage of this compressor is its compact size and lighter weight.

The amount of compression needed depends on the refrigerant used. For example, most centrifugal compressors used for chillers use two-pass evaporators and two-pass condensers. If the evaporator has a 7°F approach, the evaporator will boil the refrigerant at 40°F for a chiller with 47°F leaving water.

Note: We are using 47°F water in order to get an evaporator temperature of 40°F for use on a typical pressure/temperature chart. If the condenser has a 10°F approach, is receiving 85°F water from the tower, and has 95°F leaving water, the condenser should be condensing the refrigerant at about 105°F. The compressor must overcome the following pressures to meet the requirements for this system:

Refrigerant	Evaporator Pressure (40°F)	Condenser Pressure (105°F)
R-113	24.4 in. Hg vac.	6.4 in. Hg vac.
R-11	15.6 in. Hg vac.	11.1 psig
R-123	18.1 in. Hg vac.	8.1 psig
R-114	.44 psig	35.3 psig
R-500	46 psig	152.2 psig
R-502	80.6 psig	231.7 psig
R-12	37 psig	126.4 psig
R-134A	35 psig	134.9 psig
R-22	68.5 psig	210.8 psig

Notice that the R-113 system has only to raise the pressure from 24.4 in. Hg vac. to 6.4 in. Hg vac.; this is only 18 in. Hg vac. or 18 ÷ 2.036 = 8.84 psig. A system that uses R-502 would have to raise the pressure from 80.6 psig to 231.7 psig or 151.1 psig, and a different compressor would have to be used. These two compressors would not raise the pressure the same amount nor would they operate in the same pressure ranges. One compressor operates in a vacuum on both the high-pressure and low-pressure sides of the system and the other operates under a discharge pressure of 231.7 psig. The physical thickness of the shell of the two compressors would also be very different. The above comparison will give you some idea of the different applications for centrifugal compressors.

Figure 29-35 shows a cutaway of a high-pressure chiller shell. This is typical of most high-pressure chillers. Many of the operating features of low-pressure centrifugal chillers are the same or similar to those found on high-pressure units.

The evaporator on a low-pressure system has one working pressure for the refrigerant circuit and another for the water circuit. Low-pressure chillers must be able to accommodate the refrigerant used (typically R-113, R-11, or R-123). Since these are low-pressure chillers, the refrigerant side of the shell does not have to be as strong as with high-pressure chillers. These shells can be much lighter weight because of the pressures. The low pressure chiller shell may have a working pressure as low as 15 psig. The refrigerant safety device for the system is called a *rupture disk* (Figure 29-36). Its relief pressure is 15 psig and it is located in the low-pressure side of the system, either on the evaporator section or the suction line. This device prevents excess pressure on the shell.

Condensers Used in Low-Pressure Chillers

The condenser is the component in the system that removes the heat from the system. The condensers used in low-pressure chillers are water-cooled condensers. As with high-pressure systems, the heat is usually transferred to the atmosphere. In some manufacturing processes, the heat may be recovered and used for other applications.

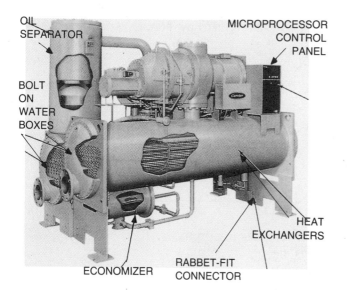

Figure 29-35. Cutaway of high-pressure chiller. *Reproduced courtesy of Carrier Corporation.*

Figure 29-36. Rupture disk. *Courtesy York International.*

Water-cooled condensers used in low-pressure chillers are the shell-and-tube type, with the water circulating in the tubes and the refrigerant flowing around the tubes. The shell must have an adequate working pressure to accommodate the refrigerant used and a water-side working pressure of 150 to 300 psig. The hot discharge gas is discharged into the top of the condenser.

Since the condenser rejects heat from the system, its capacity to reject heat determines the head pressure for the refrigeration process. The condenser is always located above the evaporator in low-pressure chillers. The compressor lifts the refrigerant to the condenser level in the vapor state. When it is condensed into a liquid, it flows via gravity with very little pressure drop to the evaporator through the metering device. The head pressure must be controlled so that it does not drop too far below the design pressure drop.

Water-cooled condensers may also have a subcooling circuit. The purpose of the subcooling circuit is to lower the liquid refrigerant to a temperature below the condensing temperature. It is accomplished by means of a separate chamber in the bottom of the condenser where the entering condenser water (at about 85°F) can remove heat from the condensed liquid refrigerant (at about 105°F). This provides additional system efficiency.

Metering Devices Used in Low-Pressure Chillers

The metering device releases the liquid refrigerant into the evaporator at the correct rate. Two types of metering devices are typically used in low-pressure chillers: the orifice and the high-side or low-side float. These were discussed earlier for high-pressure chillers.

Purge Devices Used in Low-Pressure Chillers

When a centrifugal chiller uses a low-pressure refrigerant, the low-pressure side of the system is always in a vacuum during system operation. If the system uses R-113, the complete system is in a vacuum. Any time the system has a leak and it is in a vacuum, air will enter the system. Air may cause several problems. It contains oxygen and moisture and will mix with the refrigerant, which will create acids that will eventually attack the motor windings and cause damage or motor burnout. Air will also collect in the condenser while the machine is running but it will not condense. This will cause the head pressure to rise. If too much air enters the system, the head pressure will rise to the point that it will either cause a compressor surge or it will shut the system down on a pressure overload.

When the air moves to the condenser, it will collect at the top and will take up condenser space that should be used for condensing refrigerant. This air can be collected and removed using a purge system. The purge is a device that often uses a separate compressor to collect a sample of whatever product is in the top of the condenser. If it can condense the sample, it is considered refrigerant and it will be returned to the system at the evaporator level (Figure 29-37). If it cannot condense the sample, the purge pressure will rise and a relief valve will allow the sample to be exhausted to the atmosphere.

There are several types of purge systems, and they do not all operate at the same efficiency. Some of them use the pressure difference between the high-side and the low-side to create the same pressure difference that a purge compressor will create and a condenser is used to condense any refrigerant from the sample.

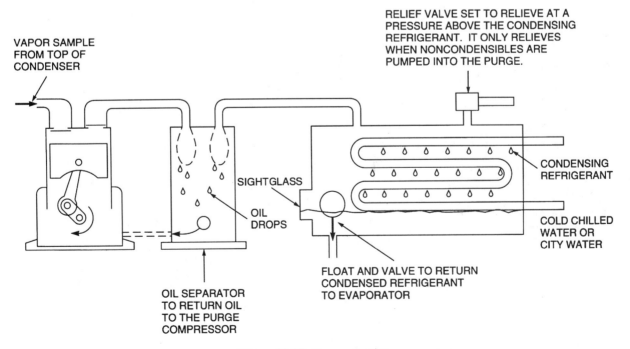

Figure 29-37. Purge operation.

Older purge devices were not very efficient and refrigerant constituted a high percentage of the samples released to the atmosphere. This is considered poor practice today due to environmental concerns and the price of lost refrigerant. Modern day purge devices must be very efficient and allow only 1% of loss during the relieving process. Figure 29-38 shows a modern purge system that releases only a very small amount of refrigerant when relieving the purge pressure. The vapor is relieved to the atmosphere only when air is present. A leak-free system will contain no air and the purge system will not have to function.

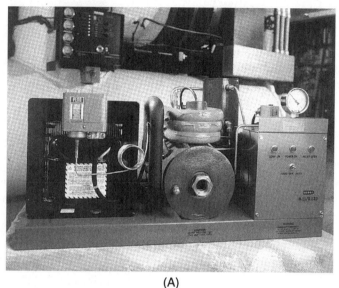
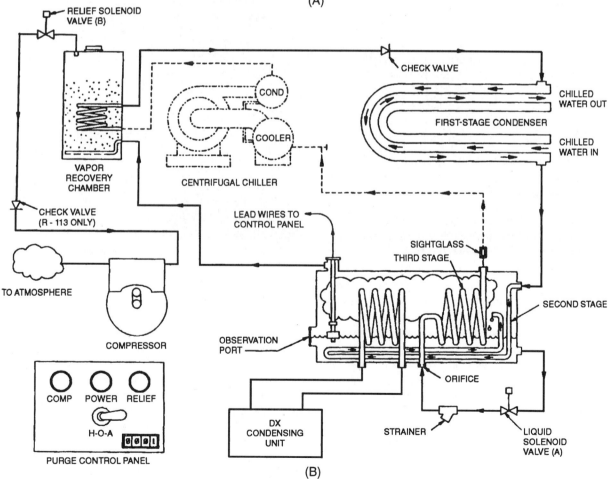

Figure 29-38. Modern purge systems include features that reduce the amount of refrigerant that escapes with the non-condensibles. *Courtesy Carolina Products Inc.*

Some manufacturers relieve the purge pressure using a pressure switch and solenoid valve. A counter is used to register the number of times that the purge functions. The operator can use this to tell when a system has a leak and air is entering. When too many purge reliefs occur, the system is considered to have a leak.

29.4 ABSORPTION AIR CONDITIONING CHILLERS

Absorption refrigeration uses a process that is considerably different from the compression refrigeration process just described. Absorption refrigeration uses heat as the driving force rather than a compressor. When heat is plentiful or economical, or when it is a byproduct of another process, absorption cooling can be a very attractive option. For example, in a manufacturing process where steam is used, it is often used at 100 psig and must be condensed before it can be reintroduced to the boiler. Condensing steam in the absorption system process is a natural choice for chilling water for air conditioning. In this case, chilled water for air conditioning can be furnished at a very economical cost, basically for the cost of operating all related pumps since no compressor is involved.

In some areas, natural gas is an inexpensive fuel in the summer. Gas companies often look for ways to market gas in the summer because their winter quota may depend on how much gas they can sell in the summer. Some gas companies may offer incentives to install gas appliances that consume fuel in the off-peak summer months.

Absorption refrigeration equipment operates in a vacuum and contains non-CFC refrigerants which are not considered a threat to the environment.

Warning: Wear gloves and use caution when working around hot steam pipes and other heated components.

The absorption system equipment looks much like a boiler except it has chilled-water piping and condenser water piping routed to it, in addition to the piping for steam or hot water. Oil or gas burners are part of the system if it is a direct-fired chiller. Figure 29-39 shows two absorption systems by different manufacturers. Small absorption chillers are considered package units, meaning that they are completely assembled and tested at the factory and shipped as one component. They are contained in a single shell or a series of chambers, so the range of equipment sizes may be limited as it must be moved as a single component. These units are rated in tons like other refrigeration systems. They range in size from approximately 100 tons up to to 1700 tons. A 1700 ton system is very large. Large systems may be manufactured in sections for ease of rigging and moving into equipment rooms. These systems are typically assembled in the factory for proper alignment, except for the welds that must be accomplished in the field.

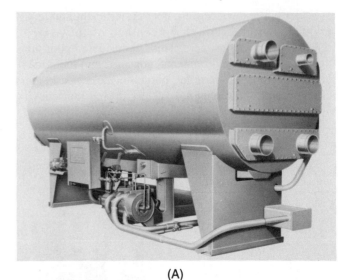

(A)

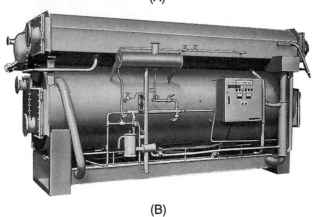

(B)

Figure 29-39. Absorption systems. *(A) Courtesy Trane Company, (B) Reproduced courtesy of Carrier Corporation.*

The absorption process is very different from the compression cycle in certain respects but is also similar in others. It will be helpful to fully to understand the compression cycle before trying to understand the absorption cycle. The boiling temperature for the refrigerant in compression cycle chillers is controlled by controlling the pressure above the boiling liquid. This boiling pressure is controlled in the compression cycle by the compressor. Figure 29-40 shows a refrigeration system using water as the refrigerant. The water pressure must be boiled at .248 in. Hg absolute or .122 psia in order to boil the water at 40°F. Figure 29-41 shows a limited pressure/temperature chart for water. This refrigeration cycle would be workable using a compressor except that the volume of vapor rising from the boiling water would be excessive. The compressor would have to remove 2444 cubic feet of water for every pound of water boiled at 40°F. Figure 29-42 shows a limited specific volume chart for water at different temperatures. A compressor is impractical for a refrigeration system using water as the refrigerant.

ABSORPTION AIR CONDITIONING CHILLERS 535

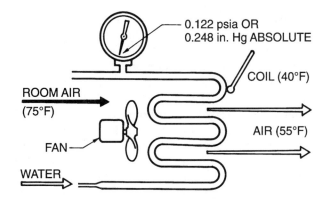

Figure 29-40. Using water as a refrigerant.

TEMPERATURE		SPECIFIC VOLUME OF WATER VAPOR	ABSOLUTE PRESSURE		
°C	°F	ft³/lb	lb/in.²	kPa	in. Hg
−12.2	10	9054	0.031	0.214	0.063
−6.7	20	5657	0.050	0.345	0.103
−1.1	30	3606	0.081	0.558	0.165
0.0	32	3302	0.089	0.613	0.180
1.1	34	3059	0.096	0.661	0.195
2.2	36	2837	0.104	0.717	0.212
3.3	38	2632	0.112	0.772	0.229
4.4	40	2444	0.122	0.841	0.248
5.6	42	2270	0.131	0.903	0.268
6.7	44	2111	0.142	0.978	0.289
7.8	46	1964	0.153	1.054	0.312
8.9	48	1828	0.165	1.137	0.336
10.0	50	1702	0.178	1.266	0.362
15.6	60	1206	0.256	1.764	0.522
21.1	70	867	0.363	2.501	0.739
26.7	80	633	0.507	3.493	1.032
32.2	90	468	0.698	4.809	1.422
37.8	100	350	0.950	6.546	1.933
43.3	110	265	1.275	8.785	2.597
48.9	120	203	1.693	11.665	3.448
54.4	130	157	2.224	15.323	4.527
60.0	140	123	2.890	19.912	5.881
65.6	150	97	3.719	25.624	7.573
71.1	160	77	4.742	32.672	9.656
76.7	170	62	5.994	41.299	12.203
82.2	180	50	7.512	51.758	15.295
87.8	190	41	9.340	64.353	19.017
93.3	200	34	11.526	79.414	23.468
98.9	210	28	14.123	97.307	28.754
100.0	212	27	14.696	101.255	29.921

Figure 29-42. Specific volume chart for water.

TEMPERATURE °F	ABSOLUTE PRESSURE	
	lb/in.²	in. Hg
10	0.031	0.063
20	0.050	0.103
30	0.081	0.165
32	0.089	0.180
34	0.096	0.195
36	0.104	0.212
38	0.112	0.229
40	0.122	0.248
42	0.131	0.268
44	0.142	0.289
46	0.153	0.312
48	0.165	0.336
50	0.178	0.362
60	0.256	0.522
70	0.363	0.739
80	0.507	1.032
90	0.698	1.422
100	0.950	1.933
110	1.275	2.597
120	1.693	3.448
130	2.224	4.527
140	2.890	5.881
150	3.719	7.573
160	4.742	9.656
170	5.994	12.203
180	7.512	15.295
190	9.340	19.017
200	11.526	23.468
210	14.123	28.754
212	14.696	29.921

Figure 29-41. Pressure/temperature chart for water.

The absorption system uses water for the refrigerant but does not use a compressor to create the pressure difference. Instead, it uses the fact that certain salt solutions are high hygroscopic; that is, they have enough attraction for water that they may be used to create the pressure difference. This attraction of salt to water is the process that clogs up your salt shaker during humid weather. This same concept is used in an absorption refrigeration system to reduce the pressure of the water to a point where it will boil at a lower temperature. A type of salt solution called *lithium bromide* is used as the absorbent (attractant) for the water. The liquid solution of lithium bromide is diluted with distilled water. It is actually a mixture of about 60% lithium bromide and 40% water.

A simplified absorption system would be similar to the one shown in Figure 29-43. This is not a complete cycle, but is simplified for the purpose of explanation. A brief description of the absorption cycle is provided below.

1. Figure 29-44 shows the equivalent of the evaporator in the absorption system. Refrigerant (water) is metered into the evaporator section through a restriction (orifice).

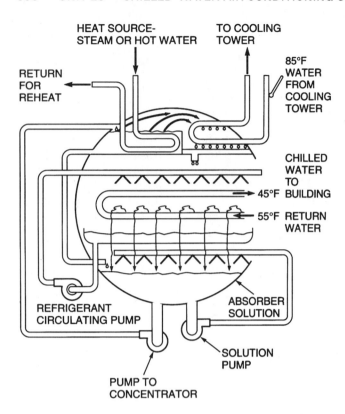

Figure 29-43. Simplified absorption refrigeration system.

heads to be sprayed over the evaporator tube bundle. This wets the tube bundle, through which the circulating water from the system passes. The heat from the system water evaporates the refrigerant. Water is constantly being evaporated and must be made up through the orifice at the top of the unit.

2. Figure 29-45 shows the absorber section which would be the equivalent of the suction side of the compressor in a compression cycle system. The salt solution spray has a low-pressure attraction for the evaporated water vapor, so it readily absorbs into the solution. The solution is recirculated through the spray heads to give the solution more surface area to attract the water. As the solution absorbs the water, it becomes diluted by the water. If this water is not removed, the solution will become so diluted that it will no longer have any attraction and the process will stop, so another pump constantly removes some of the solution and pumps it to the next step, called the *concentrator*. The solution that is pumped to the concentrator is called the *dilute* or *weak solution* because it contains water absorbed from the evaporator.

3. Figure 29-46 shows the concentrator and condenser sections. The weak solution is pumped to the concentrator where it is boiled. This boiling action changes the water into a vapor; it then leaves the solution and is attracted to the condenser coils. The water is condensed into a liquid. It collects and is metered to the evaporator section through the orifice. In this example, the heat source to boil the water could be either steam or hot water. Direct-fired machines are also available and will be discussed later. The concentrated solution is drained back to the absorber area for circulation by the absorber pump.

Figure 29-44. Evaporator section of an absorption system.

It is warm until it passes through the orifice where it flashes to a low-pressure area (about .248 in. Hg absolute or .122 psia). The reduction in pressure also reduces the temperature of the water. The water drops to the pan below the evaporator tube bundle. A refrigerant circulating pump circulates the water through spray

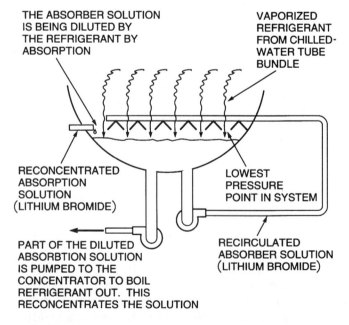

Figure 29-45. Absorber section of an absorption system.

ABSORPTION AIR CONDITIONING CHILLERS 537

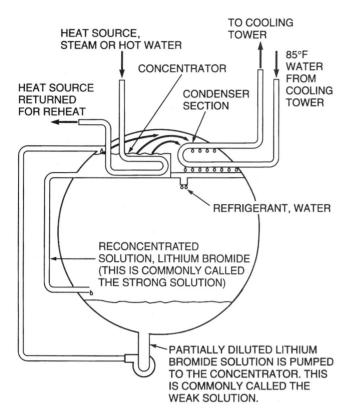

Figure 29-46. Concentrator and condenser section of an absorption system.

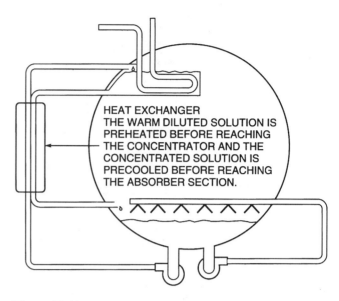

Figure 29-48. A heat exchanger between the solutions can also be used to increase efficiency.

As you can see from the above description, the absorption process is not very complicated. The only moving parts are the pump motors and the impellers.

Certain features may be added to the cycle to make it more efficient. One method is shown in Figure 29-47.

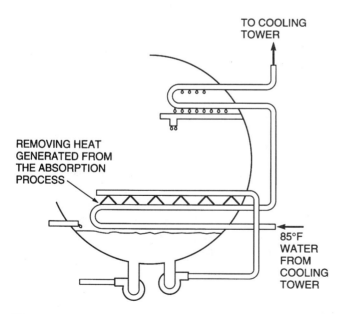

Figure 29-47. A heat exchanger in the absorber can be used to increase efficiency.

It is a heat exchanger between the cooling tower water and the absorber solution in the absorber. This heat exchange removes heat that is generated when the water vapor is absorbed into the absorber solution.

Figure 29-48 shows a heat exchanger between the weak solution and the concentrated solution. This heat exchanger serves two purposes: it preheats the weak solution before it enters the concentrator and precools the concentrated solution before it enters the absorber. It is much like the heat exchange that occurs between the suction line and capillary tube on a window air conditioner. Without this heat exchange, the machine would be less efficient.

Some manufacturers have developed a two-stage absorption machine that uses a higher pressure steam or hot water (Figure 29-49). The steam pressure may be 115 psig (346°F). These machines are more efficient than the single-stage machine discussed above.

Solution Strength

The concentration of the solution determines the ability of the unit to perform. The wider the spread between the weak solution and the strong solution, the more capacity the system has to lower the pressure to absorb the water from the evaporator. Adjusting the strengths of these solutions is the job of the start-up technician. Some units are shipped with no charge and the charge is added in the field. The lithium bromide is shipped in steel drums to be added to the absorption system. The estimated amount of lithium bromide is added first. Distilled water is used as the refrigerant charge and is added next. When the charge is adjusted, it adjusts the solution strength, and is called the *trim*. You charge a compression cycle system and trim an absorption system.

538 UNIT 29 CHILLED-WATER AIR CONDITIONING SYSTEMS

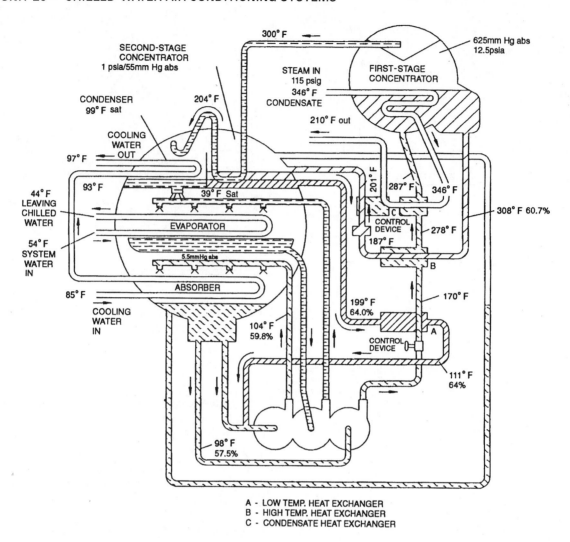

Figure 29-49. Two-stage absorption machine. *Courtesy Trane Company.*

After adding the approximate amount of lithium bromide and water, the machine can be started. The machine is gradually run up to full load. Full load may be determined by the temperature drop across the evaporator and full steam pressure, which is normally 12 to 14 psig (hot water or direct-fired equipment would be similar). When full load is obtained, pull a sample of the weak solution and a sample of the strong solution and measure the specific gravity of each using a hydrometer (Figure 29-50). Remember, specific gravity is the weight of a substance compared to the weight of an equal volume of water. You would know what these specific gravity readings should be from the manufacturer's literature or a lithium bromide pressure/temperature chart (Figure 29-51). Water is either added or removed to obtain the correct specific gravity.

This may be a relatively long process in some cases because you cannot always obtain full load operation at start-up. The machine may have to be trimmed to an approximate level and finished later when full load can be achieved. Some manufacturers furnish a trim chart for a partial load.

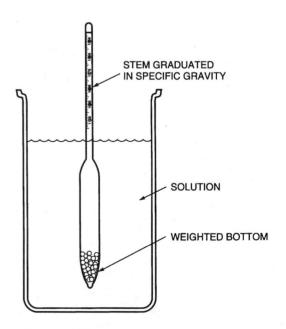

Figure 29-50. Hydrometer used for measuring the specific gravity of absorption solutions.

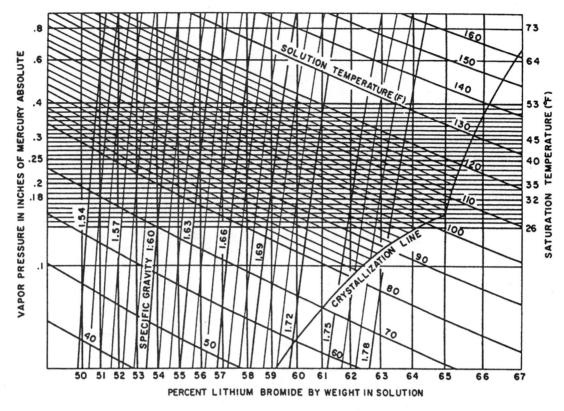

Figure 29-51. Lithium bromide pressure/temperature chart. *Reproduced courtesy of Carrier Corporation.*

System Maintenance

The solutions inside the absorption system are very corrosive. Rust is oxidation and takes place where corrosive materials and oxygen are present. The absorption machine is manufactured with materials such as steel and contains a salt water solution that will corrode the steel if air is present. It is next to impossible to manufacture and keep a system from ever letting air be exposed to the working parts.

To prevent corrosion, these solutions must be kept as clean as possible while circulating through the system. Some of the passages that the solution must circulate through are very small, such as the spray heads in the absorber. Manufacturers use different methods for removing solid materials that may clog any small passages. Filters are used to stop solid particles and magnetic attraction devices are used to attract steel particles that may be in circulation.

Circulating Pumps for Absorption Chillers

The solutions must be circulated throughout the various parts of the absorption system. There are two distinct fluid flows; the solution flow and the refrigerant flow. Some absorption machines use two circulating pumps (Figure 29-52) and others use three, depending on the manufacturer. One manufacturer uses one motor with three pumps on the end of the shaft (Figure 29-53).

Whatever the pump type, they are all similar. The pumps are centrifugal pumps and the pump impeller and shaft must be made of a material that will not corrode. The pumps must be driven with a motor and special care must be taken not to let the atmosphere enter during the pumping process. Because of this it is typical for the motor to be hermetically sealed and to operate only within the system atmosphere. The pump is submerged in the actual solution to be pumped, but the motor windings are sealed in their own atmosphere where no system solution can enter.

These motors must be cooled so cold refrigerant water from the evaporator or normal supply water is used in a closed circuit for this purpose (Figure 29-54).

Motors must also be serviced over a period of time. Many years may pass before servicing is recommended by the manufacturer, but it will eventually be required as these pumps contain moving parts with bearings. Different manufacturers require different procedures for motor servicing, but it is much easier to service the motor and drive if the solution does not have to be removed. However, if the solution must be removed, it involves pressurizing the system with dry nitrogen to above-atmospheric pressure and pushing the solution out. Once this is accomplished and the service completed, the nitrogen must be removed. This is a time-consuming process. Then the system must be recharged.

540 UNIT 29 CHILLED-WATER AIR CONDITIONING SYSTEMS

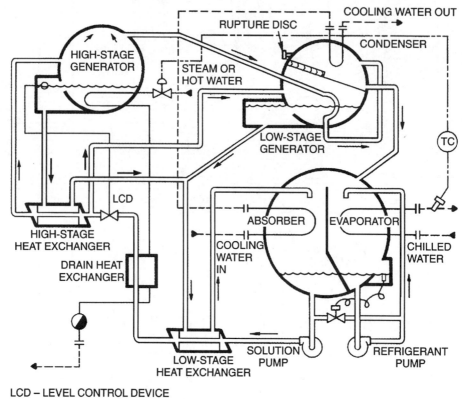

Figure 29-52. Two-pump absorption machine application. *Reproduced courtesy of Carrier Corporation.*

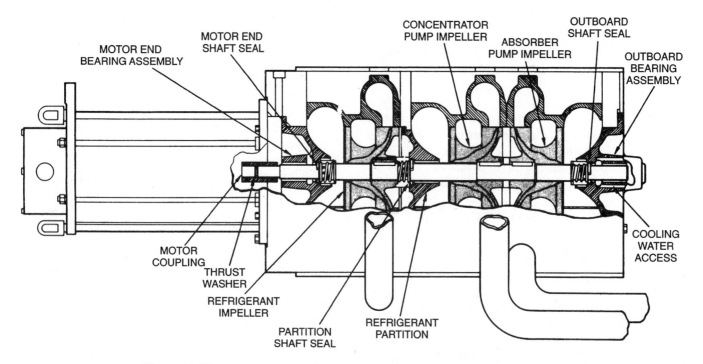

Figure 29-53. Operating three pumps with a single motor. *Courtesy Trane Company.*

ABSORPTION AIR CONDITIONING CHILLERS 541

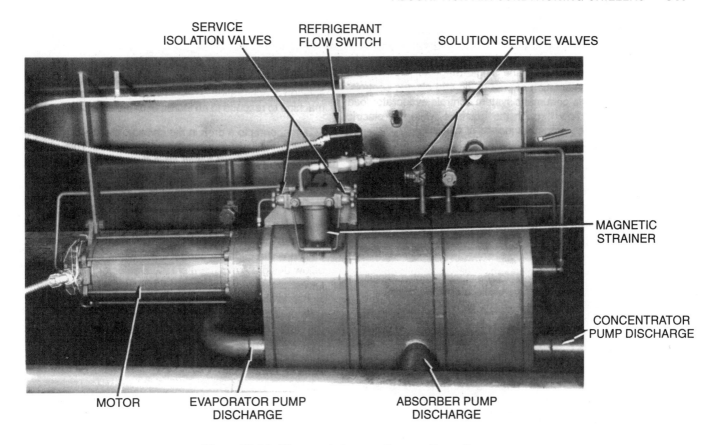

Figure 29-54. Water-cooled motor. *Courtesy Trane Company.*

Capacity Control

The capacity can be controlled for a typical absorption system by throttling the supply of heat to the concentrator. A typical system that operates on steam uses 12 to 14 psig of steam at full load and 6 psig at 50% of capacity. The steam can be controlled using a modulating valve called a *chilled-water controller* (Figure 29-55). Hot water and direct-fired machines can be controlled in a similar manner. Other manufacturers may have different methods for capacity control. These may involve controlling the internal fluid flow in the machine. For example, controlling the flow of weak solution to the concentrator is practical because at reduced load, less solution needs to be concentrated.

Crystallization

The use of a salt solution for absorption cooling creates the possibility of the solution becoming over-concentrated and actually turning back into rock salt. When this happens, it is called *crystallization*. It may occur if the machine is operated under the wrong conditions.

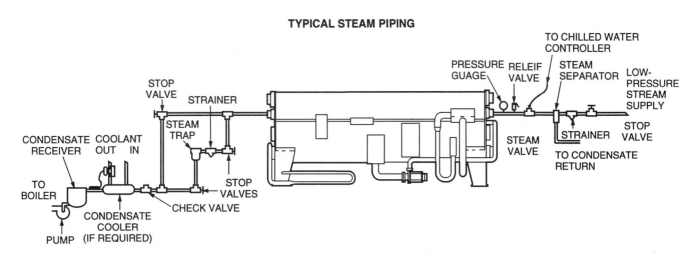

Figure 29-55. Capacity control using a modulating steam valve called a *chilled-water controller. Courtesy York International.*

For example, if the cooling tower water is allowed to become too cold while operating at full load with some systems, the condenser will become too efficient and remove too much water from the concentrate. This will result in a strong solution that has too little water. When this solution passes through the heat exchanger, it will crystallize and restrict the flow of the solution. If this is not corrected, a complete blockage will occur and the machine will stop cooling. Because this is a difficult problem to correct, manufacturers have developed various methods to help prevent this condition. One manufacturer uses a pressure drop in the strong solution across the heat exchanger as a key to the problem. The action taken may be to open a valve between the refrigerant circuit and the absorber fluid circuit to make the weak solution even weaker for a certain period of time to relieve the problem. When the situation is corrected, the valve is closed and the system resumes normal operation. Another manufacturer may shut the machine down for a dilution cycle when over-concentration occurs. Manufacturers use various other methods to correct this problem.

Crystallization can also occur following a unit a shutdown due to power failure while operating at full load. An orderly shutdown calls for the solution pumps to operate for several minutes after shutdown to dilute the strong solution. With a power failure, there is no orderly shutdown and crystallization can occur.

Since the machine operates in a vacuum, atmospheric air can be pulled into the machine at any point where a leak occurs. Air in the system can also cause crystallization.

Purge Systems

The purge system removes non-condensibles from the absorption machine during the operating cycle.

All absorption systems operate in a vacuum. If there are even tiny cracks or openings in the system, the atmosphere will enter. As little as eight ounces of air in a 500 ton machine will affect the capacity. The air will expand greatly when pulled into the unit vacuum. These machines must be kept absolutely leak-free. All piping should have factory-welded connections wherever possible. When the typical package absorption machine is assembled at the factory, it is put through a very rigid leak test known as a *mass spectrum analysis*. For this test, the machine is surrounded by an envelope of helium. The system is pulled into a deep vacuum and the exhaust of the vacuum pump is analyzed using a spectrum analyzer (spectrometer) to determine whether it contains any helium (Figure 29-56). Helium is a gas with very small molecules that will leak in through even the smallest opening.

Even with rigid leak check and welding procedures, the absorption system is subject to leaks developing during shipment to the job site. Leaks may also develop after years of operation and maintenance. When a leak occurs, the non-condensibles must be removed.

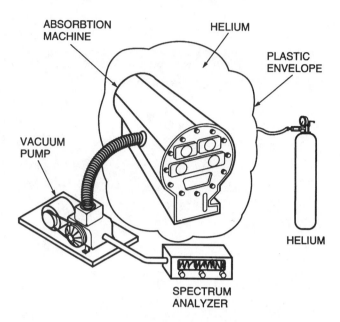

Figure 29-56. Leak checking an absorption machine in the factory.

Absorption systems also generate small amounts of non-condensibles during normal operation. A byproduct of the internal process causes hydrogen gas and some other non-condensibles to form inside the system during normal operation. These are held to a minimum with the use of additives, but they still occur and it is the responsibility of both the purge system and the operator to keep the machine free of these non-condensibles. There are two kinds of purge systems used with typical absorption cooling equipment: the non-motorized purge and the motorized purge. The non-motorized purge uses the system pumps to move non-condensible products toward a chamber where they are collected and then bled off by the machine operator (Figure 29-57). This purge operates well while the machine is in operation but may not be of much use if the machine requires service and is pressurized to atmospheric pressure using nitrogen. This particular purge manufacturer offers an optional motorized purge for removal of large amounts of non-condensibles.

The motorized purge is essentially a two-stage vacuum pump that removes a sample of whatever gas is in the absorber and pumps it into the atmosphere. These vapors are not harmful to the atmosphere. Since the absorber operates under such a low vacuum, this gas sample would only contain non-condensibles or water vapor. The motorized purge requires some maintenance because of the corrosive nature of the salt vapor that will be pulled through the vacuum pump. The vacuum pump oil must be changed on a regular basis to prevent vacuum pump failure. Vacuum pumps are very expensive and must be properly maintained.

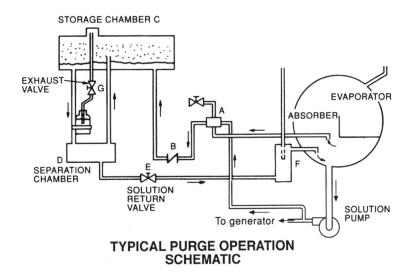

Figure 29-57. Storage chamber for non-condensible gases. *Reproduced courtesy of Carrier Corporation.*

Absorption System Heat Exchangers

The absorption machine uses heat exchangers as in the compression cycle chillers. There are actually more heat exchangers in an absorption chiller: the evaporator, absorber, concentrator, condenser, and the first-stage heat exchanger for two-stage systems. They are similar to compression cycle heat exchangers but have some significant differences. They also contain tubes. These tubes are either copper or cupro-nickel (a tube made of copper and nickel).

The chilled water heat exchanger tubes remove heat from the building water and add it to the refrigerant water. The chilled water circuit is usually a closed circuit, so the tubes rarely need maintenance except for water treatment. A flange at the end of the tube bundle allows access. These machines have either marine or standard water boxes to provide access. This section of the absorption system must have some means of preventing heat from the equipment room from transferring into the cold machine parts. Some refrigeration systems are insulated in this section while others have a double-wall construction to prevent heat exchange. Like the compression cycle chillers, there is an approach temperature between the boiling refrigerant and the leaving chilled water. This approach temperature is likely to be only 2°F or 3°F for an absorption chiller as the rate of heat exchange is very good (Figure 29-58).

The absorption heat exchanger exchanges heat between the absorber solution and the water returning from the cooling tower. The design cooling tower water temperature under full load in hot weather is 85°F. The water from this heat exchange continues on to the condenser where the refrigerant vapor is condensed for reuse in the evaporator. The leaving water is likely to be between 95°F and 103°F on a hot day at full load (Figure 29-59).

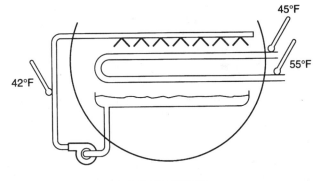

Figure 29-58. Chilled water-to-refrigerant approach temperature.

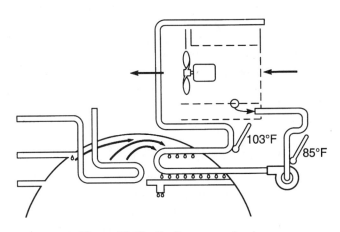

Figure 29-59. Cooling tower circuit.

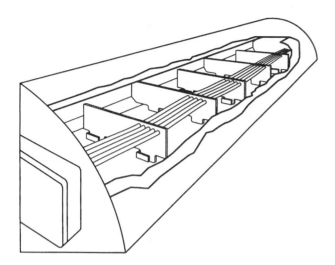

Figure 29-60. Expansion of tubes in an absorption machine. *Courtesy Trane Company.*

The other source of heat exchange in standard equipment is between the steam, hot water, or flue gases and the refrigerant. This tube bundle may have some very high temperature differences when you consider it may be at room temperature until the system is started and hot water or steam is turned into the bundle. A great deal of stress in the form of tube expansion takes place at this time. The manufacturers claim that a tube can grow up to $1/4$ inch in length during this process. Different manufacturers deal with this stress in different ways. Figure 29-60 shows a tube bundle that has the ability to float in order to minimize problems.

Some absorption equipment manufacturers provide thermometer wells located at strategic spots to check the solution temperatures for the purpose of determining machine performance and to determine if the tubes are dirty. These manufacturers also provide charts listing the correct temperatures expected for their systems.

Direct-Fired Systems

Some manufacturers furnish direct-fired equipment (Figure 29-61). These systems use gas or oil as the heat source. The unit can be furnished as a dual-fuel system for applications where gas demand may require a second fuel.

These machines range in capacities from approximately 100 tons to 1500 tons. They also may have the ability to either heat or cool by furnishing either hot or chilled water. In some cases, they may be used in older installations to replace both a boiler and an absorption machine with a single piece of new equipment

29.5 MOTORS AND DRIVES

All of the motors that drive both high-pressure and low-pressure chillers are highly efficient three-phase motors. These motors all give off some heat during normal operation. This heat must be removed from the motor or it will become overheated and burn out. Some manufacturers

Figure 29-61. Combustion and control components of a direct-fired absorption machine. *Courtesy Trane Company.*

choose to keep the motor in the atmosphere for cooling purposes. These motors are air-cooled and release heat to the equipment room (Figure 29-62). This heat must be removed or the room will overheat. The heat is typically removed using an exhaust fan system. This compressor must have a shaft seal to prevent refrigerant from leaking into the atmosphere (Figure 29-63). In the past, this has been a source of leaks, but modern seals and correct shaft alignment have improved the success of open-drive compressors.

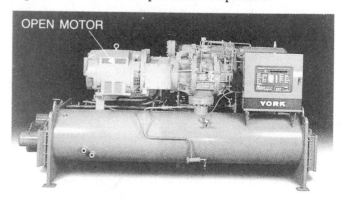

Figure 29-62. Air-cooled, direct-drive compressor motor. *Courtesy York International.*

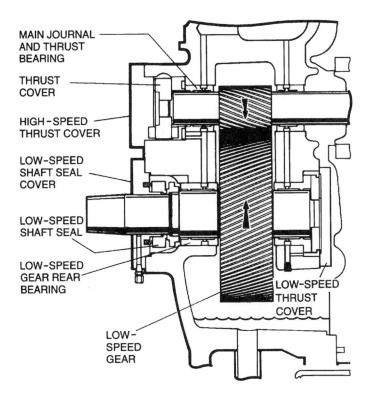

Figure 29-63. Shaft seal for direct-drive compressor. *Courtesy York International.*

Compressor motors may also be cooled using the vapor refrigerant leaving the evaporator. These compressors are called *suction-cooled compressors* and are either hermetic or semi-hermetic compressors. Fully-hermetic compressors are available in the smaller sizes and may be returned to the factory for rebuild. Semi-hermetic compressors may have the motor or compressor rebuilt in the field.

Compressor motors may also be cooled using liquid refrigerant by allowing liquid to enter the motor housing around the motor. The liquid does not actually touch the motor or windings; it is contained in a jacket around the motor.

Large amounts of electrical power must be furnished to the motor through a leak-free connection. This power is often furnished through a non-conducting terminal block made of some form of phenolic or plastic with the motor terminal protruding through both sides (Figure 29-64). The motor terminals are fastened on one side and the field connections are fastened on the other side using O rings for the terminal seals. The terminal board has a gasket to seal it to the compressor housing. This terminal block is normally checked periodically for tightness of the field connections. A loose connection may melt the board and cause a leak to develop.

The motors used in all types of chillers are very expensive and the manufacturer goes to great lengths to protect the motor by designing it to operate in a protective atmosphere. Part of achieving long motor life is to use a careful start-up procedure for the motor.

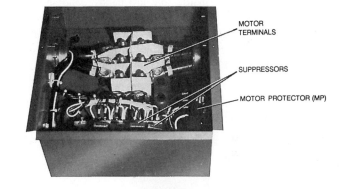

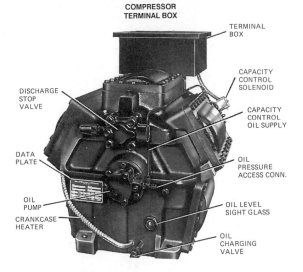

Figure 29-64. Terminal block for hermetic compressor motor. *Courtesy York International.*

Motor Starters

A large motor is usually started using a group of components called a *starter*. There are several types of starters. A motor draws about five times as much current at locked rotor amperage on start-up than it does at full load. For example, if a motor draws 200 amperes at full load, it would draw approximately 1000 amperes on start-up. This inrush current can cause problems in the electrical service so manufacturers use several different methods to start motors to minimize the inrush current and power line fluctuations. Common methods include: part winding, auto transformer, wye-delta (often called *star-delta*), and electronic starters.

Warning: Follow all recommended safety procedures when working around electrical circuits.

Part Winding Start When compressor motors reach about 25 horsepower, manufacturers often use part-winding start motors. These motors are very versatile and normally have nine leads. The same motor can be used for two different voltages. For example, the compressor manufacturer can use the same compressor for 208/230 and 460 volt applications by changing the motor terminal arrangement (Figure 29-65). This motor is actually two motors in one. For example, a 100 horsepower compressor has two motors inside that are each 50 horsepower. When connected for 208/230 V applications, the motors are wired in parallel and each motor is started separately; first one, then the other. This is done using two motor starters separated by a time delay of about one second (Figure 29-66). When the first motor is started, it starts

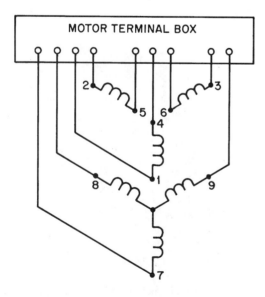

Figure 29-65. Nine-lead, dual-voltage motor connections.

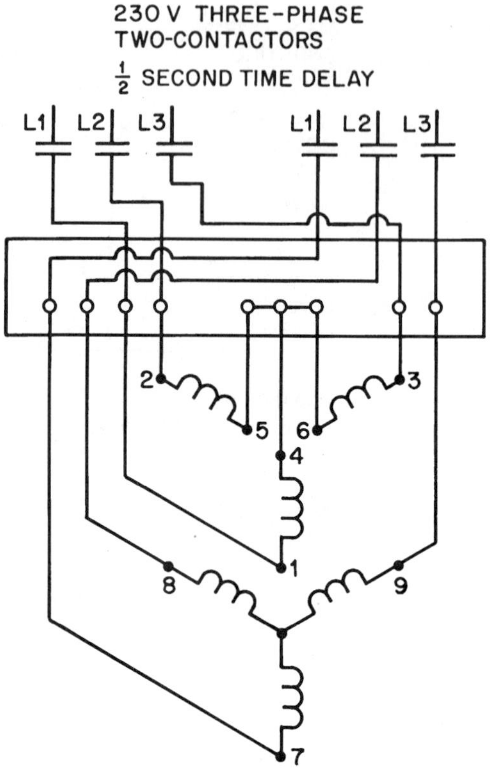

Figure 29-66. Part-winding start for a motor.

the motor shaft turning. The second motor then starts to bring the compressor up to full speed, about 1800 rpm or 3600 rpm, depending on the motor's rated speed. This only imposes the inrush current of a 50 hp motor on the line because when the second motor is energized, the shaft of the first motor is already turning and the inrush current has dissipated.

When the motor is used for 460 V applications, the motors are wired in series and are started as a single motor across the line (Figure 29-67). The higher voltage has a much lower inrush amperage on start-up. These motors are found on compressors up to about 150 hp.

Autotransformer Start An autotransformer start installation is actually a reduced voltage start. A transformer-like coil is placed between the motor and starter contacts and the voltage to the motor is supplied through the transformer during start-up. This reduces the voltage to the motor until the motor is up to speed (Figure 29-68). When the motor is up to speed, a set of contacts close that short around the transformer to run the motor at full voltage. The contacts that direct power through the transformer will then open so that current does not go through the transformer when the motor is running. This minimizes heat in the area of the transformer.

This motor has very little starting torque because it is starting at reduced voltage, so autotransformer start is used only in special applications. This type of motor

MOTORS AND DRIVES 547

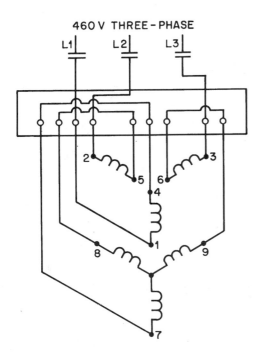

Figure 29-67. These motors are wired in series for a 460 V application.

start-up is common in large compressors that need very little starting torque, such as centrifugal compressors where the pressures are completely equalized during the off cycle and compression does not start until the compressor is up to speed and the vanes start to open.

Wye-Delta Start Wye-delta start is often called star-delta start and is used with large motors with six leads and a single voltage. The motor typically draws less amperage when operating in the wye or star circuit so the motor is started in wye or star (Figure 29-69). After the motor is up to speed, a transition to a delta connection is made where the motor pulls the load and does the work (Figure 29-70). The transition from star to delta is accomplished using a starting sequence that has three different contactors that are electrically and mechanically interlocked; two are energized in wye for start-up then one is dropped out and the other engaged for delta operation. The amount of time that the motor operates in the wye connection depends on how long it takes the compressor to get up to speed. Large centrifugal compressors may take a minute or more to get up to speed in wye, then the transition is made to delta.

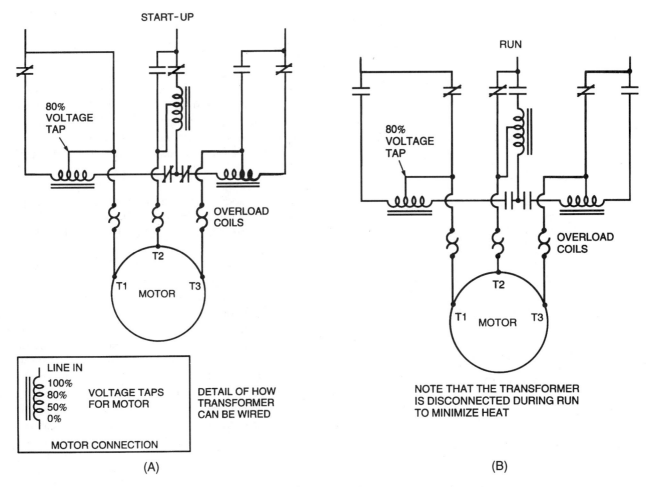

Figure 29-68. Autotransformer start.

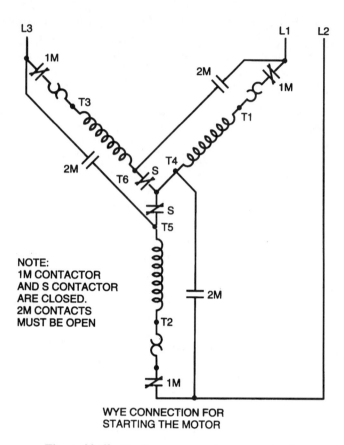

Figure 29-69. Starting a motor using a wye circuit.

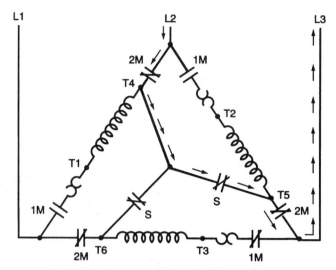

Figure 29-71. Dead short for wye-delta start sequence.

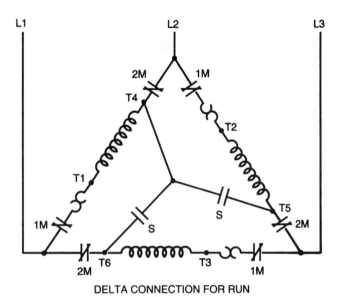

Figure 29-70. Delta connection for a motor.

When the motor starter switches from wye to delta, the wye connection is actually disconnected electrically by means of a contactor. This disconnect is proven both electrically by the timing of the auxiliary contacts and mechanically by means of a set of levers that interlock the contactors. Then the delta connection is made. The electrical and mechanical interlock feature is necessary because if the wye and delta connections were made at the same time, a dead short from phase to phase would occur and likely destroy the starter components (Figure 29-71).

When the delta connection is made, there is a large amperage draw as the motor is totally disconnected and then reconnected. This amperage spike could cause voltage problems in the vicinity of the motor (e.g., a computer room located on the same service). Many manufacturers furnish a motor starter with a set of resistors that are electrically connected as the load while the transition is being made from wye to delta. This is known as a *wye-delta closed transition starter*. The starter would be a wye-delta open transition without the resistors. The larger the motor to be started, the more likely it is to have a closed transition starter.

All of the above starters use open contactors to start and stop the motors. When contacts of this nature are brought together to start a motor, they are under a great deal of stress with the inrush current. When the motor is disconnected from the line, an electrical arc tries to maintain the electrical path. This is much like an electrical welding arc and damage to the contactor is caused each time the load is disconnected. The contacts become pitted due to this arc. The contacts will only be able to interrupt the load of a motor for a limited number of times before replacement is required.

The contacts in the starter will also have an arc shield to prevent the arc that occurs when the contacts open from spreading to the next phase of contacts. This arc shield must be in place when starting and stopping the motor or severe damage may occur.

Warning: Starting and stopping a large motor using a starter requires a lot of electrical energy to be expended inside the starter cabinet. Never start and stop a motor with the starter door open. If something should happen inside the starter, particularly during start-up, most of this energy may be converted into heat energy and the result may be molten metal blown toward the door. Don't take the chance—shut the door.

Electronic Starters As mentioned before for other starters, the inrush current can be five times the full load current for a motor. This inrush current can be greatly reduced by using electronic starters. They are also called *soft starters*. These starters use electronic circuits to reduce the voltage and vary the frequency to the motor at start-up. The result can allow a motor that would typically draw 1000 amperes at locked rotor current to be able to start with 100 amperes at locked rotor current. The motor speed is then accelerated up to full speed. This start-up is much easier on all components and reduces power line voltage fluctuations.

Unlike the rest of the starters we have discussed, the electronic starter also has the advantaged of not having any open contacts.

Motor Protection

The motor that drives the compressor is often the most expensive component in the system. The larger the compressor motor the more expensive it is and the more protection it should be afforded. Small hermetic compressors in room air conditioners have very little protection. Motors used in small chillers will have motor protection like that covered in Unit 15. Here we will only discuss the types of motor protection used with large chiller motors such as screw or centrifugal compressors.

The type of protection depends on the size of the system and the type of equipment used. The advances made in electronic devices for monitoring voltage, heat, and amperage have improved the protection offered today compared to the protection available several years ago. However, older motor protectors are still in use and will be for many years to come.

Load-Limiting Devices Motors used on screw and centrifugal chillers may use a load-limiting device to control the motor amperage. The load limiting device monitors the current the motor is drawing while operating and throttles the refrigerant to the suction intake of the compressor to prevent the motor from operating at an amperage that is higher than the full load amperage. The load-limiting device controls the slide valve on a screw

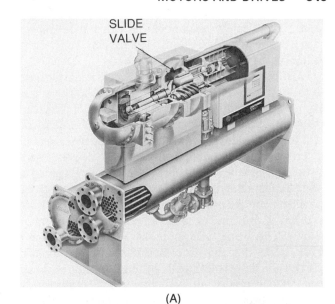

Figure 29-72. Slide capacity control and prerotation vane capacity control. *Courtesy of: (A) Trane Company, (B) York International.*

compressor or the prerotation vanes on a centrifugal compressor (Figure 29-72). This control is a precision device that is adjusted at start-up and should not present any problems during operation. It is the first line of defense in preventing motor overload and is set at exactly the full load amperage.

Note: The full load amperage may be derated to a lower value for a system that has high voltage so you cannot always go by the nameplate amperage. Look at the start-up log to find the derated amperage.

The load limiter may also have a feature that allows the operator to manually operate the chiller at reduced load. Typically, for screw chillers this load may be adjusted between 10% and 100%. For centrifugal chillers, it may be from 20% to 100%.

Most buildings are charged for electrical power based on the highest current draw for each month. The current normally has to spend only a limited amount of time at the excessive level, typically 15 to 30 minutes. This is known as *billing by demand charge* and is measured by a demand meter. Operating the equipment below the demand charge is desirable whenever possible. For example, suppose an office building did not require cooling until the last day of the month and the operator started the chiller and allowed it to run at full load to reduce the building temperature quickly. This would take more than the required time to drive up the demand meter and the building would be charged for the whole month at a higher rate, as though they had operated at full load all month. This excess cost could have been saved by an operator who started the system and kept it operating at part load, allowing it to take longer to reduce the building temperature. For example, the chiller may have a feature that would allow it to run at 40%, which would take longer to accomplish the task but would draw a much lower current.

The load limiter is set to the motor full load amperage. For example, assume a motor has a full load amperage of 200 amperes. The load-limiting device will be set so the system would not exceed 200 amperes.

Mechanical-Electrical Motor Overload Protection
All air conditioning motors must have some means of overload protection. A motor must not be allowed to operate at above the full load amperage for very long or damage will occur to the motor windings due to the excess heat. Different motors have different types of protection. Overload devices for smaller motors, typically up to about 100 hp, were discussed in Unit 15.

The mechanical-electrical overload protection for large screw or centrifugal compressors may be a simple dash-pot type of overload device. This device operates on the electromagnetic theory that when a coil of wire is wound around a core of iron, the core of iron will move when current flows. With the dash-pot, either the current from the motor or a branch of the current of the motor passing through a coil around an iron core will cause the iron core to rise (Figure 29-73).

Large motor overload protectors are typically rated to stop the motor should the current rise to 105% of the motor full load amperage. For example, assume a motor has a rating of 200 amperes. The overload device would be set to trip at 210 amperes (200 × 1.05 = 210). The motor should be allowed to run at this amperage for a few minutes to prevent unwanted overload trips before stopping the motor. Remember, these motors have a load-limiting device to limit the motor amperage, so if this device is functioning correctly, the amperage should not be excessive for longer than it takes the load limiter to react. At this point, the amperage should be no more than full load.

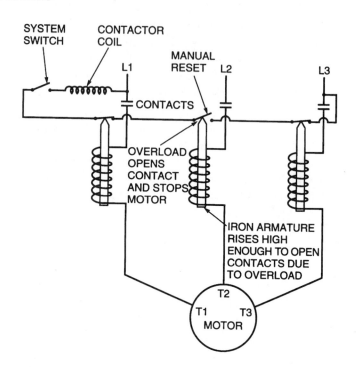

Figure 29-73. Iron core and current flow in a compressor overload.

The reason this overload device is called a *dash-pot* is because it contains a time delay that allows the motor to start without having the inrush current trip the device. The time delay in the dash-pot is accomplished using a piston and a thick fluid that the piston must rise up through before tripping (Figure 29-74). The overload mechanism is wired into the main control circuit to stop the motor when it is tripped. Overload protection for large motors is a manual reset type so the operator will be aware of any problem.

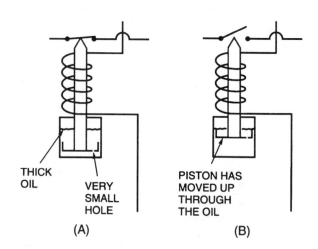

Figure 29-74. A dash-pot of oil provides the overload with the time delay required for motor start-up.

Electronic Solid-State Overload Device Protection

These protective devices are wired into the control circuit like the dash-pot but act and react as an electronic control. They have the ability to allow the motor to be started and also monitor the full load amperage very closely. They are typically located in the starter and are installed around the motor leads for accurate current monitoring.

Anti-Recycle Control All large motors should be protected from starting too often over a specified period of time. The anti-recycle timer is a device that prevents the motor from restarting unless it has had enough running time or off time to dissipate the heat from the last start-up. Manufacturers have different ideas as to how long this time period should be. Typically, the larger the motor, the more time required. Many centrifugals have a 30-minute start delay. If the motor has not been started or tried to start for 30 minutes, it is ready for a start. Always check the manufacturer's literature for the time delay for a specific chiller.

Phase Failure Protection All large motors use three-phase power. The typical voltages are 208, 230, 480, and 575 V. Higher voltages are used for some applications, such as 4160 V or 13,000 V.

Whatever the voltage, all three phases must be furnished or the motor will overload immediately. Electronic phase protection monitors the power and ensures that all three phases are present.

Checking for Voltage Imbalance

The voltage to a compressor must be balanced within certain limits. Most manufacturers use a maximum voltage imbalance of ±2%.

$$\text{max. voltage} = \frac{\text{max. deviation from average voltage}}{\text{average voltage}}$$

For example, suppose the voltages measured on a nominal 460 volt system are as follows:

Phase 1 to Phase 2	475 V
Phase 1 to Phase 3	448 V
Phase 2 to Phase 3	461 V

The average voltage is $(475 + 448 + 461) \div 3 = 461.3$ V. The maximum deviation from the average is $475 - 461.3 = 13.7$ V. The voltage imbalance is:

$$13.7 \div 461.3 = .0297 \text{ or } 2.97\%$$

This exceeds the maximum imbalance allowed by many manufacturers. You should keep an eye on voltage imbalance as it will cause the motor to overheat. Some sophisticated electronic systems include voltage imbalance protection as part of their features, otherwise; it is up to you to monitor the system.

Phase imbalance can be caused by an imbalanced building electrical load or when the power company supply voltage is out of balance.

Phase Reversal

Three-phase motors turn in the direction in which they are wired to run. If the phases are reversed, the motor rotation will also be reversed. This can be very detrimental to many compressors. Reciprocating compressors will normally perform in either direction because of bidirectional oil pumps. Scroll, screw, and centrifugal compressors must turn in the correct direction. Phase protection is often part of the control package for these compressors. Any compressor with a separate oil pump, such as the centrifugal compressor, will not start under these conditions because the oil pump will not pump when reversed. Scroll and screw compressors will start without protection and incur serious damage if operated in reverse rotation.

SUMMARY

- A chiller refrigerates circulating water.
- There are two basic categories of chillers: the compression cycle chiller and the absorption chiller.
- The compression cycle chiller has the same four basic components as other refrigeration systems: the compressor, evaporator, condenser, and metering device.
- Compression cycle chillers may be classified as either high-pressure or low-pressure systems.
- R-22 is the most popular refrigerant used in reciprocating compressor chillers.
- Cylinder unloading is used to control the capacity of a reciprocating compressor. This is accomplished using either blocked suction unloading or suction valve unloading.
- Scroll compressors used in chiller systems are usually in the 10 to 15 ton range.
- Screw compressors are used in larger chillers using high-pressure refrigerants.
- Centrifugal compressors may be used in high-pressure, single-stage chiller systems. These compressors may operate at speeds up to 30,000 rpm using a gear box. These compressors are lubricated using an oil pump with a separate motor.
- The evaporators used in high-pressure chillers are either direct expansion or flooded evaporators.
- The refrigerant absorbs heat from the cooling water and is about 7°F cooler than the leaving water in a two-pass chiller. This is called the *approach temperature*.
- The condensers used in high-pressure chillers may be either water-cooled or air-cooled condensers.
- Many condensers have subcooling circuits that reduce the liquid temperature.
- Like the evaporator, the condenser has a relationship between the refrigerant condensing temperature and the leaving water temperature.

- Air-cooled condensers eliminate the need for using water towers.
- Any of the following metering devices may be used in high-pressure chillers: thermostatic expansion valves, orifices, high-side or low-side floats, and electronic expansion valves.
- Low-pressure chillers use centrifugal compressors which may turn at very high speeds.
- These centrifugal compressors may be manufactured so that they can be operated in series, which is called *multi-stage operation*.
- The oil pump on a centrifugal compressor is energized before the compressor is started to provide prelubrication of the bearings.
- The orifice and the high-side or low-side float metering devices are typically used in low-pressure chillers.
- When a centrifugal compressor uses a low-pressure refrigerant, the low-pressure side is always in a vacuum. If there is a leak, air will enter the system. A purge unit will separate the refrigerant from the non-condensibles and these non-condensibles will be released to the atmosphere through a relief valve.
- Absorption refrigeration is a process that uses heat as the driving force rather than a compressor.
- The absorption chiller uses water as the refrigerant.
- The absorption chiller uses a salt solution consisting of lithium bromide as the attractant in the refrigeration process. The strength of this solution is very important. When the solution strength is adjusted, it is known as *trimming the system*.
- These solutions are very corrosive and it is vital that air be kept out of the system.
- Absorption chillers also have a purge system.
- The motors that drive large chillers are all three-phase motors.
- One of several start devices may be used to prevent excess current on start-up of large motors: part-winding start, autotransformer start, wye-delta start, or electronic starters.
- There are many motor protection controls, including motor overload protection devices, load-limiting devices, anti-recycle control, phase failure protection, voltage imbalance protection, and phase reversal protection.

REVIEW QUESTIONS

1. What medium is used for circulating the cooling throughout the building when a chiller is used?
2. What are the two basic categories of chillers?
3. What types of compressors are used with compression chilled water systems?
4. What is the most popular refrigerant used with reciprocating compressor chillers?
5. How is the capacity of a reciprocating compressor reduced?
6. Describe what is meant by the term *blocked suction*.
7. Describe what is meant by the term *suction valve unloading*.
8. Describe how capacity control is achieved with a rotary compressor.
9. At what approximate maximum speed does a centrifugal compressor turn in a single-stage, high-pressure system?
10. What is the purpose of multiple passes in chillers?
11. What is meant by the term *approach temperature*?
12. What type of water-cooled condenser is generally used in high-pressure chillers?
13. What is meant by the term *condenser subcooling*?
14. List the types of metering devices used in high-pressure chiller systems.
15. What type of compressor is used in low-pressure chiller systems?
16. What causes a surge?
17. Describe how a centrifugal compressor is lubricated.
18. Describe the difference between a marine water box and a standard water box.
19. What types of metering devices are generally used in low-pressure chiller systems?
20. Why is a purge unit used in a low-pressure system?
21. In an absorption chiller, what is used instead of a compressor to drive the system?
22. What is the compound used in most salt solutions in absorption chillers?
23. What term is used to describe the adjustment of the salt solution in an absorption chiller?
24. What refrigerant is generally used in absorption systems?
25. How can capacity control be achieved in an absorption chiller?
26. Why is a purge system needed in absorption chillers?
27. What types of fuels are most common in direct-fired systems?
28. What types of electric motors are typically used in chilled-water systems?
29. List two ways in which compressor motors are cooled.
30. List four start-assist devices.

COOLING TOWERS AND PUMPS 30

OBJECTIVES

After studying this unit, you should be able to

- Describe the purpose of the cooling towers used with chilled-water systems.
- State the relationship between the cooling capacity of the water tower and the wet bulb temperature of the outside air.
- State the means by which the cooling tower reduces the water temperature.
- Describe two types of cooling towers.
- Describe the various types of fill material in cooling water towers.
- State the two types of fans used in cooling towers.
- Explain the purpose of the water tower sump.
- Explain the purpose of makeup water.
- Describe a centrifugal pump.
- Describe water vortexing and explain how it may be avoided.
- Explain two types of motor pump alignment.

30.1 COOLING TOWER FUNCTIONS

The cooling tower is the component in the water-cooled system that rejects the heat from the conditioned space to the atmosphere. The water pump moves the water that contains the heat to the cooling tower. Figure 30-1 shows a basic cooling tower system.

The cooling tower must reject more heat than the chiller absorbs from the structure. The chiller absorbs the heat from the chilled-water circuit and in compression systems, the compressor adds the heat of compression to the hot gas pumped to the condenser (Figure 30-2). The compressor adds about 25% additional heat so the cooling tower must reject about 25% more heat than the capacity of the chiller. For example, a 1000 ton chiller would need a cooling tower that could reject about 1250 tons of heat.

The condenser must be furnished with water within the design limits of the system or the system will not perform adequately. The design temperature for the water leaving the cooling tower for most refrigeration systems,

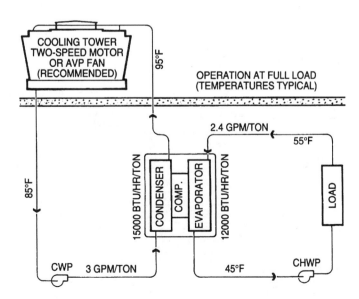

Figure 30-1. Example of a condensing water circuit with a cooling tower. *Courtesy Marley Cooling Tower Company.*

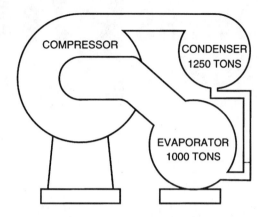

Figure 30-2. Heat of compression and the cooling tower.

including the absorption chiller, is 85°F. Using a design temperature of 95°F dry bulb and 78°F wet bulb for a typical southern section of the country, the cooling tower can lower the water temperature through the tower to within 7°F of the wet bulb temperature of the outside air. This is known as the cooling tower *approach temperature*. When the wet bulb temperature is 78°F, the water temperature leaving the tower would be 85°F (Figure 30-3). This occurs even though the air temperature is 95°F.

All cooling towers work in a similar manner; they reduce the temperature of the water in the tower by means of evaporation. As the water moves through the tower, the surface area of the water is increased to enhance the evaporation process. The different methods used to increase the surface area of the water is part of what makes one water tower different from another.

When the water is spread out, there is more surface area, so the water will evaporate faster and may be cooled to a temperature that is closer to the wet bulb temperature of the air. The water cannot be cooled below the temperature of the cooling medium, which is the wet bulb temperature. An ideal cooling tower could reduce the temperature to the entering wet bulb temperature; however, this tower would be tremendous in size. The manufacturers choose to make towers that approach the wet bulb temperature by 7°F for typical applications. Towers are also manufactured to meet a 5°F approach temperature, but they are not typically used for air conditioning applications.

30.2 TYPES OF COOLING TOWERS

There are two types of cooling towers in common use. These towers are categorized according to whether or not they use fans to move the air over the tower. Towers that do not use fans include the natural-draft tower and the spray pond, while towers that use fans include the forced-draft and induced-draft towers. The larger the installation, the more elaborate the tower.

Natural-Draft Towers and Spray Ponds

The natural-draft tower may be anything from a spray pond in front of a building to a tower on top of a building (Figure 30-4). Both the spray pond and the natural-draft cooling tower rely on the prevailing winds and will cool the water to a lower temperature when the prevailing winds are greater. Actually, the natural-draft cooling tower is a spray pond contained with sides to help hold the water in the tower. Natural-draft tower and spray pond applications usually have an approach temperature of about 10°F because they rely on only the wind speed and duration to achieve cooling.

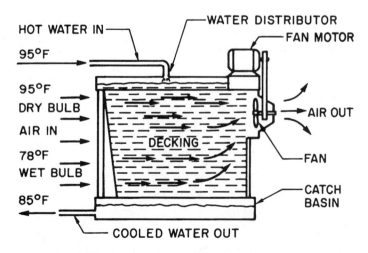

Figure 30-3. Typical cooling tower approach temperature.

TYPES OF COOLING TOWERS 555

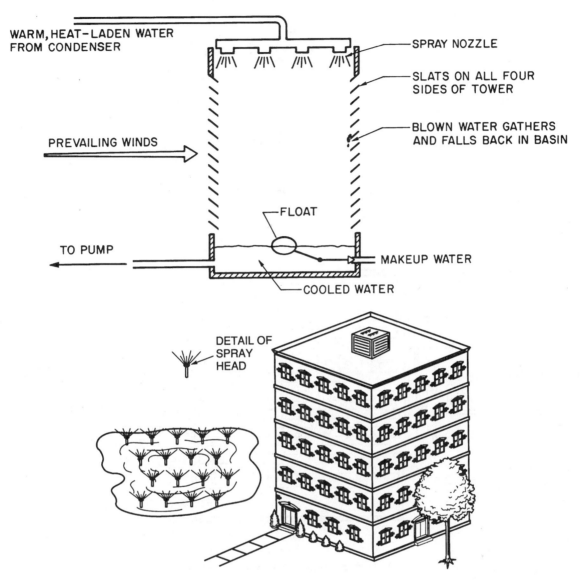

Figure 30-4. Natural-draft cooling towers.

Regardless of whether a spray pond or tower is used, the spray must be contained in the area of the pond. The wind may blow the spray water out of the pond and annoy people or cause water damage. Spray ponds on rooftops have made somewhat of a comeback in the last few years because they serve the purpose of cooling the roof and reducing the solar load. If the rooftop can be cooled to 85°F, a large reduction in the cooling load can be realized.

The spray pond and natural-draft cooling tower both use pump pressure to increase the area of the water by atomizing it into droplets. Spray heads are located at strategic locations within the pond or tower and provide a spray pattern for the water. The spray nozzles are a source of restriction and must be kept clean or they will not atomize the water. Poor atomization will result in the use of excess pump horsepower, which may be considered an unacceptable expense because of the large volume of water used in these applications.

Forced-Draft and Induced-Draft Towers

Forced-draft and induced-draft towers use a fan to move the air through the tower (Figure 30-5). These towers are popular because the efficiency does not vary, as with natural-draft cooling towers. Centrifugal fans are used to move the air in some towers. For example, when the tower air must be ducted to the outside of the building, a centrifugal fan should be used because it is able to overcome the static pressure in the ductwork. Towers with centrifugal fans can be more compact and are therefore more desirable in certain applications. These towers are available in capacities up to about 500 tons of cooling.

556 UNIT 30 COOLING TOWERS AND PUMPS

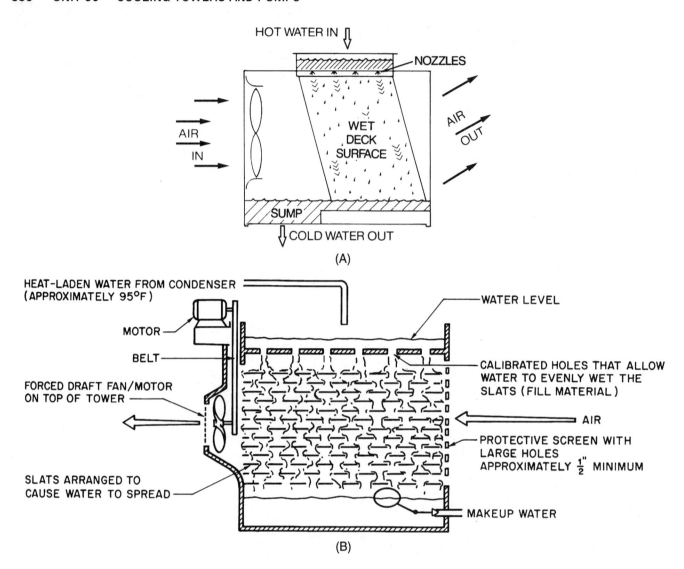

Figure 30-5. Forced-draft and induced-draft cooling towers.

Larger towers use propeller-type fans. These fans may use either belt drives or gear drives. Belt-drive fans will require belt maintenance on a regular basis. Gear-drive fans have a transmission that will only require lubrication.

30.3 FLOW PATTERNS

There are two distinct airflow patterns in cooling towers: crossflow and counterflow (Figure 30-6).

Crossflow Towers

The crossflow tower introduces the air from the side and usually pulls the air to the top of the tower for exhaust. In smaller crossflow towers, the fan is on the side of the tower and the exhaust is out the side. Care must be taken that the moisture-laden air is not exhausted to a place where it will cause problems, such as a walkway or parking lot where cars may be spotted with the water. In the tower, the water is moving downward and the air is moving at right angles to the water.

Counterflow Towers

The counterflow tower introduces the air at the bottom of the tower and exhausts the air out the top. The water is moving downward as the air moves up through it.

The water in most towers, specifically spray-type towers, consist of many small droplets of water suspended in the air. These droplets are subject to being blown out of the tower by the prevailing winds. This loss of water can be expensive, and is also a nuisance for any surrounding areas, especially if the spray contains chemicals. Water drift is minimized using eliminators. Eliminators cause the spray to change direction and rub against a solid surface, where the water deposits and runs back to the tower basin. The eliminators may be louvers on the side of the tower or they may be part of the fill material in some newer towers.

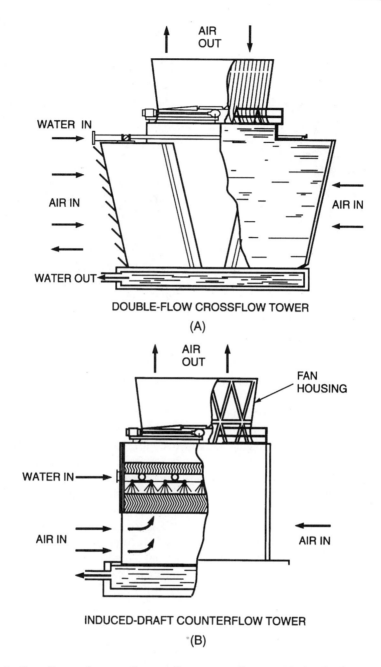

Figure 30-6. Crossflow and counterflow cooling towers. *Courtesy Marley Cooling Tower Company.*

30.4 TOWER CONSTRUCTION

The materials from which cooling towers are constructed must be able to withstand the environment in which the tower operates. There are towers all over the world and each may be located in a different chemical environment, so the materials are varied. The typical tower must be able to withstand wind, the weight of the tower components and related water, sun, cold, freezing weather (including any snow or ice that may accumulate), and vibration from the fan and drive mechanisms. Towers must be carefully designed and many different materials may be used depending on the type and location of the tower. Typically, smaller package towers are made of galvanized steel (for rust protection), fiberglass, or fiber reinforced plastic (FRP). These towers are manufactured as a complete assembly and shipped to the job (Figure 30-7).

Larger towers may have a concrete base and sump to hold the water and the sides may be made of other materials, such as corrugated asbestos-cement panels, wood (either treated or redwood), fiberglass, or corrugated FRP. Fire prevention must be considered in the selection of materials in many cases.

Figure 30-7. Package cooling tower. *Courtesy Baltimore Aircoil Company, Inc.*

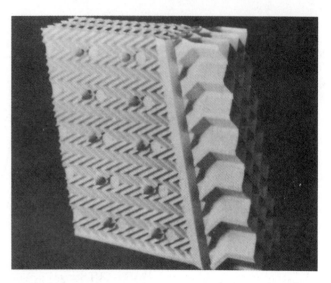

Figure 30-8. Wetted-surface tower fill material. *Courtesy Marley Cooling Tower Company.*

Fire Protection

A cooling tower contains combustible materials that may ignite during the off season when the tower components are dry. Some fire codes or insurance companies may require that the tower be manufactured of fireproof materials or that a tower wetting system be installed. A tower wetting system may consist of sprinkler heads such as those required in buildings. Some towers are controlled in such a manner that the system pump starts up on a timer system to wet the tower down from time to time during the off cycle. Another type of wetting system is an auxiliary pumping system that keeps the tower wet all the time as long as the outdoor temperature is above freezing. This type of system may also prevent expansion and contraction of any wood construction in the tower because it is wet all the time. Local codes and insurance requirements will dictate which method should be used to protect the tower from fire.

Fill Materials

The cooling tower is designed to keep the water in contact with the air for as long as possible. Manufacturers use various methods to slow the water as it trickles down through the tower, with the air moving up through the tower. There are two methods used to evaporate the water in forced-air and induced-draft towers: the splash method and the film or wetted surface method. Both methods use special fill materials in the construction of the tower. This material may consist of wood slats where the water drips down through layers of wood. Other types of materials are also used in the splash method, such as PVC plastic or FRP plastic. These materials have a slow burn rate and should be acceptable for most applications.

Towers that use the splash system have a framework that supports the slats and is designed to keep them at the correct angle for proper wetting of all slats from the top to the bottom.

The film or wetted surface type of tower uses a fill material that may be some form of plastic or fiberglass. The water is spread out on the surface while air is passed over it (Figure 30-8). This type of fill may have small passages for the air to travel through and is not used where particles from the surroundings can contaminate the tower and restrict the passages.

Both types of fill material rely on water to run down and across the fill material, which should be kept at the angle recommended by the manufacturer or the water will not take the proper path. Water running to one side or the other will cause the capacity of the tower to be reduced. If the fill material is removed for maintenance, care should be taken to replace it in the correct manner for correct water movement through the tower.

Tower Access

All cooling towers will require service on a regular basis. The tower must have access doors to the fill material for cleaning and possible removal. The tower basin must also be accessible for cleaning as large amounts of sludge will accumulate. The cooling tower becomes a large filter for airborne contaminants such as dirt, pollution, leaves, feathers, plastic wrappers, and other debris. Insects and birds may also fly into the tower. These materials will collect in the sump and must be removed. There should be an adequate water supply in the vicinity of the tower for the connection of a water hose for flushing the tower sump.

When the tower is tall, it will usually have a stairway or ladder to the top for servicing the fan and drive components (Figure 30-9). During tower construction, provisions should be made to lift any components that may need to be removed later for service. These parts

WATER CONTROL AND DISTRIBUTION 559

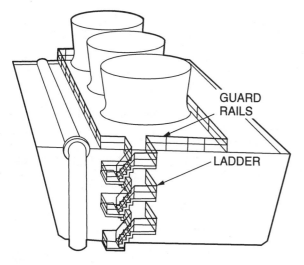

Figure 30-9. Stairway and guard rails for a large tower. *Courtesy Marley Cooling Tower Company.*

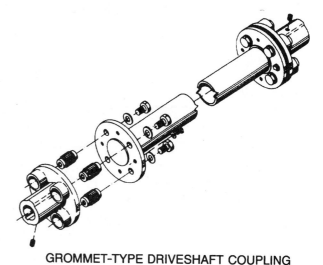

GROMMET-TYPE DRIVESHAFT COUPLING

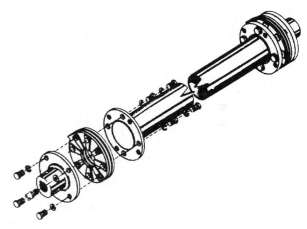

DISC-TYPE DRIVESHAFT COUPLING

Figure 30-10. Coupling and shaft for gear reduction box. *Courtesy Marley Cooling Tower Company.*

may need to be lowered to the ground using a hoist or pulley system. Any stairway or ladder installed for service must meet local codes for safety to include proper handrails and barriers to prevent anyone from stepping over the edge.

Warning: Be careful of your footing and balance when climbing up to or down from a cooling tower location. Do not move in a way that will cause a loss of balance when working around a tower at a high elevation.

Caution: When lifting a motor or pump, lift with your back straight. Wear a back belt support if recommended by your employer or insurance carrier.

30.5 FAN SECTION

In all forced-draft towers, the motor must have some method of turning the fan. There are two different types of drives used in tower fan applications: the belt-drive and the gear box (transmission) drive. Belt-drive fans are normally used in smaller towers and include an adjustable motor mount. This mechanism will incur the greatest wear of any component in the cooling tower and will require the most maintenance, so it should be located where service can be performed with as little effort as possible.

Fans for larger towers have the motor mounted out to the side and a gear box or transmission is used to change the direction of the motor drive shaft by 90°. These units may also be designed to change the motor-to-fan shaft speed relationship. Typical motors will turn at 1800 rpm or 3600 rpm and the speed of the fan shaft will turn considerably slower, depending on the gear reduction.

The motor is connected to the gear box using a coupling and shaft (Figure 30-10). The motor, gear box, and fan bearings must be accessible for service purposes.

The propeller fan blades are enclosed in a fan housing that improves the efficiency of the fan blades. This fan blade location is critical in most towers. For best performance, the fan must be located at the correct distance from the top and sides of the tower.

30.6 WATER CONTROL AND DISTRIBUTION

Various devices and methods are used to control and distribute the water in a cooling tower. These include the use of tower sumps, makeup water, blowdown, and distribution pans.

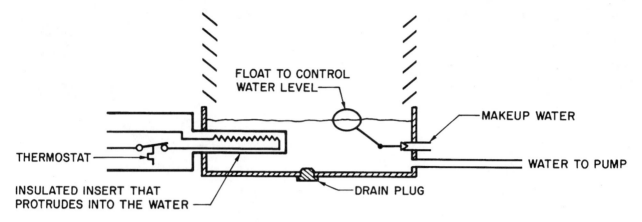

Figure 30-11. Freeze protection for cooling tower sump.

Tower Sumps

All cooling towers must have some sort of sump to collect the water. The sump on small cooling towers may consist of a metal pan that drains to a lower point to gather the water. When this sump contains water and is located outside, it must have some means to prevent freezing such as a thermostatically-operated sump heater (Figure 30-11). These sump heaters are used only in small installations. Larger installations require more heat and may have a circulating hot water coil or a method of using low-pressure steam to heat the tower basin water in cold weather.

Many cooling towers use underground sumps made of concrete. These sumps will not freeze in southern climates but must be protected from the cold in northern climates unless the refrigeration system is operating and adding heat to the sump. Some sumps are located in the heated space of a building to prevent freezing.

Wherever the sump is located, it is the collection point for all debris in the system and must be accessible for cleaning. Because of this, a bypass filter system is often used to sweep the bottom of the sump and carry a portion of the water through the filter system for cleaning. The sump usually contains a coarse screen to protect the pump (Figure 30-12).

Makeup Water

Since the cooling tower operates using the principle of evaporation, water is continually being removed from the system. There are several devices that can be used to replace this water, with the float valve being one of the most common. When the water level drops, the float ball drops with it and opens the valve to the makeup water supply. Usually, this makeup water is from the normal municipal supply or other similar water supply. Another method is to use float switches that fall with the water

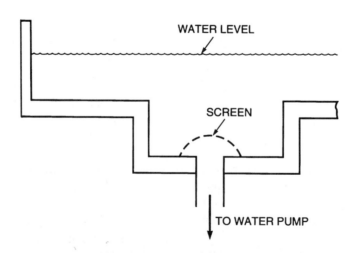

Figure 30-12. Filter system for cooling tower sump.

level and energize a solenoid valve to allow water to fill the sump. When the water reaches the proper level, the float rises and shuts off the solenoid valve. Still another method uses electrodes protruding into the water that sense the water level. When the electrodes sense that the water level is too low, a solenoid valve is opened and makeup water is allowed to enter. As the water level rises, an upper electrode senses the water level and shuts off the water. Figure 30-13 shows all of the above control methods.

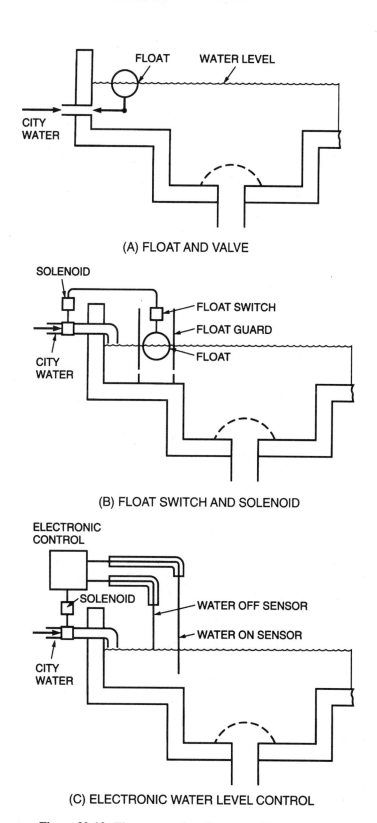

Figure 30-13. Three types of cooling tower fill methods.

Blowdown or Bleed Water

The water that evaporates leaves behind any solid materials that may have been in the water. These include dust particles, minerals, and algae, a low form of plant life. As the water evaporates, the water that is left behind becomes more and more concentrated with these particles. If this continues, the particles will drop out of the water and deposit on the surfaces of the cooling tower. These deposits are very hard to remove as they turn into a substance that is similar to cement. Even more important than what is happening to the cooling tower is what will happen to the condenser. Some of these particles will deposit on the hottest part of the condenser (the outlet). These deposits will act as an insulator to the heat exchange process. This will cause the head pressure to rise, and the condenser approach temperature will begin to spread. As mentioned in the previous unit, the original operating log will give the technician a clue as to when this is happening.

Blowdown or bleed water will solve some of these problems. It is merely a bleedoff of a portion of the water that is being circulated. When new water is added, less sediment is present as some has been removed with the water that was bled off. It is generally recognized that three gallons per minute per ton is the amount of water circulated in a water-cooled condenser used for air conditioning with a 10°F temperature rise through the condenser. Figure 30-14 shows a 30 ton application and what would happen if the cooling tower water did not have blowdown as part of the piping system. Figure 30-15 shows four common methods that are used for obtaining the correct blowdown. Often, building management personnel have a difficult time understanding the purpose of blowdown. They don't understand why water is being allowed to run down the drain and must be educated as to the purpose of blowdown. Proper blowdown is essential, and it must be managed correctly because expensive water treatment chemicals are released down the drain along with the water.

Distribution Pans

Even distribution of the water through the tower is vital to correct system operation. If a tower has two cells where the water is distributed, the same amount of water must be fed to each cell or the tower will not perform as designed. Many towers use a distribution pan at the top to distribute the water over the fill deck. These pans often have a series of calibrated holes drilled into the pan (Figure 30-16).

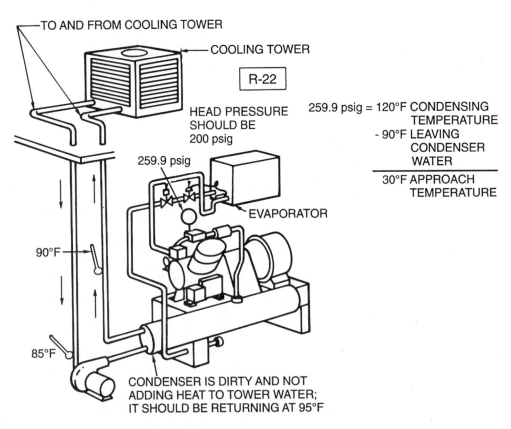

Figure 30-14. Flow rate through a cooling tower that does not have a blowdown system.

WATER CONTROL AND DISTRIBUTION 563

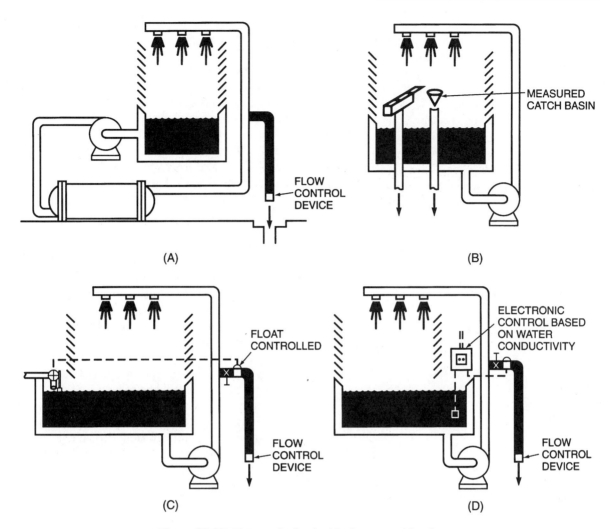

Figure 30-15. Four methods of achieving proper blowdown.

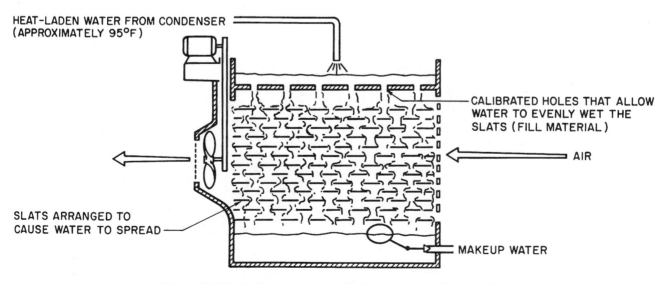

Figure 30-16. Cooling tower water distribution using calibrated holes.

Warm return water from the condenser is poured into the pan by the pump discharge. This discharge may be split to each side of the tower. A balancing valve or valves must be calibrated to obtain the correct flow (Figure 30-17). The calibrated holes in the top of the tower must be clean and correctly sized. Figure 30-18 shows a pan in which the holes are rusted out and most of the water is flowing to one side of the tower, so it is not wetting the entire fill deck. The tower will not perform correctly when this happens. For example, the tower may have a design water temperature of 85°F and it may be returning water at 90°F. This will result in high head pressures and excess operating costs.

30.7 WATER PUMPS USED IN COOLING TOWERS

The condenser water pump is the device that moves the water through the condenser and cooling tower circuit. The cooling tower pump is normally located where it takes water from the cooling tower sump into the pump inlet (Figure 30-19). The water pump is the heart of the cooling tower system because it pumps the heat-laden water. This water must be pumped at the correct rate and delivered to the cooling tower at the correct pressure.

The condenser water pump is normally a centrifugal pump. It uses centrifugal action to impart velocity to the water which is then converted into pressure. Pressure may be expressed in pounds per square inch gage (psig) or feet of head. One foot of head of water is equal to 0.433 psig; in other words, a column of water that is one foot high will cause a pressure gage to read 0.433 psig. A column of water that is 2.31 feet high (27.7 inches) will cause a pressure gage to read 1 psig (Figure 30-20). Pump capacities are discussed in feet of head. Centrifugal pump action is discussed in more detail in Unit 31; a study of that unit should be of help.

There are several types of condenser water pumps used in large condenser water systems. The close-coupled pump has the pump located very close to the motor, with the pump impeller actually mounted on the end of the motor shaft (Figure 30-21). These pumps are used in smaller applications. Note that all of the water enters the side of the pump housing. The pump shaft must have a seal to prevent water from leaking to the atmosphere or to prevent the atmosphere from leaking into the system should this portion of the piping be in a vacuum.

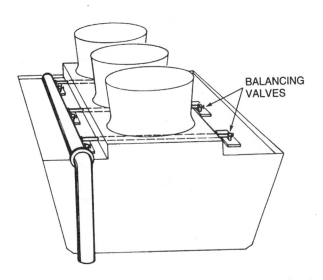

Figure 30-17. Balancing valves for balancing the water flow in the cooling tower. *Courtesy Marley Cooling Tower Company.*

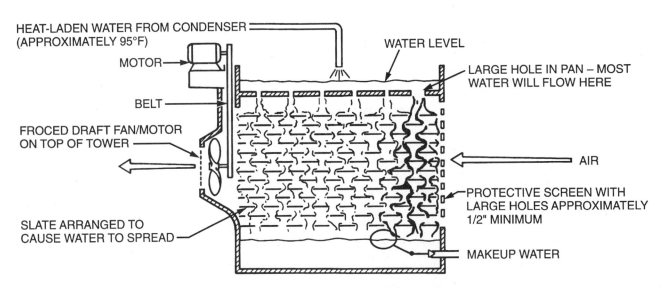

Figure 30-18. One of the calibrated holes used to distribute water in the tower is enlarged due to rust. Most of the water is running down the right side of the tower.

WATER PUMPS USED IN COOLING TOWERS 565

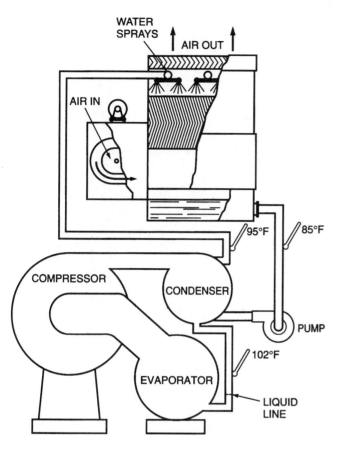

Figure 30-19. Purpose of the pump in the cooling tower system.

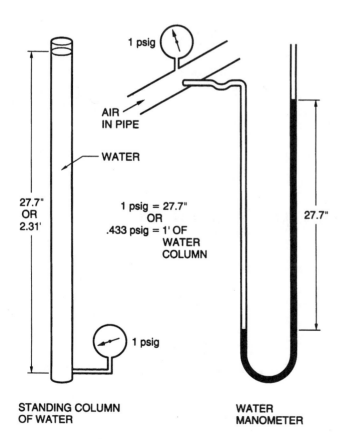

Figure 30-20. Standing column of water.

Figure 30-21. Close-coupled pump assembly. *Courtesy ITT Bell and Gossett.*

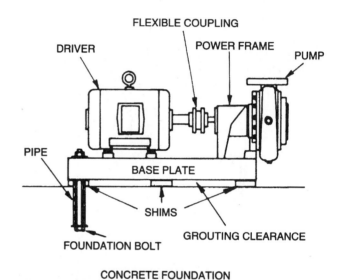

Figure 30-22. Flexible coupling between the pump and motor. *Courtesy Amtrol, Inc.*

The base-mounted pump comes as an assembly with the pump mounted at the end of the pump shaft and a flexible coupling between the driver (motor) and the pump (Figure 30-22). This pump may have a single-sided or double-sided impeller. The base is usually made of steel or cast iron and holds the pump and motor steady during operation. The pump base is usually fastened to the floor using cement as a filler in the open portion of the pump base. This is called *grouting*. The purpose is to make a firmer foundation that will last for many years. The pump and motor can still be removed from the steel or cast iron base. One of the advantages of this pump is that the motor and pump shafts are aligned at the factory and alignment is not required in the field. These pumps are used in larger applications.

As pumps become larger, the need for different designs becomes apparent. Some pumps have single-inlet impellers with all of the water entering the pump impeller on one side (Figure 30-23). The double-inlet impeller allows water to flow into both sides (Figure 30-24).

566 UNIT 30 COOLING TOWERS AND PUMPS

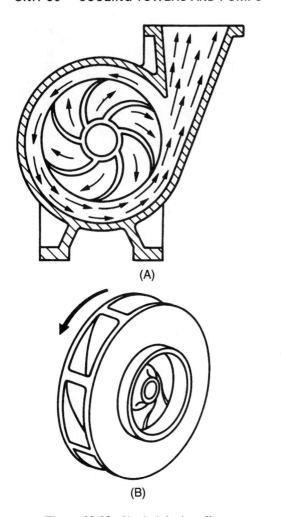

Figure 30-23. Single-inlet impeller.

Figure 30-25. Horizontal split-case pump. *Courtesy AC Pump ITT.*

The pump housing must be designed differently to allow water to enter this pump. The housing at the inlet must be split so water can flow into both sides of the impeller. There are two styles of double-inlet impeller pumps. With one type, the pump is disassembled from the ends of the pump as with the base-mounted pump, and the other is a split-case pump where the pump is disassembled by removing the top of the pump (Figure 30-25).

These pumps must also have a shaft seal. There are two kinds of seals that may be used for these pumps: one is the stuffing box seal and the other is a mechanical-type seal. The stuffing box seal is a packing gland seal that must be hand tightened using a wrench. When the nut on the seal is tightened, it squeezes a packing around the shaft to minimize leakage (Figure 30-26). Stuffing box seals are typically used for pump pressures up to about 150 psig. These seals require regular maintenance.

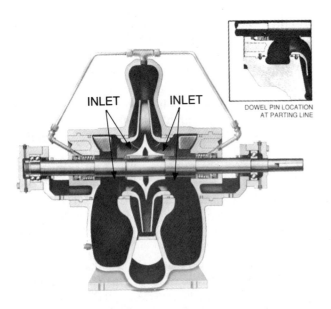

Figure 30-24. Double-inlet impeller pump. *Courtesy AC Pump ITT.*

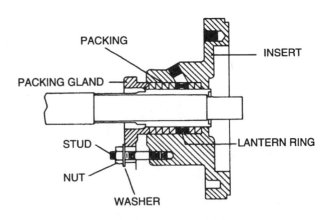

Figure 30-26. Stuffing box shaft seal.

WATER PUMPS USED IN COOLING TOWERS 567

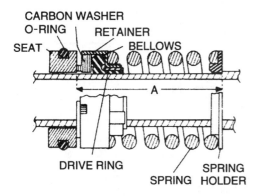

Figure 30-27. Mechanical shaft seal. *Courtesy Amtrol, Inc.*

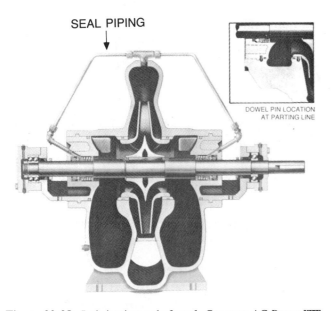

Figure 30-28. Lubricating a shaft seal. *Courtesy AC Pump ITT.*

The mechanical-type seal may be used for higher pressures (up to about 300 psig) and usually have a carbon washer mounted on the end of a bellows that turns with the shaft (Figure 30-27). The bellows is sealed to the shaft using an O-ring seal. The carbon washer rubs against a stationary ceramic drive ring while the shaft turns and seals water in the pump housing.

Both types of seals are lubricated using the circulating water. Split-case pumps may have a special piping system that injects pump discharge water into the seal (Figure 30-28).

Pumps may also be manufactured in a vertical configuration with the motor on top. These pumps are mounted on top of the water sump and protrude down into the sump for pump pickup (Figure 30-29).

Materials Used in Pump Construction

Most pumps for cooling towers are manufactured of cast iron. They are very heavy and may be used for pressures up to 300 psig with the correct piping flanges and arrangements. Cast iron has the capacity to last many years without deterioration due to rust or corrosion.

The impeller in the typical centrifugal pump is made of brass and is often mounted on a stainless steel shaft. This can be very important when the time comes to remove the impeller from the shaft. This makes it easier to remove after many years of operation. Impellers are supplied from the factory at a particular diameter for each pump in order to furnish a specific water flow against a known pressure drop. Often, the pump may furnish too much water flow to the point that a smaller impeller is required. These impellers are often adjusted in the field to new specifications for the actual water flow and pressure requirements of the system. The manufacturer can be contacted when it is necessary to make any modifications to a specific pump.

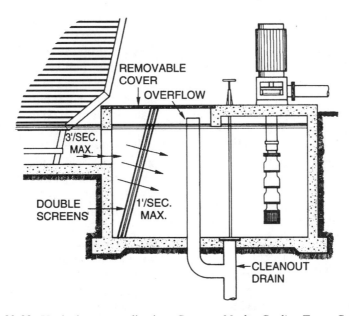

Figure 30-29. Vertical pump application. *Courtesy Marley Cooling Tower Company.*

Pump Location

The location of the condenser water pump is very important because these pumps are normally centrifugal pumps and they must be furnished with water in such a manner that the eye of the impeller is under water during start-up. Otherwise, the pump will not move any water. The pump should be located below the tower water with nothing in the pump inlet piping that will impede flow. A free flow of water to the centrifugal pump inlet is necessary. There are several things that can affect the water flow to the pump inlet, including sump location, vortexing in the tower, clogged screens, and the cooling tower bypass.

Sump Location There are times when the sump to the cooling tower is located below the pump. When this is the case, special care should be taken during start-up because only air will be in the pump casing. The pump should never be started with only air inside the casing because the pump seal will have no water for lubrication. When the sump is below the pump, a foot or check valve must be located in the line to prevent the water from flowing back into the sump at shutdown (Figure 30-30). Water must be added to the pump inlet side until the pump casing is completely full before the pump is started. This can be done by connecting a water hose to the pump inlet side and filling it until water escapes the bleed port on top of the pump. If the pump inlet cannot be filled, the foot or check valve is leaking and must be repaired.

Vortexing Vortexing in the tower is actually a whirlpool action. This vortex interferes with pump operation because it introduces air to the pump inlet. The pump is designed to pump only water, and air will cause problems with not only the pump, but also with the condenser at the chiller. Poor tower construction or piping practices may be the cause of vortexing. The sump may not be deep enough or an anti-vortexing design may not be in place. Vortexing can often be eliminated by means

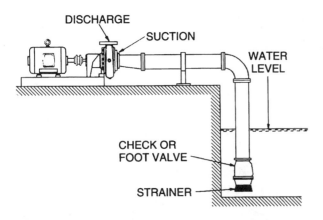

Figure 30-30. The cooling tower sump is located below the pump.

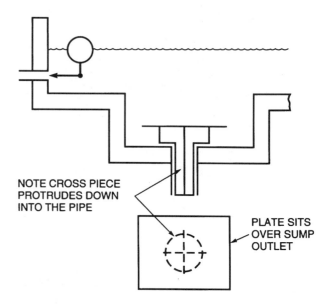

Figure 30-31. Vortex protection for a cooling tower.

of a device placed in the sump outlet that breaks up the vortex. This may be a cross-type configuration with a plate on top. The cross configuration creates four outlets from the sump and the plate causes the water to be pulled from a further distance and from the sides (Figure 30-31). Vortexing in multi-story buildings with the condenser water pump well below the tower has been cured using the above method.

Clogged Screens There must be some sort of strainer located between the tower water and the pump inlet to prevent debris from the tower from entering the pumping circuit. Typically, there is a screen-type strainer in the cooling tower exit to the sump. This may be a coarse screen. Often, a fine mesh screen is located before the pump inlet. If this mesh screen is too fine, problems from pressure drop will occur as the screen becomes restricted. Pressure drop causes the pump inlet to operate in a vacuum. It may pull in air while operating in a vacuum if the pump seals leak. Pressure drop can also cause pump cavitation (water turning to a vapor) if the cooling tower water is warm enough and the pressure is low enough. Cavitation or air at the pump inlet will be evident from noise at the pump. It will often sound like it is pumping small rocks when this occurs. If a fine screen must be used, it is better to locate it at the pump outlet (Figure 30-32). It does less harm to throttle the pump discharge with pressure drop than the pump inlet.

Bypass Valves The tower bypass valve helps to maintain the correct tower water temperature at the condenser during start-up and at low ambient conditions. The tower bypass circuit allows water from the pump outlet to be recirculated to the pump inlet so that the cold cooling tower water will not reach the condenser. In compression

WATER PUMPS USED IN COOLING TOWERS 569

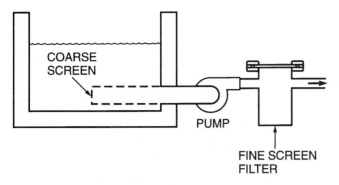

Figure 30-32. The best location for a fine screen filter is in the pump outlet.

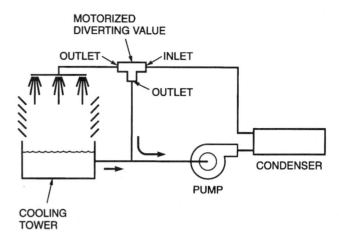

Figure 30-34. Diverting valve application.

cycle systems, condenser water that is too cold will reduce the head pressure to the point where the condenser will become so efficient that too much refrigerant will be held in the condenser and it will starve the evaporator. Cooling tower water that is too cold for an absorption machine will often cause the salt solution to crystallize. There are two types of three-way bypass valves that may be used for tower bypass. These are the mixing valve and the diverting valve. The mixing valve has two inlets and one outlet and needs to be located in the pump suction line between the tower and the pump inlet (Figure 30-33). This valve can cause a pressure drop in the pump suction line that should be avoided if possible. The diverting valve has one inlet and two outlets. It can be located in the pump discharge and piped to the cooling tower or the pump inlet (Figure 30-34). These valves do not come in large sizes (they are normally a maximum of 4 inches) so other arrangements may need to be made if a larger size is required. A straight-through valve can be used in the configuration shown in Figure 30-35. Both the diverting valve and the straight-through valve allow some cooling tower water from the basin to be introduced to the pump inlet but may cause problems at start-up until the tower water reaches the correct temperature.

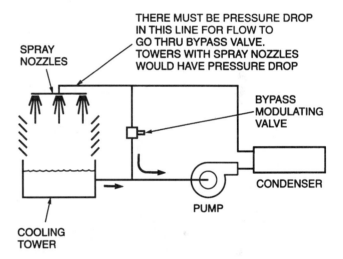

Figure 30-35. Straight-through valve for tower bypass.

All pumps must have bearings that support the turning shaft while under load. Sleeve bearings are used for small pumps, and ball or roller bearings are used for larger pumps. These bearings must be lubricated on a regular basis to provide satisfactory service. The bearings are located outside the pump on a split-case pump (Figure 30-36).

Alignment

For most applications, the pump must be fastened to the motor shaft in such a manner that it is within specified alignment tolerances. A flexible coupling may be installed between the pump and motor shaft to adjust for minor misalignment (Figure 30-37). Since this coupling can only handle slight misalignment, the shafts must be aligned carefully.

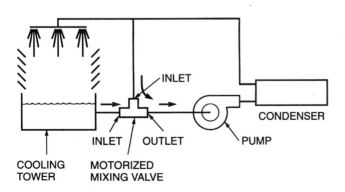

Figure 30-33. Mixing valve application.

570 UNIT 30 COOLING TOWERS AND PUMPS

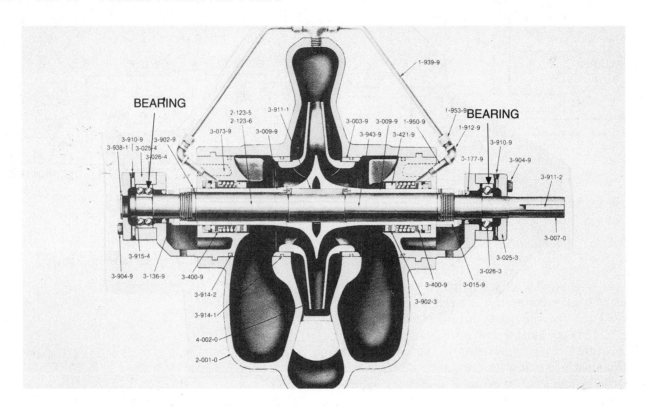

Figure 30-36. Bearings on a horizontal split-case pump. *Courtesy AC Pump ITT.*

Figure 30-37. Flexible coupling. *Courtesy TB Woods & Sons.*

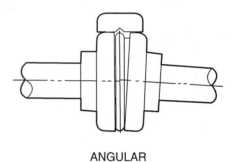

Figure 30-38. Angular coupling misalignment. *Courtesy Amtrol, Inc.*

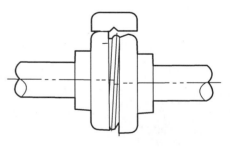

Figure 30-39. Parallel coupling misalignment. *Courtesy Amtrol, Inc.*

There are two planes of alignment that must be considered: angular and parallel. Angular alignment ensures that both shafts are at the same angle with each other (Figure 30-38). Parallel alignment ensures that the shafts are aligned end to end (Figure 30-39). Correct alignment is accomplished using a dial indicator that reads to $1/1000$ of an inch. It is mounted on the shaft and the shafts are rotated together through the full rotation (Figure 30-40). Shims (very thin sheets of steel) are placed under the pump and motor until both angular and parallel alignment is achieved to within the manufacturer's specifications. After alignment is attained, the pump and motor are tightened down and alignment is checked again. Then the base of both are drilled and tapered dowel pins are driven in the base (Figure 30-41). When done correctly, the motor and pump should provide years of service without requiring realignment.

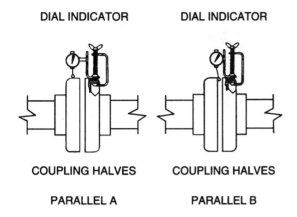

Figure 30-40. Dial indicators used to align couplings and shafts. *Courtesy Amtrol, Inc.*

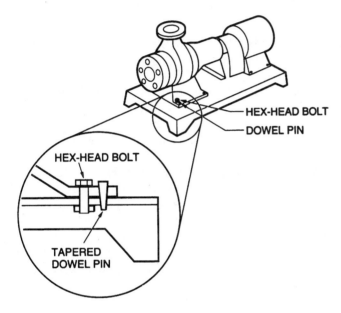

Figure 30-41. Pump and motor fastened to base using tapered dowel pins.

SUMMARY

- The cooling tower rejects heat that has been absorbed into the chilled-water system.
- Cooling towers lower the temperature of the water in a tower by means of evaporation.
- There are both natural-draft and forced-draft cooling towers.
- Cooling towers should be protected from fire. They may be manufactured from fireproof materials or may use a wetting system. All applicable codes and insurance requirements should be followed.
- Either the splash method or the film or wetted surface method may be used to spread the water out to allow additional exposure to the air to provide adequate evaporation.
- The design of a cooling tower provides for distinct airflow patterns.
- Fan drives for forced-draft towers may be either belt-drive or gear-drive units.
- The cooling tower sump needs to be flushed periodically.
- Sumps may have heaters to protect them from freezing.
- Cooling towers must have makeup water systems.
- The water flow through the tower must follow the pattern designed by the manufacturer.
- The condenser water pump is normally a centrifugal pump.
- Larger pumps may have single-inlet or double-inlet impellers.
- Most pumps for cooling towers are manufactured of cast iron.
- Vortexing (whirlpooling) may occur in the sump. This is not desirable as it may introduce air into the pump. Anti-vortexing designs or devices can be used to eliminate this problem.
- A tower bypass valve helps to maintain the correct tower water temperature at the condenser.

REVIEW QUESTIONS

1. What is the purpose of the cooling tower in a chilled-water system?
2. What is the typical temperature difference between the water at the outlet of the cooling tower and the outside air wet bulb temperature?
3. What is the purpose of spreading the water out as it moves through the cooling tower?
4. Name two types of water cooling towers?
5. What two types of fans are used in water towers?
6. Why must a cooling tower have specific fire protection or be constructed of fireproof or fire-retardant materials?
7. What two methods may be used to spread the water out and/or slow it down as it moves through the tower?
8. What are the two types of airflow patterns in forced-draft cooling towers?
9. What is the purpose of the makeup water system in cooling towers?
10. What is meant by the term *blowdown*?
11. What type of pump is normally used for the condenser water and the cooling tower water?
12. What is meant by the term *vortexing*?
13. What is the purpose of the tower bypass valve?
14. What will happen if the condenser water is too cold in a compression system chiller?
15. Describe the two planes of pump motor alignment.

31 CHILLER OPERATION, MAINTENANCE, AND TROUBLESHOOTING

OBJECTIVES

Upon completion of this unit, you should be able to

- Discuss the general start-up procedures for a chilled-water air conditioning system.
- Describe specific start-up procedures for chilled-water systems using a scroll, reciprocating, screw, and centrifugal compressors.
- Describe operating and monitoring procedures for scroll and reciprocating chilled-water systems.
- List preventive and periodic maintenance procedures that should be performed on water-cooled chiller systems.
- Describe the general start-up procedures for an absorption system.
- List preventive and periodic maintenance procedures for an absorption chiller, including those required for the purge system.
- Wear gloves and use caution when working around hot steam pipes and other heated components.
- Use caution when working around high-pressure systems. Do not attempt to tighten or loosen fittings or connections when a system is pressurized. Follow recommended procedures when using nitrogen.
- Never start a motor with the door to the starter components open.
- Check pressures regularly as indicated by the manufacturer of the system to avoid excessive pressures.

31.1 CHILLER SYSTEMS

Proper operation of chillers involves starting, running, and stopping the system in an orderly manner. You must become familiar with the chiller system before attempting to operate or monitoring it. Most chiller systems are almost fool-proof, but there is no need to take chances with this very expensive equipment. There should be product data at the site for any major equipment installation. Study the manufacturer's literature. It is good practice to become familiar with one piece of equipment at a time. Many times, the job literature is taken home for future study and is never returned. Make it a practice to never let it leave the job. Manufacturers may be called upon to furnish product data, if needed, but it may not be available in older systems.

Chiller systems encompass both compression cycle chilled-water systems and absorption chillers. These systems have some similarities, but also have enough differences that they will be discussed separately in this text.

31.2 START-UP PROCEDURES FOR CHILLED-WATER SYSTEMS

The first step in starting up a chilled-water system is to establish chilled-water flow. This is true whether it is for the first time each season or the first time each day. The next step for a water-cooled chiller is to establish condenser water flow. Air-cooled chillers have fans at the condensers and do not need the same amount of attention as water-cooled chillers. Both the chilled-water circuit and the condenser water circuit will have some means, such as a flow switch, of verifying water flow. The flow switch is often a paddle that protrudes into the water and when water passes, it moves and operates a switch. Some chiller manufacturers use pressure drop controls to establish water flow. If this is the case, these are built into the water circuit by the manufacturer. You should be aware of the check points to make sure these switches are functioning. Since these paddles are in the water stream, it is not unusual for them to break off. If so, water flow may be established and not be proven by the flow switch. Water flow may be verified using pressure gages. The starting point for verifying correct water flow through a heat exchanger is knowing the expected pressure drop.

There should be pressure test points at the inlet and outlet of any water heat exchanger, chiller, or condenser. Figure 31-1(A) shows two gages being used to check the pressures. A better arrangement is to use one gage with two pressure connections, as shown in Figure 31-1(B). If two gages are used and they are at different heights, an error is automatically built in. For example, if one gage is 2.31 feet lower than the other, there will be an error of 1 psig because of the difference in height. A standing column of water that is 2.31 feet high will have a pressure of 1 psig at the bottom (Figure 31-2). Also, if the gages have a built-in error because they are not calibrated, they cannot be expected to provide correct readings.

START-UP PROCEDURES FOR CHILLED-WATER SYSTEMS 573

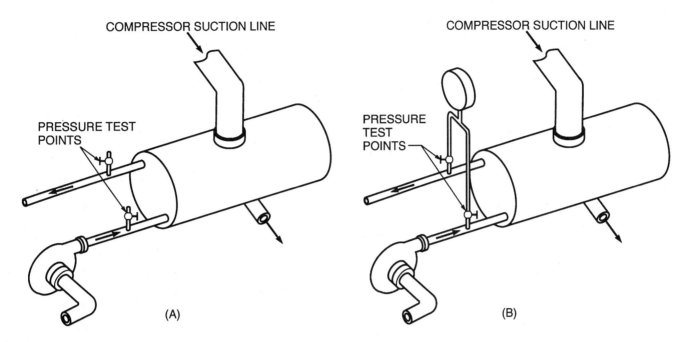

Figure 31-1. (A) Pressure test points at the inlet and outlet of a heat exchanger. (B) Using one gage and two ports for checking the pressure drop through a heat exchanger.

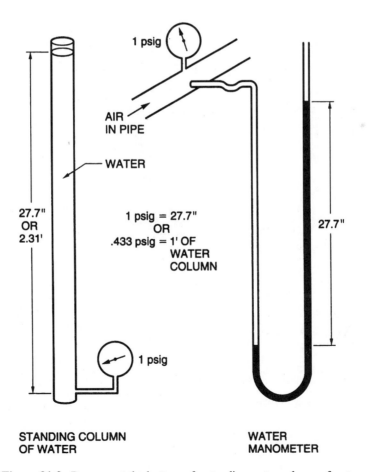

Figure 31-2. Pressure at the bottom of a standing water column of water.

574 UNIT 31 CHILLER OPERATION, MAINTENANCE, AND TROUBLESHOOTING

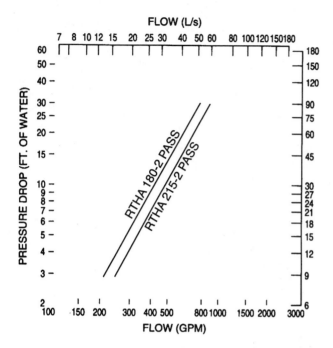

Figure 31-3. Pressure drop chart for a heat exchanger. *Courtesy Trane Company.*

For example, if one gage reads 1 psig when open to the atmosphere while the other gage reads 0 psig, there is a built-in error of 1 psig in the pressure difference. Actually, all you want to know is the pressure difference at the two points. Using one gage that is not accurate at 0 psig may still yield the correct pressure difference.

What you need to know from the pressure readings is that there is a pressure drop across the heat exchanger. As water flows through the heat exchanger, the pressure drops a specific amount. The heat exchanger is a calibrated, pressure-drop monitoring device that may be used to determine water flow. A pressure drop chart for the heat exchanger can be used to determine the gallons per minute (GPM) of water flow (Figure 31-3). The original operating log for the installation will show the initial pressure drops at start-up. This baseline information is helpful for future system servicing. If the original log cannot be found, the manufacturer may be contacted for a copy of the log sheet or a pressure drop chart for the chiller.

You should also be aware that the interlock circuit through the contactors must be satisfied before start-up. The starting sequence for most chiller systems is to start the chilled-water pump first. A set of auxiliary contacts in the chilled-water starter then starts the condenser water pump. When the condenser water pump starts, another set of auxiliary contacts make and pass power to the cooling tower fan and the chiller circuit (Figure 31-4).

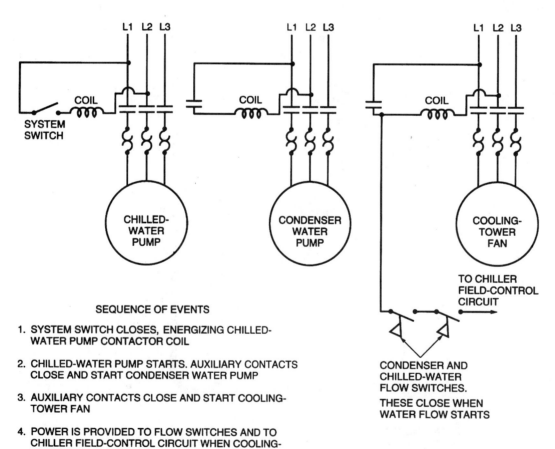

Figure 31-4. Control circuit for a typical chiller.

When the signal is received at the chiller control circuit, the compressor should start. This is often called the *field control circuit* because it is the circuit furnished by the contractor that connects to the manufacturer's control circuit. If a chiller does not start, the first thing to do is to check the field circuit to see if it is satisfied. You should know where the field circuit is wired into the chiller circuit for each chiller in the system (some large systems contain several chillers). A quick check at this point will tell whether there is a field circuit problem (flow switches, pump interlock circuit, outdoor thermostat and often, the main controller) or a chiller problem. If there is no field control circuit power, there is no need to check the chiller until there is power. Many chillers have a ready light that indicates when the field control circuit is energized.

Different chillers have different starting sequences, usually depending on the type of compressor lubrication. When the chiller uses a positive displacement compressor (scroll, reciprocating, or screw), the compressor will start soon after the field control circuit is satisfied. Some equipment has a time delay before the compressor starts after the field circuit is satisfied. Look for a time delay circuit and make a note of the time delay for each particular system. You may think there is a problem only to have the chiller start unexpectedly after a planned time delay. Reciprocating, scroll, and screw compressors are lubricated from within and do not have a separate oil pump that must be started first. Both reciprocating and screw compressors will start unloaded and will begin to load when oil pressure is developed. Scroll compressors do not use compressor unloading and will start up under full load.

Scroll Chillers

Scroll chillers can be either air-cooled or water-cooled systems. Regardless of the type, before attempting to start the chiller, make sure the water for the chilled-water circuit is at the correct level. The system must be full. Then operate the chilled-water pump and verify the water flow using the pressure drop across the chiller. Make sure that all valves in the refrigerant circuit are in the back-seated position. Visually check the system for leaks. Oil on the external portion of a fitting or valve is a sure sign of a leak as oil is entrained in the refrigerant. The compressor may have crankcase heat and it must be energized for the length of time recommended by the manufacturer, usually 24 hours.

Caution: Do not start the compressor without adequate crankcase heat (when furnished) or compressor damage may occur.

Air-cooled chillers are located outside. The chiller barrel will have a heater to prevent it from freezing in cold weather. Be sure this heater is wired correctly and is operable. If you don't do this at start-up, you may forget it when winter arrives. This heater strip is thermostatically controlled and will shut off automatically.

With an air-cooled unit, make sure that the condenser fans are free to turn.

With a water-cooled unit, the water-cooled condenser portion must be in operation. The cooling tower must be full of fresh water and the condenser water pump must be started. Make sure that water flow is established. You can look at the tower and verify this or you can check the pressure drop across the condenser.

Check to see that the field control circuit is calling for cooling. If all of the above requirements are met, the chiller should be ready to start.

When the chiller is started, observe the following:

1. Suction pressure
2. Discharge pressure
3. Check the compressor for liquid floodback
4. Check the entering and leaving water temperatures on the chiller and the condenser entering and leaving water temperatures for water-cooled condensers

When the chiller is operating normally, it can usually be left unattended except for the water tower used with water-cooled systems. The water tower requires regular observation and maintenance.

Reciprocating Chillers

The reciprocating chiller may be either an air-cooled or water-cooled system. Air-cooled chillers are located outside and a heater strip is applied to the chiller barrel to prevent freezing in cold weather. The heater strip is thermostatically controlled and only operates when needed.

Before starting a reciprocating chiller, make sure the oil level in the sightglass is correct, usually from $1/4$ to $1/2$ level in the sightglass. Reciprocating chillers will have crankcase heaters to prevent refrigerant migration to the crankcase during the off cycle. When the crankcase heater is on, the compressor will be warm to the touch. The crankcase heater must be energized for the prescribed length of time before start-up to prevent damage to the compressor. Refrigerant in the oil will dilute the lubrication quality of the oil and can easily cause bearing damage may does not show up immediately. Generally, the crankcase heater is energized for 24 hours prior to start-up unless the manufacturer states otherwise. It is poor practice to turn the crankcase heater off, even during winter shutdown, as refrigerant will migrate to the crankcase and it will be very difficult to boil it out to the extent that the oil returns to the correct consistency for proper lubrication.

When the compressor is started, the pressures can be observed using panel gages, if provided. Typically, chillers over 25 tons have gages permanently installed by the manufacturer. These gages may be isolated by means of a service valve to prevent damage to the gages during long periods of running time so these valves may need to be opened to read the gages. It is poor practice to leave the gage valves open all the time because the gage

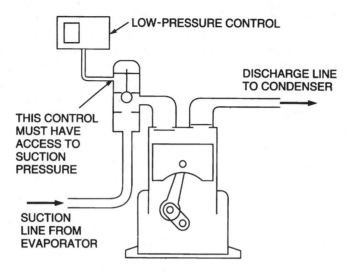

Figure 31-5. Service valve positions.

mechanism will experience considerable wear during normal operation from reciprocating compressor gas pulsations. These gages are for intermittent testing purposes. You should check these gages for accuracy from time to time as they cannot be relied upon during the entire life of the chiller.

If a chiller has been secured for winter, the shutdown technician may have pumped the refrigerant into the receiver and the valves may need to be repositioned to the running position, which is back-seated for any valve that does not have a control operating off the back-seat. Figure 31-5 shows an example of a valve where the low-pressure control is fastened to the gage port on the back seat of the valve. If the valve is back seated, the low pressure control will not be functional because it is isolated.

When the compressor is started, it will start up unloaded until the oil pressure builds up and loads the compressor. An ammeter can be applied to the motor leads to determine the load level on the compressor. When it is fully loaded, the amperage should be very close to the expected full load amperage.

As the system temperature begins to pull down, you may notice that the oil level in the compressor rises and foams. This is normal for many systems, but the oil level should stabilize after a short period of operation (about 15 minutes). This is a suction-cooled compressor, so you should verify that liquid refrigerant is not flooding back into the compressor during the running cycle. This can be determined by feeling the compressor housing. The compressor motor should be cold where the suction line enters, then it should gradually become warm where the motor housing reaches the compressor housing. The compressor crankcase should never be cool to the touch after 30 minutes of running time. It it is, liquid refrigerant is probably flooding back to the compressor (Figure 31-6).

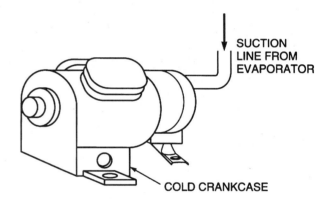

Figure 31-6. This compressor crankcase is cold because of liquid refrigerant returning to the compressor.

Screw Chillers

The rotary (screw) chiller may be either an air-cooled or water-cooled system. Air-cooled systems are located outdoors and the heat for the chiller barrel should be verified. Again, it is thermostatically controlled and will only operate when needed.

The compressor must have crankcase heat before start-up. Again, most manufacturers require that this heat be energized for 24 hours before start-up. Start-up without crankcase heat can cause compressor damage.

The water flow must be verified through the chiller and the condenser in water-cooled systems. The best way to verify this is to use pressure gages.

All valves must be in the correct position before start-up. Check the manufacturer's literature for valve positions. If the chiller was pumped down for winter, the correct procedures must be followed to let the refrigerant back into the evaporator before start-up.

Check the field control circuit to make sure it is calling for cooling.

When all of the above checks are completed, the chiller is ready to start.

Start the compressor and watch for it to load up and start cooling the water. Observe the following:

1. Suction pressure
2. Discharge pressure
3. Entering and leaving water temperatures (both chiller and condenser for water-cooled systems)
4. Look for liquid floodback to the compressor

When the chiller is operating normally, it can usually be left unattended except for the water tower in water-cooled systems. The water tower requires regular observation and maintenance.

Centrifugal Chillers

Centrifugal chillers often need to be watched for the first few minutes of operation because they have a separate oil sump system. The oil sump should also be checked before start-up. After start-up, observe the following:

1. Correct oil sump temperature (from 135°F to 165°F, depending on the manufacturer).

Warning: Do not attempt to start a compressor unless the oil sump temperature is within the range recommended by the compressor manufacturer.

After the machine has been operating for some time, check the bearing oil temperature to be sure that it is not overheating. If overheating occurs, check the oil cooling medium (usually water).

2. Correct oil level in the oil sump. If the oil level is above the glass, it may be full of liquid refrigerant. Ensure that it is at the correct temperature. You may start the oil pump in manual mode and observe what happens. If in doubt, call the manufacturer for recommendations. Unless you have experience with this particular problem, do not try to start the compressor, as marginal oil pressure may cause bearing damage.
3. Start the oil pump manually and verify the oil pressure before starting the compressor, then switch it to automatic operation before starting the compressor.

When the chiller is started, the compressor oil pump will start and build oil pressure first. When satisfactory oil pressure is established, the compressor will start and run up to speed then switch over to the run winding configuration (either autotransformer or wye delta). The centrifugal compressor starts up unloaded and only begins to load after the motor is up to speed. This is accomplished by the prerotation guide vanes mentioned earlier. When the chiller is up to speed, the prerotation vanes begin to open and will open to full load unless the machine demand limit control stops the vanes at a lower percent of full load.

When the chiller compressor begins to pull full or part load amperage, observe the pressure gages. Look for the following problems:

1. The suction pressure should be correct. Mark on the gage front or a note pad the expected operating suction pressure for normal operation, which would be at design water temperature (typically, 45°F leaving water).
2. The net oil pressure should be correct. This should be noted nearby for ready reference. If the machine has been off for a long period of time, observe the oil level in the oil sump as the compressor starts to accept the load, and watch for oil foaming. If the oil starts to foam, reduce the load on the compressor. This will raise the oil sump pressure and boil the refrigerant out of the oil at a slower rate. If you do not reduce the load, it is likely that the oil pressure will start to drop and the machine will shut off because of low oil pressure. If the compressor has an anti-recycle timer, you will not be able to restart the compressor until the timer completes its cycle and that may be 30 minutes. It is better to avoid allowing the compressor to shut off by watching the oil pressure.
3. The discharge pressure should be correct. This should be noted nearby for ready reference or marked on the pressure gage for typical operating conditions.
4. The compressor amperage should be correct and marked on the ammeter on the starter. The ammeter should be marked for all operating percentages (typically 40%, 6%0, 80%, and 100% load) so you should be aware at all times at what load the system is operating. The operating arm for the prerotation vanes should be marked so the travel can be measured; this is closely correlated with the amperage.

Chillers manufactured in the last few years all have electronic controls and the start-up sequence should be studied for the each particular system. They may use everything from default lights on the control panel to electronic readouts for troubleshooting purposes. For example, a unit may have a sequence of light emitting diodes (LEDs) on the circuit board that will be lit if a particular problem has occurred. The technician must use the manufacturer's troubleshooting chart to find the suspected problem. Some manufacturers use a specific LED flashing sequence to indicate each problem. Other use an electronic readout that explains the problem in words or code numbers. These control sequences can be quite lengthy and it is beyond the scope of this text to describe them. Again, the most complete source of system information is the manufacturer's instructions. One of the purposes of these electronic controls is to make troubleshooting easier.

31.3 OPERATING PROCEDURES FOR CHILLED-WATER SYSTEMS

Once the chiller has been started, observe it from time to time to make sure it is operating correctly.

Small Scroll and Reciprocating Chillers

Small air-cooled chillers may be located in remote locations and may not be observed on a regular basis. These chillers are usually operated unattended. There is not much that can go wrong if the chilled-water supply is maintained.

Water-Cooled Chillers

All water-cooled chillers require special attention to the water circuit and cooling tower. It is important to check any piece of operating equipment on a periodic basis, but it is also expensive, so most close observation is reserved for large systems. With large systems, the cooling tower water should be checked several times per day to make sure that the water level is correct. Often, a partial loss of water pressure will cause the tower water to drop due to the fact that the water is evaporating faster than the water supply can make it up. With careful observation, you can catch this problem before it shuts the unit off and a complaint occurs. If there is no regular technician on the job (e.g., in a small office building), someone such as management or building maintenance can make this check with very little instruction. These chillers are often on top of buildings where no one goes on a regular basis. Someone should be assigned to check these installations every week.

Water-cooled chillers must have water treatment to prevent minerals and algae from forming in the water circuit. A qualified water treatment specialist is necessary for all water tower applications or problems will occur. Review the text on water towers for the types of treatment required. All towers must have blowdown (bleed water), which is a percentage of the circulated water passing down the drain to prevent the tower water from becoming over-concentrated with minerals due to evaporation. Blowdown for the cooling tower is necessary, but often management has a problem watching water that looks clean flow down the drain. You should take the time to explain the purpose of blowdown.

Large Positive-Displacement Chillers

Large chillers that use reciprocating compressors are available in sizes up to about 150 tons, and screw compressor chillers are available in sizes of over 1000 tons. Chillers of this size are very expensive and must be observed on a regular basis or problems may occur. These problems can often be very expensive to correct. These large chillers are very reliable but regular observation is important. Look for any potential problems, such as those explained in the previous paragraphs. These chillers may also have pressure gages that are active all the time (except if they are valved off). If so, these may be observed as often as once per hour in some critical applications. With large systems, a regular operating log should be maintained. The frequency of recording in the operating log depends on the importance of the job. If it is a chiller used for a critical manufacturing process, hourly entries may be required. If it is an office building, the log may be maintained on a daily or weekly basis.

Centrifugal Chillers

Centrifugal chillers were the largest available chillers for many years. Typically, when the requirements were above 100 tons, a centrifugal chiller was considered or multiple reciprocating chillers were used. Today, reciprocating and screw compressor chillers may be used in applications up to and beyond 1000 tons.

Centrifugal chillers range in size from about 100 tons to about 10,000 tons for a single chiller. These are the most expensive of all chillers and also the most reliable, so much attention has been given to the observation and maintenance of these chillers. For years, it has been customary for the operation of these chillers to be observed systematically and an operating log maintained on a regular basis (usually daily). An operating log is valuable because it can track problems as they worsen over a period of time. For example, if the cooling tower water treatment system has not been working and the condenser water tubes are beginning to suffer from mineral buildup, the condenser water approach temperature will begin to spread. An alert operator will notice this and take corrective action. This action may be to alter the water treatment system and certainly to clean the water tubes as the heat exchange has been reduced.

31.4 MAINTENANCE PROCEDURES FOR CHILLED-WATER SYSTEMS

The maintenance required varies greatly depending on whether the system is an air-cooled or water-cooled chiller.

Air-Cooled Chillers

There is very little maintenance required for air-cooled chillers. This is one reason why they are so popular. The fan section of these chillers may require lubrication of the fan motors or the motors may have permanently-lubricated bearings. It is typical for these chillers to have multiple fans and motors. The motor horsepower can be reduced in this manner and the direct-drive fans can be used. This eliminates the extra maintenance required with belt-drive systems including replacing worn belts and routine lubrication. Also, if one fan fails, the chiller can continue to operate.

The following electrical and mechanical maintenance should be performed annually for typical air-cooled systems:

1. Inspect the complete power wiring circuit because this is where the most current is drawn. Look for places where hot spots have occurred. For example, the wires on the compressor contactor should show no signs of heat. The insulation should not appear discolored or melted. If so, it must be repaired. This can be done by cutting the wire back to clean copper. If the lead is not long enough, a splice of new wire may be required or the wire may need to be replaced back to the next junction.
2. Inspect the motor terminal connections at the compressor. These should not show any discoloration or a repair should be made. A hot terminal block can cause refrigerant leaks, so the wiring should be inspected carefully.
3. The contacts in all contactors should be inspected and replaced if pitting is excessive. If excess pitting begins to occur, it is likely that the contacts will weld shut in the future; if this happens, the motor cannot be stopped using the contactor. This can lead to single-phasing of a wye-wound motor if any two contacts weld shut. A motor burnout will then occur as there is nothing to shut the motor off except the breaker (Figure 31-7).
4. The compressor motor should be checked for internal grounds (leakage) using an ohmmeter that provides ohm readings in millions of ohms. This meter is called a *megger* (megohmmeter). Figure 31-8 shows an example of a basic megger. This particular instrument uses about 50 volts to check for grounded circuits; others may use much higher voltages and have a crank-type generator to achieve the high voltage. The motor manufacturer's recommendations should be used to determine the allowable leakage of the circuit. Typically, a motor should not have less than a 100 megohm circuit to ground or to another winding in the case of a delta-type motor or dual-voltage motor (Figure 31-9).

Figure 31-8. Megger used for checking internal circuits to ground. *Photo by Bill Johnson.*

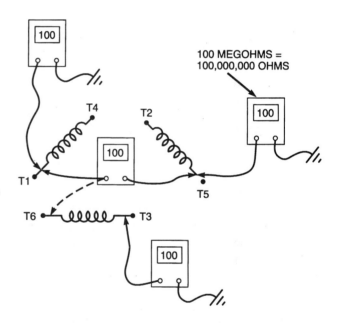

Figure 31-9. Using a megger to check a motor.

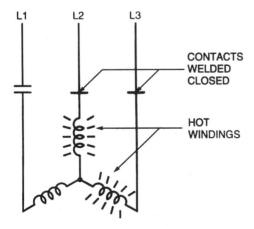

Figure 31-7. Welded contacts on a wye-wound motor.

The required reading depends on the motor temperature because the temperature changes the requirements. Consult the manufacturer if proper guidelines are not available. A reliable local motor shop can also provide you with general guidelines. It is important to start the process of megging a motor when it is new and repeat the test every year. Keep good records. If the motor megohm value begins to drop, it is a sign that moisture or some other foreign matter may be getting into the system.

5. An oil sample from the compressor crankcase can be sent to an oil laboratory to determine the condition of the compressor. This test, like the megohm test, should

be started when the chiller is new and performed every year. If certain elements start appearing in greater quantities in the oil, it is evident that problems are occurring. For example, if the bearings are made of babbitt and the babbitt content increases each year, it will be evident that the bearings need to be inspected. If small amounts of water appear, the tubes need to be inspected for leaks. This will only occur if the water pressure is greater than the refrigerant pressure. For high-pressure chillers this can occur if the chiller is located at the bottom of a high column of water.

6. Inspect the condenser for deposits on the coil surface, such as lint, dust, or dirt. These must be removed as they block airflow. It is good practice to clean all air-cooled condensers once a year, or more often if needed. If the operating log shows an increase in head pressure compared to the entering air temperature, the coil is becoming dirty. Air-cooled condensers can be very deceptive in that they can be dirty and you may not notice it because the dirt may be imbedded in the interior of the coil. This cannot be seen until the coil is cleaned. A coil should be cleaned by saturating the coil with an approved detergent and then rinsing the coil opposite to the direction of airflow. This is called *back washing*. An apparently clean coil may yield a large volume of dirt from the interior of the coil.

Water-Cooled Chillers

Routine maintenance involves cleaning the equipment room and keeping all pipes and pumps in good working order. It is always easier to keep a room clean than to have to clean it when the chiller breaks down and needs to be disassembled.

The cooling tower should be checked regularly to ensure that the strainer is not restricted. Examine the water for signs of rust, dirt, and floating debris and clean if necessary. The water treatment system should be checked on a regular basis. Some systems require daily water analysis to check for mineral content. When the water treatment is out of balance, either make the necessary adjustments or call the water treatment company. Someone must be in charge of keeping the water chemicals at the correct level for best performance.

Annual maintenance involves checking the complete system and chiller to prepare it for a season of routine maintenance. Some chillers operate year-round and are only shut down for annual maintenance. During this maintenance, it is often wise to bring in the factory maintenance representatives. These representatives are privileged to know what is happening across the nation or around the world with regard to any service or failure problems with their equipment. They are often the best source of technical knowledge for the equipment. Factory representatives may also perform a complete start-up procedure for your equipment just as though it were being started for the first time. All controls may be checked at this time to ensure they will perform correctly. It is not uncommon for a system to operate for years without having a control check. This is a poor practice because the entire system mail fail due to a single failed control. A control check once a year can be good insurance against many failures.

During annual maintenance, all electrical connections should be checked. Refer to the electrical maintenance mentioned previously for air-cooled systems. Water-cooled equipment has the same requirements.

The inside of a water-cooled condenser should be inspected every year by draining the water from the condenser and removing the heads. The tubes should be clean. You can tell that a tube is clean when you take a light and shine inside the tube and all you see is a copper tube with a dull copper finish. A pen light is good for making this check because it can be inserted inside the tube. Any film on the inside of the tube must be removed for proper heat exchange. Many technicians believe that a tube is clean if it is open enough to allow a good water flow. This is not true; the tube must be absolutely clean for proper efficiency.

When it is discovered that the tubes are dirty, there are several approaches that may be used. The tubes may be brushed using a nylon brush that is the correct size for the inside of the tube. This is customarily done using a machine that turns the brush while flushing the tube with water. Some manufacturers recommend using a brush with fine brass bristles if a nylon brush will not remove the scale from the tubes. However, check with the manufacturer because some do not recommend this practice. Always keep the tubes wet until it is time to brush them or the scale will harden. It is a good idea to be prepared to brush the tubes before removing the heads so the tubes will stay wet until they are brushed.

If brushing the tubes will not work, the tubes may be chemically cleaned using the recommended chemical solution for the application (Figure 31-10).

Caution: Tube cleaning is a very delicate process as the tubes may be easily damaged. Consult the chemical manufacturer for specific cleaning instructions.

It may be good practice to let the chemical manufacturer clean the tubes, then if damage occurs, they are responsible.

Caution: If chemicals are used to clean the tubes, make sure that the chemicals are neutralized or damage to the tubes may occur. This damage may be severe because the chemicals may eat away the tube material and leaks between the water and refrigerant may occur.

If the cleaning chemicals only soften the scale, the tubes may need brushing to remove the scale.

The water pump should be checked to make sure the coupling is in good condition.

Warning: Before checking the water pump, turn off the power, lock the disconnect, and remove the cover to the pump coupling.

MAINTENANCE PROCEDURES FOR CHILLED-WATER SYSTEMS 581

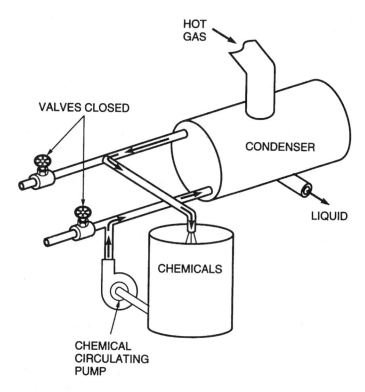

Figure 31-10. Chemically cleaning condenser tubes.

If any foreign material is found inside the coupling housing, deterioration of the coupling should be suspected. Some couplings are made of rubber and others are made of steel. Look for the flexible material that takes up the slack to see if it is wearing. Also check for metal filings that would indicate wear in a steel coupling.

After the condenser has been cleaned, the condenser tubes may be checked for defects such as stress cracks, erosion, corrosion, and wear on the outside. This can be accomplished using an eddy current check. This test should also be performed on the evaporator tubes. One common problem is wear on the outer surface of the tube where it is supported by the tube support sheets. Each set of tubes has a different type of stress applied to the outside. In the condenser, the hot discharge gas is pulsating as it enters the condenser and has a tendency to shake the tubes. In the evaporator, the boiling refrigerant shakes the tubes. The tube support sheets are sheets of steel that the tubes pass through to help prevent this action. They are evenly spaced between the end sheets. Support sheets are used in both the condenser and the evaporator. When the evaporator or condenser shell is tubed, the tubes are guided through the support sheets. When the tube is in place, a roll mechanism is inserted into the tube and expanded to hold the tube tight in the support sheet (Figure 31-11). The tube roller may miss a tube and this tube can shake or vibrate until the tube sheet wears the tube. If this wear continues, the tube will rupture and there will be a leak between the water and refrigerant circuits. In the worst cases, this can flood the chiller on the refrigerant side with water. This is more likely to occur with low-pressure chillers. The eddy current test can be used to find potential tube failures and these tubes may be pulled and replaced or plugged if there are not too many.

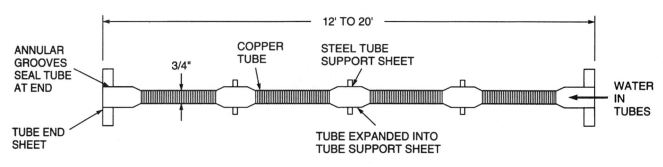

Figure 31-11. Tube assembly for a heat exchanger.

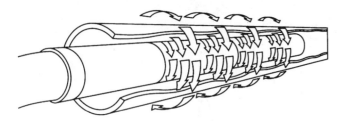

Figure 31-12. Probe for checking tubes using an eddy current and monitor.

The eddy current test instrument has a probe that can be pushed through the clean tube while an operator watches a test monitor (Figure 31-12). As the probe passes through the tube, the probe sends out a magnetic current signal that reacts with the tube and shows a profile on the monitor. When unusual profiles are noticed, the tube is marked for further study, which may involve comparing the profile to that of a known good tube. A decision may need to be made to pull or plug the tube. A tube failure can take days to repair, so it is a good idea to look for these problems before the start of the cooling season.

Remove and clean any strainers that may be in the piping. Some systems have a strainer before or after the pump. This should be checked and cleaned, if necessary.

The cooling tower should be scrubbed to remove any residue that may have accumulated in the basin.

If the chiller and water pumps will stand idle during the winter months, all precautions should be taken to prevent freezing of any components such as piping, sumps, pumps, and the chiller itself if it is located outdoors. Even if the chiller is in a building, it is good practice to drain it during the winter because if the building loses power and the interior freezes, the chiller tubes may also freeze and will be difficult to repair. It is better to simply drain it and not take a chance.

31.5 PROCEDURES FOR ABSORPTION SYSTEM START-UP

The absorption system start-up procedure is similar to that for water-cooled chillers; both chilled-water flow and condenser water flow must be established before the chiller is started. Ensure that the cooling tower water temperature is within the range specified by the chiller manufacturer. If the absorption machine is started with the cooling tower water too cold, the lithium bromide may crystallize, causing a serious problem. In addition to these checks, the heat source must be verified, whether it is steam, hot water, natural gas, or oil. Most absorption chillers are steam-operated so the correct steam pressure must be available. If the chiller is operated by a dedicated boiler, it must be started and operated until it is up to the correct temperature.

Warning: Wear gloves and exercise caution when working around hot steam pipes and other heated components.

If the chiller has been off for a long period of time, it is good practice to perform a vacuum pump purge for several hours prior to starting the machine. The purge discharge may be monitored by placing the vacuum pump exhaust into a glass of water and look for bubbles (Figure 31-13). If there are bubbles, it is a sure sign that non-condensibles are in the system. It is pointless to start the chiller until the bubbles stop. Let the vacuum pump run even when the chiller is started (unless the manufacturer recommends not to) because non-condensibles will often migrate to the purge pickup point after start-up.

When all systems are ready, the chiller may be started. The heat source should be up to the required temperature. When the chiller starts, the fluid pumps will start to circulate the lithium bromide and the refrigerant. The chiller will begin to cool very quickly; this can be verified by checking the temperature drop across the chilled-water circuit. When absorption chillers are refrigerating, they make a sound like ice cracking. When this noise is heard, the chilled-water temperature should begin to drop.

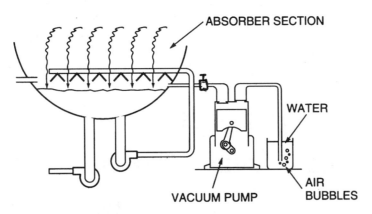

Figure 31-13. The vacuum pump exhaust is placed in a glass of water. When bubbles are observed, the vacuum pump is exhausting the non-condensibles.

31.6 OPERATION AND MAINTENANCE PROCEDURES FOR ABSORPTION SYSTEMS

Absorption chillers must be observed more closely than some compression cycle chillers. Chiller manufacturers have different check points to determine chiller operation. This involves monitoring the relationships between the refrigerant temperature, absorption fluid temperatures, and leaving chilled-water temperatures, to name a few. The manufacturer's literature must be consulted for the specific check points and procedures. The maintenance of an operating log is highly recommended for absorption installations.

When the chiller is operating, pay close attention to the purge operation. If the chiller is requiring excessive purge operation, the system may have leaks or it may be manufacturing excess hydrogen internally and may need special chemical additives. The machine must be kept free of non-condensibles.

The heat source also requires maintenance. If it is steam, the steam valve should be checked for leaks and operating problems. When steam is used, the condensate trap should also be checked to make sure it is operating properly. The condensate trap ensures that only water (condensed steam) returns to the boiler. A typical condensate trap may contain a float. When the water level rises, the float also rises and allows only water to move into the condensate return line (Figure 31-14). There are several types of condensate traps available; check the manufacturer's data for each particular system.

Condensate traps are often checked using an infrared testing device that measures the temperature on both sides of the trap.

When the chiller is operated using hot water, the hot water valve should be observed for proper operation and to ensure that there are no leaks.

Gas or oil units will require the typical maintenance for these fuels. Oil systems require special maintenance of the filter system and nozzle.

The purge system may require the most maintenance. If there is a vacuum pump used with the chiller, regular oil changes will help to make it last longer. Lithium bromide is salt and will corrode the vacuum pump on the inside.

Caution: It is not recommended that the vacuum pump used for a compression cycle system be used on an absorption system because of the corrosive nature of lithium bromide. If the same pump must be used for both systems, extra care should be taken to change the oil immediately after use.

Any lithium bromide that is spilled in the equipment room will corrode any metal it comes in contact with. To prevent problems, it is wise to regularly wash the area where lithium bromide is handled. The chiller and other equipment may need to be painted periodically for additional protection in these areas.

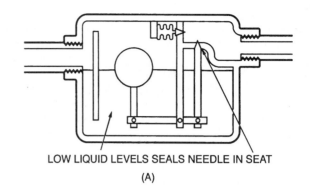

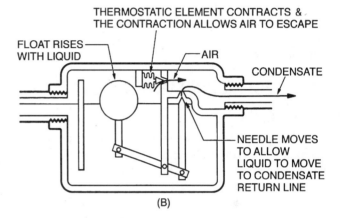

Figure 31-14. Condensate float trap for a steam coil.

The water pump and system solution pump or pumps require periodic maintenance. A check with the manufacturer will provide details regarding the type and frequency of this maintenance.

The equipment room must be kept above freezing or the refrigerant (water) inside the unit will freeze and cause tube cracks or ruptures. Unlike other chillers, this water cannot be drained, so the system must be protected against cold weather.

The steam or hot water valve and condensate trap should be inspected regularly. A strainer will be located in the vicinity of the condensate trap and should be cleaned during the off season.

The cooling tower must receive regular cleaning and water treatment. Don't forget blowdown water. Absorption chillers operate at much higher temperatures than compression cycle chillers and good water treatment is essential because the tubes become fouled much quicker.

Absorption chillers may need to have the tubes probed with an eddy current probe on a more regular basis than compression cycle chillers because of the higher temperatures in the condenser section. The tubes receive stress when the heat source is applied and they expand. They then contract when cooled during the off cycle.

These temperature variations place a large amount of stress on the tubes. Check with the manufacturer for specific maintenance requirements.

31.7 GENERAL MAINTENANCE TIPS FOR ALL CHILLERS

If you intend to maintain chillers, you must be well qualified and should stay in contact with the chiller manufacturers for the latest information. Manufacturers receive feedback from all over the world on their equipment and know about global system problems, such as premature failures. They will normally share any potential problems with you to prevent future problems with their equipment. You should also attend factory seminars and schools and make a point of becoming acquainted with the factory technicians.

31.8 TECHNICIAN SERVICE CALLS

Service Call 1

A building manager calls and reports that the building is becoming hot. The problem is that the cooling tower fan has a broken belt and the fan is off. The reciprocating chiller is off because of high head pressure.

The technician arrives and goes to the basement where the chiller is located. It is a reciprocating water-cooled chiller. The chiller is off and the technician looks around for the problem. The control panel is opened and it is noticed that the reset button on the high-pressure control is out, signaling that the control needs to be reset. The pressure drop across the condenser is normal and the water temperature is also normal (because the chiller has been off, the tower is able to lower the water temperature without the fan and with the compressor off). The technician resets the button and the compressor starts. As the technician watches, the compressor starts to load up. Everything runs normally for about 10 minutes, and then the compressor sounds as though it is straining. The technician feels the condenser water line entering the condenser. It is warm compared to hand temperature, which is typically 90°F. Remember, the cooling tower water should be 85°F. Therefore, the returning cooling tower water is too hot. The technician shuts the compressor off rather than letting it shut itself off with the high-pressure control.

The technician goes to the roof to look at the cooling tower, fully expecting to see it full of algae as the management does not perform regular maintenance. On the roof, it is noticed that the fan belt is broken on the cooling tower fan. The disconnect to the fan is shut off and locked out for safety. The numbers are written down from the old belt and the technician goes to the truck for a new one. While at the truck, the technician gets a grease gun for greasing the fan and motor bearings at the cooling tower. The belt is replaced and the tower fan is turned on. It starts to turn and the technician can see from the moist air leaving the tower that the water is loaded with heat.

The technician goes back to the basement and touches the water line leading from the tower to the condenser. The line now feel cooler than hand temperature. The compressor is restarted. The technician stays with the compressor for about 30 minutes to make sure that it will stay on and it does. This time is spent lubricating the pump and fan motors and looking for any other potential problems, such as debris in the cooling tower.

Service Call 2

A building owner reports that the building is hot and the chiller is off. The chiller is a low-pressure R-11 chiller that has a known air leak. The building maintenance personnel have been operating the purge for several days to remove air that is leaking in, and were waiting for the weekend to leak test the machine. The purge water has been valved off and the purge condenser is not operable, causing excess purge pressure and releasing refrigerant.

The technician arrives and goes to the equipment room where the chiller is located. The signal light indicates that the chiller is off because of freeze protection. The technician looks around and notices that the purge pressure is very high as the purge is still running on manual. Upon inspection, it is noticed that the purge relief valve is seeping refrigerant. The technician then checks the water circuit to the purge and discovers that there is no water cooling the condenser in the purge. The valve is opened to the purge water condenser. This system uses chilled water, which is now warmed up to 80°F, but will still condense the refrigerant in the purge system.

The technician starts the compressor in the 40% mode and watches the suction pressure. It is not low, so the load is advanced to 60%. The suction pressure begins to drop. The technician checks the refrigerant level in the sightglass at the evaporator and it is low. There is refrigerant available at the site so the technician gets set up to add refrigerant. The refrigerant drum is piped to the charging valve and the valve is opened. Meanwhile, the technician places a thermometer in the well in the evaporator to check the refrigerant temperature. The chilled-water temperature leaving the chiller is down to 55°F and the thermometer in the refrigerant reads 38°F. This is a 17°F approach temperature and it should be 7°F according to the log sheet that was left at start-up. Two hundred pounds of refrigerant are added before the compressor is operated at full load and the refrigerant approach temperature is lowered to 7°F. The technician is satisfied with the system charge. The chilled-water temperature pulls down to 45°F and the compressor begins to unload in about 30 minutes. The purge pressure is normal and the purge system is no longer relieving continuously. The technician leaves, but will return for the leak check over the weekend.

Service Call 3

A building manager reports that the centrifugal chiller is off and the building is hot. The problem is that the flow switch in the condenser water circuit has a broken paddle and the field control circuit is not passing power to the chiller.

The technician arrives and goes straight to the chiller. This chiller has a ready light on the control panel which indicates when the field circuit is calling for cooling. The light is off. The technician first checks the chilled-water temperature. It is 80°F and the controller is calling for cooling. The technician then checks the water gages that register the pressure drop across the chiller and the condenser. The correct pressure drop registers across both so there is water flow. The technician then goes to the wiring diagram for the system and looks to see what other controls are in this circuit. The flow switches and pump interlocks are all in the same circuit. It is now a matter of tracing the wiring to see what has caused the chiller to shut down.

The technician realizes that the flow switches are likely to be a problem because there is a lot of movement in these devices since they are located in a moving water stream. The compressor switch is turned off so the compressor will not try to start during the control testing. The technician removes the cover from the chilled-water flow switch and checks the voltage across it. There is no voltage.

Note: If the switch were open, there would be voltage.

The cover is replaced and the same test is performed on the condenser water flow switch and there is voltage. This flow switch is open. The technician jumpers the switch and the ready light on the chiller panel lights up. This is the problem.

The technician leaves the jumper on the flow switch and starts the chiller. Flow in the condenser water circuit is not as critical as it would be in a chilled-water circuit. If the water flow were to stop in the condenser circuit, the compressor would shut down because of high pressure.

Caution: The chilled-water flow switch should not be jumpered with the compressor running as it may freeze the chiller.

The technician explains to the building manager that the flow switch paddle must be replaced. The manager requests that this be done after hours, so the technician leaves and will return at 5:00 pm with the parts to repair the switch.

At 5:00 pm, the technician returns to the equipment room to make the repair. The chiller is off for the day because of the energy management time clock. The condenser water pump is turned off and locked. The valves on either side of the condenser system are valved off and the water is drained from the condenser. The flow switch is removed and replaced. The water valves are turned on and the condenser system is allowed to fill. Air is vented out of the water system at the high places in the piping. The technician then turns the water pump on and verifies that the flow switch is passing power. By leaving the meter leads on the flow switch and shutting off the water, it is verified that the flow switch will also shut the system off if the water flow is stopped. The system is made ready to run the next morning when the time clock calls for cooling, and the technician leaves.

Service Call 4

A building manager telephones and indicates that the centrifugal chiller in the basement of the building is making a screaming noise. The manager went to the door and was afraid to go in. The problem is that the cooling tower strainer is clogged and restricting the water flow to the condenser. The centrifugal has a high head pressure, causing a surge which produces a loud screaming sound.

The technician arrives and can hear the compressor screaming from the parking lot. When the technician enters the room, the machine capacity control is turned down to 40% to reduce the load. The machine has a high head pressure.

A check of the pressure gages across the condenser shows that there is reduced water flow, so the technician proceeds to the cooling tower on the roof.

A look at the cooling tower basin shows what has happened—there are leaves in the cooling tower basin. The technician removes the cover from the cooling tower basin and removes the leaves from the strainer. The water immediately starts flowing faster.

The technician replaces the cover to the tower basin, then goes back to the basement and turns the capacity control up to 100%.

The technician then talks to the manager about returning to clean the cooling tower after hours when the building air conditioning is off.

Late that afternoon, the technician returns. The system is off because of the time clock in the energy management system. There are many small leaves in the tower and it is thoroughly cleaned. The technician then goes to the basement and prepares to check the condenser for leaves. The condenser circuit is valved off and drained. Then the technician removes the marine water box cover on the piping inlet end of the condenser where any trash would accumulate. There are small leaves in the condenser that have passed through the cooling tower strainer. These are cleaned out. While the water box cover is removed, the technician uses a small pen-type flashlight to examine the tubes. There is no scale or fouling of the tubes. This is good news.

The water box cover is replaced and the valves are opened to the condenser. The technician waits a few minutes for the tower to refill with water after the condenser has been filled, then starts the condenser water pump in the manual mode and verifies that there is water flow. This is a necessary step if the condenser is large enough to drain most of the water from the tower; the tower must have time to refill before starting the pump. Now the technician knows that when the machine starts in the morning, there should be no problems.

Service Call 5

A maintenance technician calls and says that the building centrifugal chiller is off. A light is lit that reads FREEZE CONTROL. The maintenance technician wants to know if it should be reset. It is a mild day without much need for air conditioning. The dispatcher says not to reset the switch; a technician will be sent over within half an hour. The problem is that the main controller is not calibrated and is drifting, causing the water temperature to fall too low.

The technician arrives and goes to the chiller, which is on the roof in an equipment penthouse. The technician checks the water flow through the chiller by comparing the inlet and outlet pressures to those recorded on the log sheet. The pressures are correct. The freeze control is reset and the chiller compressor starts. When it starts to lower the water temperature, the technician watches the controller. It is a pneumatic controller and as the water pressure drops, the air pressure is supposed to drop from the 15 psig furnished to a lower value. When the water temperature reaches 43°F, the pneumatic air pressure should be 7 psig. When the water temperature reaches 41°F, the air pressure should be 3 psig, and the chiller should shut down. This is not what happens; the chiller continues to run and the water temperature reaches 39°F without shutting off. The technician then adjusts the controller to the correct calibration. The chiller shuts down.

The technician waits for the chiller to restart when the building water warms up. This takes about 30 minutes. When it is established that the chiller is operating correctly, the technician leaves.

Service Call 6

A building manager calls and reports that the chiller is off and the building is getting hot. This is a centrifugal chiller. The problem is that the building technician has been adjusting the controls and has adjusted the load-limiting control while trying to increase the motor capacity to reduce the building temperature faster. The compressor was pulling too much amperage, which tripped the overload device.

The technician arrives and goes to the equipment room with the building technician. The building technician does not tell the technician that the load-limiting control has been adjusted. The technician checks the field control circuit and discovers that all field controls are calling for cooling. A further check shows that an overload device has tripped. The technician knows that something has overloaded and caution should be used.

The technician shuts the power off to the compressor starter off and checks the motor windings for a ground or short circuit from winding to winding. The motor seems fine, so power is restored. There is not much else to do except to start the motor and observe.

The technician starts the compressor in the 40% mode. When the compressor starts to load up, the amperage is observed. The start-up data shows that the compressor should draw 600 amps at full load and 240 amps at 40%. The compressor is pulling 264 amps; this is excessive, so the technician turns the load limiter back to 240 amps and then turns the limiter up to 100%. The compressor begins to load up and reaches 600 amps and throttles to hold this amperage. The technician then turns to the building maintenance technician and asks if anyone has adjusted the load-limiting control. The building technician admits to making the adjustment. The building technician is told that these controls should not be adjusted as they are very reliable. The technician then uses enamel to mark the setting on the control so it will be apparent if it is ever adjusted again.

The technician understands that the controls belong to the building owner, but also knows that they should only be adjusted by a qualified technician.

Service Call 7

Early one Tuesday morning, a customer calls and says that the building is becoming warm. The customer says that water is still flowing over the cooling tower, but there is no cooling. The unit is a 300 ton centrifugal with a hermetic compressor. The problem is that the compressor motor is burned out and will need to be replaced or rewound.

The technician goes straight to the equipment room, which is on the 10th floor of the office building. It is noticed that the building is hot, so there is obviously a problem with the chiller.

The technician checks the field control circuit and discovers that all signals are correct and it is calling for cooling. The technician then goes to the starter and discovers that there is no line voltage to the starter. The technician finds the main power supply and discovers that the breaker is tripped. There must be something very wrong.

The technician locks out the breaker and uses a megger to check the motor leads in the starter and discovers a circuit to ground. There is always the possibility that the problem may be in the wiring between the starter and the motor terminals on the motor, so the technician removes the leads from the motor terminals and disconnects them. The motor is checked at the terminals without the wiring and it shows a circuit to ground.

The technician goes to the building management for advice as to the next step. It is explained that the motor will have to be removed from the equipment room for replacement or rebuild. Since the motor is hermetically-sealed, the refrigerant will have to be recovered. The technician explains that there are two possible ways of making the repair. The motor can be replaced with a new motor, which may take as long as two weeks for delivery from the factory, or the motor may be rebuilt at a local rebuild shop that is approved to work with hermetic motors. The building management gives permission to proceed with the rebuild to save time as the building must have air conditioning. Like many modern buildings, there are no windows that open and there are computers that rely on the air conditioning.

The technician calls the home office and requests backup help and the recommended tools for the job. This includes a hoist to lift the motor and a dolly to transport it onto the elevator and out to the truck, which is equipped with a lift gate.

While the helper is getting the tools together, the technician prepares to pull the motor. The refrigerant is recovered from the unit using the recovery system brought by the helper. Water in the chilled-water circuit and condenser water are circulated through the system during recovery to prevent the tubes from freezing and to aid in boiling the refrigerant out of the system. The refrigerant is stored in refrigerant drums. The system refrigerant (R-11) will be checked for acid content and recycled for reuse if needed. When the refrigerant is recovered down to 20 in. Hg vac. in the system, the technician breaks the vacuum with dry nitrogen that has been brought by the helper. Breaking the vacuum with dry nitrogen will minimize oxidation inside the unit when it is opened to the atmosphere. The compressor is made of steel inside and it will rust quickly when exposed to the air.

The technician then removes the wires from the motor terminal box and starts to disconnect the compressor from the motor. When the motor is free to move, the hoist is set up and the motor is placed on a rolling dolly to transport it to the truck. The motor is then rolled up the lift gate into the back of the truck and secured. The helper takes the motor to the motor shop for rebuild. This will take about 2 days. It is now 2:30 on Tuesday.

The technician goes back to the equipment room and prepares the equipment for when the motor is returned. A sample of the refrigerant and a sample of the refrigerant oil from the compressor sump are pulled and shipped by next-day delivery to a chemical analysis firm. They will be analyzed and the report can be called in by Wednesday afternoon to determine what must be done about the refrigerant. The refrigerant and oil samples do not smell acidic, but the analysis will reveal the true extent of the damage.

The technician checks the contacts in the starter because there must have been quite an amperage surge when the motor grounded. The contacts are good and do not require replacement.

The oil heater is turned off and locked out. The technician then cleans all gasket sealant from all compressor flanges and is now ready for the motor to be returned. Time has been saved by doing all of this before the motor is brought back.

On Thursday morning, the technician checks with the rebuild shop and is told that the motor will be ready by 11:30 am. The technician arranges for the helper to pick up the motor and a compressor gasket kit. They meet at the building at 1:00 pm and proceed to install the rebuilt motor. The motor is moved to the equipment room and lifted back into place. All gaskets are installed and everything is tightened. This system has a history of being very tight, so the only leak checking that needs to be performed is on the fittings, including the gasketed flanges that were removed and in the area where the work was done.

When everything is ready, the system is pressurized with nitrogen to 10 psig with a trace of R-22 refrigerant, which is cheaper than R-11 and is commonly used to detect leaks. The machine is leak checked and found to be good. The nitrogen and refrigerant are exhausted and it is time to pull the vacuum. The motor leads are connected to the motor terminals.

The vacuum pump is installed and started. It will take about five hours to pull the vacuum down to 0 in. Hg on a mercury manometer and it is now 5:30 pm. The technician leaves the job and decides to come back after 10:30 pm to check the vacuum. If the vacuum can be verified this evening and the vacuum pump shut off for eight hours, the machine can be started in the morning if all is well. The technician checks the vacuum at about 11:00 pm and the manometer reads 0 in. Hg. The vacuum pump is valved off and the technician leaves.

When the technician returns at 7:30 am Friday morning, the mercury manometer is still reading 0 in. Hg, so the machine is leak free and it is time to charge the system. The refrigerant and oil report came back okay so no extra precautions need to be taken with the refrigerant. The technician starts the chilled-water pump and the condenser water pump in preparation for start-up. A drum of refrigerant is connected to the machine at the drum vapor valve. The valve is slowly opened to the system and vapor is allowed to enter the machine until the pressure is above that corresponding to freezing. This is important because if liquid refrigerant is allowed to enter the system at such a low pressure, the tubes may freeze. The chilled water is circulating, but if there is one tube that does not have circulation, this tube may freeze. The pressure in the machine must be above 17 in. Hg vac. to prevent freezing while charging liquid.

When the machine pressure reaches 17 in. Hg vac., the technician then starts charging liquid refrigerant. While the technician is charging refrigerant, the helper pulls the oil in the oil sump using the system vacuum. When the oil is charged, the oil heater is turned on to bring the oil up to temperature.

The technician is not able to charge all of the refrigerant into the system using the vacuum because the chilled-water circuit warms the refrigerant and the pressure soon rises. The technician then changes over and pushes the remainder of the refrigerant into the system by pressurizing the drum with nitrogen on top of the liquid and drawing liquid from the bottom of the drum (Figure 31-15).

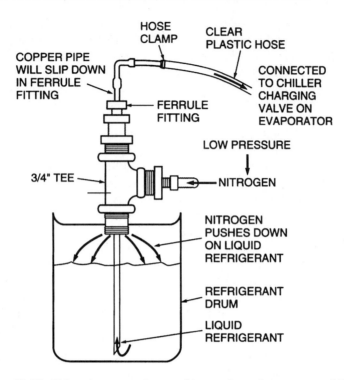

Figure 31-15. Using nitrogen to charge refrigerant into a low-pressure chiller.

The technician starts the purge pump to pull any nitrogen that may have entered the system while charging, even though the technician was very careful and could see no nitrogen in the clear charging hose while charging. The purge pressure does not rise, so the system should be free of nitrogen. The technician has closed the main breaker and is ready to start the system. The system is started and quickly brought up to 100% to reduce the building temperature as quickly as possible. There is no apparent reason why the motor developed a circuit to ground, and the motor rebuild shop could not find any problem, so the motor failure is accounted for as a random failure.

The chiller is quickly pulling the building water down to 45°F leaving water temperature. The technician stays with the chiller for two hours and completes entries in an operating log, including the motor amperage. When it is determined that the system is functioning properly, the technician leaves.

SUMMARY

- The operation of a chilled-water air conditioning system involves starting, running, and stopping the chiller system in an orderly manner.
- When starting a chilled-water system, first check for sufficient chilled-water flow and if it is a water-cooled check for condenser water flow. The interlock circuit through the contactors must be satisfied; the cooling tower fan should start, followed by the compressor. There may be a time delay before the compressor starts.
- Reciprocating, scroll, and screw compressors may be either air-cooled or water-cooled units.
- The centrifugal chiller will have a separate oil lubrication system. Check that the lubrication system is functioning satisfactorily before starting the compressor.
- After the chiller is on and operating, observe the operation for a period of time to ensure that it is working correctly.
- Water-cooled chillers must have water treatment to prevent minerals and algae from collecting in the water tower.
- Inspecting and cleaning the water tower should be done on a regular basis.
- Water-cooled condenser tubes should be checked at least annually. If they are found to be dirty or have scale buildup, they must be cleaned using a brush and/or chemicals. These chemicals must be applied carefully and rinsed well or they will cause damage to the tubes.
- Condenser and evaporator tubes may be checked for defects using an eddy current test instrument.
- Absorption chilled-water system start-up is similar to and compression cycle system start-up in many respects. Condenser and chilled-water flow must be established before the chiller is started. The cooling tower water must not be too cold or the lithium bromide may crystallize. If the chiller has been off for a long period of time, it is good practice to operate the vacuum pump purge system for several hours before starting the machine.

- Absorption chillers must be observed for proper operation on a more regular basis than most compression cycle chillers because the lithium bromide solution and the water (refrigerant) require monitoring.
- The purge system may require more maintenance than other devices in the absorption chiller. The vacuum pump oil should be changed regularly.
- Proper chiller maintenance requires ongoing contact with the manufacturers of the equipment to receive the latest information, advice, and schedules for service schools. Factory schools and seminars should be attended whenever possible.

REVIEW QUESTIONS

1. What is the first step that should be taken when starting up a chilled-water air conditioning system?
2. How does the lubrication system differ between a reciprocating compressor and a centrifugal compressor?
3. Why do reciprocating compressors have crankcase heaters?
4. Does the chiller reciprocating compressor start up fully loaded?
5. Does the rotary screw chiller compressor have crankcase heat?
6. Does the chiller centrifugal compressor start up fully loaded?
7. What component must start before the centrifugal compressor?
8. Why is the data in an operating log important?
9. Why should condenser tubes be kept wet until they are brushed or otherwise cleaned?
10. What is the name of the test used to determine the condition of the interior of the condenser and evaporator tubes?
11. If the water tower temperature is too cold when an absorption chilled-water system is started, what may happen to the lithium bromide?

APPENDICES

Appendix I	TEMPERATURE CONVERSION TABLE	591
Appendix II	DESIGN CONDITIONS TABLES	593
	Table 1. United States	593
	Table 2. Canada	605
	Table 3. Other Countries	607
Appendix III	ELECTRICAL SYMBOLS	612

APPENDIX I—TEMPERATURE CONVERSION TABLE

TEMPERATURE CONVERSION TABLE

°F	Temperature to be Converted	°C	°F	Temperature to be Converted	°C
−76.0	−60	−51.1	23.0	−5	−20.6
−74.2	−59	−50.6	24.8	−4	−20.0
−72.4	−58	−50.0	26.6	−3	−19.4
−70.6	−57	−49.4	28.4	−2	−18.9
−68.8	−56	−48.9	30.2	−1	−18.3
−67.0	−55	−48.3	32.0	0	−17.8
−65.2	−54	−47.8	33.8	1	−17.2
−63.4	−53	−47.2	35.6	2	−16.7
−61.6	−52	−46.7	37.4	3	−16.1
−59.8	−51	−46.1	39.2	4	−15.6
−58.0	−50	−45.6	41.0	5	−15.0
−56.2	−49	−45.0	42.8	6	−14.4
−54.4	−48	−44.4	44.6	7	−13.9
−52.6	−47	−43.9	46.4	8	−13.3
−50.8	−46	−43.3	48.2	9	−12.8
−49.0	−45	−42.8	50.0	10	−12.2
−47.2	−44	−42.2	51.8	11	−11.7
−45.4	−43	−41.7	53.6	12	−11.1
−43.6	−42	−41.1	55.4	13	−10.6
−41.8	−41	−40.6	57.2	14	−10.0
−40.0	−40	−40.0	59.0	15	−9.4
−38.2	−39	−39.4	60.8	16	−8.9
−36.4	−38	−38.9	62.6	17	−8.3
−34.6	−37	−38.3	64.4	18	−7.8
−32.8	−36	−37.8	66.2	19	−7.2
−31.0	−35	−37.2	68.0	20	−6.7
−29.2	−34	−36.7	69.8	21	−6.1
−27.4	−33	−36.1	71.6	22	−5.6
−25.6	−32	−35.6	73.4	23	−5.0
−23.8	−31	−35.0	75.2	24	−4.4
−22.0	−30	−34.4	77.0	25	−3.9
−20.2	−29	−33.9	78.8	26	−3.3
−18.4	−28	−33.3	80.6	27	−2.8
−16.6	−27	−32.8	82.4	28	−2.2
−14.8	−26	−32.2	84.2	29	−1.7
−13.0	−25	−31.7	86.0	30	−1.1
−11.2	−24	−31.1	87.8	31	−0.6
−9.4	−23	−30.6	89.6	32	0.0
−7.6	−22	−30.0	91.4	33	0.6
−5.8	−21	−29.4	93.2	34	1.1
−4.0	−20	−28.9	95.0	35	1.7
−2.2	−19	−28.3	96.8	36	2.2
−0.4	−18	−27.8	98.6	37	2.8
1.4	−17	−27.2	100.4	38	3.3
3.2	−16	−26.7	102.2	39	3.9
5.0	−15	−26.1	104.0	40	4.4
6.8	−14	−25.6	105.8	41	5.0
8.6	−13	−25.0	107.6	42	5.6
10.4	−12	−24.4	109.4	43	6.1
12.2	−11	−23.9	111.2	44	6.7
14.0	−10	−23.3	113.0	45	7.2
15.8	−9	−22.8	114.8	46	7.8
17.6	−8	−22.2	116.6	47	8.3
19.4	−7	−21.7	118.4	48	8.9
21.2	−6	−21.1	120.2	49	9.4

°F	Temperature to be Converted	°C	°F	Temperature to be Converted	°C
122.0	50	10.0	208.4	98	36.7
123.8	51	10.6	210.2	99	37.2
125.6	52	11.1	212.0	100	37.8
127.4	53	11.7	213.8	101	38.3
129.2	54	12.2	215.6	102	38.9
131.0	55	12.8	217.4	103	39.4
132.8	56	13.3	219.2	104	40.0
134.6	57	13.9	221.0	105	40.6
136.4	58	14.4	222.8	106	41.1
138.2	59	15.0	224.6	107	41.7
140.0	60	15.6	226.4	108	42.2
141.8	61	16.1	228.2	109	42.8
143.6	62	16.7	230.0	110	43.3
145.4	63	17.2	231.8	111	43.9
147.2	64	17.8	233.6	112	44.4
149.0	65	18.3	235.4	113	45.0
150.8	66	18.9	237.2	114	45.6
152.6	67	19.4	239.0	115	46.1
154.4	68	20.0	240.8	116	46.6
156.2	69	20.6	242.6	117	47.2
158.0	70	21.1	244.4	118	47.7
159.8	71	21.7	246.2	119	48.3
161.8	72	22.2	248.0	120	48.8
163.4	73	22.8	257.0	125	51.7
165.2	74	23.3	266.0	130	54.4
167.0	75	23.9	275.0	135	57.2
168.8	76	24.4	284.0	140	60.0
170.6	77	25.0	293.0	145	62.8
172.4	78	25.6	302.0	150	65.6
174.2	79	26.1	311.0	155	68.3
176.0	80	26.7	320.0	160	71.1
177.8	81	27.2	329.0	165	73.9
179.6	82	27.8	338.0	170	76.7
181.4	83	28.3	347.0	175	79.4
183.2	84	28.9	356.0	180	82.2
185.0	85	29.4	365.0	185	85.0
186.8	86	30.0	374.0	190	87.8
188.6	87	30.6	383.0	195	90.6
190.4	88	31.1	392.0	200	93.3
192.2	89	31.7	401.0	205	96.1
194.0	90	32.2	410.0	210	98.9
195.8	91	32.8	413.6	212	100.0
197.6	92	33.3	428.0	220	104.4
199.4	93	33.9	446.0	230	110.0
201.2	94	34.4	464.0	240	115.6
203.0	95	35.0	482.0	250	121.1
204.8	96	35.6	500.0	260	126.7
206.6	97	36.1			

APPENDIX II—DESIGN CONDITIONS TABLES

Table 1 Climatic Conditions for the United States

Col. 1	Col. 2		Col. 3		Col. 4	Col. 5 Winter,[b] °F Design Dry-Bulb		Col. 6 Summer,[c] °F Design Dry-Bulb and Mean Coincident Wet-Bulb			Col. 7 Mean Daily Range	Col. 8 Design Wet-Bulb			Col. 9 Prevailing Wind				Col. 10 Temp., °F Median of Annual Extr.	
State and Station[a]	Lat. °	'N	Long. °	'W	Elev. Feet	99%	97.5%	1%	2.5%	5%		1%	2.5%	5%	Winter	Knots[d]	Summer		Max.	Min.
ALABAMA																				
Alexander City	32	57	85	57	660	18	22	96/77	93/76	91/76	21	79	78	78						
Anniston AP	33	35	85	51	599	18	22	97/77	94/76	92/76	21	79	78	78	SW	5	SW		98.4	12.4
Auburn	32	36	85	30	652	18	22	96/77	93/76	91/76	21	79	78	78					99.8	14.6
Birmingham AP	33	34	86	45	620	17	21	96/74	94/75	92/74	21	78	77	76	NNW	8	WNW		98.5	12.9
Decatur	34	37	86	59	580	11	16	95/75	93/74	91/74	22	78	77	76						
Dothan AP	31	19	85	27	374	23	27	94/76	92/76	91/76	20	80	79	78						
Florence AP	34	48	87	40	581	17	21	97/74	94/74	92/74	22	78	77	76	NW	7	NW			
Gadsden	34	01	86	00	554	16	20	96/76	94/75	92/74	22	78	77	76	NNW	8	WNW			
Huntsville AP	34	42	86	35	606	11	16	95/75	93/74	91/74	23	78	77	76	N	9	SW			
Mobile AP	30	41	88	15	211	25	29	95/77	93/77	91/76	18	80	79	78	N	10	N			
Mobile Co	30	40	88	15	211	25	29	95/77	93/77	91/76	16	80	79	78					97.9	22.3
Montgomery AP	32	23	86	22	169	22	25	96/76	95/76	93/76	21	79	79	78	NW	7	W		98.9	18.2
Selma, Craig AFB	32	20	87	59	166	22	26	97/77	95/77	93/77	21	81	80	79	N	9	SW		100.1	17.6
Talladega	33	27	86	06	565	18	22	97/77	94/76	92/76	21	79	78	78					99.6	11.2
Tuscaloosa AP	33	13	87	37	169	20	23	98/75	96/76	94/76	22	79	78	77	N	5	WNW			
ALASKA																				
Anchorage AP	61	10	150	01	114	−23	−18	71/59	68/58	66/56	15	60	59	57	SE	3	WNW			
Barrow	71	18	156	47	31	−45	−41	57/53	53/50	49/47	12	54	50	47	SW	8	SE			
Fairbanks AP	64	49	147	52	436	−51	−47	82/62	78/60	75/59	24	64	62	60	N	5	S			
Juneau AP	58	22	134	35	12	−4	1	74/60	70/58	67/57	15	61	59	58	N	7	W			
Kodiak	57	45	152	29	73	10	13	69/58	65/56	62/55	10	60	58	56	WNW	14	NW			
Nome AP	64	30	165	26	13	−31	−27	66/57	62/55	59/54	10	58	56	55	N	4	W			
ARIZONA																				
Douglas AP	31	27	109	36	4098	27	31	98/63	95/63	93/63	31	70	69	68					104.4	14.0
Flagstaff AP	35	08	111	40	7006	−2	4	84/55	82/55	80/54	31	61	60	59	NE	5	SW		90.0	−11.6
Fort Huachuca AP	31	35	110	20	4664	24	28	95/62	93/62	90/62	27	69	68	67	SW	5	W			
Kingman AP	35	12	114	01	3539	18	25	103/65	100/64	97/64	30	70	69	69						
Nogales	31	21	110	55	3800	28	32	99/64	96/64	94/64	31	71	70	69	SW	5	W			
Phoenix AP	33	26	112	01	1112	31	34	109/71	107/71	105/71	27	76	75	75	E	4	W		112.8	26.7
Prescott AP	34	39	112	26	5010	4	9	96/61	94/60	92/60	30	66	65	64						
Tucson AP	32	07	110	56	2558	28	32	104/66	102/66	100/66	26	72	71	71	SE	6	WNW		108.9	0.3
Winslow AP	35	01	110	44	4895	5	10	97/61	95/60	93/60	32	66	65	64	SW	6	WSW		102.7	−0.4
Yuma AP	32	39	114	37	213	36	39	111/72	109/72	107/71	27	79	78	77	NNE	6	WSW		114.8	30.8
ARKANSAS																				
Blytheville AFB	35	57	89	57	264	10	15	96/78	94/77	91/76	21	81	80	78	N	8	SSW			
Camden	33	36	92	49	116	18	23	98/76	96/76	94/76	21	80	79	78						
El Dorado AP	33	13	92	49	277	18	23	98/76	96/76	94/76	21	80	79	78	S	6	SE		101.0	13.9
Fayetteville AP	36	00	94	10	1251	7	12	97/72	94/73	92/73	23	77	76	75	NE	9	SSW		99.4	−0.4
Fort Smith AP	35	20	94	22	463	12	17	101/75	98/76	95/76	24	80	79	78	NW	8	SW		101.9	7.0
Hot Springs	34	29	93	06	535	17	23	101/77	97/77	94/77	22	80	79	78	N	8	SW		103.0	10.6
Jonesboro	35	50	90	42	345	10	15	96/78	94/77	91/76	21	81	80	78					101.7	7.3
Little Rock AP	34	44	92	14	257	15	20	99/76	96/77	94/77	22	80	79	78	N	9	SSW		99.0	11.2
Pine Bluff AP	34	18	92	05	241	16	22	100/78	97/77	95/78	22	81	80	80	N	7	SW		102.2	13.1
Texarkana AP	33	27	93	59	389	18	23	98/76	96/77	93/76	21	80	79	78	WNW	9	SSW		104.8	14.0
CALIFORNIA																				
Bakersfield AP	35	25	119	03	475	30	32	104/70	101/69	98/68	32	73	71	70	ENE	5	WNW		109.8	25.3
Barstow AP	34	51	116	47	1927	26	29	106/68	104/68	102/67	37	73	71	70	WNW	7	W		110.4	17.4
Blythe AP	33	37	114	43	395	30	33	112/71	110/71	108/70	28	75	75	74					116.8	24.1
Burbank AP	34	12	118	21	775	37	39	95/68	91/68	88/67	25	71	70	69	NW	3	S			
Chico	39	48	121	51	238	28	30	103/69	101/68	98/67	36	71	69	68	NW	5	SSE		109.0	22.6
Concord	37	58	121	59	200	24	27	100/69	97/68	94/67	32	71	70	68	WNW	5	NW			
Covina	34	05	117	52	575	32	35	98/69	95/68	92/67	31	73	71	70						
Crescent City AP	41	46	124	12	40	31	33	68/60	65/59	63/58	18	62	60	59						
Downey	33	56	118	08	116	37	40	93/70	89/70	86/69	22	72	71	70						
El Cajon	32	49	116	58	367	42	44	83/69	80/69	78/68	30	71	70	68						
El Centro AP	32	49	115	40	−43	35	38	112/74	110/74	108/74	34	81	80	78	W	6	SE			
Escondido	33	07	117	05	660	39	41	89/68	85/68	82/68	30	71	70	69						
Eureka, Arcata AP	40	59	124	06	218	31	33	68/60	65/59	63/58	11	62	60	59	E	5	NW		75.8	29.7
Fairfield, Travis AFB	38	16	121	56	62	29	32	99/68	95/67	91/66	34	70	68	67	N	5	WSW			
Fresno AP	36	46	119	43	328	28	30	102/70	100/69	97/68	34	72	71	70	E	4	WNW		108.7	25.8
Hamilton AFB	38	04	122	30	3	30	32	89/68	84/66	80/65	28	72	69	67	N	4	SE			
Laguna Beach	33	33	117	47	35	41	43	83/68	80/68	77/67	18	70	69	68						
Livermore	37	42	121	57	545	24	27	100/69	97/68	93/67	24	71	70	68	WNW	4	NW			
Lompoc, Vandenberg AFB	34	43	120	34	368	35	38	75/61	70/61	67/60	20	63	61	60	ESE	5	NW			
Long Beach AP	33	49	118	09	30	41	43	83/68	80/68	77/67	22	70	69	68	NW	4	WNW			

[a] AP, AFB, following the station name designates airport or military airbase temperature observations. Co designates office locations within an urban area that are affected by the surrounding area. Undersigned stations are semirural and may be compared to airport data.

[b] Winter design data are based on the 3-month period, December through February.

[c] Summer design data are based on the 4-month period, June through September.

[d] Mean wind speeds occurring coincidentally with the 99.5% dry-bulb winter design temperature.

Reproduced by permission of American Society of Heating, Refrigerating and Air-Conditioning Engineers.

Table 1 Climatic Conditions for the United States (*Continued*)

Col. 1	Col. 2	Col. 3	Col. 4	Col. 5 Winter,[b] °F Design Dry-Bulb		Col. 6 Summer,[c] °F Design Dry-Bulb and Coincident Wet-Bulb			Col. 7 Mean Daily Range	Col. 8 Design Wet-Bulb			Col. 9 Prevailing Wind Knots[d]				Col. 10 Temp., °F Median of Annual Extr.	
State and Station[a]	Lat. ° ′N	Long. ° ′W	Elev. Feet	99%	97.5%	1%	2.5%	5%		1%	2.5%	5%	Winter		Summer		Max.	Min.
Los Angeles AP	33 56	118 24	97	41	43	83/68	80/68	77/67	15	70	69	68	E	4	WSW			
Los Angeles Co	34 03	118 14	270	37	40	93/70	89/70	86/69	20	72	71	70	NW	4	NW		98.1	35.9
Merced, Castle AFB	37 23	120 34	188	29	31	102/70	99/69	96/68	36	72	71	70	ESE	4	NW			
Modesto	37 39	121 00	91	28	30	101/69	98/68	95/67	36	71	70	69					105.8	26.2
Monterey	36 36	121 54	39	35	38	75/63	71/61	68/61	20	64	62	61	SE	4	NW			
Napa	38 13	122 17	56	30	32	100/69	96/68	92/67	30	71	69	68					103.1	25.8
Needles AP	34 46	114 37	913	30	33	112/71	110/71	108/70	27	75	75	74					116.4	26.7
Oakland AP	37 49	122 19	5	34	36	85/64	80/63	75/62	19	66	64	63	E	5	WNW		93.0	31.8
Oceanside	33 14	117 25	26	41	43	83/68	80/68	77/67	13	70	69	68						
Ontario	34 03	117 36	952	31	33	102/70	99/69	96/67	36	74	72	71	E	4	WSW			
Oxnard	34 12	119 11	49	34	36	83/66	80/64	77/63	19	70	68	67						
Palmdale AP	34 38	118 06	2542	18	22	103/65	101/65	98/64	35	69	67	66	SW	5	WSW			
Palm Springs	33 49	116 32	411	33	35	112/71	110/70	108/70	35	76	74	73						
Pasadena	34 09	118 09	864	32	35	98/69	95/68	92/67	29	73	71	70					102.8	30.4
Petaluma	38 14	122 38	16	26	29	94/68	90/66	87/65	31	72	70	68					102.0	24.2
Pomona Co	34 03	117 45	934	28	30	102/70	99/69	95/68	36	74	72	71	E	4	W		105.7	26.2
Redding AP	40 31	122 18	495	29	31	105/68	102/67	100/66	32	71	69	68					109.2	26.0
Redlands	34 03	117 11	1318	31	33	102/70	99/69	96/68	33	74	72	71					106.7	27.1
Richmond	37 56	122 21	55	34	36	85/64	80/63	75/62	17	66	64	63						
Riverside, March AFB	33 54	117 15	1532	29	32	100/68	98/68	95/67	37	72	71	70	N	4	NW		107.6	26.6
Sacramento AP	38 31	121 30	17	30	32	101/70	98/70	94/69	36	72	71	70	NNW	6	SW		105.1	27.6
Salinas AP	36 40	121 36	75	30	32	74/64	70/60	67/59	24	62	61	59						
San Bernardino, Norton AFB	34 08	117 16	1125	31	33	102/70	99/69	96/68	38	74	72	71	E	3	W		109.3	25.3
San Diego AP	32 44	117 10	13	42	44	83/69	80/69	78/68	12	71	70	68	NE	3	WNW		91.2	37.4
San Fernando	34 17	118 28	965	37	39	95/68	91/68	88/67	38	71	70	69						
San Francisco AP	37 37	122 23	8	35	38	82/64	77/63	73/62	20	65	64	62	S	5	NW			
San Francisco Co	37 46	122 26	72	38	40	74/63	71/62	69/61	14	64	62	61	W	5	W		91.3	35.9
San Jose AP	37 22	121 56	56	34	36	85/66	81/65	77/64	26	68	67	65	SE	4	NNW		98.6	28.2
San Luis Obispo	35 20	120 43	250	33	35	92/69	88/70	84/69	26	73	71	70	E	4	W		99.8	29.3
Santa Ana AP	33 45	117 52	115	37	39	89/69	85/68	82/68	28	71	70	69	E	3	SW		101.0	29.9
Santa Barbara AP	34 26	119 50	10	34	36	81/67	77/66	75/65	24	68	67	66	NE	3	SW		97.1	31.7
Santa Cruz	36 59	122 01	125	35	38	75/63	71/61	68/61	28	64	62	61					97.5	26.8
Santa Maria AP	34 54	120 27	236	31	33	81/64	76/63	73/62	23	65	64	63	E	4	WNW			
Santa Monica Co	34 01	118 29	64	41	43	83/68	80/68	77/67	16	70	69	68						
Santa Paula	34 21	119 05	263	33	35	90/68	86/67	84/66	36	71	69	68						
Santa Rosa	38 31	122 49	125	27	29	99/68	95/67	91/66	34	70	68	67	N	5	SE		102.5	23.4
Stockton AP	37 54	121 15	22	28	30	100/69	97/68	94/67	37	71	70	68	WNW	4	NW		104.1	24.5
Ukiah	39 09	123 12	623	27	29	99/69	95/68	91/67	40	70	68	67					108.1	21.6
Visalia	36 20	119 18	325	28	30	102/70	100/69	97/68	38	72	71	70					108.4	25.1
Yreka	41 43	122 38	2625	13	17	95/65	92/64	89/63	38	67	65	64					102.8	7.1
Yuba City	39 08	121 36	80	29	31	104/68	101/67	99/66	36	71	69	68						
COLORADO																		
Alamosa AP	37 27	105 52	7537	−21	−16	84/57	82/57	80/57	35	62	61	60						
Boulder	40 00	105 16	5445	2	8	93/59	91/59	89/59	27	64	63	62					96.0	−8.4
Colorado Springs AP	38 49	104 43	6145	−3	2	91/58	88/57	86/57	30	63	62	61	N	9	S		92.3	−12.1
Denver AP	39 45	104 52	5283	−5	1	93/59	91/59	89/59	28	64	63	62	S	8	SE		96.8	−10.4
Durango	37 17	107 53	6550	−1	4	89/59	87/59	85/59	30	64	63	62					92.4	−11.2
Fort Collins	40 35	105 05	4999	−10	−4	93/59	91/59	89/59	28	64	63	62					95.2	−18.1
Grand Junction AP	39 07	108 32	4843	2	7	96/59	94/59	92/59	29	64	63	62	ESE	5	WNW		99.9	−3.4
Greeley	40 26	104 38	4648	−11	−5	96/60	94/60	92/60	29	65	64	63						
Lajunta AP	38 03	103 30	4160	−3	3	100/68	98/68	95/67	31	72	70	69	W	8	S			
Leadville	39 15	106 18	10155	−8	−4	84/52	81/51	78/50	30	56	55	54					79.7	−17.8
Pueblo AP	38 18	104 29	4641	−7	0	97/61	95/61	92/61	31	67	66	65	W	5	SE		100.5	−12.2
Sterling	40 37	103 12	3939	−7	−2	95/62	93/62	90/62	30	67	66	65					100.3	−15.4
Trinidad AP	37 15	104 20	5740	−2	3	93/61	91/61	89/61	32	66	65	64	W	7	WSW		96.8	−10.5
CONNECTICUT																		
Bridgeport AP	41 11	73 11	25	6	9	86/73	84/71	81/70	18	75	74	73	NNW	13	WSW			
Hartford, Brainard Field	41 44	72 39	19	3	7	91/74	88/73	85/72	22	77	75	74	N	5	SSW		95.7	−4.4
New Haven AP	41 19	73 55	6	3	7	88/75	84/73	82/72	17	76	75	74	NNE	7	SW		93.0	−0.2
New London	41 21	72 06	59	5	9	88/73	85/72	83/71	16	76	75	74						
Norwalk	41 07	73 25	37	6	9	86/73	84/71	81/70	19	75	74	73						
Norwich	41 32	72 04	20	3	7	89/75	86/73	83/72	18	76	75	74						
Waterbury	41 35	73 04	843	−4	2	88/83	85/71	82/70	21	75	74	72	N	8	SW			
Windsor Locks, Bradley Fld	41 56	72 41	169	0	4	91/74	88/72	85/71	22	76	75	73	N	8	SW			
DELAWARE																		
Dover AFB	39 08	75 28	28	11	15	92/75	90/75	87/74	18	79	77	76	W	9	SW		97.0	7.0
Wilmington AP	39 40	75 36	74	10	14	92/74	89/74	87/73	20	77	76	75	WNW	9	WSW		95.4	4.9
DISTRICT OF COLUMBIA																		
Andrews AFB	38 5	76 5	279	10	14	92/75	90/74	87/73	18	78	76	75						
Washington, National AP	38 51	77 02	14	14	17	93/75	91/74	89/74	18	78	77	76	WNW	11	S		97.6	7.4

APPENDIX II—DESIGN CONDITIONS TABLES

Table 1 Climatic Conditions for the United States (*Continued*)

	Col. 1	Col. 2		Col. 3		Col. 4	Col. 5		Col. 6			Col. 7	Col. 8			Col. 9		Col. 10	
		Lat.		Long.		Elev.	Winter,[b] °F Design Dry-Bulb		Summer,[c] °F Design Dry-Bulb and Mean Coincident Wet-Bulb			Mean Daily Range	Design Wet-Bulb			Prevailing Wind Knots[d]		Temp., °F Median of Annual Extr.	
State and Station[a]		°	'N	°	'W	Feet	99%	97.5%	1%	2.5%	5%		1%	2.5%	5%	Winter	Summer	Max.	Min.
FLORIDA																			
Belle Glade		26	39	80	39	16	41	44	92/76	91/76	89/76	16	79	78	78			94.7	30.9
Cape Kennedy AP		28	29	80	34	16	35	38	90/78	88/78	87/78	15	80	79	79				
Daytona Beach AP		29	11	81	03	31	32	35	92/78	90/77	88/77	15	80	79	78	NW 8			
E Fort Lauderdale		26	04	80	09	10	42	46	92/78	91/78	90/78	15	80	79	79	NW 9	ESE		
Fort Myers AP		26	35	81	52	15	41	44	93/78	92/78	91/77	18	80	79	79	NNE 7	W	94.9	34.9
Fort Pierce		27	28	80	21	25	38	42	91/78	90/78	89/78	15	80	79	79			96.1	34.0
Gainesville AP		29	41	82	16	152	28	31	95/77	93/77	92/77	18	80	79	78	W 6	W	97.8	23.3
Jacksonville AP		30	30	81	42	26	29	32	96/77	94/77	92/76	19	79	79	78	NW 7	SW	97.5	25.4
Key West AP		24	33	81	45	4	55	57	90/78	90/78	89/78	09	80	79	79	NNE 12	SE	92.0	51.5
Lakeland Co		28	02	81	57	214	39	41	93/76	91/76	89/76	17	79	78	78	NNW 9	SSW		
Miami AP		25	48	80	16	7	44	47	91/77	90/77	89/77	15	79	79	78	NNW 8	SE	92.5	39.0
Miami Beach Co		25	47	80	17	10	45	48	90/77	89/77	88/77	10	79	79	78				
Ocala		29	11	82	08	89	31	34	95/77	93/77	92/76	18	80	79	78			98.6	24.8
Orlando AP		28	33	81	23	100	35	38	94/76	93/76	91/76	17	79	78	78	NNW 9	SSW		
Panama City, Tyndall AFB		30	04	85	35	18	29	33	92/78	90/78	89/77	14	81	80	79	N 8	WSW		
Pensacola Co		30	25	87	13	56	25	29	94/77	93/77	91/77	14	80	79	79	NNE 7	SW	96.3	23.3
St. Augustine		29	58	81	20	10	31	35	92/78	89/78	87/78	16	80	79	78	NW 7	W	97.6	25.8
St. Petersburg		27	46	82	80	35	36	40	92/77	91/77	90/76	16	79	79	78	N 8	W	94.8	35.6
Sanford		28	46	81	17	89	35	38	94/76	93/76	91/76	17	79	78	78				
Sarasota		27	23	82	33	26	39	42	93/77	92/77	90/76	17	79	79	78				
Tallahassee AP		30	23	84	22	55	27	30	94/77	92/76	90/76	19	79	78	78	NW 6	NW	97.6	20.9
Tampa AP		27	58	82	32	19	36	40	92/77	91/77	90/76	17	79	79	78	N 8	W	95.0	31.5
West Palm Beach AP		26	41	80	06	15	41	45	92/78	91/78	90/78	16	80	79	79	NW 9	ESE		
GEORGIA																			
Albany, Turner AFB		31	36	84	05	223	25	29	97/77	95/76	93/76	20	80	79	78	N 7	W	100.6	19.9
Americus		32	03	84	14	456	21	25	97/77	94/77	92/75	20	79	78	77			100.4	16.5
Athens		33	57	83	19	802	18	22	94/74	92/74	90/74	21	78	77	76	NW 9	WNW	98.7	13.5
Atlanta AP		33	39	84	26	1010	17	22	94/74	92/74	90/73	19	77	76	75	NW 11	NW	95.7	11.9
Augusta AP		33	22	81	58	145	20	23	97/76	95/76	93/76	19	80	79	78	W 4	WSW	99.0	17.5
Brunswick		31	15	81	29	25	29	32	92/78	89/78	87/78	18	80	79	79			99.3	24.7
Columbus, Lawson AFB		32	31	84	56	242	21	24	95/76	93/76	91/75	21	79	78	77	NW 8	W		
Dalton		34	34	84	57	720	17	22	94/76	93/76	91/76	22	79	78	77				
Dublin		32	20	82	54	215	21	25	96/77	94/77	92/77	20	79	78	77			101.0	16.7
Gainesville		34	11	83	41	50	24	27	96/77	93/77	91/77	20	80	79	78	WNW 7	SW	98.7	21.9
Griffin		33	13	84	16	981	18	22	93/76	90/75	88/74	21	78	77	76				
LaGrange		33	01	85	04	709	19	23	94/76	91/75	89/74	21	78	77	76			97.7	11.9
Macon AP		32	42	83	39	354	21	25	96/77	93/76	91/75	22	79	78	77	NW 8	WNW	99.6	16.9
Marietta, Dobbins AFB		33	55	84	31	1068	17	21	94/74	92/74	90/74	21	78	77	76	NNW 12	NW		
Savannah		32	08	81	12	50	24	27	96/77	93/77	91/77	20	80	79	78	WNW 7	SW	98.7	21.9
Valdosta-Moody AFB		30	58	83	12	233	28	31	96/77	94/77	92/76	20	80	79	78	WNW 6	W		
Waycross		31	15	82	24	148	26	29	96/77	94/77	91/76	20	80	79	78			100.0	19.5
HAWAII																			
Hilo AP		19	43	155	05	36	61	62	84/73	83/72	82/72	15	75	74	74	SW 6	NE		
Honolulu AP		21	20	157	55	13	62	63	87/73	86/73	85/72	12	76	75	74	ENE 12	ENE		
Kaneohe Bay MCAS		21	27	157	46	18	65	66	85/75	84/74	83/74	12	76	76	75	NNE 9	NE		
Wahiawa		21	03	158	02	900	58	59	86/73	85/72	84/72	14	75	74	73	WNW 5	E		
IDAHO																			
Boise AP		43	34	116	13	2838	3	10	96/65	94/64	91/64	31	68	66	65	SE 6	NW	103.2	0.6
Burley		42	32	113	46	4156	−3	2	99/62	95/61	92/62	35	64	63	61			98.6	−8.3
Coeur D'Alene AP		47	46	116	49	2972	−8	−1	89/62	86/61	83/60	31	64	63	61			99.9	−4.5
Idaho Falls AP		43	31	112	04	4741	−11	−6	89/61	87/61	84/59	38	65	63	61	N 9	S	96.2	−16.0
Lewiston AP		46	23	117	01	1413	−1	6	96/65	93/64	90/63	32	67	66	64	W 3	WNW	105.9	2.7
Moscow		46	44	116	58	2660	−7	0	90/63	87/62	84/61	32	65	64	62			98.0	−5.9
Mountain Home AFB		43	02	115	54	2996	6	12	99/64	97/63	94/62	36	66	65	63	ESE 7	NW	103.2	−6.5
Pocatello AP		42	55	112	36	4454	−8	−1	94/61	91/60	89/59	35	64	63	61	NE 5	W	97.9	−11.4
Twin Falls AP		42	29	114	29	4150	−3	2	99/62	95/61	92/60	34	64	63	61	SE 6	NW	100.9	−5.1
ILLINOIS																			
Aurora		41	45	88	20	744	−6	−1	93/76	91/76	88/75	20	79	78	76			96.7	−13.0
Belleville, Scott AFB		38	33	89	51	453	1	6	94/76	92/76	89/75	21	79	78	76	WNW 8	S		
Bloomington		40	29	88	57	876	−6	−2	92/75	90/74	88/73	21	78	76	75			98.4	−9.6
Carbondale		37	47	89	15	417	2	7	95/77	93/77	90/76	21	80	79	77			100.9	−0.8
Champaign/Urbana		40	02	88	17	777	−3	2	95/75	92/74	90/73	21	78	77	75				
Chicago, Midway AP		41	47	87	45	607	−5	0	94/74	91/73	88/72	20	77	75	74	NW 11	SW		
Chicago, O'Hare AP		41	59	87	54	658	−8	−4	91/74	89/74	86/72	20	77	76	74	WNW 9	SW		
Chicago Co		41	53	87	38	590	−3	2	94/75	91/74	88/73	15	79	77	75			96.1	−8.3
Danville		40	12	87	36	695	−4	1	93/75	90/74	88/73	21	78	77	75	W 10	SSW	98.2	−8.4
Decatur		39	50	88	52	679	−3	2	94/75	91/74	88/73	21	78	77	75	NW 10	SW	99.0	−8.1
Dixon		41	50	89	29	696	−7	−2	93/75	90/74	88/73	23	78	77	75			97.5	−13.5
Elgin		42	02	88	16	758	−7	−2	91/75	88/75	86/73	21	78	77	75				
Freeport		42	18	89	37	780	−9	−4	91/74	89/73	87/72	24	77	76	74				
Galesburg		40	56	90	26	764	−7	−2	93/75	91/75	88/74	22	78	77	75	WNW 8	SW		

Table 1 Climatic Conditions for the United States (Continued)

				Winter,[b] °F		Summer,[c] °F							Prevailing Wind		Temp., °F	
Col. 1	Col. 2	Col. 3	Col. 4	Col. 5		Col. 6			Col. 7	Col. 8			Col. 9		Col. 10	
	Lat.	Long.	Elev.	Design Dry-Bulb		Design Dry-Bulb and Mean Coincident Wet-Bulb			Mean Daily	Design Wet-Bulb			Winter	Summer	Median of Annual Extr.	
State and Station[a]	° 'N	° 'W	Feet	99%	97.5%	1%	2.5%	5%	Range	1%	2.5%	5%	Knots[d]		Max.	Min.
Greenville	38 53	89 24	563	−1	4	94/76	92/75	89/74	21	79	78	76				
Joliet	41 31	88 10	582	−5	0	93/75	90/74	88/73	20	78	77	75	NW 11	SW		
Kankakee	41 05	87 55	625	−4	1	93/75	90/74	88/73	21	78	77	75				
La Salle/Peru	41 19	89 06	520	−7	−2	93/75	91/75	88/74	22	78	77	75				
Macomb	40 28	90 40	702	−5	0	95/76	92/75	89/75	22	79	78	76				
Moline AP	41 27	90 31	582	−9	−4	93/75	91/75	88/74	23	78	77	75	WNW 8	SW	96.8	−12.7
Mt Vernon	38 19	88 52	479	0	5	95/76	92/75	89/74	21	79	78	76			100.5	−2.9
Peoria AP	40 40	89 41	652	−8	−4	91/75	89/74	87/73	22	78	76	75	WNW 8	SW	98.0	−10.9
Quincy AP	39 57	91 12	769	−2	3	96/76	93/76	90/76	22	80	78	77	NW 11	SSW	101.1	−6.7
Rantoul, Chanute AFB	40 18	88 08	753	−4	1	94/75	91/74	89/73	21	78	77	75	W 10	SSW		
Rockford	42 21	89 03	741	−9	−4	91/74	89/73	87/72	24	77	76	74			97.4	−13.8
Springfield AP	39 50	89 40	588	−3	2	94/75	92/75	89/74	21	79	77	76	NW 10	SW	98.1	−7.2
Waukegan	42 21	87 53	700	−6	−3	92/76	89/75	87/73	21	78	76	75			96.5	−10.6
INDIANA																
Anderson	40 06	85 37	919	0	6	95/76	92/75	89/74	22	79	78	76	W 9	SW	95.1	−6.0
Bedford	38 51	86 30	670	0	5	95/76	92/75	89/74	22	79	78	76			97.5	−4.4
Bloomington	39 08	86 37	847	0	5	95/76	92/75	89/74	22	79	78	76	W 9	SW	97.8	−4.6
Columbus, Bakalar AFB	39 16	85 54	651	3	7	95/76	92/75	90/74	22	79	78	76	W 9	SW	98.3	−6.4
Crawfordsville	40 03	86 54	679	−2	3	94/75	91/75	88/73	22	79	77	76			98.4	−7.6
Evansville AP	38 03	87 32	381	4	9	95/76	93/75	91/75	22	79	78	77	NW 9	SW	98.2	0.2
Fort Wayne AP	41 00	85 12	791	−4	1	92/73	89/72	87/72	24	77	75	74	WSW 10	SW		
Goshen AP	41 32	85 48	827	−3	1	91/73	89/73	86/72	23	77	75	74			96.8	−10.5
Hobart	41 32	87 15	600	−4	2	91/73	88/73	85/72	21	77	75	74			98.5	−8.5
Huntington	40 53	85 30	802	−4	1	92/73	89/72	87/72	23	77	75	74			96.9	−8.1
Indianapolis AP	39 44	86 17	792	−2	2	92/74	90/74	87/73	22	78	76	75	WNW 10	SW	96	−7
Jeffersonville	38 17	85 45	455	5	10	95/74	93/74	90/74	23	79	77	76			98	2
Kokomo	40 25	86 03	855	−4	0	91/74	90/73	88/73	22	77	75	74			98.2	−7.5
Lafayette	40 2	86 5	600	−3	3	94/74	91/73	88/73	22	78	76	75				
La Porte	41 36	86 43	810	−3	3	93/74	90/74	88/73	22	78	76	75			98.1	−10.5
Marion	40 29	85 41	859	−4	0	91/74	90/73	88/73	23	77	75	74			97.0	−8.6
Muncie	40 11	85 21	957	−3	2	92/73	90/74	87/73	22	76	76	75				
Peru, Grissom AFB	40 39	86 09	813	−6	−1	90/74	88/73	86/73	22	77	75	74	W 10	SW		
Richmond AP	39 46	84 50	1141	−2	2	92/74	90/74	87/73	22	78	76	75			94.8	−8.5
Shelbyville	39 31	85 47	750	−1	3	93/74	91/74	88/73	22	78	76	75			97.7	−6.0
South Bend AP	41 42	86 19	773	−3	1	91/73	89/73	86/72	22	77	75	74	SW 11	SSW	96.2	−9.2
Terre Haute AP	39 27	87 18	585	−2	4	95/75	92/74	89/73	22	79	77	76	NNW 7	SSW	98.3	−4.9
Valparaiso	41 31	87 02	801	−3	3	93/74	90/74	87/73	22	78	76	75			95.5	−11.0
Vincennes	38 41	87 32	420	1	6	95/75	92/74	90/73	22	79	77	76			100.3	−2.8
IOWA																
Ames	42 02	93 48	1099	−11	−6	93/75	90/74	87/73	23	78	76	75			97.4	−17.8
Burlington AP	40 47	91 07	692	−7	−3	94/74	91/74	88/73	22	78	77	75	NW 9	SSW	98.6	−11.0
Cedar Rapids AP	41 53	91 42	863	−10	−5	91/76	88/75	86/74	23	78	77	75	NW 9	S	97.7	−15.6
Clinton	41 50	90 13	595	−8	−3	92/75	90/75	87/74	23	78	77	75			97.5	−13.8
Council Bluffs	41 20	95 49	1210	−8	−3	94/76	91/75	88/74	22	78	77	75				
Des Moines AP	41 32	93 39	938	−10	−5	94/75	91/74	88/73	23	78	77	75	NW 11	S	98.2	−14.2
Dubuque	42 24	90 42	1056	−12	−7	90/74	88/73	86/72	22	77	75	74	N 10	SSW	95.2	−15.0
Fort Dodge	42 33	94 11	1162	−12	−7	91/74	88/74	86/72	23	77	75	74	NW 11		98.5	−19.1
Iowa City	41 38	91 33	661	−11	−6	92/76	89/76	87/74	22	80	78	76	NW 9	SSW	97.4	−15.2
Keokuk	40 24	91 24	574	−5	0	95/75	92/75	89/74	22	79	77	76			98.4	−8.8
Marshalltown	42 04	92 56	898	−12	−7	92/76	90/75	88/74	23	78	77	75			98.5	−13.4
Mason City AP	43 09	93 20	1213	−15	−11	90/74	88/74	85/72	24	77	75	74	NW 11	S	96.5	−21.7
Newton	41 41	93 02	936	−10	−5	94/75	91/74	88/73	23	78	77	75			98.2	−14.7
Ottumwa AP	41 06	92 27	840	−8	−4	94/75	91/74	88/73	22	78	77	75			99.1	−12.0
Sioux City AP	42 24	96 23	1095	−11	−7	95/74	92/74	89/73	24	78	77	75	NNW 9	S	99.9	−17.7
Waterloo	42 33	92 24	868	−15	−10	91/76	89/75	86/74	23	78	77	75	NW 9	S	97.7	−19.8
KANSAS																
Atchison	39 34	95 07	945	−2	2	96/77	93/76	91/76	23	81	79	77			100.5	−8.8
Chanute AP	37 40	95 29	981	3	7	100/74	97/74	94/74	23	78	77	76	NNW 11	SSW	102.8	−2.8
Dodge City AP	37 46	99 58	2582	0	5	100/69	97/69	95/69	25	74	73	71	N 12	SSW	102.9	−7.0
El Dorado	37 49	96 50	1282	3	7	101/72	98/73	96/73	24	77	76	75			103.5	−5.0
Emporia	38 20	96 12	1210	1	5	100/74	97/74	94/73	25	78	77	76			102.4	−6.4
Garden City AP	37 56	100 44	2880	−1	4	99/69	96/69	94/69	28	74	73	71				
Goodland AP	39 22	101 42	3654	−5	0	99/66	96/65	93/66	31	71	70	68	WSW 10	S	103.2	−10.4
Great Bend	38 21	98 52	1889	0	4	101/73	98/73	95/73	28	78	76	75			105.3	−6.1
Hutchinson AP	38 04	97 52	1542	4	8	102/72	99/72	97/72	28	77	75	74	N 14	S	105.8	−3.8
Liberal	37 03	100 58	2870	2	7	99/68	96/68	94/68	28	73	72	71				
Manhattan, Ft Riley	39 03	96 46	1065	−1	3	99/75	95/75	92/74	24	78	77	76	NNE 8	S	104.5	−8.6
Parsons	37 20	95 31	899	5	9	100/74	97/74	94/74	23	79	77	76	NNW 11	SSW		
Russell AP	38 52	98 49	1866	0	4	101/73	98/73	95/73	29	78	76	75				
Salina	38 48	97 39	1272	0	5	103/74	100/74	97/73	26	78	77	75	N 8	SSW		
Topeka AP	39 04	95 38	877	0	4	99/75	96/75	93/74	24	79	78	76	NNW 10	S	101.8	−6.4
Wichita AP	37 39	97 25	1321	3	7	101/72	98/73	96/73	23	77	76	75	NNW 12	SSW	102.5	−2.8

APPENDIX II—DESIGN CONDITIONS TABLES

Table 1 Climatic Conditions for the United States (*Continued*)

				Winter,[b] °F		Summer,[c] °F							Prevailing Wind				Temp., °F	
Col. 1	Col. 2	Col. 3	Col. 4	Col. 5		Col. 6			Col. 7	Col. 8			Col. 9				Col. 10	
				Design Dry-Bulb		Design Dry-Bulb and Mean Coincident Wet-Bulb			Mean Daily Range	Design Wet-Bulb			Winter		Summer		Median of Annual Extr.	
State and Station[a]	Lat. ° 'N	Long. ° 'W	Elev. Feet	99%	97.5%	1%	2.5%	5%		1%	2.5%	5%		Knots[d]			Max.	Min.
KENTUCKY																		
Ashland	38 33	82 44	546	5	10	94/76	91/74	89/73	22	78	77	75	W	6	SW		97.4	0.8
Bowling Green AP	35 58	86 28	535	4	10	94/77	92/75	89/74	21	79	77	76					99.9	1.2
Corbin AP	36 57	84 06	1174	4	9	94/73	92/73	89/72	23	77	76	75						
Covington AP	39 03	84 40	869	1	6	92/73	90/72	88/72	22	77	75	74	W	9	SW			
Hopkinsville, Ft Campbell	36 40	87 29	571	4	10	94/77	92/75	89/74	21	79	77	76	N	6	W		100.1	−0.4
Lexington AP	38 02	84 36	966	3	8	93/73	91/73	88/72	22	77	76	75	WNW	9	SW		95.3	−0.5
Louisville AP	38 11	85 44	477	5	10	95/74	93/74	90/74	23	79	77	76	NW	8	SW		97.4	1.2
Madisonville	37 19	87 29	439	5	10	96/76	93/75	90/75	22	79	78	77						
Owensboro	37 45	87 10	407	5	10	97/76	94/75	91/75	23	79	78	77	NW	9	SW		98.0	−0.2
Paducah AP	37 04	88 46	413	7	12	98/76	95/75	92/75	20	79	78	77						
LOUISIANA																		
Alexandria AP	31 24	92 18	92	23	27	95/77	94/77	92/77	20	80	79	78	N	7	S		100.1	−5.7
Baton Rouge AP	30 32	91 09	64	25	29	95/77	93/77	92/77	19	80	80	79	ENE	8	W		98.0	21.4
Bogalusa	30 47	89 52	103	24	28	95/77	93/77	92/77	19	80	80	79					99.3	20.2
Houma	29 31	90 40	13	31	35	95/78	93/78	92/77	15	81	80	79					97.2	22.5
Lafayette AP	30 12	92 00	42	26	30	95/78	94/78	92/78	18	81	80	79	N	8	SW		98.2	22.6
Lake Charles AP	30 07	93 13	9	27	31	95/77	93/77	92/77	17	80	79	79	N	9	SSW		99.2	20.5
Minden	32 36	93 18	250	20	25	99/77	96/76	94/76	20	79	79	78					101.7	−4.9
Monroe AP	32 31	92 02	79	20	25	99/77	96/76	94/76	20	79	79	78	N	9	S		101.1	−5.9
Natchitoches	31 46	93 05	130	22	26	97/77	95/77	93/77	20	80	79	78						
New Orleans AP	29 59	90 15	4	29	33	93/78	92/78	90/77	16	81	80	79	NNE	9	SSW		96.3	27.7
Shreveport AP	32 28	93 49	254	20	25	99/77	96/76	94/76	20	79	79	78	N	9	S			
MAINE																		
Augusta AP	44 19	69 48	353	−7	−3	88/73	85/70	82/68	22	74	72	70	NNE	10	WNW			
Bangor, Dow AFB	44 48	68 50	192	−11	−6	86/70	83/68	80/67	22	73	71	69	WNW	7	S			
Caribou AP	46 52	68 01	624	−18	−13	84/69	81/67	78/66	21	71	69	67	WSW	10	SW			
Lewiston	44 02	70 15	200	−7	−2	88/73	85/70	82/68	22	74	72	70					94.0	−13.7
Millinocket AP	45 39	68 42	413	−13	−9	87/69	83/68	80/66	22	72	70	68	WNW	11	WNW		92.4	−23.0
Portland	43 39	70 19	43	−6	−1	87/72	84/71	81/69	22	74	72	70	W	7	S		93.5	−9.9
Waterville	44 32	69 40	302	−8	−4	87/72	84/69	81/68	22	74	72	70						
MARYLAND																		
Baltimore AP	39 11	76 40	148	10	13	94/75	91/75	89/74	21	78	77	76	W	9	WSW			
Baltimore Co	39 20	76 25	20	14	17	92/75	89/76	87/75	17	80	78	76	WNW	9	S		97.9	7.2
Cumberland	39 37	78 46	790	6	10	92/75	89/74	87/74	22	77	76	75	WNW	10	W			
Frederick AP	39 27	77 25	313	8	12	94/76	91/75	88/74	22	78	77	76	N	9	WNW			
Hagerstown	39 42	77 44	704	8	12	94/75	91/74	89/74	22	77	76	75	WNW	10	W			
Salisbury	38 20	75 30	59	12	16	93/75	91/75	88/74	18	79	77	76					96.8	7.4
MASSACHUSETTS																		
Boston AP	42 22	71 02	15	6	9	91/73	88/71	85/70	16	75	74	72	WNW	16	SW		95.7	−1.2
Clinton	42 24	71 41	398	−2	2	90/72	87/71	84/69	17	75	73	72					91.7	−8.5
Fall River	41 43	71 08	190	5	9	87/72	84/71	81/69	18	74	73	72	NW	10	SW		92.1	−1.0
Framingham	42 17	71 25	170	3	6	89/72	86/71	83/69	17	74	73	71					96.0	−7.7
Gloucester	42 35	70 41	10	2	5	89/73	86/71	83/70	15	75	74	72						
Greenfield	42 3	72 4	205	−7	−2	88/72	85/71	82/69	23	74	73	71						
Lawrence	42 42	71 10	57	−6	0	90/73	87/72	84/70	22	76	74	73	NW	8	WSW		95.2	−9.0
Lowell	42 39	71 19	90	−4	1	91/74	88/72	85/70	21	76	74	73					95.1	−8.5
New Bedford	41 41	70 58	79	5	9	85/72	82/71	80/69	19	74	73	72	NW	10	SW		91.4	2.2
Pittsfield AP	42 26	73 18	1194	−8	−3	87/71	84/70	81/68	23	73	72	70	NW	12	SW			
Springfield, Westover AFB	42 12	72 32	245	−5	0	90/72	87/71	84/69	19	75	73	72	N	8	SSW		95.7	−4.7
Taunton	41 54	71 04	20	5	9	89/73	86/72	83/70	18	75	74	73					92.9	−9.8
Worcester AP	42 16	71 52	986	0	4	87/71	84/70	81/68	18	73	72	70	W	14	W			
MICHIGAN																		
Adrian	41 55	84 01	754	−1	3	91/73	88/72	85/71	23	76	75	73					97.2	−7.0
Alpena AP	45 04	83 26	610	−11	−6	89/73	85/70	83/69	27	73	72	70	W	5	SW		93.9	−14.8
Battle Creek AP	42 19	85 15	941	1	5	92/74	88/72	85/70	23	76	74	73	SW	8	SW			
Benton Harbor AP	42 08	86 26	643	1	5	91/72	88/72	85/70	20	75	74	72	SSW	8	WSW			
Detroit	42 25	83 01	619	3	6	91/73	88/72	86/71	20	76	74	73	W	11	SW		95.1	−2.6
Escanaba	45 44	87 05	607	−11	−7	87/70	83/69	80/68	17	73	71	69					88.8	−16.1
Flint AP	42 58	83 44	771	−4	1	90/73	87/72	85/71	25	76	74	72	SW	8	SW		95.3	−9.9
Grand Rapids AP	42 53	85 31	784	1	5	91/72	88/72	85/70	24	75	74	72	WNW	8	WSW		95.4	−5.6
Holland	42 42	86 06	678	2		88/72	86/71	83/70	22	75	73	72					94.1	−6.8
Jackson AP	42 16	84 28	1020	1	5	92/74	88/72	85/70	23	76	74	73					96.5	−7.8
Kalamazoo	42 17	85 36	955	1	5	92/74	88/72	85/70	23	76	74	73					95.9	−6.7
Lansing AP	42 47	84 36	873	−3	1	90/73	87/72	84/70	24	75	74	72	SW	12	W		94.6	−11.0
Marquette Co	46 34	87 24	735	−12	−8	84/70	81/69	77/66	18	72	70	68					94.5	−11.8
Mt Pleasant	43 35	84 46	796	0	4	91/73	87/72	84/71	24	76	74	72					95.4	−11.1
Muskegon AP	43 10	86 14	625	2	6	86/72	84/70	82/70	21	75	73	72	E	8	SW			
Pontiac	42 40	83 25	981	0	4	90/73	87/72	85/71	21	76	74	73					95.0	−6.8

Table 1 Climatic Conditions for the United States (Continued)

Col. 1	Col. 2		Col. 3		Col. 4	Col. 5		Col. 6			Col. 7	Col. 8			Col. 9		Col. 10	
	Lat.		Long.		Elev.	Winter,[b] °F Design Dry-Bulb		Summer,[c] °F Design Dry-Bulb and Mean Coincident Wet-Bulb			Mean Daily Range	Design Wet-Bulb			Prevailing Wind Knots[d]		Temp., °F Median of Annual Extr.	
State and Station[a]	°'N		°'W		Feet	99%	97.5%	1%	2.5%	5%		1%	2.5%	5%	Winter	Summer	Max.	Min.
Port Huron	42	59	82	25	586	0	4	90/73	87/72	83/71	21	76	74	73	W 8	S		
Saginaw AP	43	32	84	05	667	0	4	91/73	87/72	84/71	23	76	74	72	WSW 7	SW	96.1	−7.6
Sault Ste. Marie AP	46	28	84	22	721	−12	−8	84/70	81/69	77/66	23	72	70	68	E 7	SW	89.8	−21.0
Traverse City AP	44	45	85	35	624	−3	1	89/72	86/71	83/69	22	75	73	71	SSW 9	SW	95.4	−10.7
Ypsilanti	42	14	83	32	716	1	5	92/72	89/71	86/70	22	75	74	72	SW 10	SW		
MINNESOTA																		
Albert Lea	43	39	93	21	1220	−17	−12	90/74	87/72	84/71	24	77	75	73				
Alexandria AP	45	52	95	23	1430	−22	−16	91/72	88/72	85/70	24	76	74	72			95.1	−28.0
Bemidji AP	47	31	94	56	1389	−31	−26	88/69	85/69	81/67	24	73	71	69	N 8	S	94.5	−36.9
Brainerd	46	24	94	08	1227	−20	−16	90/73	87/71	84/69	24	75	73	71				
Duluth AP	46	50	92	11	1428	−21	−16	85/70	82/68	79/66	22	72	70	68	WNW 12	WSW	90.9	−27.4
Fairbault	44	18	93	16	940	17	−12	91/74	88/72	85/71	24	77	75	73			95.8	−24.3
Fergus Falls	46	16	96	04	1210	−21	−17	91/72	88/72	85/70	24	76	74	72			96.9	−27.8
International Falls AP	48	34	93	23	1179	−29	−25	85/68	83/68	80/66	26	71	70	68	N 9	S	93.4	−36.5
Mankato	44	09	93	59	1004	−17	−12	91/72	88/72	85/70	24	77	75	73				
Minneapolis/St. Paul AP	44	53	93	13	834	−16	−12	92/75	89/73	86/71	22	77	75	73	NW 8	S	96.5	−22.0
Rochester AP	43	55	92	30	1297	−17	−12	90/74	87/72	84/71	24	77	75	73	NW 9	SSW		
St. Cloud AP	45	35	94	11	1043	−15	−11	91/74	88/72	85/70	24	76	74	72				
Virginia	47	30	92	33	1435	−25	−21	85/69	83/68	80/66	23	71	70	68			92.6	−33.0
Willmar	45	07	95	05	1128	−15	−11	91/74	88/72	85/71	24	76	74	72			96.8	−24.3
Winona	44	03	91	38	652	−14	−10	91/75	88/73	85/72	24	77	75	74				
MISSISSIPPI																		
Biloxi, Keesler AFB	30	25	88	55	26	28	31	94/79	92/79	90/78	16	82	81	80	N 8	S	98.0	23.0
Clarksdale	34	12	90	34	178	14	19	96/77	94/77	92/76	21	80	79	78			100.9	13.2
Columbus AFB	33	39	88	27	219	15	20	95/77	93/77	91/76	22	80	79	78	N 7	W	101.6	12.7
Greenville AFB	33	29	90	59	138	15	20	95/77	93/77	91/76	21	80	79	78			99.5	14.9
Greenwood	33	30	90	05	148	15	20	95/77	93/77	91/76	21	80	79	78			100.6	15.3
Hattiesburg	31	16	89	15	148	24	27	96/78	94/77	92/77	21	81	80	79			99.9	18.2
Jackson AP	32	19	90	05	310	21	25	97/76	95/76	93/76	21	79	78	78	NNW 6	NW	99.8	16.0
Laurel	31	40	89	10	236	24	27	96/78	94/77	92/77	21	81	80	79			99.7	17.8
McComb AP	31	15	90	28	469	21	26	96/77	94/76	92/76	18	80	79	78				
Meridian AP	32	20	88	45	290	19	23	97/77	95/76	93/76	22	80	79	78	N 6	WSW	98.3	15.7
Natchez	31	33	91	23	195	23	27	96/78	94/78	92/77	21	81	80	79			98.4	18.4
Tupelo	34	16	88	46	361	14	19	96/77	94/77	92/76	22	80	79	78			100.7	11.8
Vicksburg Co	32	24	90	47	262	22	26	97/78	95/78	93/77	21	81	80	79			96.9	18.0
MISSOURI																		
Cape Girardeau	37	14	89	35	351	8	13	98/76	95/75	92/75	21	79	78	77				
Columbia AP	38	58	92	22	778	−1	4	97/74	94/74	91/73	22	78	77	76	WNW 9	WSW	99.5	−6.2
Farmington AP	37	46	90	24	928	3	8	96/76	93/75	90/74	22	78	77	75			99.9	−2.1
Hannibal	39	42	91	21	489	−2	3	96/76	93/76	90/76	22	80	78	77	NNW 11	SSW	98.4	−7.6
Jefferson City	38	34	92	11	640	2	7	98/75	95/75	92/74	23	78	77	76			101.2	−6.1
Joplin AP	37	09	94	30	980	6	10	100/73	97/73	94/73	24	78	77	76	NNW 12	SSW		
Kansas City AP	39	07	94	35	791	2	6	99/75	96/74	93/74	20	78	77	76	NW 9	S	100.2	−4.3
Kirksville AP	40	06	92	33	964	−5	0	96/74	93/74	90/73	24	78	77	76			98.3	−10.8
Mexico	39	11	91	54	775	−1	4	97/74	94/74	91/73	22	78	77	76			101.2	−8.0
Moberly	39	24	92	26	850	−2	3	97/74	94/74	91/73	23	78	77	76				
Poplar Bluff	36	46	90	25	380	11	16	98/78	95/76	92/76	22	81	79	78				
Rolla	37	59	91	43	1204	3	9	94/77	91/75	89/74	22	78	77	76			99.4	−3.1
St. Joseph AP	39	46	94	55	825	−3	2	96/77	93/76	91/76	23	81	79	77	NNW 9	S	100.6	−8.0
St. Louis AP	38	45	90	23	535	2	6	97/75	94/75	91/74	21	78	77	76	NW 9	WSW		
St. Louis Co	38	39	90	38	462	3	8	98/75	94/75	91/74	18	78	77	76	NW 6	S	99.1	−2.7
Sikeston	36	53	89	36	325	9	15	98/77	95/76	92/75	21	80	78	77				
Sedalia, Whiteman AFB	38	43	93	33	869	−1	4	95/76	92/75	90/75	22	79	78	76	NNW 7	SSW	100.0	−5.1
Sikeston	36	53	89	36	325	9	15	98/77	95/76	92/75	21	80	78	77				
Springfield AP	37	14	93	23	1268	3	9	96/73	93/74	91/74	23	78	77	75	NNW 10	S	97.2	−2.4
MONTANA																		
Billings AP	45	48	108	32	3567	−15	−10	94/64	91/64	88/63	31	67	66	64	NE 9	SW	100.5	−19.1
Bozeman	45	47	111	09	4448	−20	−14	90/61	87/60	84/59	32	63	62	60			92.2	−23.2
Butte AP	45	57	112	30	5553	−24	−17	86/58	83/56	80/56	35	60	58	57	S 5	NW	91.8	−26.3
Cut Bank AP	48	37	112	22	3838	−25	−20	88/61	85/61	82/60	35	64	62	61			94.7	−30.9
Glasgow AP	48	25	106	32	2760	−22	−18	92/64	89/63	85/62	29	68	66	64	E 8	S		
Glendive	47	08	104	48	2476	−18	−13	95/66	92/64	89/62	29	69	67	65			103.3	−29.8
Great Falls AP	47	29	111	22	3662	−21	−15	91/60	88/60	85/59	28	64	62	60	SW 7	WSW	98.0	−25.1
Havre	48	34	109	40	2492	−18	−11	94/65	90/64	87/63	33	68	66	65			99.7	−31.3
Helena AP	46	36	112	00	3828	−21	−16	91/60	88/60	85/59	32	64	62	61	N 12	WNW	95.6	−23.7
Kalispell AP	48	18	114	16	2974	−14	−7	91/62	87/61	84/60	34	65	63	62			94.4	−16.8
Lewistown AP	47	04	109	27	4122	−22	−16	90/62	87/61	83/60	30	65	63	62	NW 9	NW	96.2	−27.7
Livingstown AP	45	42	110	26	4618	−20	−14	90/61	87/60	84/59	32	63	62	60			97.2	−21.2
Miles City AP	46	26	105	52	2634	−20	−15	98/66	95/66	92/65	30	70	68	67	NW 7	SE	103.6	−27.7
Missoula AP	46	55	114	05	3190	−13	−6	92/62	88/61	85/60	36	65	63	62	ESE 7	NW	98.6	−13.9

Table 1 Climatic Conditions for the United States (Continued)

Col. 1	Col. 2		Col. 3		Col. 4	Winter,[b] °F Col. 5		Summer,[c] °F Col. 6			Col. 7	Col. 8			Prevailing Wind Col. 9				Temp., °F Col. 10	
	Lat.		Long.		Elev.	Design Dry-Bulb		Design Dry-Bulb and Mean Coincident Wet-Bulb			Mean Daily Range	Design Wet-Bulb			Winter		Summer		Median of Annual Extr.	
State and Station[a]	°N	′	°W	′	Feet	99%	97.5%	1%	2.5%	5%		1%	2.5%	5%	Knots[d]				Max.	Min.
NEBRASKA																				
Beatrice	40	16	96	45	1235	−5	−2	99/75	95/74	92/74	24	78	77	76					103.1	−11.3
Chadron AP	42	50	103	05	3313	−8	−3	97/66	94/65	91/65	30	71	69	68						
Columbus	41	28	97	20	1450	−6	−2	98/74	95/73	92/73	25	77	76	75						
Fremont	41	26	96	29	1200	−6	−2	98/75	95/74	92/74	22	78	77	76						
Grand Island AP	40	59	98	19	1860	−8	−3	97/72	94/71	91/71	28	75	74	73	NNW	10		S	103.3	−14.2
Hastings	40	36	98	26	1954	−7	−3	97/72	94/71	91/71	27	75	74	73	NNW	10		S	103.5	−10.7
Kearney	40	44	99	01	2132	−9	−4	96/72	93/70	90/70	28	74	73	72					102.9	−13.7
Lincoln Co	40	51	96	45	1180	−5	−2	99/75	95/74	92/74	24	78	77	76	N	8		S	102.0	−12.4
McCook	40	12	100	38	2768	−6	−2	98/69	95/69	91/69	28	74	72	71						
Norfolk	41	59	97	26	1551	−8	−4	97/74	93/74	90/73	30	78	77	75					102.0	−20.0
North Platte AP	41	08	100	41	2779	−8	−4	97/69	94/69	90/69	28	74	72	71	NW	9		SSE	100.8	−15.8
Omaha AP	41	18	95	54	977	−8	−3	94/76	91/75	88/74	22	78	77	75	NW	8		S	100.2	−13.2
Scottsbluff AP	41	52	103	36	3958	−8	−3	95/65	92/65	90/64	31	70	68	67	NW	9		SE	101.6	−18.9
Sidney AP	41	13	103	06	4399	−8	−3	95/65	92/65	90/64	31	70	68	67						
NEVADA																				
Carson City	39	10	119	46	4675	4	9	94/60	91/59	89/58	42	63	61	60	SSW	3		WNW	99.2	−5.0
Elko AP	40	50	115	47	5050	−8	−2	94/59	92/59	90/58	42	63	62	60	E	4		SW		
Ely AP	39	17	114	51	6253	−10	−4	89/57	87/56	85/55	39	60	59	58	S	9		SSW		
Las Vegas AP	36	05	115	10	2178	25	28	108/66	106/65	104/65	30	71	70	69	ENE	7		SW		
Lovelock AP	40	04	118	33	3903	8	12	98/63	96/63	93/62	42	66	65	64					103.0	−1.0
Reno AP	39	30	119	47	4404	5	10	95/61	92/60	90/59	45	64	62	61	SSW	3		WNW		
Reno Co	39	30	119	47	4408	6	11	96/61	93/60	91/59	45	64	62	61					98.9	0.2
Tonopah AP	38	04	117	05	5426	5	10	94/60	92/59	90/58	40	64	62	61	N	8		S		
Winnemucca AP	40	54	117	48	4301	−1	3	96/60	94/60	92/60	42	64	62	61	SE	10		W	100.1	−8.1
NEW HAMPSHIRE																				
Berlin	44	03	71	01	1110	−14	−9	87/71	84/69	81/68	22	73	71	70					93.2	−24.7
Claremont	43	02	72	02	420	−9	−4	89/72	86/70	83/69	24	74	73	71						
Concord AP	43	12	71	30	342	−8	−3	90/72	87/70	84/69	26	74	73	71	NW	7		SW	94.8	−16.0
Keene	42	55	72	17	490	−12	−7	90/72	87/70	83/69	24	74	73	71					94.6	−18.9
Laconia	43	03	71	03	505	−10	−5	89/72	86/70	83/69	25	74	73	71						
Manchester, Grenier AFB	42	56	71	26	233	−8	−3	91/72	88/71	85/70	24	75	74	72	N	11		SW	93.7	−12.6
Portsmouth, Pease AFB	43	04	70	49	101	−2	2	89/73	85/71	83/70	22	75	74	72	W	8		W		
NEW JERSEY																				
Atlantic City Co	39	23	74	26	11	10	13	92/74	89/74	86/72	18	78	77	75	NW	11		WSW	93.0	7.5
Long Branch	40	19	74	01	15	10	13	93/74	90/73	87/72	18	78	77	75					95.9	4.3
Newark AP	40	42	74	10	7	10	14	94/74	91/73	88/72	20	77	76	75	WNW	11		WSW		
New Brunswick	40	29	74	26	125	6	10	92/74	89/73	86/72	19	77	76	75						
Paterson	40	54	74	09	100	6	10	94/74	91/73	88/72	21	77	76	75						
Phillipsburg	40	41	75	11	180	1	6	92/73	89/72	86/71	21	76	75	74					97.4	−0.7
Trenton Co	40	13	74	46	56	11	14	91/75	88/74	85/73	19	78	76	75	W	9		SW	96.2	4.2
Vineland	39	29	75	00	112	8	11	91/75	89/74	86/73	19	78	76	75						
NEW MEXICO																				
Alamagordo, Holloman AFB	32	51	106	06	4093	14	19	98/64	96/64	94/64	30	69	68	67						
Albuquerque AP	35	03	106	37	5311	12	16	96/61	94/61	92/61	27	66	65	64	N	7		W	98.1	5.1
Artesia	32	46	104	23	3320	13	19	103/67	100/67	97/67	30	72	71	70					105.5	3.7
Carlsbad AP	32	20	104	16	3293	13	19	103/67	100/67	97/67	28	72	71	70	N	6		SSE		
Clovis AP	34	23	103	19	4294	8	13	95/65	93/65	91/65	28	69	68	67					102.0	2.5
Farmington AP	36	44	108	14	5503	1	6	95/63	93/62	91/61	30	67	65	64	ENE	5		SW		
Gallup	35	31	108	47	6465	0	5	90/59	89/58	86/58	32	64	62	61						
Grants	35	10	107	54	6524	−1	4	89/59	88/58	85/57	32	64	62	61						
Hobbs AP	32	45	103	13	3690	13	18	101/66	99/66	97/66	29	71	70	69						
Las Cruces	32	18	106	55	4544	15	20	99/64	96/64	94/64	30	69	68	67	SE	5		SE		
Los Alamos	35	52	106	19	7410	5	9	89/60	87/60	85/60	32	62	61	60					89.8	−2.3
Raton AP	36	45	104	30	6373	−4	1	91/60	89/60	87/60	34	65	64	63						
Roswell, Walker AFB	33	18	104	32	3676	13	18	100/66	98/66	96/66	33	71	70	69	N	6		SSE	103.0	2.7
Santa Fe Co	35	37	106	05	6307	6	10	90/61	88/61	86/61	28	63	62	61					90.1	−1.2
Silver City AP	32	38	108	10	5442	5	10	95/61	94/60	91/60	30	66	64	63						
Socorro AP	34	03	106	53	4624	13	17	97/62	95/62	93/62	30	67	66	65						
Tucumcari AP	35	11	103	36	4039	8	13	99/66	97/66	95/65	28	70	69	68	NE	8		SW	102.7	1.1

Table 1 Climatic Conditions for the United States (Continued)

Col. 1	Col. 2		Col. 3		Col. 4	Col. 5 Winter,[b] °F Design Dry-Bulb		Col. 6 Summer,[c] °F Design Dry-Bulb and Coincident Wet-Bulb			Col. 7 Mean Daily Range	Col. 8 Design Wet-Bulb			Col. 9 Prevailing Wind				Col. 10 Temp., °F Median of Annual Extr.	
State and Station[a]	Lat.		Long.		Elev.										Winter		Summer			
	°	'N	°	'W	Feet	99%	97.5%	1%	2.5%	5%	Range	1%	2.5%	5%	Knots[d]				Max.	Min.
NEW YORK																				
Albany AP	42	45	73	48	275	−6	−1	91/73	88/72	85/70	23	75	74	72	WNW	8		S		
Albany Co	42	39	73	45	19	−4	1	91/73	88/72	85/70	20	75	74	72					95.2	−11.4
Auburn	42	54	76	32	715	−3	2	90/73	87/71	84/70	22	75	73	72					92.4	−9.5
Batavia	43	00	78	11	922	1	5	90/72	87/71	84/70	22	75	73	72					92.2	−7.5
Binghamton AP	42	13	75	59	1590	−2	1	86/71	83/69	81/68	20	73	72	70	WSW	10		WSW	92.9	−9.3
Buffalo AP	42	56	78	44	705	2	6	88/71	85/70	83/69	21	74	73	72	W	10		SW	90.0	−3.2
Cortland	42	36	76	11	1129	−5	0	88/71	85/71	82/70	23	74	73	71					93.8	−11.2
Dunkirk	42	29	79	16	692	4	9	88/73	85/72	83/71	18	75	74	72	SSW	10		WSW		
Elmira AP	42	10	76	54	955	−4	1	89/71	86/71	83/70	24	74	73	71					96.2	−6.7
Geneva	42	45	76	54	613	−3	2	90/73	87/71	84/70	22	75	73	72					96.1	−6.5
Glens Falls	43	20	73	37	328	−11	−5	88/72	85/71	82/69	23	74	73	71	NNW	6		S		
Gloversville	43	02	74	21	760	−8	−2	89/72	86/71	83/69	23	75	74	72					93.2	−14.6
Hornell	42	21	77	42	1325	−4	0	88/71	85/70	82/69	24	74	73	72						
Ithaca	42	27	76	29	928	−5	0	88/71	85/71	82/70	24	74	73	71	W	6		SW		
Jamestown	42	07	79	14	1390	−1	3	88/70	86/70	83/69	20	74	72	71	WSW	9		WSW		
Kingston	41	56	74	00	279	−3	2	91/73	88/72	85/70	22	76	74	73						
Lockport	43	09	79	15	638	4	7	89/74	86/72	84/71	21	76	74	73	N	9		SW	92.2	−4.8
Massena AP	44	56	74	51	207	−13	−8	86/70	83/69	80/68	20	73	72	70						
Newburgh, Stewart AFB	41	30	74	06	471	−1	4	90/73	88/72	85/70	21	76	74	73	W	10		W		
NYC-Central Park	40	47	73	58	157	11	15	92/74	89/73	87/72	17	76	75	74					94.9	3.8
NYC-Kennedy AP	40	39	3	47	13	12	15	90/73	87/72	84/71	16	76	75	74	WNW	4		SSW		
NYC-La Guardia AP	40	46	73	54	11	11	15	92/74	89/73	87/72	16	76	75	74	WNW	15		SW		
Niagara Falls AP	43	06	79	57	590	4	7	89/74	86/72	84/71	20	76	74	73	W	9		SW		
Olean	42	14	78	22	2119	−2	2	87/71	84/71	81/70	23	74	73	71						
Oneonta	42	31	75	04	1775	−7	−4	86/71	83/69	80/68	24	73	72	70						
Oswego Co	43	28	76	33	300	1	7	86/73	83/71	80/70	20	75	73	72	E	7		WSW	91.3	−7.4
Plattsburg AFB	44	39	73	28	235	−13	−8	86/70	83/69	80/68	22	73	72	70	NW	6		SE		
Poughkeepsie	41	38	73	55	165	0	6	92/74	89/74	86/72	21	77	75	74	NNE	6		SSW	98.1	−5.6
Rochester AP	43	07	77	40	547	1	5	91/73	88/71	85/70	22	75	73	72	WSW	11		WSW		
Rome, Griffiss AFB	43	14	75	25	514	−11	−5	88/71	85/70	83/69	22	75	73	71	NW	5		W		
Schenectady	42	51	73	57	377	−4	1	90/73	87/72	84/70	22	75	74	72	WNW	8		S		
Suffolk County AFB	40	51	72	38	67	7	10	86/72	83/71	80/70	16	76	74	73	NW	9		SW		
Syracuse AP	43	07	76	07	410	−3	2	90/73	87/71	84/70	20	75	73	72	N	7		WNW	93.	−10.0
Utica	43	09	75	23	714	−12	−6	88/73	85/71	82/70	22	75	73	71	NW	12		W		
Watertown	43	59	76	01	325	−11	−6	86/73	83/71	81/70	20	75	73	72	E	7		WSW	91.7	−19.6
NORTH CAROLINA																				
Asheville AP	35	26	82	32	2140	10	14	89/73	87/72	85/71	21	75	74	72	NNW	12		NNW	91.9	5.8
Charlotte AP	35	13	80	56	736	18	22	95/74	93/74	91/74	20	77	76	76	NNW	6		SW	97.8	12.6
Durham	35	52	78	47	434	16	20	94/75	92/75	90/75	20	78	77	76					98.9	9.6
Elizabeth City AP	36	16	76	11	12	12	19	93/78	91/77	89/76	18	80	78	78	NW	8		SW		
Fayetteville, Pope AFB	35	10	79	01	218	17	20	95/76	92/76	90/75	20	79	78	77	N	6		SSW	99.1	13.1
Goldsboro	35	20	77	58	109	18	21	94/77	91/76	89/75	18	79	78	77	N	8		SW	99.8	13.0
Greensboro AP	36	05	79	57	897	14	18	93/74	91/73	89/73	21	77	76	75	NE	7		SW	97.7	9.7
Greenville	35	37	77	25	75	18	21	93/76	91/76	89/75	19	79	78	77						
Henderson	36	22	78	25	480	12	15	95/77	92/76	90/76	20	79	78	77						
Hickory	35	45	81	23	1187	14	18	92/73	90/72	88/72	21	75	74	73					96.5	9.6
Jacksonville	34	50	77	37	95	20	24	92/78	90/78	88/77	18	80	79	78						
Lumberton	34	37	79	04	129	18	21	95/76	92/76	90/75	20	79	78	77						
New Bern AP	35	05	77	03	20	20	24	92/78	90/78	88/77	18	80	79	78					98.2	15.1
Raleigh/Durham AP	35	52	78	47	434	16	20	94/75	92/75	90/75	20	78	77	76	N	7		SW	97.7	12.2
Rocky Mount	35	58	77	48	121	18	21	94/77	91/76	89/75	19	79	78	77						
Wilmington AP	34	16	77	55	28	23	26	93/79	91/78	89/77	18	81	80	79	N	8		SW	96.9	18.2
Winston-Salem AP	36	08	80	13	969	16	20	94/74	91/73	89/73	20	76	75	74	NW	8		WSW		
NORTH DAKOTA																				
Bismarck AP	46	46	100	45	1647	−23	−19	95/68	91/68	88/67	27	73	71	70	WNW	7		S	100.3	−31.5
Devils Lake	48	07	98	54	1450	−25	−21	91/69	88/68	85/66	25	73	71	69					97.5	−30.4
Dickinson AP	46	48	102	48	2585	−21	−17	94/68	90/66	87/65	25	71	69	68	WNW	12		SSE	101.0	−31.3
Fargo AP	46	54	96	48	896	−22	−18	92/73	89/71	85/69	25	76	74	72	SSE	11		S	97.3	−29.7
Grand Forks AP	47	57	97	24	911	−26	−22	91/70	87/70	84/68	25	74	72	70	N	8		S	97.6	−29.0
Jamestown AP	46	55	98	41	1492	−22	−18	94/70	90/69	87/68	26	74	74	71					101.3	−27.9
Minot AP	48	25	101	21	1668	−24	−20	92/68	89/67	86/65	25	72	70	68	WSW	10		S		
Williston	48	09	103	35	1876	−25	−21	91/68	88/67	85/65	25	72	70	68					99.7	−32.9

APPENDIX II—DESIGN CONDITIONS TABLES

Table 1 Climatic Conditions for the United States (Continued)

Col. 1	Col. 2		Col. 3		Col. 4	Col. 5 Winter,[b] °F		Col. 6 Summer,[c] °F			Col. 7	Col. 8			Col. 9 Prevailing Wind		Col. 10 Temp., °F	
	Lat.		Long.		Elev.	Design Dry-Bulb		Design Dry-Bulb and Coincident Wet-Bulb			Mean Daily Range	Design Wet-Bulb			Winter	Summer	Median of Annual Extr.	
State and Station[a]	°N	'N	°W	'W	Feet	99%	97.5%	Mean 1%	2.5%	5%		1%	2.5%	5%	Knots[d]		Max.	Min.
OHIO																		
Akron, Canton AP	40	55	81	26	1208	1	6	89/72	86/71	84/70	21	75	73	72	SW 9	SW	94.4	−4.6
Ashtabula	41	51	80	48	690	4	9	88/73	85/72	83/71	18	75	74	72				
Athens	39	20	82	06	700	0	6	95/75	92/74	90/73	22	78	76	74				
Bowling Green	41	23	83	38	675	−2	2	92/73	89/73	86/71	23	76	75	73			96.7	−7.3
Cambridge	40	04	81	35	807	1	7	93/75	90/74	87/73	23	78	76	75				
Chillicothe	39	21	83	00	640	0	6	95/75	92/74	90/73	22	78	76	74	W 8	WSW	98.2	−2.1
Cincinnati Co	39	09	84	31	758	1	6	92/73	90/72	88/72	21	77	75	74	W 9	SW	97.2	−0.2
Cleveland AP	41	24	81	51	777	1	5	91/73	88/72	86/71	22	76	74	73	SW 12	N	94.7	−3.1
Columbus AP	40	00	82	53	812	0	5	92/73	90/73	87/72	24	77	75	74	W 8	SSW	96.0	−3.4
Dayton AP	39	54	84	13	1002	−1	4	91/73	89/72	86/71	20	76	75	73	WNW 11	SW	96.6	−4.5
Defiance	41	17	84	23	700	−1	4	94/74	91/73	88/72	24	77	76	74				
Findlay AP	41	01	83	40	804	2	3	92/74	90/73	87/72	24	77	76	74			97.4	−7.4
Fremont	41	20	83	07	600	−3	1	90/74	88/73	85/71	24	76	75	73				
Hamilton	39	24	84	35	650	0	5	92/73	90/73	87/71	22	76	75	73			98.2	−2.8
Lancaster	39	44	82	38	860	0	5	93/74	91/73	88/72	23	77	75	74				
Lima	40	42	84	02	975	−1	4	94/74	91/73	88/72	24	77	76	74	WNW 11	SW	96.0	−6.5
Mansfield AP	40	49	82	31	1295	0	5	90/73	87/72	85/72	22	76	74	73	W 8	SW	93.8	−10.7
Marion	40	36	83	10	920	0	5	93/74	91/73	88/72	23	77	76	74				
Middletown	39	31	84	25	635	0	5	92/73	90/72	87/71	22	76	75	73				
Newark	40	01	82	28	880	−1	5	94/74	92/73	89/72	23	77	75	74	W 8	SSW	95.8	−6.8
Norwalk	41	16	82	37	670	−3	1	90/73	88/73	85/71	22	76	75	73			97.3	−8.3
Portsmouth	38	45	82	55	540	5	10	95/76	92/74	89/73	22	78	77	75	W 8	SW	97.9	1.0
Sandusky Co	41	27	82	43	606	1	6	93/73	91/72	88/71	21	76	74	73			96.7	−1.9
Springfield	39	50	83	50	1052	−1	3	91/74	89/73	87/72	21	77	76	74	W 7	W		
Steubenville	40	23	80	38	992	1	5	89/72	86/71	84/70	22	74	73	72				
Toledo AP	41	36	83	48	669	−3	1	90/73	88/73	85/71	25	76	75	73	WSW 8	SW	95.4	−5.2
Warren	41	20	80	51	928	0	5	89/71	87/71	85/70	23	74	73	71				
Wooster	40	47	81	55	1020	1	6	89/72	86/71	84/70	22	75	73	72			94.0	−7.7
Youngstown AP	41	16	80	40	1178	−1	4	88/71	86/71	84/70	23	74	73	71	SW 10	SW		
Zanesville AP	39	57	81	54	900	1	7	93/75	90/74	87/73	23	78	76	75	W 6	WSW		
OKLAHOMA																		
Ada	34	47	96	41	1015	10	14	100/74	97/74	95/74	23	77	76	75				
Altus AFB	34	39	99	16	1378	11	16	102/73	100/73	98/73	25	77	76	75	N 10	S		
Ardmore	34	18	97	01	771	13	17	100/74	98/74	95/74	23	77	77	76				
Bartlesville	36	45	96	00	715	6	10	101/73	98/73	95/74	23	77	77	76				
Chickasha	35	03	97	55	1085	10	14	101/74	98/74	95/74	24	78	77	76				
Enid, Vance AFB	36	21	97	55	1307	9	13	103/74	100/74	97/74	24	79	77	76				
Lawton AP	34	34	98	25	1096	12	16	101/74	99/74	96/74	24	78	77	76				
McAlester	34	50	95	55	776	14	19	99/74	96/74	93/74	23	77	76	75	N 10	S		
Muskogee AP	35	40	95	22	610	10	15	101/74	98/75	95/74	23	79	78	77				
Norman	35	15	97	29	1181	9	13	99/74	96/74	94/74	24	77	76	75				
Oklahoma City AP	35	24	97	36	1285	9	13	100/74	97/74	95/73	23	78	77	76	N 14	SSW		
Ponca City	36	44	97	06	997	5	9	100/74	97/74	94/74	24	77	76					
Seminole	35	14	96	40	865	11	15	99/74	96/74	94/73	23	77	76	75				
Stillwater	36	10	97	05	984	8	13	100/74	96/74	93/74	24	77	76	75	N 12	SSW	103.7	1.6
Tulsa AP	36	12	95	54	650	8	13	101/74	98/75	95/75	22	79	78	77	N 11	SSW		
Woodward	36	36	99	31	2165	6	10	100/73	97/73	94/73	26	78	76	75			107.1	−1.3
OREGON																		
Albany	44	38	123	07	230	18	22	92/67	89/66	86/65	31	69	67	66			97.5	16.6
Astoria AP	46	09	123	53	8	25	29	75/65	71/62	68/61	16	65	63	62	ESE 7	NNW		
Baker AP	44	50	117	49	3372	−1	6	92/63	89/61	86/60	30	65	63	61			97.5	−6.8
Bend	44	04	121	19	3595	−3	4	90/62	87/60	84/59	33	64	62	60			96.4	−5.8
Corvallis	44	30	123	17	246	18	22	92/67	89/66	86/65	31	69	67	66	N 6	N	98.5	17.1
Eugene AP	44	07	123	13	359	17	22	92/67	89/66	86/65	31	69	67	66	N 7	N		
Grants Pass	42	26	123	19	925	20	24	99/69	96/68	93/67	33	71	69	68	N 5	N	103.6	16.4
Klamath Falls AP	42	09	121	44	4092	4	9	90/61	87/60	84/59	36	63	61	60	N 4	W	96.3	0.9
Medford AP	42	22	122	52	1298	19	23	98/68	94/67	91/66	35	70	68	67	S 4	WMW	103.8	15.0
Pendleton AP	45	41	118	51	1482	−2	5	97/65	93/64	90/62	29	66	65	63	NNW 6	WNW		
Portland AP	45	36	122	36	21	17	23	89/68	85/67	81/65	23	69	68	66	ESE 12	NW	96.6	18.3
Portland Co	45	32	122	40	75	18	24	90/68	86/67	82/65	21	69	67	66			97.6	20.5
Roseburg AP	43	14	123	22	525	18	23	93/67	90/66	87/65	30	69	67	66			99.6	19.5
Salem AP	44	55	123	01	196	18	23	92/68	88/66	84/65	31	69	68	66	N 6	N	98.9	15.9
The Dalles	45	36	121	12	100	13	19	93/69	89/68	85/66	28	70	68	67			105.1	7.9

Table 1 Climatic Conditions for the United States (Continued)

Col. 1	Col. 2		Col. 3		Col. 4	Winter,[b] °F Col. 5		Summer,[c] °F Col. 6			Col. 7	Col. 8			Prevailing Wind Col. 9				Temp., °F Col. 10	
State and Station[a]	Lat.		Long.		Elev.	Design Dry-Bulb		Design Dry-Bulb and Mean Coincident Wet-Bulb			Mean Daily Range	Design Wet-Bulb			Winter		Summer		Median of Annual Extr.	
	°	'N	°	'W	Feet	99%	97.5%	1%	2.5%	5%		1%	2.5%	5%	Knots[d]				Max.	Min.
PENNSYLVANIA																				
Allentown AP	40	39	75	26	387	4	9	92/73	88/72	86/72	22	76	75	73	W	11	SW			
Altoona Co	40	18	78	19	1504	0	5	90/72	87/71	84/70	23	74	73	72	WNW	11	WSW		93.7	−5.2
Butler	40	52	79	54	1100	1	6	90/73	87/72	85/71	22	75	74	73						
Chambersburg	39	56	77	38	640	4	8	93/75	90/74	87/73	23	77	76	75					97.4	−0.3
Erie AP	42	05	80	11	731	4	9	88/73	85/72	83/71	18	75	74	72	SSW	0	WSW		91.3	−2.2
Harrisburg AP	40	12	76	46	308	7	11	94/75	91/74	88/73	21	77	76	75	NW	11	WSW		96.5	3.7
Johnstown	40	19	78	50	2284	−3	2	86/70	83/70	80/68	23	72	71	70	WNW	8	WSW		96.4	−1.8
Lancaster	40	07	76	18	403	4	8	93/75	90/74	87/73	22	77	76	75	NW	11	WSW			
Meadville	41	38	80	10	1065	0	4	88/71	85/70	83/69	21	73	72	71					93.2	−8.5
New Castle	41	01	80	22	825	2	7	91/73	88/72	86/71	23	75	74	73	WSW	10	WSW		94.7	−6.4
Philadelphia AP	39	53	75	15	5	10	14	93/75	90/74	87/72	21	77	76	75	WNW	10	WSW		96.4	5.9
Pittsburgh AP	40	30	80	13	1137	1	5	89/72	86/71	84/70	22	74	73	72	WSW	10	WSW			
Pittsburgh Co	40	27	80	00	1017	3	7	91/72	88/71	86/70	19	74	73	72					94.6	−1.1
Reading Co	40	20	75	38	266	9	13	92/73	89/72	86/72	19	76	75	73	W	11	SW		97.0	3.6
Scranton/Wilkes-Barre	41	20	75	44	930	1	5	90/72	87/71	84/70	19	74	73	72	SW	8	WSW		94.8	−2.2
State College	40	48	77	52	1175	3	7	90/72	87/71	84/70	23	74	73	72	NNW	8	WSW		93.2	−3.6
Sunbury	40	53	76	46	446	2	7	92/73	89/72	86/72	22	75	74	73						
Uniontown	39	55	79	43	956	5	9	91/74	88/73	85/72	22	76	75	74					93.9	−2.5
Warren	41	51	79	08	1280	−2	4	89/71	86/71	83/70	24	74	73	72					93.3	−10.7
West Chester	39	58	75	38	450	9	13	92/75	89/74	86/72	20	77	76	75						
Williamsport AP	41	15	76	55	524	2	7	92/73	89/72	86/70	23	75	74	73	W	9	WSW		95.5	−4.6
York	39	55	76	45	390	8	12	94/75	91/74	88/73	22	77	76	75					97.0	−2.4
RHODE ISLAND																				
Newport	41	30	71	20	10	5	9	88/73	85/72	82/70	16	76	75	73	WNW	10	SW			
Providence AP	41	44	71	26	51	5	9	89/73	86/72	83/70	19	75	74	73	WNW	11	SW		94.6	−0.5
SOUTH CAROLINA																				
Anderson	34	30	82	43	774	19	23	94/74	92/74	90/74	21	77	76	75					99.5	13.3
Charleston AFB	32	54	80	02	45	24	27	93/78	91/78	89/77	18	81	80	79	NNE	8	SW			
Charleston Co	32	54	79	58	3	25	28	94/78	92/78	90/77	13	81	80	79					97.8	21.4
Columbia AP	33	57	81	07	213	20	24	97/76	95/75	93/75	22	79	78	77	W	6	SW		100.6	16.2
Florence AP	34	11	79	43	147	22	25	94/77	92/77	90/76	21	80	79	78	N	7	SW		99.5	16.5
Georgetown	33	23	79	17	14	23	26	92/79	90/78	88/77	18	81	80	79	N	7	SSW		98.2	19.1
Greenville AP	34	54	82	13	957	18	22	93/74	91/74	89/74	21	77	76	75	NW	8	SW		97.3	12.6
Greenwood	34	10	82	07	620	18	22	95/75	93/74	91/74	21	78	77	76					99.5	14.1
Orangeburg	33	30	80	52	260	20	24	97/76	95/75	93/75	20	79	78	77					101.2	18.0
Rock Hill	34	59	80	58	470	19	23	96/75	94/74	92/74	20	78	77	76						
Spartanburg AP	34	58	82	00	823	18	22	93/74	91/74	89/74	20	77	76	75					99.5	13.9
Sumter, Shaw AFB	33	54	80	22	169	22	25	95/77	92/76	90/75	21	79	78	77	NNE	6	W		100.0	15.4
SOUTH DAKOTA																				
Aberdeen AP	45	27	98	26	1296	−19	−15	94/73	91/72	88/70	27	77	75	73	NNW	8	S		102.3	−28.1
Brookings	44	18	96	48	1637	−17	−13	95/73	92/72	89/71	25	77	75	73					97.8	−26.5
Huron AP	44	23	98	13	1281	−18	−14	96/73	93/72	90/71	28	77	75	73	NNW	8	S		101.5	−25.8
Mitchell	43	41	98	01	1346	−15	−10	96/72	93/71	90/70	28	76	75	73					103.0	−22.7
Pierre AP	44	23	100	17	1742	−15	−10	99/71	95/71	92/69	29	75	74	72	NW	11	SSE		105.7	−20.6
Rapid City AP	44	03	103	04	3162	−11	−7	95/66	92/65	89/65	28	71	69	67	NNW	10	SSE		100.9	−19.0
Sioux Falls AP	43	34	96	44	1418	−15	−11	94/73	91/72	88/71	24	76	75	73	NW	8	S			
Watertown AP	44	55	97	09	1738	−19	−15	94/73	91/72	88/71	26	76	75	73					97.8	−26.5
Yankton	42	55	97	23	1302	−13	−7	94/73	91/72	88/71	25	77	76	74					100.8	−19.1
TENNESSEE																				
Athens	35	26	84	35	940	13	18	95/74	92/73	90/73	22	77	76	75						
Bristol-Tri City AP	36	29	82	24	1507	9	14	91/72	89/72	87/71	22	75	75	73	WNW	6	SW			
Chattanooga AP	35	02	85	12	665	13	18	96/75	93/74	91/74	22	78	77	76	NNW	8	WSW		97.2	9.8
Clarksville	36	33	87	22	382	6	12	95/76	93/74	90/74	21	78	77	76					99.8	3.7
Columbia	35	38	87	02	690	10	15	97/75	94/74	91/74	21	78	77	76						
Dyersburg	36	01	89	24	344	10	15	96/78	94/77	91/76	21	81	80	78						
Greeneville	36	04	82	50	1319	11	16	92/73	90/72	88/72	22	76	75	74					95.6	0.8
Jackson AP	35	36	88	55	423	11	16	98/76	95/75	92/75	21	79	78	77					99.2	6.6
Knoxville AP	35	49	83	59	980	13	19	94/74	92/73	90/73	21	77	76	75	NE	8	W		96.0	7.0
Memphis AP	35	03	90	00	258	13	18	98/77	95/76	93/76	21	80	79	78	N	10	SW		97.9	10.4
Murfreesboro	34	55	86	28	600	9	14	97/75	94/74	91/74	22	78	77	76					97.7	4.5
Nashville AP	36	07	86	41	590	9	14	97/75	94/74	91/74	21	78	77	76	NW	8	WSW			
Tullahoma	35	23	86	05	1067	8	13	96/74	93/73	91/73	22	77	76	75	NW	9	WSW		96.7	3.7

APPENDIX II—DESIGN CONDITIONS TABLES

Table 1 Climatic Conditions for the United States (*Continued*)

Col. 1	Col. 2	Col. 3	Col. 4	Col. 5 Winter,[b] °F Design Dry-Bulb		Col. 6 Summer,[c] °F Design Dry-Bulb and Mean Coincident Wet-Bulb			Col. 7 Mean Daily Range	Col. 8 Design Wet-Bulb			Col. 9 Prevailing Wind Knots[d]		Col. 10 Temp., °F Median of Annual Extr.	
State and Station[a]	Lat. ° 'N	Long. ° 'W	Elev. Feet	99%	97.5%	1%	2.5%	5%		1%	2.5%	5%	Winter	Summer	Max.	Min.
TEXAS																
Abilene AP	32 25	99 41	1784	15	20	101/71	99/71	97/71	22	75	74	74	N 12	SSE	103.6	10.4
Alice AP	27 44	98 02	180	31	34	100/78	98/77	95/77	20	82	81	79			104.9	24.8
Amarillo AP	35 14	101 42	3604	6	11	98/67	95/67	93/67	26	71	70	70	N 11	S	100.8	0.9
Austin AP	30 18	97 42	597	24	28	100/74	98/74	97/74	22	78	77	77	N 11	S	101.6	19.7
Bay City	29 00	95 58	50	29	33	96/77	94/77	92/77	16	80	79	79				
Beaumont	29 57	94 01	16	27	31	95/79	93/78	91/78	19	81	80	80			99.7	23.5
Beeville	28 22	97 40	190	30	33	99/78	97/77	95/77	18	82	81	79	N 9	SSE	103.1	22.5
Big Spring AP	32 18	101 27	2598	16	20	100/69	97/69	95/69	26	74	73	72			105.3	10.7
Brownsville AP	25 54	97 26	19	35	39	94/77	93/77	92/77	18	80	79	79	NNW 13	SE	98.1	30.1
Brownwood	31 48	98 57	1386	18	22	101/73	99/73	96/73	22	77	76	75	N 9	S	105.3	13.0
Bryan AP	30 40	96 33	276	24	29	98/76	96/76	94/76	20	79	78	78				
Corpus Christi AP	27 46	97 30	41	31	35	95/78	94/78	92/78	19	80	80	79	N 12	SSE	97.0	27.2
Corsicana	32 05	96 28	425	20	25	100/75	98/75	96/75	21	79	78	77			104.2	15.2
Dallas AP	32 51	96 51	481	18	22	102/75	100/75	97/75	20	78	77	77	N 11	S		
Del Rio, Laughlin AFB	29 22	100 47	1081	26	31	100/73	98/73	97/73	24	79	77	76			103.8	23.0
Denton	33 12	97 06	630	17	22	101/74	99/74	97/74	22	78	77	76			104.5	11.8
Eagle Pass	28 52	100 32	884	27	32	101/73	99/73	98/73	24	78	78	77	NNW 9	ESE	107.7	22.1
El Paso AP	31 48	106 24	3918	20	24	100/64	98/64	96/64	27	69	68	68	N 7	S	103.0	15.7
Fort Worth AP	32 50	97 03	537	17	22	101/74	99/74	97/74	22	78	77	76	NW 11	S	103.2	13.5
Galveston AP	29 18	94 48	7	31	36	90/79	89/79	88/78	10	81	80	80	N 15	S	93.9	27.5
Greenville	33 04	96 03	535	17	22	101/74	99/74	97/74	21	78	77	76			103.6	11.7
Harlingen	26 14	97 39	35	35	39	96/77	94/77	93/77	19	80	79	79	NNW 10	SSE	102.3	29.3
Houston AP	29 58	95 21	96	27	32	96/77	94/77	92/77	18	80	79	79	NNW 11	S		
Houston Co	29 59	95 22	108	28	33	97/77	95/77	93/77	18	80	79	79			99.0	23.5
Huntsville	30 43	95 33	494	22	27	100/75	98/75	96/75	20	78	78	77			100.8	18.7
Killeen, Robert Gray AAF	31 05	97 41	850	20	25	99/73	97/73	95/73	22	77	76	75				
Lamesa	32 42	101 56	2965	13	17	99/69	96/69	94/69	26	73	72	71			105.5	8.9
Laredo AFB	27 32	99 27	512	32	36	102/73	101/73	99/74	23	78	78	77	N 8	SE		
Longview	32 28	94 44	330	19	24	99/76	97/76	95/76	20	80	79	78				
Lubbock AP	33 39	101 49	3254	10	15	98/69	96/69	94/69	26	73	72	71	NNE 10	SSE		
Lufkin AP	31 25	94 48	277	25	29	99/76	97/76	94/76	20	80	79	78	NNW 12	S		
McAllen	26 12	98 13	122	35	39	97/77	95/77	94/77	21	80	79	79				
Midland AP	31 57	102 11	2851	16	21	100/69	98/69	96/69	26	73	72	71	NE 9	SSE	103.6	10.8
Mineral Wells AP	32 47	98 04	930	17	22	101/74	99/74	97/74	22	78	77	76				
Palestine Co	31 47	95 38	600	23	27	100/76	98/76	96/76	20	79	79	78			101.2	16.3
Pampa	35 32	100 59	3250	7	12	99/67	96/67	94/67	26	71	70	70				
Pecos	31 25	103 30	2610	16	21	100/69	98/69	96/69	27	73	72	71				
Plainview	34 11	101 42	3370	8	13	98/68	96/68	94/68	26	72	71	70			102.7	3.1
Port Arthur AP	29 57	94 01	16	27	31	95/79	93/78	91/78	19	81	80	80	N 9	S	97.7	24.0
San Angelo, Goodfellow AFB	31 26	100 24	1877	18	22	101/71	99/71	97/70	24	75	74	73	NNE 10	SSE		
San Antonio AP	29 32	98 28	788	25	30	99/72	97/73	96/73	19	77	76	76	N 8	SSE	101.3	21.1
Sherman, Perrin AFB	33 43	96 40	763	15	20	100/75	98/75	95/74	22	78	77	76	N 10	S	103.0	11.9
Snyder	32 43	100 55	2325	13	18	100/70	98/70	96/70	26	74	73	72				
Temple	31 06	97 21	700	22	27	100/74	99/74	97/74	22	78	77	77				
Tyler AP	32 21	95 16	530	19	24	99/76	97/76	95/76	21	80	79	78	NNE 23	S		
Vernon	34 10	99 18	1212	13	17	102/73	100/73	97/73	24	77	76	75				
Victoria AP	28 51	96 55	104	29	32	98/78	96/77	94/77	18	82	81	79			101.4	23.4
Waco AP	31 37	97 13	500	21	26	101/75	99/75	97/75	22	78	78	77				
Wichita Falls AP	33 58	98 29	994	14	18	103/73	101/73	98/73	24	77	76	75	NNW 12	S		
UTAH																
Cedar City AP	37 42	113 06	5617	−2	5	93/60	91/60	89/59	32	65	63	62	SE 5	SW		
Logan	41 45	111 49	4785	−3	2	93/62	91/61	88/60	33	65	64	63			95.5	−7.8
Moab	38 36	109 36	3965	6	11	100/60	98/60	96/60	30	65	64	63				
Ogden AP	41 12	112 01	4455	1	5	93/63	91/61	88/61	33	66	65	64	S 6	SW	99.5	−3.9
Price	39 37	110 50	5580	−2	5	93/60	91/60	89/59	33	65	63	62				
Provo	40 13	111 43	4448	1	6	98/62	96/62	94/61	32	66	65	64	SE 5	SW		
Richfield	38 46	112 05	5270	−2	5	93/60	91/60	89/59	34	65	63	62			98.1	−10.5
St George Co	37 02	113 31	2900	14	21	103/65	101/65	99/64	33	70	68	67			109.3	11.1
Salt Lake City AP	40 46	111 58	4220	3	8	97/62	95/62	92/61	32	66	65	64	SSE 6	N	99.4	−0.1
Vernal AP	40 27	109 31	5280	−5	0	91/61	89/60	86/59	32	64	63	62				
VERMONT																
Barre	44 12	72 31	600	−16	−11	84/71	81/69	78/68	23	73	71	70				
Burlington AP	44 28	73 09	332	−12	−7	88/72	85/70	82/69	23	74	72	71	E 7	SSW	92.4	−16.9
Rutland	43 36	72 58	620	−13	−8	87/72	84/70	81/69	23	74	72	71			92.5	−17.5

Table 1 Climatic Conditions for the United States (*Concluded*)

Col. 1	Col. 2	Col. 3	Col. 4	Col. 5 Winter,[b] °F Design Dry-Bulb		Col. 6 Summer,[c] °F Design Dry-Bulb and Coincident Wet-Bulb			Col. 7	Col. 8 Design Wet-Bulb			Col. 9 Prevailing Wind Knots[d]		Col. 10 Temp., °F Median of Annual Extr.	
State and Station[a]	Lat. ° 'N	Long. ° 'W	Elev. Feet	99%	97.5%	Mean 1%	Coincident 2.5%	Wet-Bulb 5%	Mean Daily Range	1%	2.5%	5%	Winter	Summer	Max.	Min.
VIRGINIA																
Charlottesville	38 02	78 31	870	14	18	94/74	91/74	88/73	23	77	76	75	NE 7	SW	97.4	8.0
Danville AP	36 34	79 20	590	14	16	94/74	92/73	90/73	21	77	76	75			100.1	9.2
Fredericksburg	38 18	77 28	100	10	14	96/76	93/75	90/74	21	78	77	76				
Harrisonburg	38 27	78 54	1370	12	16	93/72	91/72	88/71	23	75	74	73				
Lynchburg AP	37 20	79 12	916	12	16	93/74	90/74	88/73	21	77	76	75	NE 7	SW	97.2	7.6
Norfolk AP	36 54	76 12	22	20	22	93/77	91/76	89/76	18	79	78	77	NW 10	SW	97.2	15.3
Petersburg	37 11	77 31	194	14	17	95/76	92/76	90/75	20	79	78	77				
Richmond AP	37 30	77 20	164	14	17	95/76	92/76	90/75	21	79	78	77	N 6	SW	97.9	9.6
Roanoke AP	37 19	79 58	1193	12	16	93/72	91/72	88/71	23	75	74	73	NW 9	SW		
Staunton	38 16	78 54	1201	12	16	93/72	91/72	88/71	23	75	74	73	NW 9	SW	95.9	2.5
Winchester	39 12	78 10	760	6	10	93/75	90/74	88/74	21	77	76	75			97.3	3.7
WASHINGTON																
Aberdeen	46 59	123 49	12	25	28	80/65	77/62	73/61	16	65	63	62	ESE 6	NNW	91.9	19.3
Bellingham AP	48 48	122 32	158	10	15	81/67	77/65	74/63	19	68	65	63	NNE 15	WSW	87.4	10.3
Bremerton	47 34	122 40	162	21	25	82/65	78/64	75/62	20	66	64	63	E 8	N		
Ellensburg AP	47 02	120 31	1735	2	6	94/65	91/64	87/62	34	66	65	63				
Everett, Paine AFB	47 55	122 17	596	21	25	80/65	76/64	73/62	20	67	64	63	ESE 6	NNW	84.9	15.2
Kennewick	46 13	119 08	392	5	11	99/66	95/67	92/66	30	70	68	67			103.4	2.0
Longview	46 10	122 56	12	19	24	88/68	85/67	81/65	30	69	67	66	ESE 9	NW	96.0	14.8
Moses Lake, Larson AFB	47 12	119 19	1185	1	7	97/66	94/65	90/63	32	67	66	64	N 8	SSW		
Olympia AP	46 58	122 54	215	16	22	87/66	83/65	79/64	32	67	66	64	NE 4	NE		
Port Angeles	48 07	123 26	99	24	27	72/62	69/61	67/60	18	64	62	61			83.5	19.4
Seattle-Boeing Field	47 32	122 18	23	21	26	84/68	81/66	77/65	24	69	67	65				
Seattle Co	47 39	122 18	20	22	27	85/68	82/66	78/65	19	69	67	65	N 7	N	90.2	22.0
Seattle-Tacoma AP	47 27	122 18	400	21	26	84/65	80/64	76/62	22	66	64	63	E 9	N	90.1	19.9
Spokane AP	47 38	117 31	2357	−6	2	93/64	90/63	87/62	28	65	64	62	NE 6	SW	98.8	−4.9
Tacoma, McChord AFB	47 15	122 30	100	19	24	86/66	82/65	79/63	22	68	66	64	S 5	NNE	89.4	18.8
Walla Walla AP	46 06	118 17	1206	0	7	97/67	94/66	90/65	27	69	67	66	W 5	W	103.0	3.8
Wenatchee	47 25	120 19	632	7	11	99/67	96/66	92/64	32	68	67	65			101.1	1.0
Yakima AP	46 34	120 32	1052	−2	5	96/65	93/65	89/63	36	68	66	65	W 5	NW		
WEST VIRGINIA																
Beckley	37 47	81 07	2504	−2	4	83/71	81/69	79/69	22	73	71	70	WNW 9	WNW		
Bluefield AP	37 18	81 13	2867	−2	4	83/71	81/69	79/69	22	73	71	70				
Charleston AP	38 22	81 36	939	7	11	92/74	90/73	87/72	20	76	75	74	SW 8	SW	97.2	2.9
Clarksburg	39 16	80 21	977	6	10	92/74	90/73	87/72	21	76	75	74				
Elkins AP	38 53	79 51	1948	1	6	86/72	84/70	82/70	22	74	72	71	WNW 9	WNW	90.6	−7.3
Huntington Co	38 25	82 30	565	5	10	94/76	91/74	89/73	22	78	77	75	W 6	SW	97.1	2.1
Martinsburg AP	39 24	77 59	556	6	10	93/75	90/74	88/74	21	77	76	75	WNW 10	W	99.0	1.1
Morgantown AP	39 39	79 55	1240	4	8	90/74	87/73	85/73	22	76	75	74				
Parkersburg Co	39 16	81 34	615	7	11	93/75	90/74	88/73	21	77	76	75	WSW 7	WSW	95.9	0.7
Wheeling	40 07	80 42	665	1	5	89/72	86/71	84/70	21	74	73	72	WSW 10	WSW	97.5	−0.6
WISCONSIN																
Appleton	44 15	88 23	730	−14	−9	89/74	86/72	83/71	23	76	74	72			94.6	−16.2
Ashland	46 34	90 58	650	−21	−16	85/70	82/68	79/66	23	72	70	68			94.1	−26.8
Beloit	42 30	89 02	780	−7	−3	92/75	90/75	88/74	24	78	77	75				
Eau Claire AP	44 52	91 29	888	−15	−11	92/75	89/73	86/71	23	77	75	73				
Fond Du Lac	43 48	88 27	760	−12	−8	89/74	86/72	84/71	23	76	74	72			96.0	−17.7
Green Bay AP	44 29	88 08	682	−13	−9	88/74	85/72	83/71	23	76	74	72	W 8	SW	94.3	−17.9
La Crosse AP	43 52	91 15	651	−13	−9	91/75	88/73	85/72	22	77	75	74	NW 10	S	95.7	−21.3
Madison AP	43 08	89 20	858	−11	−7	91/74	88/73	85/72	22	77	75	73	NW 8	SW	93.6	−16.8
Manitowoc	44 06	87 41	660	−11	−7	89/74	86/72	83/71	21	76	74	72			94.1	−13.7
Marinette	45 06	87 38	605	−15	−11	87/73	84/71	82/70	20	75	73	71			95.9	−15.8
Milwaukee AP	42 57	87 54	672	−8	−4	90/74	87/73	84/71	21	76	74	73	WNW 10	SSW		
Racine	42 43	87 51	730	−6	−2	91/75	88/73	85/72	21	77	75	74				
Sheboygan	43 45	87 43	648	−10	−6	89/75	86/73	83/72	20	77	75	74			97.0	−12.4
Stevens Point	44 30	89 34	1079	−15	−11	92/75	89/73	86/71	23	77	75	73			95.3	−24.1
Waukesha	43 01	88 14	860	−9	−5	90/74	87/73	84/71	22	76	74	73			95.7	−14.3
Wausau AP	44 55	89 37	1196	−16	−12	91/74	88/72	85/70	23	76	74	72				
WYOMING																
Casper AP	42 55	106 28	5338	−11	−5	92/58	90/57	87/57	31	63	61	60	NE 10	SW	97.3	−20.9
Cheyenne	41 09	104 49	6126	−9	−1	89/58	86/58	84/57	30	63	62	60	N 11	WNW	92.5	−15.9
Cody AP	44 33	109 04	4990	−19	−13	89/60	86/60	83/59	32	64	63	61			97.4	−21.9
Evanston	41 16	110 57	6780	−9	−3	86/55	84/55	82/54	32	59	58	57			89.2	−21.2
Lander AP	42 49	108 44	5563	−16	−11	91/61	88/61	85/60	32	64	63	61	E 5	NW	94.9	−22.6
Laramie AP	41 19	105 41	7266	−14	−6	84/56	81/56	79/55	28	61	60	59				
Newcastle	43 51	104 13	4265	−17	−12	91/64	87/63	84/63	30	69	68	66			99.4	−19.0
Rawlins	41 48	107 12	6740	−12	−4	86/57	83/57	81/56	40	62	61	60				
Rock Springs AP	41 36	109 04	6745	−9	−3	86/55	84/55	82/54	32	59	58	57	WSW 10	W		
Sheridan AP	44 46	106 58	3964	−14	−8	94/62	91/62	88/61	32	66	65	63	NW 7	N	99.8	−23.6
Torrington	42 05	104 13	4098	−14	−8	94/62	91/62	88/61	30	66	65	63			101.1	−20.7

APPENDIX II—DESIGN CONDITIONS TABLES

Table 2 Climatic Conditions for Canada

Col. 1	Col. 2		Col. 3		Col. 4	Col. 5		Col. 6			Col. 7	Col. 8			Col. 9	
						Winter,[b] °F		Summer,[c] °F							Prevailing Wind	
	Lat.		Long.		Elev.	Design Dry-Bulb		Design Dry-Bulb and Mean Coincident Wet-Bulb			Mean Daily Range	Design Wet-Bulb			Winter	Summer
State and Station[a]	°	′N	°	′W	Feet	99%	97.5%	1%	2.5%	5%		1%	2.5%	5%	Knots[d]	
ALBERTA																
Calgary AP	51	06	114	01	3540	−27	−23	84/63	81/61	79/60	25	65	63	62	NNW 8	SE
Edmonton AP	53	34	113	31	2219	−29	−25	85/66	82/65	79/63	23	68	66	65	E 9	SE
Grande Prairie AP	55	11	118	53	2190	−39	−33	83/64	80/63	78/61	23	66	64	62		
Jasper	52	53	118	04	3480	−31	−26	83/64	80/62	77/61	28	66	64	63		
Lethbridge AP	49	38	112	48	3018	−27	−22	90/65	87/64	84/63	28	68	66	65		
McMurray AP	56	39	111	13	1216	−41	−38	86/67	82/65	79/64	26	69	67	65		
Medicine Hat AP	50	01	110	43	2365	−29	−24	93/66	90/65	87/64	28	70	68	66		
Red Deer AP	52	11	113	54	2965	−31	−26	84/65	81/64	78/62	25	67	66	64		
BRITISH COLUMBIA																
Dawson Creek	55	44	120	11	2164	−37	−33	82/64	79/63	76/61	26	66	64	62		
Fort Nelson AP	58	50	122	35	1230	−43	−40	84/64	81/63	78/62	23	67	65	64		
Kamloops Co	50	43	120	25	1133	−21	−15	94/66	91/65	88/64	29	68	66	65		
Nanaimo	49	11	123	58	230	16	20	83/67	80/65	77/64	21	68	66	65		
New Westminster	49	13	122	54	50	14	18	84/68	81/67	78/66	19	69	68	66		
Penticton AP	49	28	119	36	1121	0	4	92/68	89/67	87/66	31	70	68	67		
Prince George AP	53	53	122	41	2218	−33	−28	84/64	80/62	77/61	26	66	64	62	N 11	N
Prince Rupert Co	54	17	130	23	170	−2	2	64/59	63/57	61/56	12	60	58	57		
Trail	49	08	117	44	1400	−5	0	92/66	89/65	86/64	33	68	67	65		
Vancouver AP	49	11	123	10	16	15	19	79/67	77/66	74/65	17	68	67	66	E 6	WNW
Victoria Co	48	25	123	19	228	20	23	77/64	73/62	70/60	16	64	62	60		
MANITOBA																
Brandon	49	52	99	59	1200	−30	−27	89/72	86/70	83/68	25	74	72	70		
Churchill AP	58	45	94	04	155	−41	−39	81/66	77/64	74/62	18	67	65	63	SE 11	S
Dauphin AP	51	06	100	03	999	−31	−28	87/71	84/70	81/68	23	74	72	70		
Flin Flon	54	46	101	51	1098	−41	−37	84/68	81/66	79/65	19	70	68	67		
Portage La Prairie AP	49	54	98	16	867	−28	−24	88/73	86/72	83/70	22	76	74	71		
The Pas AP	53	58	101	06	894	−37	−33	85/68	82/67	79/66	20	71	69	68	W 8	W
Winnipeg AP	49	54	97	14	786	−30	−27	89/73	86/71	84/70	22	75	73	71	W 8	S
NEW BRUNSWICK																
Campbellton Co	48	00	66	40	25	−18	−14	85/68	82/67	79/66	21	72	70	68		
Chatham AP	47	01	65	27	112	15	−10	89/69	85/68	82/67	22	72	71	69		
Edmundston Co	47	22	68	20	500	−21	−16	87/70	83/68	80/67	21	73	71	69		
Fredericton AP	45	52	66	32	74	−16	−11	89/71	85/69	82/68	23	73	71	70		
Moncton AP	46	07	64	41	248	−12	−8	85/70	82/69	79/67	23	72	71	69		
Saint John AP	45	19	65	53	352	−12	−8	80/67	77/65	75/64	19	70	68	66		
NEWFOUNDLAND																
Corner Brook	48	58	57	57	15	−5	0	76/64	73/63	71/62	17	67	66	65		
Gander AP	48	57	54	34	482	−5	−1	82/66	79/65	77/64	19	69	67	66	WNW 11	SW
Goose Bay AP	53	19	60	25	144	−27	−24	85/66	81/64	77/63	19	68	66	64	N 9	SW
St John's AP	47	37	52	45	463	3	7	77/66	75/65	73/64	18	69	67	66	N 20	WSW
Stephenville AP	48	32	58	33	44	−3	4	76/65	74/64	71/63	14	67	66	65	WNW 10	S
NORTHWEST TERRITORIES																
Fort Smith AP	60	01	111	58	665	−49	−45	85/66	81/64	78/63	24	68	66	65	NW 4	S
Frobisher AP	63	45	68	33	68	−43	−41	66/53	63/51	59/50	14	54	52	51	NNW 9	NW
Inuvik	68	18	133	29	200	−56	−53	79/62	77/60	75/59	21	64	62	61		
Resolute AP	74	43	94	59	209	−50	−47	57/48	54/46	51/45	10	50	48	46		
Yellowknife AP	62	28	114	27	682	−49	−46	79/62	77/61	74/60	16	64	63	62	SSE 7	S
NOVA SCOTIA																
Amherst	45	49	64	13	65	−11	−6	84/69	81/68	79/67	21	72	70	68		
Halifax AP	44	39	63	34	83	1	5	79/66	76/65	74/64	16	69	67	66		
Kentville	45	03	64	36	40	−3	1	85/69	83/68	80/67	22	72	71	69		
New Glasgow	45	37	62	37	317	−9	−5	81/69	79/68	77/67	20	72	70	69		
Sydney AP	46	10	60	03	197	−1	3	82/69	80/68	77/66	19	71	70	68		
Truro Co	45	22	63	16	131	−8	−5	82/70	80/69	78/68	22	73	71	70		
Yarmouth AP	43	50	66	05	136	5	9	74/65	72/64	70/63	15	68	66	65	NW 11	S

Reproduced by permission of American Society of Heating, Refrigerating and Air-Conditioning Engineers.

Table 2 Climatic Conditions for Canada (*Concluded*)

Col. 1	Col. 2	Col. 3	Col. 4	Col. 5 Winter,[b] °F Design Dry-Bulb		Col. 6 Summer,[c] °F Design Dry-Bulb and Mean Coincident Wet-Bulb			Col. 7 Mean Daily Range	Col. 8 Design Wet-Bulb			Col. 9 Prevailing Wind	
State and Station[a]	Lat. ° 'N	Long. ° 'W	Elev. Feet	99%	97.5%	1%	2.5%	5%		1%	2.5%	5%	Winter Knots[d]	Summer
ONTARIO														
Belleville	44 09	77 24	250	−11	−7	86/73	84/72	82/71	20	75	74	73		
Chatham	42 24	82 12	600	0	3	89/74	87/73	85/72	19	76	75	74		
Cornwall	45 01	74 45	210	−13	−9	89/73	87/72	84/71	21	75	74	72		
Hamilton	43 16	79 54	303	−3	1	88/73	86/72	83/71	21	76	74	73		
Kapuskasing AP	49 25	82 28	752	−31	−28	86/70	83/69	80/67	23	72	70	69		
Kenora AP	49 48	94 22	1345	−32	−28	84/70	82/69	80/68	19	73	71	70		
Kingston	44 16	76 30	300	−11	−7	87/73	84/72	82/71	20	75	74	73		
Kitchener	43 26	80 30	1125	−6	−2	88/73	85/72	83/71	23	75	74	72		
London AP	43 02	81 09	912	−4	0	87/74	85/73	83/72	21	76	74	73		
North Bay AP	46 22	79 25	1210	−22	−18	84/68	81/67	79/66	20	71	70	68		
Oshawa	43 54	78 52	370	−6	−3	88/73	86/72	84/71	20	75	74	73		
Ottawa AP	45 19	75 40	413	−17	−13	90/72	87/71	84/70	21	75	73	72		
Owen Sound	44 34	80 55	597	−6	−2	84/71	82/70	80/69	21	73	72	70		
Peterborough	44 17	78 19	635	−13	−9	87/72	85/71	83/70	21	75	73	72		
St Catharines	43 11	79 14	325	−1	3	87/73	85/72	83/71	20	76	74	73		
Sarnia	42 58	82 22	625	0	3	88/73	86/72	84/71	19	76	74	73		
Sault Ste Marie AP	46 32	84 30	675	−7	−13	85/71	82/69	79/68	22	73	71	70		
Sudbury AP	46 37	80 48	1121	−22	−19	86/69	83/67	81/66	22	72	70	68		
Thunder Bay AP	48 22	89 19	644	−27	−24	85/70	83/68	80/67	24	72	70	68	W 8	W
Timmins AP	48 34	81 22	965	−33	−29	87/69	84/68	81/66	25	72	70	68		
Toronto AP	43 41	79 38	578	−5	−1	90/73	87/72	85/71	20	75	74	73	N 10	SW
Windsor AP	42 16	82 58	637	0	4	90/74	88/73	86/72	20	77	75	74		
PRINCE EDWARD ISLAND														
Charlottetown AP	46 17	63 08	186	−7	−4	80/69	78/68	76/67	16	71	70	68		
Summerside AP	46 26	63 50	78	−8	−4	81/69	79/68	77/67	16	72	70	68		
QUEBEC														
Bagotville AP	48 20	71 00	536	−28	−23	87/70	83/68	80/67	21	72	70	68		
Chicoutimi	48 25	71 05	150	−26	−22	86/70	83/68	80/67	20	72	70	68		
Drummondville	45 53	72 29	270	−18	−14	88/72	85/71	82/69	21	75	73	71		
Granby	45 23	72 42	550	−19	−14	88/72	85/71	83/70	21	75	73	72		
Hull	45 26	75 44	200	−18	−14	90/72	87/71	84/70	21	75	73	72		
Megantic AP	45 35	70 52	1362	−20	−16	86/71	83/70	81/69	20	74	72	71		
Montreal AP	45 28	73 45	98	−16	−10	88/73	85/72	83/71	17	75	74	72		
Quebec AP	46 48	71 23	245	−19	−14	87/72	84/70	81/68	20	74	72	70		
Rimouski	48 27	68 32	117	−16	−12	83/68	79/66	76/65	18	71	69	67		
St Jean	45 18	73 16	129	−15	−11	88/73	86/72	84/71	20	75	74	72		
St Jerome	45 48	74 01	556	−17	−13	88/72	86/71	83/70	23	75	73	72		
Sept. Iles AP	50 13	66 16	190	−26	−21	76/63	73/61	70/60	17	67	65	63		
Shawinigan	46 34	72 43	306	−18	−14	86/72	84/70	82/69	21	74	72	71		
Sherbrooke Co	45 24	71 54	595	−25	−21	86/72	84/71	81/69	20	74	73	71		
Thetford Mines	46 04	71 19	1020	−19	−14	87/71	84/70	81/69	21	74	72	71		
Trois Rivieres	46 21	72 35	50	−17	−13	88/72	85/70	82/69	23	74	72	71		
Val D'or AP	48 03	77 47	1108	−32	−27	85/70	83/68	80/67	22	72	70	68		
Valleyfield	45 16	74 06	150	−14	−10	89/73	86/72	84/71	20	75	74	72		
SASKATCHEWAN														
Estevan AP	49 04	103 00	1884	−30	−25	92/70	89/68	86/67	26	72	70	69		
Moose Jaw AP	50 20	105 33	1857	−29	−25	93/69	89/67	86/66	27	71	69	68		
North Battleford AP	52 46	108 15	1796	−33	−30	88/67	85/66	82/65	23	69	68	66		
Prince Albert AP	53 13	105 41	1414	−42	−35	87/67	84/66	81/65	25	70	68	67		
Regina AP	50 26	104 40	1884	−33	−29	91/69	88/68	84/67	26	72	70	68		
Saskatoon AP	52 10	106 41	1645	−35	−31	89/68	86/66	83/65	26	70	68	67		
Swift Current AP	50 17	107 41	2677	−28	−25	93/68	90/66	87/65	25	70	69	67		
Yorkton AP	51 16	102 28	1653	−35	−30	87/69	84/68	80/66	23	72	70	68		
YUKON TERRITORY														
Whitehorse AP	60 43	135 04	2289	−46	−43	80/59	77/58	74/56	22	61	59	58	NW 5	SE

[a] AP following the station name designates airport temperature observations. Co designates office locations within an urban area that are affected by the surrounding area. Undesignated stations are semirural and may be compared to airport data.
[b] Winter design data are based on the month of January only.
[c] Summer design data are based on the month of July only. See also Boughner (1960).
[d] Mean wind speeds occurring coincidentally with the 99.5% dry-bulb winter design temperature.

APPENDIX II—DESIGN CONDITIONS TABLES

Table 3 Climatic Conditions for Other Countries

Col. 1 Country and Station	Col. 2 Lat. ° 'N		Col. 2 Long. ° 'W		Col. 3 Elevation, ft	Winter, °F Col. 4 Mean of Annual Extremes	Col. 4 99%	Col. 4 97.5%	Summer, °F Col. 5 Design Dry-Bulb 1%	Col. 5 2.5%	Col. 5 5%	Col. 6 Mean Daily Range	Col. 7 Design Wet-Bulb 1%	Col. 7 2.5%	Col. 7 5%	Prevailing Wind Winter Knots		Prevailing Wind Summer Knots
AFGHANISTAN																		
Kabul	34	35N	69	12E	5955	2	6	9	98	96	93	32	66	65	64	N	4	N
ALGERIA																		
Algiers	36	46N	3	03E	194	38	43	45	95	92	89	14	77	76	75			
ARGENTINA																		
Buenos Aires	34	35S	58	29W	89	27	32	34	91	89	86	22	77	76	75	SW	9	NNE
Cordoba	31	22S	64	15W	1388	21	28	32	100	96	93	27	76	75	74			
Tucuman	26	50S	65	10W	1401	24	32	36	102	99	96	23	76	75	74			
AUSTRALIA																		
Adelaide	34	56S	138	35E	140	36	38	40	98	94	91	25	72	70	68	NE	5	NW
Alice Springs	23	48S	133	53E	1795	28	34	37	104	102	100	27	75	74	72	N	6	SE
Brisbane	27	28S	153	02E	137	39	44	47	91	88	86	18	77	76	75	N	7	NNE
Darwin	12	28S	130	51E	88	60	64	66	94	93	91	16	82	81	81	E	10	WNW
Melbourne	37	49S	144	58E	114	31	35	38	95	91	86	21	71	69	68			
Perth	31	57S	115	51E	210	38	40	42	100	96	93	22	76	74	73	N	6	E
Sydney	33	52S	151	12E	138	38	40	42	89	84	80	13	74	73	72	N	8	NE
AUSTRIA																		
Vienna	48	15N	16	22E	644	−2	6	11	88	86	83	16	71	69	67	W	13	SSE
AZORES																		
Lajes	38	45N	27	05W	170	42	46	49	80	78	77	11	73	72	71	W	9	NW
BAHAMAS																		
Nassau	25	05N	77	21W	11	55	61	63	90	89	88	13	80	80	79			
BANGLADESH																		
Chittagong	22	21N	91	50E	87	48	52	54	93	91	89	20	82	81	81			
BELGIUM																		
Brussels	50	48N	4	21E	328	13	15	19	83	79	77	19	70	68	67	NE	8	ENE
BELIZE																		
Belize	17	31N	88	11W	17	55	60	62	90	90	89	13	82	82	81			
BERMUDA																		
Kindley AFB	33	22N	64	41W	129	47	53	55	87	86	85	12	79	78	78	NW	16	S
BOLIVIA																		
La Paz	16	30S	68	09W	12001	28	31	33	71	69	68	24	58	57	56			
BRAZIL																		
Belem	1	27S	48	29W	42	67	70	71	90	89	87	19	80	79	78	SE	5	E
Belo Horizonte	19	56S	43	57W	3002	42	47	50	86	84	83	18	76	75	75			
Brasilia	15	53S	47	56W	3481	46	52	54	86	84	84	23	71	70	70	W	1	E
Campinas	23	01S	47	08W	2169	—	52	54	91	90	88	23	76	75	74	ESE	7	N
Congonhas	23	38S	46	39W	2635	—	46	49	88	86	84	21	73	72	71	S	3	NNW
Cirotoba	25	25S	49	17W	3114	28	34	37	86	84	82	21	75	74	74			
Fortaleza	3	46S	38	33W	89	66	69	70	91	90	89	17	79	78	78			
Galeao	22	50S	43	15W	20	—	61	63	97	95	92	21	80	79	78	NW	2	SSE
Porto Alegre	30	02S	51	13W	33	32	37	40	95	92	89	20	76	76	75			
Recife	8	04S	34	53W	97	67	69	70	88	87	86	10	78	77	77	S	7	ESE
Rio de Janeiro	22	55S	43	12W	201	56	58	60	94	92	90	11	80	79	78	N	5	S
Salvador	13	00S	38	30W	154	65	67	68	88	87	86	12	79	79	78			
Sao Paulo	23	31S	46	37W	2598	36	47	50	89	87	85	14	74	73	72	E	6	NW
BULGARIA																		
Sofia	42	42N	23	20E	1805	−2	3	8	89	86	84	26	71	70	69			
CAMBODIA																		
Phnom Penh	11	33N	104	51E	36	62	66	68	98	96	94	19	83	82	82	N	4	W
CHILE																		
Concepcion	36	47S	73	04W	30	—	41	41	75	73	72	22	64	63	62	S	4	SW
Punta Arenas	53	10S	70	54W	26	22	25	27	68	66	64	14	56	55	54			
Santiago	33	24S	70	47W	1555	27	30	32	90	88	86	37	68	67	66	NE	1	SW
Valparaiso	33	01S	71	38W	135	39	43	46	81	79	77	16	67	66	65			
CHINA																		
Chungking	29	33N	106	33E	755	34	37	39	99	97	95	18	81	80	79			
Shanghai	31	12N	121	26E	23	16	23	26	94	92	90	16	81	81	80	WNW	6	S
COLOMBIA																		
Baranquilla	10	59N	74	48W	44	66	70	72	95	94	93	17	83	82	82			
Bogota	4	36N	74	05W	8406	42	45	46	72	70	69	19	60	59	58	E	8	E
Cali	3	25N	76	30W	3189	53	57	58	84	82	79	15	70	69	68			
Medellin	6	13N	75	36W	4650	48	53	55	87	85	84	25	73	72	72			

Reproduced by permission of American Society of Heating, Refrigerating and Air-Conditioning Engineers.

Table 3 Climatic Conditions for Other Countries (*Continued*)

Col. 1 Country and Station	Col. 2 Lat. ° 'N		Col. 2 Long. ° 'W		Col. 3 Elevation, ft	Winter, °F Col. 4 Mean of Annual Extremes	Col. 4 99%	Col. 4 97.5%	Summer, °F Col. 5 Design Dry-Bulb 1%	Col. 5 2.5%	Col. 5 5%	Col. 6 Mean Daily Range	Col. 7 Design Wet-Bulb 1%	Col. 7 2.5%	Col. 7 5%	Prevailing Wind Winter		Prevailing Wind Summer
COMMONWEALTH OF INDEPENDENT STATES (formerly SOVIET UNION)																		
Alma Ata	43	14N	76	53E	2543	−18	−10	−6	88	86	83	21	69	68	67			
Archangel	64	33N	40	32E	22	−29	−23	−18	75	71	68	13	60	58	57			
Ekaterinburg (Sverdlousk)	56	49N	60	38E	894	−34	−25	−20	80	76	72	16	63	62	60			
Kaliningrad	54	43N	20	30E	23	−3	1	6	83	80	77	17	67	66	65			
Krasnoyarsk	56	01N	92	57E	498	−41	−23	−27	84	80	76	12	64	62	60			
Kiev	50	27N	30	30E	600	−12	−5	1	87	84	81	22	69	68	67			
Kharkov	50	00N	36	14E	472	−19	−10	−3	87	84	82	23	69	68	67			
Minsk	53	54N	27	33E	738	−19	−11	−4	80	77	74	16	67	66	65			
Moscow	55	46N	37	40E	505	−19	−11	−6	84	81	78	21	69	67	65	SW	11	S
Odessa	46	29N	30	44E	214	−1	4	8	87	84	82	14	70	69	68			
Petropavlovsk	52	53N	158	42E	286	−9	−3	0	70	68	65	13	58	57	56			
Rostov on Don	47	13N	39	43E	159	−9	−2	4	90	87	84	20	70	69	68			
Samara	53	11N	50	06E	190	−23	−19	−13	89	85	81	20	69	67	66			
St. Petersburg (Leningrad)	59	56N	30	16E	16	−14	−9	−5	78	75	72	15	65	64	63			
Tashkent	41	20N	69	18E	1569	−4	3	8	95	93	90	29	71	70	69			
Tbilisi	41	43N	44	48E	1325	12	18	22	87	85	83	18	68	67	66			
Vladivostok	43	07N	131	55E	94	−15	−10	−7	80	77	74	11	70	69	68			
Volgograd	48	42N	44	31E	136	−21	−13	−7	93	89	86	19	71	70	69			
CONGO																		
Brazzaville	4	15S	15	15E	1043	54	60	62	93	92	91	21	81	81	80			
CUBA																		
Guantanamo Bay	19	54N	75	09W	21	60	64	66	94	93	92	16	82	81	80	N	6	ESE
Havana	23	08N	82	21W	80	54	59	62	92	91	89	14	81	81	80	N	11	E
CYPRUS																		
Akrotiri	34	36N	32	59E	75	—	41	43	91	90	87	17	78	77	76	NNW	4	SW
Lainaca	34	53N	33	39E	7	—	37	41	93	91	90	21	78	77	76	NW	6	SSW
Paphos	34	43N	32	30E	26	—	42	45	88	86	86	13	79	79	78	NE	7	W
CZECHOSLOVAKIA																		
Prague	50	05N	14	25E	662	3	4	9	88	85	83	16	66	65	64			
DENMARK																		
Copenhagen	55	41N	12	33E	43	11	16	19	79	76	74	17	68	66	64	NE	11	N
DOMINICAN REPUBLIC																		
Santo Domingo	18	29N	69	54W	57	61	63	65	92	90	88	16	81	80	80	NNE	6	SE
EQUADOR																		
Guayaquil	2	10S	79	53W	20	61	64	65	92	91	89	20	80	80	79			
Quito	0	13S	78	32W	9446	30	36	39	73	72	71	32	63	62	62	N	3	N
EGYPT																		
Cairo	29	52N	31	20E	381	39	45	46	102	100	98	26	76	75	74	N	9	NNW
Luxor	25	40N	32	43E	289	—	38	41	109	108	106	31	76	74	73	E	1	N
EL SALVADOR																		
San Salvador	13	42N	89	13W	2238	51	54	56	98	96	95	32	77	76	75	N	7	S
ETHIOPIA																		
Addis Ababa	9	02N	38	45E	7753	35	39	41	84	82	81	28	66	65	64	E	10	S
Asmara	15	17N	38	55E	7628	36	40	42	83	81	80	27	65	64	63	E	9	WNW
FINLAND																		
Helsinki	60	10N	24	57E	30	−11	−7	−1	77	74	72	14	66	65	63	E	4	S
FRANCE																		
Lyon	45	42N	4	47E	938	−1	10	14	91	89	86	23	71	70	69	N	7	S
Marseilles	43	18N	5	23E	246	23	25	28	90	87	84	22	72	71	69	SE	14	W
Nantes	47	15N	1	34W	121	17	22	26	86	83	80	21	70	69	67	NNE	6	E
Nice	43	42N	7	16E	39	31	34	37	87	85	83	15	73	72	72			
Paris	48	49N	2	29E	164	16	22	25	89	86	83	21	70	68	67	NE	7	E
Strasbourg	48	35N	7	46E	465	9	11	16	86	83	80	20	70	69	67			
FRENCH GUIANA																		
Cayenne	4	56N	52	27W	20	69	71	72	92	91	90	17	83	83	82	ENE	5	E
GERMANY																		
Berlin	52	27N	13	18E	187	6	7	12	84	81	78	19	68	67	66	E	6	E
Hamburg	53	33N	9	58E	66	10	12	16	80	76	73	13	68	66	65			
Hannover	52	24N	9	40E	561	7	16	20	82	78	75	17	68	67	65	E	8	E
Mannheim	49	34N	8	28E	359	2	8	11	87	85	82	18	71	69	68	N	5	S
Munich	48	09N	11	34E	1729	−1	5	9	86	83	80	18	68	66	64	S	4	N
GHANA																		
Accra	5	33N	0	12W	88	65	68	69	91	90	89	13	80	79	79	WSW	5	SW
GIBRALTAR																		
Gibraltar	36	09N	5	22W	11	38	42	45	92	89	86	14	76	75	74			
GREECE																		
Athens	37	58N	23	43E	351	29	33	36	96	93	91	18	72	71	71	N	9	NNE
Souda	35	32N	24	09E	479	—	38	41	95	91	90	21	76	74	72	NNW	4	WNW
Thessaloniki	40	37N	22	57E	78	23	28	32	95	93	91	20	77	76	75			

Table 3 Climatic Conditions for Other Countries (*Continued*)

Col. 1 Country and Station	Col. 2 Lat. ° 'N	Col. 2 Long. ° 'W	Col. 3 Elevation, ft	Winter, °F Col. 4 Mean of Annual Extremes	Winter, °F Col. 4 99%	Winter, °F Col. 4 97.5%	Summer, °F Col. 5 Design Dry-Bulb 1%	Summer, °F Col. 5 Design Dry-Bulb 2.5%	Summer, °F Col. 5 Design Dry-Bulb 5%	Summer, °F Col. 6 Mean Daily Range	Summer, °F Col. 7 Design Wet-Bulb 1%	Summer, °F Col. 7 Design Wet-Bulb 2.5%	Summer, °F Col. 7 Design Wet-Bulb 5%	Prevailing Wind Winter (Knots)		Prevailing Wind Summer
GREENLAND																
Narsarssuaq	61 11N	45 25W	85	−23	−12	−8	66	63	61	20	56	54	52			
GUATEMALA																
Guatemala City	14 37N	90 31W	4855	45	48	51	83	82	81	24	69	68	67	N	9	S
GUYANA																
Georgetown	6 50N	58 12W	6	70	72	73	89	88	87	11	80	79	79			
HAITI																
Port au Prince	18 33N	72 20W	121	63	65	67	97	95	93	20	82	81	80	N	6	ESE
HONDURAS																
Tegucigalpa	14 06N	87 13W	3094	44	47	50	89	87	85	28	73	72	71	N	8	E
HONG KONG																
Hong Kong	22 18N	114 10E	109	43	48	50	92	91	90	10	81	80	80	N	9	W
HUNGARY																
Budapest	47 31N	19 02E	394	8	10	14	90	86	84	21	72	71	70	N	5	S
ICELAND																
Reykjavik	64 08N	21 56E	59	8	14	17	59	58	56	16	54	53	53	E	12	E
INDIA																
Ahmenabad	23 02N	72 35E	163	49	53	56	109	107	105	28	80	79	78			
Bangalore	12 57N	77 37E	3021	53	56	58	96	94	93	26	75	74	74			
Bombay	18 54N	72 49E	37	62	65	67	96	94	92	13	82	81	81	NW		NW
Calcutta	22 32N	88 20E	21	49	52	54	98	97	96	22	83	82	82	N	4	S
Madras	13 04N	80 15E	51	61	64	66	104	102	101	19	84	83	83	W	3	W
Nagpur	21 09N	79 07E	1017	45	51	54	110	108	107	30	79	79	78			
New Delhi	28 35N	77 12E	703	35	39	41	110	107	105	26	83	82	82	N	6	NW
INDONESIA																
Djakarta	6 11S	106 50E	26	69	71	72	90	89	88	14	80	79	78	N	11	N
Kupang	10 10S	123 34E	148	63	66	68	94	93	92	20	81	80	80			
Makassar	5 08S	119 28E	61	64	66	68	90	89	88	17	80	80	79			
Medan	3 35N	98 41E	77	66	69	71	92	91	90	17	81	80	79			
Palembang	3 00S	104 46E	20	67	70	71	92	91	90	17	80	79	79			
Surabaya	7 13S	112 43E	10	64	66	68	91	90	89	18	80	79	79			
IRAN																
Abadan	30 21N	48 16E	7	32	39	41	116	113	110	32	82	81	81	W	6	WNW
Meshed	36 17N	59 36E	3104	3	10	14	99	96	93	29	68	67	66			
Tehran	35 41N	51 25E	4002	15	20	24	102	100	98	27	75	74	73	W	5	SE
IRAQ																
Baghdad	33 20N	44 24E	111	27	32	35	113	111	108	34	73	72	72	WNW	5	WNW
Mosul	36 19N	43 09E	730	23	29	32	114	112	110	40	73	72	72			
IRELAND																
Dublin	53 22N	6 21W	155	19	24	27	74	72	70	16	65	64	62	W	9	SW
Shannon	52 41N	8 55W	8	19	25	28	76	73	71	14	65	64	63	SE	4	W
IRIAN BARAT																
Manokwari	0 52S	134 05E	62	70	71	72	89	88	87	12	82	81	81			
ISRAEL																
Jerusalem	31 47N	35 13E	2485	31	36	38	95	94	92	24	70	69	69	W	12	NW
Tel Aviv	32 06N	34 47E	36	33	39	41	96	93	91	16	74	73	72	N	8	W
ITALY																
Milan	45 27N	9 17E	341	12	18	22	89	87	84	20	76	75	74	W	4	SW
Naples	40 53N	14 18E	220	28	34	36	91	88	86	19	74	73	72	N	6	SSW
Rome	41 48N	12 36E	377	25	30	33	94	92	89	24	74	73	72	E	6	WSW
IVORY COAST																
Abidjan	5 19N	4 01W	65	64	67	69	91	90	88	15	83	82	81	WSW	5	SW
JAMAICA																
Kingston	17 56N	76 47W	46	—	71	72	93	91	89	10	81	80	80	N	10	ESE
Montego	18 30N	77 55W	10	—	68	69	89	89	88	11	80	79	79	ESE	10	ENE
JAPAN																
Fukuoka	33 35N	130 27E	22	26	29	31	92	90	89	20	82	80	79			
Sapporo	43 04N	141 21E	56	−7	1	5	86	83	80	20	76	74	72	SE	3	SE
Tokyo	35 41N	139 46E	19	21	26	28	91	89	87	14	81	80	79	SW	10	S
JOHNSTON ISLAND	16 44N	169 31W	16	—	72	73	87	86	85	6	80	80	79	NE	15	E
JORDAN																
Amman	31 57N	35 57E	2548	29	33	36	97	94	92	25	70	69	68	N	6	NNW
KENYA																
Nairobi	1 16S	36 48E	5971	45	48	50	81	80	78	24	66	65	65	E	13	ENE
KOREA																
Pyongyang	39 02N	125 41E	186	−10	−2	3	89	87	85	21	77	76	76			
Seoul	37 34N	126 58E	285	−1	7	9	91	89	87	16	81	79	78	NW	7	W
LEBANON																
Beirut	33 54N	35 28E	111	40	42	45	93	91	90	15	78	77	76	N	7	SW
LIBERIA																
Monrovia	6 18N	10 48W	75	64	68	69	90	89	88	19	82	82	81	E	3	WSW
LIBYA																
Benghazi	32 06N	20 04E	82	41	46	48	97	94	91	13	77	76	75	SSE	8	S
MADAGASCAR																
Tananarive	18 55S	47 33E	4531	39	43	46	86	84	83	23	73	72	71			

Table 3 Climatic Conditions for Other Countries (*Continued*)

Col. 1 Country and Station	Col. 2 Lat. ° 'N		Col. 2 Long. ° 'W		Col. 3 Elevation, ft	Winter, °F Col. 4 Mean of Annual Extremes	Winter, °F Col. 4 99%	Winter, °F Col. 4 97.5%	Summer, °F Col. 5 Design Dry-Bulb 1%	Summer, °F Col. 5 Design Dry-Bulb 2.5%	Summer, °F Col. 5 Design Dry-Bulb 5%	Col. 6 Mean Daily Range	Col. 7 Design Wet-Bulb 1%	Col. 7 Design Wet-Bulb 2.5%	Col. 7 Design Wet-Bulb 5%	Prevailing Wind Winter Knots		Prevailing Wind Summer Knots
MALAYSIA																		
Kuala Lumpur	3	07N	101	42E	127	67	70	71	94	93	92	20	82	82	81	N	4	W
Penang	5	25N	100	19E	17	69	72	73	93	93	92	18	82	81	80			
MARTINIQUE																		
Fort de France	14	37N	61	05W	13	62	64	66	90	89	88	14	81	81	80			
MEXICO																		
Guadalajara	20	41N	103	20W	5105	35	39	42	93	91	89	29	68	67	66	N	7	W
Merida	20	58N	89	38W	72	56	59	61	97	95	94	21	80	79	77	E	11	E
Mexico City	19	24N	99	12W	7575	33	37	39	83	81	79	25	61	60	59	N	8	N
Monterrey	25	40N	100	18W	1732	31	38	41	98	95	93	20	79	78	77			
Tampico	22	17N	97	52W	79	—	48	50	91	89	89	10	83	81	81	N	11	E
Vera Cruz	19	12N	96	08W	184	55	60	62	91	89	88	12	83	83	82			
MIDWAY ISLAND	28	13N	177	23W	10	—	56	58	87	86	86	9	78	77	76	NNW	9	E
MOROCCO																		
Casablanca	33	35N	7	39W	164	36	40	42	94	90	86	50	73	72	70			
MYANMAR																		
Mandalay	21	59N	96	06E	252	50	54	56	104	102	101	30	81	80	80			
Rangoon	16	47N	96	09E	18	59	62	63	100	98	95	25	83	82	82	W	6	W
NEPAL																		
Katmandu	27	42N	85	12E	4388	30	33	35	89	87	86	25	78	77	76	W	4	NW
NETHERLANDS																		
Amsterdam	52	23N	4	55E	5	17	20	23	79	76	73	10	65	64	63	S	8	E
NEW ZEALAND																		
Auckland	36	51S	174	46E	140	37	40	42	78	77	76	14	67	66	65			
Christ Church	43	32S	172	37E	32	25	28	31	82	79	76	17	68	67	66	W	4	NNW
Wellington	41	17S	174	46E	394	32	35	37	76	74	72	14	66	65	64	NE	6	NNE
NICARAGUA																		
Managua	12	10N	86	15W	135	62	65	67	94	93	92	21	81	80	79	E	9	E
NIGERIA																		
Lagos	6	27N	3	24E	10	67	70	71	92	91	90	12	82	82	81	WSW	8	S
NORWAY																		
Bergen	60	24N	5	19E	141	14	17	20	75	74	73	21	67	66	65			
Oslo	59	56N	10	44E	308	−2	0	4	79	77	74	17	67	66	64	N	10	S
PAKISTAN																		
Karachi	24	48N	66	59E	13	45	49	51	100	98	95	14	82	82	81	N	4	SSW
Lahore	31	35N	74	20E	702	32	35	37	109	107	105	27	83	82	81	NW	3	SE
Peshwar	34	01N	71	35E	1164	31	35	37	109	106	103	29	81	80	79	W	5	NE
PANAMA AND CANAL ZONE																		
Panama City	8	58N	79	33W	21	69	72	73	93	92	91	18	81	81	80			
PAPUA NEW GUINEA																		
Port Moresby	9	29S	147	09E	126	62	67	69	92	91	90	14	80	80	79			
PARAGUAY																		
Ascuncion	25	17S	57	30W	456	35	43	46	100	98	96	24	81	81	80	NE	7	NE
PERU																		
Lima	12	05S	77	03W	394	51	53	55	86	85	84	17	76	75	74	N	10	S
San Juan de Marcona	15	24S	75	10W	197	—	55	57	82	81	79	12	75	73	72	S	10	S
Talara	4	35S	81	15W	282	—	59	61	90	88	86	15	79	78	76	SSE	18	S
PHILIPPINES																		
Manila	14	35N	120	59E	47	69	73	74	94	92	91	20	82	81	81	N	3	ESE
POLAND																		
Krakow	50	04N	19	57E	723	−2	2	6	84	81	78	19	68	67	66			
Warsaw	52	13N	21	02E	394	−3	3	8	84	81	78	19	71	70	68	E	7	SE
PORTUGAL																		
Lisbon	38	43N	9	08W	313	32	37	39	89	86	83	16	69	68	67	ENE	5	N
PUERTO RICO																		
San Juan	18	29N	66	07W	82	65	67	68	89	88	87	11	81	80	79	ENE	10	E
RUMANIA																		
Bucharest	44	25N	26	06E	269	−2	3	8	93	91	89	26	72	71	70			
SAUDI ARABIA																		
Dhahran	26	17N	50	09E	80	39	45	48	111	110	108	32	86	85	84	N	8	N
Jedda	21	28N	39	10E	20	52	57	60	106	103	100	22	85	84	83			
Riyadh	24	39N	46	42E	1938	29	37	40	110	108	106	32	78	77	76	N	8	N
SENEGAL																		
Dakar	14	42N	17	29W	131	58	61	62	95	93	91	13	81	80	80	N	8	NW
SINGAPORE																		
Singapore	1	18N	103	50E	33	69	71	72	92	91	90	14	82	81	80	N	4	SE
SOMALIA																		
Mogadiscio	2	02N	49	19E	39	67	69	70	91	90	89	12	82	82	81	SSW	16	E
SOUTH AFRICA																		
Cape Town	33	56S	18	29E	55	36	40	42	93	90	86	20	72	71	70			
Johannesburg	26	11S	78	03E	5463	26	31	34	85	83	81	24	70	69	69			
Pretoria	25	45S	28	14E	4491	27	32	35	90	87	85	23	70	69	68	N	4	W
SOUTH YEMEN																		
Aden	12	50N	45	02E	10	63	68	70	102	100	98	11	83	82	82			

Table 3 Climatic Conditions for Other Countries (*Concluded*)

Col. 1 Country and Station	Col. 2 Lat. °N		Col. 2 Long. °W		Col. 3 Elevation, ft	Winter, °F Col. 4 Mean of Annual Extremes	Winter, °F Col. 4 99%	Winter, °F Col. 4 97.5%	Summer, °F Col. 5 Design Dry-Bulb 1%	Summer, °F Col. 5 Design Dry-Bulb 2.5%	Summer, °F Col. 5 Design Dry-Bulb 5%	Col. 6 Mean Daily Range	Col. 7 Design Wet-Bulb 1%	Col. 7 Design Wet-Bulb 2.5%	Col. 7 Design Wet-Bulb 5%	Prevailing Wind Winter Knots		Prevailing Wind Summer Knots
SPAIN																		
Barcelona	41	24N	2	09E	312	31	33	36	88	86	84	13	75	74	73	N	10	S
Madrid	40	25N	3	41W	2188	22	25	28	93	91	89	25	71	69	67	NNE	5	W
Valencia	39	28N	0	23W	79	31	33	37	92	90	88	14	75	74	73	W	7	ESE
SRI LANKA																		
Colombo	6	54N	79	52E	24	65	69	70	90	89	88	15	81	80	80	W		6 W
SUDAN																		
Khartoum	15	37N	32	33E	1279	47	53	56	109	107	104	30	77	76	75	N	6	NW
SURINAM																		
Paramaribo	5	49N	55	09W	12	66	68	70	93	92	90	18	82	82	81	NE	9	E
SWEDEN																		
Stockholm	59	21N	18	04E	146	3	5	8	78	74	72	15	64	62	60	W	4	S
SWITZERLAND																		
Zurich	47	23N	8	33E	1617	4	9	14	84	81	78	21	68	67	66			
SYRIA																		
Damascus	33	30N	36	20E	2362	25	29	32	102	100	98	35	72	71	70			
TAIWAN																		
Tainan	22	57N	120	12E	70	40	46	49	92	91	90	14	84	83	82	N	10	W
Taipei	25	02N	121	31E	30	41	44	47	94	92	90	16	83	82	81	E	7	E
TANZANIA																		
Dar es Salaam	6	50S	39	18E	47	62	64	65	90	89	88	13	82	81	81			
THAILAND																		
Bangkok	13	44N	100	30E	39	57	61	63	97	95	93	18	82	82	81	N	4	S
TRINIDAD																		
Port of Spain	10	40N	61	31W	67	61	64	66	91	90	89	16	80	80	79			
TUNISIA																		
Tunis	36	47N	10	12E	217	35	39	41	102	99	96	22	77	76	74	W	10	E
TURKEY																		
Adana	36	59N	35	18E	82	25	33	35	100	97	95	22	79	78	77			
Ankara	39	57N	32	53E	2825	2	9	12	94	92	89	28	68	67	66	N	8	W
Istanbul	40	58N	28	50E	59	23	28	30	91	88	86	16	75	74	73	N	10	NE
Izmir	38	26N	27	10E	16	24	27	29	98	96	94	23	75	74	73	NNE	8	N
UNITED KINGDOM																		
Belfast	54	36N	5	55W	24	19	23	26	74	72	69	16	65	64	62			
Birmingham	52	29N	1	56W	535	21	24	27	79	76	73	15	66	64	63			
Cardiff	51	28N	3	10W	203	21	24	27	79	76	73	14	64	63	62			
Edinburgh	55	55N	3	11W	441	22	25	28	73	70	68	13	64	62	61	WSW	6	WSW
Glasgow	55	52N	4	17W	85	17	21	24	74	71	68	13	64	63	61			
London	51	29N	0	00	149	20	24	26	82	79	76	16	68	66	65	W	7	E
URUGUAY																		
Montevideo	34	51S	56	13W	72	34	37	39	90	88	85	21	73	72	71	N	11	NNE
VENEZUELA																		
Caracas	10	30N	66	56W	3418	49	52	54	84	83	81	21	70	69	69	E	8	ENE
Maracaibo	10	39N	71	36W	20	69	72	73	97	96	95	17	84	83	83			
VIETNAM																		
Da Nang	16	04N	108	13E	23	56	60	62	97	95	93	14	86	86	85	NW	5	N
Hanoi	21	02N	105	52E	53	46	50	53	99	97	95	16	85	85	84			
Ho Chi Minh City	10	47N	106	42E	30	62	65	67	93	91	89	16	85	84	83			
YUGOSLAVIA																		
Belgrade	44	48N	20	28E	453	4	9	13	92	89	86	23	74	73	72	ESE	9	SE
ZAIRE																		
Kinshasa	4	20S	15	18E	1066	54	60	62	92	91	90	19	81	80	80	NNW	7	W
Kisangani	0	26S	15	14E	1370	65	67	68	92	91	90	19	81	80	80			

APPENDIX III—ELECTRICAL SYMBOLS

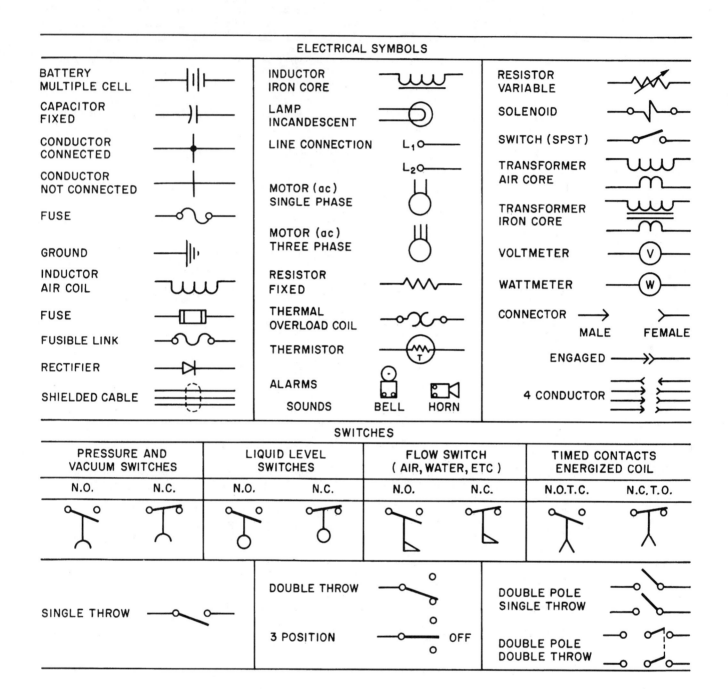

GLOSSARY

Because the HVAC industry encompasses not only air conditioning but heating and ventilation as well, we have included a complete HVAC glossary of terms for your convenience.

Note: English term appears boldface followed by Spanish term set in Italic.

ABS pipe. Acrylonitrilebutadiene styrene plastic pipe used for water, drains, waste, and venting.

Tubo de acronitrilo-butadieno-estireno. Tubo plástico de acronitrilo-butadieno-estireno utilizado para el agua, los drenajes, los desperdicios y la ventilación.

Absolute pressure. Gage pressure plus the pressure of the atmosphere, normally 14.696 at sea level at 68°F.

Presión absoluta. La presión del calibrador más la presión de la atmósfera, que generalmente es 14,696 al nivel del mar a 68°F (20°C).

Absolute zero temperature. The lowest obtainable temperature where molecular motion stops, −460°F and −273°C.

Temperatura del cero absoluto. La temperatura más baja obtenible donde se detiene el movimiento molecular, −460°F and −273°C.

Absorbent (attractant). The salt solution used to attract water in an absorption chiller.

Hidrófilo. Solución salina utilizada para atraer el agua en un enfriador por absorción.

Absorber. That part of an absorption chiller where the water is absorbed by the salt solution.

Absorbedor. El lugar en el enfriador por absorción donde la solución salina absorbe el agua.

Absorption air conditioning chiller. A system using a salt substance, water, and heat to provide cooling for an air conditioning system.

Enfriador por absorción para acondicionamiento de aire. Sistema que utiliza una sustancia salina, agua y calor para proveer enfriamiento en un sistema de acondicionamiento de aire.

Accumulator. A storage tank located in the suction line of a compressor. It allows small amounts of liquid refrigerant to boil away before entering the compressor. Sometimes used to store excess refrigerant in heat pump systems during the winter cycle.

Acumulador. Tanque de almacenaje ubicado en el conducto de aspiración de un compresor. Permite que pequeñas cantidades de refrigerante líquido se evaporen antes de entrar al compresor. Algunas veces se utiliza para almacenar exceso de refrigerante en sistemas de bombas de calor durante el ciclo de invierno.

Acid-contaminated system. A refrigeration system that contains acid due to contamination.

Sistema contaminado de ácido. Sistema de refrigeración que, debido a la contaminación, contiene ácido.

ACR tubing. Air Conditioning and Refrigeration tubing that is very clean, dry, and normally charged with dry nitrogen. The tubing is sealed at the ends to contain the nitrogen.

Tubería ACR. Tubería para el acondicionamiento de aire y la refrigeración que es muy limpia y seca, y que por lo general está cargada de nitrógeno seco. La tubería se sella en ambos extremos para contener el nitrógeno.

Activated alumina. A chemical desiccant used in refrigerant driers.

Alúmina activada. Disecante químico utilizado en secadores de refrigerantes.

Active solar system. A system that uses electrical and/or mechanical devices to help collect, store, and distribute the sun's energy.

Sistema solar activo. Sistema que utiliza dispositivos eléctricos y/o mecánicos para ayudar a acumular, almacenar y distribuir la energía del sol.

Air-acetylene. A mixture of air and acetylene gas that when ignited is used for soldering, brazing, and other applications.

Aire-acetilénico. Mezcla de aire y de gas acetileno que se utiliza en la soldadura, la broncesoldadura y otras aplicaciones al ser encendida.

Air heat exchanger. A device used to exchange heat between air and another medium at different temperature levels, such as air-to-air, air-to-water, or air-to-refrigerant.

Intercambiador de aire y calor. Dispositivo utilizado para intercambiar el calor entre el aire y otro medio, como por ejemplo aire y aire, aire y agua o aire y refrigerante, a diferentes niveles de temperatura.

Air conditioner. Equipment that conditions air by cleaning, cooling, heating, humidifying, or dehumidifying it. A term often applied to comfort cooling equipment.

Acondicionador de aire. Equipo que acondiciona el aire limpiándolo, enfriándolo, calentándolo, humidificándolo o deshumidificándolo. Término comúnmente aplicado al equipo de enfriamiento para comodidad.

Air conditioning. A process that maintains comfort conditions in a defined area.

Acondicionamiento de aire. Proceso que mantiene condiciones agradables en un área definida.

Air-cooled condenser. One of the four main components of an air-cooled refrigeration system. It receives hot gas from the compressor and rejects it to a place where it makes no difference.

Condensador enfriado por aire. Uno de los cuatro componentes principales de un sistema de refrigeración enfriado por aire. Recibe el gas caliente del compresor y lo dirige a un lugar donde no afecte la temperatura.

Air gap. The clearance between the rotating rotor and the stationary winding on an open motor. Known as a vapor gap in a hermetically sealed compressor motor.

614 GLOSSARY

Espacio de aire. Espacio libre entre el rotor giratorio y el devandado fijo en un motor abierto. Conocido como espacio de vapor en un motor de compresor sellado herméticamente.

Air handler. The device that moves the air across the heat exchanger in a forced-air system—normally considered to be the fan and its housing.

Tratante de aire. Dispositivo que dirige el aire a través del intercambiador de calor en un sistema de aire forzado—considerado generalmente como el abanico y su alojamiento.

Air pressure control (switch). Used to detect air pressure drop across the coil in a heat pump outdoor unit due to ice buildup.

Regulador de la presión de aire (conmutador). Utilizado para detectar una caída en la presión del aire a través de la bobina en una unidad de bomba de calor para exteriores debido a la acumulación de hielo.

Air sensor. A device that registers changes in air conditions such as pressure, velocity, temperature, or moisture content.

Sensor de aire. Dispositivo que registra los cambios en las condiciones del aire, como por ejemplo cambios en presión, velocidad, temperatura o contenido de humedad.

Air, standard. Dry air at 70°F and 14.696 psi at which it has a mass density of 0.075 lb/ft^3 and a specific volume of 13.33 ft^3/lb, ASHRAE 1986.

Aire, estándar. Aire seco a 70°F (21.11°C) y 14,696 psi [libra por pulgada cuadrada]; a dicha temperatura tiene una densidad de masa de 0,075 pies/libras^3 y un volumen específico de 13,33 pies^3/libras, ASHRAE 1986.

Air vent. A fitting used to vent air manually or automatically from a system.

Válvula de aire. Accesorio utilizado para darle al aire salida manual o automática de un sistema.

Algae. A form of green or black, slimy plant life that grows in water systems.

Alga. Tipo de planta legamosa de color verde o negro que crece en sistemas acuáticos.

Allen head. A recessed hex head in a fastener.

Cabeza allen. Cabeza de concavidad hexagonal en un asegurador.

Alternating current. An electric current that reverses its direction at regular intervals.

Corriente alterna. Corriente eléctrica que invierte su dirección a intervalos regulares.

Altitude adjustment. An adjustment to a refrigerator thermostat to account for a lower than normal atmospheric pressure such as may be found at a high altitude.

Ajuste para elevación. Ajuste al termóstato de un refrigerador para regular una presión atmosférica más baja que la normal, como la que se encuentra en elevaciones altas.

Ambient temperature. The surrounding air temperature.

Temperatura ambiente. Temperatura del aire circundante.

American standard pipe thread. Standard thread used on pipe to prevent leaks.

Rosca estándar estadounidense para tubos. Rosca estándar utilizada en tubos para evitar fugas.

Ammeter. A meter used to measure current flow in an electrical circuit.

Amperímetro. Instrumento utilizado para medir el flujo de corriente en un circuito eléctrico.

Amperage. Amount of electron or current flow (the number of electrons passing a point in a given time) in an electrical circuit.

Amperaje. Cantidad de flujo de electrones o de corriente (el número de electrones que sobrepasa un punto específico en un tiempo fijo) en un circuito eléctrico.

Ampere. Unit of current flow.

Amperio. Unidad de flujo de corriente.

Anemometer. An instrument used to measure the velocity of air.

Anemómetro. Instrumento utilizado para medir la velocidad del aire.

Angle valve. Valve with one opening at a 90° angle from the other opening.

Válvula en ángulo. Válvula con una abertura a un ángulo de 90° con respecto a la otra abertura.

Anode. A terminal or connection point on a semiconductor.

Ánodo. Punto de conexión o terminal en un semiconductor.

Approach temperature. The difference in temperature between the refrigerant and the leaving water in a chilled-water system.

Temperatura de acercamiento. Diferencia en temperatura entre el refrigerante y el agua de salida en un sistema de agua enfriada.

A.S.A. Abbreviation for the American Standards Association [now known as American National Standards Institute (ANSI)].

A.S.A. Abreviatura de Asociación Estadounidense de Normas [conocida ahora como Instituto Nacional Estadounidense de Normas (ANSI)].

ASHRAE. Abbreviation for the American Society of Heating, Refrigerating, and Air Conditioning Engineers.

ASHRAE. Abreviatura de Sociedad Estadounidense de Ingenieros de Calefacción, Refrigeración y Acondicionamiento de Aire.

ASME. Abbreviation for the American Society of Mechanical Engineers.

ASME. Abreviatura de Sociedad Estadounidense de Ingenieros Mecánicos.

Aspect ratio. The ratio of the length to width of a component.

Coeficiente de alargamiento. Relación del largo al ancho de un componente.

Atmospheric pressure. The weight of the atmosphere's gases pressing down on the earth. Equal to 14.696 psi at sea level and 70°F.

Presión atmosférica. El peso de la presión ejercida por los gases de la atmósfera sobre la tierra, equivalente a 14,696 psi al nivel del mar a 70°F.

Atom. The smallest particle of an element.

Átomo. Partícula más pequeña de un elemento.

Atomize. Using pressure to change liquid to small particles of vapor.

Atomizar. Utilizar la presión para cambiar un líquido a partículas pequeñas de vapor.

Automatic combination gas valve. A gas valve for gas furnaces that incorporates a manual control, gas supply for the pilot, adjustment and safety features for the pilot, pressure regulator, the controls for and the main gas valve.

Válvula de gas de combinación automática. Válvula de gas para hornos de gas que incorpora un regulador manual, suministro de gas para la llama piloto, ajuste y dispositivos de seguridad, regulador de presión, la válvula de gas principal y los reguladores de la válvula.

Automatic control. Controls that react to a change in conditions to cause the condition to stabilize.

Regulador automático. Reguladores que reaccionan a un cambio en las condiciones para provocar la estabilidad de dicha condición.

Automatic defrost. Using automatic means to remove ice from a refrigeration coil.

Desempañador automático. La utilización de medios automáticos para remover el hielo de una bobina de refrigeración.

Automatic expansion valve. A refrigerant control valve that maintains a constant pressure in an evaporator.

Válvula de expansión automática. Válvula de regulación del refrigerante que mantiene una presión constante en un evaporador.

Back pressure. The pressure on the low-pressure side of a refrigeration system (also known as suction pressure).

Contrapresión. La presión en el lado de baja presión de un sistema de refrigeración (conocido también como presión de aspiración).

Back seat. The position of a refrigeration service valve when the stem is turned away from the valve body and seated.

Asiento trasero. Posición de una válvula de servicio de refrigeración cuando el vástago está orientado fuera del cuerpo de la válvula y aplicado sobre su asiento.

Baffle. A plate used to keep fluids from moving back and forth at will in a container.

Deflector. Placa utilizada para evitar el libre movimiento de líquidos en un recipiente.

Balanced port TXV. A valve that will meter refrigerant at the same rate when the condenser head pressure is low.

Válvula electrónica de expansión con conducto equilibrado. Válvula que medirá el refrigerante a la misma proporción cuando la presión en la cabeza del condensador sea baja.

Ball check valve. A valve with a ball-shaped internal assembly that only allows fluid flow in one direction.

Válvula de retención de bolas. Válvula con un conjunto interior en forma de bola que permite el flujo de fluido en una sola dirección.

Barometer. A device used to measure atmospheric pressure that is commonly calibrated in inches or millimeters of mercury. There are two types: mercury column and aneroid.

Barómetro. Dispositivo comúnmente calibrado en pulgadas o en milímetros de mercurio que se utiliza para medir la presión atmosférica. Existen dos tipos: columna de mercurio y aneroide.

Base. A terminal on a semiconductor.

Base. Punto terminal en un semiconductor.

Battery. A device that produces electricity from the interaction of metals and acid.

Pila. Dispositivo que genera electricidad de la interacción entre metales y el ácido.

Bearing. A device that surrounds a rotating shaft and provides a low-friction contact surface to reduce wear from the rotating shaft.

Cojinete. Dispositivo que rodea un árbol giratorio y provee una superficie de contacto de baja fricción para disminuir el desgaste de dicho árbol.

Bellows. An accordion-like device that expands and contracts when internal pressure changes.

Fuelles. Dispositivo en forma de acordeón con pliegues que se expanden y contraen cuando la presión interna sufre cambios.

Bellows seal. A method of sealing a rotating shaft or valve stem that allows rotary movement of the shaft or stem without leaking.

Cierre hermético de fuelles. Método de sellar un árbol giratorio o el vástago de una válvula que permite el movimiento giratorio del árbol o del vástago sin producir fugas.

Bending spring. A coil spring that can be fitted inside or outside a piece of tubing to prevent its walls from collapsing when being formed.

Muelle de flexión. Muelle helicoidal que puede acomodarse dentro o fuera de una pieza de tubería para evitar que sus paredes se doblen al ser formadas.

Bimetal. Two dissimilar metals fastened together to create a distortion of the assembly with temperature changes.

Bimetal. Dos metales distintos fijados entre sí para producir una distorción del conjunto al ocurrir cambios de temperatura.

Bimetal strip. Two dissimilar metal strips fastened back to back.

Banda bimetálica. Dos bandas de metales distintos fijadas entre sí en su parte posterior.

Bleeding. Allowing pressure to move from one pressure level to another very slowly.

Sangradura. Proceso a través del cual se permite el movimiento de presión de un nivel a otro de manera muy lenta.

Bleed valve. A valve with a small port usually used to bleed pressure from a vessel to the atmosphere.

Válvula de descarga. Válvula con un conducto pequeño utilizado normalmente para purgar la presión de un depósito a la atmósfera.

Blocked suction. A method of cylinder unloading. The suction line passage to a cylinder in a reciprocating compressor is blocked, thus causing that cylinder to stop pumping.

Aspiración obturada. Método de descarga de un cilindro. El paso del conducto de aspiración a un cilindro en un compresor alternativo se obtura, provocando así que el cilindro deje de bombear.

Blowdown. A system in a cooling tower whereby some of the circulating water is bled off and replaced with fresh water to dilute the sediment in the sump.

Vaciado. Sistema en una torre de refrigeración por medio del cual se purga parte del agua circulante y se reemplaza con agua fresca para diluir el sedimento en el sumidero.

Boiler. A container in which a liquid may be heated using any heat source. When the liquid is heated to the point that vapor forms and is used as the circulating medium, it is called a steam boiler.

Caldera. Recipiente en el que se puede calentar un líquido utilizando cualquier fuente de calor. Cuando se calienta el líquido al punto en que se produce vapor y se utiliza éste como el medio para la circulación, se llama caldera de vapor.

Boiling point. The temperature level of a liquid at which it begins to change to a vapor. The boiling temperature is controlled by the vapor pressure above the liquid.

Punto de ebullición. El nivel de temperatura de un líquido al que el líquido empieza a convertirse en vapor. La temperatura de ebullición se regula por medio de la presión del vapor sobre líquido.

Bore. The inside diameter of a cylinder.

Calibre. Diámetro interior de un cilindro.

Bourdon tube. C-shaped tube manufactured of thin metal and closed on one end. When pressure is increased inside, it tends to straighten. It is used in a gage to indicate pressure.

Tubo Bourdon. Tubo en forma de C fabricado de metal delgado y cerrado en uno de los extremos. Al aumentarse la presión en su interior, el tubo tiende a enderezarse. Se utiliza dentro de un calibrador para indicar la presión.

Brazing. High-temperature (above 800°F) soldering of two metals.

Broncesoldadura. Soldadura de dos metales a temperaturas altas (sobre los 800°F ó 430°C).

Breaker. A heat-activated electrical device used to open an electrical circuit to protect it from excessive current flow.

Interruptor. Dispositivo eléctrico activado por el calor que se utiliza para abrir un circuito eléctrico a fin protegerlo de un flujo excesivo de corriente.

British thermal unit. The amount (quantity) of heat required to raise the temperature of 1 lb of water 1°F.

Unidad térmica británica. Cantidad de calor necesario para elevar en 1°F (-17.56°C) la temperatura de una libra inglesa de agua.

Btu. Abbreviation for British thermal unit.

BTU. Abreviatura de unidad térmica británica.

Bulb, sensor. The part of a sealed automatic control used to sense temperature.

Bombilla sensora. Pieza de un regulador automático sellado que se utiliza para advertir la temperatura.

Burner. A device used to prepare and burn fuel.

Quemador. Dispositivo utilizado para la preparación y la quema de combustible.

Burr. Excess material squeezed into the end of tubing after a cut has been made. This burr must be removed.

Rebaba. Exceso de material introducido por fuerza en el extremo de una tubería después de hacerse un corte. Esta rebaba debe removerse.

Butane gas. A liquefied petroleum gas burned for heat.

Gas butano. Gas licuado derivado del petróleo que se quema para producir calor.

Cad cell. A device used to prove the flame in an oil burning furnace containing cadmium sulfide.

Celda de cadmio. Dispositivo utilizado para probar la llama en un horno de aceite pesado que contiene sulfuro de cadmio.

Calibrate. To adjust instruments or gages to the correct setting for conditions.

Calibrar. Ajustar instrumentos o calibradores en posición correcta para su operación.

Capacitance. The term used to describe the electrical storage ability of a capacitor.

Capacitancia. Término utilizado para describir la capacidad de almacenamiento eléctrico de un capacitador.

Capacitor. An electrical storage device used to start motors (start capacitor) and to improve the efficiency of motors (run capacitor).

Capacitador. Dispositivo de almacenamiento eléctrico utilizado para arrancar motores (capacitador de arranque) y para mejorar el rendimiento de motores (capacitador de funcionamiento).

Capacity. The rating system of equipment used to heat or cool substances.

Capacidad. Sistema de clasificación de equipo utilizado para calentar o enfriar sustancias.

Capillary attraction. The attraction of a liquid material between two pieces of material such as two pieces of copper or copper and brass. For instance, in a joint made up of copper tubing and a brass fitting, the solder filler material has a greater attraction to the copper and brass than to itself and is drawn into the space between them.

Atracción capilar. Atracción de un material líquido entre dos piezas de material, como por ejemplo dos piezas de cobre o cobre y latón. Por ejemplo, en una junta fabricada de tubería de cobre y un accesorio de latón, el material de relleno de la soldadura tiene mayor atracción al cobre y al latón que a sí mismo y es arrastrado hacia el espacio entre éstos.

Capillary tube. A fixed-bore metering device. This is a small diameter tube that can vary in length from a few inches to several feet. The amount of refrigerant flow needed is predetermined and the length and diameter of the capillary tube is sized accordingly.

Tubo capilar. Dispositivo de medición de calibre fijo. Este es un tubo de diámetro pequeño cuyo largo puede oscilar entre unas cuantas pulgadas a varios pies. La cantidad de flujo de refrigerante requerida es predeterminada y, de acuerdo a esto, se fijan el largo y el diámetro del tubo capilar.

Carbon dioxide. A by-product of natural gas combustion that is not harmful.

Bióxido de carbono. Subproducto de la combustión del gas natural que no es nocivo.

Carbon monoxide. A poisonous, colorless, odorless, tasteless gas generated by incomplete combustion.

Monóxido de carbono. Gas mortífero, inodoro, incoloro e insípido que se desprende en la combustión incompleta del carbono.

Catalytic combustor stove. A stove that contains a cell-like structure consisting of a substrate, washcoat, and catalyst producing a chemical reaction causing pollutants to be burned at much lower temperatures.

Estufa de combustor catalítico. Estufa con una estructura en forma de celda compuesta de una subestructura, una capa brochada y un catalizador que produce una reacción química. Esta reacción provoca la quema de contaminantes a temperaturas mucho más bajas.

Cathode. A terminal or connection point on a semiconductor.

Cátodo. Punto de conexión o terminal en un semiconductor.

Cavitation. A vapor formed due to a drop in pressure in a pumping system. Air at a pump inlet may be caused at a cooling tower if the pressure is low and water is turned to vapor.

Cavitación. Vapor producido como consecuencia de una caída de presión en un sistema de bombeo. El aire a la entrada de una bomba puede ser producido en una torre de refrigeración si la presión es baja y el agua se convierte en vapor.

Celsius scale. A temperature scale with 1200 graduations between water freezing (0°C) and water boiling (100°C).

Escala Celsio. Escala dividida en cien grados, con el cero marcado a la temperatura de fusión del hielo (0°C) y el cien a la de ebullición del agua (100°C).

Centigrade scale. See Celsius.

Centígrado. Véase escala Celsio.

Centrifugal compressor. A compressor used for large refrigeration systems. It is not positive displacement, but it is similar to a blower.

Compresor centrífugo. Compresor utilizado en sistemas grandes de refrigeración. No es desplazamiento positivo, pero es similar a un soplador.

Centrifugal switch. A switch that uses a centrifugal action to disconnect the start windings from the circuit.

Conmutador centrífugo. Conmutador que utiliza una acción centrífuga para desconectar los devanados de arranque del circuito.

Change of state. The condition that occurs when a substance changes from one physical state to another, such as ice to water and water to steam.

Cambio de estado. Condición que ocurre cuando una sustancia cambia de un estado físico a otro, como por ejemplo el hielo a agua y el agua a vapor.

Charge. The quantity of refrigerant in a system.

Carga. Cantidad de refrigerante en un sistema.

Charging cylinder. A device that allows the technician to accurately charge a refrigeration system with refrigerant.

Cilindro cargador. Dispositivo que le permite al mecánico cargar correctamente un sistema de refrigeración con refrigerante.

Check valve. A device that permits fluid flow in one direction only.

Válvula de retención. Dispositivo que permite el flujo de fluido en una sola dirección.

Chill factor. A factor or number that is a combination of temperature, humidity, and wind velocity that is used to compare a relative condition to a known condition.

Factor de frío. Factor o número que es una combinación de la temperatura, la humedad y la velocidad del viento utilizado para comparar una condición relativa a una condición conocida.

Chilled-water system. An air conditioning system that circulates refrigerated water to the area to be cooled. The refrigerated water picks up heat from the area, thus cooling the area.

Sistema de agua enfriada. Sistema de acondicionamiento de aire que hace circular agua refrigerada al área que será enfriada. El agua refrigerada atrapa el calor del área y la enfria.

Chiller purge unit. A system that removes air from a low-pressure chiller.

Unidad enfriadora de purga. Sistema que remueve el aire de un enfriador de baja presión.

Chimney. A vertical shaft used to convey flue gases above the rooftop.

Chimenea. Cañón vertical utilizado para conducir los gases de combustión por encima del techo.

Chimney effect. A term used to describe air or gas when it expands and rises when heated.

Efecto de chimenea. Término utilizado para describir el aire o el gas cuando se expande y sube al calentarse.

Chlorofluocarbons (CFC). Those refrigerants thought to contribute to the depletion of the ozone layer.

Cloroflurocarburos. Líquidos refrigerantes que, según algunos, han contribuido a la reducción de la capa de ozono.

Circuit. An electron or fluid-flow path that makes a complete loop.

Circuito. Electrón o trayectoria del flujo de fluido que hace un ciclo completo.

Circuit breaker. A device that opens an electric circuit when an overload occurs.

Interruptor para circuitos. Dispositivo que abre un circuito eléctrico cuando ocurre una sobrecarga.

Clamp-on ammeter. An instrument that can be clamped around one conductor in an electrical circuit and measure the current.

Amperímetro fijado con abrazadera. Instrumento que puede fijarse con una abrazadera a un conductor en un circuito eléctrico y medir la corriente.

Clearance volume. The volume at the top of the stroke in a compressor cylinder between the top of the piston and the valve plate.

Volumen de holgura. Volumen en la parte superior de una carrera en el cilindro de un compresor entre la parte superior del pistón y la placa de una válvula.

Closed circuit. A complete path for electrons to flow on.

Circuito cerrado. Circuito de trayectoria ininterrumpida que permite un flujo continuo de electrones.

Closed loop. Piping circuit that is complete and not open to the atmosphere.

Ciclo cerrado. Circuito de tubería completo y no abierto a la atmósfera.

Code. The local, state, or national rules that govern safe installation and service of systems and equipment for the purpose of safety of the public and trade personnel.

Código. Reglamentos locales, estaduales o federales que rigen la instalación segura y el servicio de sistemas y equipo con el propósito de garantizar la seguridad del personal público y profesional.

Coefficient of performance (COP). The ratio of usable output energy divided by input energy.

Coeficiente de rendimiento. Relación de la de energía de salida utilizable dividida por la energía de entrada.

CO_2 indicator. An instrument used to detect the quantity of carbon dioxide in flue gas for efficiency purposes.

Indicador del CO_2. Instrumento utilizado para detectar la cantidad de bióxido de carbono en el gas de combustión a fin de lograr un mejor rendimiento.

Cold. The word used to describe heat at lower levels of intensity.

Frío. Término utilizado para describir el calor a niveles de intensidad más bajos.

Cold anticipator. A device that anticipates a need for cooling and starts the cooling system early enough for it to reach capacity when it is needed.

Anticipador de frío. Dispositivo que anticipa la necesidad de enfriamiento y pone en marcha el sistema de enfriamiento con suficiente anticipación para que éste alcance su máxima capacidad cuando vaya a ser utilizado.

Cold junction. The opposite junction to the hot junction in a thermocouple.

Empalme frío. El empalme opuesto al empalme caliente en un termopar.

Cold trap. A device to help trap moisture in a refrigeration system.

Trampa del frío. Dispositivo utilizado para ayudar a atrapar la humedad en un sistema de refrigeración.

Cold wall. The term used in comfort heating to describe a cold outside wall and its effect on human comfort.

Pared fría. Término utilizado en la calefacción para comodidad que describe una pared exterior fría y sus efectos en la comodidad de una persona.

Collector. A terminal on a semiconductor.

Colector. Punto terminal en un semiconductor.

Combustion. A reaction called rapid oxidation or burning produced with the right combination of a fuel, oxygen, and heat.

Combustión. Reacción conocida como oxidación rápida o quema producida con la combinación correcta de combustible, oxígeno y calor.

Comfort chart. A chart used to compare the relative comfort of one temperature and humidity condition to another condition.

Esquema de comodidad. Esquema utilizado para comparar la comodidad relativa de una condición de temperatura y humedad a otra condición.

Compound gage. A gage used to measure the pressure above and below the atmosphere's standard pressure. It is a Bourdon tube sensing device and can be found on all gage manifolds used for air conditioning and refrigeration service work.

Calibrador compuesto. Calibrador utilizado para medir la presión mayor y menor que la presión estándar de la atmósfera. Es un dispositivo sensor de tubo Bourdon que puede encontrarse en todos los distribuidores de calibrador utilizados para el servicio de sistemas de acondicionamiento de aire y de refrigeración.

Compression. A term used to describe a vapor when pressure is applied and the molecules are compacted closer together.

Compresión. Término utilizado para describir un vapor cuando se aplica presión y se compactan las moléculas.

Compression ratio. A term used with compressors to describe the actual difference in the low- and high-pressure sides of the compression cycle. It is absolute discharge pressure divided by absolute suction pressure.

Relación de compresión. Término utilizado con compresores para describir la diferencia real en los lados de baja y alta presión del ciclo de compresión. Es la presión absoluta de descarga dividida por la presión absoluta de aspiración.

Compressor. A vapor pump that pumps vapor (refrigerant or air) from one pressure level to a higher pressure level.

Compresor. Bomba de vapor que bombea el vapor (refrigerante o aire) de un nivel de presión a un nivel de presión más alto.

Compressor displacement. The internal volume of a compressor, used to calculate the pumping capacity of the compressor.

Desplazamiento del compresor. Volumen interno de un compresor, utilizado para calcular la capacidad de bombeo del mismo.

Compressor shaft seal. The seal that prevents refrigerant inside the compressor from leaking around the rotating shaft.

Junta de estanqueidad del árbol del compresor. La junta de estanqueidadque evita la fuga, alrededor del árbol giratorio, del refrigerante en el interior del compresor.

Concentrator. That part of an absorption chiller where the dilute salt solution is boiled to release the water.

Concentrador. El lugar en el enfriador por absorción donde se hierve la solución salina diluida para liberar el agua.

Condensate. The moisture collected on an evaporator coil.

Condensado. Humedad acumulada en la bobina de un evaporador.

Condensate pump. A small pump used to pump condensate to a higher level.

Bomba para condensado. Bomba pequeña utilizada para bombear el condensado a un nivel más alto.

Condensation. Liquid formed when a vapor condenses.

Condensación. El líquido formado cuando se condensa un vapor.

Condense. Changing a vapor to a liquid at a particular pressure.

Condensar. Convertir un vapor en líquido a una presión específica.

Condenser. The component in a refrigeration system that transfers heat from the system by condensing refrigerant.

Condensador. Componente en un sistema de refrigeración que transmite el calor del sistema al condensar el refrigerante.

Condenser flooding. A method of maintaining a correct head pressure by adding liquid refrigerant to the condenser from a receiver to increase the head pressure.

Inundación del condensador. Método de mantener una presión correcta en la cabeza agregando refrigerante líquido al condensador de un receptor para aumentar la presión en la cabeza.

Condensing-gas furnace. A furnace with a condensing heat exchanger that condenses moisture from the flue gases resulting in greater efficiency.

Horno para condensación de gas. Horno con un intercambiador de calor para condensación que condensa la humedad de los gases de combustión. El resultado será un mayor rendimiento.

Condensing pressure. The pressure that corresponds to the condensing temperature in a refrigeration system.

Presión para condensación. La presión que corresponde a la temperatura de condensación en un sistema de refrigeración.

Condensing temperature. The temperature at which a vapor changes to a liquid.

Temperatura de condensación. Temperatura a la que un vapor se convierte en líquido.

Condensing unit. A complete unit that includes the compressor and the condensing coil.

Conjunto del condensador. Unidad completa que incluye el compresor y la bobina condensadora.

Conduction. Heat transfer from one molecule to another within a substance or from one substance to another.

Conducción. Transmisión de calor de una molécula a otra dentro de una sustancia o de una sustancia a otra.

Conductivity. The ability of a substance to conduct electricity or heat.

Conductividad. Capacidad de una sustancia de conducir electricidad o calor.

Conductor. A path for electrical energy to flow on.

Conductor. Trayectoria que permite un flujo continuo de energía eléctrica.

Contactor. A larger version of the relay. It can be repaired or rebuilt and has moveable and stationary contacts.

Contactador. Versión más grande del relé. Puede ser reparado o reconstruido. Tiene contactos móviles y fijos.

Contaminant. Any substance in a refrigeration system that is foreign to the system, particularly if it causes damage.

Contaminante. Cualquier sustancia en un sistema de refrigeración extraña a éste, principalmente si causa averías.

Control. A device to stop, start, or modulate flow of electricity or fluid to maintain a preset condition.

Regulador. Dispositivo para detener, poner en marcha o modular el flujo de electricidad o de fluido a fin de mantener una condición establecida con anticipación.

Control system. A network of controls to maintain desired conditions in a system or space.

Sistema de regulación. Red de reguladores que mantienen las condiciones deseadas en un sistema o un espacio.

Convection. Heat transfer from one place to another using a fluid.

Convección. Transmisión de calor de un lugar a otro por medio de un fluido.

Conversion factor. A number used to convert from one equivalent value to another.

Factor de conversión. Número utilizado en la conversión de un valor equivalente a otro.

Cooler. A walk-in or reach-in refrigerated box.

Nevera. Caja refrigerada donde se puede entrar o introducir la mano.

Cooling tower. The final device in many water-cooled systems, which rejects heat from the system into the atmosphere by evaporation of water.

Torre de refrigeración. Dispositivo final en muchos sistemas enfriados por agua, que dirige el calor del sistema a la atmósfera por medio de la evaporación de agua.

Copper plating. Small amounts of copper are removed by electrolysis and deposited on the ferrous metal parts in a compressor.

Encobrado. Remoción de pequeñas cantidades de cobre por medio de electrólisis que luego se colocan en las piezas de metal férreo en un compresor.

Corrosion. A chemical action that eats into or wears away material from a substance.

Corrosión. Acción química que carcome o desgasta el material de una sustancia.

Counter EMF. Voltage generated or induced above the applied voltage in a single-phase motor.

Contra EMF. Tensión generada o inducida sobre la tensión aplicada en un motor unifásico.

Counterflow. Two fluids flowing in opposite directions.

Contraflujo. Dos fluidos que fluyen en direcciones opuestas.

Coupling. A device for joining two fluid-flow lines. Also the device connecting a motor drive shaft to the driven shaft in a direct-drive system.

Acoplamiento. Dispositivo utilizado para la conexión de dos conductos de flujo de fluido. Es también el dispositivo que conecta un árbol de mando del motor al árbol accionado en un sistema de mando directo.

CPVC (Chlorinated polyvinyl chloride). Plastic pipe similar to PVC except that it can be used with temperatures up to 180°F at 100 psig.

CPVC (Cloruro de polivinilo clorado). Tubo plástico similar al PVC, pero que puede utilizarse a temperaturas de hasta 180°F (82°C) a 100 psig [indicador de libras por pulgada cuadrada].

Crackage. Small spaces in a structure that allow air to infiltrate the structure.

Formación de grietas. Espacios pequeños en una estructura que permiten la infiltración del aire dentro de la misma.

Crankcase heat. Heat provided to the compressor crankcase.

Calor para el cárter del cigüeñal. Calor suministrado al cárter del cigüeñal del compresor.

Crankcase pressure regulator (CPR). A valve installed in the suction line, usually close to the compressor. It is used to keep a low-temperature compressor from overloading on a hot pull down.

Regulador de la presión del cárter del cigüeñal. Válvula instalada en el conducto de aspiración, normalmente cerca del compresor. Se utiliza para evitar la sobrecarga en un compresor de temperatura baja durante un arrastre caliente hacia abajo.

Crankshaft seal. Same as the compressor shaft seal.

Junta de estanqueidad del árbol del cigüeñal. Exactamente igual que la junta de estanqueidad del árbol del compresor.

Crankshaft throw. The off-center portion of a crankshaft that changes rotating motion to reciprocating motion.

Excentricidad del cigüeñal. Porción descentrada de un cigüeñal que cambia el movimiento giratorio a un movimiento alternativo.

Creosote. A mixture of unburned organic material found in the smoke from a wood-burning fire.

Creosota. Mezcla del material orgánico no quemado que se encuentra en el humo proveniente de un incendio de madera.

Crisper. A refrigerated compartment that maintains a high humidity and a low temperature.

Encrespador. Compartimiento refrigerado que mantiene una humedad alta y una temperatura baja.

Cross charge. A control with a sealed bulb that contains two different fluids that work together for a common specific condition.

Carga transversal. Regulador con una bombilla sellada compuesta de dos fluidos diferentes que pueden funcionar juntos para una condición común específica.

Cross liquid charge bulb. A type of charge in the sensing bulb of the TXV that has different characteristics from the system refrigerant. This is designed to help prevent liquid refrigerant from flooding to the compressor at startup.

Bombilla de carga del líquido transversal. Tipo de carga en la bombilla sensora de la válvula electrónica de expansión que tiene características diferentes a las del refrigerante del sistema. La carga está diseñada para ayudar a evitar que el refrigerante líquido se derrame dentro del compresor durante la puesta en marcha.

Cross vapor charge bulb. Similar to the vapor charge bulb but contains a fluid different from the system refrigerant. This is a special-type charge and produces a different pressure/temperature relationship under different conditions.

Bombilla de carga del vapor transversal. Similar a la bombilla de carga del vapor pero contiene un fluido diferente al del refrigerante del sistema. Esta es una carga de tipo especial y produce una relación diferente entre la presión y la temperatura bajo condiciones diferentes.

Crystallization. When a salt solution becomes too concentrated and part of the solution turns to salt.

Cristalización. Condición que ocurre cuando una solución salina se concentra demasiado y una parte de la solución se convierte en sal.

Current, electrical. Electrons flowing along a conductor.

Corriente eléctrica. Electrones que fluyen a través de un conductor.

Current relay. An electrical device activated by a change in current flow.

Relé para corriente. Dispositivo eléctrico accionado por un cambio en el flujo de corriente.

Cut-in and cut-out. The two points at which a control opens or closes its contacts based on the condition it is supposed to maintain.

Puntos de conexión y desconexión. Los dos puntos en los que un regulador abre o cierra sus contactos según las condiciones que debe mantener.

Cycle. A complete sequence of events (from start to finish) in a system.

Ciclo. Secuencia completa de eventos, de comienzo a fin, que ocurre en un sistema.

Cylinder. A circular container with straight sides used to contain fluids or to contain the compression process (the piston movement) in a compressor.

Cilindro. Recipiente circular con lados rectos, utilizado para contener fluidos o el proceso de compresión (movimiento del pistón) en un compresor.

Cylinder, compressor. The part of the compressor that contains the piston and its travel.

Cilindro del compresor. Pieza del compresor que contiene el pistón y su movimiento.

Cylinder head, compressor. The top to the cylinder on the high-pressure side of the compressor.

Culata del cilindro del compresor. Tapa del cilindro en el lado de alta presión del compresor.

Cylinder, refrigerant. The container that holds refrigerant.

Cilindro del refrigerante. El recipiente que contiene el refrigerante.

Cylinder unloading. A method of providing capacity control by causing a cylinder in a reciprocating compressor to stop pumping.

Descarga del cilindro. Método de suministrar regulación de capacidad provocando que el cilindro en un compresor alternativo deje de bombear.

Damper. A component in an air distribution system that restricts airflow for the purpose of air balance.

Desviador. Componente en un sistema de distribución de aire que limita el flujo de aire para mantener un equilibrio de aire.

Declination angle. The angle of the tilt of the earth on its axis.

Ángulo de declinación. Ángulo de inclinación de la Tierra en su eje.

Defrost. Melting of ice.

Descongelar. Convertir hielo en líquido.

Defrost cycle. The portion of the refrigeration cycle that melts the ice off the evaporator.

Ciclo de descongelación. Parte del ciclo de refrigeración que derrite el hielo del evaporador.

Defrost timer. A timer used to start and stop the defrost cycle.

Temporizador de descongelación. Temporizador utilizado para poner en marcha y detener el ciclo de descongelación.

Degreaser. A cleaning solution used to remove grease from parts and coils.

Desengrasador. Solución limpiadora utilizada para remover la grasa de piezas y bobinas.

Dehumidify. To remove moisture from air.

Deshumidificar. Remover la humedad del aire.

Dehydrate. To remove moisture from a sealed system or a product.

Deshidratar. Remover la humedad de un sistema sellado o un producto.

Density. The weight per unit of volume of a substance.

Densidad. Relación entre el peso de una sustancia y su volumen.

Desiccant. Substance in a refrigeration system drier that collects moisture.

Disecante. Sustancia en el secador de un sistema de refrigeración que acumula la humedad.

Design pressure. The pressure at which the system is designed to operate under normal conditions.

Presión de diseño. Presión a la que el sistema ha sido diseñado para funcionar bajo condiciones normales.

De-superheating. Removing heat from the superheated hot refrigerant gas down to the condensing temperature.

De sobrecalentamiento. Reducir el calor del gas caliente del refrigerante sobrecalentado hasta alcanzar la temperatura de condensación.

Detector. A device to search and find.

Detector. Dispositivo de búsqueda y detección.

Dew. Moisture droplets that form on a cool surface.

Rocío. Gotitas de humedad que se forman en una superficie fría.

Dew-point. The exact temperature at which moisture begins to form.

Punto de rocío. Temperatura exacta a la que la humedad comienza a formarse.

DIAC. A semiconductor often used as a voltage-sensitive switching device.

DIAC. Semiconductor utilizado frecuentemente como dispositivo de conmutación sensible a la tensión.

Diaphragm. A thin flexible material (metal, rubber, or plastic) that separates two pressure differences.

Diafragma. Material delgado y flexible, como por ejemplo el metal, el caucho o el plástico, que separa dos presiones diferentes.

Die. A tool used to make an external thread such as on the end of a piece of pipe.

Troquel. Herramienta utilizada para formar un filete externo, como por ejemplo, en el extremo de un tubo.

Differential. The difference in the cut-in and cut-out points of a control, pressure, time, temperature, or level.

Diferencial. Diferencia entre los puntos de conexión y desconexión de un regulador, una presión, un intervalo de tiempo, una temperatura o un nivel.

Diffuser. The terminal or end device in an air distribution system that directs air in a specific direction using louvers.

Placa difusora. Punto o dispositivo terminal en un sistema de distribución de aire que dirige el aire a una dirección específica, utilizando aberturas tipo celosía.

Diode. A solid state device composed of both P-type and N-type material. When connected in a circuit one way, current will flow. When the diode is reversed, current will not flow.

Diodo. Dispositivo de estado sólido compuesto de material P y de material N. Cuando se conecta a un circuito de una manera, la corriente fluye. Cuando la dirección del diodo cambia, la corriente deja de fluir.

Direct current. Electricity in which all electron flow is continuously in one direction.

Corriente continua. Electricidad en la que todos los electrones fluyen continuamente en una sola dirección.

Direct expansion. The term used to describe an evaporator with an expansion device other than a low-side float type.

Expansión directa. Término utilizado para describir un evaporador con un dispositivo de expansión diferente al tipo de dispositivo flotador de lado bajo.

Direct-spark ignition (DSI). A system that provides direct ignition to the main burner.

Encendido de chispa directa. Sistema que le provee un encendido directo al quemador principal.

Discus compressor. A reciprocating compressor distinguished by its disc-type valve system.

Compresor de disco. Compresor alternativo caracterizado por su sistema de válvulas de tipo disco.

Discus valve. A reciprocating compressor valve design with a low clearance volume and larger bore.

Válvula de disco. Válvula de compresor alternativo diseñada con un volumen de holgura bajo y un calibre más grande.

Distributor. A component installed at the outlet of the expansion valve that distributes the refrigerant to each evaporator circuit.

Distribuidor. Componente instalado a la salida de la váluva de expansión que distribuye el refrigerante a cada circuito del evaporador.

Doping. Adding an impurity to a semiconductor to produce a desired charge.

Impurificación. La adición de una impureza para producir una carga deseada.

Double flare. A connection used on copper, aluminum, or steel tubing that folds tubing wall to a double thickness.

Abocinado doble. Conexión utilizada en tuberías de cobre, aluminio, o acero que pliega la pared de la tubería y crea un espesor doble.

Dowel pin. A pin, which may or may not be tapered, used to align and fasten two parts.

Pasador de espiga. Pasador, que puede o no ser cónico, utilizado para alinear y fijar dos piezas.

Draft gage. A gage used to measure very small pressures (above and below atmospheric) and compare them to the atmosphere's pressure. Used to determine the flow of flue gas in a chimney or vent.

Calibrador de tiro. Calibrador utilizado para medir presiones sumamente pequeñas, (mayores o menores que la atmosférica), y compararlas con la presión de la atmósfera. Utilizado para determinar el flujo de gas de combustión en una chimenea o válvula.

Drier. A device used in a refrigerant line to remove moisture.

Secador. Dispositivo utilizado en un conducto de refrigerante para remover la humedad.

Drip pan. A pan shaped to collect moisture condensing on an evaporator coil in an air conditioning or refrigeration system.

Colector de goteo. Un colector formado para acumular la humedad que se condensa en la bobina de un evaporador en un sistema de acondicionamiento de aire o de refrigeración.

Dry-bulb temperature. The temperature measured using a plain thermometer.

Temperatura de bombilla seca. Temperatura que se mide con un termómetro sencillo.

Duct. A sealed channel used to convey air from the system to and from the point of utilization.

Conducto. Canal sellado que se emplea para dirigir el aire del sistema hacia y desde el punto de utilización.

Eccentric. An off-center device that rotates in a circle around a shaft.

Excéntrico. Dispositivo descentrado que gira en un círculo alrededor de un árbol.

Eddy current test. A test with an instrument to find potential failures in evaporator or condenser tubes.

Prueba para la corriente de Foucault. Prueba que se realiza con un instrumento para detectar posibles fallas en los tubos del evaporador o del condensador.

Effective temperature. Different combinations of temperature and humidity that provide the same comfort level.

Temperatura efectiva. Diferentes combinaciones de temperatura y humedad que proveen el mismo nivel de comodidad.

Electric heat. The process of converting electrical energy, using resistance, into heat.

Calor eléctrico. Proceso de convertir energía eléctrica en calor a través de la resistencia.

Electrical power. Electrical power is measured in watts. One watt is equal to one ampere flowing with a potential of one volt. Watts = Volts × Amperes (P = E × I)

Potencia eléctrica. La potencia eléctrica se mide en watios. Un watio equivale a un amperio que fluye con una potencia de un voltio. Watios = voltios × amperios P = E × I)

Electrical shock. When an electrical current travels through a human body.

Sacudida eléctrica. Paso brusco de una corriente eléctrica a través del cuerpo humano.

Electromagnet. A coil of wire wrapped around a soft iron core that creates a magnet.

Electroimán. Bobina de alambre devanado alrededor de un núcleo de hierro blando que crea un imán.

Electron. The smallest portion of an atom that carries a negative charge.

Electrón. Partícula más pequeña de un átomo que tiene carga negativa.

Electronic air filter. A filter that charges dust particles using high-voltage direct current and then collects these particles on a plate of an opposite charge.

Filtro de aire electrónico. Filtro que carga partículas de polvo utilizando una corriente continua de alta tensión y luego las acumula en una placa de carga opuesta.

Electronic charging scale. An electronically operated scale used to accurately charge refrigeration systems by weight.

Escala electrónica para carga. Escala accionada electrónicamente que se utiliza para cargar correctamente sistemas de refrigeración por peso.

Electronic expansion valve (TXV). A metering valve that uses a thermistor as a temperature-sensing element that varies the voltage to a heat motor-operated valve.

Válvula electrónica de expansión. Válvula de medición que utiliza un termistor como elemento sensor de temperatura para variar la tensión a una válvula de calor accionada por motor.

Electronic leak detector. An instrument used to detect gases in very small portions by using electronic sensors and circuits.

Detector electrónico de fugas. Instrumento que se emplea para detectar cantidades de gases sumamente pequeñas utilizando sensores y circuitos electrónicos.

Electronics. The use of electron flow in conductors, semiconductors, and other devices.

Electrónica. La utilización del flujo de electrones en conductores, semiconductores y otros dispositivos.

Emitter. A terminal on a semiconductor.

Emisor. Punto terminal en un semiconductor.

End bell. The end structure of an electric motor that normally contains the bearings and lubrication system.

Extremo acampanado. Estructura terminal de un motor eléctrico que generalmente contiene los cojinetes y el sistema de lubrificación.

End play. The amount of lateral travel in a motor or pump shaft.

Holgadura. Amplitud de movimiento lateral en un motor o en el árbol de una bomba.

Energy. The capacity for doing work.

Energía. Capacidad para realizar un trabajo.

Energy efficiency ratio (EER). An equipment efficiency rating that is determined by dividing the output in Btuh by the input in watts. This does not take into account the startup and shutdown for each cycle.

Relación del rendimiento de engería. Clasificación del rendimiento de un equipo que se determina al dividir la salida en Btuh por la entrada en watios. Esto no toma en cuenta la puesta en marcha y la parada de cada ciclo.

Enthalpy. The amount of heat a substance contains determined from a predetermined base or point.

Entalpía. Cantidad de calor que contiene una sustancia, establecida desde una base o un punto predeterminado.

Environment. Our surroundings, including the atmosphere.

Medio ambiente. Nuestros alrededores, incluyendo la atmósfera.

Ethane gas. The fossil fuel, natural gas, used for heat.

Gas etano. Combustible fósil, gas natural, utilizados para generar calor.

Evacuation. The removal of any gases not characteristic to a system or vessel.

Evacuación. Remoción de los gases no característicos de un sistema o depósito.

Evaporation. The condition that occurs when heat is absorbed by liquid and it changes to vapor.

Evaporación. Condición que ocurre cuando un líquido absorbe calor y se convierte en vapor.

Evaporator. The component in a refrigeration system that absorbs heat into the system and evaporates the liquid refrigerant.

Evaporador. El componente en un sistema de refrigeración que absorbe el calor hacia el sistema y evapora el refrigerante líquido.

Evaporator fan. A forced convector used to improve the efficiency of an evaporator by air movement over the coil.

Abanico del evaporador. Convector forzado que se utiliza para mejorar el rendimiento de un evaporador por medio del movimiento de aire a través de la bobina.

Evaporator pressure regulator (EPR). A mechanical control installed in the suction line at the evaporator outlet that keeps the evaporator pressure from dropping below a certain point.

Regulador de presión del evaporador. Regulador mecánico instalado en el conducto de aspiración de la salida del evaporador; evita que la presión del evaporador caiga hasta alcanzar un nivel por debajo del nivel específico.

Evaporator types. Flooded—an evaporator where the liquid refrigerant level is maintained to the top of the heat exchange coil. Dry type—an evaporator coil that achieves the heat exchange process with a minimum of refrigerant charge.

Clases de evaporadores. Inundado—un evaporador en el que se mantiene el nivel del refrigerante líquido en la parte superior de la bobina de intercambio de calor. Seco—una bobina de evaporador que logra el proceso de intercambio de calor con una mínima cantidad de carga de refrigerante.

Exhaust valve. The movable component in a refrigeration compressor that allows hot gas to flow to the condenser and prevents it from refilling the cylinder on the downstroke.

Válvula de escape. Componente móvil en un compresor de refrigeración que permite el flujo de gas caliente al condensador y evita que este gas rellene el cilindro durante la carrera descendente.

Expansion (metering) device. The component between the high-pressure liquid line and the evaporator that feeds the liquid refrigerant into the evaporator.

Dispositivo de (medición) de expansión. Componente entre el conducto de líquido de alta presión y el evaporador que alimenta el refrigerante líquido hacia el evaporador.

Expansion joint. A flexible portion of a piping system or building structure that allows for expansion of the materials due to temperature changes.

Junta de expansión. Parte flexible de un sistema de tubería o de la estructura de un edificio que permite la expanión de los materiales debido a cambios de temperatura.

External drive. An external type of compressor motor drive, as opposed to a hermetic compressor.

Motor externo. Motor tipo externo de un compresor, en comparación con un compresor hermético.

External equalizer. The connection from the evaporator outlet to the bottom of the diaphragm on a thermostatic expansion valve.

Equilibrador externo. Conexión de la salida del evaporador a la parte inferior del diafragma en una válvula de expansión termostática.

Fahrenheit scale. The temperature scale that places the boiling point of water at 212°F and the freezing point at 32°F.

Escala Fahrenheit. Escala de temperatura en la que el punto de ebullición del agua se encuentra a 212°F y el punto de fusión del hielo a 32°F.

Fan. A device that produces a pressure difference in air to move it.

Abanico. Dispositivo que produce una diferencia de presión en el aire para moverlo.

Fan cycling. The use of a pressure control to turn a condenser fan on and off to maintain a correct pressure within the system.

Funcionamiento cíclico. La utilización de un regulador de presión para poner en marcha y detener el abanico de un condensador a fin de mantener una presión correcta dentro del sistema.

Fan relay coil. A magnetic coil that controls the starting and stopping of a fan.

Bobina de relé del abanico. Bobina magnética que regula la puesta en marcha y la parada de un abanico.

Farad. The unit of capacity of a capacitor. Capacitors in our industry are rated in microfarads.

Faradio. Unidad de capacidad de un capacitador. En nuestro medio, los capacitadores se clasifican en microfaradios.

Female thread. The internal thread in a fitting.

Filete hembra. Filete interno en un accesorio.

Fill or wetted-surface method. Water in a cooling tower is spread out over a wetted surface while air is passed over it to enhance evaporation.

Método de relleno o de superficie mojada. El agua en una torre de refrigeración se extiende sobre una superficie mojada mientras el aire se dirige por encima de la misma para facilitar la evaporación.

Film factor. The relationship between the medium giving up heat and the heat exchange surface (evaporator). This relates to the velocity of the medium passing over the evaporator. When the velocity is too slow, the film between the air and the evaporator becomes greater and becomes an insulator, which slows the heat exchange.

Factor de película. Relación entre el medio que emite calor y la superficie del intercambiador de calor (evaporador). Esto se refiere a la velocidad del medio que pasa sobre el evaporador. Cuando la velocidad es demasiado lenta, la película entre el aire y el evaporador se expande y se convierte en un aislador, disminuyendo así la velocidad del intercambio del calor.

Filter. A fine mesh or porous material that removes particles from passing fluids.

Filtro. Malla fina o material poroso que remueve partículas de los fluidos que pasan por él.

Fin comb. A hand tool used to straighten the fins on an air-cooled condenser.

Herramienta para aletas. Herramienta manual utilizada para enderezar las aletas en un condensador enfriado por aire.

Fixed resistor. A nonadjustable resistor. The resistance cannot be changed.

Resistor fijo. Resistor no ajustable. La resistencia no se puede cambiar.

Fixed-bore device. An expansion device with a fixed diameter that does not adjust to varying load conditions.

Dispositivo de calibre fijo. Dispositivo de expansión con un diámetro fijo que no se ajusta a las condiciones de carga variables.

Flapper valve. See reed valve.

Chapaleta. Véase válvula de lámina.

Flare. The angle that may be fashioned at the end of a piece of tubing to match a fitting and create a leak-free connection.

Abocinado. Ángulo que puede formarse en el extremo de una pieza de tubería para emparejar un accesorio y crear una conexión libre de fugas.

Flare nut. A connector used in a flare assembly for tubing.

Tuerca abocinada. Conector utilizado en un conjunto abocinado para tuberías.

Flash gas. A term used to describe the pressure drop in an expansion device when some of the liquid passing through the valve is changed quickly to a gas and cools the remaining liquid to the corresponding temperature.

Gas instantáneo. Término utilizado para describir la caída de la presión en un dispositivo de expansión cuando una parte del líquido que pasa a través de la válvula se convierte rápidamente en gas y enfria el líquido restante a la temperatura correspondiente.

Float, valve or switch. An assembly used to maintain or monitor a liquid level.

Válvula o conmutador de flotador. Conjunto utilizado para mantener o controlar el nivel de un líquido.

Flooded system. A refrigeration system operated with the liquid refrigerant level very close to the outlet of the evaporator coil for improved heat exchange.

Sistema inundado. Sistema de refrigeración que funciona con el nivel del refrigerante líquido bastante próximo a la salida de la bobina del evaporador para mejorar el intercambio de calor.

Flooding. The term applied to a refrigeration system when the liquid refrigerant reaches the compressor.

Inundación. Término aplicado a un sistema de refrigeración cuando el nivel del refrigerante líquido llega al compresor.

Flue. The duct that carries the products of combustion out of a structure for a fossil- or a solid-fuel system.

Conducto de humo. Conducto que extrae los productos de combustión de una estructura en sistemas de combustible fósil o sólido.

Flue-gas analysis instruments. Instruments used to analyze the operation of fossil—fuel-burning equipment such as oil and gas furnaces by analyzing the flue gases.

Instrumentos para el análisis del gas de combustión. Instrumentos utilizados para llevar a cabo un análisis del funcionamiento de los quemadores de combustible fósil, como por ejemplo hornos de aceite pesado o gas, a través del estudio de los gases de combustión.

Fluid. The state of matter of liquids and gases.

Fluido. Estado de la materia de líquidos y gases.

Fluid expansion device. Using a bulb or sensor, tube, and diaphragm filled with fluid, this device will produce movement at the diaphragm when the fluid is heated or cooled. A bellows may be added to produce more movement. These devices may contain vapor and liquid.

Dispositivo para la expansión del fluido. Utilizando una bombilla o sensor, un tubo y un diafragma lleno de fluido, este dispositivo generará movimiento en el diafragma cuando se caliente o enfríe el fluido. Se le puede agregar un fuelle para generar aún más movimiento. Dichos dispositivos pueden contener vapor y líquido.

Flush. The process of using a fluid to push contaminants from a system.

Descarga. Proceso de utilizar un fluido para remover los contaminantes de un sistema.

Flux. A substance applied to soldered and brazed connections to prevent oxidation during the heating process.

Fundente. Sustancia aplicada a conexiones soldadas y broncesoldadas para evitar la oxidación durante el proceso de calentamiento.

Foaming. A term used to describe oil when it has liquid refrigerant boiling out of it.

Espumación. Término utilizado para describir el aceite cuando el refrigerante líquido se derrama del mismo.

Foot-pound. The amount of work accomplished by lifting 1 lb of weight 1 ft; a unit of energy.

Libra-pie. Medida de la cantidad de energía o fuerza que se requiere para levantar una libra a una distancia de un pie; unidad de energía.

Force. Energy exerted.

Fuerza. Energía ejercida sobre un objeto.

Forced convection. The movement of fluid by mechanical means.

Convección forzada. Movimiento de fluido por medios mecánicos.

Fossil fuels. Natural gas, oil, and coal formed millions of years ago from dead plants and animals.

Combustibles fósiles. El gas natural, el petroleo y el carbón que se formaron hace millones de años de plantas y animales muertos.

Four-way valve. The valve in a heat pump system that changes the direction of the refrigerant flow between the heating and cooling cycles.

Válvula con cuatro vías. Válvula en un sistema de bomba de calor que cambia la dirección del flujo de refrigerante entre los ciclos de calentamiento y enfriamiento.

Freezer burn. The term applied to frozen food when it becomes dry and hard from dehydration due to poor packaging.

Quemadura del congelador. Término aplicado a la comida congelada cuando se seca y endurece debido a la deshidratación ocacionada por el empaque de calidad inferior.

Freeze up. Excess ice or frost accumulation on an evaporator to the point that airflow may be affected.

Congelación. Acumulación excesiva de hielo o congelación en un evaporador a tal extremo que el flujo de aire puede ser afectado.

Freezing. The change of state of water from a liquid to a solid.

Congelamiento. Cambio de estado del agua de líquido a sólido.

Freon. The trade name for refrigerants manufactured by E. I. du Pont de Nemours & Co., Inc.

Freón. Marca registrada para refrigerantes fabricados por la compañía E. I. du Pont de Nemours, S.A.

Frequency. The cycles per second (cps) of the electrical current supplied by the power company. This is normally 60 cps in the United States.

Frecuencia. Ciclos por segundo (cps), generalmente 60 cps en los Estados Unidos, de la corriente eléctrica suministrada por la empresa de fuerza motriz.

Front seated. A position on a valve that will not allow refrigerant flow in one direction.

Sentado delante. Posición en una válvula que no permite el flujo de refrigerante en una dirección.

Frost back. A condition of frost on the suction line and even the compressor body usually due to liquid refrigerant in the suction line.

Obturación por congelación. Condición de congelación que ocurre en el conducto de aspiración e inclusive en el cuerpo del compresor, normalmente debido a la presencia de refrigerante líquido en el conducto de aspiración.

Frostbite. When skin freezes.

Quemadura por frío. Congelación de la piel.

Frozen. The term used to describe water in the solid state; also used to describe a rotating shaft that will not turn.

Congelado. Término utilizado para describir el agua en un estado sólido; utilizado también para describir un árbol giratorio que no gira.

Fuel oil. The fossil fuel used for heating; a petroleum distillate.

Aceite pesado. Combustible fósil utilizado para calentar; un destilado de petróleo.

Full-load amperage (FLA). The current an electric motor draws while operating under a full-load condition. Also called the run-load amperage.

Amperaje de carga total. Corriente que un motor eléctrico consume mientras funciona en una condición de carga completa. Conocido también como amperaje de carga de funcionamiento.

Furnace. Equipment used to convert heating energy, such as fuel oil, gas, or electricity, to usable heat. It usually contains a heat exchanger, a blower, and the controls to operate the system.

Horno. Equipo utilizado para la conversión de energía calórica, como por ejemplo el aceite pesado, el gas o la electricidad, en calor utilizable. Normalmente contiene un intercambiador de calor, un soplador y los reguladores para accionar el sistema.

Fuse. A safety device used in electrical circuits for the protection of the circuit conductor and components.

Fusible. Dispositivo de segurididad utilizado en circuitos eléctricos para la protección del conductor y de los componentes del circuito.

Fusible link. An electrical safety device normally located in a furnace that burns and opens the circuit during an overheat situation.

Cartucho de fusible. Dispositivo eléctrico de seguridad ubicado por lo general en un horno, que quema y abre el circuito en caso de sobrecalentamiento.

Fusible plug. A device (made of low-melting temperature metal) used in pressure vessels that is sensitive to low temperatures and relieves the vessel contents in an overheating situation.

Tapón de fusible. Dispositivo utilizado en depósitos en presión, hecho de un metal que tiene una temperatura de fusión baja. Este dispositivo es sensible a temperaturas bajas y alivia el contenido del depósito en caso de sobrecalentamiento.

Gage. An instrument used to detect pressure.

Calibrador. Instrumento utilizado para detectar presión.

Gage manifold. A tool that may have more than one gage with a valve arrangement to control fluid flow.

Distribuidor de calibrador. Herramienta que puede tener más de un calibrador con las válvulas arregladas a fin de regular el flujo de fluido.

Gage port. The service port used to attach a gage for service procedures.

Orificio de calibrador. Orificio de servicio utilizado con el propósito de fijar un calibrador para procedimientos de servicio.

Gas. The vapor state of matter.

Gas. Estado de vapor de una materia.

Gas-pressure switch. Used to detect gas pressure before gas burners are allowed to ignite.

Conmutador de presión del gas. Utilizado para detectar la presión del gas antes de que los quemadores de gas puedan encenderse.

Gas valve. A value used to stop, start, or modulate the flow of natural gas.

Válvula de gas. Válvula utilizada para detener, poner en marcha o modular el flujo de gas natural.

Gasket. A thin piece of flexible material used between two metal plates to prevent leakage.

Guarnición. Pieza delgada de material flexible utilizada entre dos placas de metal para evitar fugas.

Gate. A terminal on a semiconductor.

Compuerta. Punto terminal en un semiconductor.

Germanium. A substance from which many semiconductors are made.

Germanio. Sustancia de la que se fabrican muchos semiconductores.

Glow coil. A device that automatically reignites a pilot light if it goes out.

Bobina encendedora. Dispositivo que automáticamente vuelve a encender la llama piloto si ésta se apaga.

Graduated cylinder. A cylinder with a visible column of liquid refrigerant used to measure the refrigerant charged into a system. Refrigerant temperatures can be dialed on the graduated cylinder.

Cilindro graduado. Cilindro con una columna visible de refrigerante líquido utilizado para medir el refrigerante inyectado al sistema. Las temperaturas del refrigerante pueden marcarse en el cilindro graduado.

Grain. Unit of measure. One pound = 7000 grains.

Grano. Unidad de medida. Una libra equivale a 7000 granos.

Gram. Metric measurement term used to express weight.

Gramo. Término utilizado para referirse a la unidad básica de peso en el sistema métrico.

Grille. A louvered, often decorative, component in an air system at the inlet or the outlet of the airflow.

Rejilla. Componente con celosías, comúnmente decorativo, en un sistema de aire que se encuentra a la entrada o a la salida del flujo de aire.

Grommet. A rubber, plastic, or metal protector usually used where wire or pipe goes through a metal panel.

Guardaojal. Protector de caucho, plástico o metal normalmente utilizado donde un alambre o un tubo pasa a través de una base de metal.

Ground, electrical. A circuit or path for electron flow to the earth ground.

Tierra eléctrica. Circuito o trayectoria para el flujo de electrones a la puesta a tierra.

Ground wire. A wire from the frame of an electrical device to be wired to the earth ground.

Alambre a tierra. Alambre que va desde el armazón de un dispositivo eléctrico para ser conectado a la puesta a tierra.

Guide vanes. Vanes used to produce capacity control in a centrifugal compressor. Also called prerotation guide vanes.

Paletas directrices. Paletas utilizadas para producir la regulación de capacidad en un compresor centrífugo. Conocidas también como paletas directrices para prerotación.

Halide refrigerants. Refrigerants that contain halogen chemicals; R-12, R-22, R-500, and R-502 are among them.

Refrigerantes de hálido. Refrigerantes que contienen productos químicos de halógeno; entre ellos se encuentran el R-12, R-22, R-500 y R-502.

Halide torch. A torch-type leak detector used to detect the halogen refrigerants.

Soplete de hálido. Detector de fugas de tipo soplete utilizado para detectar los refrigerantes de halógeno.

Halogens. Chemical substances found in many refrigerants containing chlorine, bromine, iodine, and fluorine.

Halógenos. Sustancias químicas presentes en muchos refrigerantes que contienen cloro, bromo, yodo y flúor.

Hand truck. A two-wheeled piece of equipment that can be used for moving heavy objects.

Vagoneta para mano. Equipo con dos ruedas que puede utilizarse para transportar objetos pesados.

Hanger. A device used to support tubing, pipe, duct, or other components of a system.

Soporte. Dispositivo utilizado para apoyar tuberías, tubos, conductos u otros componentes de un sistema.

Head. Another term for pressure, usually referring to gas or liquid.

Carga. Otro término para presión, refiriéndose normalmente a gas o líquido.

Head pressure control. A control that regulates the head pressure in a refrigeration or air conditioning system.

Regulador de la presión de la carga. Regulador que controla la presión de la carga en un sistema de refrigeración o de acondicionamiento de aire.

Header. A pipe or containment to which other pipe lines are connected.

Conductor principal. Tubo o conducto al que se conectan otras conexiones.

Heat. Energy that causes molecules to be in motion and to raise the temperature of a substance.

Calor. Energía que ocasiona el movimiento de las moléculas provocando un aumento de temperatura en una sustancia.

Heat anticipator. A device that anticipates the need for cutting off the heating system prematurely so the fan can cool the furnace.

Anticipador de calor. Dispositivo que anticipa la necesidad de detener la marcha del sistema de calentamiento para que el abanico pueda enfriar el horno.

Heat coil. A device made of tubing or pipe designed to transfer heat to a cooler substance by using fluids.

Bobina de calor. Dispositivo hecho de tubos, diseñado para transmitir calor a una sustancia más fría por medio de fluidos.

Heat exchanger. A device that transfers heat from one substance to another.

Intercambiador de calor. Dispositivo que transmite calor de una sustancia a otra.

Heat of compression. That part of the energy from the pressurization of a gas or a liquid converted to heat.

Calor de compresión. La parte de la energía generada de la presurización de un gas o un líquido que se ha convertido en calor.

Heat of fusion. The heat released when a substance is changing from a liquid to a solid.

Calor de fusión. Calor liberado cuando una sustancia se convierte de líquido a sólido.

Heat of respiration. When oxygen and carbon hydrates are taken in by a substance or when carbon dioxide and water are given off. Associated with fresh fruits and vegetables during their aging process while stored.

Calor de respiración. Cuando se admiten oxígeno e hidratos de carbono en una sustancia o cuando se emiten bióxido de carbono y agua. Se asocia con el proceso de maduración de frutas y legumbres frescas durante su almacenamiento.

Heat pump. A refrigeration system used to supply heat or cooling using valves to reverse the refrigerant gas flow.

Bomba de calor. Sistema de refrigeración utilizado para suministrar calor o frío mediante válvulas que cambian la dirección del flujo de gas del refrigerante.

Heat reclaim. Using heat from a condenser for purposes such as space and domestic water heating.

Reclamación de calor. La utilización del calor de un condensador para propósitos tales como la calefacción de espacio y el calentamiento doméstico de agua.

Heat sink. A low-temperature surface to which heat can transfer.

Fuente fría. Superficie de temperatura baja a la que puede transmitírsele calor.

Heat transfer. The transfer of heat from a warmer to a colder substance.

Transmisión de calor. Cuando se transmite calor de una sustancia más caliente a una más fría.

Helix coil. A bimetal formed into a helix-shaped coil that provides longer travel when heated.

Bobina en forma de hélice. Bimetal encofrado en una bobina en forma de hélice que provee mayor movimiento al ser calentado.

Hermetic system. A totally enclosed refrigeration system where the motor and compressor are sealed within the same system with the refrigerant.

Sistema hermético. Sistema de refrigeración completamente cerrado donde el motor y el compresor se obturan dentro del mismo sistema con el refrigerante.

Hertz. Cycles per second.

Hertz. Ciclos por segundo.

Hg. Abbreviation for the element mercury.

Hg. Abreviatura del elemento mercurio.

High-pressure control. A control that stops a boiler heating device or a compressor when the pressure becomes too high.

Regulador de alta presión. Regulador que detiene la marcha del dispositivo de calentamiento de una caldera o de un compresor cuando la presión alcanza un nivel demasiado alto.

High side. A term used to indicate the high-pressure or condensing side of the refrigeration system.

Lado de alta presión. Término utilizado para indicar el lado de alta presión o de condensación del sistema de refrigeración.

High-temperature refrigeration. A refrigeration temperature range starting with evaporator temperatures no lower than 35°F, a range usually used in air conditioning (cooling).

Refrigeración a temperatura alta. Margen de la temperatura de refrigeración que comienza con temperaturas de evaporadores no menores de 35°F (2°C). Este margen se utiliza normalmente en el acondicionamiento de aire (enfriamiento).

High-vacuum pump. A pump that can produce a vacuum in the low micron range.

Bomba de vacío alto. Bomba que puede generar un vacío dentro del margen de micrón bajo.

Horsepower. A unit equal to 33,000 ft-lb of work per minute.

Potencia en caballos. Unidad equivalente a 33.000 libras-pies de trabajo por minuto.

Hot gas. The refrigerant vapor as it leaves the compressor. This is often used to defrost evaporators.

Gas caliente. El vapor del refrigerante al salir del compresor. Esto se utiliza con frecuencia para descongelar evaporadores.

Hot gas bypass. Piping that allows hot refrigerant gas into the cooler low-pressure side of a refrigeration system usually for system capacity control.

Desviación de gas caliente. Tubería que permite la entrada de gas caliente del refrigerante en el lado más frío de baja presión de un sistema de refrigeración, normalmente para la regulación de la capacidad del sistema.

Hot gas defrost. A system where the hot refrigerant gases are passed through the evaporator to defrost it.

Descongelación con gas caliente. Sistema en el que los gases calientes del refrigerante se pasan a través del evaporador para descongelarlo.

Hot gas line. The tubing between the compressor and condenser.

Conducto de gas caliente. Tubería entre el compresor y el condensador.

Hot junction. That part of a thermocouple or thermopile where heat is applied.

Empalme caliente. El lugar en un termopar o pila termoeléctrica donde se aplica el calor.

Hot pull down. The process of lowering the refrigerated space to the design temperature after it has been allowed to warm up considerably over this temperature.

Descenso caliente. Proceso de bajar la temperatura del espacio refrigerado a la temperatura de diseño luego de habérsele permitido calentarse a un punto sumamente superior a esta temperatura.

Hot water heat. A heating system using hot water to distribute the heat.

Calor de agua caliente. Sistema de calefacción que utiliza agua caliente para la distribución del calor.

Hot wire. The wire in an electrical circuit that has a voltage potential between it and another electrical source or between it and ground.

Conductor electrizado. Conductor en un circuito eléctrico a través del cual fluye la tensión entre éste y otra fuente de electricidad o entre éste y la tierra.

Humidifier. A device used to add moisture to the air.

Humedecedor. Dispositivo utilizado para agregarle humedad al aire.

Humidistat. A control operated by a change in humidity.

Humidistato. Regulador activado por un cambio en la humedad.

Humidity. Moisture in the air.

Humedad. Vapor de agua existente en el ambiente.

Hydraulics. Producing mechanical motion by using liquids under pressure.

Hidráulico. Generación de movimiento mecánico por medio de líquidos bajo presión.

Hydrocarbons. Organic compounds containing hydrogen and carbon found in many heating fuels.

Hidrocarburos. Compuestos orgánicos que contienen el hidrógeno y el carbón presentes en muchos combustibles de calentamiento.

Hydrochlorofluorocarbons (HCFC). Refrigerants thought to contribute to the depletion of the ozone layer although not to the extent of chlorofluorocarbons.

Hidroclorofluorocarburos. Líquidos refrigerantes que, según algunos, han contribuido a la reducción de la capa de ozono aunque no en tal grado como los cloroflurocarburos.

Hydrometer. An instrument used to measure the specific gravity of a liquid.

Hidrómetro. Instrumento utilizado para medir la gravedad específica de un líquido.

Hydronic. Usually refers to a hot water heating system.

Hidrónico. Normalmente se refiere a un sistema de calefacción de agua caliente.

Hygrometer. An instrument used to measure the amount of moisture in the air.

Higrómetro. Instrumento utilizado para medir la cantidad de humedad en el aire.

Idler. A pulley on which a belt rides. It does not transfer power but is used to provide tension or reduce vibration.

Polea tensora. Polea sobre la que se mueve una correa. No sirve para transmitir potencia, pero se utiliza para proveer tensión o disminuir la vibración.

Ignition transformer. Provides a high-voltage current, usually to produce a spark to ignite a furnace fuel, either gas or oil.

Transformador para encendido. Provee una corriente de alta tensión, normalmente para generar una chispa a fin de encender el combustible de un horno, sea gas o aceite pesado.

Impedance. A form of resistance in an alternating current circuit.

Impedancia. Forma de resistencia en un circuito de corriente alterna.

Impeller. The rotating part of a pump that causes the centrifugal force to develop fluid flow and pressure difference.

Impulsor. Pieza giratoria de una bomba que hace que la fuerza centrífuga desarrolle flujo de fluido y una diferencia en presión.

Impingement. The condition in a gas or oil furnace when the flame strikes the sides of the combustion chamber, resulting in poor combustion efficiency.

Golpeo. Condición que ocurre en un horno de gas o de aceite pesado cuando la llama golpea los lados de la cámara de combustión. Esta condición trae como resultado un rendimiento de combustión pobre.

Inclined water manometer. Indicates air pressures in very low-pressure systems.

Manómetro de agua inclinada. Señala las presiones de aire en sistemas de muy baja presión.

Induced magnetism. Magnetism produced, usually in a metal, from another magnetic field.

Magnetismo inducido. Magnetismo generado, normalmente en un metal, desde otro campo magnético.

Inductance. An induced voltage producing a resistance in an alternating current circuit.

Inductancia. Tensión inducida que genera una resistencia en un circuito de corriente alterna.

Induction motor. An alternating current motor where the rotor turns from induced magnetism from the field windings.

Motor inductor. Motor de corriente alterna donde el rotor gira debido al magnetismo inducido desde los devanados inductores.

Inductive reactance. A resistance to the flow of an alternating current produced by an electromagnetic induction.

Reactancia inductiva. Resistencia al flujo de una corriente alterna generada por una inducción electromagnética.

Inert gas. A gas that will not support most chemical reactions, particularly oxidation.

Gas inerte. Gas incapaz de resistir la mayoría de las reacciones químicas, especialmente la oxidación.

Infiltration. Air that leaks into a structure through cracks, windows, doors, or other openings due to less pressure inside the structure than outside the structure.

Infiltración. Penetración de aire en una estructura a través de grietas, ventanas, puertas u otras aberturas debido a que la presión en el interior de la estructura es menor que en el exterior.

Infrared rays. The rays that transfer heat by radiation.

Rayos infrarrojos. Rayos que transmiten calor por medio de la radiación.

In-phase. When two or more alternating current circuits have the same polarity at all times.

En fase. Cuando dos o más circuitos de corriente alterna tienen siempre la misma polaridad.

Insulation, electric. A substance that is a poor conductor of electricity.

Aislamiento eléctrico. Sustancia que es un conductor pobre de electricidad.

Insulation, thermal. A substance that is a poor conductor of the flow of heat.

Aislamiento térmico. Sustancia que es un conductor pobre de flujo de calor.

Intermittent ignition. Ignition system for a gas furnace that operates only when needed or when the furnace is operating.

Encendido interrumpido. Sistema de encendido para un horno de gas que funciona solamente cuando es necesario o cuando el horno está trabajando.

Isolation relays. Components used to prevent stray unwanted electrical feedback that can cause erratic operation.

Relés de aislación. Componentes utilizados para evitar la realimentación eléctrica dispersa no deseada que puede ocasionar un funcionamiento errático.

Joule. Metric measurement term used to express the quantity of heat.

Joule. Término utilizado para referirse a la unidad básica de cantidad de calor en el sistema métrico.

Junction box. A metal or plastic box within which electrical connections are made.

Caja de empalme. Caja metálica o plástica dentro de la cual se hacen conexiones eléctricas.

Kelvin. A temperature scale where absolute 0 equals 0 or where molecular motion stops at 0. It has the same graduations per degree of change as the Celsius scale.

Escala absoluta. Escala de temperaturas donde el cero absoluto equivale a 0 ó donde el movimiento molecular se detiene en 0. Tiene las mismas graduaciones por grado de cambio que la escala Celsio.

Kilopascal. A metric unit of measurement for pressure used in the air conditioning, heating, and refrigeration field. There are 6.89 kilopascals in 1 psi.

Kilopascal. Unidad métrica de medida de presión utilizada en el ramo del acondicionamiento de aire, calefacción y refrigeración. 6,89 kilopascales equivalen a 1 psi.

Kilowatt. A unit of electrical power equal to 1000 watts.

Kilowatio. Unidad eléctrica de potencia equivalente a 1000 watios.

Kilowatt-hour. 1 kilowatt (1000 watts) of energy used for 1 hour.

Kilowatio hora. Unidad de energía equivalente a la que produce un kilowatio durante una hora.

King valve. A service valve at the liquid receiver.

Válvula maestra. Válvula de servicio ubicada en el receptor del líquido.

Latent heat. Heat energy absorbed or rejected when a substance is changing state and there is no change in temperature.

Calor latente. Energía calórica absorbida o rechazada cuando un sustancia cambia de estado y no se experimentan cambios de temperatura.

Leak detector. Any device used to detect leaks in a pressurized system.

Detector de fugas. Cualquier dispositivo utilizado para detectar fugas en un sistema presurizado.

Lever truck. A long-handled, two-wheeled device that can be used to lift and assist in moving heavy objects.

Vagoneta con palanca. Dispositivo con dos ruedas y una manivela larga que puede utilizarse para levantar y ayudar a transportar objetos pesados.

Limit control. A control used to make a change in a system, usually to stop it when predetermined limits of pressure or temperature are reached.

Regulador de límite. Regulador utilizado para realizar un cambio en un sistema, normalmente para detener su marcha cuando se alcanzan niveles predeterminados de presión o de temperatura.

Line set. A term used for tubing sets furnished by the manufacturer.

Juego de conductos. Término utilizado para referise a los juegos de tubería suministrados por el fabricante.

Liquified petroleum. Liquified propane, butane, or a combination of these gases. The gas is kept as a liquid under pressure until ready to use.

Petróleo licuado. Propano o butano licuados, o una combinación de estos gases. El gas se mantiene en estado líquido bajo presión hasta que se encuentre listo para usar.

Liquid. A substance where molecules push outward and downward and seek a uniform level.

Líquido. Sustancia donde las moléculas empujan hacia afuera y hacia abajo y buscan un nivel uniforme.

Liquid charge bulb. A type of charge in the sensing bulb of the thermostatic expansion valve. This charge is characteristic of the refrigerant in the system and contains enough liquid so that it will not totally boil away.

Bombilla de carga líquida. Tipo de carga en la bombilla sensora de la válvula de expansión termostática. Esta carga es característica del refrigerante en el sistema y contiene suficiente líquido para que el mismo no se evapore completamente.

Liquid line. A term applied in the industry to refer to the tubing or piping from the condenser to the expansion device.

Conducto de líquido. Término aplicado en nuestro medio para referirse a la tubería que va del condensador al dispositivo de expansión.

Liquid nitrogen. Nitrogen in liquid form.

Nitrógeno líquido. Nitrógeno en forma líquida.

Liquid receiver. A container in the refrigeration system where liquid refrigerant is stored.

Receptor del líquido. Recipiente en el sistema de refrigeración donde se almacena el refrigerante líquido.

Liquid refrigerant charging. The process of allowing liquid refrigerant to enter the refrigeration system through the liquid line to the condenser and evaporator.

Carga para refrigerante líquido. Proceso de permitir la entrada del refrigerante líquido al condensador y al evaporador en el sistema de refrigeración a través del conducto de líquido.

Liquid slugging. A large amount of liquid refrigerant in the compressor cylinder, usually causing immediate damage.

Relleno de líquido. Acumulación de una gran cantidad de refrigerante líquido en el cilindro del compresor, que normalmente provoca una avería inmediata.

Lithium-bromide. A type of salt solution used in an absorption chiller.

Bromuro de litio. Tipo de solución salina utilizada en un enfriador por absorción.

Locked-rotor amperage (LRA). The current an electric motor draws when it is first turned on. This is normally five times the full-load amperage.

Amperaje de rotor bloqueado. Corriente que un motor eléctrico consume al ser encendido, la cual generalmente es cinco veces mayor que el amperaje de carga completa.

Low-pressure control. A pressure switch that can provide low charge protection by shutting down the system on low pressure. It can also be used to control space temperature.

Regulador de baja presión. Conmutador de presión que puede proveer protección contra una carga baja al detener el sistema si éste alcanza una presión demasiado baja. Puede utilizarse también para regular la temperatura de un espacio.

Low side. A term used to refer to that part of the refrigeration system that operates at the lowest pressure, between the expansion device and the compressor.

Lado bajo. Término utilizado para referirse a la parte del sistema de refrigeración que funciona a niveles de presión más baja, entre el dispositivo de expansión y el compresor.

Low-temperature refrigeration. A refrigeration temperature range starting with evaporator temperatures no higher than 0°F for storing frozen food.

Refrigeración a temperatura baja. Margen de la temperatura de refrigeración que comienza con temperaturas de evaporadores no mayores de 0°F (−18°C) para almacenar comida congelada.

LP fuel. Liquefied petroleum. A substance used as a gas for fuel. It is transported and stored in the liquid state.

Combustible PL. Petróleo licuado. Sustancia utilizada como gas para combustible. El petróleo licuado se transporta y almacena en estado líquido.

Magnetic field. A field or space where magnetic lines of force exist.

Campo magnético. Campo o espacio donde existen líneas de fuerza magnética.

Magnetism. A force causing a magnetic field to attract ferrous metals, or where like poles of a magnet repel and unlike poles attract each other.

Magnetismo. Fuerza que hace que un campo magnético atraiga metales férreos, o cuando los polos iguales de un imán se rechazan y los opuestos se atraen.

Male thread. A thread on the outside of a pipe, fitting, or cylinder; an external thread.

Filete macho. Filete en la parte exterior de un tubo, accesorio o cilindro; filete externo.

Manometer. An instrument used to check low vapor pressures. The pressures may be checked against a column of mercury or water.

Manómetro. Instrumento utilizado para revisar las presiones bajas de vapor. Las presiones pueden revisarse comparándolas con una columna de mercurio o de agua.

Mapp gas. A composite gas similar to propane that may be used with air.

Gas Mapp. Gas compuesto similar al propano que puede utilizarse con aire.

Marine water box. A water box with a removable cover.

Caja marina para agua. Caja para agua con un tapón desmontable.

Mass. Matter held together to the extent that it is considered one body.

Masa. Materia compacta que se considera un solo cuerpo.

Mass spectrum analysis. An absorption machine factory leak test performed using helium.

Análisis del límite de masa. Prueba para fugas y absorción llevada a cabo en la fábrica utilizando helio.

Matter. A substance that takes up space and has weight.

Materia. Sustancia que ocupa espacio y tiene peso.

Medium-temperature refrigeration. Refrigeration where evaporator temperatures are 32°F or below, normally used for preserving fresh food.

Refrigeración a temperatura media. Refrigeración, donde las temperaturas del evaporador son 32°F (0°C) o menos, utilizada generalmente para preservar comida fresca.

Megger. An instrument (megohmmeter) that can detect very high resistances, in millions of ohms. Megger relates to megohm or 1,000,000 ohms.

Megóhmetro. Instrumento que puede detectar resistencias sumamente altas, en millones de ohmios. Este término está relacionado al megohmio o 1.000.000 de ohmios.

Megohm. A measure of electrical resistance equal to 1,000,000 ohms.

Megohmio. Medida de resistencia eléctrica equivalente a 1.000.000 de ohmios.

Melting point. The temperature at which a substance will change from a solid to a liquid.

Punto de fusión. Temperatura a la que una sustancia se convierte de sólido a líquido.

Mercury bulb. A glass bulb containing a small amount of mercury and electrical contacts used to make and break the electrical circuit in a low-voltage thermostat.

Bombilla de mercurio. Bombilla de cristal que contiene una pequeña cantidad de mercurio y contactos eléctricos, utilizada para conectar y desconectar el circuito eléctrico en un termostato de baja tensión.

Metering device. A valve or small fixed-size tubing or orifice that meters liquid refrigerant into the evaporator.

Dispositivo de medida. Válvula o tubería pequeña u orificio que mide la cantidad de refrigerante líquido que entra en el evaporador.

Methane. Natural gas is composed of 90% to 95% methane, a combustible hydrocarbon.

Metano. El gas natural se compone de un 90% a un 95% de metano, un hidrocarburo combustible.

Metric system. System International (SI)—system of measurement used by most countries in the world.

Sistema métrico. Sistema internacional; el sistema de medida utilizado por la mayoría de los países del mundo.

Micro. A prefix meaning 1/1,000,000.

Micro. Prefijo que significa una parte de un millón.

Microfarad. Capacitor capacity equal to 1/1,000,000 of a farad.

Microfaradio. Capacidad de un capacitador equivalente a 1/1.000.000 de un faradio.

Micrometer. A precision measuring instrument.

Micrómetro Instrumento de precisión utilizado para medir.

Micron. A unit of length equal to 1/1000 of a millimeter, 1/1,000,000 of a meter.

Micrón. Unidad de largo equivalente a 1/1000 de un milímetro, o 1/1.000.000 de un metro.

Micron gage. A gage used when it is necessary to measure pressure close to a perfect vacuum.

Calibrador de micrón. Calibrador utilizado cuando es necesario medir la presión de un vacío casi perfecto.

Midseated (cracked). A position on a valve that allows refrigerant flow in all directions.

Sentado en el medio (agrietado). Posición en una válvula que permite el flujo de refrigerante en cualquier dirección.

Milli. A prefix meaning 1/1000.

Mili. Prefijo que significa una parte de mil.

Modulator. A device that adjusts by small increments or changes.

Modulador. Dispositivo que se ajusta por medio de incrementos o cambios pequeños.

Moisture indicator. A device for determining moisture in a refrigerant.

Indicador de humedad. Dispositivo utilizado para determinar la humedad en un refrigerante.

Molecule. The smallest particle that a substance can be broken into and still retain its chemical identity.

Molécula. La partícula más pequeña en la que una sustancia puede dividirse y aún conservar sus propias características.

Molecular motion. The movement of molecules within a substance.

Movimiento molecular. Movimiento de moléculas dentro de una sustancia.

Monochlorodifluoromethane. The refrigerant R-22.

Monoclorodiflorometano. El refrigerante R-22.

Motor service factor. A factor above an electric motor's normal operating design parameters, indicated on the nameplate, under which it can operate.

Factor de servicio del motor. Factor superior a los parámetros de diseño normales de funcionamiento de un motor eléctrico, indicados en el marbete; este factor indica su nivel de funcionamiento.

Motor starter. Electromagnetic contactors that contain motor protection and are used for switching electric motors on and off.

Arrancador de motor. Contactadores electromagnéticos que contienen protección para el motor y se utilizan para arrancar y detener motores eléctricos.

Muffler compressor. Sound absorber at the compressor.

Silenciador del compresor. Absorbedor de sonido ubicado en el compresor.

Mullion. Stationary frame between two doors.

Parteluz. Armazón fijo entre dos puertas.

Mullion heater. Heating element mounted in mullion of a refrigerator to keep moisture from forming on it.

Calentador del parteluz. Elemento de calentamiento montado en el parteluz de un refrigerador para evitar la formación de humedad en el mismo.

Multimeter. An instrument that will measure voltage, resistance, and milliamperes.

Multímetro. Instrumento que mide la tensión, la resistencia y los miliamperios.

Multiple evacuation. A procedure for removing the refrigerant from a system. A vacuum is pulled, a small amount of refrigerant allowed into the system, and the procedure duplicated. This is often done three times.

Evacuación múltiple. Procedimiento para remover el refrigerante de un sistema. Se crea un vacío, se permite la entrada de una pequeña cantidad de refrigerante al sistema, y se repite el procedimiento. Con frecuencia esto se lleva a cabo tres veces.

National electrical code (NEC). A publication that sets the standards for all electrical installations, including motor overload protection.

Código estadounidense de electricidad. Publicación que establece las normas para todas las instalaciones eléctricas, incluyendo la protección contra la sobrecarga de un motor.

National pipe taper (NPT). The standard designation for a standard tapered pipe thread.

Cono estadounidense para tubos. Designación estándar para una rosca cónica para tubos estándar.

Natural convection. The natural movement of a gas or fluid caused by differences in temperature.

Convección natural. Movimiento natural de un gas o fluido ocacionado por diferencias en temperatura.

Natural gas. A fossil fuel formed over millions of years from dead vegetation and animals that were deposited or washed deep into the earth.

Gas natural. Combustible fósil formado a través de millones de años de la vegetación y los animales muertos que fueron depositados o arrastrados a una gran profundidad dentro la tierra.

Needlepoint valve. A device having a needle and a very small orifice for controlling the flow of a fluid.

Válvula de aguja. Dispositivo que tiene una aguja y un orificio bastante pequeño para regular el flujo de un fluido.

Negative electrical charge. An atom or component that has an excess of electrons.

Carga eléctrica negativa. Átomo o componente que tiene un exceso de electrones.

Neoprene. Synthetic flexible material used for gaskets and seals.

Neopreno. Material sintético flexible utilizado en guarniciones y juntas de estanqueidad.

Net oil pressure. Difference in the suction pressure and the compressor oil pump outlet pressure.

Presión neta del aceite. Diferencia en la presión de aspiración y la presión a la salida de la bomba de aceite del compresor.

Neutralizer. A substance used to counteract acids.

Neutralizador. Sustancia utilizada para contrarrestar ácidos.

Newton/meter^2. Metric unit of measurement for pressure. Also called a pascal.

Metro-Newton^2. Unidad métrica de medida de presión. Conocido también como pascal.

Nichrome. A metal made of nickle chromium that when formed into a wire is used as a resistance heating element in electric heaters and furnaces.

Níquel-cromio. Metal fabricado de níquel-cromio que al ser convertido en alambre, se utiliza como un elemento de calentamiento de resistencia en calentadores y hornos eléctricos.

Nitrogen. An inert gas often used to "sweep" a refrigeration system to help ensure that all refrigerant and contaminants have been removed.

Nitrógeno. Gas inerte utilizado con frecuencia para purgar un sistema de refrigeración. Esta gas ayuda a asegurar la remoción de todo el refrigerante y los contaminantes del sistema.

Nominal. A rounded-off stated size. The nominal size is the closest rounded-off size.

Nominal. Tamaño redondeado establecido. El tamaño nominal es el tamaño redondeado más cercano.

Noncondensable gas. A gas that does not change into a liquid under normal operating conditions.

Gas no condensable. Gas que no se convierte en líquido bajo condiciones de funcionamiento normales.

Nonferrous. Metals containing no iron.

No férreos. Metales que no contienen hierro.

North pole, magnetic. One end of a magnet.

Polo norte magnético. El extremo de un imán.

Nut driver. These tools have a socket head used primarily to drive hex head screws on air conditioning, heating, and refrigeration cabinets.

Extractor de tuercas. Estas herramientas tienen una cabeza hueca utilizada principalmente para darles vueltas a tornillos de cabeza hexagonal en gabinetes de acondicionamiento de aire, de calefacción y de refrigeración.

Off cycle. A period when a system is not operating.

Ciclo de apagado. Período de tiempo cuando un sistema no está en funcionamiento.

Ohm. A unit of measurement of electrical resistance.

Ohmio. Unidad de medida de la resistencia eléctrica.

Ohmmeter. A meter that measures electrical resistance.

Ohmiómetro. Instrumento que mide la resistencia eléctrica.

Ohm's law. A law involving electrical relationships discovered by Georg Ohm: $E = I \times R$.

Ley de ohm. Ley que define las relaciones eléctricas, descubierta por Georg Ohm: $E = I \times R$.

Oil-pressure safety control (switch). A control used to ensure that a compressor has adequate oil lubricating pressure.

Regulador de seguridad para la presión de aceite (conmutador). Regulador utilizado para asegurar que un compresor tenga la presión de lubrificación de aceite adecuada.

Oil, refrigeration. Oil used in refrigeration systems.

Aceite de refrigeración. Aceite utilizado en sistemas de refrigeración.

Oil separator. Apparatus that removes oil from a gaseous refrigerant.

Separador de aceite. Aparato que remueve el aceite de un refrigerante gaseoso.

Open compressor. A compressor with an external drive.

Compresor abierto. Compresor con un motor externo.

Operating pressure. The actual pressure under operating conditions.

Presión de funcionamiento. La presión real bajo las condiciones de funcionamiento.

Organic. Materials formed from living organisms.

Orgánico. Materiales formados de organismos vivos.

Orifice. A small opening through which fluid flows.

Orificio. Pequeña abertura a través de la cual fluye un fluido.

Overload protection. A system or device that will shut down a system if an overcurrent condition exists.

Protección contra sobrecarga. Sistema o dispositivo que detendrá la marcha de un sistema si existe una condición de sobreintensidad.

Oxidation. The combining of a material with oxygen to form a different substance. This results in the deterioration of the original substance.

Oxidación. La combinación de un material con oxígeno para formar una sustancia diferente, lo que ocasiona el deterioro de la sustancia original.

Ozone. A form of oxygen (O_3). A layer of ozone in the stratosphere that protects the earth from certain of the sun's ultraviolet wave lengths.

Ozono. Forma de oxígeno (O_3). La capa de ozono en la estratósfera que protege la tierra de ciertos rayos ultravioletas del sol.

Package unit. A refrigerating system where all major components are located in one cabinet.

Unidad completa. Sistema de refrigeración donde todos los componentes principales se encuentran en un solo gabinete.

Packing. A soft material that can be shaped and compressed to provide a seal. It is commonly applied around valve stems.

Empaquetadura. Material blando que puede formarse y comprimirse para proveer una junta de estanqueidad. Comúnmente se aplica alrededor de los vástagos de válvulas.

Parallel circuit. An electrical or fluid circuit where the current or fluid takes more than one path at a junction.

Circuito paralelo. Corriente eléctrica o fluida donde la corriente o el fluido siguen más de una trayectoria en un empalme.

Pascal. A metric unit of measurement of pressure.

Pascal. Unidad métrica de medida de presión.

Passive solar design. The use of nonmoving parts of a building to provide heat or cooling, or to eliminate certain parts of a building that cause inefficient heating or cooling.

Diseño solar pasivo. La utilización de piezas fijas de un edificio para proveer calefacción o enfriamiento, o para eliminar ciertas piezas de un edificio que causan calefacción o enfriamiento ineficientes.

PE (polyethylene). Plastic pipe used for water, gas, and irrigation systems.

Polietileno. Tubo plástico utilizado en sistemas de agua, de gas y de irrigación.

Permanent magnet. An object that has its own permanent magnetic field.

Imán permanente. Objeto que tiene su propio campo magnético permanente.

Permanent split-capacitor motor (PSC). A split-phase motor with a run capacitor only. It has a very low starting torque.

Motor permanente de capacitador separado. Motor de fase separada que sólo tiene un capacitador de funcionamiento. Su par de arranque es sumamente bajo.

Phase. One distinct part of a cycle.

Fase. Una parte específica de un ciclo.

Pilot light. The flame that ignites the main burner on a gas furnace.

Llama piloto. Llama que enciende el quemador principal en un horno de gas.

Piston. The part that moves up and down in a cylinder.

Pistón. La pieza que asciende y desciende dentro de un cilindro.

Piston displacement. The volume within the cylinder that is displaced with the movement of the piston from top to bottom.

Desplazamiento del pistón. Volumen dentro del cilindro que se desplaza de arriba a abajo con el movimiento del pistón.

Pitot tube. Part of an instrument for measuring air velocities.

Tubo Pitot. Pieza de un instrumento para medir velocidades de aire.

Planned defrost. Shutting the compressor off with a timer so that the space temperature can provide the defrost.

Descongelación proyectada. Detención de la marcha de un compresor con un temporizador para que la temperatura del espacio lleve a cabo la descongelación.

Plenum. A sealed chamber at the inlet or outlet of an air handler. The duct attaches to the plenum.

Plenum. Cámara sellada a la entrada o a la salida de un tratante de aire. El conducto se fija al plenum.

Polycyclic organic matter. By-products of wood combustion found in smoke and considered to be health hazards.

Materia orgánica policíclica. Subproductos de la combustión de madera presentes en el humo y considerados nocivos para la salud.

Polyphase. Three or more phases.

Polifase. Tres o más fases.

Porcelain. A ceramic material.

Porcelana. Material cerámico.

Portable dolly. A small platform with four wheels on which heavy objects can be placed and moved.

Carretilla portátil. Plataforma pequeña con cuatro ruedas sobre la que pueden colocarse y transportarse objetos pesados.

Positive displacement. A term used with a pumping device such as a compressor that is designed to move all matter from a volume such as a cylinder or it will stall, possibly causing failure of a part.

Desplazamiento positivo. Término utilizado con un dispositivo de bombeo, como por ejemplo un compresor, diseñado para mover toda la materia de un volumen, como un cilindro o se bloqueará, posiblemente causándole fallas a una pieza.

Positive electrical charge. An atom or component that has a shortage of electrons.

Carga eléctrica positiva. Átomo o componente que tiene una insuficiencia de electrones.

Positive temperature coefficient start device. A thermistor used to provide start assistance to a permanent split-capacitor motor.

Dispositivo de arranque de coeficiente de temperatura positiva. Termistor utilizado para ayudar a arrancar un motor permanente de capacitador separado.

Potential relay. A switching device used with hermetic motors that breaks the circuit to the start windings after the motor has reached approximately 75% of its running speed.

Relé de potencial. Dispositivo de conmutación utilizado con motores herméticos que interrupe el circuito de los devandos de arranque antes de que el motor haya alcanzado aproximadamente un 75% de su velocidad de marcha.

Potentiometer. An instrument that controls electrical current.

Potenciómetro. Instrumento que regula corriente eléctrica.

Power. The rate at which work is done.

Potencia. Velocidad a la que se realiza un trabajo.

Pressure. Force per unit of area.

Presión. Fuerza por unidad de área.

Pressure drop. The difference in pressure between two points.

Caída de presión. Diferencia en presión entre dos puntos.

Pressure/enthalpy diagram. A chart indicating the pressure and heat content of a refrigerant and the extent to which the refrigerant is a liquid and vapor.

Diagrama de presión y entalpía. Esquema que indica la presión y el contenido de calor de un refrigerante y el punto en que el refrigerante es líquido y vapor.

Pressure limiter. A device that opens when a certain pressure is reached.

Dispositivo limitador de presión. Dispositivo que se abre cuando se alcanza una presión específica.

Pressure-limiting TXV. A valve designed to allow the evaporator to build only to a predetermined temperature when the valve will shut off the flow of refrigerant.

Válvula electrónica de expansión limitadora de presión. Válvula diseñada para permitir que la temperatura del evaporador alcance un límite predeterminado cuando la válvula detenga el flujo de refrigerante.

Pressure regulator. A valve capable of maintaining a constant outlet pressure when a variable inlet pressure occurs. Used for regulating fluid flow such as natural gas, refrigerant, and water.

Regulador de presión. Válvula capaz de mantener una presión constante a la salida cuando ocurre una presión variable a la entrada. Utilizado para regular el flujo de fluidos, como por ejemplo el gas natural, el refrigerante y el agua.

Pressure switch. A switch operated by a change in pressure.

Conmutador accionado por presión. Conmutador accionado por un cambio de presión.

Pressure/temperature relationship. This refers to the pressure/temperature relationship of a liquid and vapor in a closed container. If the temperature increases, the pressure will also increase. If the temperature is lowered, the pressure will decrease.

Relación entre presión y temperatura. Se refiere a la relación entre la presión y la temperatura de un líquido y un vapor en un recipiente cerrado. Si la temperatura aumenta, la presión también aumentará. Si la temperatura baja, habrá una caída de presión.

Pressure vessels and piping. Piping, tubing, cylinders, drums, and other containers that have pressurized contents.

Depósitos y tubería con presión. Tubería, cilindros, tambores y otros recipientes que tienen un contenido presurizado.

Primary control. Controlling device for an oil burner to ensure ignition within a specific time span, usually 90 seconds.

Regulador principal. Dispositivo de regulación para un quemador de aceite pesado. El regulador principal asegura el encendido dentro de un período de tiempo específico, normalmente 90 segundos.

Propane. An LP gas used for heat.

Propano. Gas de petróleo licuado que se utiliza para producir calor.

Proton. That part of an atom having a positive charge.

Protón. Parte de un átomo que tiene carga positiva.

psi. Abbreviation for pounds per square inch.

psi. Abreviatura de libras por pulgada cuadrada.

psia. Abbreviation for pounds per square inch absolute.

psia. Abreviatura de libras por pulgada cuadrada absoluta.

psig. Abbreviation for pounds per square inch gage.

psig. Abreviatura de indicador de libras por pulgada cuadrada.

Psychrometer. An instrument for determining relative humidity.

Sicrómetro. Instrumento para medir la humedad relativa.

Psychrometric chart. A chart that shows the relationship of temperature, pressure, and humidity in the air.

Esquema sicrométrico. Esquema que indica la relación entre la temperatura, la presión y la humedad en el aire.

Pump. A device that forces fluids through a system.

Bomba. Dispositivo que introduce fluidos por fuerza a través de un sistema.

Pump down. To use a compressor to pump the refrigerant charge into the condenser and/or receiver.

Extraer con bomba. Utilizar un compresor para bombear la carga del refrigerante dentro del condensador y/o receptor.

Purge. To remove or release fluid from a system.

Purga. Remover o liberar el fluido de un sistema.

PVC (Polyvinyl chloride). Plastic pipe used in pressure applications for water and gas as well as for sewage and certain industrial applications.

Cloruro de polivinilo (PVC). Tubo plástico utilizado tanto en aplicaciones de presión para agua y gas, como en ciertas aplicaciones industriales y de aguas negras.

Quench. To submerge a hot object in a fluid for cooling.

Enfriamiento por inmersión. Sumersión de un objeto caliente en un fluido para enfriarlo.

Quick-connect coupling. A device designed for easy connecting or disconnecting of fluid lines.

Acoplamiento de conexión rápida. Dispositivo diseñado para facilitar la conexión o desconexión de conductos de fluido.

R-12. Dichlorodifluoromethane, a popular refrigerant for refrigeration systems.

R-12. Diclorodiflorometano, refrigerante muy utilizado en sistemas de refrigeración.

R-22. Monochlorodifluoromethane, a popular refrigerant for air conditioning systems.

R-22. Monoclorodiflorometano, refrigerante muy utilizado en sistemas de acondicionamiento de aire.

R-123. Dichlorotrifluoroethane, a refrigerant developed for low-pressure application.

R-123. Diclorotrifloroetano, refrigerante elaborado para aplicaciones de baja presión.

R-134a. Tetrafluoroethane, a refrigerant developed for refrigeration systems and as a possible replacement for R-12.

R-134a. Tetrafloroetano, refrigerante elaborado para sistemas de refrigeración y como posible sustituto del R-12.

R-502. An azeotropic mixture of R-22 and R-115, a popular refrigerant for low-temperature refrigeration systems.

R-502. Mezcla azeotrópica de R-22 y R-115, refrigerante muy utilizado en sistemas de refrigeración de temperatura baja.

Radiant heat. Heat that passes through air, heating solid objects that in turn heat the surrounding area.

Calor radiante. Calor que pasa a través del aire y calienta objetos sólidos que a su vez calientan el ambiente.

Radiation. Heat transfer. See radiant heat.

Radiación. Transferencia de calor. Véase calor radiante.

Random or off-cycle defrost. Defrost provided by the space temperature during the normal off cycle.

Descongelación variable o de ciclo apagado. Descongelación llevada a cabo por la temperatura del espacio durante el ciclo normal de apagado.

Rankine. The absolute Fahrenheit scale with 0 at the point where all molecular motion stops.

Rankine. Escala absoluta de Fahrenheit con el 0 al punto donde se detiene todo movimiento molecular.

Reactance. A type of resistance in an alternating current circuit.

Reactancia. Tipo de resistencia en un circuito de corriente alterna.

Reamer. Tool to remove burrs from inside a pipe after it has been cut.

Escariador. Herramienta utilizada para remover las rebabas de un tubo después de haber sido cortado.

Receiver-drier. A component in a refrigeration system for storing and drying refrigerant.

Receptor-secador. Componente en un sistema de refrigeración que almacena y seca el refrigerante.

Reciprocating. Back-and-forth motion.

Movimiento alternativo. Movimiento de atrás para adelante.

Reciprocating compressor. A compressor that uses a piston in a cylinder and a back-and-forth motion to compress vapor.

Compresor alternativo. Compresor que utiliza un pistón en un cilindro y un movimiento de atrás para adelante a fin de comprir el vapor.

Rectifier. A device for changing alternating current to direct current.

Rectificador. Dispositivo utilizado para convertir corriente alterna en corriente continua.

Reed valve. A thin steel plate used as a valve in a compressor.

Válvula con lámina. Placa delgada de acero utilizada como una válvula en un compresor.

Refrigerant. The fluid in a refrigeration system that changes from a liquid to a vapor and back to a liquid at practical pressures.

Refrigerante. Fluido en un sistema de refrigeración que se convierte de líquido en vapor y nuevamente en líquido a presiones prácticas.

Refrigerant reclaim. Recovering the refrigerant and processing it so that it can be reused.

Recuperación del refrigerante. La recuperación del refrigerante y su procesamiento para que pueda ser utilizado de nuevo.

Refrigerant reclaim. "To process refrigerant to new product specifications by means which may include distillation. It will require chemical analysis of the refrigerant to determine that appropriate product specifications are met. This term usually implies the use of processes or procedures available only at a reprocessing or manufacturing facility."

Recuperación del refrigerante. "Procesar refrigerante según nuevas especificaciones para productos a través de métodos que pueden incluir la destilación. Se requiere un análisis químico del refrigerante para asegurar el cumplimiento de las especificaciones para productos adecuadas. Por lo general este término supone la utilización de procesos o de procedimientos disponibles solamente en fábricas de reprocesamiento o manufactura."

Refrigerant recovery. "To remove refrigerant in any condition from a system and store it in an external container without necessarily testing or processing it in any way."

Recobrar refrigerante líquido. "Remover refrigerante en cualquier estado de un sistema y almacenarlo en un recipiente externo sin ponerlo a prueba o elaborarlo de ninguna manera."

Refrigerant recycling. "To clean the refrigerant by oil separation and single or multiple passes through devices, such as replaceable core filter-driers, which reduce moisture, acidity and particulate matter. This term usually applies to procedures implemented at the job site or at a local service shop."

Recirculación de refrigerante. "Limpieza del refrigerante por medio de la separación del aceite y pasadas sencillas o múltiples a través de dispositivos, como por ejemplo secadores filtros con núcleos reemplazables que disminuyen la humedad, la acidez y las partículas. Por lo general este término se aplica a los procedimientos utilizados en el lugar del trabajo o en un taller de servicio local."

Refrigeration. The process of removing heat from a place where it is not wanted and transferring that heat to a place where it makes little or no difference.

Refrigeración. Proceso de remover el calor de un lugar donde no es deseado y transferirlo a un lugar donde no afecte la temperatura.

Register. A terminal device on an air distribution system that directs air but also has a damper to adjust airflow.

Registro. Dispositivo terminal en un sistema de distribución de aire que dirige el aire y además tiene un desviador para ajustar su flujo.

Relative humidity. The amount of moisture contained in the air as compared to the amount the air could hold at that temperature.

Humedad relativa. Cantidad de humedad presente en el aire, comparada con la cantidad de humedad que el aire pueda contener a dicha temperatura.

Relay. A small electromagnetic device to control a switch, motor, or valve.

Relé. Pequeño dispositivo electromagnético utilizado para regular un conmutador, un motor o una válvula.

Relief valve. A valve designed to open and release liquids at a certain pressure.

Válvula para alivio. Válvula diseñada para abrir y liberar líquidos a una presión específica.

Remote system. Often called a split system where the condenser is located away from the evaporator and/or other parts of the system.

Sistema remoto. Llamado muchas veces sistema separado donde el condensador se coloca lejos del evaporador y/u otras piezas del sistema.

Resistance. The opposition to the flow of an electrical current or a fluid.

Resistencia. Oposición al flujo de una corriente eléctrica o de un fluido.

Resistor. An electrical or electronic component with a specific opposition to electron flow. It is used to create voltage drop or heat.

Resistor. Componente eléctrico o eletrónico con una oposición específica al flujo de electrones; se utiliza para producir una caída de tensión o calor.

Restrictor. A device used to create a planned resistance to fluid flow.

Limitador. Dispositivo utilizado para producir una resistencia proyectada al flujo de fluido.

Reverse cycle. The ability to direct the hot gas flow into the indoor or the outdoor coil in a heat pump to control the system for heating or cooling purposes.

Ciclo invertido. Capacidad de dirigir el flujo de gas caliente dentro de la bobina interior o exterior en una bomba de calor a fin de regular el sistema para propósitos de calentamiento o enfriamiento.

Rod and tube. The rod and tube are each made of a different metal. The tube has a high expansion rate and the rod a low expansion rate.

Varilla y tubo. La varilla y el tubo se fabrican de un metal diferente. El tubo tiene una tasa de expansión alta y la varilla una tasa de expansión baja.

Rotary compressor. A compressor that uses rotary motion to pump fluids. It is a positive-displacement pump.

Compresor giratorio. Compresor que utiliza un movimiento giratorio para bombear fluidos. Es una bomba de desplazmiento positivo.

Rotor. The rotating or moving component of a motor, including the shaft.

Rotor. Componente giratorio o en movimiento de un motor, incluyendo el árbol.

Running time. The time a unit operates. Also called the on time.

Período de funcionamiento. El período de tiempo en que funciona una unidad. Conocido también como período de conexión.

Run winding. The electrical winding in a motor that draws current during the entire running cycle.

Devanado de funcionamiento. Devanado eléctrico en un motor que consume corriente durante todo el ciclo de funcionamiento.

Rupture disk. Safety device for a centrifugal low-pressure chiller.

Disco de ruptura. Dispositivo de seguridad para un enfriador centrífugo de baja presión.

Saddle valve. A valve that straddles a fluid line and is fastened by solder or screws. It normally contains a device to puncture the line for pressure readings.

Válvula de silleta. Válvula que está sentada a horcajadas en un conducto de fluido y se fija por medio de la soldadura o tornillos. Por lo general contiene un dispositivo para agujerear el conducto a fin de que se puedan tomar lecturas de presión.

Safety control. An electrical, mechanical, or electromechanical control to protect the equipment or public from harm.

Regulador de seguridad. Regulador eléctrico, mecánico o electromecánico para proteger al equipo de posibles averías o al público de sufrir alguna lesión.

Safety plug. A fusible plug.

Tapón de seguridad. Tapón fusible.

Sail switch. A safety switch with a lightweight sensitive sail that operates by sensing an airflow.

Conmutador con vela. Conmutador de seguridad con una vela liviana sensible que funciona al advertir el flujo de aire.

Saturated vapor. The refrigerant when all of the liquid has changed to a vapor.

Vapor saturado. El refrigerante cuando todo el líquido se ha convertido en vapor.

Saturation. A term used to describe a substance when it contains all of another substance it can hold.

Saturación. Término utilizado para describir una sustancia cuando contiene lo más que puede de otra sustancia.

Scavenger pump. A pump used to remove the fluid from a sump.

Bomba de barrido. Bomba utilizada para remover el fluido de un sumidero.

Schrader valve. A valve similar to the valve on an auto tire that allows refrigerant to be charged or discharged from the system.

Válvula Schrader. Válvula similar a la válvula del neumático de un automóvil que permite la entrada o la salida de refrigerante del sistema.

Scotch yoke. A mechanism used to create reciprocating motion from the electric motor drive in very small compressors.

Yugo escocés. Mecanismo utilizado para producir movimiento alternativo del accionador del motor eléctrico en compresores bastante pequeños.

Screw compressor. A form of positive-displacement compressor that squeezes fluid from a low-pressure area to a high-pressure area, using screw-type mechanisms.

Compresor de tornillo. Forma de compresor de desplazamiento positivo que introduce por fuerza el fluido de un área de baja presión a un área de alta presión, a través de mecanismos de tipo de tornillo.

Scroll compressor. A compressor that uses two scroll-type components to compress vapor.

Compresor espiral. Compresor que utiliza dos componentes de tipo espiral para comprimir el vapor.

Sealed unit. The term used to describe a refrigeration system, including the compressor, that is completely welded closed. The pressures can be accessed by saddle valves.

Unidad sellada. Término utilizado para describir un sistema de refrigeración, incluyendo el compresor, que es soldado completamente cerrado. Las presiones son accesibles por medio de válvulas de silleta.

Seasonal energy efficiency ratio (SEER). An equipment efficiency rating that takes into account the startup and shutdown for each cycle.

Relación del rendimiento de energía temporal. Clasificación del rendimiento de un equipo que toma en cuenta la puesta en marcha y la parada de cada ciclo.

Seat. The stationary part of a valve that the moving part of the valve presses against for shutoff.

Asiento. Pieza fija de una válvula contra la que la pieza en movimiento de la válvula presiona para cerrarla.

Semiconductor. A component in an electronic system that is considered neither an insulator nor a conductor but a partial conductor.

Semiconductor. Componente en un sistema eléctrico que no se considera ni aislante ni conductor, sino conductor parcial.

Semihermetic compressor. A motor compressor that can be opened or disassembled by removing bolts and flanges. Also known as a serviceable hermetic.

Compresor semihermético. Compresor de un motor que puede abrirse o desmontarse al removerle los pernos y bridas. Conocido también como hermético utilizable.

Sensible heat. Heat that causes a change in the level of a thermometer.

Calor sensible. Calor que produce un cambio en el nivel de un termómetro.

Sensor. A component for detection that changes shape, form, or resistance when a condition changes.

Sensor. Componente para la deteción que cambia de forma o de resistencia cuando cambia una condición.

Sequencer. A control that causes a staging of events, such as a sequencer between stages of electric heat.

Regulador de secuencia. Regulador que produce una sucesión de acontecimientos, como por ejemplo etapas sucesivas de calor eléctrico.

Series circuit. An electrical or piping circuit where all of the current or fluid flows through the entire circuit.

Circuito en serie. Circuito eléctrico o de tubería donde toda la corriente o todo el fluido fluye a través de todo el circuito.

Service valve. A manually operated valve in a refrigeration system used for various service procedures.

Válvula de servicio. Válvula de un sistema de refrigeración accionada manualmente que se utiliza en varios procedimientos de servicio.

Serviceable hermetic. See semihermetic compressor.

Compresor hermético utilizable. Véase compresor semihermético.

Shaded-pole motor. An alternating current motor used for very light loads.

Motor polar en sombra. Motor de corriente alterna utilizado en cargas sumamente livianas.

Shell and coil. A vessel with a coil of tubing inside that is used as a heat exchanger.

Coraza y bobina. Depósito con una bobina de tubería en su interior que se utiliza como intercambiador de calor.

Shell and tube. A heat exchanger with straight tubes in a shell that can normally be mechanically cleaned.

Coraza y tubo. Intercambiador de calor con tubos rectos en una coraza que por lo general puede limpiarse mecánicamente.

Short circuit. A circuit that does not have the correct measurable resistance; too much current flows and will overload the conductors.

Cortocircuito. Corriente que no tiene la resistencia medible correcta; un exceso de corriente fluye a través del circuito provocando una sobrecarga de los conductores.

Short cycle. The term used to describe the running time (on time) of a unit when it is not running long enough.

Ciclo corto. Término utilizado para describir el período de funcionamiento (de encendido) de una unidad cuando no funciona por un período de tiempo suficiente.

Shroud. A fan housing that ensures maximum airflow through the coil.

Bóveda. Alojamiento del abanico que asegura un flujo máximo de aire a través de la bobina.

Sight glass. A clear window in a fluid line.

Mirilla para observación. Ventana clara en un conducto de fluido.

Silica gel. A chemical compound often used in refrigerant driers to remove moisture from the refrigerant.

Gel silíceo. Compuesto químico utilizado a menudo en secadores de refrigerantes para remover la humedad del refrigerante.

Silicon. A substance from which many semiconductors are made.

Silicio. Sustancia de la cual se fabrican muchos semiconductores.

Silicon-controlled rectifier (SCR). A semiconductor control device.

Rectificador controlado por silicio. Dispositivo para regular un semiconductor.

Silver brazing. A high-temperature (above 800°F) brazing process for bonding metals.

Soldadura con plata. Soldadura a temperatura alta (sobre los 800°F ó 430°C) para unir metales.

Sine wave. The graph or curve used to describe the characteristics of alternating current voltage.

Onda sinusoidal. Gráfica o curva utilizada para describir las características de tensión de corriente alterna.

Single phase. The electrical power supplied to equipment or small motors, normally under 7½ hp.

Monofásico. Potencia eléctrica suministrada a equipos o motores pequeños, por lo general menor de 7½ hp.

Single phasing. The condition in a three-phase motor when one phase of the power supply is open.

Fasaje sencillo. Condición en un motor trifásico cuando una fase de la fuente de alimentación está abierta.

Sling psychrometer. A device with two thermometers, one a wet bulb and one a dry bulb, used for checking air conditions, temperature, and humidity.

Sicrómetro con eslinga. Dispositivo con dos termómetros, uno con una bombilla húmeda y otro con una bombilla seca, utilizados para revisar las condiciones del aire, de la temperatura y de la humedad.

Slip. The difference in the rated rpm of a motor and the actual operating rpm.

Deslizamiento. Diferencia entre las rpm nominales de un motor y las rpm de funcionamiento reales.

Slugging. A term used to describe the condition when large amounts of liquid enter a pumping compressor cylinder.

Relleno. Término utilizado para describir la condición donde grandes cantidades de líquido entran en el cilindro de un compresor de bombeo.

Smoke test. A test performed to determine the amount of unburned fuel in an oil burner flue-gas sample.

Prueba de humo. Prueba llevada a cabo para determinar la cantidad de combustible no quemado en una muestra de gas de combustión que se obtiene de un quemador de aceite pesado.

Snap-disc. An application of the bimetal. Two different metals fastened together in the form of a disc that provides a warping condition when heated. This also provides a snap action that is beneficial in controls that start and stop current flow in electrical circuits.

Disco de acción rápida. Aplicación del bimetal. Dos metales diferentes fijados entre sí en forma de un disco que provee una deformación al ser calentado. Esto provee también una acción rápida, ventajosa para reguladores que ponen en marcha y detienen el flujo de corriente en circuitos eléctricos.

Solar collectors. Components of a solar system designed to collect the heat from the sun, using air, a liquid, or refrigerant as the medium.

Colectores solares. Componentes de un sistema solar diseñados para acumular el calor emitido por el sol, utilizando el aire, un líquido o un refrigerante como el medio.

Solar heat. Heat from the sun's rays.

Calor solar. Calor emitido por los rayos del sol.

Soldering. Fastening two base metals together by using a third, filler metal that melts at a temperature below 800°F.

Soldadura. La fijación entre sí de dos metales bases utilizando un tercer metal de relleno que se funde a una temperatura menor de 800°F (430°C).

Solder pot. A device using a low-melting solder and an overload heater sized for the amperage of the motor it is protecting. The solder will melt, opening the circuit when there is an overload. It can be reset.

Olla para soldadura. Dispositivo que utiliza una soldadura con un punto de fusión bajo y un calentador de sobrecarga diseñado para el amperaje del motor al que provee protección. La soldadura se fundirá, abriendo así el circuito cuando ocurra una sobrecarga. Puede ser reconectado.

Solenoid. A coil of wire designed to carry an electrical current producing a magnetic field.

Solenoide. Bobina de alambre diseñada para conducir una corriente eléctrica generando un campo magnético.

Soild. Molecules of a solid are highly attracted to each other forming a mass that exerts all of its weight downward.

Sólido. Las moléculas de un sólido se atraen entre sí y forman una masa que ejerce todo su peso hacia abajo.

Specific gravity. The weight of a substance compared to the weight of an equal volume of water.

Gravedad específica. El peso de una sustancia comparada con el peso de un volumen igual de agua.

Specific heat. The amount of heat required to raise the temperature of 1 lb of a substance 1°F.

Calor específico. La cantidad de calor requerido para elevar la temperatura de una libra de una sustancia 1°F (-17°C).

Specific volume. The volume occupied by 1 lb of a fluid.

Volumen específico. Volumen que ocupa una libra de fluido.

Splash lubrication system. A system of furnishing lubrication to a compressor by agitating the oil.

Sistema de lubrificación por salpicadura. Método de proveerle lubrificación a un compresor agitando el aceite.

Splash method. A method of water dropping from a higher level in a cooling tower and splashing on slots with air passing through for more efficient evaporation.

Método de salpicaduras. Método de dejar caer agua desde un nivel más alto en una torre de refrigeración y salpicándola en ranuras, mientras el aire pasa a través de las mismas con el propósito de lograr una evaporación más eficaz.

Split-phase motor. A motor with run and start windings.

Motor de fase separada. Motor con devandos de funcionamiento y de arranque.

Split system. A refrigeration or air conditioning system that has the condensing unit remote from the indoor (evaporator) coil.

Sistema separado. Sistema de refrigeración o de acondicionamiento de aire cuya unidad de condensación se encuentra en un sitio alejado de la bobina interior del evaporador.

Spray pond. A pond with spray heads used for cooling water in water-cooled air conditioning or refrigeration systems.

Tanque de rociado. Tanque con una cabeza rociadora utilizada para enfriar el agua en sistemas de acondicionamiento de aire o de refrigeración enfriados por agua.

Squirrel cage fan. A fan assembly used to move air.

Abanico con jaula de ardilla. Conjunto de abanico utilizado para mover el aire.

Standard atmosphere or standard conditions. Air at sea level at 70°F when the atmosphere's pressure is 14.696 psia (29.92 in. Hg). Air at this condition has a volume of 13.33 ft^3/lb.

Atmósfera estándar o condiciones estándares. El aire al nivel del mar a una temperatura de 70°F (21°C) cuando la presión de la atmósfera es 14,696 psia (29,92 pulgadas Hg). Bajo esta condición, el aire tiene un volumen de 13,33 ft^3/lb (libras/pies).

Standing pilot. Pilot flame that remains burning continuously.

Piloto constante. Llama piloto que se quema de manera continua.

Start capacitor. A capacitor used to help an electric motor start.

Capacitador de arranque. Capacitador utilizado para ayudar en el arranque de un motor eléctrico.

Starting relay. An electrical relay used to disconnect the start winding in a hermetic compressor.

Relé de arranque. Relé eléctrico utilizado para desconectar el devanado de arranque en un compresor hermético.

Starting winding. The winding in a motor used primarily to give the motor extra starting torque.

Devanado de arranque. Devanado en un motor utilizado principalmente para proveerle al motor mayor par de arranque.

Starved coil. The condition in an evaporator when the metering device is not feeding enough refrigerant to the evaporator.

Bobina estrangulada. Condición que ocurre en un evaporador cuando el dispositivo de medida no le suministra suficiente refrigerante al evaporador.

Stator. The component in a motor that contains the windings; it does not turn.

Estátor. Componente en un motor que contiene los devanados y que no gira.

Steam. The vapor state of water.

Vapor. Estado de vapor del agua.

Strainer. A fine-mesh device that allows fluid flow and holds back solid particles.

Colador. Dispositivo de malla fina que permite el flujo de fluido a través de él y atrapa partículas sólidas.

Stratification. The condition where a fluid appears in layers.

Estratificación. Condición que ocurre cuando un fluido aparece en capas.

Stress crack. A crack in piping or other component caused by age or abnormal conditions such as vibration.

Grieta por tensión. Grieta que aparece en una tubería u otro componente ocasionada por envejecimiento o condiciones anormales, como por ejemplo vibración.

Subbase. The part of a space temperature thermostat that is mounted on the wall and to which the interconnecting wiring is attached.

Subbase. Pieza de un termóstato que mide la temperatura de un espacio que se monta sobre la pared y a la que se fijan los conductores eléctricos interconectados.

Subcooling. The temperature of a liquid when it is cooled below its condensing temperature.

Subenfriamiento. La temperatura de un líquido cuando se enfría a una temperatura menor que su temperatura de condensación.

Sublimation. When a substance changes from the solid state to the vapor state without going through the liquid state.

Sublimación. Cuando una sustancia cambia de sólido a vapor sin convertirse primero en líquido.

Suction gas. The refrigerant vapor in an operating refrigeration system found in the tubing from the evaporator to the compressor and in the compressor shell.

Gas de aspiración. El vapor del refrigerante en un sistema de refrigeración en funcionamiento presente en la tubería que va del evaporador al compresor y en la coraza del compresor.

Suction line. The pipe that carries the heat-laden refrigerant gas from the evaporator to the compressor.

Conducto de aspiración. Tubo que conduce el gas de refrigerante lleno de calor del evaporador al compresor.

Suction service valve. A manually operated valve with front and back seats located at the compressor.

Válvula de aspiración para servicio. Válvula accionada manualmente que tiene asientos delanteros y traseros ubicados en el compresor.

Suction valve lift unloading. The suction valve in a reciprocating compressor cylinder is lifted, causing that cylinder to stop pumping.

Descarga por levantamiento de la válvula de aspiración. La válvula de aspiración en el cilindro de un compresor alternativo se levanta, provocando que el cilindro deje de bombear.

Sump. A reservoir at the bottom of a cooling tower to collect the water that has passed through the tower.

Sumidero. Tanque que se encuentra en el fondo de una torre de refrigeración para acumular el agua que ha pasado a través de la torre.

Superheat. The temperature of vapor refrigerant above its saturation change of state temperature.

Sobrecalor. Temperatura del refrigerante de vapor mayor que su temperatura de cambio de estado de saturación.

Surge. When the head pressure becomes too great or the evaporator pressure too low, refrigerant will flow from the high- to the low-pressure side of a centrifugal compressor system, making a loud sound.

Movimiento repentino. Cuando la presión en la cabeza aumenta demasiado o la presión en el evaporador es demasiado baja, el refrigerante fluye del lado de alta presión al lado de baja presión de un sistema de compresor centrífugo. Este movimiento produce un sonido fuerte.

Swaged joint. The joining of two pieces of copper tubing by expanding or stretching the end of one piece of tubing to fit over the other piece.

Junta estampada. La conexión de dos piezas de tubería de cobre dilatando o alargando el extremo de una pieza de tubería para ajustarla sobre otra.

Swaging tool. A tool used to enlarge a piece of tubing for a solder or braze connection.

Herramienta de estampado. Herramienta utilizada para agrandar una pieza de tubería a utilizarse en una conexión soldada o broncesoldada.

Swamp cooler. A slang term used to describe an evaporative cooler.

Nevera pantanoso. Término del argot utilizado para describir una nevera de evaporación.

Sweating. A word used to describe moisture collection on a line or coil that is operating below the dew point temperature of the air.

Exudación. Término utilizado para describir la acumulación de humedad en un conducto o una bobina que está funcionando a una temperatura menor que la del punto de rocío del aire.

Tank. A closed vessel used to contain a fluid.

Tanque. Depósito cerrado utilizado para contener un fluido.

Tap. A tool used to cut internal threads in a fastener or fitting.

Macho de roscar. Herramienta utilizada para cortar filetes internos en un aparto fijador o en un accesorio.

Temperature. A word used to describe the level of heat or molecular activity, expressed in Fahrenheit, Rankine, Celsius, or Kelvin units.

Temperatura. Término utilizado para describir el nivel de calor o actividad molecular, expresado en unidades Fahrenheit, Rankine, Celsio o Kelvin.

Test light. A light bulb arrangement used to prove the presence of electrical power in a circuit.

Luz de prueba. Arreglo de bombillas utilizado para probar la presencia de fuerza eléctrica en un circuito.

Therm. Quantity of heat, 100,000 Btu.

Therm. Cantidad de calor, mil unidades térmicas inglesas.

Thermistor. A semiconductor electronic device that changes resistance with a change in temperature.

Termistor. Dispositivo eléctrico semiconductor que cambia su resistencia cuando se produce un cambio en temperatura.

Thermocouple. A device made of two unlike metals that generates electricity when there is a difference in temperature from one end to the other. Thermocouples have a hot and cold junction.

Thermopar. Dispositivo hecho de dos metales distintos que genera electricidad cuando hay una diferencia en temperatura de un extremo al otro. Los termopares tienen un empalme caliente y uno frío.

Thermometer. An instrument used to detect differences in the level of heat.

Termómetro. Instrumento utilizado para detectar diferencias en el nivel de calor.

Thermopile. A group of thermocouples connected in series to increase voltage output.

Pila termoeléctrica. Grupo de termopares conectados en serie para aumentar la salida de tensión.

Thermostat. A device that senses temperature change and changes some dimension or condition within to control an operating device.

Termostato. Dispositivo que advierte un cambio en temperatura y cambia alguna dimensión o condición dentro de sí para regular un dispositivo en funcionamiento.

Thermostatic expansion valve (TXV). A valve used in refrigeration systems to control the superheat in an evaporator by metering the correct refrigerant flow to the evaporator.

Válvula de gobierno termostático para expansión. Válvula utilizada en sistemas de refrigeración para regular el sobrecalor en un evaporador midiendo el flujo correcto de refrigerante al evaporador.

Three-phase power. A type of power supply usually used for operating heavy loads. It consists of three sine waves that are out of phase with each other.

Potencia trifásica. Tipo de fuente de alimentación normalmente utilizada en el funcionamiento de cargas pesadas. Consiste de tres ondas sinusoidales que no están en fase la una con la otra.

Throttling. Creating a restriction in a fluid line.

Estrangulamiento. Que ocasiona una restricción en un conducto de fluido.

Timers. Clock-operated devices used to time various sequences of events in circuits.

Temporizadores. Dispositivos accionados por un reloj utilizados para medir el tiempo de varias secuencias de eventos en circuitos.

Ton of refrigeration. The amount of heat required to melt a ton (2000 lb) of ice at 32°F, 288,000 Btu/24 h, 12,000 Btu/h, or 200 Btu/min.

Tonelada de refrigeración. Cantidad de calor necesario para fundir una tonelada (2000 libras) de hielo a 32°F (0°C), 288.000 Btu/24 h, 12.000 Btu/h, o 200 Btu/min.

Torque. The twisting force often applied to the starting power of a motor.

Par de torsión. Fuerza de torsión aplicada con frecuencia a la fuerza de arranque de un motor.

Torque wrench. A wrench used to apply a prescribed amount of torque or tightening to a connector.

Llave de torsión. Llave utilizada para aplicar una cantidad específica de torsión o de apriete a un conector.

Total heat. The total amount of sensible heat and latent heat contained in a substance from a reference point.

Calor total. Cantidad total de calor sensible o de calor latente presente en una sustancia desde un punto de referencia.

Transformer. A coil of wire wrapped around an iron core that induces a current to another coil of wire wrapped around the same iron core. Note: A transformer can have an air core.

Transformador. Bobina de alambre devanado alrededor de un núcleo de hierro que induce una corriente a otra bobina de alambre devanado alrededor del mismo núcleo de hierro. Nota: Un transformador puede tener un núcleo de aire.

Transistor. A semiconductor often used as a switch or amplifier.

Transistor. Semiconductor que suele utilizarse como conmutador o amplificador.

TRIAC. A semiconductor switching device.

TRIAC. Dispositivo de conmutación para semiconductores.

Tube within a tube coil. A coil used for heat transfer that has a pipe in a pipe and is fastened together so that the outer tube becomes one circuit and the inner tube another.

Bobina de tubo dentro de un tubo. Bobina utilizada en la transferencia de calor que tiene un tubo dentro de otro y se sujeta de manera que el tubo exterior se convierte en un circuito y el tubo interior en otro circuito.

Tubing. Pipe with a thin wall used to carry fluids.

Tubería. Tubo que tiene una pared delgada utilizado para conducir fluidos.

Two-temperature valve. A valve used in systems with multiple evaporators to control the evaporator pressures and maintain different temperatures in each evaporator. Sometimes called a <u>holdback valve</u>.

Válvula de dos temperaturas. Válvula utilizada en sistemas con evaporadores múltiples para regular las presiones de los evaporadores y mantener temperaturas diferentes en cada uno de ellos. Conocida también como <u>válvula de retención</u>.

Ultraviolet. Light waves that can only be seen under a special lamp.

Ultravioleta. Ondas de luz que pueden observarse solamente utilizando una lámpara especial.

Urethane foam. A foam that can be applied between two walls for insulation.

Espuma de uretano. Espuma que puede aplicarse entre dos paredes para crear un aislamiento.

U-Tube mercury manometer. A U-tube containing mercury, which indicates the level of vacuum while evacuating a refrigeration system.

Manómetro de mercurio de tubo en U. Tubo en U que contiene mercurio y que indica el nivel del vacío mientras vacía un sistema de refrigeración.

U-Tube water manometer. Indicates natural gas and propane gas pressures. It is usually calibrated in inches of water.

Manómetro de agua de tubo en U. Indica las presiones del gas natural y del propano. Se calibra normalmente en pulgadas de agua.

Vacuum. The pressure range between the earth's atmosphere and no pressure, normally expressed in inches of mercury (in. Hg) vacuum.

Vacío. Margen de presión entre la atmósfera de la Tierra y cero presión, por lo general expresado en pulgadas de mercurio (pulgadas Hg) en vacío.

Vacuum pump. A pump used to remove some fluids such as air and moisture from a system at a pressure below the earth's atmosphere.

Bomba de vacío. Bomba utilizada para remover algunos fluidos, como por ejemplo aire y humedad de un sistema a una presión menor que la de la atmósfera de la Tierra.

Valve. A device used to control fluid flow.

Válvula. Dispositivo utilizado para regular el flujo de fluido.

Valve plate. A plate of steel bolted between the head and the body of a compressor that contains the suction and discharge reed or flapper valves.

Placa de válvula. Placa de acero empernado entre la cabeza y el cuerpo de un compresor que contiene la lámina de aspiración y de descarga o las chapaletas.

Valve seat. That part of a valve that is usually stationary. The movable part comes in contact with the valve seat to stop the flow of fluids.

Asiento de la válvula. Pieza de una válvula que es normalmente fija. La pieza móvil entra en contacto con el asiento de la válvula para detener el flujo de fluidos.

Vapor. The gaseous state of a substance.

Vapor. Estado gaseoso de una sustancia.

Vapor barrier. A thin film used in construction to keep moisture from migrating through building materials.

Película impermeable. Película delgada utilizada en construcciones para evitar que la humeded penetre a través de los materiales de construcción.

Vapor charge valve. A charge in a thermostatic expansion valve bulb that boils to a complete vapor. When this point is reached, an increase in temperature will not produce an increase in pressure.

Válvula para la carga de vapor. Carga en la bombilla de una válvula de expansión termostática que hierve a un vapor completo. Al llegar a este punto, un aumento en temperatura no produce un aumento en presión.

Vapor lock. A condition where vapor is trapped in a liquid line and impedes liquid flow.

Bolsa de vapor. Condición que ocurre cuando el vapor queda atrapado en el conducto de líquido e impide el flujo de líquido.

Vapor pump. Another term for compressor.

Bomba de vapor. Otro término para compresor.

Vapor refrigerant charging. Adding refrigerant to a system by allowing vapor to move out of the vapor space of a refrigerant cylinder and into the low-pressure side of the refrigeration system.

Carga del refrigerante de vapor. Agregarle refrigerante a un sistema permitiendo que el vapor salga del espacio de vapor de un cilindro de refrigerante y que entre en el lado de baja presión del sistema de refrigeración.

Vaporization. The changing of a liquid to a gas or vapor.

Vaporización. Cuando un líquido se convierte en gas o vapor.

Variable pitch pulley. A pulley whose diameter can be adjusted.

Polea de paso variable. Polea cuyo diámetro puede ajustarse.

Variable resistor. A type of resistor where the resistance can be varied.

Resistor variable. Tipo de resistor donde la resistencia puede variarse.

V belt. A belt that has a V-shaped contact surface and is used to drive compressors, fans, or pumps.

Correa en V. Correa que tiene una superficie de contacto en forma de V y se utiliza para accionar compresores, abanicos o bombas.

Velocity meter. A meter used to detect the velocity of fluids, air, or water.

Velocímetro. Instrumento utilizado para medir la velocidad de fluidos, aire o agua.

Velocity. The speed at which a substance passes a point.

Velocidad. Rapidez a la que una sustancia sobrepasa un punto.

Volt-ohm-milliammeter (VOM). A multimeter that measures voltage, resistance, and current in milliamperes.

Voltio-ohmio-miliamperímetro. Multímetro que mide tensión, resistencia y corriente en miliamperios.

Voltage. The potential electrical difference for electron flow from one line to another in an electrical circuit.

Tensión. Diferencia de potencial eléctrico del flujo de electrones de un conducto a otro en un circuito eléctrico.

Voltmeter. An instrument used for checking electrical potential.

Voltímetro. Instrumento utilizado para revisar la potencia eléctrica.

Volumetric efficiency. The pumping efficiency of a compressor or vacuum pump that describes the pumping capacity in relationship to the actual volume of the pump.

Rendimiento volumétrico. Rendimiento de bombeo de un compresor o de una bomba de vacío que describe la capacidad de bombeo con relación al volumen real de la bomba.

Vortexing. A whirlpool action in the sump of a cooling tower.

Acción de vórtice. Torbellino en el sumidero de una torre de refrigeración.

Walk-in cooler. A large refrigerated space used for storage of refrigerated products.

Nevera con acceso al interior. Espacio refrigerado grande utilizado para almacenar productos refrigerados.

Water box. A container or reservoir at the end of a chiller where water is introduced and contained.

Caja de agua. Recipiente o depósito al extremo de un enfriador por donde entra y se retiene el agua.

Water column (WC). The pressure it takes to push a column of water up vertically. One inch of water column is the amount of pressure it would take to push a column of water in a tube up one inch.

Columna de agua. Presión necesaria para levantar una columna de agua verticalmente. Una pulgada de columna de agua es la cantidad de presión necesaria para levantar una columna de agua a una distancia de una pulgada en un tubo.

Water-cooled condenser. A condenser used to reject heat from a refrigeration system into water.

Condensador enfriado por agua. Condensador utilizado para dirigir el calor de un sistema de refrigeración al agua.

Water-regulating valve. An operating control regulating the flow of water.

Válvula reguladora de agua. Regulador de mando que controla el flujo de agua.

Watt. A unit of power applied to electron flow. One watt equals 3.414 Btu.

Watio. Unidad de potencia eléctrica aplicada al flujo de electrones. Un watio equivale a 3,414 Btu.

Watt-hour. The unit of power that takes into consideration the time of consumption. It is the equivalent of a 1-watt bulb burning for 1 hour.

Watio hora. Unidad de potencia eléctrica que toma en cuenta la duración de consumo. Es el equivalente de una bombilla de 1 watio encendida por espacio de una hora.

Wet-bulb temperature. A wet-bulb temperature of air is used to evaluate the humidity in the air. It is obtained with a wet thermometer bulb to record the evaporation rate with an airstream passing over the bulb to help in evaporation.

Temperatura de una bombilla húmeda. La temperatura de una bombilla húmeda se utiliza para evaluar la humedad presente en el aire. Se obtiene con la bombilla húmeda de un termómetro para registrar el margen de evaporación con un flujo de aire circulando sobre la bombilla para ayudar en evaporar el agua.

Wet heat. A heating system using steam or hot water as the heating medium.

Calor húmedo. Sistema de calentamiento que utiliza vapor o agua caliente como medio de calentamiento.

Window unit. An air conditioner installed in a window that rejects the heat outside the structure.

Acondicionador de aire para la ventana. Acondicionador de aire instalado en una ventana que desvía el calor proveniente del exterior de la estructura.

Work. A force moving an object in the direction of the force. Work = Force × Distance.

Trabajo. Fuerza que mueve un objeto en la dirección de la fuerza. Trabajo = Fuerza × Distancia.

INDEX

ABS (acrylonitrilebutadiene styrene), 65–66
   joining ABS piping, 66
Absorption air conditioning chillers
   absorption system heat exchangers, 543–544
   capacity control, 541
   circulating pumps for absorption chillers, 539–541
   crystallization, 541–542
   direct-fired systems, 544
   purge systems, 542–543
   solution strength, 537–539
   system maintenance, 539
Absorption system heat exchangers, 543–544
Acetylene cylinders, 3
Adjustment and packing gland, 333
Air circulation over the evaporator coil, 281
Air conditioning and heating with room units, 251
Air conditioning equipment, 376–410
   installation, 458
   piping practices, 386–388
   refrigerant line sizing, 388–403
   special-application cooling equipment, 403–410
   types of equipment, 377–386
Air conditioning system start-up, 475–476
Air distribution and balance, 411–457
   blending the conditioned air with room air, 428–430
   combination duct systems, 426–427
   commercial duct applications, 443–455
   duct materials, 422–426
   duct system pressures, 413–416
   duct system standards, 422
   forced-air systems, 411–413
   friction loss, 432–434
   installation considerations, 427–428
   measuring air movement for balancing, 434–443
   return air duct system, 430–432
   supply duct system, 419–422
   types of fans, 416–419
Air friction charts, 435–443
Air measuring instruments, 416
Air movement through fittings, 427
Air quantity in forced-air systems, 411–413
Air system installation
   air movement through fittings, 427
   balancing dampers, 428
   duct insulation, 428

Air-cooled chillers, maintenance procedures, 578–580
Air-cooled condenser subcooling, 527
Air-cooled condensers, high-pressure chillers, 526–527
Air-cooled package systems, 378–382
Airstream, sensing temperature in, 145
Altering line set lengths, 473–474
Alternating current, 107
Amperage ratings, 216
Anti-recycle (time delay), 175–176, 551
Application of air conditioning equipment, 376–377
Atmospheric pressure, 17–18
Automatic controls, 129–153
   electromechanical controls, 151
   introduction to, 129–135
   maintenance of controls, 151
   pressure-sensitive devices, 145–150
   recognition of control components, 136
   technician service call, 151–152
   temperature controls, 136–140
   temperature measurements, 140–145
   temperature-sensitive devices, 129–134
   types of, 129
   used in air conditioning systems, 129
Autotransformer start, 546–547

Balanced port expansion valves, 339
Balancing dampers, 428
Ball bearings, 208
Basic controls, troubleshooting, 154–169
   introduction to, 154–155
   pictorial and line diagrams, 161–164
   technician service calls, 164–168
   troubleshooting a simple circuit, 155–156
   troubleshooting a thermostat, 157–159
   troubleshooting amperage in low-voltage circuits, 159–160
   troubleshooting an air conditioning circuit, 156–157
   troubleshooting voltage in low-voltage circuits, 160–161
Basic electricity, 105–128
   alternating current, 107
   capacitance, 114–115
   circuit protection devices, 120–121
   conductors, 106
   direct current, 107

641

electrical circuit, 107–108
electrical measuring instruments, 115–119
electrical power, 111
electrical units of measurement, 107
electricity produced from magnetism, 106–107
impedance, 115
inductance, 112
insulators, 106
magnetism, 111–112
making electrical measurements, 108–109
movement of electrons, 106
Ohm's Law, 109–110
parallel circuits, 110–111
semiconductors, 122–127
series circuits, 110
structure of matter, 105
transformers, 112–114
wire sizes, 119
Bearings
ball bearings, 208
sleeve bearings, 207–208
Bellyband-mount motors, 210
Belt-driven motors
applications, 211–212
belt tension, 227
compressors, 301–303
Bending tubing, 51–52
Bimetal devices, 130–131
Bleed water, 323, 562
Blocked suction unloading, 306
Blowdown (See bleed water)
Boyle's Law, 21
Brazing, 52–53
cleaning and fluxing, 56
heating and applying solder, 57
Bulb charges
cross liquid charge bulb, 334
cross vapor charge bulb, 335
liquid charge bulb, 334
vapor charge bulb, 334
Bypass dampers, 453
Bypass valves, 568–569

Capacitance, 114–115
Capacitor identification, 232
Capacitor-start motors, 191–192
Capacitor-start, capacitor-run motors, 191–192
Capacity checks, 249–250
Capacity control, 541
Capillary tubes, 341–342
restrictions in, 268
Cartridge fuses, 121
Centrifugal chillers, 577–578
start-up procedures, 577
Centrifugal compressors, high-pressure chillers, 515–519
Centrifugal fans, 416–418
Centrifugal switches, 196–197

CFC refrigerants, 86–87
Charging a refrigeration system, 98
Charging procedures
field charging systems with fixed-bore metering devices, 499–500
field charging systems with thermostatic expansion valves, 500–501
Charles' Law, 21–22
Checking capacitors, 231–232
Checking for voltage imbalance, 551
Chemical additives, 322–323
Chilled-water systems, 511–552
absorption air conditioning chillers, 534–544
applications, 511–512
high-pressure compression cycle chillers, 512–529
low-pressure compression cycle chillers, 529–534
motors and drives, 544–551
Chiller systems, 572–590
general maintenance tips for all chillers, 584
maintenance procedures for chilled-water systems, 578–582
operating procedures for chilled-water systems, 578
operation and maintenance procedures for absorption systems, 583–584
procedures for absorption system start-up, 582
start-up procedures for chilled-water systems, 572–577
technician service calls, 584–588
Circuit breakers, 121
Circuit protection devices
circuit breakers, 121
fuses, 120–121
ground fault circuit interrupters, 121
Circulating pumps for absorption chillers, 539–541
Clamp-on ammeter usage, 117–118
Cleanup after compressor burnout, 82–84
Clogged screens, 568
Cold traps, 78
Combination cooling/heating units, 256–257
Combination duct systems, 426–427
Comfort and psychrometrics
comfort, 235
food energy and the body, 235–236
psychrometric chart, 238–244
psychrometrics, 236–238
ventilation and infiltration, 244–250
Commercial building ventilation, 246–249, 369–370
Commercial duct applications, 443–455
dual-duct applications, 450–452
multi-zone applications, 446–450
single-duct applications, 445–446
types of fans used in commercial systems, 454–455
variable air volume (VAV) systems, 450, 452–453
Compressor capacity control
blocked suction unloading, 306
cylinder unloading, 304–306
gas diversion unloading, 306

INDEX

Compressors, high-pressure chillers
    centrifugal, 515–519
    reciprocating, 512–514
    rotary (screw), 514–516
    scroll, 514–515
Compressors, low-pressure chillers, 529–531
Compressors, 31–33, 312–313
    capacity control, 304–306
    electrical checkup, 503
    functions, 289
    hermetic, 298
    moisture removal using a vacuum, 78–79
    motors for hermetic and semi-hermetic, 309–311
    open-drive, 299–303
    overload problems, 503
    reciprocating efficiency, 303–304
    reciprocating, 289–298, 578
    rotary, 306–307
    scroll, 308–309, 578
    semi-hermetic, 298–299
    two-speed, 198
    used in high-pressure chillers, 512–519
    used in low-pressure chillers, 529–531
Condensate
    drain piping, 469–470
    evaporator, 286
Condensers, high-pressure chillers
    air-cooled, 526–527
    water-cooled, 524–526
Condensers, low-pressure chillers, 531–532
Condensers, 33–35, 314–326
    condenser functions, 314–315
    refrigerant-to-air heat exchangers, 323–325
    refrigerant-to-earth heat exchangers, 316
    refrigerant-to-water heat exchangers, 316–323
    shell-and-coil, 317–318
    shell-and-tube, 318–319
    used in high-pressure chillers, 524–527
    used in low-pressure chillers, 531–532
Condensing unit, 471
Conduction, 14–15
Conductors, 106
Contactors, 217–218
Convection, 15
Cooling towers
    construction, 557–559
    fan section, 559
    flow patterns, 556–557
    functions, 553–554
    types of, 554–556
    water control and distribution, 559–564
    water pumps used in, 564–571
    water supply systems, 321–322
Cooling towers, types of
    forced-draft and induced-draft towers, 555–556
    natural-draft towers and spray ponds, 554–555
Cooling-only refrigeration cycle, 252–256

Cooling-only units, 251–256
    refrigeration cycle, 252–256
    controls for, 262–264
Cooling/heating units, controls for, 264
Counterflow towers, 556–557
CPVC (chlorinated polyvinyl chloride), 65–66
    joining CPVC piping, 66
Cradle-mount motors, 209–210
Cross liquid charge bulb, 334
Cross vapor charge bulb, 335
Crossflow towers, 556–557
Crystallization, 541–542
Current
    capacity, 137, 203–204
    monitoring, 174–175
    relays, 197–198
Cutting tubing and piping
    steel pipe, 64
    tubing, 50–51
Cylinder unloading, 304–306, 513–514

Dalton's Law, 23
Dehumidification, 281–282
Density, 20
    and resistance, 236
Design conditions tables
    Canada, 605–606
    other countries, 607–611
    United States, 593–604
Design conditions, 346–354
Dew-point temperature, 237–238
Diacs, 126
Diagnostic thermostats, 179
Diaphragm, 330
Diodes, 122–123
    checking a diode, 123
Direct current, 107
Direct expansion evaporators, high-pressure chillers, 519–520
Direct-driven motors
    applications, 212–213
    compressors, 300–301
Direct-fired systems, 544
Discharge line temperatures, 498
Distribution pans, 562–564
Double-thickness flare, 59–60
Drive systems
    belt, 211–212, 301–303
    direct, 212–213, 300–301
    open, 299–303
    removal of, 226
Dry bulb temperatures, 237
Dual-duct applications, 450–452
Dual-element plug fuses, 121
Duct
    fiberglass duct, 423, 425, 461
    flexible duct, 423, 426, 462

644 INDEX

galvanized steel duct, 422–424, 458–461
installation, 422, 428, 458–462
insulation, 428
return, 430-432
spiral metal duct, 423
supply, 419–422
Duct system pressures, 413–416
    air measuring instruments, 416
    identifying system pressures, 414–416
Duct system standards, 422
Ductless air conditioning, 409

EER (Energy Efficiency Ratio), 488
Electric motors
    and magnetism, 187
    applications, 186
    capacity control, 198–200
    circuit control devices, 196–198
    components and operation, 187–190
    cooling, 200–201
    used in air conditioning applications, 190–196
Electrical circuit, 107–108
Electrical hazards, 3–6
    electrical burns, 6
    electrical shock, 4–6
    nonconducting ladders, 6
Electrical installation, 462–464
Electrical measurements, 108–119
    use in troubleshooting, 118–119
    using a clamp-on ammeter, 117–118
    using a VOM to measure AC voltage, 116–117
    using a VOM to measure DC voltage, 116–117
    using a VOM to measure resistance, 117
Electrical operating conditions, 488–490
    finding a point of reference for an unknown motor rating, 489–490
    matching the unit to the correct power supply, 489
    starting the equipment, 489
Electrical power, 111
    watts, 24
Electrical service, 269–271
Electrical symbols chart, 612
Electrical troubleshooting, 501–504
    checking capacitors, 231–232
    compressor electrical checkup, 503
    compressor overload problems, 503
    fuses and circuit breakers, 504
    hermetic motors, 232–233
    mechanical motor problems, 225–227
    motor starting problems, 230
    open windings, 227
    short circuit to ground (frame), 228–230
    shorted motor windings, 227–228
    technician service calls, 233–234
    using electric instruments for, 118–119
    wiring and connectors, 232
Electrical units of measurement, 107

Electricity produced from magnetism, 106–107
Electromechanical controls, 151
Electronic and programmable controls, 170–185
    air conditioning applications, 72–79, 172–179
    anti-recycle, 175–176
    current monitoring, 174–175
    diagnostic thermostats, 179
    electronic thermostats, 179
    phase protection, 174–175
    pressures and temperatures, 176–178
    programmed operation, 178
    refrigerant flow control, 178–179
    technician service calls, 182–184
    troubleshooting, 179–182
    voltage monitoring, 172–174
Electronic expansion valves, high-pressure chillers, 528–529
Electronic relays, 197
Electronic solid-state overload device protection, 551
Electronic starters, 549
Electronic thermostats, 179
Electrons, movement of, 106
End-mount motors, 210
Energy, 23–24
EPA evacuation requirements, 90–91
    for residential and commercial refrigeration systems, 91
    for small appliances, 91
Equivalent lengths, 390
Equipment
    efficiency ratings, 488
    grades, 482
    loads, 372–374
    sizing, 346
Establishing a reference point on unknown equipment, 484–488
    operating conditions near design conditions for high-efficiency equipment, 486–487
    operating conditions near design conditions for standard equipment, 485
    other-than-design conditions for high-efficiency equipment, 487–488
    space temperature higher than normal for standard equipment, 485–486
Evacuation (*See* System evacuation)
Evaporative cooling, 403–408
    maintenance, 405
    operation, 404–406
    sizing, 407–408
Evaporator construction and operation, 278–284
    air circulation over the evaporator coil, 281
    dehumidification, 281–282
    routing of the evaporator piping, 282
    troubleshooting, 282–284
Evaporators, high-pressure chillers
    direct expansion evaporators, 519–520
    flooded evaporators, 520–524

Evaporators, 30–31, 277–288
    checking airflow in, 287
    condensate, 286
    construction and operation, 278–284
    functions, 277–278
    operating conditions, 479–481
    outlet temperatures, 496
    piping, 282
    sensible and latent heat removal, 286–287
    types of evaporator coils, 284–285
    used in high-pressure chillers, 519–524
Expansion (metering) devices, 35–36, 327–344
    balanced port expansion valves, 339
    bulb charges, 334–335
    capillary tubes, 268, 341–342
    fixed-bore metering devices, 341–343
    float type, 528
    functions, 327–328
    receivers, 341
    solid-state controlled expansion valves, 339–340
    thermostatic expansion valves, 330–333, 527
Extended plenum system, 421
External equalizer, 337–338
External motor protection, 221

Fan section, 559
Fans
    centrifugal fans, 416–418
    drives, 418–419
    outlet dampers, 453
    propeller fans, 416–417
    removing, 268–269
    used in commercial systems, 454–455
Federal excise tax, 88
Fiberglass duct, 423, 425, 461
Field charging systems
    with fixed-bore metering devices, 499–500
    with thermostatic expansion valves, 500–501
Fill materials, 558
Finding a point of reference for an unknown motor rating
    determining the compressor running amperage, 489–490
    determining the fan motor amperage, 489
Fire protection in water towers, 558
Fixed-bore metering devices
    capillary tubes, 341–342
    orifice metering devices, 342, 527
    operating charge for, 342–343
Flare joints
    flare fittings, 60
    making a double-thickness flare, 59–60
    making a single-thickness flare, 59
Flexible duct, 423, 426, 462
Float-type metering devices, high-pressure chillers, 528
Flooded evaporators, high-pressure chillers, 520–524

Flow patterns
    counterflow towers, 556–557
    crossflow towers, 556–557
Fluid expansion devices, 131–134
Fluorescent lighting, 371–372
Fluxing, 55–56
Food energy and the body, 235–236
Forced-air systems
    air quantity, 411–413
    components, 413
Forced-draft and induced-draft towers, 555–556
Fouling, 320
Frequency, 203
Friction loss, 432–434
Fuses
    cartridge fuses, 121
    dual-element plug fuses, 121
    plug fuses, 120

Gage manifolds, 492–495
    high-side gage readings, 495
    hose leaks, 81
    low-side gage readings, 493, 495
    selecting and using, 80–81
    when to connect the gages, 493–494
Galvanized steel duct, 422–424
Gas diversion unloading, 306
Gas laws, 20–23
    Boyle's Law, 21
    Charles' Law, 21–22
    Dalton's Law, 23
    General Law of Perfect Gas, 22
Gas line sizing, 393–403
General evacuation tips
    cold traps, 78
    gage manifold hose leaks, 81
    removing moisture from the compressor, 78–79
    selecting and using a gage manifold, 80–81
    system valves, 81–82
    using dry nitrogen, 82
General Law of Perfect Gas, 22
Ground fault circuit interrupters, 121

Hard-drawn copper tubing, 50
HCFC refrigerants, 87
Heat exchangers
    refrigerant-to-air, 323–325
    refrigerant-to-earth, 316
    refrigerant-to-water, 316–323
Heat gain calculations, 345–375
    design conditions, 346–354
    heat gain due to infiltration and ventilation, 367–370
    heat gain through ductwork, 370
    heat transfer through building materials, 354–363
    internal heat gains, 370–374
    solar heat gain, 363–367
    system design, 345–346

Heat pump refrigeration cycle, 256–257
Heat sinks, 126–127
Heat sources for soldering and brazing, 53–55
Heat transfer
    conduction, 14–15
    convection, 15
    due to infiltration and ventilation, 367–370
    radiation, 15
    through building materials, 354–363
    through ductwork, 370
    through internal heat gains, 370–374
    through solar heat gain, 363–367
Heat, 6–7
    latent heat, 15–16
    sensible heat, 15
    removal, 286–287
Hermetic compressors, 298
HFC refrigerants, 87
High-pressure compression cycle chillers
    compressors used in, 512–519
    condensers used in, 524–527
    evaporators used in, 519–524
    metering devices used in, 527–529
High-pressure controls, 148
High-side gage readings, 495
High-temperature soldering (brazing), 57
High-voltage space temperature controls, 139–140
    cover, 140
    sensing element, 140
    subbase, 140
    switching mechanism, 140
    thermostat assembly, 140
Horsepower, 24
Humidity control, 27

Impedance, 115
Incandescent lighting, 371
Inductance, 112
Inherent motor protection, 221
Inlet air temperatures, 496
Installation, 458–477
    air system, 427–428
    duct, 428, 458–462
    electrical system, 462–464
    equipment start-up, 475–476
    package systems, 464–467
    room air conditioners, 257–261
    split systems, 467–475
    steel pipe, 65
Insulators, 106
Internal equalizer, 335–337
Internal heat gains
    equipment loads, 372–374
    people loads, 370–371
    lighting loads, 371–372
    room-by-room load calculations, 374

Joining steel pipe
    cutting, 64
    pipe fittings, 63
    pipe threads, 62
    reaming, 64
    threading, 64-65
    tools, 63–64

Large positive-displacement chillers, 578
Latent heat, 15–16
    removal, 286–287
Leak detection, 266
    while in a vacuum, 74
    standing pressure test, 75
Lighting loads
    fluorescent lighting, 371–372
    incandescent lighting, 371
Line sets, 50
    altering, 473–474
Liquid charge bulb, 334
Liquid line
    sizing, 390–393
    temperatures, 498
Liquid refrigerant charging
    empty system, 100
    operating system, 100–101
Load-limiting devices, 549–550
Low-pressure compression cycle chillers
    compressors used in, 529–531
    condensers used in, 531–532
    metering devices used in, 532
    purge devices used in, 532–534
Low-pressure controls, 148
Low-side gage readings, 493, 495
Low-temperature soldering, 57
Low-voltage space temperature controls
    cold anticipator, 138
    cover, 139
    electrical contacts, 137–138
    subbase, 139
    thermostat assembly, 139

Magnetic overload devices, 222–223
Magnetism, 111–112
Maintenance
    chilled-water systems, 578–582
    controls, 151
    room air conditioners, 265–271
Makeup water, 560–561
Mass, 20
Matching the unit to the correct power supply, 489
Matter, 19–20
Measuring a vacuum, 69–71
Measuring air movement for balancing, 434–443
    air friction charts, 435–443

Measuring operating conditions, 478
Measuring refrigerant
    measuring refrigerant using charging charts, 104
    measuring refrigerant using charging cylinders, 103
    refrigerant chargers, 104
    weighing refrigerant, 102–103
Mechanical equipment, 7–9
Mechanical motor problems, 225–227
    belt tension, 227
    pulley alignment, 227
    removing drive assemblies, 226
Mechanical operating conditions, 478–481
    evaporator operating conditions, 479–481
    system component relationships under load changes, 479
Mechanical recovery systems, 92–96
Mechanical troubleshooting, 491
Mechanical-electrical motor overload protection, 550
Metal duct
    insulation, 459, 461
    round, 422–424, 458–459, 461
    square and rectangular, 422–424, 458–460
Metering devices used in high-pressure chillers
    electronic expansion valves, 528–529
    float-type metering devices, 528
    orifice metering devices, 527
    thermostatic expansion valves, 527
Metering devices used in low-pressure chillers, 532
Metering devices (See Expansion devices)
Moisture removal using a vacuum, 75–77
Motor capacity control
    special-application motors, 198–200
    two-speed compressor motors, 198
    variable speed motors, 199–200
Motor circuit control devices, 215
    centrifugal switches, 196–197
    current relays, 197–198
    electronic relays, 197
    positive temperature coefficient (PTC) devices, 198
    potential relays, 197
Motor drives
    belt-drive applications, 211–212
    direct-drive applications, 212–213
Motor mounting characteristics
    bellyband-mount motors, 210
    cradle-mount motors, 209–210
    end-mount motors, 210
    rigid-mount motors, 210
Motor protection, 220–221
    anti-recycle control, 551
    electronic solid-state overload devices, 551
    external, 221
    for hermetic and semi-hermetic compressors, 310–311
    inherent, 221
    load-limiting devices, 549–550
    mechanical-electrical overload devices, 550
    phase failure protection, 551
    temperature controls, 140–143
Motor starters, 218–220
    autotransformer, 546–547
    electronic, 549
    part winding, 546–547
    wye-delta, 547–549
Motor starting, 215–224
    amperage ratings, 216
    contactors, 217–218
    motor control devices, 215
    motor protection, 220–221
    motor starters, 218–220, 546–549
    problems, 230
    relays, 216–217
    service factor, 221–223
Motors
    and magnetism, 187
    applications, 202–214
    bearings, 207–208
    capacitor-start, 191–192
    capacitor-start, capacitor-run, 191–192
    capacity control, 198–200
    checking for voltage imbalance, 551
    circuit control devices, 196–198, 215
    components and operation, 187–190
    cooling, 200–201
    determining motor speed, 188
    drives, 211–213
    hermetic and semi-hermetic compressors, 309–311
    mounting characteristics, 209–210
    permanent split-capacitor, 192–193
    phase reversal, 551
    power supplies for, 189–190, 202–205
    protection, 140–143, 220–221, 310–311, 549–551
    run winding, 187–188
    selection, 202
    shaded-pole, 193
    single-phase hermetic, 194–195
    split-phase open, 190–191
    start winding, 188
    starters, 218–220, 546–549
    starting and running characteristics, 188–189
    starting, 215–224, 230
    temperature classifications, 206
    three-phase, 193–196
    working conditions, 206
Moving heavy objects, 8–10
Multi-zone applications, 446–450
Multiple evacuation, 73–74

National Electrical Code (NEC) standards, 221
Natural-draft towers, 554–555
Needle and seat, 330–332
Nitrogen cylinders, 2–3

Ohm's Law, 109–110
Oil pressure safety controls, 148–150
Oil return and piping practices, 387–388
Oil separators, 386–387
Open windings, 227
Open-drive compressors, 299–303
    belt-drive, 301–303
    direct-drive, 300–301
    refrigerant shaft seal in, 302–303
Operating charge for fixed-bore devices, 342–343
Operating conditions, typical
    equipment efficiency ratings, 488
    equipment grades, 482
    establishing a reference point on unknown equipment, 484–488
    high-efficiency equipment, 486–487
    manufacturer's literature, 482–484
    measuring operating conditions, 478
    mechanical operating conditions, 478–481
    standard equipment, 485
    typical electrical operating conditions, 488–490
Operating procedures for chilled-water systems
    centrifugal chillers, 578
    large positive-displacement chillers, 578
    small scroll and reciprocating compressors, 578
    water-cooled chillers, 578
Operation and maintenance procedures for absorption systems, 583–584
Orifice (fixed-bore) metering devices, 342
    high-pressure chillers, 527
Other-than-design conditions
    for high-efficiency equipment, 487–488
    for standard equipment, 485–486
Oxygen cylinders, 2–3
Ozination, 323
Ozone depletion potential, 87

Package systems, 464–467
    duct connections for package equipment, 466–467
    vibration isolation, 465–466
Parallel circuits, 110–111
Part winding start, 546–547
PE (polyethylene), 65–66
    joining PE piping, 66
People loads, 370–371
Perimeter loop system, 422
Permanent split-capacitor motors, 192–193
Phase protection, 174–175, 551
Phase reversal, 551
Phase, 204–205
Pictorial and line diagrams, 161–164
Pipe fittings, 63
Pipe threads, 62
Pipe-in-pipe condensers, 317–318
Piping (*See* Tubing)
Piping practices, 474–475

oil separators, 386–387
oil return and piping practices, 387–388
Plastic pipe
    ABS (acrylonitrilebutadiene styrene), 65
    CPVC (chlorinated polyvinyl chloride), 65
    Joining ABS piping, 66
    Joining PE piping, 66
    Joining PVC or CPVC piping, 66
    PE (polyethylene), 65
    PVC (polyvinyl chloride), 65
Plenum system, 419–421
Plotting the refrigeration cycle, 39–48
Plug fuses, 120
Portable air conditioning, 409–410
Positive temperature coefficient (PTC) devices, 198
Potential relays, 197
Power outages and the electronic thermostat, 182
Power supplies
    current capacity, 203–204
    for electric motors, 189–190
    frequency, 203
    phase, 204–205
    voltage, 202–203
Practical soldering and brazing tips
    different joints, 57
    general soldering and brazing tips, 58–59
    heat sources for soldering and brazing, 57–58
    high-temperature soldering (brazing), 57
    low-temperature soldering, 57
Precharged line sets (quick-connect line sets), 474
Pressure vessels and piping
    acetylene cylinders, 3
    nitrogen and oxygen cylinders, 2–3
    refrigerant cylinders and piping, 1–2
Pressure
    atmospheric, 17–18
    controls, 176–178
    gages, 18–19
    measured in metric terms, 19
    relief valves, 150
    vessels and piping, 1–3
Pressure-sensitive devices, 145–150
    high-pressure controls, 148
    low-pressure controls, 148
    oil pressure safety controls, 148–150
    pressure relief valves, 150
    water pressure regulators, 150
Pressure/temperature relationship, 27–30
Preventive maintenance, 504–505
Propeller fans, 416–417
Psychrometrics
    density and resistance, 236
    dew-point temperature, 237–238
    dry bulb and wet bulb temperatures, 237
    superheated gases in air, 236

INDEX

using the psychrometric chart, 238–244
water vapor in air, 236–237
Pulley alignment, 227
Purge systems, 542–543
    for low-pressure chillers, 532–534
PVC (polyvinyl chloride), 65–66
    joining PVC piping, 66

Radiation, 15
Rating air conditioning equipment, 27
Reaming, 64
Receivers, 341
Reciprocating chillers, start-up procedures, 575–576
Reciprocating compressors, 289–298
    components of, 292–298
    efficiency, 303–304
    for high-pressure chillers, 512–514
    small, 578
Reclaiming refrigerant, 39, 90, 96
Recovering refrigerant, 39, 71
    and refrigerant oil, 89
    cylinders for, 89
    determining when it is necessary, 88–89
    from window air conditioners, 96
    selecting equipment for, 95–96
Rectifiers, 123–124
Recycling, 39, 90
Reducing plenum system, 421
Refrigerant, 10
    blends, 87–88
    charge, 472
    chargers, 104
    classifications, 86–88
    cylinders and piping, 1–2
    flow control, 178–179
    heat exchangers, 316–325
    line sizing, 388–403
    management, 39, 71, 86–97
    Refrigerant-22 (R-22), 37–39
    shaft seal in open-drive compressors, 302–303
Refrigerant classifications
    CFC refrigerants, 86–87
    HCFC refrigerants, 87
    HFC refrigerants, 87
    ozone depletion potential, 87
    refrigerant blends, 87–88
Refrigerant line sizing, 388–403
    resistance and equivalent lengths, 390
    sizing the gas lines, 393–403
    sizing the liquid line, 390–393
Refrigerant management, 86–97
    classifications, 86–88
    environmental concerns, 86
    epa evacuation requirements, 90–91
    mechanical recovery systems, 92–96

reclaiming, 39, 90, 96
recovering, 39, 71, 89, 95–96
recycling, 39, 90
regulations, 88
technician certification, 91
technician information, 96
Refrigerant-to-air heat exchangers, 323–325
Refrigerant-to-earth heat exchangers, 316
Refrigerant-to-water heat exchangers, 316–323
    fouling, 320
    pipe-in-pipe condensers, 317–318
    shell-and-coil condensers, 317–318
    shell-and-tube condensers, 318–319
    water supply systems, 320–322
    water treatment, 322–323
Refrigeration system, 26–48
    and the refrigeration cycle, 36–37, 39–48
    compressors, 31–33, 198, 289–313, 503, 512–519, 529–531, 578
    condensers, 33–35, 314–326, 524–527, 531–532
    evaporators, 30–31, 277–288, 479–481, 496, 519–524
    expansion devices, 35–36, 268, 327–344, 527–528
    installation, 464–475
    pressure/temperature relationship, 27–30
    rating, 27, 488
Refrigeration system installation, 458–477
    air system, 427–428
    duct, 428, 458–462
    electrical system, 462–464
    equipment start-up, 475–476
    package systems, 464–467
    room air conditioners, 257–261
    split systems, 467–475
    steel pipe, 65
Relays, 216–217
Residential infiltration and ventilation, 368–369
    ventilation calculations, 245–246
Resistance, air system, 390
Restarting, motor, 222–223
Return air duct system, 430–432
Rigid-mount motors, 210
Room air conditioners, 251–276
    air conditioning and heating with room units, 251
    combination cooling/heating units, 256–257
    cooling-only units, 251–256
    controls for cooling-only units, 262–264
    controls for cooling/heating units, 264
    installation, 257–261
    maintenance and service, 265–271
    technician service calls, 271–275
Room-by-room load calculations, 374
Rotary (screw) compressors
    for high-pressure chillers, 514–516
    rotary vane compressors, 306–307
    stationary vane compressors, 306–307

Round metal duct, 422–424, 458–461
Run winding, 187–188

Safety practices, 1–11
   chemicals, 10
   cold, 7
   electrical hazards, 3–6
   heat, 6–7
   mechanical equipment, 7–9
   moving heavy objects, 8–10
   pressure vessels and piping, 1–3
   refrigerants, 10
Screw chillers, start-up procedures, 576–577
Scroll chillers, start-up procedures, 575
Scroll compressors, 308–309
   for high-pressure chillers, 514–515
   small, 578
SEER (Seasonal Energy Efficiency Ratio), 488
Semi-hermetic compressors, 298–299
Semiconductors
   diacs, 126
   diodes, 122–123
   heat sinks, 126–127
   rectifiers, 123–124
   silicon-controlled rectifiers, 124–125
   thermistors, 125–126
   transistors, 125
   triacs, 126
Sensible heat, 15
   removal, 286–287
Sensing bulb and transmission tube, 333
Series circuits, 110
Service factor
   magnetic overload devices, 222–223
   National Electrical Code (NEC) standards, 221
   restarting the motor, 222–223
   temperature-sensing devices, 221–222
Shaded-pole motors, 193
Shell-and-coil condensers, 317–318
Shell-and-tube condensers, 318–319
Short circuit to ground (frame), 228–230
Shorted motor windings, 227–228
Silicon-controlled rectifiers, 124–125
Single-duct applications, 445–446
Single-phase hermetic motors, 194–195
Single-thickness flare, 59
Sleeve bearings, 207–208
Soft copper tubing, 50
Solar heat gain, 363–367
Solar influence, outdoor unit, 471
Soldering, 52–53
   assembly, 56
   cleaning, 55
   fluxing, 55–56
   heating and applying solder, 56
   wiping, 56

Solid-state controlled expansion valves, 339–340
Solution strength, 537–539
Special-application cooling equipment
   ductless air conditioning, 409
   evaporative cooling, 403–408
   portable air conditioning, 409–410
Special-application motors, 198–200
Specific gravity, 20
Specific heat, 16–17
Specific volume, 20
Spiral metal duct, 423
Split systems, 382
   non-precharged line sets, 384–386
   precharged line sets, 384–385
   evaporator section, 467–471
   installing refrigerant piping, 471–475
Split-phase open motors, 190–191
Spray ponds, 554–555
Spring, 332–333
Square and rectangular metal duct, 422–424, 458–461
Start winding, 188
Start-up procedures for absorption systems, 582
Start-up procedures for chilled-water
      systems, 572–577
   centrifugal chillers, 577
   screw chillers, 576–577
   scroll chillers, 575
Starting and running characteristics, 188–189
Starting the equipment, 489
Stationary vane compressors, 306–307
Steel and wrought iron pipe, 62
Structure of matter, 105
Suction line temperatures, 496–498
Sump location, 568
Superheated gases in air, 236
Supply duct system
   extended plenum system, 421
   perimeter loop system, 422
   plenum system, 419–421
   reducing plenum system, 421
Swaging, 60–61
System charging
   adjusting the system charge, 266–268
   charging a refrigeration system, 98
   liquid refrigerant charging, 100–101
   measuring refrigerant, 102–104
   vapor refrigerant charging, 98–100
System component relationships under load
      changes, 479
System dehydration using a vacuum pump, 71–72
System design, 345–346
   equipment sizing, 346
System evacuation, 68–85
   achieving a deep vacuum, 72–73
   cleanup after compressor burnout, 82–84
   evacuating air conditioning systems, 68

general evacuation tips, 78–82
leak detection while in a vacuum, 74
leak detection—standing pressure test, 75
measuring a vacuum, 69–71
multiple evacuation, 73–74
recovering refrigerant, 71
removing moisture with a vacuum, 75–77
removing moisture from the compressor, 78–79
understanding evacuation, 68–69
vacuum pump oil, 72
vacuum pumps, 71–72
System evaluation using a bench test, 265–266
System maintenance, 539
System pressures, identifying 414–416
System valves, 81–82

Technician certification
    Type I certification, 91
    Type II certification, 91
    Type III certification, 91
    Universal certification, 91
Technician information, 96
Technician service calls
    for automatic control components, 151–152
    for basic controls, 164–168
    for chillers, 584–588
    for electric motors, 233–234
    for electronic and programmable controls, 182–184
    for general troubleshooting, 505–509
    for room air conditioners, 271–275
Temperature, 12–13
    classifications, 206
    control, 129
    conversions, 13, 591–592
Temperature controls, 176–178
    current-carrying capacity, 137
    high-voltage, 139–140
    low-voltage, 137–139
Temperature measurements, 495–498
    discharge line, 498
    evaporator outlet, 496
    fluids, 144–145
    in an airstream, 145
    inlet air, 496
    liquid line, 498
    motor temperature controls, 140–143
    sensing devices, 129–134, 221–222
    solids, 143–144
    suction line, 496–498
Temperature-sensitive devices
    bimetal devices, 130–131
    fluid expansion devices, 131–134
    thermistors, 134
    types of, 129
Thermistors, 125–126, 134
Thermostatic expansion valves
    adjustment and packing gland, 333
    diaphragm, 330
    for high-pressure chillers, 527
    installation and service considerations, 338–339
    needle and seat, 330–332
    operation, 335–338
    sensing bulb and transmission tube, 333
    spring, 332–333
    valve body, 328–330
Threading, 64–65
Three-phase motors, 193–196
    hermetic motors, 195–196
Towers
    access, 558–559
    construction, 557–559
    fill materials, 558
    fire protection, 558
    sumps, 560
Transformers, 112–114
Transistors, 125
Triacs, 126
Troubleshooting, 491–510
    air conditioning circuits, 156–157
    air conditioning systems, 491
    amperage in low-voltage circuits, 159–160
    basic controls, 154–169
    charging procedures, 498–501
    electric motors, 225–234
    electrical circuits, 501–504
    electronic controls, 179–182
    electronic thermostats, 180–182
    evaporators, 282–284
    fuses and circuit breakers, 504
    gage manifold usage, 492–495
    hermetic motors, 232–233
    mechanical, 491
    preventive maintenance, 504–505
    simple circuits, 155–156
    technician service calls, 505–509
    temperature readings, 495–498
    thermostat, 157–159
    voltage in low-voltage circuits, 160–161
Troubleshooting electronic controls, 179–182
    power outages and the electronic thermostat, 182
    troubleshooting the electronic thermostat, 180–182
Tubing, 49–67
    bending, 51–52
    brazing techniques, 56–57
    cutting, 50–51
    heat sources for soldering and brazing, 53–55
    installing steel pipe, 65
    insulation, 50
    joining steel pipe, 62–65
    line sets, 50
    making flare joints, 59–60
    plastic pipe, 65–66
    practical soldering and brazing tips, 57–59
    purpose in air conditioning systems, 49

652 INDEX

soldering and brazing processes, 52–53
soldering techniques, 55–56
steel and wrought iron pipe, 62
swaging, 60–61
types and sizes, 49–50
leak test and evacuation, 472–473
Tubing, types and sizes, 49–50
hard-drawn copper tubing, 50
soft copper tubing, 50
Two-speed compressor motors, 198
TXV (*See* Thermostatic expansion valves)
Type I certification, 91
Type II certification, 91
Type III certification, 91

Universal certification, 91

Vacuum (*See* System evacuation)
Vacuum pumps, 71–72
Valves
balanced port expansion, 339
bypass, 568–569
electronic expansion for high-pressure chillers, 528–529
pressure relief, 150
solid-state controlled expansion, 339–340
system, 81–82
thermostatic expansion valves, 330–333, 527
Vapor charge bulb, 334
Vapor refrigerant charging, 98–100
vapor charging an empty system, 98
vapor charging an operating system, 98–99
Variable air volume (VAV) systems, 450, 452–453
bypass dampers, 453
fan outlet dampers, 453
variable fan inlet vanes, 453
variable fan speeds, 453
Variable fan inlet vanes, 453
Variable fan speeds, 453
Variable speed motors, 199–200
Ventilation and infiltration, 244–250
calculating ventilation air for residences, 245–246
capacity checks, 249–250
commercial ventilation requirements, 246–249
Vibration isolation, 465–466
Voltage monitoring, 172–174
Voltage, 202–203

VOM usage
to measure AC voltage, 116–117
to measure DC voltage, 116–117
to measure resistance, 117
Vortexing, 568
Wastewater supply systems, 320–321
Water control and distribution, 559–564
blowdown or bleed water, 562
distribution pans, 562–564
makeup water, 560–561
tower sumps, 560
Water pressure regulators, 150
Water pump location
bypass valves, 568–569
clogged screens, 568
sump location, 568
vortexing, 568
Water pumps used in cooling towers, 564–571
alignment, 569–571
materials used in pump construction, 567
pump location, 568–569
Water supply systems
cooling tower systems, 321–322
wastewater systems, 320–321
Water treatment
bleed water, 323
chemical additives, 322–323
ozination, 323
Water vapor in air, 236–237
Water-cooled chillers, 578
maintenance procedures, 580–582
Water-cooled condenser subcooling, 524–526
Water-cooled condensers, high-pressure chillers, 524–526
Water-cooled package systems, 382–383
Weighing refrigerant, 102–103
Wet bulb temperatures, 237
Winding
open, 227
part, 546–547
run, 187–188
shorted, 227–228
start, 188
Wire sizes, 119
Wiring and connectors, 232
Working conditions, 206
Wye-delta start, 547–549